# Cancer-Causing Viruses and Their Inhibitors

Edited by **Satya Prakash Gupta**
National Institute of Technical Teachers' Training and Research (NITTTR), Bhopal, India

# Cancer-Causing Viruses and Their Inhibitors

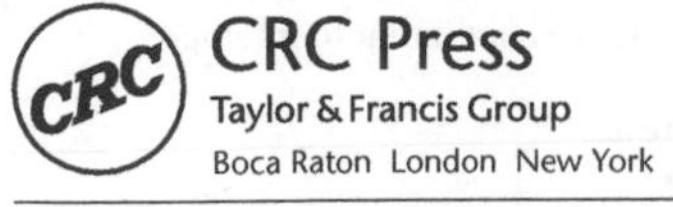


CRC Press is an imprint of the
Taylor & Francis Group, an **informa** business

CRC Press
Taylor & Francis Group
6000 Broken Sound Parkway NW, Suite 300
Boca Raton, FL 33487-2742

First issued in paperback 2018



ISBN-13: 978-1-4665-8977-3 (hbk)
ISBN-13: 978-1-138-37483-6 (pbk)



**Visit the Taylor & Francis Web site at**
**http://www.taylorandfrancis.com**

**and the CRC Press Web site at**
**http://www.crcpress.com**

# Contents

# Preface

Cancer-causing viruses are also called oncoviruses or tumor viruses. Their infection in the human body leads to cancer. According to the World Health Organization Cancer Report 2008, cancer-causing viruses are responsible for up to 20% of cancer deaths all over the world; approximately 20%–30% of all cancers worldwide can be linked to viral infections. Cancer develops when cells start to divide uncontrollably because the "cell cycle machinery" that regulates this process stops working properly. These cancer cells can then invade other tissues. Cancer development is a complex process involving a series of genetic changes that disrupt the cell cycle machinery, interfering with cellular functions such as cell growth. Cancer-causing viruses play a key role in the development of certain cancers by contributing to these genetic changes. Although cancer itself is not an infectious disease, a significant number of human viruses have been linked to specific cancers, such as adult T-cell leukemia (ATL), hepatocellular carcinoma (HCC), Burkitt's lymphoma (BL), nasopharyngeal carcinoma (NPC), posttransplant lymphomas, Hodgkin's disease, cervical cancer, skin cancer in patients with epidermodysplasia verruciformis (EV), head and neck cancers, primary effusion lymphoma, and Castleman's disease. The involvement of these viruses in human cancer development means that the frequency of these cancers can be reduced either prophylactically by vaccinating against the viruses or therapeutically by treating the infections. Extensive studies of these viruses are currently taking place with respect to their life cycle, their mechanism of infection, the inhibition of their growth, and the development of vaccines to treat the cancer caused by them. This book will cover in detail all of these aspects of oncoviruses, and it may be of great interest to medicinal and pharmaceutical chemists as well as to those working in the area of biotechnology. Cancer researchers may also find it of immense value.

There are a total of fourteen chapters in this volume. In the first chapter, Gupta and Gautam discuss the cancer-causing viruses discovered to date and include a brief description of their structure, genotypes, replication, and mechanisms of infection leading to cancers. The second chapter, by Yasunaga, presents the pathogenesis of human T-cell leukemia virus type 1 (HTLV-1) as well as current and upcoming therapeutic strategies against HTLV-1-mediated diseases. Chapter 3, by Deval and Huber, discusses the hepatitis C virus (HCV)—which leads to hepatocellular carcinoma—and its inhibitors. The number of HCV infection cases has now reached more than 170 million worldwide; one of the main protein targets for the generation of HCV therapeutics is the viral RNA-dependent RNA polymerase (RdRp) NS5B. This chapter summarizes what is known to date about the structure and function(s) of NS5B, with a particular focus on its role in HCV replication, and it reviews the recent advances in the development of nucleoside and nucleotide analogs as inhibitors of NS5B. Combination therapy utilizing interferon-alpha (IFN-α), ribavirin (RBV), and a protease inhibitor is the current standard of care for HCV genotype 1 infection. In Chapter 4, Dash et al. discuss the hypothesis that a better understanding of interferon's and ribavirin's HCV clearance mechanisms will lead to improvement in treatment response and will reduce the burden of liver cirrhosis and primary liver cancer. They present an overview of HCV infection clearance by treatment with type I and type III interferon (IFN), ribavirin, and other new emerging antivirals. Dash et al. also review novel antiviral strategies that aim to eradicate HCV infection and reduce the burden of liver cirrhosis and primary liver cancer. Like HCV, hepatitis B virus (HBV) is also a leading risk factor for hepatocellular carcinoma (HCC), with more than 80% of HCC cases occurring in regions that are also endemic for hepatitis B. In Chapter 5, Maiwall and Sharma discuss the factors associated with increased risk of developing HCC in chronic HBV-infected subjects and the preventive strategies for HBV-related HCC including the role of vaccination, nucleoside and nucleotide analogs, and interferon.

Epstein–Barr virus (EBV) is a ubiquitous pathogen that has adopted a unique and effective strategy for infection, persistence, and spread in its only biological host—human beings. It infects more than 90% of the human population worldwide and is associated with a diverse array of proliferative diseases in different parts of the world. It leads to BL, nasopharyngeal carcinoma (NPC), posttransplant lymphomas, and Hodgkin's disease. In Chapter 6, Jha et al. present the biology and pathophysiology of EBV and discuss currently available techniques for preventing EBV infection as well as for treating EBV-mediated malignancies.

More than four decades have elapsed since the initial hypothesis regarding the possible role of human papillomavirus (HPV) in the genesis of human cancers was postulated. It is now a matter of common knowledge that HPV infection accounts for almost 100% of cervical cancers and that HPV is responsible for 5.2% of all cancers including 25% of head and neck cancers. In Chapter 7, Chaudhury et al. present the structure and properties of HPV, its association with cervical and neck and head cancers, and the prevention and therapeutics of these cancers. Several recent studies have reported that high-risk HPVs are also present in human breast cancer tissues. Although several investigations did not confirm this fact, in Chapter 8, Al Moustafa presents an overview of the presence and role of high-risk HPVs in human breast carcinogenesis and focuses on the function of E5, E6, and E7 oncoproteins of high-risk HPVs in the progression of breast cancer.

Genital infection by HPV is the most common sexually transmitted disease in the world. Its global prevalence among women without cervical abnormalities is 12%. Chapter 9, by Cobucci et al., presents research data on promising prospective genital HPV infection therapies along with a brief description of current and future therapies.

Another of the viruses that infect humans, human herpesvirus 8 (HHV-8)/Kaposi's sarcoma-associated herpesvirus (KSHV), is described as a novel gamma-2 herpesvirus closely related to the human gamma-1 herpesvirus, Epstein–Barr virus (EBV). In Chapter 10, El Hajj et al. discuss the pathogenesis of HHV-8 and the treatment of the diseases associated with it, such as primary effusion lymphoma and Castleman's disease. Herpes simplex virus 1 and 2 (HSV-1 and HSV-2), also known as human herpesvirus 1 and 2 (HHV-1 and HHV-2), are two members of the Herpesviridae virus family that infects humans. Although HSV-1 and HSV-2 are closely related, they differ in several aspects of their pathology. HSV-1 is primarily associated with throat and mouth diseases as well as ocular and genital infections, while HSV-2 is the leading cause of recurring genital herpes cases worldwide. HSVs currently have no cure, and some have developed drug resistance. However, in Chapter 11, Chu et al. discuss the importance of tea polyphenolic compounds, such as epigallocatechin gallate (a green tea polyphenol) and theaflavin (a black tea polyphenol), against these HSVs.

People with AIDS have a weak immune system and thus are at an increased risk of developing infections, lymphoma, and other types of cancer. Therefore, HIV is also considered a cancer-causing virus. The most common types of AIDS-related cancers are Kaposi sarcoma and non-Hodgkin lymphoma. Other AIDS-related cancers include Hodgkin's disease and cancers of the lung, mouth, cervix, and digestive system. Anti-HIV drugs target multiple pathways in the life cycle of HIV and provide a large source of potential anticancer drugs. HIV integrase (IN) is one of the major viral targets for the development of new anti-HIV drugs. In Chapter 12, Garg and Ko have presented a review of HIV-related cancers and repositioning efforts to target some HIV IN inhibitors as anticancer agents as well as quantitative structure–activity relationship (QSAR) studies on HIV IN inhibitors that might be of great help in the design of more effective HIV IN inhibitors relevant to cancer treatment.

Oncolytic (tumor-killing) viruses vis-à-vis oncoviruses represent a heterogeneous group of viruses that have been observed/engineered to preferentially replicate in and destroy tumor cells but not normal cells. In Chapter 13, Vähä-Koskela et al. present a brief history of oncolytic viruses and their mechanism of action and clinical development. The authors assess whether there is any rationale—either molecular or immunological—for deliberate targeting of oncovirus-associated disease by oncolytic viruses. In this regard, they discuss a limited but expanding number of studies that indicate that it may be feasible to deliberately infect oncovirus tumors with oncolytic viruses and that such "virus wars" may result in substantial therapeutic benefit and long-term tumor control. Whereas oncolytic viruses have shown significant promise in cancer therapy, histone deacetylase inhibitors have also shown great promise for inhibiting cell proliferation in various cancer cell lines. Therefore, current research is directed toward combined use of these approaches. A detailed description of this aspect is presented by Patil and Gupta in Chapter 14.

In this volume, an attempt has been made to present a detailed account of cancer-causing viruses—their structure, genotypes, and replication; the mechanisms of infection leading to human cancers; modes of infection prevention; and the treatment of the cancers produced by these viruses. This book will be of great value to those working on virology and cancers and in the area of biotechnology as well as to general readers interested in viral infections. As the editor of this book, I have greatly enjoyed reading all the chapters and hope the readers do as well. I gratefully acknowledge the interest and zeal of all the authors for contributing such important, timely, and useful material for this book.

**Satya P. Gupta**
*National Institute of Technical Teachers' Training and Research (NITTTR)*
*Bhopal, India*

# About the Editor

**Dr. Satya P. Gupta**, formerly a renowned professor of chemistry at Birla Institute of Technology and Science (BITS), Pilani, India, is presently a professor in the Department of Applied Science at the National Institute of Technical Teachers' Training and Research (NITTTR), Bhopal, India. Prior to this, he served as a distinguished professor in the Department of Pharmaceutical Technology at the Meerut Institute of Engineering and Technology (MIET), Meerut, India. Initially, he spent a few years at the Tata Institute of Fundamental Research (TIFR), Bombay, under Professor Girjesh Govil, where he worked on the structure and functions of biomembranes. He later moved to BITS, where he initiated work on the quantitative structure–activity relationship (QSAR), a very prominent area of medicinal chemistry. Dr. Gupta developed a deep understanding of QSAR, the modes of drug–receptor interactions, and the roles of physicochemical properties in drug–receptor interactions. Dr. Gupta earned an MSc in physical chemistry and PhD in quantum chemistry.

In 1985, Dr. Gupta was made a fellow of the National Academy of Sciences, India, and in 1989, he was awarded the coveted Ranbaxy Research Foundation Award. Dr. Gupta is one of the pioneers of QSAR. He has published approximately 200 papers in various journals of repute on the topic and has contributed dozens of authoritative reviews in prestigious journals such as *Chemical Reviews* (American Chemical Society), *Progress in Drug Research* (Birkhauser Verlag Basel), and *Current Medicinal Chemistry* (Bentham Science, the Netherlands and the United States). Dr. Gupta has also contributed several chapters in books produced by internationally reputed publishers such as J. R. Prous Science Spain, Elsevier Science BV (the Netherlands), Springer Heidelberg, and Springer Basel. He has edited several books published by Springer Heidelberg/Basel, namely, *Topics in Heterocyclic Chemistry,* Vol. 3 and 4 (2006); *Ion Channels and Their Inhibitors* (2011); *Matrix Metalloproteinase Inhibitors—Specificity of Binding and Structure–Activity Relationships* (2012); and *Hydroxamic Acids: A Unique Family of Chemicals with Multiple Biological Activities* (2013). In addition, Dr. Gupta has been the editor-in-chief of four international journals published by Bentham Science: *Cardiovascular and Hematological Agents in Medicinal Chemistry*, *Current Computer-Aided Drug Design*, *Current Enzyme Inhibition*, and *Current Bioinformatics*. Dr. Gupta also serves on the editorial boards of several journals.

In 1996, Dr. Gupta's book, *Quantum Biology* (New Age International Publishers, New Delhi), was highly acclaimed by theoretical chemists and biologists. His recent book, *QSAR and Molecular Modeling* (published in 2011 by Anamaya, New Delhi, in collaboration with Springer, the Netherlands), has become the most popular volume among those interested in QSAR and molecular modeling—the most fascinating area of drug design. This book is a result of his vast work experience in QSAR.

# Contributors

**Sandra D. Adams**
Department of Biology and Molecular Biology
Montclair State University
Montclair, New Jersey

**Nilanjan Adhikari**
Natural Science Laboratory
Division of Medicinal and Pharmaceutical Chemistry
Department of Pharmaceutical Technology
Jadavpur University
Kolkata, India

**Ala-Eddin Al Moustafa**
Department of Oncology
McGill University
Montreal, Quebec, Canada

and

Syrian Research Cancer Centre
Syrian Society against Cancer
Aleppo, Syria

**Luis A. Balart**
Department of Medicine
Tulane University Health Sciences Center
New Orleans, Louisiana

**Ali Bazarbachi**
Department of Internal Medicine
Faculty of Medicine
American University of Beirut
Beirut, Lebanon

**Vincenzo Cerullo**
Immunovirotherapy Group
Division of Biopharmaceutics and Pharmacokinetics
University of Helsinki
Helsinki, Finland

**Partha K. Chandra**
Department of Pathology
Tulane University Health Sciences Center
New Orleans, Louisiana

**Indranil Chatterjee**
Department of Surgical Oncology
Saroj Gupta Cancer Centre and Research Institute
Kolkata, India

**Susri Ray Chaudhuri (Guha)**
Drug Development, Diagnostics and Biotechnology Division
CSIR–Indian Institute of Chemical Biology
Kolkata, India

**Rahul Roy Chowdhury**
Department of Gynaecological Oncology
Saroj Gupta Cancer Centre and Research Institute
Kolkata, India

**Tin-Chun Chu**
Department of Biological Sciences
Seton Hall University
South Orange, New Jersey

**Ricardo Ney Oliveira Cobucci**
Department of Medicine
Potiguar University
Natal, Brazil

**Ana Katherine da Silveira Gonçalves**
Department of Obstetrics and Gynecology
Federal University of Rio Grande do Norte
Natal, Brazil

**Srikanta Dash**
Departments of Pathology and Medicine
Tulane University Health Sciences Center
New Orleans, Louisiana

**Janaina Cristiana de Oliveira Crispim**
Department of Pharmaceutical Sciences
Federal University of Rio Grande do Norte
Natal, Brazil

**Jerome Deval**
Alios BioPharma, Inc.
San Francisco, California

**Hiba El Hajj**
Department of Internal Medicine
Faculty of Medicine
American University of Beirut
Beirut, Lebanon

**Rajni Garg**
Computational Science Research Center
San Diego State University
San Diego, California

**Robert F. Garry**
Department of Microbiology
Tulane University Health Sciences Center
New Orleans, Louisiana

**Vertika Gautam**
Department of Chemistry
University of Malaya
Kualalumpur, Malaysia

**Satya P. Gupta**
National Institute of Technical Teachers' Training and Research (NITTTR)
Bhopal, India

**Amit Kumar Halder**
Natural Science Laboratory
Division of Medicinal and Pharmaceutical Chemistry
Department of Pharmaceutical Technology
Jadavpur University
Kolkata, India

**Martin Huber**
Humapharm
Liestal, Switzerland

**Tarun Jha**
Natural Science Laboratory
Division of Medicinal and Pharmaceutical Chemistry
Department of Pharmaceutical Technology
Jadavpur University
Kolkata, India

**Gene M. Ko**
Computational Science Research Center
San Diego State University
San Diego, California

**Fabrice Le Boeuf**
Centre for Innovative Cancer Research
Ottawa Hospital Research Institute
Ottawa, Ontario, Canada

**Lee H. Lee**
Department of Biology and Molecular Biology
Montclair State University
Montclair, New Jersey

**Rakhi Maiwall**
Assistant Professor of Hepatology
Institute of Liver and Biliary Sciences
New Delhi, India

**Raghida Abou Merhi**
Department of Internal Medicine
Faculty of Medicine
American University of Beirut
Beirut, Lebanon

**Vaishali M. Patil**
Department of Pharmacy
Bharat Institute of Technology
Meerut, India

**Kurt Ramazan**
Department of Medicine
Tulane University Health Sciences Center
New Orleans, Louisiana

**Anirban Roy**
Drug Development, Diagnostics and Biotechnology Division
CSIR–Indian Institute of Chemical Biology
Kolkata, India

**Manoj Kumar Sharma**
Associate Professor of Hepatology
Institute of Liver and Biliary Sciences
New Delhi, India

**Snehasikta Swarnakar**
Drug Development, Diagnostics and Biotechnology Division
CSIR–Indian Institute of Chemical Biology
Kolkata, India

**Markus Vähä-Koskela**
Translational Virology Group
Haartman Institute
University of Helsinki
Helsinki, Finland

**Jun-ichirou Yasunaga**
Laboratory of Virus Control
Institute for Virus Research
Kyoto University
Kyoto, Japan

# 1 Cancer-Causing Viruses
## *An Introduction*

*Satya P. Gupta and Vertika Gautam*

## 1.1 INTRODUCTION

Till the dawn of 19th century, viruses were classified as small infectious agents that, unlike cells and bacteria, pass through fine-pore filters. After a decade, Peyton Rous realized a relation between cancer and viruses. In 1966, he was awarded the Nobel Prize in recognition for his seminal discovery of tumor-inducing viruses. Later, a series of studies found viruses as physical and chemical initiators of cancer. Rapid progress in tumor virology has taken place over the last 40 years. According to the World Health Organization (WHO) cancer report 2008, some cancer-causing viruses are responsible for up to 20% of cancer deaths worldwide. Approximately 20%–30% of all cancers can be linked to viral infections (Goedert 2000).

Cancers are the result of a disruption of the normal restraints on cellular proliferation. There are several cellular genes in which mutation or some other disruption of their expression can lead to unrestrained cell growth. There are two classes of such genes in which altered expression can lead to loss of growth control:

1. Oncogenes: These genes are stimulatory for growth and cause cancer when hyperactive. Mutations in these genes will be dominant.
2. Tumor suppressor genes or anti-oncogenes: These genes inhibit cell growth and cause cancer when they are turned off. Mutations in these genes will be recessive.

Viruses are involved in cancers because they can either carry a copy of one of these genes or alter expression of the cell's copy of one of them. Such viruses are called cancer-causing viruses, oncoviruses, or tumor viruses. The majority of tumor virus-infected individuals do not develop cancer; many years may pass between initial infection and tumor appearance. Additional cofactors, such as host immunity and chronic inflammation, as well as additional host cellular mutations, play an important role in the transformation process. Actually, the long-term interactions between virus and host are the key features of the oncogenic viruses because they set the stage for a variety of molecular events that may contribute to eventual virus-mediated carcinogenesis (Cohen et al. 1999).

## 1.2 FAMILIES OF CANCER-CAUSING VIRUSES

Broadly, there are two different families of oncogenic viruses that depend on the type of nucleic acid present in the viral genome and the type of strategy to induce malignant transformation. One is the RNA virus family with two subfamilies known as *Retroviridae* and *Flaviviridae*, and the other is the DNA virus family with the subfamilies *Hepadnaviridae*, *Herpesviridae*, and *Papillomaviridae*. These two major classes of viruses utilize diverse mechanisms in cancer development but often have one aspect of their life cycle in common—the ability to integrate their own genome into that of the host cell. However, such integration is not a prerequisite for tumor formation. Moreover, tumor viruses have the ability to infect but not kill their host cell. So, unlike other disease causing viruses, they establish long-term persistent infections.

Tumor viruses induce oncogenesis through manipulating an array of different cellular pathways. These viruses initiate a series of cellular events that lead to immortalization and proliferation of the infected cells by disrupting the mitotic checkpoint upon infection of the host cell. This is often accomplished by functional inhibition or proteasomal degradation of many tumor suppressor proteins by virally encoded gene products. The virally infected cells can either be eliminated via cell-mediated apoptosis or persist in a state of chronic infection. Importantly, the chronic persistence of infection by tumor viruses can lead to oncogenesis (Saha et al. 2010). The two different families of tumor viruses or cancer-causing viruses are described as follows.

### 1.2.1 RNA Tumor Viruses/Retroviruses

RNA tumor viruses are retroviruses. A retrovirus is an RNA virus that replicates in a host cell through the process of reverse transcription. First, it uses its own reverse transcriptase enzyme to produce DNA from its RNA genome—the reverse of the usual pattern, thus, *retro* (backward). This new DNA is then incorporated into the host's genome by an enzyme called integrase. The cell then treats the viral DNA as part of its own instructions, making the proteins required to assemble new copies of the virus (Figure 1.1). Retroviruses are enveloped viruses that belong to the viral family of *Retroviridae*. The incorporation of the viral DNA into the host's genome converts proto-oncogenes—the cellular genes that code for components of the cellular machinery that regulates cell proliferation, differentiation, and death—into oncogenes in two ways: the gene sequence may be altered or truncated so that it codes for proteins with abnormal activity, or the gene may be brought under the control of powerful viral regulators that cause its product to be made in excess or in inappropriate circumstances. Retroviruses may also exert similar oncogenic effects by insertional mutation when DNA copies of the viral RNA are integrated into the host-cell genome at a site close to or even within proto-oncogenes. Cellular oncogenes are expressed by the cell at some period in the life of the cell, often when the cell is growing, replicating, and differentiating normally. They are usually proteins that are involved in growth control. Cellular oncogenes are highly conserved.

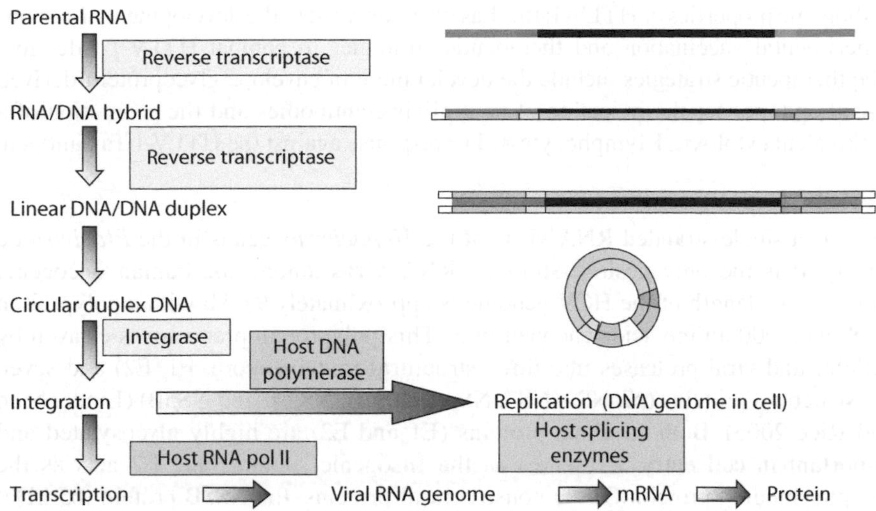


**FIGURE 1.1 (See color insert.)** The pathway of RNA virus infection.

So far only two RNA viruses have been recognized: human T-cell leukemia virus of type 1 (HTLV-1) and hepatitis C virus (HCV), described in the following sections.

#### 1.2.1.1 Human T-Cell Leukemia Virus of Type 1

Human T-cell leukemia virus of type 1 (HTLV-1) is the first human retrovirus to be discovered that is clearly associated with human malignancy (Poiesz et al. 1980; Yoshida et al. 1982). It is a delta-type complex retrovirus and is the etiologic agent of adult T-cell leukemia (ATL). It is endemic to Japan, South America, Africa, and the Caribbean (Wong-Staal and Gallo 1985; Gessian et al. 1991). Although exact cellular events involved in the leukemogenesis of ATL by HTLV-1 are not well known, a variety of steps, including virus, host cell, and immune factors, have been implicated (Matsuoka 2003). Several studies, such as Pozzatti et al. (1990), have suggested that Tax, a multifunctional viral accessory protein, is a major protein, transforming the activities of HTLV-1 to immune modulation, apoptosis regulation, proliferation, and genetic instability.

To date, four types of HTLVs (HTLV-1, HTLV-2, HTLV-3, and HTLV-4) have been identified, of which HTLV-1 is the most widely studied for its association with human malignancy. Its genome is diploid, composed of two copies of a single-stranded RNA virus whose genome is copied into a double-stranded DNA form that integrates into the host-cell genome, at which point the virus is referred to as a provirus. The association of HTLV-1 with ATL was demonstrated by the presence of extracellular type C viral particles in cells from a patient with ATL (Hinuma et al. 1981; Gallo 2005) that has been confirmed by further studies (Matsuoka and Jeang 2007; Boxus and Willems 2009). HTLV-1 proviral DNA can be detected in virtually every ATL patient tested. Attempts have been made to understand the biological and pathogenic properties of HTLV-1; this has ultimately led to the development of various experimental vaccination and therapeutic strategies to combat HTLV-1 infection. The therapeutic strategies include the development of envelope glycoprotein-derived B-cell epitopes for the induction of neutralizing antibodies and the generation of a multivalent cytotoxic T-lymphocyte (CTL) response against the HTLV-1 Tax antigen.

#### 1.2.1.2 Hepatitis C Virus

HCV is a single-stranded RNA virus of the *Hepacivirus* genus in the *Flaviviridae* family. It is the only positive-stranded RNA virus among the human oncogenic viruses. The length of the HCV genome is approximately 9.6 kb with a polyprotein of about 3,000 amino acids encoded in it. This polyprotein precursor is cleaved by cellular and viral proteases into three structural proteins (core, E1, E2) and seven nonstructural proteins (p7, NS2, NS3, NS4A, NS4B, NS5A, and NS5B) (Lindenbach and Rice 2005). Both envelope proteins (E1 and E2) are highly glycosylated and important in cell entry. E1 serves as the fusogenic subunit, and E2 acts as the receptor-binding protein. Of the nonstructural proteins, the NS5B protein (65 kDa) is the viral RNA-dependent RNA polymerase, which has the key function of replicating the HCV's viral RNA by using the viral positive RNA strand as its template. NS5B catalyzes the polymerization of ribonucleoside triphosphates (rNTP) during RNA replication (Moradpour et al. 2007; Rigat et al. 2010).

Persistent HCV infection can lead to hepatitis, hepatic steatosis, cirrhosis, and hepatocellular carcinoma (HCC) (Saito et al. 1990; Alter 1995; Lauer and Walker 2001; Poynard et al. 2003; Pawlotsky 2004). HCV infection also causes chronic inflammation and fibrosis, which can progress to cirrhosis. Though the chronic inflammation and cirrhosis are thought to play key roles in HCV-induced carcinogenesis, the exact mechanism of carcinogenesis is not well understood (Fattovich et al. 2004). However, the immune modulation by quasispecies variation, apoptosis regulation through core protein, proliferation through NS2, and genetic stability through NS5B are thought to be the major biological activities involved in the induction of carcinogenesis by HCV (McLaughlin-Drubin and Munger 2008).

### 1.2.2 The DNA Tumor Viruses

The family of DNA tumor viruses consists of papilloma, polyoma, adeno, hepadna, and herpesviruses and produces tumors of different types in various species. DNA tumor viruses are thought to play a role in the pathogenesis of about 15%–20% of human cancers. These include Burkitt's lymphoma (BL), nasopharyngeal carcinoma (NPC), immunoblastic lymphomas in immunosuppressed individuals, and a proportion of Hodgkin's lymphomas that are all associated with the Epstein–Barr virus (EBV) of the herpes family and with liver carcinoma associated with chronic hepatitis B virus (HBV) infection. Unlike retroviral oncogenes, DNA virus oncogenes are essential components of the viral genome and have no counterpart in the normal host cells. Also, they are of viral, not cellular, origin and are necessary for the replication of the virus (Damania 2007). The analysis of DNA tumor virus oncoproteins has not only elucidated the etiology of several human diseases but also has revealed the mechanisms controlling mammalian cell growth, ultimately leading to discovery of a cellular tumor suppressor gene. Most of the DNA tumor viruses, such as human papillomavirus (HPV), HBV, EBV, and Kaposi's sarcoma-associated herpesvirus (KSHV), cause malignancy in their natural hosts, but a few, such as human adenoviruses, transform cultured cells and cause tumors only in heterologous animal models.

The transforming ability of DNA tumor viruses has been traced to several viral proteins that cooperate to stimulate cell proliferation, overriding some of the normal growth control mechanisms in the infected cell and its progeny. Some of these viral proteins bind to the protein products of two key tumor suppressor genes of the host cells, the retinoblastoma and the p53, deactivating them and thereby permitting the cell to replicate its DNA and divide. Studies on small DNA viruses, such as adenoviruses, polyomaviruses, and papillomaviruses, have helped in elucidating the underlying molecular mechanisms of virus-induced cell transformation. Although these small viruses differ in their evolution, they have striking similarity in their transforming functions, which emphasize their mutual need to utilize the host cell's replication machinery for efficient viral replication (McLaughlin-Drubin and Munger 2008). The following different types of DNA tumor viruses have been recognized.

#### 1.2.2.1 Epstein–Barr Virus

EBV or human herpesvirus 4 is a B-lymphotropic gammaherpesvirus that belongs to a subfamily of *Lymphocryptovirus* (LCV) genus that infects more than 95% of the world's population (Evans and Niederman 1989). It was the first human virus to be classified as carcinogenic. The majority of EBV infection occurs during childhood without causing overt symptoms. The most common manifestation of primary infection with this organism is acute infectious mononucleosis, a self-limiting lymphoproliferative disease that most frequently affects adolescents and young adults. Classic symptoms include sore throat, fever, and lymphadenopathy that may or may not be accompanied by a faint, generalized maculopapular rash. Humans are the only known reservoir of EBV. The relatively low incidence of EBV-related tumors compared with this prevalence of infection shows that there are definitely many other factors that contribute to tumor development.

After exposure, EBV establishes a persistent infection in the host and is intermittently shed in oropharyngeal secretions. In uncomplicated EBV infections, the main target cells are B lymphocytes that express CD21, which along with HLA class II molecules serve as the viral entry receptor and coreceptor, respectively. Infected B lymphocytes resemble antigen-activated B cells. The virus can also infect epithelial cells and may initially replicate in the oropharynx before infecting B cells. Naive B cells become infected in local lymphoid tissue, most often Waldeyer's tonsillar ring, and establish latent infected memory B-cell pools that migrate to other lymphoid tissues. Periodic reactivations of latently infected memory B cells are thought to facilitate infection of epithelial cells in the oropharynx that permits shedding of virus into saliva for transmission to new hosts. Latently infected memory B cells are long-lived and, due to low levels of expression of viral proteins, they escape detection by CD8+ cytotoxic T cells (Kutok and Wang 2006); thus, persistence of EBV in the body is lifelong. EBV gene expression in these cells is limited to a B-cell growth program, termed Latency III, that includes nine latency-associated viral proteins, including six nuclear antigens (Epstein–Barr nuclear antigen [EBNA]-1, -2, -3A, -3B, -3C, and -LP) and three membrane proteins (latent membrane protein [LMP]-1, -2A, and -2B) that include LMP-1 and LMP-2a/b; EBNA-1, -2, -3a, -3b, -3c, and -LP; miRNAs; BamHI A Rightward Transcripts, BamHI: a type II restriction endonuclease from *Bacillus amyloliquefaciens*, having the capacity for recognizing short sequences (6 b.p.) of DNA and specifically cleaving them at a target site (BARTs); and EBV-encoded RNAs (EBERs) (Gruhne et al. 2009). These cells are eliminated by a robust immune response to EBNA-3 proteins, resulting in Latency I, a reservoir of latently infected resting memory B cells expressing only EBNA-1 and LMP-2. The differentiation of memory cells to plasma cells results in reactivation of the replication phase of the viral life cycle that includes expression of Latency III gene products. In addition, there is likely another amplification step by reinfection of the epithelial cells followed by shedding virus in the saliva to the next host (McLaughlin-Drubin and Munger 2008). Apart from B cells, which are known to be primarily infected by EBV, other non-B-cell NHLs associated with EBV infection include T-cell lymphoproliferative disorders that include a subset of peripheral T-cell lymphomas such as extranodal nasal-type NK/T-cell lymphoma.

All phases of the EBV life cycle are associated with human disease (Ocheni et al. 2010). In immunocompetent individuals, infected cells increase in number, and eventually B-cell growth control pathways are activated, inducing transformation and leading to malignancies such as BL, NPC, and Hodgkin's lymphoma. Other types of malignancies include immunoblastic lymphomas, such as X-linked and acquired immunodeficiency syndrome (AIDS)-associated lymphoproliferative disorders, and posttransplant lymphoproliferative disorders (PTLDS).

#### 1.2.2.2 Hepatitis B Virus

HBV is a small DNA virus that is maintained in circular conformation within the virions. HBV replicates via a reverse transcription of an RNA pregenome and predominantly infects human hepatocytes in the liver. This marked hepatotropism is a prominent feature of hepadnaviruses. The majority of individuals that are chronically infected with HBV ultimately experience severe liver disease and are at a high risk of developing HCC (Beasley et al. 1981). HBV is supposed to be a major etiological factor in the development of HCC. HBV-infected individuals develop chronic active hepatitis (CAH) that can in turn lead to cirrhosis, liver failure, or HCC (Feitelson 1999; Lok 2004). Patients with cirrhosis are more prone to develop HCC than those without cirrhosis (Brechot 2004; Tanaka et al. 2005). It is suggested that there is a direct oncogenic contribution of HBV to the carcinogenic process. The cause of HBV-associated HCC is supposed to be a combination of HBV encoded oncogenic activities along with the synergistic effects of chronic inflammation. It is estimated that more than 400 million people worldwide are infected with HBV.

In the life cycle of HBV, the replication of HBV is dependent on reverse transcription, which is a trait of retroviruses. However, unlike retroviruses, integration of the viral genome into the host chromosome is not necessary for viral replication but does allow for persistence of the viral genome (Feitelson and Lee 2007). The 3.2 kb HBV DNA genome codes for four genes named C, S, P, and X. The C gene codes for the core protein and the serum e antigen; the S gene codes for three related viral envelope proteins known as surface antigen; the P gene codes for the viral DNA polymerase; and the X gene codes for a 16.5 kDa protein. The viral transcription is governed by four promoters and two enhancer elements that are located in the HBV genome. The compact nature of this small genome necessitates an extensive overlapping arrangement of the genetic information. Therefore, it is likely that a number of complex mechanisms regulate the temporal and differential expression of the viral RNAs. Among the proteins encoded by the HBV genome, the X gene product, termed HBx, which is essential for productive infection of the mammalians HBV (Chen et al. 1993; Zoulim et al. 1994), has drawn considerable attention due to its pleiotropic functions. HBx does not directly bind with DNA but functions via protein–protein interaction (Maguire et al. 1991). It can activate cellular signaling pathways including mitogen-activated protein kinase (MAPK), c-Jun N-terminal kinase (JNK), and Src tyrosine kinases (Benn and Schneider 1994; Klein and Schneider 1997). It also regulates proteasomal function (Huang et al. 1996; Hu et al. 1999; Zhang et al. 2000), affects mitochondrial function (Rahmani et al. 2000; Huh and Siddiqui 2002), and modulates calcium homeostasis (Bouchard et al. 2001; Chami et al. 2003), and its expression causes genomic instability (Forgues et al. 2003).

HBV-associated liver carcinogenesis is thought to be a multifactorial process in which the mutations in core protein preS of HBV lead to immune modulation; HBx leads to apoptosis regulation and proliferation; and preS mutation followed by HBx action leads to genetic instability. Further studies are, however, necessary to find the molecular mechanisms for the development of HBV-associated HCC.

#### 1.2.2.3 Human Papillomavirus

HPVs belong to a family of small DNA viruses. Their circular double-stranded DNA genomes of less than 8000 base pairs are packaged into nonenveloped particles of approximately 55 nm. They show a marked preference for infection of epithelial cells and have been isolated from a wide range of hosts, including humans. More than 200 different HPV types have been described to date (de Villiers et al. 2004), and they can be classified into two groups: mucosal and cutaneous. Of these, mucosal HPVs are further designated into low-risk HPVs (HPV-6 and HPV-11) that can cause benign hyperplastic lesions (warts) that have little or no probability for malignant progression and high-risk HPVs (HPV-16 and HPV-18) that are associated with the vast majority of cervical carcinoma and a pathologically distinct group of oropharyngeal tumors.

HPV infects squamous epithelial cells by interacting with putative host-cell surface receptors such as heparan sulfate proteoglycans (Patterson et al. 2005) and alpha 6 integrins (Doorbar 2006). Depending on the HPV type, the virus enters the cell via either clathrin-mediated (Day et al. 2003) or caveolin-mediated endocytosis (Bousarghin et al. 2003). After uncoating, the virus delivers its genome of eight genes to the host-cell nucleus, which is expressed as autonomous replicating episomal or extrachromosomal elements. Viral DNA is replicated using host-cell machinery and then segregated into progeny cells. Afterward, one of the daughter cells remains in the basal layer to serve as a reservoir for the replication of more viral DNA, while the other daughter cell migrates suprabasally; HPV inhibits the cell from its normal exit from the cell cycle. Differentiation of infected cells is necessary for productive infection of the host cell and is coupled with expression of late viral capsid genes for completion of the viral life cycle (Doorbar 2005).

The HPV genome is comprised of eight genes with multiple promoters and a number of splice variants that are expressed either early (E) or late (L) in the HPV life cycle. The early genes encode nonstructural proteins that participate in DNA replication, transcriptional regulation, cell transformation, and viral assembly and release. The late (L) genes, L1 and L2, encode viral capsid proteins. An additional 1000 base pairs noncoding region of the 8000 base pairs HPV genome contains transcriptional regulatory sequences and the viral origin of replication (Bernard 2002). All these E and L genes and their functions are shown in Table 1.1.

Papillomaviruses are supposed to have two modes of replication: stable and vegetative. The former mode exists as a circular episomal genome in the cells of the basal and parabasal layers of low-risk HPV-infected squamous epithelia, while the latter form occurs in highly differentiated cells at the epithelial surface. E6 and E7, the first viral genes to be expressed, are involved in cell transformation in HPV high-risk types. E6 complexes with the cellular ubiquitin ligase. The E6AP and the E6AP/E6 complex act as a p53-specific ubiquitin protein ligase to increase the rate of p53 degradation, thereby negating the antiproliferative and pro-apoptotic functions of p53 in cases of DNA damage and cellular stress (Scheffner et al. 1993). E7, similar to E6, is also an oncoprotein. E7 complexes with hypophosphorylated pocket proteins pRb, p107, or p130, facilitating the release of the transcription factor E2F, which then constitutively activates DNA synthesis and cell proliferation (Munger and

**TABLE 1.1**
**HPV Genes and Their Functions**

| Gene | Function |
|---|---|
| E1 | DNA-dependent ATPase, ATP-dependent helicase: Allow unwinding of the viral genome and act as an elongation factor for DNA replication. |
| E2 | Responsible for recognition and binding of origin of replication. Exists in two forms: full length (transcriptional transactivator) and truncated (transcriptional repressor). The ratio of these found in the heterotrimeric complex formed before complexing with E1 regulates transcription of viral genome. |
| E3 | ??? |
| E4 | Late expression: C terminal binds intermediate filament, allowing release of virus-like particles. Also involved in transformation of host cell by deregulation of host-cell mitogenic signaling pathway. |
| E5 | Obstruction of growth suppression mechanisms, for example, EGF receptor; activation of mitogenic signaling pathways via transcription factors: c-Jun and c-Fos (important in ubiquitin pathway degradation of p53 complex by E6). Inactivation of p21 (p53-induced expression halts cell cycle until DNA is proofread for mutations). |
| E6 | E6 transformation of host cell by binding p53 tumor suppressor protein. |
| E7 | Transforming protein: Binds to pRB/p107. |
| E8 | ??? |
| L1 | Major capsid protein: Can form virus-like particles. |
| L2 | Minor capsid protein: Possible DNA-packaging protein. |

*Source:* www.microbiologybytes.com/virology/Papillomaviruses.html, accessed on April 8, 2009.

Howley 2002). The Rb tumor suppressor protein normally binds to and inactivates E2F. E7 has also been shown to inactivate the cyclin-dependent kinase (Cdk) inhibitors, p21/Cipl and p27/Kipl, and to complex with cyclins A and E (Duensing and Munger 2004). However, in low-risk HPV types, such as HPV-6 and HPV-11, E6 may or may not bind to p53. In either case, it does not stimulate p53 degradation, and E7 does not bind pRb with high affinity. El and E2 are both DNA-binding proteins. El is a DNA helicase/ATPase involved in viral genome replication, while E2 is a transcription factor that binds to four sites within the viral noncoding region and recruits El to the viral origin of replication (Hughes and Romanos 1993). El and E2 have been shown to be sufficient for transient replication of the HPV genome.

In addition, increased levels of active El and E2 have been correlated with increased viral copy number (Del Vecchio et al. 1992). E2 also binds to the early promoter and decreases the expression of E6 and E7. Therefore, loss of E2 during integration of viral DNA into the host genome is the first stage in transformation (Talora et al. 2002). E2 exists in a full-length form that acts as a transcriptional activator of the early promoter at low levels of E2 expression and a truncated form at high levels of E2 expression that acts as a transcriptional repressor of the early promoter production of oncoproteins E6 and E7 (Bernard et al. 1989; Ushikai et al. 1994). At higher concentrations of E2, displacement of basal transcription factors is thought to be responsible for the inactivation of the promoter and for the repression of E6 and E7 expression (Doorbar 2006). Additionally, a switch in promoter usage to the differentiation-dependent late promoter (which is not repressible by E2) during differentiation is due to changes in cellular signaling and not to genome amplification. This switch results in increased levels of El and E2, and subsequent E4 gene sequences are not highly conserved among the various HPV types. Although E4 has been shown to associate with cytokeratins to destabilize the cytokeratin network and to aid in viral release at the epithelial cell layer surface, its primary function is still unknown (Sterling et al. 1993). However, E4 has been shown to cause G2 cell cycle arrest and to antagonize E7-mediated cell proliferation. Researchers also have hypothesized that E4 associates with E2 (Doorbar 2006). E5 proteins are hydrophobic transmembrane proteins that seem to be weakly involved in cellular transformation and that modulate the activity of the epidermal growth factor (EGF) receptor tyrosine kinase signaling pathway (Dimaio and Mattoon 2001). E5 is present on plasma and on Golgi and endoplasmic reticulum (ER) membranes.

Researchers also believe that E5 associates with the vacuolar ATPase and delays endosomal acidification in addition to inhibiting gap junction mediated cell–cell communication in keratinocytes. This results in an increase in EGF receptor recycling to the cell membrane and an inability to respond to paracrine growth inhibitory signals (Thomsen et al. 2000; Doorbar 2006). E5 expression in high-risk HPV types is thought to occur early during carcinogenesis because expression is inhibited by incorporation of the viral genome during progression to cervical cancer (Oelze et al. 1995). The functions of E3 and E8 are still unknown. As infected host cells become terminally differentiated, the late genes, Ll and L2, which encode the major and minor structural capsid proteins, are expressed until the viral copy number increases enough that viral particles are sloughed off at the skin surface to infect other cells and hosts. L2 expression precedes that of L1, and the presence of E2 is thought to increase the efficiency of this process (Zhao et al. 2000). The lytic release of viral progeny requires terminal differentiation of stratified epithelium, a prerequisite for the activation of differentiation-dependent late viral promoters. Most HPV infections are latent or asymptomatic in immunocompetent hosts but can become reactivated in cases of cell-mediated immunosuppression replication of the viral genome (Doorbar 2006).

Progression to cancer with many HPV types results after years of active infection. Although it has been shown that E6 and E7 expression is not sufficient to transform human keratinocytes in culture, E6 inactivation of the p53 response to DNA damage and cellular stress and repeated E7-mediated S phase entry and cell proliferation do result in the accumulation of multiple mutations in the host genome (Hawley-Nelson et al. 1989; Doorbar 2006). However, cases of rapid onset of carcinogenesis after initial infection have also been noted (Woodman et al. 2001). The highest incidences of HPV-related cancers have been found to be associated with cancer of the cervix in females and of the anus in homosexual men (Bosch and de Sanjosé 2003). Research has suggested that HPV can more easily access the basal epithelial cells at both of these sites due to the presence of reserve cells in the transformation zone that will eventually take on the basal cell phenotype (Doorbar 2006). Integration of HPV DNA at specific sites in the host-cell genome has been suggested to be an early and critical step in cancer progression (Bosch and de Sanjosé 2003). The E6 and E7 oncoproteins of HPV are supposed to mechanistically contribute to initiation and progression of cervical cancer. The participation of certain flanking sequences in the integration of viral DNA has also been suggested to be a step in carcinogenesis (Yu et al. 2005).

#### 1.2.2.4 Kaposi's Sarcoma-Associated Herpesvirus/Human Herpesvirus 8

KSHV is the eighth human herpesvirus, the formal name of which according to the International Committee on Taxonomy of Viruses (ITCV) is human herpesvirus 8 (HHV-8). This virus causes Kaposi's sarcoma, a cancer commonly occurring in patients with AIDS (Boshoff and Weiss 2002), as well as primary effusion lymphoma (PEL) (Cesarman et al. 1995) and some types of multicentric Castleman's disease. In the 1980s, frequent occurrence of aggressive Kaposi's sarcoma (KS) was noted in AIDS patients and, subsequently, an increased incidence rate of KS was noticed specifically among homosexual human immunodeficiency virus (HIV)-infected individuals (Sarid and Gao 2011). This led to the suggestion that KS might be caused by a sexually transmitted agent distinct from HIV (Beral et al. 1990, 1992; Cohen et al. 2005). The use of advanced molecular techniques led Chang et al. (1994) to identify the infectious agent responsible for KS, which was named as KSHV. Shortly after its discovery, KSHV was shown to be associated with PEL (a rare subgroup of B-cell non-Hodgkin lymphoma) and a subset of multicentric Castleman's disease (Cesarman et al. 1995; Soulier et al. 1995). Among patients with AIDS, KSHV infection was shown to precede KS onset, corroborating a causative role of KSHV in KS (Whitby et al. 1995; Gao et al. 1996; Moore et al. 1996).

KSHV is a large double-stranded DNA virus with a protein that packages its nucleic acids, called capsid, which is then surrounded by an amorphous protein layer called tegument, and is finally enclosed in a lipid envelop derived in part from the cell membrane. KSHV has a genome of about 165,000 nucleic acid bases in length. KSHV contains several proteins that promote cell proliferation and survival and thus contribute to cellular transformation. These proteins include open reading frames 73-71 (Orf 73-71), locus-encoded latency-associated antigen (LANA, Orf 73), viral cyclin (v-cyclin, Orf 72), viral FLICE inhibitory protein (vFLIP, Orf 71), viral interferon regulatory factor 1 (vIRF-1), and the kaposin/K12 gene. Among the other proteins that have been mentioned as contributing to the transforming activities of KSHV/HHV-8 are modulator of immune response (MIR), complement control protein homolog (CCPH), viral G-protein-coupled receptor (vGPCR), latency-associated membrane protein (LAMP), and viral B-cell lymphoma 2 (vBlc-2) (McLaughlin-Drubin and Munger 2008).

### 1.2.3 Miscellaneous

In addition to the aforementioned RNA and DNA viruses, there are some oncoviruses that have been identified, but their association with human cancer is yet to be firmly established. Some of them are described here.

#### 1.2.3.1 Polyomaviruses

Polyomaviruses belong to *Polyomaviridae* family, which is a group of nonenveloped, small, double-stranded DNA viruses. They are potentially oncogenic, and the first polyomavirus, known as mouse polyomavirus (Gross 1953; Stewart et al. 1958), is one of the most aggressive oncogenic viruses. It causes tumors in almost every tissue of susceptible mouse strains (Eddy 1969; Gross 1983). The name polyoma refers to the ability of viruses to produce multiple (poly-) tumors (-oma). Polyomaviruses have approximately 5000 base pairs, a circular genome and icosahedral shape, and do not have a lipoprotein envelope.

Several polyomaviruses have been found in humans. Four of these viruses, John Cunningham (JC), BK, Karolinska Institute (KI), and Washington University (WU), are closely related to simian virus 40 (SV40), the second polyomavirus discovered after the mouse polyoma and isolated as a contaminant of the early batches of the polio vaccine, which was produced in African green monkey kidney cells (Sweet and Hilleman 1960). Therefore, infection of these four viruses can be confused with SV40 infection (zur Hausen 2003; Poulin and DeCaprio 2006). Infection with BK and JC viruses is widespread in humans and is usually subclinical. However, in immune-suppressed patients, the former is associated with renal nephropathy (Randhawa and Demetris 2000; Hirsch et al. 2002) and the latter can cause progressive multifocal leukoencephalopathy (Safak and Khalili 2003; Berger and Koralnik 2005). However, no conclusive proof exists whether either of them causes or acts as a cofactor in human cancer.

The Merkel cell carcinoma (MCC) was the first human cancer that was found to harbor an integrated polyomavirus genome. Merkel cell polyomavirus (MCPyV) is highly divergent from the other human polyomaviruses and is most closely related to murine polyomavirus. After it was cloned, several studies suggested a casual role of this virus in MCC.

Polyomaviruses replicate in the nucleus of the host. Because their genomic structure is homologous to that of the mammalian host, they utilize the host machinery for replication. Their replication occurs in two distinct phases: early and late gene expression, separated by genome replication. The early gene expression is responsible for the synthesis of nonstructural proteins, whereas the late gene expression synthesizes the structural proteins, which are required for the viral particle composition. Prior to genome replication, there occur the processes of viral attachment, entry, and uncoating. Cellular receptors for polyomaviruses are sialic acid residues of gangliosides. The attachment of polyomaviruses to host cells is mediated by viral protein 1 (VP1) via the sialic acid attachment region. Then polyomavirus virions are endocytosed and transported first to the endoplasmatic reticulum where a conformational change occurs revealing VP2. After this, the virus is exported to the nucleus for which the mechanism is not known.

#### 1.2.3.2 Adenoviruses

Adenoviruses belong to the family of *Adenoviridae*. They are so named because they were initially isolated from human adenoids (in 1953). They are medium-sized (90–100 nm), nonenveloped (without an outer lipid bilayer) viruses with an icosahedral nucleocapsid containing a double-stranded DNA genome. In humans, more than 50 different adenovirus serotypes have been described. Adenoviruses are able to replicate in the nucleus of mammalian cells using the host's replication machinery. Once the virus has successfully gained entry into the host cell, the endosome acidifies, altering virus topology by causing capsid components to disassociate. These changes as well as the toxic nature of the pentons result in the release of the virions into the cytoplasm. They are then transported to the nuclear pore complex with the help of cellular microtubules, whereby the adenovirus particles disassemble. Viral DNA is subsequently released, which can enter the nucleus via the nuclear pore (Meier and Greber 2004). After this, the DNA associates with histone molecules. Thus, viral gene expression can occur, and new virus particles can be generated.

The transforming and oncogenic potential of adenoviruses has been ascribed to E1A and E1B oncoproteins, where the former is similar to SV40 TAg and the latter to high-risk HPV E7/E6 target pRb and p53. However, the possible role of adenoviruses in producing any kind of tumor is to be carefully ascertained. Adenoviruses typically cause respiratory illnesses such as the common cold, conjunctivitis (an infection in the eye), croup, bronchitis, or pneumonia. In children, adenoviruses usually cause infections in the respiratory tract and the intestinal tract.

#### 1.2.3.3 Torque Teno Virus

Torque teno virus (TTV) is a nonenveloped small virus containing a circular, single-stranded, negative-sense DNA genome. It belongs to the *Circoviridae* family. It was first found in a patient with non-A-E-hepatitis (Nishizawa et al. 1997). Within a few years after its discovery, TTV was found to have several genotypes (Niel et al. 2000) and to be highly prevalent globally among healthy individuals; up to 94% of subjects are viremic. Persistent infections and coinfections with several genotypes occur frequently. However, the pathogenicity of TTV is at present unclear. A study had suggested that a high TTV load may have prognostic significance in HCV-associated liver disease, but whether the high TTV load mediates HCV-associated disease progression remains to be ascertained. However, a relatively high prevalence of

TTV-related DNA sequences in human cancers, specifically gastrointestinal cancer, lung cancer, breast cancer, and myeloma, has been observed (de Villiers et al. 2002).

#### 1.2.3.4 Human Mammary Tumor Virus/Pogo Virus

The inspiration for searching for human mammary tumor virus (HMTV) came from the existence of mouse mammary tumor virus (MMTV). MMTV is a milk-transmitted retrovirus such as HTLV, HIV, and BLV (bovine leukemia virus). It belongs to the genus *Betaretrovirus* and was formerly known as Bittner virus because it was demonstrated by Bittner that MMTV can cause breast cancer in mice (Bittner 1936). Previously, MMTV was also known as "milk factor" because Bittner established the theory that it could be transmitted by cancerous mothers to young mice from a virus in their mother's milk (Bittner 1936). The majority of mammary tumors in mice are caused by MMTV. From several studies, it has been suggested that an MMTV-related virus, HMTV, sometimes also referred to as Pogo virus, could be associated with human breast cancer (Holland and Pogo 2004). Primary breast cancer cells have been found to produce HMTV particles in vitro (Melana et al. 2007). In fresh breast cancers, the entire HMTV genome, which contains hormone response elements, has also been observed (Liu et al. 2001). HMTV particles in primary breast cancer cells were reported to have morphogenic and molecular characteristics similar to MMTV (Melana et al. 2007). However, further studies failed to corroborate this (Bindra et al. 2007). Also, it could not be well established that HMTV makes any major contribution to the development of breast cancer.

#### 1.2.3.5 Human Endogenous Retrovirus

Human endogenous retrovirus (HERV) is an HIV-like retrovirus, but where HIV is exogenous (meaning that it is originated outside the organism's body), HERV is endogenous (meaning that it is the part of human genome). There are many families of endogenous retroviruses that exist throughout the human genome. Unlike HIV, these retroviruses purportedly integrated themselves into the genome long ago and have since accumulated mutations that have rendered them unable to produce infectious, exogenous viruses. HERVs are supposed to be "junk DNA." The regulatory role of HERVs has been demonstrated in the liver, placenta, colon, and other locations (Bannert and Kurth 2004). One of the families of HERVs, HERV-K, is proposed to be fairly young (less than five million years old) because it still contains a complete set of genes (albeit with mutations) necessary for a retrovirus to produce infectious viruses (Bannert and Kurth 2004; Dewanneiux et al. 2006). Some HERVs have been implicated in human malignancy because of increased expression of HERV messenger RNA (mRNA) (Andersson et al. 1998), functional protein (Sauter et al. 1995), and retrovirus-like particles (Lower et al. 1993) in certain cancers. Studies are being made on the association of HERV-K with a number of cancers such as germ cell tumors like seminomas (Sauter et al. 1995, 1996), breast cancer, myeloproliferative disease, ovarian cancer (Wang-Johanning et al. 2007a, b), melanoma (Muster et al. 2003; Buscher et al. 2005), and prostate cancer (Tomlins et al. 2007). However, it could not be firmly established so far that HERV-K expression contributes to the development of any kind of cancer (McLaughlin-Drubin and Munger 2008).

#### 1.2.3.6 Xenotropic Murine Leukemia Virus-Related Virus

Xenotropic murine leukemia virus-related virus (XMRV) is a chimeric virus that arose from the recombination of two endogenous mouse retroviruses during the mid-1990s. First described in 2006 as an apparently novel retrovirus and potential human pathogen, it was linked to prostate cancer and later to chronic fatigue syndrome (CFS). However, no evidence gathered so far proves that XMRV can infect humans, nor has it been demonstrated that XMRV is associated with or causes any human disease.

XMRV is a murine leukemia virus (MLV) that formed through the recombination of the genomes of two parent MLVs known as preXMRV-1 and preXMRV-2 (Cingöz et al. 2012). MLV is a gammaretrovirus, a genus of the *Retroviridae* family, and has a single-stranded RNA genome that replicates through a DNA intermediate. The name XMRV was given because the discoverers of the virus initially thought that it was a novel potential human pathogen that was related to but distinct from MLVs. XMRV has now been established as a laboratory contaminant. This contamination hypothesis has been supported by several studies (Smith 2010).

## 1.3 CONCLUSION

In conclusion, we find that currently acknowledged viruses that are associated with human malignancies include two RNA viruses—namely HTLV-1 and HCV—and four DNA viruses—namely EBV, HBV, HPV, and KSHV/HHV-8. Of the two RNA viruses, HTLV-1 causes ATL (Poiesz et al. 1980; Giam and Jeang 2007). HCV and HBV, which belongs to the class of DNA viruses, lead to HCC (Dane et al. 1970; Choo et al. 1989; Levrero 2006; Jin 2007). EBV (of the DNA class) produces BL, NPC, posttransplant lymphomas, and Hodgkin's disease (Epstein et al. 1964; Pagano 1999; Tao et al. 2006; Klein et al. 2007). Of the remaining two DNA viruses, HPV leads to cervical cancer, skin cancer in patients with epidermodysplasia verruciformis, head and neck cancers, and other anogenital cancers (Durst et al. 1983; Boshart et al. 1984; Hausen 1996; Munger et al. 2004), and KSHV/HHV-8 causes PEL and Castleman's disease (Chang et al. 1994; Cathomas 2003). The oncogenic roles of other viruses discovered are not well established.

All tumor viruses induce oncogenesis through manipulating an array of different cellular pathways. These viruses initiate a series of cellular events, which lead to immortalization and proliferation of the infected cells by disrupting the mitotic checkpoint upon infection of the host cell. This is often accomplished by functional inhibition or proteasomal degradation of many tumor suppressor proteins by virally encoded gene products. The virally infected cells can either be eliminated via cell-mediated apoptosis or persist in a state of chronic infection. Importantly, the chronic persistence of infection by tumor viruses can lead to oncogenesis (Saha et al. 2010).

## REFERENCES

Alter MJ. (1995). Epidemiology of hepatitis C in the West. *Semin Liver Dis* 15: 5–14.

Andersson AC, Svensson AC, Rolny C, et al. (1998). Expression of human endogenous retrovirus ERV3 (HERV-R) mRNA in normal and neoplastic tissues. *Int J Oncol* 12: 309–313.

Bannert N, Kurth R. (2004). Retro elements and the human genome: New perspectives on an old relation. *Proc Natl Acad Sci U S A* 101: 14572–14579.

Beasley RP, Lin CC, Hwang LY, et al. (1981). Hepatocellular carcinoma and hepatitis B virus. *Lancet* 2: 1129–1133.

Benn J, Schneider RJ. (1994). Hepatitis B virus HBx protein activates Ras-GTP complex formation and establishes a Ras, Raf, MAP Kinase signaling 19 cascade. *Proc Natl Acad Sci U S A* 91: 10350–10354.

Beral V, Bull D, Darby S, et al. (1992). Risk of Kaposi's sarcoma and sexual practices associated with faecal contact in homosexual or bisexual men with AIDS. *Lancet* 339: 632–635.

Beral V, Peterman RL, Jaffe HW. (1990). Kaposi's sarcoma among persons with AIDS: A sexually transmitted infection? *Lancet* 335: 123–128.

Berger JR, Koralnik IJ. (2005). Progressive multifocal leukoencephalopathy and natalizumab-unforeseen consequences. *N Engl J Med* 353: 414–416.

Bernard BA, Bailly C, Lenoir MC. (1989). The human papillomavirus type 18 (HPV18) E2 gene product is a repressor of the HPV18 regulatory region in human keratinocytes. *J Virol* 63: 4317–4432.

Bernard HU. (2002). Gene expression of genital human papillomaviruses and considerations on potential antiviral approaches. *Antivir Ther* 7: 219–237.

Bindra A, Muradrasoli S, Kisekka R, et al. (2007). Search for DNA of exogenous mouse mammary tumor virus-related virus in human breast cancer samples. *J Gen Virol* 88: 1806–1809.

Bittner JJ. (1936). Some possible effects of nursing on the mammary gland tumor incidence in mice. *Science* 84: 162.

Bosch FX, de Sanjosé S (2003). Chapter 1: Human papillomavirus and cervical cancer—burden and assessment of causality. *J Natl Cancer Inst Monogr* 31: 3–13.

Boshart M, Gissmann L, Ikenberg H. (1984). A new type of papillomavirus DNA, its presence in genital cancer biopsies and in cell lines derived from cervical cancer. *EMBO J* 3: 1151–1157.

Boshoff C, Weiss R. (2002). AIDS-related malignancies. *Nat Rev Cancer* 2: 373–382.

Bouchard MJ, Wang LH, Schneider RJ. (2001). Calcium signaling by HBx protein in hepatitis B virus DNA replication. *Science* 294: 2376–2378.

Bousarghin L, Touzé A, Sizaret PY, et al. (2003). Human papillomavirus types 16, 31, and 58 use different endocytosis pathways to enter cells. *J Virol* 77: 3846–3850.

Boxus M, Willems L. (2009). Mechanisms of HTLV-1 persistence and transformation. *Brit J Cancer* 101: 1497–1501.

Brechot C. (2004). Pathogenesis of hepatitis B virus-related hepatocellular carcinoma: Old and new paradigms. *Gastroenterology* 127: S56–S61.

Buscher K, Trefzer U, Hofmann M, et al. (2005). Expression of human endogenous retrovirus K in melanomas and melanoma cell lines. *Cancer Res* 65: 4172–4180.

Cathomas G. (2003). Kaposi's sarcoma-associated herpesvirus (KSHV)/human herpesvirus 8 (HHV-8) as a tumour virus. *Herpes* 10: 72–77.

Cesarman E, Chang Y, Moore PS, et al. (1995). Kaposi's sarcoma- associated herpesvirus-like DNA sequences in AIDS-related body-cavity-based lymphomas. *N Engl J Med* 332: 1186–1191.

Chami M, Ferrari D, Nicotera P, et al. (2003). Caspase-dependent alterations of $Ca^{2+}$ singaling in the induction of apoptosis by hepatitis B virus X protein. *J Biol Chem* 278: 31745–31755.

Chang Y, Cesarman E, Pessin MS, et al. (1994). Identification of herpesvirus-like DNA sequences in AIDS-associated Kaposi's sarcoma. *Science* 226: 1865–1869.

Chen HS, Kaneko R, Girones RW, et al. (1993). The woodchuck hepatitis virus X gene is important for establishment of virus infection in woodchucks. *J Virol* 67: 1218–1226.

Choo QL, Kuo G, Weiner AJ, et al. (1989). Isolation of a cDNA clone derived from a blood-borne non-A, non-B viral hepatitis genome. *Science* 244: 359–362.

Cingöz O, Paprotka T, Delviks-Frankenberry KA, et al. (2012). Characterization, mapping, and distribution of the two XMRV parental proviruses. *J Virol* 86: 328–338.

Cohen A, Wolf DG, Guttman-Yassky E, et al. (2005). Kaposi's sarcoma associated herpesvirus: Clinical, diagnostic and epidemiological aspects. *Crit Rev Clin Lab Sci* 42: 101–153.

Cohen SM, Parsonnet J, Horning S, et al. (1999). Infection, cell proliferation, and malignancy. In: Parsonnet J, Horning S (eds.), *Microbes and Malignancy: Infection as a Cause of Cancer*. Oxford University Press, New York, pp. 89–106.

Damania B. (2007). DNA tumor viruses and human cancer. *Trends Microbiol* 15: 38–44.

Dane DS, Cameron CH, Briggs M. (1970). Virus-like particles in serum of patients with Australia-antigen-associated hepatitis. *Lancet* 295: 695–698.

Day PM, Lowy DR, Schiller JT. (2003). Papillomaviruses infect cells via a clathrin-dependent pathway. *Virology* 307: 1–11.

de Villiers EM, Fauquet C, Broker TR, et al. (2004). Classification of papillomaviruses. *Virology* 324: 17–27.

de Villiers EM, Schmidt R, Delius H, et al. (2002). Heterogeneity of TT virus related sequences isolated from human tumour biopsy specimens. *J Mol Med* 80: 44–50.

Del Vecchio AM, Romanczuk H, Howley PM, et al. (1992). Transient replication of human papillomavirus DNAs. *J Virol* 66: 5949–5958.

Dewanneiux M, Harper F, Richaud A, et al. (2006). Identification of an infectious progenitor for the multiple-copy HERV-K human endogenous retroelements. *Genome Res* 16: 1548–1556.

DiMaio D, Mattoon D. (2001). Mechanisms of cell transformation by papillomavirus E5 proteins. *Oncogene* 20: 7866–7873.

Doorbar J. (2005). The papillomavirus life cycle. *J Clin Virol* 32: S7–S15.

Doorbar J. (2006). Molecular biology of human papillomavirus infection and cervical cancer. *Clin Sci (Lond)* 110: 525–541.

Duensing S, Münger K. (2004). Mechanisms of genomic instability in human cancer: Insights from studies with human papillomavirus oncoproteins. *Int J Cancer* 109: 157–162.

Durst M, Gissmann L, Ikenberg H. (1983). A papillomavirus DNA from a cervical carcinoma and its prevalence in cancer biopsy samples from different geographic regions. *Proc Natl Acad Sci U S A* 80: 3812–3815.

Eddy BE. (1969). In: Gard CHS, Meyers KF (eds.), Polyoma virus. *Virology Monographs*. Springer-Verlag, New York, pp. 1–114.

Epstein MA, Achong BG, Barr YM. (1964). Virus particles in cultured lymphoblasts from Burkitt's lymphoma. *Lancet* 1: 702–703.

Evans AS, Niederman JC. (1989). *Viral Infections of Humans*. Plenum Press, New York.

Fattovich G, Stroffolini T, Zagni I, et al. (2004). Hepatocellular carcinoma in cirrhosis: Incidence and risk factors. *Gastroenterology* 127: S35–S50.

Feitelson MA. (1999). Hepatitis B virus in hepatocarcinogenesis. *J Cell Physiol* 181: 188–202.

Feitelson MA, Lee J. (2007). Hepatitis B virus integration, fragile sites, and hepatocarcinogenesis. *Cancer Lett* 252: 157–170.

Forgues M, Difilippantonio MJ, Linke SP, et al. (2003). Involvement of Crm1 in hepatitis B virus X protein-induced aberrant centriole replication and abnormal mitotic spindles. *Mol Cell Biol* 23: 5282–5292.

Gallo RC. (2005). History of the discoveries of the first human retroviruses: HTLV-1 and HTLV-2. *Oncogene* 24: 5926–5930.

Gao SJ, Kingsley L, Li M, et al. (1996). KSHV antibodies among Americans, Italians and Ugandans with and without Kaposi's sarcoma. *Nat Med* 2: 925–928.

Gessian A, Yanagihara R, Franchini G, et al. (1991). Highly divergent molecular variants of human T-lymphotropic virus type I from isolated populations in Papua New Guinea and the Solomon Islands. *Proc Natl Acad Sci U S A* 88: 7694–7698.

Giam CZ, Jeang KT. (2007). HTLV-1 Tax and adult T-cell leukemia. *Front Biosci* 12: 1496–1507.

Goedert JJ (ed.). (2000). *Infectious Causes of Cancer: Targets for Intervention*. Springer Link, Germany.

Gross L. (1953). A filterable agent, recovered from Ak leukemic extracts, causing salivary gland carcinomas in C3H mice. *Proc Soc Exp Biol Med* 83: 414–421.

Gross L. (1983). *Oncogenic Viruses*. Pergamon Press, Oxford.

Gruhne B, Sompallae R, Masucci MG. (2009). Three Epstein-Barr virus latency proteins independently promote genomic instability by inducing DNA damage, inhibiting DNA repair and inactivating cell cycle checkpoints. *Oncogene* 28: 3999–4008.

Hausen HZ. (1996). Papillomavirus infections – a major cause of human cancers. *Biochim Biophys Acta* 1288: F55–F78.

Hawley-Nelson P, Vousden KH, Hubbert NL, Lowy DR, Schiller JT. (1989). HPV16 E6 and E7 proteins cooperate to immortalize human foreskin keratinocytes. *EMBO J* 8: 3905–3910.

Hinuma Y, Nagata K, Hanaoka M, et al. (1981). Adult T-cell leukemia: Antigen in an ATL cell line and detection of antibodies to the antigen in human sera. *Proc Natl Acad Sci U S A* 78: 6476–6480.

Hirsch HH, Knowles W, Dickenmann M, et al. (2002). Prospective study of polyomavirus type BK replication and nephropathy in renal-transplant recipients. *N Engl J Med* 347: 488–496.

Holland JF, Pogo BGT. (2004). Mouse mammary tumor virus-like infection and human breast cancer. *Clin Cancer Res* 10: 5647–5649.

Hu Z, Zhang Z, Doo E, et al. (1999). Hepatitis B virus X protein is both a substrate and a potential inhibitor of the proteasome complex. *J Virol* 73: 7231–7240.

Huang J, Kwong J, Sun EC, et al. (1996). Proteasome complex as a potential cellular target of hepatitis B Virus X protein. *J Virol* 70: 5582–5591.

Hughes FJ, Romanos MA. (1993). E1 protein of human papillomavirus is a DNA helicase/ ATPase. *Nucleic Acids Res* 21: 5817–5823.

Huh KW, Siddiqui A. (2002). Characterization of the mitochondrial association of hepatitis B virus X protein, H Bx. *Mitochodrion* 1: 349–359.

Jin DY. (2007). Molecular pathogenesis of hepatitis C virus-associated hepatocellular carcinoma. *Front Biosci* 12: 222–233.

Klein E, Kis LL, Klein G. (2007). Epstein–Barr virus infection in humans: From harmless to life endangering virus–lymphocyte interactions. *Oncogene* 26: 1297–1305.

Klein NP, Schneider RJ. (1997). Activation of Src family of kinase by HBV HBx protein, and coupled signaling to Ras. *Mol Cell Biol* 17: 6427–6436.

Kutok JL, Wang FW. (2006). Spectrum of Epstein-Barr virus-associated diseases. *Annu Rev Pathol Mech Dis* 1: 375–404.

Lauer GM, Walker BD. (2001). Hepatitis C virus infection. *N Engl J Med* 345: 41–52.

Levrero M. (2006). Viral hepatitis and liver cancer: The case of hepatitis C. *Oncogene* 25: 3834–3847.

Lindenbach BD, Rice CM. (2005). Unravelling hepatitis C virus replication from genome to function. *Nature* 436: 933–938.

Liu B, Wang Y, Melana SM, et al. (2001). Identification of a proviral structure in human breast cancer. *Cancer Res* 61: 1754–1759.

Lok AS. (2004). Prevention of hepatitis B virus-related hepatocellular carcinoma. *Gastroenterology* 127: S303–S309.

Lower R, Lower J, Koch CT, et al. (1993). A general method for the identification of transcribed retrovirus sequences (R-U5 PCR) reveals the expression of the human endogenous retrovirus loci HERV-H and HERV-K in teratocarcinoma cells. *Virology* 192: 501–511.

Maguire HF, Hoeffler JP, Siddiqui A. (1991). HBV X protein alters the DNA binding specificity of CREB and ATF-2 by protein-protein interactions. *Science* 252: 842–844.

Matsuoka M. (2003). Human T-cell leukemia virus type I and adult T-cell leukemia. *Oncogene* 22: 5131–5140.

Matsuoka M, Jeang KT. (2007). Human T-cell leukemia virus type 1 (HTLV-1) infectivity and cellular transformation. *Nat Rev Cancer* 7: 270–280.

McLaughlin-Drubin ME, Munger K. (2008). Viruses associated with human cancer. *Biochim Biophys Acta* 1782: 127–150.

Meier O, Greber UF. (2004). Adenovirus endocytosis. *J Gene Med* 6: S152–S163.

Melana SM, Nepomnaschy I, Sakalian M, et al. (2007). Characterization of viral particles isolated from primary cultures of human breast cancer cells. *Cancer Res* 67: 8960–8965.

Moore PS, Kingsley LA, Holmberg SD, et al. (1996). Kaposi's sarcoma-associated herpesvirus infection prior to onset of Kaposi's sarcoma. *AIDS* 10: 175–180.

Moradpour D, Penin F, Rice CM. (2007). Replication of hepatitis C virus. *Nat Rev Microbiol* 5: 453–463.

Munger K, Baldwin A, Edwards KM, et al. (2004). Mechanisms of human papillomavirus-induced oncogenesis. *J Virol* 78: 11451–11460.

Munger K, Howley PM. (2002). Human papillomavirus immortalization and transformation functions. *Virus Res* 89: 213–228.

Muster T, Waltenberger A, Grassauer A, et al. (2003). An endogenous retrovirus derived from human melanoma cells. *Cancer Res* 63: 8735–8741.

Niel C, Saback FL, Lampe E. (2000). Coinfection with multiple TT virus strains belonging to different genotypes is a common event in healthy Brazilian adults. *J Clin Microbiol* 38: 1926–1930.

Nishizawa T, Okamoto H, Konishi K, et al. (1997). A novel DNA virus (TTV) associated with elevated transaminase levels in posttransfusion hepatitis of unknown etiology. *Biochem Biophys Res Commun* 241: 92–97.

Ocheni S, Olusina DB, Oyekunle AA, et al. (2010). EBV-associated malignancies. *Open Infect Dis J* 4: 101–112.

Oelze I, Kartenbeck J, Crusius K, et al. (1995). Human papillomavirus type 16 E5 protein affects cell-cell communication in an epithelial cell line. *J Virol* 69: 4489–4494.

Pagano JS. (1999). Epstein–Barr virus: The first human tumor virus and its role in cancer. *Proc Assoc Am Physicians* 111: 573–580.

Patterson NA, Smith, JL, Ozbun MA. (2005). Human papillomavirus type 31b infection of human keratinocytes does not require heparan sulfate. *J Virol* 79: 6838–6847.

Pawlotsky JM. (2004). Pathophysiology of hepatitis C virus infection and related liver disease. *Trends Microbiol* 12: 96–102.

Poiesz BJ, Ruscetti FW, Gazdar AF, et al. (1980). Detection and isolation of type C retrovirus particles from fresh and cultured lymphocytes of a patient with cutaneous T-cell lymphoma. *Proc Natl Acad Sci U S A* 77: 7415–7419.

Poulin DL, DeCaprio JA. (2006). Is there a role for SV40 in human cancer? *J Clin Oncol* 24: 4356–4365.

Poynard T, Yuen MF, Ratziu V, Lai CL. (2003). Viral hepatitis C. *Lancet* 362: 2095–2100.

Pozzatti R, Vogel J, Jay G. (1990). The human T-lymphotropic virus type 1 tax gene can cooperate with the Ras oncogene to induce neoplastic transformation of cells. *Mol Cell Biol* 10: 413–417.

Rahmani Z, Huh KW, Lasher R, et al. (2000). Hepatitis B virus X protein colocalizes to mitochondria with a human voltage-dependent anion channel, HVDAC3, and alters its transmembrane potential. *J Virol* 74: 2840–2846.

Randhawa PS, Demetris AJ. (2000). Nephropathy due to polyomavirus type BK. *N Engl J Med* 342: 1361–1363.

Rigat K, Wang Y, Hudyma TW, et al. (2010). Ligand-induced changes in hepatitis C virus NS5B polymerase structure. *Antiviral Res* 88: 197–206.

Safak M, Khalili K. (2003). An overview: Human polyomavirus JC virus and its associated disorders. *J Neurovirol* 9 (Suppl 1): 3–9.

Saha A, Kaul R, Murakami M, et al. (2010). Tumor viruses and cancer biology: Modulating signaling pathways for therapeutic intervention. *Cancer Biol Ther* 10: 961–978.

Saito I, Miyamura T, Ohbayashi A, et al. (1990). Hepatitis C virus infection is associated with the development of hepatocellular carcinoma. *Proc Natl Acad Sci U S A* 87: 6547–6549.

Sarid R, Gao SJ. (2011). Viruses and human cancer: From detection to causality. *Cancer Lett* 305: 218–227.

Sauter M, Roemer K, Best B, et al. (1996). Specificity of antibodies directed against Env protein of human endogenous retroviruses in patients with germ cell tumors. *Cancer Res* 56: 4362–4365.

Sauter M, Schommer S, Kremmer E, et al. (1995). Human endogenous retrovirus K10: Expression of Gag protein and detection of antibodies in patients with seminomas. *J Virol* 69: 414–421.

Scheffner M, Huibregtse JM, Vierstra RD, et al. (1993). The HPV-16 E6 and E6-AP complex functions as a ubiquitin-protein ligase in the ubiquitination of p53. *Cell* 75: 495–505.

Smith RA. (2010). Contamination of clinical specimens with MLV-encoding nucleic acids: Implications for XMRV and other candidate human retroviruses. *Retrovirology* 7: 112.

Soulier J, Grollet L, Oksenhendler E, et al. (1995). Kaposi's sarcoma-associated herpesvirus-like DNA sequences in multicentric Castelmen's disease. *Blood* 86: 1276–1280.

Sterling JC, Skepper JN, Stanley MA. (1993). Immunoelectron microscopical localization of human papillomavirus type 16 L1 and E4 proteins in cervical keratinocytes cultured in vivo. *J Invest Dermatol* 100: 154–158.

Stewart SE, Eddy BE, Borgese N. (1958). Neoplasms in mice inoculated with a tumor agent carried in tissue culture. *J Natl Cancer Inst* 20: 1223–1243.

Sweet BH, Hilleman MR. (1960). The vacuolating virus SV 40. *Proc Soc Exp Biol Med* 105: 420–427.

Talora C, Sgroi DC, Crum CP, et al. (2002). Specific down-modulation of Notch1 signaling in cervical cancer cells is required for sustained HPV-E6/E7 expression and late steps of malignant transformation. *Genes Dev* 16: 2252–2263.

Tanaka H, Iwasaki Y, Nousa K, et al. (2005). Possible contribution of prior hepatitis B virus infection to the development of hepatocellular carcinoma. *J Gastroenterol Hepatol* 20: 850–856.

Tao Q, Young LS, Woodman CB, et al. (2006). Epstein–Barr virus (EBV) and its associated human cancers—Genetics, epigenetics, pathobiology and novel therapeutics. *Front Biosci* 11: 2672–2713.

Thomsen P, van Deurs B, Norrild B, et al. (2000). The HPV16 E5 oncogene inhibits endocytic trafficking. *Oncogene* 19: 6023–6032.

Tomlins SA, Laxman B, Dhanasekaran SM, et al. (2007). Distinct classes of chromosomal rearrangements create oncogenic ETS gene fusions in prostate cancer. *Nature* 448: 595–599.

Ushikai M, Lace MJ, Yamakawa Y, et al. (1994). Trans activation by the full-length E2 proteins of human papillomavirus type 16 and bovine papillomavirus type 1 in vitro and in vivo: Cooperation with activation domains of cellular transcription factors. *J Virol* 68: 6655–6666.

Wang-Johanning F, Huang M, Liu J, et al. (2007a). Sheep stromal-epithelial cell interactions and ovarian tumor progression. *Int J Cancer* 10: 2346–2354.

Wang-Johanning F, Liu J, Rycaj K, et al. (2007b). Expression of multiple human endogenous retrovirus surface envelope proteins in ovarian cancer. *Int J Cancer* 120: 81–90.

Whitby D, Howard MR, Tenant-Flowers M, et al. (1995). Detection of Kaposi's sarcoma associated herpesvirus in peripheral blood of HIV-infected individuals and progression to Kaposi's sarcoma. *Lancet* 346: 799–802.

Wong-Staal F, Gallo RC. (1985). Human T-lymphotropic retroviruses. *Nature* 317: 395–403.

Woodman CB, Collins S, Winter H, et al. (2001). Natural history of cervical human papillomavirus infection in young women: A longitudinal cohort study. *Lancet* 357: 1831–1836.

Yoshida M, Miyoshi I, Hinuma Y. (1982). Isolation and characterisation of retrovirus from cell lines of human adult T-cell leukemia and its implication in the disease. *Proc Natl Acad Sci U S A* 79: 2031–2035.

Yu T, Ferber MJ, Cheung TH, et al. (2005). The role of viral integration in the development of cervical cancer. *Cancer Genet Cytogenet* 158: 27–34.

Zhang Z, Torii N, Furusaka N, et al. (2000). Structural and functional characterization of interaction between hepatitis B virus X protein and the proteasome complex. *J Biol Chem* 275: 15157–15165.

Zhao KN, Hengst K, Liu WJ, et al. (2000). BPV1 E2 protein enhances packaging of full-length plasmid DNA in BPV1 pseudovirions. *J Virol* 272: 382–393.

Zoulim F, Saputelli J, Seeger C. (1994). Woodchuck hepatitis B virus X gene is important for viral infection in vivo. *J Virol* 68: 2026–2030.

zur Hausen H. (2003). SV40 in human cancers—An endless tale? *Int J Cancer* 107: 687–687.

# 2 Targeting Human T-Cell Leukemia Virus Type 1

## *Pathogenesis and Treatment Strategies*

*Jun-ichirou Yasunaga*

## 2.1 INTRODUCTION

Human T-cell leukemia virus type 1 (HTLV-1) was the first retrovirus that was identified as a causative agent of human diseases (Gallo 2005; Takatsuki 2005; Matsuoka and Jeang 2007). HTLV-1 induces malignancy of CD4+ CD25+ T-lymphocytes, adult T-cell leukemia (ATL), and several inflammatory diseases, such as HTLV-1-associated myelopathy (HAM)/tropical spastic paraparesis (TSP) and HTLV-1 uveitis (HU), in a small fraction of the HTLV-1-infected individuals (Matsuoka and Jeang 2007; Yasunaga and Matsuoka 2011).

At present, about 10–20 million people worldwide are infected with HTLV-1 (Yasunaga and Matsuoka 2007). HTLV-1 is endemic in Japan, the Caribbean islands, the areas surrounding the Caribbean basin, Central Africa, South America, Papua New Guinea, and the Solomon islands. HTLV-1 belongs to the delta-type retroviruses, which also include bovine leukemia virus (BLV), human T-cell leukemia virus type 2 (HTLV-2), and simian T-cell leukemia virus. Like other retroviruses, the HTLV-1 proviral genome has the structural genes, *gag*, *pol*, and *env*, which are flanked by long terminal repeats (LTRs) at both ends. A unique structure, termed the pX region, is found between *env* and the 3′-LTR (Figure 2.1). The plus strand of the pX region encodes the regulatory proteins $p40^{tax}$ (Tax), $p27^{rex}$ (Rex), p12, p13, p30, and p21 in four different open reading frames (ORFs) (Franchini et al. 2003; Matsuoka 2003). The HTLV-1 bZIP factor (HBZ) is encoded in the minus strand of the provirus, and alternative splicing generates the isoforms of HBZ, the spliced and unspliced HBZ forms (Gaudray et al. 2002; Cavanagh et al. 2006; Murata et al. 2006; Satou et al. 2006). The major host of HTLV-1 is CD4+ T lymphocytes in HTLV-1 carriers, although it is known that HTLV-1 can infect many types of cells in vitro (Koyanagi et al. 1993; Yasunaga et al. 2001; Satou et al. 2012).

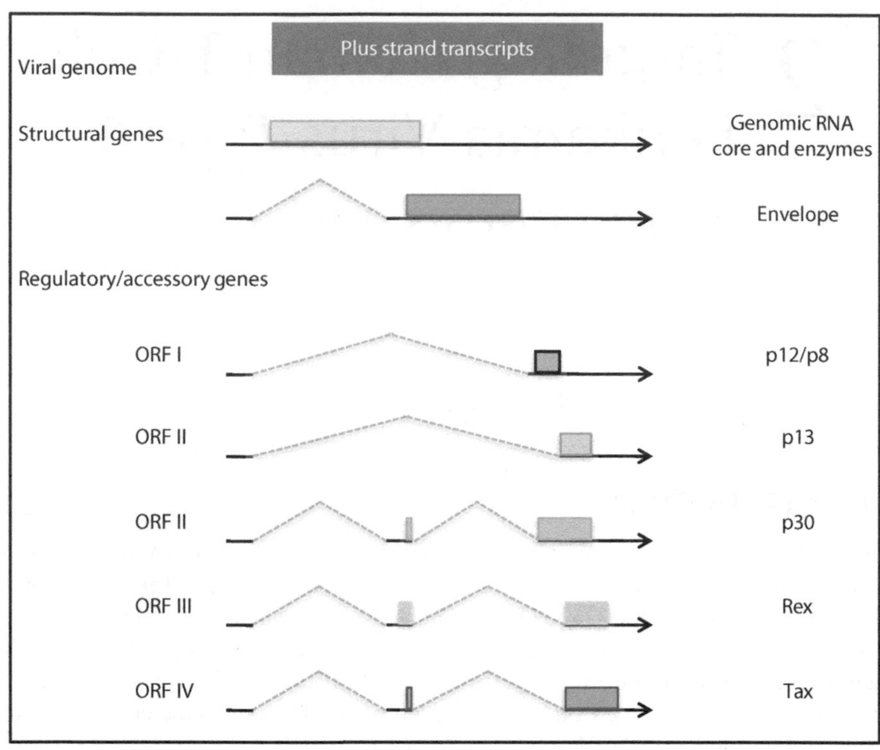


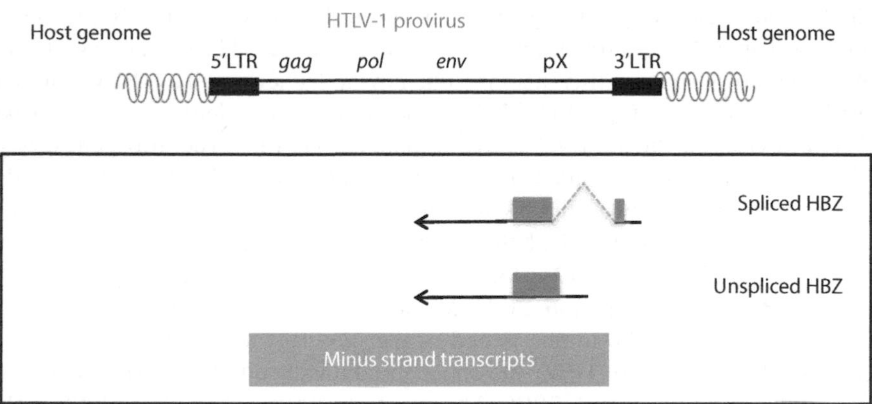


**FIGURE 2.1 (See color insert.)** HTLV-1-encoded genes. The HTLV-1 proviral genome consists of *gag*, *pol,* and *env* genes flanked by long terminal repeat (LTR) sequences at both ends. The pX region encodes several regulatory/accessory proteins in four open reading frames (ORFs) and the minus strand.

## 2.2 LIFE CYCLE OF HTLV-1

### 2.2.1 Viral Transfer via Cell-to-Cell Contact

It is well known that infection by free HTLV-1 virions is very inefficient and that HTLV-1 viral particles are hardly detectable in serum of the HTLV-1 carriers (Matsuoka and Jeang 2007). Indeed, HTLV-1 replication is not active in vivo; this fact indicates that the mechanism of HTLV-1 transmission is quite different from that of another human retrovirus, human immunodeficiency virus 1 (HIV-1), which vigorously replicates in vivo to promote efficient transmission through cell-free virions. Instead, HTLV-1 is transmitted from infected cells to uninfected cells via cell-to-cell contact (Igakura et al. 2003; Pais-Correia et al. 2010; Van Prooyen et al. 2010). Therefore, living HTLV-1-infected cells are required for efficient transmission. There are three major routes of HTLV-1 infection: mother-to-infant transmission (mainly through breast milk), parenteral transmission, and sexual transmission (Matsuoka 2003). Infected lymphocytes in the fluids are the main source of HTLV-1. Because the replication level of HTLV-1 in infected cells is generally low, efficient HTLV-1 viral transmission is established directly between cells.

### 2.2.2 Viral Entry into Target Cells

HTLV-1 is an enveloped virus. After transfer of the virions from infected to uninfected cells, viral particles attach to the receptor complex on the cell surface and enter target cells by fusion of the viral membrane with the cellular membrane (Jones et al. 2011). The receptor complex consists of at least three molecules: a glucose transporter (GLUT1) (Manel et al. 2003), heparan sulfate proteoglycan (HSPG) (Jones et al. 2005), and neuropilin 1 (Ghez et al. 2006), which are important for the interaction between the HTLV-1 envelope and the cell membrane. It has been suggested that the virus may first contact HSPG and then form complexes with neuropilin 1, followed by association with the GLUT1 on the cell surface prior to membrane fusion and entry into the cell.

### 2.2.3 Reverse Transcription

After entry into target cells and uncoating of the virus, HTLV-1 genomic RNA is reverse transcribed into double-stranded DNA by reverse transcriptase (RT). RT and integrase (IN) are incorporated in virions and released along with the viral genome after viral entry.

### 2.2.4 Integration

Viral IN mediates integration of viral double-stranded DNA into the cellular genome; the integrated viral DNA is called the provirus. Integration of HTLV-1 has been shown to occur at random sites in the host genome, whereas proviruses in leukemic cells tend to be integrated near the transcriptional start sites of cellular genes, indicating positive selection of cells with this feature during leukemogenesis (Doi et al. 2005; Gillet et al. 2011).

### 2.2.5 Viral Transcription

To produce HTLV-1 progeny, viral transcripts are expressed from the provirus. The LTRs at each end of the provirus are used as promoters: the 5′-LTR and 3′-LTR control the transcription of the viral genes encoded in the plus and minus strands of the provirus, respectively (Figure 2.1). Because the plus strand of the provirus encodes all the structural proteins and the viral genomic RNA, 5′-LTR-mediated transcription is required for viral replication and transmission. In contrast, the 3′-LTR drives the transcription of the spliced form of HBZ, which is constitutively expressed in HTLV-1-infected cells. The regulatory genes (*tax* and *rex*) and accessory genes (*p12*, *p13*, *p30*, and *HBZ*) encoded in the pX region regulate viral replication and the proliferation of infected cells. Among the products of these genes, Tax is a critical activator of viral transcription through the 5′-LTR. Tax does not bind to DNA, but rather activates the transcription of target genes by recruiting various transcription factors and modifying the epigenetic status of promoter regions (Kashanchi and Brady 2005). The association between Tax and cAMP response element-binding protein (CREB) is required for viral gene transcription. There are three 21-bp repeat elements, called Tax-responsive elements (TREs), located in the 5′-LTR, and the Tax-CREB complex recruits several histone acetyl transferases, including CREB-binding protein (CBP), p300, and p300/CBP-associated factor, to the LTR, resulting in induction of viral expression. Another regulatory protein, Rex, is an RNA-binding protein and a positive post-transcriptional regulator of virus production (Inoue et al. 1986; Hidaka et al. 1988). Rex increases nuclear export of viral genomic RNA and singly spliced messenger RNA (mRNA) encoding viral structural genes, such as *gag*, *pol*, and *env*, which are required for the production of an infectious virus. In contrast, p30 and HBZ negatively regulate 5′-LTR-mediated transcription (Gaudray et al. 2002; Nicot et al. 2004; Yoshida et al. 2008). Because viral antigens, especially Tax, are the targets of cytotoxic T-lymphocytes (CTLs) (Kannagi et al. 2004, 2011), silencing of viral expression by p30 and HBZ supports the escape of HTLV-1-infected cells from host immune surveillance. Transcription from the 5′-LTR is transiently induced and complicatedly regulated by many factors, whereas 3′-LTR-driven transcription is constitutively activated by a ubiquitously expressed transcription factor, Sp1 (Yoshida et al. 2008; Gazon et al. 2012). These data suggest that HBZ has important roles not only in viral latency but also in the proliferation of infected cells.

### 2.2.6 Assembly and Budding

Translation of unspliced viral RNA produces Gag, Gag-Protease (Pro), and Gag-Pro–Polymerase (Pol) polyproteins, and mature structural proteins and enzymes are generated by cleavage with Pro (Nam et al. 1988; Trentin et al. 1998; Kadas et al. 2004). The singly spliced mRNA encodes the envelope protein that is cleaved into the surface glycoprotein gp46 and the transmembrane glycoprotein gp21 (Pique et al. 1990). HTLV-1 particles are assembled at the host cellular membrane and released from cells by budding (Poiesz et al. 1980).

## 2.3 CLONAL EXPANSION OF HTLV-1-INFECTED CELLS

Replication of HTLV-1 is known to be generally suppressed in vivo (Hanon et al. 2000). Because Tax is a highly immunogenic protein, cells expressing Tax protein are eliminated by the host immune system. Therefore, vigorous replication of HTLV-1 viruses is not observed in HTLV-1-infected individuals. Instead, HTLV-1 increases its copy number by promoting proliferation of the infected cells; HTLV-1-infected T-cell clones can proliferate and survive for a long time (Wattel et al. 1995; Etoh et al. 1997). The number of HTLV-1-infected cells has been shown to be almost constant (Okayama et al. 2004; Tanaka et al. 2005; Umeki et al. 2009), and some HTLV-1-infected clones persist for many years in the same HTLV-I carriers (Etoh et al. 1997). Among the viral proteins, HBZ is thought to be important for clonal proliferation of HTLV-1-infected cells. HBZ is the only viral protein that is conserved and constitutively expressed in HTLV-1-infected cells (Satou et al. 2006; Miyazaki et al. 2007; Fan et al. 2010) because nonsense mutations were reported in all other viral genes except the HBZ gene, even in asymptomatic HTLV-1 carriers. In addition, knockdown of HBZ inhibits growth of ATL and HTLV-1-infected cell lines (Satou et al. 2006), indicating that HBZ is critical for the expansion of HTLV-1-infected cells.

## 2.4 HTLV-1-INDUCED DISEASES

Most HTLV-1-infected individuals are asymptomatic and do not develop any HTLV-1-mediated diseases during their lifetime. In a small fraction of carriers, HTLV-1 induces a malignant disease, ATL, and several inflammatory diseases, such as HAM/TSP and HU. It is also known that HTLV-1 carriers have mild immunodeficiency and develop some opportunistic infections. These diseases are pictorially represented and briefly described in Figure 2.2.

### 2.4.1 Adult T-Cell Leukemia

ATL is a fatal T-cell malignancy induced by HTLV-1 and an outcome of the malignant transformation of one HTLV-1-infected T-cell clone. Among individuals infected with HTLV-1, a small proportion of carriers (6% of males and 2% of females in Japan) develop ATL (Matsuoka and Jeang 2007). Most HTLV-1 carriers do not develop any HTLV-1-associated diseases. There is a long latency period from the initial infection until the onset of ATL—about 60 years in Japan (Iwanaga et al. 2012) and 40 years in Jamaica (Hanchard 1996). These findings suggest that multistep leukemogenic mechanisms are involved in the genesis of ATL. The diagnostic criteria for ATL have been defined as (Takatsuki 2005):

1. Presence of a morphologically proven lymphoid malignancy expressing T-cell surface antigens (typically, CD4+ and CD25+). These abnormal T lymphocytes have hyper-lobulated nuclei in acute ATL and are called "flower cells."
2. Presence of antibodies to HTLV-1 in serum.
3. Demonstration of monoclonal integration of HTLV-1 provirus in tumor cells by Southern blotting.

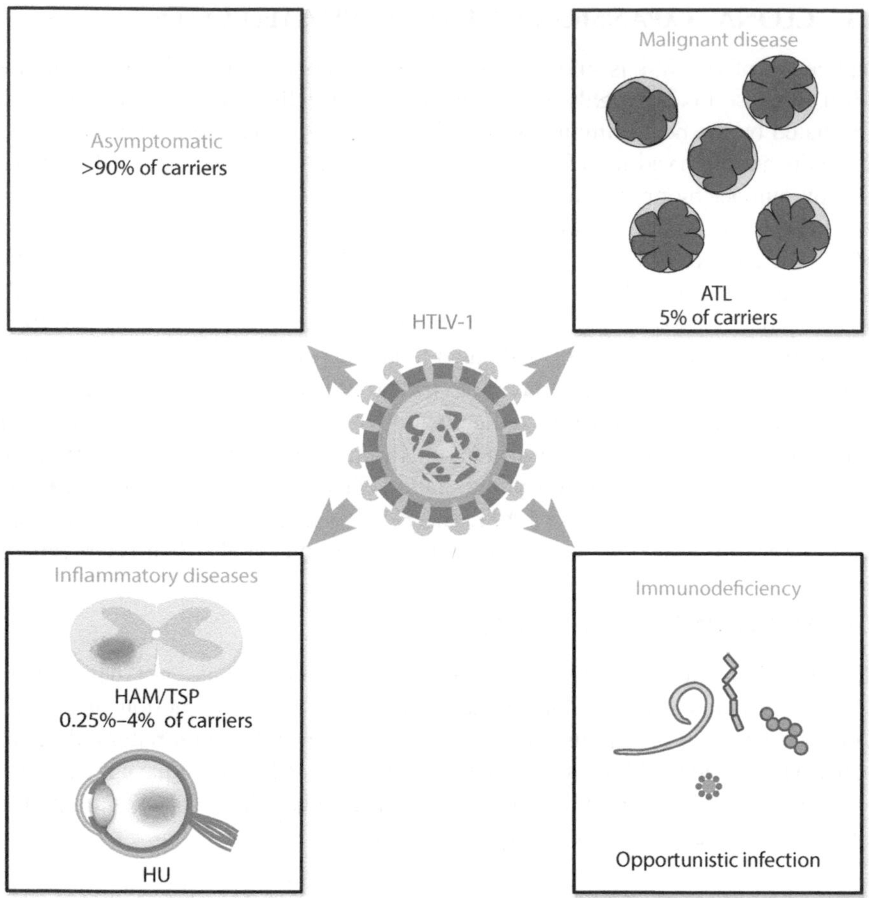


**FIGURE 2.2 (See color insert.)** A pictorial representation of HTLV-1-associated diseases.

Patients are subdivided into four clinical subtypes according to the clinical features (Shimoyama 1991): acute, lymphoma, chronic, and smoldering types. The acute type is the prototypic ATL in which patients exhibit an increased number of ATL cells, frequent skin lesions, systemic lymphadenopathy, and hepatosplenomegaly. Lymphoma-type ATL is characterized by prominent systemic lymphadenopathy, with few abnormal cells in the peripheral blood. In chronic ATL, the white blood cell count is somewhat higher, and skin lesions, lymphadenopathy, and hepatosplenomegaly are sometimes exhibited. Smoldering ATL is characterized by the presence of a few ATL cells in which monoclonal integration of HTLV-1 provirus can be confirmed in peripheral blood. Acute- and lymphoma-type ATL are very aggressive, and patients with these types are resistant to intensive chemotherapy. Chronic and smoldering types are indolent, although these types progress to the aggressive types within a few years.

### 2.4.2 HTLV-1-Associated Myelopathy/Tropic Spastic Paraparesis

HAM/TSP is a chronic progressive neurological disorder that is characterized by spasticity and/or hyperreflexia, urinary bladder disturbances, lower extremity muscle weakness, and sensory disturbance (Osame et al. 1990). The lifetime risk of developing HAM/TSP is different among different ethnic groups and ranges between 0.25% and 3.8% (Saito and Bangham 2012). The annual incidence of HAM/TSP is higher among Jamaican subjects than among Japanese subjects (20 vs. 3 cases per 100,000), with about twice the risk for women in both populations (Hisada et al. 2004). Symptom progression appears to be more rapid in blood transfusion-associated HAM/TSP than in the cases of mother-to-child transmission (Gout et al. 1990; Osame et al. 1990). Pathologic examinations indicate that the thoracic spinal cord is predominantly involved, and mononuclear cell infiltration, destruction of myelin and axon, and astrocytic gliosis are observed in the lesions.

### 2.4.3 Other HTLV-1-Associated Diseases

HTLV-1 uveitis (HU) is defined as an intraocular inflammation developed in HTLV-1-infected subjects (Mochizuki 2010). A diagnosis of HU is obtained when the following two criteria are met: (1) patients must be seropositive to HTLV-1 and (2) routine systemic and ophthalmic examinations exclude all other defined entities of uveitis. HU is characterized by moderate to severe vitritis accompanied by mild iritis and mild retinal vasculitis (Yoshimura et al. 1993; Takahashi et al. 2000).

Because HTLV-1 induces mild cellular immunodeficiency, opportunistic infections often occur in HTLV-1 carriers (Yasunaga et al. 2001). Among the infectious diseases developed in HTLV-1 carriers, HTLV-1-associated infective dermatitis (HAID) or infective dermatitis associated with HTLV-1 (IDH) and strongyloidiasis are associated with high proviral loads (Nakada et al. 1987; Gabet et al. 2000; Primo et al. 2009). HAID/IDH is a pediatric skin disease prevalent in the Caribbean and Brazil (McGill et al. 2012). It is characterized by a chronic exudative eczematous eruption and persistent infection with *Staphylococcus aureus* and beta-hemolytic streptococci. Concurrent infection with an intestinal nematode, *Strongyloides stercoralis*, in HTLV-1-infected individuals is strongly associated with parasite dissemination and development of severe strongyloidiasis, which results in a high death rate (Carvalho and Da Fonseca Porto 2004). A recent study suggests that strongyloidiasis and HAID/IDH increase the risk of development of HTLV-1-associated diseases by altering the clonality of HTLV-1-infected clones in vivo (Gillet et al. 2013).

## 2.5 PATHOGENESIS OF HTLV-1

HTLV-1 induces malignancy and several inflammatory diseases in some of the infected carriers. Although the precise mechanisms of HTLV-1-mediated pathogenesis are not completely understood, it has been suggested that not only viral factors but also host factors are involved in the development of HTLV-1-associated disorders (Yasunaga and Matsuoka 2007). Because a high number of HTLV-1-infected cells in the peripheral blood—a high proviral load—is a risk factor of both ATL and HTLV-1-associated inflammatory diseases (Kannagi et al. 2011), HTLV-1-induced cellular proliferation seems to be important in HTLV-1-mediated pathogenesis.

Among the viral proteins encoded in HTLV-1, Tax and HBZ are thought to be the major players in the development of HTLV-1-associated diseases (Matsuoka and Jeang 2007). Using transgenic mouse models, the oncogenic properties of both of these proteins have been demonstrated (Grossman et al. 1995; Hasegawa et al. 2006; Satou et al. 2011). Tax is a potent transactivator of viral transcription and is critical in viral replication (Kashanchi and Brady 2005). Tax also affects the transcription of numerous cellular genes and modulates the function of host proteins through physical interactions (Boxus et al. 2008). In addition, Tax induces oncogenic stress in primary cells and inhibits DNA repair machineries (Matsuoka and Jeang 2007; Kinjo et al. 2010). On the other hand, Tax is a major target of host CTLs. To escape from CTLs, ATL cells frequently lose the expression of Tax through several mechanisms. The 5′-LTR is a viral promoter for the transcription of viral genes, including the *tax* gene. The 5′-LTR is reported to be lost in 40% of clinical ATL cases, indicating that ATL cells with a provirus lacking an intact 5′-LTR can no longer produce Tax (Takeda et al. 2004). The second mechanism is the nonsense or missense mutation of the *tax* gene in ATL cells. Interestingly, in some cases, ATL cells have mutations in the class I major histocompatibility complex recognition site of the Tax protein, resulting in an escape from immune recognition (Furukawa et al. 2001). The third mechanism is an epigenetic change in the 5′-LTR, whereby DNA hypermethylation and histone modification silence the transcription of viral genes (Koiwa et al. 2002; Taniguchi et al. 2005). ATL cells can escape from the host immune system by suppressing Tax expression, indicating that Tax is dispensable at least for the final step of leukemogenesis.

In contrast, HBZ is constitutively expressed in all ATL cells and is required for proliferation of HTLV-1-infected cells. HBZ has a pleiotropic function just like Tax (Satou and Matsuoka 2012). Intriguingly, HBZ and Tax have opposite functions in several key signaling pathways related to oncogenesis (Yasunaga and Matsuoka 2011). The mechanism(s) through which they influence cellular transformation and interfere with each other's function is yet to be elucidated.

Tax is critical for viral replication and de novo infection, while HBZ is indispensable in clonal proliferation of HTLV-1-infected cells. During the carrier state, Tax and HBZ might collaborate to promote persistence of infection in vivo. Because cells expressing HTLV-1 antigens are eliminated by CTLs, the number of HTLV-1-infected cells is kept stable by the balance between expansion of HTLV-1 and host immunity. During this period, HTLV-1-infected cells might induce inflammatory diseases in subjects with high proviral loads. During the long latency period, the genetic and epigenetic alterations accumulate in the host genomes (Yasunaga and Matsuoka 2007). Finally, a fully transformed HTLV-1-infected clone, which no longer requires Tax, might emerge, resulting in development of ATL. HBZ appears to be required in all leukemogenic processes. Interestingly, it has been reported that HBZ expression levels correlate positively with disease severity in HAM/TSP patients, suggesting the importance of HBZ in the pathogenesis of this neurological disorder (Saito and Bangham 2012).

## 2.6 OUTLINE OF TREATMENT STRATEGIES

Because HTLV-1 is generally latent in infected cells and most of the infected individuals are asymptomatic, no treatment is required. Once aggressive forms of ATL occur, intensive therapies are mandatory. In addition to traditional combination chemotherapy regimens, several options, such as anti-CCR4 antibody, hematopoietic stem cell transplantation (HSCT), and antiviral therapy, were recently proven to be effective in ATL; related clinical trials addressing the utility of such therapies in ATL management are ongoing. On the other hand, patients with indolent types of ATL need not be treated immediately and only require careful monitoring. However, most of the indolent types progress to aggressive types of ATL within a few years, and it is reported that the five-year overall survival (OS) is 47% in Japan (Takasaki et al. 2010). To compare the outcome of careful monitoring and waiting versus an antiviral therapy in indolent ATL, the Japanese lymphoma study group (Japan Clinical Oncology Group, JCOG) is currently carrying out a randomized clinical trial (Fields and Taylor 2012).

Oral prednisone and interferon-alpha (IFN-α) are often administrated to patients with HAM/TSP, and this treatment provides a transient beneficial effect (Osame et al. 1987; Izumo et al. 1996; Nakagawa et al. 1996). A scheme of natural history of HTLV-1 infection and treatment strategies are summarized in Figure 2.3, which shows that initial infection of HTLV-1 is established by cell-to-cell transmission of the virus. In carrier state, HTLV-1 increases its progenies by de novo infection and clonal expansion of infected cells. After a long latent period,

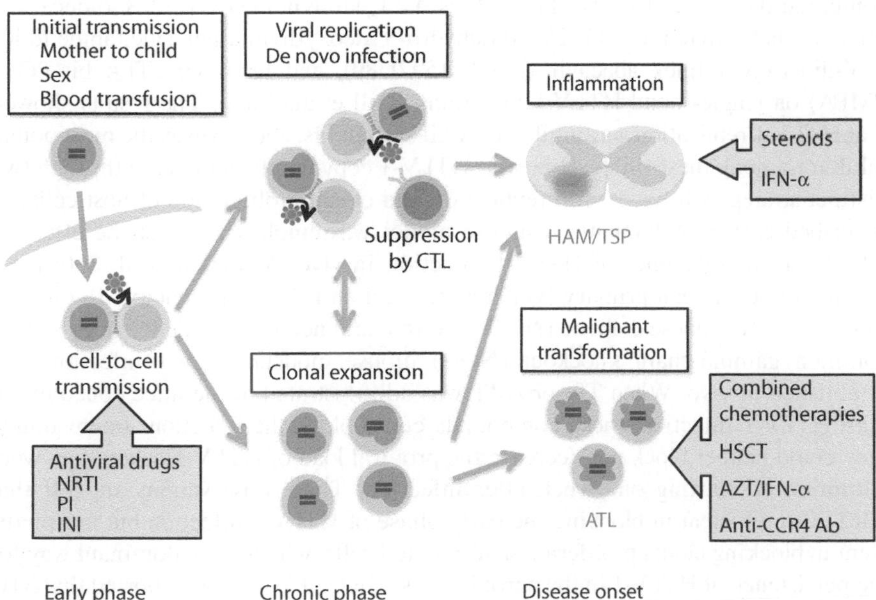


**FIGURE 2.3 (See color insert.)** A pictorial representation of HTLV-1 infection and its treatment strategies.

HTLV-1-associated diseases develop in a part of the carriers. Antiviral therapies targeting viral proteins, such as RT, protease, and IN, might be effective only in early phases of infection. Because ATL is a very aggressive disease, intensive therapies are required.

## 2.7 ANTIVIRAL THERAPIES

To reduce the viral burden, several antiviral compounds targeting viral enzymes have been evaluated. In general, these compounds were effective in inhibiting HTLV-1 transmission and replication in vitro, whereas their effects on viral load in chronic infections were not promising (Miyazato et al. 2006). However, the effectiveness of a combination therapy with an RT inhibitor, zidovudine (ZDV or AZT; 3′-azido-3′-deoxythymidine), and IFN-α in ATL deserves special mention, although its mode of action is not yet clear (Bazarbachi et al. 2011; Fields and Taylor 2012).

### 2.7.1 Reverse Transcriptase Inhibitors

After entry of HTLV-1 into cells, viral genomic RNA is reverse transcribed by viral RT. Because nucleoside analog RT inhibitors (NRTIs) are quite effective in inhibiting HIV replication, they were also expected to exert inhibitory effects on HTLV-1 replication. Several early studies tested the efficacy of AZT on HTLV-1 transmission in vitro and showed its inhibitory effect on HTLV-1 replication in cell culture systems (Matsushita et al. 1987; Macchi et al. 1997; Derse et al. 2001). Another study compared the effect of six NRTIs such as AZT, lamivudine (3TC; 2′,3′-dideoxy-3′-thiacytidine), stavudine (d4T; 2′,3′-didehydro-3′-deoxythymidine), zalcitabine (ddC; 2′,3′-dideoxycytidine), abacavir (ABC; 1592U89), and tenofovir (Tfv; bis-POC-PMPA) on single-round HTLV-1 infections (Hill et al. 2003). The results showed that HTLV-1 replication was inhibited by all six NRTIs, and Tfv was the most potent inhibitor among these drugs. However, HTLV-1 copy numbers increase through two distinct strategies in vivo: viral replication and clonal proliferation of host cells, as described earlier. A humanized mouse model was employed to test the effect of NRTIs on the dynamics of HTLV-1 infection in vivo (Miyazato et al. 2006). By inoculation of human primary lymphocytes and an HTLV-1-producing T-cell line, MT-2, into a nonobese diabetic (NOD)-severe combined immunodeficiency (SCID)/common gamma-chain knockout (NOG) mouse model, HTLV-1 infection was established in vivo. When Tfv or AZT was administrated to the mice concomitant with HTLV-1 infection, these compounds could block the infection. Interestingly, they could neither block nor decrease the proviral load of HTLV-1, when they were administered starting one week after infection. These observations suggest that NRTIs are efficient in blocking the early phase of HTLV-1 infection but are inefficient in blocking clonal proliferation of infected cells, which is the dominant way for the persistence of HTLV-1 in the chronic phase. Indeed, one study reported that 3TC reduced the proviral load in HAM/TSP patients (Taylor et al. 1999), whereas another study concluded that there was no definite effect (Machuca et al. 2001); the efficacy of NRTIs in inhibition of HTLV-1 replication in vivo is still not clear.

### 2.7.2 Combination of Zidovudine and IFN-α

A combination therapy with AZT and IFN-α is now the focus as an effective regimen against ATL, although the mechanism of action of this combination remains unknown. In 1995, two independent phase 2 studies demonstrated that the combination therapy of AZT and IFN-α achieved a high response rate in patients with ATL (Gill et al. 1995; Hermine et al. 1995). This therapy was effective, especially in the patients with previously untreated acute-type ATL. So far, several clinical studies have confirmed the efficacy of this combination therapy (Bazarbachi et al. 2011). A high response rate was observed in patients with acute, chronic, and smoldering ATL, although this regimen was not effective against lymphoma-type ATL. Another study has shown that the response rate was 92% for patients who received AZT/IFN-α as the initial treatment (complete response [CR], 58%; partial response [PR], 33%) (Hermine et al. 2002). However, it has also been reported that response and survival in patients treated with AZT/IFN-α after initial chemotherapy are less impressive than in those treated with AZT/IFN-α as the first-line therapy, although the median OS is 11 months for the whole population.

The mechanism for the anti-ATL effect of AZT/IFN-α therapy has not been clarified yet. It is known that type I IFNs have antiviral effects: IFNs can induce IFN-responsive genes that inhibit viral replication and activate the immune response by increasing expression of human leukocyte antigen class I molecules on the surface of infected cells (Goodbourn et al. 2000). Indeed, IFN-α is often used to treat patients with HAM/TSP. Although AZT is an NRTI, it has been reported that it can cause telomere shortening, induce cellular senescence, and stimulate p53-dependent apoptotic signaling (Datta et al. 2006; Gomez et al. 2012). In addition, a recent study showed that AZT alone did not affect cell viability or NF-κB activities in HTLV-1-infected T cells, whereas the combination of AZT and IFN-α markedly induced cellular apoptosis that is associated with the phosphorylation of p53 and induction of p53-responsive genes (Kinpara et al. 2013). Considering that HTLV-1 replication is generally suppressed in ATL cells, it has been suggested that AZT/IFN-α might have antitumor properties rather than antiviral effects.

### 2.7.3 Protease Inhibitors

Because the HTLV-1 protease is responsible for cleavage of the Gag and Pol precursor proteins, it is a potential target for inducing the inhibition of HTLV-1 replication. Several potent inhibitors that target HIV-1 protease are used in clinical practice. Previous studies addressed the utility of HIV-1 protease inhibitors (PIs) such as indinavir, saquinavir, nelfinavir, ritonavir, and amprenavir and showed that none of the HIV-1 PIs were effective against HTLV-1 protease (Bagossi et al. 2004; Kadas et al. 2004; Maegawa et al. 2004), most likely because HTLV-1 protease is quite different from HIV-1 in size and amino acid sequence. Recently, small compounds against HTLV-1 protease were designed and shown to be effective at low concentrations, suggesting that they are good candidates for anti-HTLV-1 therapy (Nguyen et al. 2011a, 2011b). The efficacy of the HTLV-1 protease-specific inhibitors warrants further study.

### 2.7.4 Integrase Inhibitors

It is known that HTLV-1 and HIV-1 integrases (INs) share important structural properties, suggesting that the inhibitors against HIV-1 IN may also block HTLV-1 integration. Rabaaoui et al. (2008) were the first to show that two classes of HIV-1 IN inhibitors, styrylquinolines and diketo acids (DKAs), were also active against HTLV-1 IN in vitro and ex vivo. Another study demonstrated that the DKA-based compounds raltegravir and MK-2048 had potent antiviral effects in vitro in cell-free and cell-to-cell HTLV-1 transmission systems (Seegulam and Ratner 2011). However, a clinical study that examined the effect of raltegravir on five HTLV-1-infected individuals did not show significant reductions of proviral loads (Trevino et al. 2012). This finding suggests that IN inhibitors can block the early phase of HTLV-1 transmission/replication but do not impact clonal proliferation of HTLV-1-infected cells.

## 2.8 ANTI-ATL THERAPIES

### 2.8.1 Combination Chemotherapy

Intensive combination chemotherapy has been used to treat the aggressive forms of ATL. Because ATL is a neoplasm of mature T-cells, it has been treated with chemotherapies that are used for non-Hodgkin lymphoma. A combination of *c*yclophosphamide, adriamycin (*h*ydroxydoxorubicin), vincristine (*o*ncovin), and *p*rednisolone (called CHOP therapy) has been the standard regimen for aggressive types of ATL. Granulocyte colony-stimulating, factor-supported, biweekly CHOP is an intensified therapy that is also used for ATL. JCOG has reported that the CR rate with biweekly CHOP is 25% and that the median survival time (MST) is 11 months (Tsukasaki et al. 2007, 2012). It has also been reported that the more potent chemotherapy, vincristine, cyclophosphamide, adriamycin, and prednisolone (VCAP)—adriamycin, MCNU, and prednisolone (AMP)—vindesine, etoposide, carboplatin, and prednisolone (VECP) improves the prognosis of ATL (CR is 40% and MST is 13 months) (Tsukasaki et al. 2012). However, survival rates of ATL patients with these potent chemotherapy regimens are still poor.

### 2.8.2 Allogeneic HSCT

The first report of successful bone marrow transplantation for ATL was published in 1996 (Borg et al. 1996). Subsequently, HSCT is increasingly used as a curative option for ATL. A recent retrospective cohort study in Japan showed that the three-year OS was 33% with HSC, which was a surprising improvement in the prognosis of aggressive ATL (Hishizawa et al. 2010). It was also shown that the cases with no symptoms of graft-versus-host disease relapsed with ATL, suggesting that the graft-versus-leukemia effect is essential for the curative influence of HSCT (Utsunomiya et al. 2001). However, there were two reports showing that donor-cell-derived ATL occurred in recipients after allo-HSCT (Tamaki and Matsuoka 2006; Nakamizo et al. 2011). The donors in both cases were HTLV-1-infected siblings who were asymptomatic. These data indicate that the immunosuppressive status in recipients of allo-HSCT may also contribute to the development of ATL in donor-derived HTLV-1-infected T cells.

### 2.8.3 Molecular-Targeted Agents

#### 2.8.3.1 Monoclonal Antibodies

An alternative approach to ATL therapy is to target cell-surface markers on the malignant cells with monoclonal antibodies. CC chemokine receptor 4 (CCR4) is expressed on leukemic cells from most ATL cases (Yoshie et al. 2002; Ishida et al. 2003). A humanized anti-CCR4 monoclonal antibody, mogamulizumab, with a defucosylated Fc region was developed and proved to markedly enhance antibody-dependent cellular cytotoxicity in ATL cells (Ishida et al. 2012). A clinical trial of mogamulizumab demonstrated significant response in relapsed ATL patients. Now, mogamulizumab is available in Japan as one of the treatment options for refractory ATL. An anti-CD25 (anti-Tac) monoclonal antibody was first administered to ATL patients in the late 1980s (Waldmann et al. 1988) and was reported to be effective in some patients (CR, 2/19; PR, 4/19) (Waldmann et al. 1993). It was reported that the anti-CD52 monoclonal antibody Campath-1H was effective in one patient with AZT/IFN-α-refractory ATL (Mone et al. 2005). A humanized anti-CD2 antibody (MEDI-507) has also been shown to be effective in a xenograft mouse model (Zhang et al. 2003).

#### 2.8.3.2 NF-κB Inhibitors

Nuclear factor kappa-light-chain-enhancer of activated B cells (NF-κB) is a protein complex that controls the transcription of DNA. NF-κB activation has been reported in ATL cells and is strongly associated with oncogenesis; therefore, NF-κB is an attractive molecular target in ATL (Sun and Yamaoka 2005). Bortezomib is a proteasome inhibitor that blocks the degradation of IκB, resulting in inhibition of NF-κB activation. Bortezomib is highly effective against ATL cells in vitro and in vivo (Satou et al. 2004). The combination of arsenic trioxide and IFN-α has been reported to induce cell cycle arrest and apoptosis in HTLV-1-infected cells through inhibition of the NF-κB pathway (El-Sabban et al. 2000). Several other NF-κB inhibitors were also identified and shown to exhibit anti-ATL effects (Mori et al. 2002; Watanabe et al. 2005).

## 2.9 CONCLUDING REMARKS

HTLV-1 can increase its copy number in vivo by clonal proliferation of infected cells and by de novo infection from infected cells to uninfected cells. It is suggested that de novo infection is important in the early phase of viral transmission; however, clonal expansion is predominant in chronic phase and is implicated in HTLV-1-mediated pathogenesis. As a prophylaxis of HTLV-1 transmission, therefore, antiviral drugs might be effective. Because HTLV-1 is generally latent in vivo, blocking proliferation of HTLV-1-infected cells should be important to prevent development of HTLV-1-associated diseases.

HTLV-1 was first discovered in 1980, and intensive studies on viral replication and cellular transformation have revealed the molecular mechanisms of HTLV-1-induced pathogenesis. However, the numbers of HTLV-1 carriers and patients with HTLV-1-associated diseases have not decreased much over the last two decades (Watanabe 2011), and no effective therapies could be developed to combat HTLV-1 infection. However, there exist some anti-ATL therapies, such as combination chemotherapy, allogeneic HSCT, and the application of molecular-targeted agents (e.g., monoclonal antibodies and NF-κB inhibitors). Further studies are needed to develop new treatments and prophylactic strategies based on the growing knowledge of HTLV-1 biology.

## ACKNOWLEDGMENTS

This study was supported by a grant-in-aid for scientific research from the Ministry of Education, Science, Sports, and Culture of Japan (23591384).

## REFERENCES

Bagossi P, Kadas J, Miklossy G, et al. (2004). Development of a microtiter plate fluorescent assay for inhibition studies on the HTLV-1 and HIV-1 proteinases. *J Virol Methods* 119:87–93.

Bazarbachi A, Suarez F, Fields P, Hermine O. (2011). How I treat adult T-cell leukemia/lymphoma. *Blood* 118:1736–45.

Borg A, Yin JA, Johnson PR, et al. (1996). Successful treatment of HTLV-1-associated acute adult T-cell leukaemia lymphoma by allogeneic bone marrow transplantation. *Br J Haematol* 94:713–15.

Boxus M, Twizere JC, Legros S, et al. (2008). The HTLV-1 Tax interactome. *Retrovirology* 5:76.

Carvalho EM, Da Fonseca Porto A. (2004). Epidemiological and clinical interaction between HTLV-1 and *Strongyloides stercoralis*. *Parasite Immunol* 26:487–97.

Cavanagh MH, Landry S, Audet B, et al. (2006). HTLV-I antisense transcripts initiating in the 3'LTR are alternatively spliced and polyadenylated. *Retrovirology* 3:15.

Datta A, Bellon M, Sinha-Datta U, et al. (2006). Persistent inhibition of telomerase reprograms adult T-cell leukemia to p53-dependent senescence. *Blood* 108:1021–9.

Derse D, Hill SA, Lloyd PA, Chung H, Morse BA. (2001). Examining human T-lymphotropic virus type 1 infection and replication by cell-free infection with recombinant virus vectors. *J Virol* 75:8461–8.

Doi K, Wu X, Taniguchi Y, et al. (2005). Preferential selection of human T-cell leukemia virus type I provirus integration sites in leukemic versus carrier states. *Blood* 106:1048–53.

El-Sabban ME, Nasr R, Dbaibo G, et al. (2000). Arsenic-interferon-alpha-triggered apoptosis in HTLV-I transformed cells is associated with tax down-regulation and reversal of NF-kappa B activation. *Blood* 96:2849–55.

Etoh K, Tamiya S, Yamaguchi K, et al. (1997). Persistent clonal proliferation of human T-lymphotropic virus type I-infected cells in vivo. *Cancer Res* 57:4862–7.

Fan J, Ma G, Nosaka K, et al. (2010). APOBEC3G generates nonsense mutations in human T-cell leukemia virus type 1 proviral genomes in vivo. *J Virol* 84:7278–87.

Fields PA, Taylor GP. (2012)."Antivirals" in the treatment of adult T cell leukaemia-lymphoma (ATLL). *Curr Hematol Malig Rep* 7:267–75.

Franchini G, Fukumoto R, Fullen JR. (2003). T-cell control by human T-cell leukemia/lymphoma virus type 1. *Int J Hematol* 78:280–96.

Furukawa Y, Kubota R, Tara M, Izumo S, Osame M. (2001). Existence of escape mutant in HTLV-I tax during the development of adult T-cell leukemia. *Blood* 97:987–93.

Gabet AS, Mortreux F, Talarmin A, et al. (2000). High circulating proviral load with oligoclonal expansion of HTLV-1 bearing T cells in HTLV-1 carriers with strongyloidiasis. *Oncogene* 19:4954–60.

Gallo RC. (2005). The discovery of the first human retrovirus: HTLV-1 and HTLV-2. *Retrovirology* 2:17.

Gaudray G, Gachon F, Basbous J, et al. (2002). The complementary strand of the human T-cell leukemia virus type 1 RNA genome encodes a bZIP transcription factor that downregulates viral transcription. *J Virol* 76:12813–22.

Gazon H, Lemasson I, Polakowski N, et al. (2012). Human T-cell leukemia virus type 1 (HTLV-1) bZIP factor requires cellular transcription factor JunD to upregulate HTLV-1 antisense transcription from the 3′ long terminal repeat. *J Virol* 86:9070–8.

Ghez D, Lepelletier Y, Lambert S, et al. (2006). Neuropilin-1 is involved in human T-cell lymphotropic virus type 1 entry. *J Virol* 80:6844–54.

Gill PS, Harrington W, Jr, Kaplan MH, et al. (1995). Treatment of adult T-cell leukemia-lymphoma with a combination of interferon alfa and zidovudine. *N Engl J Med* 332:1744–8.

Gillet NA, Cook L, Laydon DJ, et al. (2013). Strongyloidiasis and infective dermatitis alter human T lymphotropic virus-1 clonality in vivo. *PLoS Pathog* 9:e1003263.

Gillet NA, Malani N, Melamed A, et al. (2011). The host genomic environment of the provirus determines the abundance of HTLV-1-infected T-cell clones. *Blood* 117:3113–22.

Gomez DE, Armando RG, Alonso DF. (2012). AZT as a telomerase inhibitor. *Front Oncol* 2:113.

Goodbourn S, Didcock L, Randall RE. (2000). Interferons: Cell signalling, immune modulation, antiviral response and virus countermeasures. *J Gen Virol* 81:2341–64.

Gout O, Baulac M, Gessain A, et al. (1990). Rapid development of myelopathy after HTLV-I infection acquired by transfusion during cardiac transplantation. *N Engl J Med* 322:383–8.

Grossman WJ, Kimata JT, Wong FH, et al. (1995). Development of leukemia in mice transgenic for the tax gene of human T-cell leukemia virus type I. *Proc Natl Acad Sci U S A* 92:1057–61.

Hanchard B. (1996). Adult T-cell leukemia/lymphoma in Jamaica: 1986–1995. *J Acquir Immune Defic Syndr Hum Retrovirol* 13(Suppl 1):S20–5.

Hanon E, Hall S, Taylor GP, et al. (2000). Abundant tax protein expression in CD4+ T cells infected with human T-cell lymphotropic virus type I (HTLV-I) is prevented by cytotoxic T lymphocytes. *Blood* 95:1386–92.

Hasegawa H, Sawa H, Lewis MJ, et al. (2006). Thymus-derived leukemia-lymphoma in mice transgenic for the Tax gene of human T-lymphotropic virus type I. *Nat Med* 12:466–72.

Hermine O, Allard I, Levy V, et al. (2002). A prospective phase II clinical trial with the use of zidovudine and interferon-alpha in the acute and lymphoma forms of adult T-cell leukemia/lymphoma. *Hematol J* 3:276–82.

Hermine O, Bouscary D, Gessain A, et al. (1995). Brief report: Treatment of adult T-cell leukemia-lymphoma with zidovudine and interferon alfa. *N Engl J Med* 332:1749–51.

Hidaka M, Inoue J, Yoshida M, Seiki M. (1988). Post-transcriptional regulator (rex) of HTLV-1 initiates expression of viral structural proteins but suppresses expression of regulatory proteins. *EMBO J* 7:519–23.

Hill SA, Lloyd PA, McDonald S, Wykoff J, Derse D. (2003). Susceptibility of human T cell leukemia virus type I to nucleoside reverse transcriptase inhibitors. *J Infect Dis* 188:424–7.

Hisada M, Stuver SO, Okayama A, et al. (2004). Persistent paradox of natural history of human T lymphotropic virus type I: Parallel analyses of Japanese and Jamaican carriers. *J Infect Dis* 190:1605–9.

Hishizawa M, Kanda J, Utsunomiya A, et al. (2010). Transplantation of allogeneic hematopoietic stem cells for adult T-cell leukemia: A nationwide retrospective study. *Blood* 116:1369–76.

Igakura T, Stinchcombe JC, Goon PK, et al. (2003). Spread of HTLV-I between lymphocytes by virus-induced polarization of the cytoskeleton. *Science* 299:1713–16.

Inoue J, Seiki M, Yoshida M. (1986). The second pX product p27 chi-III of HTLV-1 is required for gag gene expression. *FEBS Lett* 209:187–90.

Ishida T, Joh T, Uike N, et al. (2012). Defucosylated anti-CCR4 monoclonal antibody (KW-0761) for relapsed adult T-cell leukemia-lymphoma: A multicenter phase II study. *J Clin Oncol* 30:837–42.

Ishida T, Utsunomiya A, Iida S, et al. (2003). Clinical significance of CCR4 expression in adult T-cell leukemia/lymphoma: Its close association with skin involvement and unfavorable outcome. *Clin Cancer Res* 9:3625–34.

Iwanaga M, Watanabe T, Yamaguchi K. (2012). Adult T-cell leukemia: A review of epidemiological evidence. *Front Microbiol* 3:322.

Izumo S, Goto I, Itoyama Y, et al. (1996). Interferon-alpha is effective in HTLV-I-associated myelopathy: A multicenter, randomized, double-blind, controlled trial. *Neurology* 46:1016–21.

Jones KS, Lambert S, Bouttier M, et al. (2011). Molecular aspects of HTLV-1 entry: Functional domains of the HTLV-1 surface subunit (SU) and their relationships to the entry receptors. *Viruses* 3:794–810.

Jones KS, Petrow-Sadowski C, Bertolette DC, Huang Y, Ruscetti FW. (2005). Heparan sulfate proteoglycans mediate attachment and entry of human T-cell leukemia virus type 1 virions into CD4+ T cells. *J Virol* 79:12692–702.

Kadas J, Weber IT, Bagossi P, et al. (2004). Narrow substrate specificity and sensitivity toward ligand-binding site mutations of human T-cell leukemia virus type 1 protease. *J Biol Chem* 279:27148–57.

Kannagi M, Hasegawa A, Kinpara S, et al. (2011). Double control systems for human T-cell leukemia virus type 1 by innate and acquired immunity. *Cancer Sci* 102:670–6.

Kannagi M, Ohashi T, Harashima N, Hanabuchi S, Hasegawa A. (2004). Immunological risks of adult T-cell leukemia at primary HTLV-I infection. *Trends Microbiol* 12:346–52.

Kashanchi F, Brady JN. (2005). Transcriptional and post-transcriptional gene regulation of HTLV-1. *Oncogene* 24:5938–51.

Kinjo T, Ham-Terhune J, Peloponese JM, Jr, Jeang KT. (2010). Induction of reactive oxygen species by human T-cell leukemia virus type 1 tax correlates with DNA damage and expression of cellular senescence marker. *J Virol* 84:5431–7.

Kinpara S, Kijiyama M, Takamori A, et al. (2013). Interferon-alpha (IFN-alpha) suppresses HTLV-1 gene expression and cell cycling, while IFN-alpha combined with zidovudin induces p53 signaling and apoptosis in HTLV-1-infected cells. *Retrovirology* 10:52.

Koiwa T, Hamano-Usami A, Ishida T, et al. (2002). 5′-long terminal repeat-selective CpG methylation of latent human T-cell leukemia virus type 1 provirus in vitro and in vivo. *J Virol* 76:9389–97.

Koyanagi Y, Itoyama Y, Nakamura N, et al. (1993). In vivo infection of human T-cell leukemia virus type I in non-T cells. *Virology* 196:25–33.

Macchi B, Faraoni I, Zhang J, et al. (1997). AZT inhibits the transmission of human T cell leukaemia/lymphoma virus type I to adult peripheral blood mononuclear cells in vitro. *J Gen Virol* 78(Pt 5):1007–16.

Machuca A, Rodes B, Soriano V. (2001). The effect of antiretroviral therapy on HTLV infection. *Virus Res* 78:93–100.

Maegawa H, Kimura T, Arii Y, et al. (2004). Identification of peptidomimetic HTLV-I protease inhibitors containing hydroxymethylcarbonyl (HMC) isostere as the transition-state mimic. *Bioorg Med Chem Lett* 14:5925–9.

Manel N, Kim FJ, Kinet S, et al. (2003). The ubiquitous glucose transporter GLUT-1 is a receptor for HTLV. *Cell* 115:449–59.

Matsuoka M. (2003). Human T-cell leukemia virus type I and adult T-cell leukemia. *Oncogene* 22:5131–40.

Matsuoka M, Jeang KT. (2007). Human T-cell leukaemia virus type 1 (HTLV-1) infectivity and cellular transformation. *Nat Rev Cancer* 7:270–80.

Matsushita S, Mitsuya H, Reitz MS, Broder S. (1987). Pharmacological inhibition of in vitro infectivity of human T lymphotropic virus type I. *J Clin Invest* 80:394–400.

McGill NK, Vyas J, Shimauchi T, Tokura Y, Piguet V. (2012). HTLV-1-associated infective dermatitis: Updates on the pathogenesis. *Exp Dermatol* 21:815–21.

Miyazaki M, Yasunaga J, Taniguchi Y, et al. (2007). Preferential selection of human T-cell leukemia virus type 1 provirus lacking the 5′ long terminal repeat during oncogenesis. *J Virol* 81:5714–23.

Miyazato P, Yasunaga J, Taniguchi Y, et al. (2006). De novo human T-cell leukemia virus type 1 infection of human lymphocytes in NOD-SCID, common gamma-chain knockout mice. *J Virol* 80:10683–91.

Mochizuki M. (2010). Regional immunity of the eye. *Acta Ophthalmol* 88:292–9.

Mone A, Puhalla S, Whitman S, et al. (2005). Durable hematologic complete response and suppression of HTLV-1 viral load following alemtuzumab in zidovudine/IFN-{alpha}-refractory adult T-cell leukemia. *Blood* 106:3380–2.

Mori N, Yamada Y, Ikeda S, et al. (2002). Bay 11–7082 inhibits transcription factor NF-kappaB and induces apoptosis of HTLV-I-infected T-cell lines and primary adult T-cell leukemia cells. *Blood* 100:1828–34.

Murata K, Hayashibara T, Sugahara K, et al. (2006). A novel alternative splicing isoform of human T-cell leukemia virus type 1 bZIP factor (HBZ-SI) targets distinct subnuclear localization. *J Virol* 80:2495–505.

Nakada K, Yamaguchi K, Furugen S, et al. (1987). Monoclonal integration of HTLV-I proviral DNA in patients with strongyloidiasis. *Int J Cancer* 40:145–8.

Nakagawa M, Nakahara K, Maruyama Y, et al. (1996). Therapeutic trials in 200 patients with HTLV-I-associated myelopathy/tropical spastic paraparesis. *J Neurovirol* 2:345–55.

Nakamizo A, Akagi Y, Amano T, et al. (2011). Donor-derived adult T-cell leukaemia. *Lancet* 377:1124.

Nam SH, Kidokoro M, Shida H, Hatanaka M. (1988). Processing of gag precursor polyprotein of human T-cell leukemia virus type I by virus-encoded protease. *J Virol* 62:3718–28.

Nguyen JT, Kato K, Hidaka K, et al. (2011a). Design and synthesis of several small-size HTLV-I protease inhibitors with different hydrophilicity profiles. *Bioorg Med Chem Lett* 21:2425–9.

Nguyen JT, Kato K, Kumada HO, et al. (2011b). Maintaining potent HTLV-I protease inhibition without the P3-cap moiety in small tetrapeptidic inhibitors. *Bioorg Med Chem Lett* 21:1832–7.

Nicot C, Dundr M, Johnson JM, et al. (2004). HTLV-1-encoded p30II is a post-transcriptional negative regulator of viral replication. *Nat Med* 10:197–201.

Okayama A, Stuver S, Matsuoka M, et al. (2004). Role of HTLV-1 proviral DNA load and clonality in the development of adult T-cell leukemia/lymphoma in asymptomatic carriers. *Int J Cancer* 110:621–5.

Osame M, Janssen R, Kubota H, et al. (1990). Nationwide survey of HTLV-I-associated myelopathy in Japan: Association with blood transfusion. *Ann Neurol* 28:50–6.

Osame M, Matsumoto M, Usuku K, et al. (1987). Chronic progressive myelopathy associated with elevated antibodies to human T-lymphotropic virus type I and adult T-cell leukemialike cells. *Ann Neurol* 21:117–22.

Pais-Correia AM, Sachse M, Guadagnini S, et al. (2010). Biofilm-like extracellular viral assemblies mediate HTLV-1 cell-to-cell transmission at virological synapses. *Nat Med* 16:83–9.

Pique C, Tursz T, Dokhelar MC. (1990). Mutations introduced along the HTLV-I envelope gene result in a non-functional protein: A basis for envelope conservation? *EMBO J* 9:4243–8.

Poiesz BJ, Ruscetti FW, Gazdar AF, et al. (1980). Detection and isolation of type C retrovirus particles from fresh and cultured lymphocytes of a patient with cutaneous T-cell lymphoma. *Proc Natl Acad Sci U S A* 77:7415–19.

Primo J, Siqueira I, Nascimento MC, et al. (2009). High HTLV-1 proviral load, a marker for HTLV-1 associated myelopathy/tropical spastic paraparesis, is also detected in patients with infective dermatitis associated with HTLV-1. *Braz J Med Biol Res* 42:761–4.

Rabaaoui S, Zouhiri F, Lancon A, et al. (2008). Inhibitors of strand transfer that prevent integration and inhibit human T-cell leukemia virus type 1 early replication. *Antimicrob Agents Chemother* 52:3532–41.

Saito M, Bangham CR (2012). Immunopathogenesis of human T-cell leukemia virus type-1-associated myelopathy/tropical spastic paraparesis: Recent perspectives. *Leuk Res Treatment* 2012:259045.

Satou Y, Matsuoka M. (2012). Molecular and cellular mechanism of leukemogenesis of ATL: Emergent evidence of a significant role for HBZ in HTLV-1-induced pathogenesis. *Leuk Res Treatment* 2012:213653.

Satou Y, Nosaka K, Koya Y, et al. (2004). Proteasome inhibitor, bortezomib, potently inhibits the growth of adult T-cell leukemia cells both in vivo and in vitro. *Leukemia* 18:1357–63.

Satou Y, Utsunomiya A, Tanabe J, et al. (2012). HTLV-1 modulates the frequency and phenotype of FoxP3+CD4+ T cells in virus-infected individuals. *Retrovirology* 9:46.

Satou Y, Yasunaga J, Yoshida M, Matsuoka M. (2006). HTLV-I basic leucine zipper factor gene mRNA supports proliferation of adult T cell leukemia cells. *Proc Natl Acad Sci U S A* 103:720–5.

Satou Y, Yasunaga J, Zhao T, et al. (2011). HTLV-1 bZIP factor induces T-cell lymphoma and systemic inflammation in vivo. *PLoS Pathog* 7:e1001274.

Seegulam ME, Ratner L. (2011). Integrase inhibitors effective against human T-cell leukemia virus type 1. *Antimicrob Agents Chemother* 55:2011–17.

Shimoyama M. (1991). Diagnostic criteria and classification of clinical subtypes of adult T-cell leukaemia-lymphoma. A report from the Lymphoma Study Group (1984–87). *Br J Haematol* 79:428–37.

Sun SC, Yamaoka S. (2005). Activation of NF-kappaB by HTLV-I and implications for cell transformation. *Oncogene* 24:5952–64.

Takahashi T, Takase H, Urano T, et al. (2000). Clinical features of human T-lymphotropic virus type 1 uveitis: A long-term follow-up. *Ocul Immunol Inflamm* 8:235–41.

Takasaki Y, Iwanaga M, Imaizumi Y, et al. (2010). Long-term study of indolent adult T-cell leukemia-lymphoma. *Blood* 115:4337–43.

Takatsuki K. (2005). Discovery of adult T-cell leukemia. *Retrovirology* 2:16.

Takeda S, Maeda M, Morikawa S, et al. (2004). Genetic and epigenetic inactivation of tax gene in adult T-cell leukemia cells. *Int J Cancer* 109:559–67.

Tamaki H, Matsuoka M. (2006). Donor-derived T-cell leukemia after bone marrow transplantation. *N Engl J Med* 354:1758–9.

Tanaka G, Okayama A, Watanabe T, et al. (2005). The clonal expansion of human T lymphotropic virus type 1-infected T cells: A comparison between seroconverters and long-term carriers. *J Infect Dis* 191:1140–7.

Taniguchi Y, Nosaka K, Yasunaga J, et al. (2005). Silencing of human T-cell leukemia virus type I gene transcription by epigenetic mechanisms. *Retrovirology* 2:64.

Taylor GP, Hall SE, Navarrete S, et al. (1999). Effect of lamivudine on human T-cell leukemia virus type 1 (HTLV-1) DNA copy number, T-cell phenotype, and anti-tax cytotoxic T-cell frequency in patients with HTLV-1-associated myelopathy. *J Virol* 73:10289–95.

Trentin B, Rebeyrotte N, Mamoun RZ. (1998). Human T-cell leukemia virus type 1 reverse transcriptase (RT) originates from the pro and pol open reading frames and requires the presence of RT-RNase H (RH) and RT-RH-integrase proteins for its activity. *J Virol* 72:6504–10.

Trevino A, Parra P, Bar-Magen T, et al. (2012). Antiviral effect of raltegravir on HTLV-1 carriers. *J Antimicrob Chemother* 67:218–21.

Tsukasaki K, Tobinai K, Hotta T, Shimoyama M. (2012). Lymphoma study group of JCOG. *Jpn J Clin Oncol* 42:85–95.

Tsukasaki K, Utsunomiya A, Fukuda H, et al. (2007). VCAP-AMP-VECP compared with biweekly CHOP for adult T-cell leukemia-lymphoma: Japan Clinical Oncology Group Study JCOG9801. *J Clin Oncol* 25:5458–64.

Umeki K, Hisada M, Maloney EM, Hanchard B, Okayama A. (2009). Proviral loads and clonal expansion of HTLV-1-infected cells following vertical transmission: A 10-year follow-up of children in Jamaica. *Intervirology* 52:115–22.

Utsunomiya A, Miyazaki Y, Takatsuka Y, et al. (2001). Improved outcome of adult T cell leukemia/lymphoma with allogeneic hematopoietic stem cell transplantation. *Bone Marrow Transplant* 27:15–20.

Van Prooyen N, Gold H, Andresen V, et al. (2010). Human T-cell leukemia virus type 1 p8 protein increases cellular conduits and virus transmission. *Proc Natl Acad Sci U S A* 107:20738–43.

Waldmann TA, Goldman CK, Bongiovanni KF, et al. (1988). Therapy of patients with human T-cell lymphotrophic virus I-induced adult T-cell leukemia with anti-Tac, a monoclonal antibody to the receptor for interleukin-2. *Blood* 72:1805–16.

Waldmann TA, White JD, Goldman CK, et al. (1993). The interleukin-2 receptor: A target for monoclonal antibody treatment of human T-cell lymphotrophic virus I-induced adult T-cell leukemia. *Blood* 82:1701–12.

Watanabe M, Ohsugi T, Shoda M, et al. (2005). Dual targeting of transformed and untransformed HTLV-1-infected T cells by DHMEQ, a potent and selective inhibitor of NF-kappaB, as a strategy for chemoprevention and therapy of adult T-cell leukemia. *Blood* 106:2462–71.

Watanabe T. (2011). Current status of HTLV-1 infection. *Int J Hematol* 94:430–4.

Wattel E, Vartanian JP, Pannetier C, Wain-Hobson S. (1995). Clonal expansion of human T-cell leukemia virus type I-infected cells in asymptomatic and symptomatic carriers without malignancy. *J Virol* 69:2863–8.

Yasunaga J, Matsuoka M. (2007). Leukaemogenic mechanism of human T-cell leukaemia virus type I. *Rev Med Virol* 17:301–11.

Yasunaga J, Matsuoka M. (2011). Molecular mechanisms of HTLV-1 infection and pathogenesis. *Int J Hematol* 94:435–42.

Yasunaga J, Sakai T, Nosaka K, et al. (2001). Impaired production of naive T lymphocytes in human T-cell leukemia virus type I-infected individuals: Its implications in the immunodeficient state. *Blood* 97:3177–83.

Yoshida M, Satou Y, Yasunaga J, Fujisawa J, Matsuoka M. (2008). Transcriptional control of spliced and unspliced human T-cell leukemia virus type 1 bZIP factor (HBZ) gene. *J Virol* 82:9359–68.

Yoshie O, Fujisawa R, Nakayama T, et al. (2002). Frequent expression of CCR4 in adult T-cell leukemia and human T-cell leukemia virus type 1-transformed T cells. *Blood* 99:1505–11.

Yoshimura K, Mochizuki M, Araki S, et al. (1993). Clinical and immunologic features of human T-cell lymphotropic virus type I uveitis. *Am J Ophthalmol* 116:156–63.

Zhang Z, Zhang M, Ravetch JV, Goldman C, Waldmann TA. (2003). Effective therapy for a murine model of adult T-cell leukemia with the humanized anti-CD2 monoclonal antibody, MEDI-507. *Blood* 102:284–8.

# 3 Hepatitis C Virus and Its Inhibitors

## *The Polymerase as a Target for Nucleoside and Nucleotide Analogs*

*Jerome Deval and Martin Huber*

## 3.1 INTRODUCTION

Hepatitis C virus (HCV) is believed to have infected approximately 175 million individuals worldwide, with an estimated two to four million new infections each year (Moradpour et al. 2007). In the United States alone, it is estimated that approximately 3.2 million persons are chronically infected with HCV (according to the Centers for Disease Control). The primary mode of transmission of HCV is via exposure to infected blood, including transfusions from infected donors, and through intravenous use of illicit drugs. Although a minority of all HCV infections will spontaneously resolve without any clinical outcome, an estimated 80% of cases will progress into chronic hepatitis, leading to a significant proportion of cirrhosis and cases of hepatocellular carcinoma (Alter and Seeff 2000). This makes HCV the leading cause of liver transplantation in the United States.

HCV is a member of the *Flaviviridae* family. It contains a single- and positive-strand RNA genome of about 9.5 kilobases (kb). The viral genome encodes only one open reading frame translated to a polyprotein of approximately 3000 amino acids (Figure 3.1). The three major structural proteins core, E1 and E2, together with NS2 and p7 are involved in viral assembly and morphogenesis, followed in the C-terminus by the five nonstructural (NS) proteins NS3, NS4A, NS4B, NS5A, and NS5B that are required for HCV RNA replication.

This review will summarize HCV infection, its replication, and its nucleoside inhibitors.

### 3.1.1 HCV Replication Cycle as a Target for Therapeutic Intervention

In order to initiate the replication cycle of HCV, HCV particles attach to human hepatocytes through interactions with a number of cellular receptors including CD-81, SR-B1, and the late-stage entry protein claudin-1 (for a complete review

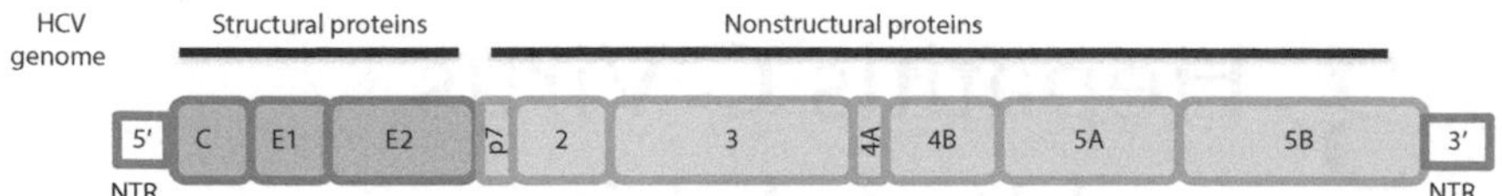


**FIGURE 3.1** Organization of the hepatitis C virus genome. Three major structural proteins—core (C), E1, and E2—are involved in viral assembly and morphogenesis. The five nonstructural proteins NS3, NS4A, NS4B, NS5A, and NS5B are required for HCV RNA replication.

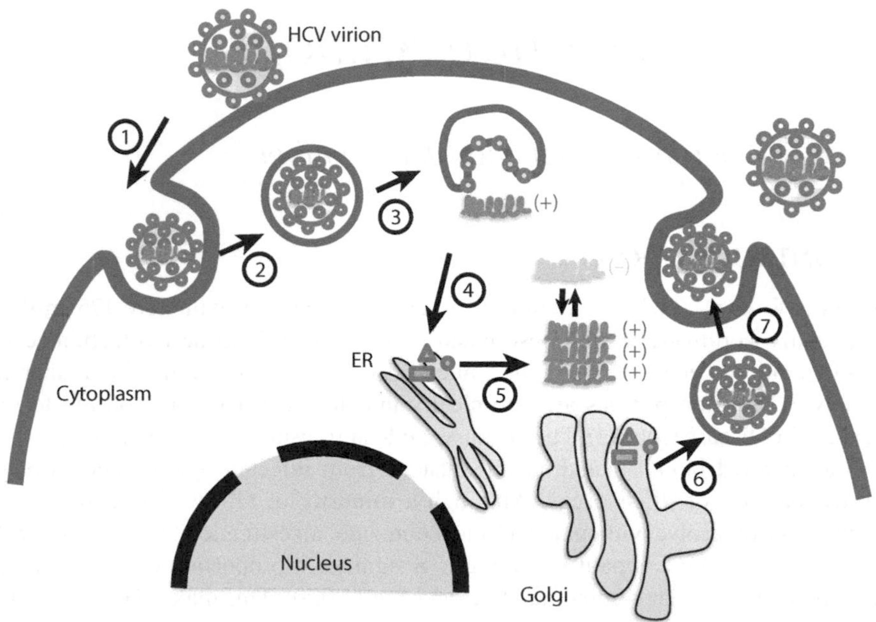


**FIGURE 3.2 (See color insert.)** Replication cycle of hepatitis C virus: (1) attachment, (2) endocytosis, (3) uncoating, (4) translation, (5) RNA replication, (6) virion assembly and maturation, and (7) exocytosis.

on the replication cycle, see Moradpour et al. 2007; Pawlotsky et al. 2007). After attachment, the HCV nucleocapsid is released in the cell cytoplasm as a result of a pH-dependent fusion process between viral and cellular membranes (Figure 3.2).

Decapsidation of viral nucleocapsids liberates free positive-strand genomic RNA in the cell cytoplasm. The viral RNA is used directly as messenger RNA (mRNA) for synthesis of the HCV polyprotein. The immature polyprotein resulting from ribosomal translation only becomes functional after proteolytic cleavage. At least two host cellular and two viral proteases are required for processing the HCV structural and nonstructural proteins. Abundance of the viral nonstructural proteins enables the replication of viral RNA by the replication complex, which causes significant rearrangements of the membranes of endoplasmic reticulum (ER). These "membranous webs" contain the RNA-bound replicase consisting of NS3 (protease and helicase),

NS4B, NS5A, and the NS5B (polymerase). The replicase machinery amplifies the formation of positive-strand RNA and the negative complementary strand used as replication and transcription intermediate. HCV core proteins can then package the nascent RNA by self-assembly to form the nucleocapsid. The newly formed viruses travel away from the ER into the inside portion of the plasma membrane, prior to budding. The plasma membrane encircles the virus and then releases it, initiating the next round of attachment and infection of neighboring liver cells.

Although nearly every single step of HCV replication has been targeted for therapeutic intervention with small molecules, three major proteins are now emerging as the most attractive targets. The NS3 protease function is the target of the two clinically approved protease inhibitors boceprevir and telaprevir. Aside from protease inhibitors, the next generation of anti-HCV small molecules target either NS5A or NS5B.

### 3.1.2 The Limitations of the Current Standard of Care

Current treatment options for HCV infections remain poor and largely suboptimal due to limited efficacy and substantial toxicity. Until 2011, the standard of care (SOC) consisted of a 48-week combination of injected pegylated interferon alpha (PEG-IFN-α) together with the nucleoside analog ribavirin (RBV) taken orally (Figure 3.3). Effective clearance of the virus was achieved in less than 50% of the cases of genotype-1 (GT1) infection, the most prevalent strain of HCV in the United States and Europe. In addition, this regimen often causes significant side effects, such as flu-like symptoms and fatigue (from PEG-IFN-α), as well as anemia (RBV). Since 2011, two inhibitors of the viral protease NS3/4A, boceprevir and telaprevir, are approved to be combined with PEG-IFN-α and RBV in the clinic. These molecules are called direct-acting antivirals (DAAs) because they were designed to specifically bind to, and inhibit, one of the viral proteins required for virus replication. Although the toxicity burden of these new treatment options remains high, the cure rate in the presence of protease inhibitors has increased from less than 50% to 70%–80% in difficult-to-treat GT1 patients (Jensen 2011; Zeuzem et al. 2011). Other DAAs designed to specifically block HCV enzymatic functions have been intensely studied over the last decade, and the polymerase function of NS5B emerges as one of the most attractive targets for the next generation of anti-HCV medicine.

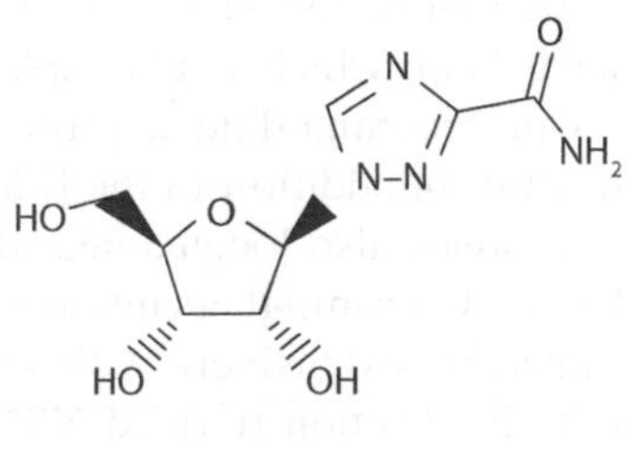

O
N
NH2
N–N
HO
O
HO
OH

**FIGURE 3.3** Chemical structure of ribavirin. Oral ribavirin was approved by the FDA as part of a combination treatment with interferon for hepatitis C in 1998.

## 3.2 STRUCTURE AND FUNCTION(S) OF HCV NS5B

The HCV NS5B protein is an RNA-dependent RNA polymerase (RdRp). The synthesis of RNA by NS5B is required for replication of the viral genome by synthesis of the minus-strand intermediate and also, at the transcription level, for synthesis of viral mRNA. The enzymatic activity of NS5B, consisting of using an RNA template to synthesize RNA product, is unique to viruses and is not found in human cells. This makes NS5B an attractive target for antiviral drug development. The following paragraph provides further details on the folding and function of NS5B.

### 3.2.1 STRUCTURAL CHARACTERIZATION OF HCV NS5B

NS5B is composed of 591 amino acids. Similar to other known RdRps, the HCV NS5B contains six conserved motifs designated A to F. The amino acids involved in the catalytic activity of NS5B are located within motifs A (aspartate at position 220) and the catalytic triad GDD at position 318–320 in motif C (Lesburg et al. 1999). The orientation of these residues in the active site of NS5B and their contribution to the catalytic activity were supported by the crystal structure of the soluble form of NS5B truncated from its C-terminal transmembrane region (delta-21) (Ago et al. 1999; Bressanelli et al. 1999; Lesburg et al. 1999). Using the polymerase right-hand analogy model, the HCV NS5B delta-21 protein also features the fingers, palm, and thumb subdomains (Figure 3.4a). Unlike the traditional open-hand conformation shared by many DNA polymerases, HCV NS5B has an encircled active site due to the extensive interactions between the fingers and thumb subdomains. These contacts are believed to restrict the flexibility of the subdomains and favor the first steps–or initiation–of RNA synthesis leading to the formation of the primer strand. Therefore, it is believed that primer extension during the elongation steps requires important structural changes involving an opening of the thumb and the fingers (Caillet-Saguy et al. 2011; Scrima et al. 2012). Another unique feature of NS5B is its β-hairpin loop that protrudes into the active site located at the base of the palm subdomain. This 12 amino acid loop located within the thumb (residues 443–453) was suggested to interfere with binding to double-stranded RNA due to steric hindrance, and its deletion allows the truncated enzyme to favor primer-dependent RNA synthesis (Hong et al. 2001; Maag et al. 2001; Ranjith-Kumar et al. 2003). Using the same truncated enzyme, it was possible to co crystallize a complex of NS5B together with a short primer-template substrate that artificially mimics the conformation of the polymerase in the elongation mode (Mosley et al. 2012). The equivalent RdRp in poliovirus lacks a similar β-loop, which could explain why polymerases from the *Picornaviridae* family can accommodate a primer-template substrate more efficiently than HCV polymerase. In addition to the β-loop, the C-terminal tail of NS5B contains hydrophobic residues also located near the active site. This region, which immediately precedes the C-terminal membrane anchorage domain, forms a hydrophobic pocket and interacts extensively with several important structural elements including the β-loop. Its deletion (termed NS5B delta-55) also results in an increased overall polymerase activity, most likely by altering the stability of the closed conformation of the enzyme that restricts the transition from de novo

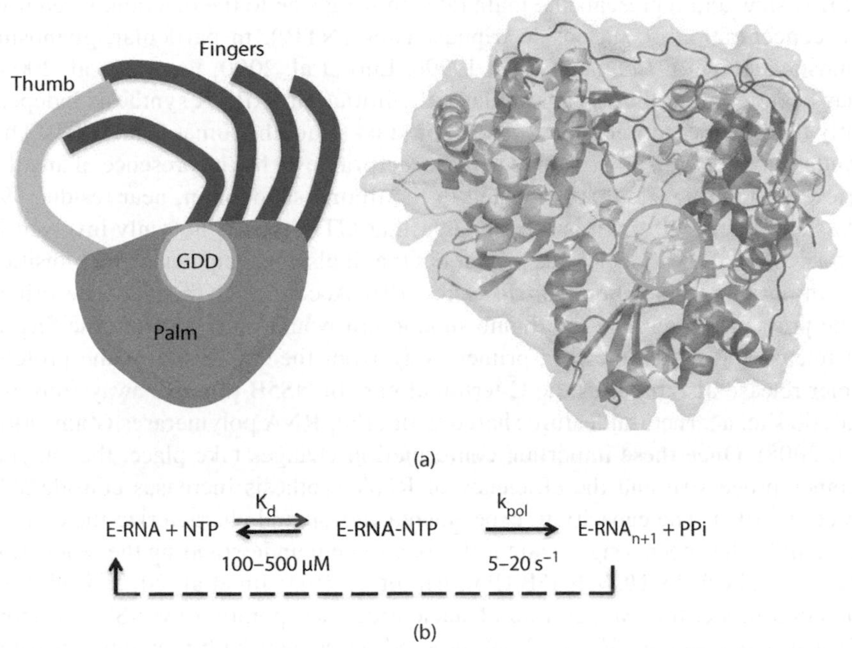


**FIGURE 3.4 (See color insert.)** Three-dimensional structure of NS5B. (a) The polymerase of HCV adopts the shape of a right hand in a closed-state, where fingers (blue) and thumb (green) subdomains form extensive interaction. The catalytic site containing the GDD motif (yellow circle) is at the base of the palm subdomain (magenta); (b) simplified scheme of the nucleotide incorporation reaction. Under the elongation mode, NS5B incorporates nucleotides at a rate of 5–20 bases per second. The enzyme is processive, which means it does not dissociate from the RNA between each cycle of the catalytic reaction.

initiation to primer-dependent elongation (Lohmann et al. 1997; Tomei et al. 2000; Adachi et al. 2002; Leveque and Wang 2002). In summary, the current structural model for NS5B is that the polymerase adopts a closed encircled shape to initiate RNA synthesis, followed by an open state that releases the primer and transitions to the elongation mode.

### 3.2.2 Role(s) of NS5B Polymerase in the Replication of Viral RNA

#### 3.2.2.1 Slow Initiation and Fast Elongation

In HCV-infected cells, the nonstructural proteins NS3 to NS5B of the replication complex are necessary and sufficient for autonomous replication of viral RNA. In this membrane-attached replicase, NS5B is the key enzyme to viral RNA replication because it acts as the motor of RNA synthesis. De novo initiation at the 3′ end of the viral positive- and negative-strand RNA is the physiological mode of initiation of RNA synthesis in infected cells for HCV as well as most other positive-strand RNA viruses. In enzymatic assays, initiation of RNA synthesis by

NS5B is slow and represents the main rate-limiting step to the reaction. It requires high concentration of nucleoside triphosphates (NTPs), in particular, guanosine triphosphate (GTP) (Lohmann et al. 1999b; Luo et al. 2000; Ferrari et al. 2008). It has been reported that GTP stimulates the initiation of RNA synthesis independently from its incorporation in the RNA product (Ranjith-Kumar et al. 2003). This stimulation effect can be explained at the structural level by the presence of an allosteric GTP-binding site that is located in the thumb subdomain, near residue 499 (Bressanelli et al. 2002). It is now thought that GTP itself is not only involved in the formation of the dinucleotide primer but that it also participates in the transition from initiation to elongation (Harrus et al. 2010). According to the model described in the previous paragraph, the thumb subdomain would open up from the fingers and release the neosynthesized primer away from the active site of the protein. Primer release also requires the C-terminal part of NS5B to move away from the catalytic site, a structural feature shared with other RNA polymerases (Zamyatkin et al. 2008). Once these important conformation changes take place, the enzyme becomes processive and the efficiency of RNA synthesis increases considerably (Jin et al. 2012). The capacity to experimentally trap and characterize the elongation complex has been a significant milestone to help understanding the replication of RNA synthesis by HCV NS5B (Powdrill et al. 2011; Jin et al. 2012). Under the elongation mode, the average rate of nucleotide incorporation by NS5B is about 5–20 bases per second (Figure 3.4b), in good agreement with previous estimated rates of 200–700 bases per minute for NS5B replicating longer RNA templates (Lohmann et al. 1998; Tomei et al. 2000). Although it is not fully understood, there have been several independent observations that HCV NS5B may form functional oligomers (Qin et al. 2002; Wang et al. 2002). These protein–protein interactions have been demonstrated physically by chromatography and covalent cross-linking and enzymatically by determining the level of cooperativity. Interestingly, one recent observation that the NS5B–RNA complex forms an active precipitate suggests that oligomerization might be important for the transition to the elongation mode (Jin et al. 2012).

#### 3.2.2.2 Other Functions of NS5B

As we just described, NS5B is capable of making intermolecular interactions in order to promote its own polymerase activity, which is also modulated by the nonstructural protein NS5A through extensive protein–protein interactions. NS5A is a multifunctional protein that is involved in the production of viral particles via its interaction with HCV core nucleocapsid and the synthesis of RNA by the replication complex. In the latter case, NS5A could function as a cofactor for NS5B by stimulating the level of RNA products from the purified polymerase (Quezada and Kane 2013), although other assay conditions could result in an inhibition effect (Shirota et al. 2002). Importantly, any disruption of the interaction between NS5B and NS5A by point mutations results in a loss of viral RNA replication in the cells (Shimakami et al. 2004). Finally, NS5B could also be involved in virus morphogenesis (Gouklani et al. 2012). In this case, mutagenesis experiments have shown that certain residues in NS5B are critical for virion assembly and/or maturation.

### 3.2.3 Functional Assays and Efficacy Models to Discover and Characterize NS5B Inhibitors

Although a nucleoside analog that may be an effective HCV inhibitor is dosed as parent molecule in a nonphosphorylated state (except for monophosphate prodrugs), the active species that inhibits NS5B polymerase inside the cell is the NTP counterpart. It is only after the nucleoside analog enters cells that it is phosphorylated in three different steps by human kinases to be converted to a mono-, di-, and finally triphosphate form (Figure 3.5). As will be described in the next sections, biochemical assays have been developed to study the interaction between NS5B and NTP analogs, while only parent nucleosides as well as monophosphate prodrugs are used in cell-based assays or in vivo studies.

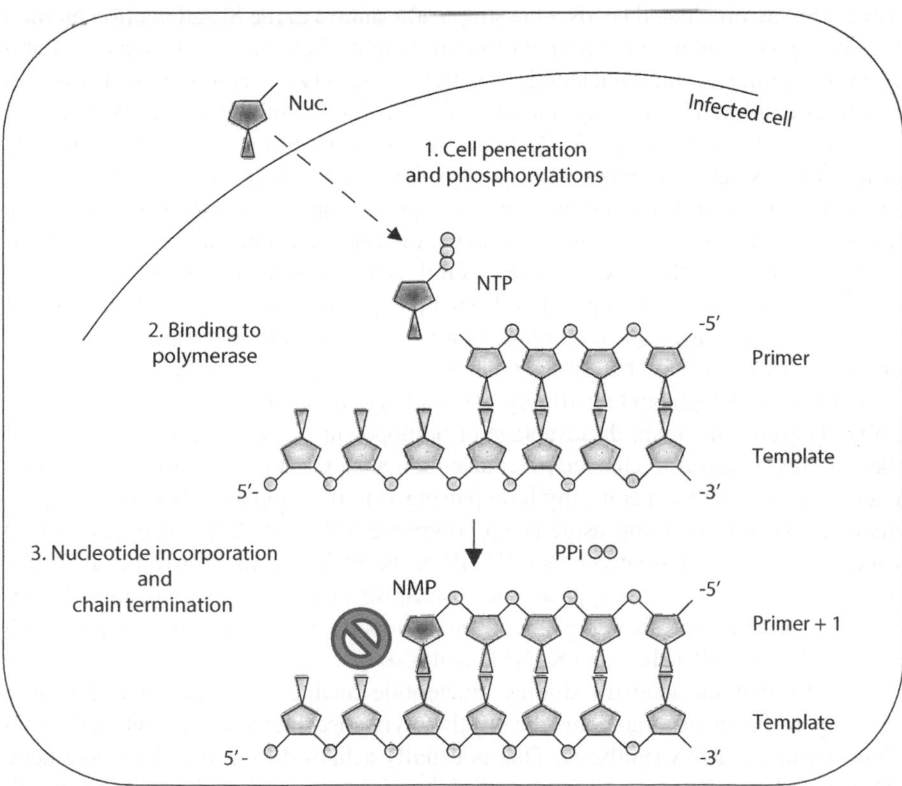


**FIGURE 3.5** Mode of action of nucleoside analogs used against HCV replication. The parent nucleoside (Nuc.) is transported through the cell membrane into the cytoplasm. It is metabolized by host kinases to a mono-, di-, and triphosphate form (NTP). The NTP is recognized as substrate by NS5B and incorporated to the nascent RNA primer strand with release of pyrophosphate anion (PPi) coproduct. In most cases, incorporation of an anti-HCV nucleotide analog results in immediate chain termination, meaning that further RNA synthesis is blocked.

#### 3.2.3.1 Biochemical Assays

Recombinant NS5B protein purified from prokaryotic expression systems has been widely used as target for drug discovery in primary screens for small molecule inhibitors. Although the hydrophobic C-terminal tail of 21 amino acids that tether NS5B onto the membranes is needed for viral replication in cells, it is dispensable for the enzymatic activity of the recombinant protein (Yamashita et al. 1998; Tomei et al. 2000; Vo et al. 2004). NS5B lacking the C-terminal 21 residues (delta-21) is widely used in biochemical, structural, and inhibitor studies because it can be purified to homogeneity and does not require detergents in the buffer. The polymerase reaction carried by the truncated NS5B protein is easily compatible with high-throughput screening of small molecule inhibitors using microplate assays. For instance, a 5′-biotinylated oligo(G) primer is annealed to a poly(rC) template such that the enzyme incorporates radiolabeled [$^3$H]GTP as substrate in the reaction. The radiolabeled product can then be traced by the scintillation proximity assay (SPA) using streptavidin-coated beads. One simple alternative to the SPA detection method is to precipitate and filter the high molecular weight RNA products to separate them from the unincorporated nucleotide substrate. These two methods have been successfully employed to identify and characterize several inhibitors of NS5B (Love et al. 2003; Tomei et al. 2003; Di Marco et al. 2005; Shi et al. 2009; Nyanguile et al. 2010). When combined with biophysical or crystallographic methods, these high-throughput HCV polymerase assays helped to rapidly identify small molecule leads. This is the reason why there are now more than 50 crystal structures of NS5B-inhibitor complexes that have been registered to the Protein Data Bank.

Another advantage of using high-throughput polymerase assays with homopolymeric RNA substrates is the requirement for very low enzyme concentration, typically around or below 20 nM. This parameter is important to accurately measure the potency of inhibitors with high binding affinity, also called tight-binding inhibitors (Hang et al. 2009). However, the main disadvantage of homopolymeric RNA-based assays is that they are not designed to characterize nucleotide analogs. Because nucleotide analogs interact with NS5B by specifically base-pairing with the template RNA, proper enzymatic assays must be setup using heteropolymeric RNA template sequences derived from the 3′- or 5′-untranslated region (UTR) of the HCV genome (Carroll et al. 2003). Although these assays are typically less sensitive and require significantly higher enzyme concentrations, they provide convenient tools to measure kinetic constants such as $K_m$(NTP), as well as $IC_{50}$ and $K_i$(NTP analogs).

For detailed mechanistic studies, nucleotide analogs are evaluated for their capacity to be incorporated by the viral polymerase into the nascent RNA and chain terminate RNA synthesis. This is usually achieved by using short, synthetic RNA templates (50-mer or less) containing single sites for nucleotide incorporation, and the short RNA products are resolved by sequencing gel analysis (Carroll et al. 2003; Olsen et al. 2004; Dutartre et al. 2006; Klumpp et al. 2006; Deval et al. 2007; Murakami et al. 2008). Finally, it should be noted that HCV polymerase activity can also be measured from native replicase complexes extracted from infected or replicon-transfected cells (Migliaccio et al. 2003; Tomassini et al. 2003). This biochemical assay has the advantage of containing all the protein constituents of the authentic replicase complex, including full-length NS5B bound to NS5A, NS3, and NS4B (Ma et al. 2005). However, this system is rarely used mainly because it is time consuming and it can measure only low levels of polymerase activity.

#### 3.2.3.2 Cell-Based Assays

The development of a broadly useful HCV subgenomic replicon system came as a major breakthrough after HCV research and drug discovery had been hampered for a long time by the inability to efficiently propagate the virus in tissue culture. The first version consisted of a bicistronic RNA design encoding neomycin resistance as a selectable marker and the nonstructural HCV proteins (NS2, or NS3 to NS5B, respectively) derived from a GT1b consensus sequence ("Con1"). Human hepatocellular carcinoma cells (Huh-7) transfected with this construct supported autonomous HCV RNA replication at high levels under G418 selective pressure (Lohmann et al. 1999a), thereby providing a model for testing compounds that target nonstructural proteins, in particular the NS3/4A protease and the NS5B RNA polymerase (Figure 3.6a). Subgenomic replicons were also extensively used to discover and characterize resistance mutations associated with inhibitor testing (Figure 3.6b). Later, replicons were constructed representing additional genotypes and incorporating reporter genes to simplify the measurement of viral RNA replication.

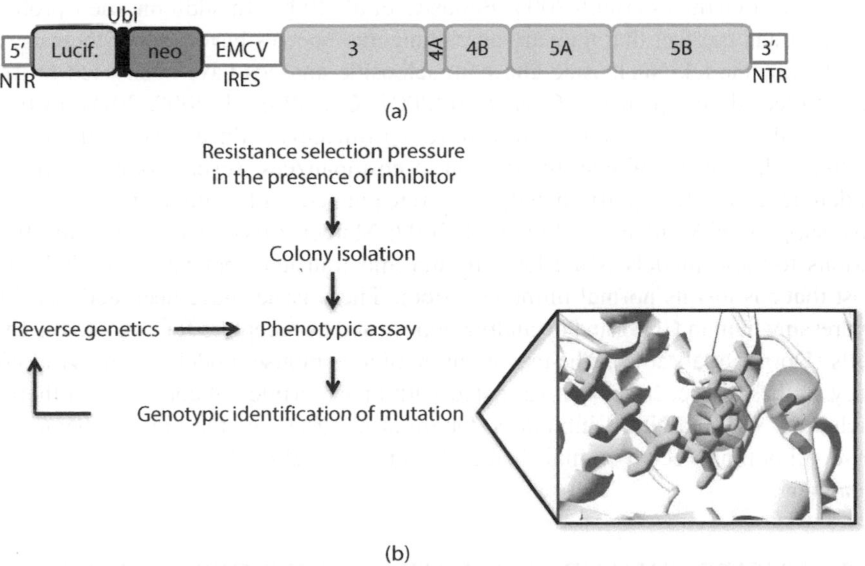


**FIGURE 3.6 (See color insert.)** Use of subgenomic replicon in cell-based assays. (a) Example of a subgenomic HCV replicon construct. In this case, the bis-cistronic replicon is first composed of the 5′-NTR, followed by the reporter gene (i.e., luciferase), and the selectable marker gene neomycin joined in the first cistron by a cleavable ubiquitin sequence. The internal ribosome entry site (IRES) enables the translation of the HCV nonstructural genes NS3 to NS5B and the construct ends with a 3′-NTR. (b) Principle of the characterization of resistance mutations.

Finally, the preparation of a full-length replicon based on GT2 isolates was reported, which was capable of complete replication in vitro including production of infectious virus particles (Lindenbach et al. 2005).

Replicon systems generally rely on transformed cell lines, which do not fully reflect host-cell-dominated processes, but HCV models using infectious particles have been developed to overcome these limitations. An infectious clone from GT2a capable of replicating in tissue culture was isolated from a Japanese patient with fulminant hepatitis (JFH-1) (Wakita et al. 2005; Lindenbach et al. 2006; Zhong et al. 2006). This clone has become widely used to study all steps of HCV replication, in particular those requiring structural proteins, such as cell entry, fusion, and assembly. However, the limited availability of other strains and genotypes has restricted the use of infectious HCV particles for drug development. In the case of nucleoside analogs, they do not provide any significant advantage over the subgenomic replicon system.

#### 3.2.3.3 Animal Efficacy Models

Animal models of viral infections have played a central role in studying various aspects of host–pathogen interaction, with applications ranging from histopathology to vaccine development. In the case of HCV, chimpanzees are the only animal species that are known to be susceptible to virus replication (Choo et al. 1989). These animals present an attenuated form of infection, manifested by low virus titers and lack of liver cirrhosis (Bukh 2004; Boonstra et al. 2009). In addition, their prohibitive cost and the fact that they are an endangered species have limited their use as an efficacy model. Nucleoside and nonnucleoside analogs have been used to treat HCV-infected chimpanzees (Chen et al. 2007; Carroll et al. 2009, 2011). In these studies, the NS5B inhibitors were able to significantly reduce viral loads by four to five orders of magnitude. In addition to chimpanzees, severe combined immunodeficiency (SCID) or irradiated mice transplanted with human liver tissue can also support HCV infection (Ilan et al. 2002; Mercer 2011). One of the main limitations to these models is the low viremia and transient replication of HCV in a host that has lost its normal immune system. These issues have been addressed by expressing human CD81 and occluding genes used as receptors for entry to the liver cells (Dorner et al. 2011). The use of "humanized" mouse models for antiviral efficacy studies has recently been validated with an experimental combination therapy including NS3 and NS5B inhibitors (Ohara et al. 2011). At the highest therapeutic dose, all of the infected animals had undetectable viral load after only one week of treatment.

## 3.3 NUCLEOSIDE AND NUCLEOTIDE INHIBITORS OF HCV RdRp

### 3.3.1 Nucleosides with Modifications in the 2′ Position

#### 3.3.1.1 Nucleosides with a Single 2′ Modification

RdRp nucleoside inhibitors with modifications in the 2′-position were disclosed in patents in 2001 and 2002 (Sommadossi and La Colla 2001a, 2001b; Carroll et al. 2002). Since then, many representatives of this class of compounds have been studied in detail (Table 3.1), and several of them eventually advanced into

**TABLE 3.1**
**Biochemical Data for 2′-Modified Nucleosides**

| | | | Cell-Based Assay | | Enzyme Assay | |
|---|---|---|---|---|---|---|
| **Compound Name** | **No.** | **Fig.** | **$EC_{50}$ [μM]** | **Selectivity[a]** | **$IC_{50}$ [μM]** | **Reference** |
| NM107, 2′-*C*-methylcytidine | **2** | 3.7 | 1.2 | >83 | 0.1–0.2 | Ali et al. (2008); McCown et al. (2008) |
| 2′-*O*-Methylcytidine | **4** | 3.8 | 21 | >4.8 | 2.7–3.8 | Carroll et al. (2003) |
| 2′-*C*-Methyladenosine | **5** | 3.8 | 0.3 | >330 | 1.9 | Carroll et al. (2003) |
| 2′-*C*-Methylguanosine | **6** | 3.8 | 3.5 | >28 | 0.13 | Eldrup et al. (2004a) |
| MK-0608, 2′-*C*-methyl-7-deazaadenosine | **7** | 3.8 | 0.25 | >400 | 0.11 | Eldrup et al. (2004b); Olsen et al. (2004) |
| 2′-*C*-Methyl-7-deazaguanosine | **8** | 3.8 | >100 | | 0.12 | Eldrup et al. (2004b); Olsen et al. (2004) |
| 2′-*C*-Methyl-7-fluoro-7-deazaadenosine | **9** | 3.8 | 0.07 | >1420 | | Eldrup et al. (2004b) |
| 2′-Deoxy-2′-fluorocytidine | **10** | 3.9 | 5[b] | >20 | 14.9 | Stuyver et al. (2004) |
| PSI-6130, 2′-deoxy-2′-fluoro-2′-*C*-methylcytidine | **11** | 3.9 | 0.5– 0.61 | >490 | 0.12 | Stuyver et al. (2006); McCown et al. (2008) |
| RO2433, PSI-6206, 2′-deoxy-2′-fluoro-2′-*C*-methyluridine | **12** | 3.9 | >100 | | 0.52 | Ma et al. (2007) |
| Mericitabine, R7128, RG7128 | **13** | 3.9 | 0.3–0.9 | | | Ali et al. (2008) |

[a] $CC_{50}/EC_{50}$.
[b] $EC_{90}$.

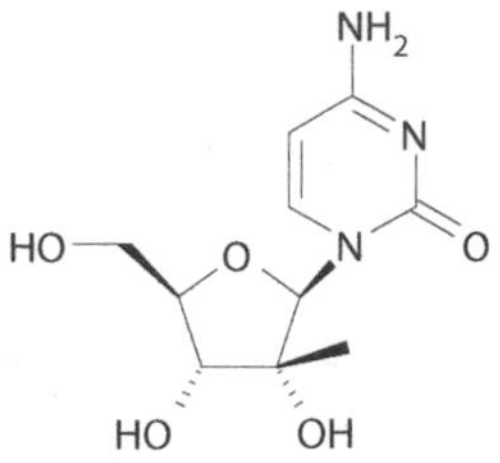

NM107
2′-C-Methylcytidine
**(2)**

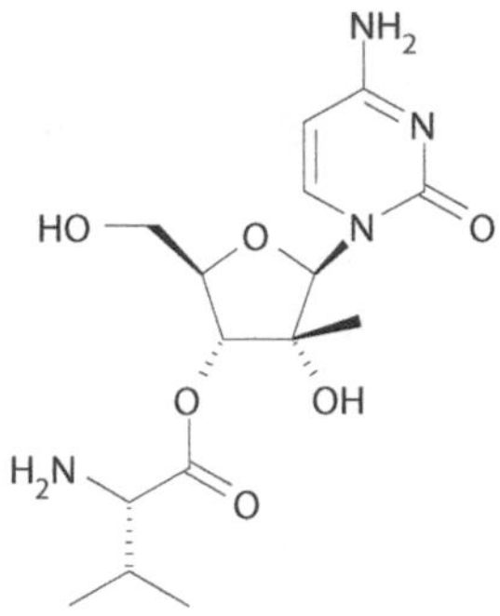

Valopicitabine
**(3)**

**FIGURE 3.7** Chemical structure of 2′-*C*-methylcytidine and its valine ester prodrug valopicitabine.

clinical trials. 2′-*C*-methylcytidine (NM107, **2**, Table 3.1) was first described as an inhibitor of bovine viral diarrhea virus (BVDV) (Sommadossi and La Colla 2001a). It is closely related to HCV and has been successfully used as a surrogate model for the discovery of HCV inhibitors (Buckwold et al. 2003). $EC_{50}$ of (Figure 3.7) **2** against BVDV in vitro was 2 μM, and the $EC_{50}$ values for HCV replicon assays were later reported to be in the range of 1.1–4.8 μM (Table 3.1) (Tomassini et al. 2005; Klumpp et al. 2006). Further studies suggested that **2** acts synergistically with IFN-α-2b in cells infected with BVDV (Bichko et al. 2005), whereas RBV antagonizes the effect of 2′-*C*-methylcytidine (Coelmont et al. 2006). This compound showed low toxicity and high selectivity for viral RdRp over human polymerases in cell culture but suffered from poor oral bioavailability. Esterification of the 3′-OH group of NM107 with L-valine led to valopicitabine (NM283, **3**), a prodrug with considerably improved pharmacokinetic properties (Pierra et al. 2006) that proved to have antiviral activity in chimpanzees chronically infected with HCV GT1 (Standring et al. 2003). In a dose-escalating phase 1/2 clinical study in patients with chronic HCV GT1 infection, monotherapy for 15 days with valopicitabine resulted in a mean viral load reduction of 1.21 $\log_{10}$ IU/mL at the highest dose of 800 mg qd (Table 3.2). Improvements ranged from 0.41 to 2.37 $\log_{10}$ IU/mL (61%–99% reduction) in individual patients in contrast to a 0.03 $\log_{10}$ IU/mL increase in HCV RNA levels observed in the placebo group (Sorbera et al. 2006). Phase 2 clinical trials with valopicitabine alone or in combination with PEG-IFN-α-2b confirmed the antiviral effect of this drug (Table 3.2). Viral breakthrough did not occur in any of these trials, and the resistance conferring S282T mutation in NS5B was not detected in viruses isolated from nine patients (Toniutto et al. 2007). However, dose-limiting gastrointestinal adverse events occurred, and further development of valopicitabine was discontinued due to an unfavorable risk/benefit profile (Poordad et al. 2007).

At the beginning of the long road to the clinic of MK-0608 (**7**), extensive structure–activity relationship (SAR) studies were made that identified six

**TABLE 3.2**
**Select Clinical Trial Data for 2′- and 4′-Modified Nucleosides**

| Drug Name | No. | Dose (mg) | Dosing | Duration | RBV | IFN | GT | Prev. SOC[a] | Mean Viral Load Reduction ($\log_{10}$ IU/mL) | RVR (%) | Status | Reference |
|---|---|---|---|---|---|---|---|---|---|---|---|---|
| NM283 | **3** | 0 | qd | 15 d | – | – | 1 | Yes (87%) | Increase of 0.03 | | T[b] | Sorbera et al. (2006) |
| | | 50 | qd | | | | | | 0.15 | | | |
| | | 100 | qd | | | | | | 0.39 | | | |
| | | 200 | qd | | | | | | 0.3 | | | |
| | | 400 | qd | | | | | | 0.67 | | | |
| | | 800 | qd | | | | | | 1.21 | | | |
| NM283 | **3** | 0 | | 24 w | + | + | 1 | Naïve | 2.3 | | T | Toniutto et al. (2007) |
| | | 800 | qd | | – | – | | | 0.5 | | | |
| | | 400 | qd | | – | + | | | 2.5 | | | |
| | | 400-800 | qd | | – | + | | | 3.0 | | | |
| | | 800 | qd | | – | + | | | 3.3 | | | |
| Mericitabine | **13** | 0 | | 14 d | – | – | 1 | Yes | 0.3 | | Phase 2 | Reddy et al. (2007) |
| | | 750 | qd | | | | | | 0.9 | | | |
| | | 1500 | qd | | | | | | 1.5 | | | |
| | | 750 | bid | | | | | | 2.1 | | | |
| | | 1500 | bid | | | | | | 2.7 | | | |
| Mericitabine | **13** | 0 | | 28 d | + | + | 2/3 | Yes | 3.7 | 60 | Phase 2 | Gane et al. (2010) |
| | | 1500 | bid | | | | | | 5.0 | 95 | | |

*Continued*

**TABLE 3.2 (*Continued*)**
**Select Clinical Trial Data for 2′- and 4′-Modified Nucleosides**

| Drug Name | No. | Dose (mg) | Dosing | Duration | RBV | IFN | GT | Prev. SOC[a] | Mean Viral Load Reduction ($log_{10}$ IU/mL) | RVR (%) | Status | Reference |
|---|---|---|---|---|---|---|---|---|---|---|---|---|
| Balapiravir | **25** | 0 | | 14 d | – | – | 1 | Naïve | 0 | | T | Roberts et al. (2008) |
| | | 500 | bid | | | | | | 0.3 | | | |
| | | 1500 | bid | | | | | | 1.2 | | | |
| | | 3000 | bid | | | | | | 2.6 | | | |
| | | 4500 | bid | | | | | | 3.7 | | | |
| Balapiravir | **25** | 0 | | 28 d | + | + | 1 | Naïve | 2.4 | 5 | T | Pockros et al. (2008) |
| | | 1500 | bid | | – | + | | | 3.6 | 29 | | |
| | | 3000 | bid | | – | + | | | 4.5 | 69 | | |
| | | 1500 | bid | | + | + | | | 5.2 | 74 | | |

[a] Patients previously treated with standard of care.
[b] Terminated.

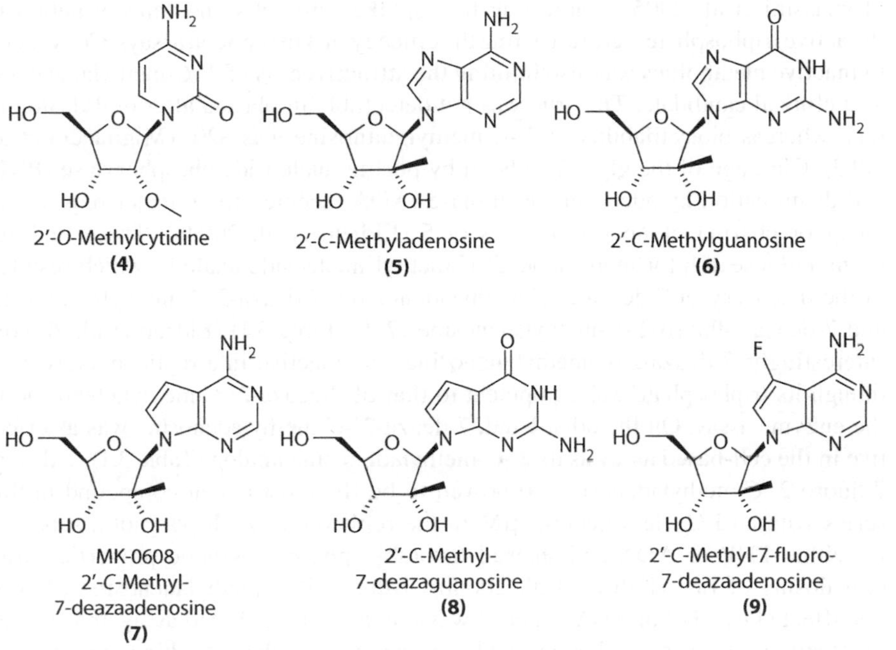

2′-*O*-Methylcytidine
**(4)**

2′-*C*-Methyladenosine
**(5)**

2′-*C*-Methylguanosine
**(6)**

MK-0608
2′-*C*-Methyl-
7-deazaadenosine
**(7)**

2′-*C*-Methyl-
7-deazaguanosine
**(8)**

2′-*C*-Methyl-7-fluoro-
7-deazaadenosine
**(9)**

**FIGURE 3.8** Chemical structures of 2′-*C*-methyl and 2′-*O*-methyl modified nucleosides.

compounds **4–9** (Figure 3.8) (Table 3.1) with promising activity in vitro (Carroll et al. 2003; Eldrup et al. 2004a, 2004b). Specificity evaluations revealed that the 2′-modifications confer a preference for viral RdRp over human DNA polymerases (Carroll et al. 2003); 7-deaza 2′-C-methyladenosine (**7**), for example, was not toxic to Jurkat cells exposed for up to 72 hours to a 100 μM concentration of this compound, whereas its analog 7-deaza adenosine (tubercidin) damaged cells with a $CC_{50}$ of 0.15 μM after 72 hours of exposure (Olsen et al. 2004). Comparing the activities of compounds **4–6** (2′-*O*-methylcytidine, 2′-*C*-methyladenosine, 2′-*C*-methylguanosine) in cell-based replicon assays to $IC_{50}$s of the respective triphosphates in enzyme assays showed a marked discrepancy (Table 3.1): 2′-*C*-methyladenosine is about tenfold more potent than 2′-*C*-methylguanosine in the replicon assay, although the numbers are reversed for the triphosphates in the enzyme assay; 2′-*O*-methylcytidine, which is seventyfold less active than 2′-*C*-methyladenosine in the replicon assay, is almost as potent as 2′-*C*-methyladenosine in the enzyme assay.

The apparent lack of consistency in this dataset could be explained by differences in cellular uptake of the compounds and efficiency of conversion to their triphosphates (Carroll et al. 2003; Tomassini et al. 2005)—most evident in the inverse correlation among intracellular triphosphate levels of **4–6** and their $EC_{50}$s in replicon assays (Migliaccio et al. 2003). Metabolic instability turned out to be another reason for the limited or even complete lack of activity of 2′-*O*-methylcytidine in cells (Carroll et al. 2003). A bis-*S*-tBu-acyl-2-thioethyl (tBuSATE) monophosphate prodrug of **4** showed sevenfold better cell-based activity than the parent compound

(Tomassini et al. 2005), confirming that uptake into cells and transformation to the active triphosphate were restricting the efficacy of **4** in replicon assays. Conversion to inactive metabolites was also limiting the attractiveness of 2′-*C*-methyladenosine as a clinical candidate. The drug was not detectable in plasma after oral dosing of rats, whereas bioavailability of 2′-*C*-methylguanosine was 82% (Migliaccio et al. 2003). Cleavage of the glycosidic bond by purine nucleoside phosphorylase (PNP) and deamination by adenosine deaminase (ADA) seemed to be major reasons for the poor pharmacokinetic properties of **5** (Eldrup et al. 2004a). These findings prompted a search for more stable 2′-*C*-methyl nucleoside analogs, which resulted in the discovery of 7-deaza-2′-*C*-methyladenosine, 7-deaza-2′-*C*-methylguanosine, and 7-deaza-7-fluoro-2′-*C*-methyladenosine (**7–9**; Table 3.1) (Eldrup et al. 2004b). Interestingly, 7-deaza-2′-*C*-methylguanosine was inactive in a replicon assay even though its triphosphate was equipotent to that of 7-deaza-2′-*C*-methyladenosine in the enzyme assay. On the other hand, 7-deaza-2′-*C*-methyladenosine was as effective in the cell-based assay as its 2′-*C*-methyladenosine analog (Table 3.1); 7-deaza-7-fluoro-2′-*C*-methyladenosine (**9**) proved to be the most potent compound in this series with an $EC_{50}$ less than 0.1 μM in the replicon assay. It was not a substrate for either PNP or ADA and showed favorable pharmacokinetic properties after oral dosing of rats (Eldrup et al. 2004b); 7-deaza-2′-*C*-methyladenosine (**7**) was not affected by PNP or ADA either. It was potent in a cell-based assay and showed low toxicity in several cell assays and in mice, and its pharmacokinetic properties upon oral dosing were very favorable in three species (Olsen et al. 2004). Taken together, these characteristics warranted advancing this drug into phase 1 clinical trials as MK-0608; however, further development was halted for unknown reasons. In vivo efficacy of MK-0608 was demonstrated in chronically HCV-infected chimpanzees. Viral loads dropped up to 5.9 $\log_{10}$ IU/mL as a result of either intravenous dosing at 2 mg/kg for seven days or oral dosing with 1 mg/kg for 37 days. Virus concentrations decreased even below the detection limit in some animals, convincingly demonstrating that the just described drug discovery efforts culminated in an efficacious 2′-*C*-methyl modified inhibitor of HCV replication in a complex animal model (Carroll et al. 2009).

#### 3.3.1.2 2′-Deoxy-2′-α-Fluoro Nucleosides

Beta-D-2′-deoxy-2′-fluorocytidine (**10**; Table 3.1) is yet another 2′-modified nucleoside with anti-HCV activity in the replicon system. However, some of its apparent antiviral potency had to be attributed to a cytostatic effect (Stuyver et al. 2004). Introduction of a methyl group in the 2′-position yielded 2′-deoxy-2′-fluoro-2′-*C*-methylcytidine (PSI-6130, **11**) (Stuyver et al. 2006), a 2′-fluoro analog of NM107 (**2**) with an $EC_{50}$ of 0.5–0.61 μM in replicon assays and an $IC_{50}$ of 0.12 μM against NS5B for the corresponding triphosphate (Table 3.1). In contrast to **10** (Figure 3.9), the new compound showed little or no toxicity against various cell types, and the no effect level in mice was found to be greater than 100 mg/kg per day. In primary human hepatocytes, the actual target cells of anti-HCV drugs, PSI-6130 is metabolized predominantly to its triphosphate and the triphosphate of the deamination product of PSI-6130 2′-deoxy-2′-fluoro-2′-*C*-methyluridine (RO2433, PSI-6206, **12**; Table 3.1). RO2433 itself has no activity in the replicon assay; however, its triphosphate is an

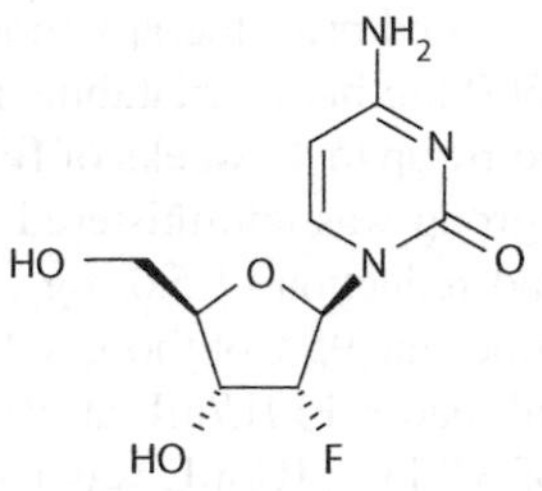


2′-Deoxy-2′-fluorocytidine
**(10)**

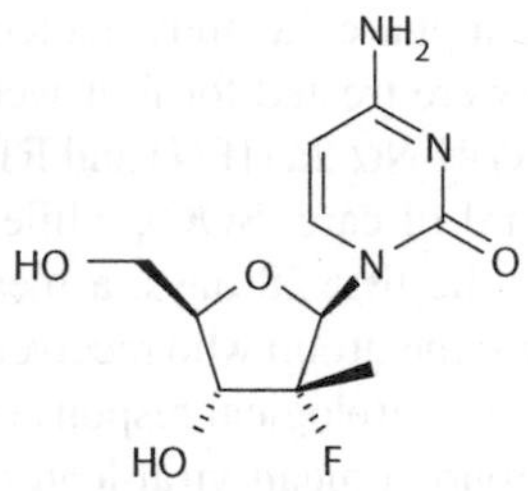


PSI-6130
2′-Deoxy-2′-fluoro-
2′-*C*-methylcytidine
**(11)**

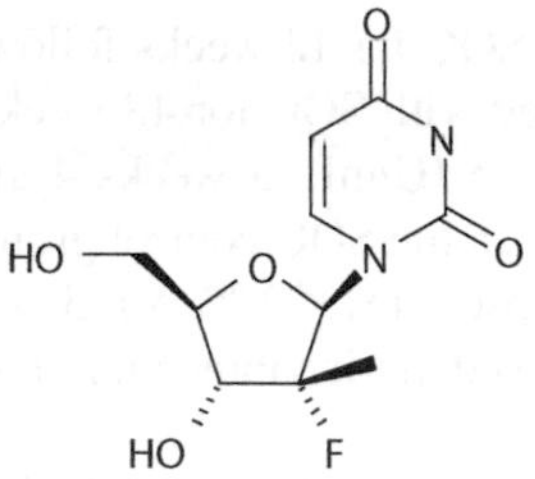


2′-Deoxy-2′-fluoro-
2′-*C*-methyluridine
**(12)**

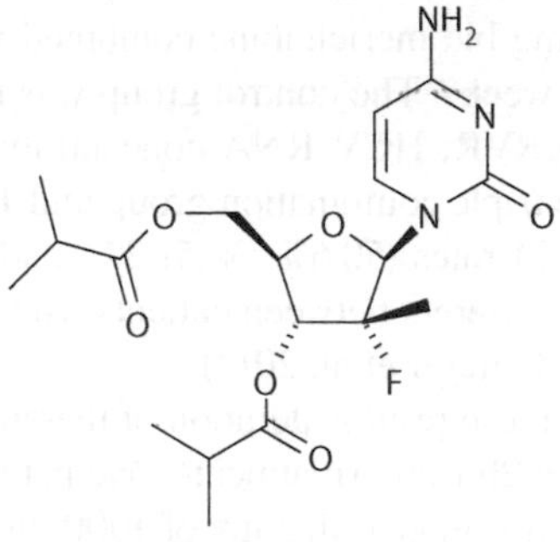


Mericitabine
**(13)**

**FIGURE 3.9** Chemical structures of 2′-deoxy-2′-α-fluorocytidine analogs.

NS5B inhibitor with an $IC_{50}$ of 0.52 μM ($K_i$ = 0.14 μM), which means that a major metabolite of **11** is also effective against HCV (Ma et al. 2007; Murakami et al. 2008); 2′-deoxy-2′-fluoro-2′-*C*-methylcytidine differs from the other 2′-modified nucleoside analogs described so far in two important aspects: It is specific for HCV and does not inhibit other flaviviruses in vitro, and it shows only a 2.5-fold reduction in potency against an S282T mutant of RdRp, which is resistant to inhibition by other 2′-*C*-methyl modified nucleosides (see Chapter 3, sec. 3.4). Taken together, these observations suggest that **11** binds to HCV NS5B in a different way than the other 2′-*C*-methyl modified compounds; 2′-deoxy-2′-fluoro-2′-*C*-methyl derivatives of adenosine and guanosine were also prepared, which turned out to be inactive or only moderately active, respectively, in the HCV replicon assay (Clark et al. 2006).

A prodrug of PSI-6130 with improved oral bioavailability was synthesized by esterification of the 3′- and 5′-OH groups with isobutyric acid. The resulting compound, mericitabine (R7128, **13**), completed phase 2 clinical development. In a dose-escalating, 14-day monotherapy phase 1b study in HCV-infected individuals, mericitabine resulted in a drop in viral RNA levels of 2.7 $\log_{10}$ IU/mL at the most effective dose of 1500 mg bid (Table 3.2); it was safe and well tolerated, and resistance selection was not detected in any of the dose groups (Reddy et al. 2007). Efficacy against HCV GT2 and GT3 was tested in a phase 2a study; patients who had failed prior therapy (nonresponders or relapsers) were treated for four weeks with 1500 mg bid mericitabine in combination with PEG-IFNα-2a (IFN) and RBV followed by up to 44 weeks of IFN and RBV alone (standard of care, SOC), while a control group was administered SOC for 48 weeks. After the first 28 days, a mean viral load reduction of 5.0 $\log_{10}$ IU/mL was determined for the group who received mericitabine, and 95% of those individuals had achieved a rapid virological response (RVR, viral load < 15 IU/mL at week 4). In the SOC-only group, a mean viral load reduction of 3.7 $\log_{10}$ IU/mL was measured and 60% of persons presented with an RVR (Table 3.2). Twelve weeks after completion of follow-up treatment, 65% of patients in the mericitabine/SOC group enjoyed a sustained virological response (SVR12), which suggested that therapy with mericitabine plus SOC is a promising combination for this patient population (Gane et al. 2010). In a phase 2b trial, treatment-naïve patients infected with GT1 or GT4 HCV were given 1000 mg bid mericitabine combined with SOC for 12 weeks followed by SOC alone for 36 weeks. The control group was treated with SOC for 48 weeks. Extended RVR rates (eRVR; HCV RNA concentration < 15 IU/mL at weeks 4 and 12) were 56.8% in the triple combination group and 17.9% in the SOC control group. However, overall SVR24 rates (50.6% vs. 51.2%) and relapse rates (29.3% vs. 31.1%) were not significantly different between patients who received mericitabine and those who got a placebo (Wedemeyer et al. 2013).

Based on these results, duration of therapy with mericitabine was doubled in a subsequent phase 2b trial: Treatment-naïve patients infected with GT1 or GT4 HCV were given a response-guided therapy of 1000 mg bid mericitabine combined with SOC for 24 weeks followed by 24 weeks of SOC alone if they did not achieve an eRVR, whereas the control group was treated with SOC for 48 weeks. Overall SVR24 rates were 56.8% for individuals receiving the triple combination compared to 36.5% for the SOC control group, and relapse rates were 27.7% and 32.0%, respectively. The most pronounced difference in relapse rates was observed for patients with a non-CC IL28B genotype (low responders to IFN) with 38.5% in mericitabine-treated individuals and 62.5% in the placebo control group. Nature and occurrence of adverse events and laboratory abnormalities in the group receiving the study drug were similar to the placebo group (Pockros et al. 2013), and mericitabine has demonstrated a high barrier to resistance in vitro (see Chapter 3, sec. 3.4) as well as in patients who were treated with the drug. Overall, more than 600 participants of clinical studies were monitored for resistance to mericitabine, and the known resistance-conferring RdRp mutation S282T was not detected nor was selection of strains resistant to the drug observed (Pawlotsky et al. 2012).

The discovery of 2′-*C*-methyl modified nucleoside analogs was a major breakthrough in the quest for direct-acting anti-HCV drugs and spurred considerable interest in this class of compounds, resulting in the identification of several promising HCV RdRp inhibitors (Smith et al. 2004; Ding et al. 2005; Koh et al. 2005; Fogt et al. 2008; Berke et al. 2011).

#### 3.3.1.3 Prodrugs of 2′-Modified Nucleoside Monophosphates

Cell-based assays do not only reflect a compound's intrinsic RdRp inhibition but also uptake into the cell and the cell's ability to convert a nucleoside into its active triphosphate. Therefore, discrepancies between the activity of the 5′-triphosphate of a particular nucleoside in an enzyme assay and the potency of the parent compound in a replicon assay can be observed, as discussed earlier (e.g., 2′-*C*-methylguanosine **6**, 2′-deoxy-2′-fluoro-2′-*C*-methyluridine **12**). The initial phosphorylation of the 5′-OH group is often rate limiting, and administering a preformed 5′-monophosphate would allow bypassing of this step. The monophosphate itself would suffer from poor absorption due to its negative charge at physiological pH, and it would be prone to enzymatic degradation by phosphatases and 5′ nucleotidases. That is why the monophosphate must be reversibly masked and delivered as a prodrug, which then will be converted to the free monophosphate and subsequently to the active triphosphate during first-pass metabolism in the liver. Several types of prodrugs have been developed, and some have been successfully harnessed to deliver 5′ monophosphates of nucleoside HCV polymerase inhibitors into cells in vitro and in vivo (for detailed review, see Madela and McGuigan 2012).

Phosphoramidate prodrugs typically involve masking the phosphate with an α-amino acid ester and an aryloxy group (Mehellou et al. 2009). An $IC_{50}$ of 0.13 μM of the triphosphate in an RdRp assay and favorable pharmacokinetic properties in rats made 2′-*C*-methylguanosine (**6**) an interesting candidate to test this prodrug approach. Extensive SAR studies on a series of 5′-phosphoramidate derivatives of 2′-*C*-methylguanosine (**14**, Figure 3.10) (McGuigan et al. 2009, 2010a) culminated in

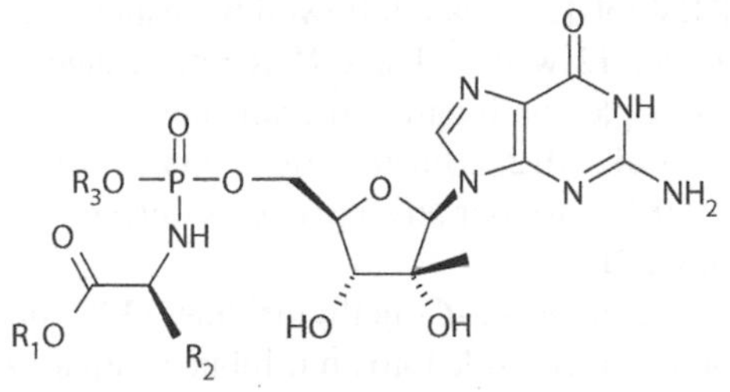


2′-*C*-Methylguanosine
phosphoramidate
**(14)**

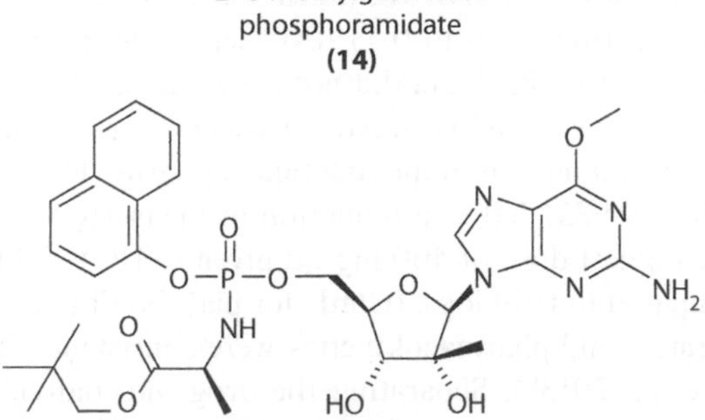


INX-08189
BMS-986094
**(15)**

**FIGURE 3.10** General structure of a 2′-*C*-methylguanonsine phosphoramidate prodrugs and the specific example INX-08189.

the synthesis of the 2′-*C*-methyl-6-methoxyguanosine phosphoramidate INX-08189 (BMS-986094, **15**), a double prodrug of 2′-*C*-methylguanosine, which was readily converted to the active 5′-triphosphate in primary human hepatocytes and in the liver of rats after oral dosing (Vernachio et al. 2011). Anti-HCV activity of **15** in a GT1 replicon assay was 10 nM (Table 3.3), a 350-fold improvement compared to the $EC_{50}$ of the parent 2′-*C*-methylguanosine, of which about a factor six is attributable to the 6-*O*-methyl group. A variety of substituents in the 6-position of the purine improved potency against HCV compared to guanosine, most likely by increasing the compounds' lipophilicity, thus improving cellular uptake (McGuigan et al. 2011b). The synthesis of phosphoramidate prodrugs resulted in a mixture of diastereomers with the phosphorus atom as an additional chiral center. The individual diastereomers of **15** showed similar $EC_{50}$s; consequently, a mixture of isomers was advanced into clinical development (McGuigan et al. 2010b). INX-08189 completed a phase 1b dose-finding study in individuals infected with GT1 HCV. The median HCV RNA reduction after seven days of monotherapy was 4.25 $\log_{10}$ IU/mL at the highest dose of 200 mg qd (Table 3.4). The synergy between **15** and RBV observed in vitro could be confirmed in the clinic: Viral load reduction after seven days of treatment with 100 mg INX-08189 qd plus RBV was 3.79 $\log_{10}$ IU/mL compared to 2.54 $\log_{10}$ IU/mL for monotherapy with the same dose (Table 3.4) (Rodriguez-Torres et al. 2011). In a subsequent phase 2 study, patients infected with GT 2/3 HCV were dosed with 25 mg, 50 mg, or 100 mg of BMS-986094 qd, or with placebo, respectively, combined with IFN and RBV for 12 weeks, followed by response-guided treatment with IFN and RBV for another 12 weeks. The eRVR rates ranged from 71% to 78% in the groups treated with triple combination therapy compared to 58% in the placebo control group (Lawitz et al. 2012a). Further development of BMS-986094 was terminated due to severe cardiovascular adverse events during a phase 2 study (Bristol-Myers Squibb Company 2012).

Converting 2′-deoxy-2′-fluoro-2′-*C*-methyluridine (**12**) into a monophosphate prodrug turned an inactive nucleoside into an inhibitor with a 1.6 μM $EC_{90}$ in a replicon assay (Furman et al. 2009). Further optimization of the phosphoramidate moiety led to an alanine isopropylester (PSI-7851, GS-9851, **16**; Table 3.3) with $EC_{50}$s in a replicon assay ranging from 0.09 to 0.43 μM, depending on virus genotype. The compound was specific for HCV and did not show any activity against the related flaviviruses yellow fever virus and West Nile virus, or against influenza A virus, hepatitis B virus (HBV), and human immunodeficiency virus (HIV) (Lam et al. 2010). In a phase 1 study, GS-9851 led to a reduction of mean HCV RNA levels of 1.95 $\log_{10}$ IU/mL at the highest dose of 400 mg qd given to HCV GT1-infected patients for three days compared to 0.09 $\log_{10}$ IU/mL for placebo (Table 3.5). The drug was safe and well tolerated, and pharmacokinetics were consistent with rapid uptake by the liver (Lawitz et al. 2013b). Separating the drug into individual diastereomers yielded two molecules with an almost 18-fold difference in $EC_{90}$ in a replicon assay (0.42 μM vs. 7.5 μM) with the more potent compound being in the *S* configuration at the phosphorus atom as x-ray crystallography revealed (Sofia et al. 2010). Consequently, the more efficacious diastereomer, sofosbuvir (PSI-7977, GS-7977, **17**; Table 3.3; Figure 3.11), was selected for further clinical development. Median viral load reduction was 4.7 $\log_{10}$ IU/mL after administering the drug at 400 mg qd for

**TABLE 3.3**
**Biochemical Data for Compounds 14–33**

| | | | Cell-Based Assay | | Enzyme Assay | |
|---|---|---|---|---|---|---|
| **Compound Name** | **No.** | **Fig.** | **$EC_{50}$ (μM)** | **Selectivity[a]** | **$IC_{50}$ (μM)** | **References** |
| 2′-*C*-Methylguanosine phosphoramidate | **14** | 3.10 | 0.05–1.7 | > 55–2000 | | McGuigan et al. (2009,2010a) |
| BMS-986094, INX-08189, 2′-*C*-methyl-6-methoxyguanosine phosphoramidate | **15** | 3.10 | 0.01 | 700 | | McGuigan et al. (2010b) |
| PSI-7851, GS-9851, 2′-deoxy-2′-fluoro-2′-*C*-methyluridine phosphoramidate | **16** | | 0.09–0.43 | >120 | 0.7–2.8 | Lam et al. (2010) |
| Sofosbuvir, PSI-7977, GS-7977, 2′-deoxy-2′-fluoro-2′-*C*-methyluridine phosphoramidate | **17** | 3.11 | 0.42[b] | 163 | | Sofia et al. (2010) |
| PSI-353661, 2′-deoxy-2′-fluoro-2′-*C*-methyl-6-methoxyguanosine phosphoramidate | **18** | 3.11 | 0.008[b] | > 12,500 | 5.9 | Chang et al. (2011) |
| 2′-*C*-Methyl-6-methoxyguanosine phosphorodiamidate | **19** | 3.12 | 0.06 | > 1600 | | McGuigan et al. (2011a) |
| IDX184, 2′-*C*-methylguanosine monophosphate prodrug | **20** | 3.12 | 0.3–0.45 | > 220 | 0.31 | Cretton-Scott et al. (2008) |
| PSI-352938, 2′-deoxy-2′-fluoro-2′-*C*-methyl-6-ethoxyguanosine cyclic monophosphate | **21** | 3.13 | 0.13–0.2 | > 500 | 1.0–4.7 | Lam et al. (2011b) |
| R1479, 4′-azidocytidine | **23** | 3.14 | 1.3 | 1500 | 0.29 | Klumpp et al. (2006) |
| 4′-Azidouridine | **24** | 3.14 | > 100 | | 0.3 | Smith et al. (2007) |
| 4′-Azidouridine phosphoramidate | **26** | | 0.6–14 | > 7–170 | 0.22 | Perrone et al. (2007b) |
| RO-9187, 4′-azidoarabinocytidine | **27** | 3.14 | 0.17–0.9 | > 1100 | 0.19–0.24 | Klumpp et al. (2008) |
| RO-0622, 4′-azido-2′-deoxy-2′-fluoroarabinocytidine | **28** | 3.14 | 0.02 | > 50,000 | 0.39–0.51 | Klumpp et al. (2008) |
| 4′-Azido-2′-*C*-methylcytidine | **29** | 3.14 | > 100 | | | Smith et al. (2009b) |
| 4′-Azido-2′-deoxy-2′-*C*-methylcytidine | **30** | 3.14 | 1.2 | > 82 | | Nilsson et al. (2012) |

*Continued*

**TABLE 3.3 (*Continued*)**
**Biochemical Data for Compounds 14–33**

| | | | Cell-Based Assay | | Enzyme Assay | |
|---|---|---|---|---|---|---|
| **Compound Name** | **No.** | **Fig.** | **$EC_{50}$ (μM)** | **Selectivity[a]** | **$IC_{50}$ (μM)** | **References** |
| β-D-$N^4$-Hydroxycytidine | **31** | 3.15 | 0.7–1.6 | > 45 | | Stuyver et al. (2003) |
| ODE-(*S*)-HPMPA | **32** | 3.15 | 1.55–1.65 | > 22 | | Wyles et al. (2009) |
| ODE-(*S*)-MPMPA | **33** | 3.15 | 1.4–2.4 | > 63 | | Valiaeva et al. (2011) |

[a] $CC_{50}/EC_{50}$.
[b] $EC_{90}$.

**TABLE 3.4**
**Select Clinical Trial Data for 2′-C-Methylguanosine Monophosphate Prodrugs**

| Drug Name | No. | Dose (mg) | Dosing | Duration | RBV | IFN | GT | Prev. SOC[a] | Mean Viral Load Reduction ($log_{10}$ IU/mL) | RVR (%) | Status | Reference |
|---|---|---|---|---|---|---|---|---|---|---|---|---|
| BMS-986094 | **15** | 0 | | 7 d | – | – | 1 | Naïve | 0.2 | | T[b] | Rodriguez-Torres et al. (2011); |
| | | 9 | qd | | | | | | 0.62 | | | Palmer (2012) |
| | | 25 | qd | | | | | | 1.02 | | | |
| | | 50 | qd | | | | | | 1.53 | | | |
| | | 100 | qd | | | | | | 2.54 | | | |
| | | 200 | qd | | | | | | 4.25 | | | |
| BMS-986094 | **15** | 0 | | 7 d | + | – | 1 | Naïve | 0.04 | | T | |
| | | 9 | qd | | | | | | 0.75 | | | Rodriguez-Torres et al. (2011); Palmer (2012) |
| | | 25 | qd | | | | | | 1.56 | | | |
| | | 100 | qd | | | | | | 3.79 | | | |
| IDX184 | **20** | 0 | | 3 d | – | – | 1 | Naïve | 0.05 | | T | Lalezari et al. (2012) |
| | | 25 | qd | | | | | | 0.5 | | | |
| | | 50 | qd | | | | | | 0.7 | | | |
| | | 75 | qd | | | | | | 0.6 | | | |
| | | 100 | qd | | | | | | 0.7 | | | |
| IDX184 | **20** | 0 | | 14 d | + | + | 1 | Naïve | 1.5 | 6 | T | Lalezari et al. (2010) |
| | | 50 | qd | | | | | | 2.7 | 13 | | |
| | | 50 | bid | | | | | | 4.0 | 50 | | |
| | | 100 | qd | | | | | | 4.2 | 50 | | |
| | | 150 | qd | | | | | | 4.1 | 40 | | |
| | | 100 | bid | | | | | | 4.3 | 29 | | |
| | | 200 | qd | | | | | | 3.7 | 25 | | |

[a] Patients previously treated with standard of care.
[b] Terminated.

**TABLE 3.5**
**Select Clinical Trial Data for 2′-Deoxy-2′-Fluoro-2′-*C*-Methyl Modified Nucleoside Monophosphate Prodrugs**

| Drug Name | No. | Dose (mg) | Dosing | Duration | RBV | IFN | GT | Prev. SOC[a] | Mean Viral Load Reduction ($\log_{10}$ IU/mL) | RVR (%) | Status | Reference |
|---|---|---|---|---|---|---|---|---|---|---|---|---|
| GS-9851 | **16** | 0 | qd | 3 d | – | – | 1 | Naïve | 0.09 | | T[b] | Lawitz et al. (2013b) |
| | | 50 | qd | | | | | | 0.49 | | | |
| | | 100 | qd | | | | | | 0.56 | | | |
| | | 200 | qd | | | | | | 1.01 | | | |
| | | 400 | qd | | | | | | 1.95 | | | |
| Sofosbuvir | **17** | 400 | qd | 7 d | – | – | 1 | Naïve | 4.7[c] | | Phase 3 | Lawitz et al. (2011b) |
| Sofosbuvir | **17** | 0 | | 28 d | + | + | 1 | Naïve | 2.8 | 21 | Phase 3 | Lawitz et al. (2010) |
| | | 100 | qd | | | | | | 5.3 | 88 | | |
| | | 200 | qd | | | | | | 5.1 | 94 | | |
| | | 400 | qd | | | | | | 5.3 | 93 | | |
| PSI-352938 | **21** | 100 | qd | 7 d | – | – | 1 | Naïve | 4.31 | 13 | T | Rodriguez-Torres et al. (2011) |
| | | 200 | qd | | | | | | 4.64 | 63 | | |
| | | 300 | qd | | | | | | 3.94 | 50 | | |
| | | 100 | bid | | | | | | 4.59 | 25 | | |
| PSI-352938 | **21** | 300 | qd | 14 d | – | – | 1 | Naïve | 5.2 | 50 | T | Lawitz et al. (2011b) |
| PSI-352938 | **21** | 300 | qd | 14 d | – | – | 1 | Naïve | 5.0 | 88 | T[d] | Lawitz et al. (2011b) |
| + sofosbuvir | **17** | 400 | qd | | | | | | | | | |

[a] Patients previously treated with standard of care.
[b] Terminated.
[c] Median viral load reduction.
[d] PSI-352938 terminated.

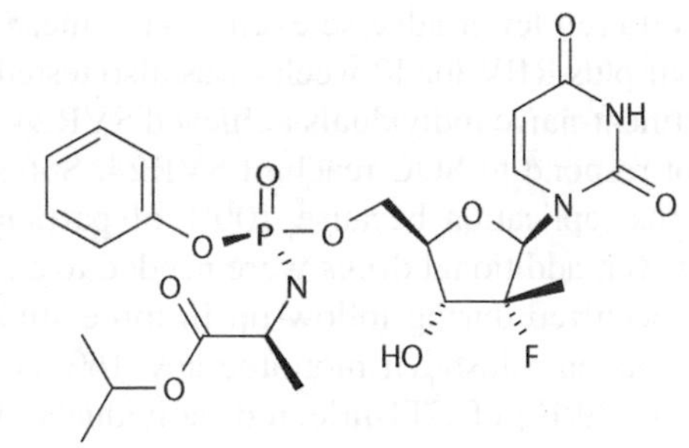


Sofosbuvir
**(17)**

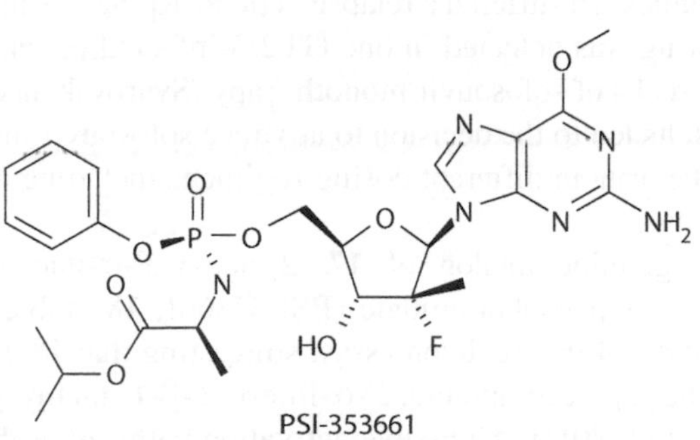


PSI-353661
2′-Deoxy-2′-fluoro-2′-C-methyl-
6-methoxyguanosine phosphoramidate
**(18)**

**FIGURE 3.11** Chemical structures of sofosbuvir and its guanosine analog PSI-353661.

seven days to HCV GT1-infected persons (Table 3.5) (Lawitz et al. 2011a). In a dose-ranging study, treatment-naïve patients infected with HCV GT1 were treated with 100, 200, or 400 mg of sofosbuvir qd combined with SOC followed by 44 weeks of SOC. The control group got 48 weeks of SOC. After 28 days, mean viral load reductions and RVR rates were similar in all three triple combination groups and markedly higher than placebo control. However, the 200 and 400 mg doses resulted in higher SVR24 rates of 80%–83% compared to 56% for the 100 mg group and 43% for the placebo group, respectively (Table 3.5) (Rodriguez-Torres et al. 2013). Subsequently, four additional phase 2 studies were conducted to determine optimum dose (Lawitz et al. 2013a), treatment duration (Kowdley et al. 2013), and combinations with RBV and PEG IFN (Gane et al. 2013; Herbst and Reddy 2013). In one of these trials, treatment-naïve patients infected with GT2/3 HCV were dosed with 400 mg qd of sofosbuvir for 12 weeks as a monotherapy, in combination with RBV, or with both RBV and 4, 8, or 12 weeks of PEG-IFN-α-2a. Monotherapy resulted in SVR24 for 60% of individuals, while 100% of study participants in all other cohorts achieved SVR24. This study demonstrated that 12 weeks of therapy are sufficient and that PEG IFN does not add any benefit in this patient population. Data evaluation further revealed that patients who were not exposed to PEG IFN suffered fewer adverse events. The interferon-free regimen of 400 mg qd of sofosbuvir plus RBV for 12 weeks was also tested in patients infected with GT1; 84% of treatment-naïve individuals achieved SVR24, and 10% of patients who previously did not respond to SOC reached SVR24. Sofosbuvir is very effective at suppressing virus replication because 100% of participants in all cohorts enjoyed an eRVR; however, additional drugs were needed to clear the virus in some populations. Relapses occurred during follow-up in three study arms: 40% (4/10) of GT2/3-infected patients on sofosbuvir monotherapy, 16% (4/25) of GT1-infected treatment-naïve, and 90% (9/10) of GT1-infected individuals, who did not respond to prior SOC therapy, did not achieve SVR12 (Gane et al. 2013). Sofosbuvir demonstrated a high barrier to resistance in vitro and in vivo. Of 572 participants in phase 2 clinical studies, 26 suffered a relapse. The RdRp S282T mutation conferring resistance to the drug was detected in one GT2/3-infected patient four weeks after completion of 12 weeks of sofosbuvir monotherapy (Svarovskaia et al. 2012). Taken together, these results led to the decision to advance sofosbuvir into phase 3 clinical trials evaluating the drug in different dosing regimens including all-oral interferon-sparing combinations.

A 6-*O*-methyl guanine analog of **17**, 2′-deoxy-2′-α-fluoro-2′-β-*C*-methyl-6-*O*-methyl guanosine 5′-phosphoramidate (PSI-353661, **18**; Table 3.3; Figure 3.11), showed an $EC_{90}$ of 8 nM in a replicon assay, suggesting that **18** is almost 8700-fold more potent than the parent compound, 2′-α-fluoro-2′-β-*C*-methyl guanosine ($EC_{90}$ = 69.2 μM) (Chang et al. 2011). A possible activation pathway of the double prodrug has been elucidated in detail, and toxicity studies revealed a favorable in vitro safety profile, moving PSI-353661 toward clinical candidate (Furman et al. 2011).

Phosphorodiamidates represent a class of phosphate prodrugs that can be designed to be symmetrical at the phosphorus center, thus potentially simplifying the development and production of the compounds. In an SAR study, 67 phosphorodiamidates of 2′-*C*-methyl-6-alkoxyguanosine were synthesized and assayed for activity against HCV. A vast majority of the compounds demonstrated submicromolar potency combined with low cytotoxicity (e.g., **19**, Table 3.3; Figure 3.12). A subset was tested in more detail and proved to be generally stable and efficiently converted to the active 2′-*C*-methylguanosine triphosphate in the liver of rats after oral dosing (McGuigan et al. 2011a).

The *S*-acyl-2-thioethyl (SATE) group was designed as an esterase-cleavable phosphate protection, and it was applied successfully to deliver monophosphates of several HIV reverse transcriptase inhibitors to cells (Peyrottes et al. 2004). IDX184 (**20**; Figure 3.12) is a prodrug of 2′-*C*-methylguanosine 5′-monophosphate employing a SATE moiety and benzylamine to mask the phosphate. Its $EC_{50}$ in a replicon assay was 0.3–0.45 μM (Table 3.3), showing that it was about tenfold more potent than the parent 2′-*C*-methylguanosine (Cretton-Scott et al. 2008). IDX184 was advanced into clinical trials as a mixture of diastereomers, demonstrating pharmacokinetic properties in healthy volunteers consistent with the intended liver targeting (Zhou et al. 2011). As a monotherapy, the highest dose of 100 mg qd for three days resulted in a mean HCV RNA decrease of 0.74 $\log_{10}$ IU/mL (Table 3.4) (Lalezari et al. 2012). In a subsequent phase 2a trial, treatment-naïve patients with chronic GT1 HCV infections were dosed for 14 days with IDX184 or a placebo in combination with

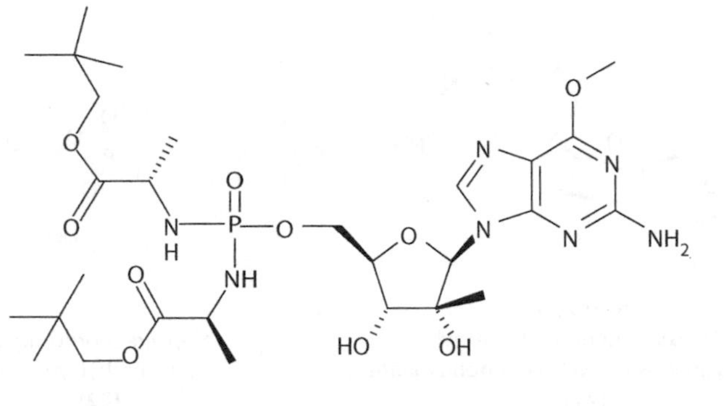

2′-*C*-Methyl-6-methoxyguanosine
phosphorodiamidate
**(19)**

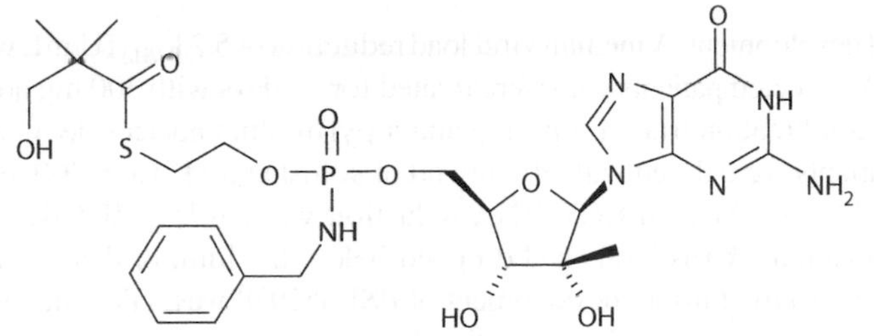

IDX184
**(20)**

**FIGURE 3.12** Chemical structures of a phosphorodiamidate and of IDX184.

PEG-IFN-α-2a and RBV. Mean viral load reduction at the end of treatment was 4.3 $\log_{10}$ IU/mL for the most effective dose of 100 mg bid compared to 1.5 $\log_{10}$ IU/mL for SOC alone (Table 3.4) (Lalezari et al. 2010). The known NS5B S282T resistance mutation was not detected in any viruses isolated from patients participating in these two studies (McCarville et al. 2011). In a further study, treatment-naïve patients infected with HCV GT1 received 50 or 100 mg IDX184 qd in combination with SOC for 12 weeks, followed by response-guided therapy with SOC for an additional 12 or 36 weeks. Complete RVR was achieved by 82% of subjects in the 100 mg qd arm (Lawitz et al. 2012b). Development of IDX184 was terminated because of serious adverse events, which occurred in a study with the related compound BMS-986094 (**15**) (Idenix Pharmaceuticals 2013).

Forming a 3′,5′-cyclic phosphate is another strategy to reduce polarity and deliver a nucleoside monophosphate into cells. Masking the remaining phosphate acid as an ester or amide was the key contribution to the several thousand-fold improved potency compared to the parent nucleosides (Gunic et al. 2007). This approach was applied to 2′-α-fluoro-2′-β-*C*-methyl guanosine, and optimizing the substituent in the 6-position of guanine and the cyclic phosphate ester led to the discovery of PSI-352938 (**21**; Table 3.3; Figure 3.13), a single Rp diastereomer (Reddy et al. 2011). The compound was advanced

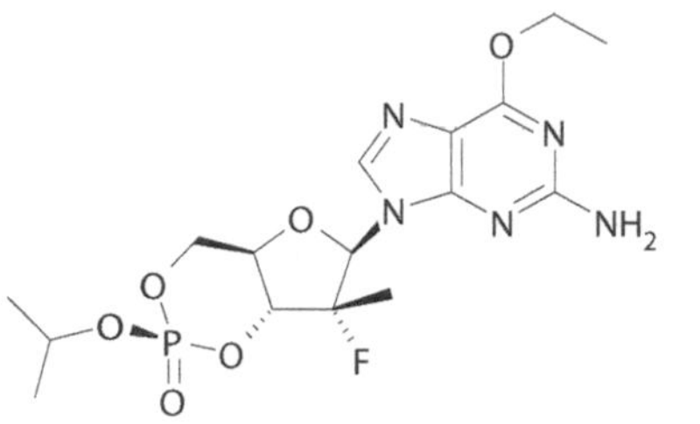

PSI-352938
2′-Deoxy-2′-fluoro-2′-C-methyl-
6-ethoxyguanosine cyclic monophosphate
**(21)**

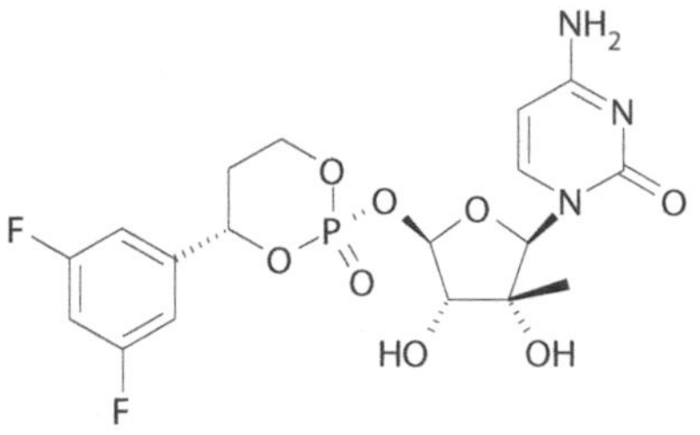

HepDirect prodrug of
2′-C-methylcytidine
**(22)**

**FIGURE 3.13** Chemical structures of cyclic phosphate prodrugs.

into clinical development. A median viral load reduction of 5.2 $\log_{10}$ IU/mL was observed in GT1 HCV-infected patients who were treated for 14 days with 300 mg qd (Table 3.5). In the first combination trial of a purine and a pyrimidine nucleoside monophosphate prodrug, patients were dosed with 400 mg qd of sofosbuvir (**17**) and 300 mg qd of PSI-352938 for 14 days. Median HCV RNA reduction was 5.0 $\log_{10}$ IU/mL, and in seven of eight individuals, virus levels had dropped below the limit of detection (Table 3.5) (Lawitz et al. 2011b). Further development of PSI-352938 was halted due to laboratory abnormalities associated with liver function found in some patients.

HepDirect is the designation for a class of cyclic 1-aryl-1,3-propanyl ester prodrugs of nucleoside monophosphates designed to target the liver. Release of the active compound is initiated by cytochrome P450 3A4-mediated oxidation at the tertiary carbon atom of the diester ring (Erion et al. 2004). A series of HepDirect prodrugs of 2′-*C*-methylcytidine were synthesized in an attempt to increase the drug's conversion to the active triphosphate in the liver. The pharmacokinetics profile of an optimized derivative (**22**; Figure 3.13) in rhesus macaques and its efficacy in HCV-infected chimpanzees revealed unsatisfactory oral bioavailability of the drug (Carroll et al. 2011).

Delivering nucleosides as 5′-monophosphate prodrugs to bypass the initial phosphorylation step can greatly enhance a compound's potency and at the same time improve its oral bioavailability. These strengths observed in vitro and in animal testing have successfully translated into the clinic as demonstrated by clinical trial data for four such prodrugs. Impressive viral load reductions, RVR, and SVR rates were achieved with convenient once a day oral dosing of 400 mg or less of the compounds. However, the toxicity-related termination of three of the programs cast a shadow on the monophosphate prodrug approach. On the other hand, a monophosphate prodrug, sofosbuvir, is currently the most advanced HCV RNA polymerase inhibitor in phase 3 clinical development, and it promises to enable all-oral IFN-α-free HCV therapy for the first time.

### 3.3.2 Nucleosides with Modifications in the 4′-Position

Deoxyribonucleoside analogs with modifications in the 4′-position were first described as inhibitors of HIV replication that act as chain terminators despite the presence of a 3′-OH group (Maag et al. 1992), suggesting that they are nonobligate chain terminators similar to the 2′-*C*-methyl nucleoside analogs discussed earlier. Therefore, 4′-modified nucleoside analogs were included in a compound library that was evaluated for inhibition of the HCV replicon system. Among the hits identified in this screen, 4′-azidocytidine (R1479, **23**; Table 3.3; Figure 3.14) showed the best inhibition to cytotoxicity selectivity, whereas its uridine analog displayed only low activity; 4′-propynyluridine and 4′-vinyl uridine and cytidine were moderately active, suggesting that 4′-alkenyl and alkynyl substituents might be worth further exploration. Interestingly, the triphosphates of 4′-azidouridine (**24**) and 4′-vinylcytidine were NS5B inhibitors with $IC_{50}$s of 0.3 and 2.7 μM, respectively,

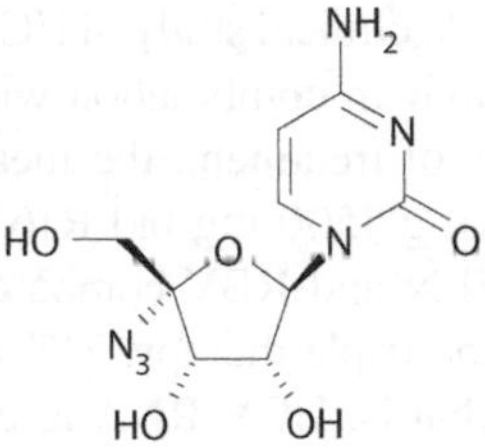

4′-Azidocytidine
**(23)**

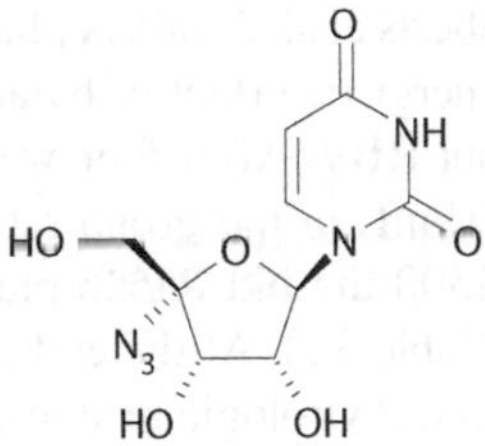

4′-Azidouridine
**(24)**

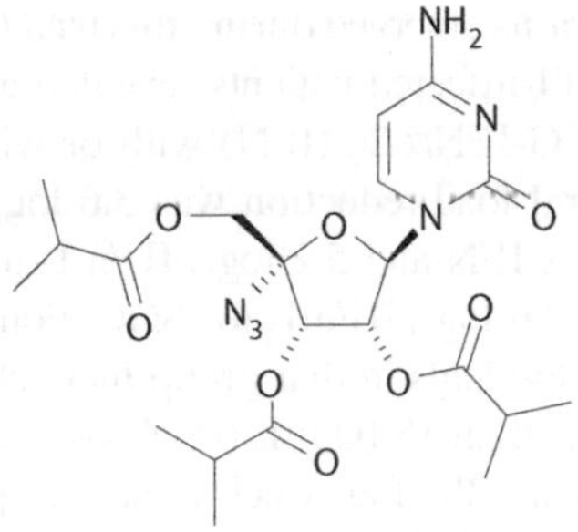

Balapiravir
**(25)**

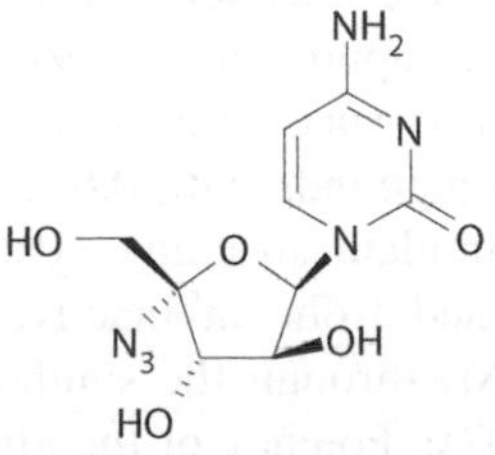

4′-Azidoarabinocytidine
**(27)**

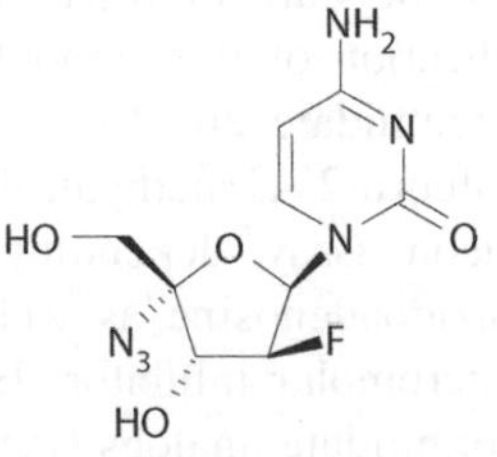

4′-Azido-2′-deoxy-
2′-fluoroarabinocytidine
**(28)**

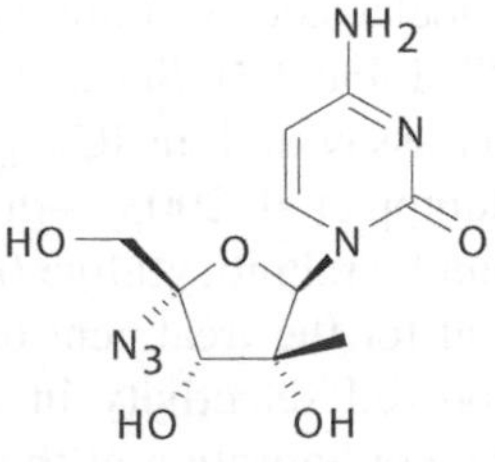

4′-Azido-2′-*C*-methylcytidine
**(29)**

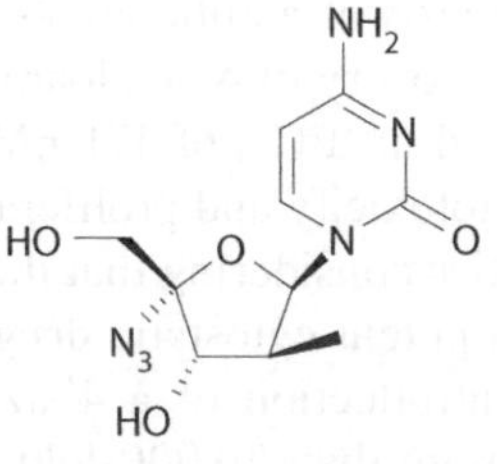

4′-Azido-2′-deoxy-
2′-*C*-methylcytidine
**(30)**

**FIGURE 3.14** Chemical structures of 4′-azido modified nucleosides.

even though their parent form showed little activity in the replicon system, demonstrating once more that intrinsic activity of the triphosphate is not sufficient for a nucleoside to show inhibition in a cell-based replicon assay (Klumpp et al. 2006; Smith et al. 2007).

Based on its potent and specific inhibition of HCV replication in vitro, 4′-azidocytidine was selected for development, and a 2′-,3′-,5′-isobutyric acid triester (balapiravir, R1626, **25**) was prepared in order to improve bioavailability (Li et al. 2008). This prodrug was efficiently converted to R1479 after uptake from the gastrointestinal tract and showed dose-dependent mean viral load reductions with a maximum of 3.7 $\log_{10}$ IU/mL at the highest dose of 4500 mg bid in a 14-day monotherapy phase 1b clinical study in treatment-naïve individuals infected with HCV GT1 (Table 3.2). No drug-resistant strains emerged during this trial (Roberts et al. 2008). A phase 2a clinical study in HCV GT1-infected patients revealed a synergistic effect of balapiravir in combination with PEG-IFNα-2a (IFN) with or without RBV. After four weeks of treatment, the mean viral load reduction was 3.6 $\log_{10}$ IU/mL in the group receiving 1500 mg bid R1626 plus IFN and 5.2 $\log_{10}$ IU/mL for 1500 mg bid R1626 plus IFN and RBV compared to 2.6 $\log_{10}$ IU/mL for SOC alone (Table 3.2). At the end of the triple therapy, 74% of individuals in that group had achieved a virologic response, that is, HCV RNA levels less than 15 IU/mL (Pockros et al. 2008). Viral resistance to balapiravir did not occur during the four-week treatment period; however, combination therapy was associated with significant hematologic changes, particularly in neutropenia. These adverse events could not be prevented by adjusting drug dosing, and the development of balapiravir was discontinued (Nelson et al. 2012).

Based on the data outlined here, 4′-azidouridine was an obvious candidate for the application of the monophosphate prodrug approach. Converting **24** into a 5′-phosphoramidate **26** (Table 3.3) was indeed as successful as in the case of 2′-deoxy-2′-fluoro-2′-*C*-methyluridine, yielding compounds with 0.6–14 μM $EC_{50}$s in the replicon assay, depending on phosphoramidate structure (Perrone et al. 2007b); 4′-azidoadenosine as well could be turned from an inactive compound into a submicromolar inhibitor ($EC_{50}$ 0.2–4.0 μM) through the synthesis of various phosphoramidate analogs (Perrone et al. 2007a). Potency of the already active 4′-azidocytidine, on the other hand, could not be enhanced significantly through application of the monophosphate prodrug approach despite extensive SAR efforts (McGuigan et al. 2009).

The synthesis of additional 4′-azido modified nucleosides to improve on R1479 led to the discovery of 4′-azidoarabinocytidine (**27**; Table 3.3) (Smith et al. 2009b), which showed an $EC_{50}$ of 171 nM in the replicon assay and an $IC_{50}$ greater than 1 mM in cytotoxicity and proliferation assays (Klumpp et al. 2008)—quite remarkable properties considering that the parent compound, arabinocytidine (AraC, cytarabine), is a potent cytostatic drug that is approved for the treatment of leukemia. Thus, the introduction of a 4′-azido moiety improved selectivity in a proliferation assay more than 30,000-fold. Inversion of the configuration at the 2′-position increased inhibition of HCV replication about sixfold (**27** vs. **23**), which was another unexpected result because **27** lacks a 2′-α OH group, a structural feature considered to be required for binding of ribonucleosides to RdRp. Replacing the 2′-OH group of AraC with fluorine yielded 2′-deoxy-2′-fluoroarabinocytidine (dFAraC), another cytostatic compound with an $IC_{50}$ of 174 nM in a cell proliferation assay. Adding a 4′-azido group turned dFAraC into a very potent nucleoside inhibitor of the HCV replicon system with an $EC_{50}$ of 24 nM (**28**; Table 3.3) and with about the same activity against S96T and S282T replicon mutants (Table 3.6), which are resistant against some of the other nucleoside inhibitors. In this case, the 4′-azido group increased the $IC_{50}$ in the proliferation assay about 600-fold (Klumpp et al. 2008; Smith et al. 2009a).

Could the attractive properties of the 4′-azido group be combined with the potential of the 2′-*C*-methyl modification to generate an HCV inhibitor embodying the strengths of both classes of compounds? To answer this question, 4′-azido-2′-*C*-methylcytidine (**29**; Table 3.3) was synthesized as a hybrid of R1479 and NM107—the two drugs that were efficacious in clinical trials. Surprisingly, **29** was devoid of activity in a replicon assay, most likely due to lack of 5′-triphosphate formation (Smith et al. 2009b); 4′-azidocytidine is a substrate of deoxycytidine kinase (dCK), whereas 2′-*C*-methylcytidine is phosphorylated by either uridine-cytidine kinase-1 or -2 (UCK1, UCK2) (Klumpp et al. 2008; Smith et al. 2009a; Golitsina et al. 2010). Combining the 4′-azido and 2′-*C*-methyl moieties in a single molecule apparently excluded **29** from the active sites of all three nucleoside kinases. A phosphoramidate prodrug of **29**, which was synthesized later, showed moderate potency in a replicon assay with an $EC_{50}$ of 4.9 μM, confirming that phosphorylation of the 5′-OH group was the limiting step in activating 4′-azido-2′-*C*-methylcytidine (Rondla et al. 2009).

Interestingly, the 2′-deoxy analog of **29**, 4′-azido-2′-deoxy-2′-*C*-methyl cytidine (**30**), is a potent HCV RdRp inhibitor with an $EC_{50}$ of 1.2 μM in a replicon assay (Table 3.3). Converting **30** into a phosphoramidate prodrug did not improve its potency, suggesting that the nucleoside is efficiently phosphorylated in cells. The 3′,5′-diisobutyryl ester was prepared (TMC649128) to improve the compound's bioavailability (Nilsson et al. 2012).

Although extensive SAR studies exemplified how modifications at the 4′-position of cytidine analogs, in particular the 4′-azido group, could modulate antiviral potency, these favorable in vitro properties did not translate into clinical success.

### 3.3.3 Other Compounds

Beta-D-$N^4$-hydroxycytidine (**31**; Table 3.3; Figure 3.15) showed an antiviral effect against BVDV and HCV with an $EC_{50}$ of 1.6 μM in an HCV replicon assay (Stuyver et al. 2003). Its triphosphate was incorporated into RNA, but it was not a chain terminator and the mechanism of action remained unclear. Metabolism studies revealed that **31** is phosphorylated to the triphosphate and also transformed into cytidine and uridine in various cell types. Overall, the stability of **31** was found to be insufficient for clinical studies (Hernandez-Santiago et al. 2004).

Several acyclic nucleoside phosphonates were found to be active in HCV replicon assays. An $EC_{50}$ of 1.6 μM was measured for ODE-(*S*)-HPMPA (**32**; Table 3.3;

**TABLE 3.6**
**Biochemical Data for Select Resistance Conferring NS5b Mutations**

| Compound Name | No. | Replicon $EC_{50}$ (μM) | | | | | | Polymerase $IC_{50}$ (μM) | | | Reference |
|---|---|---|---|---|---|---|---|---|---|---|---|
| | | wt | S96T | Shift | wt | S282T | Shift | wt | S282T | Shift | |
| NM107 | **2** | | | | 1.55 | 26.31 | 17.0 | 0.016[a] | 0.58 | 36.3 | Ali et al. (2008) |
| 2′-*O*-methylcytidine | **4** | | | | 12.6 | 9 | 0.7 | 3.8 | 16 | 4.2 | Migliaccio et al. (2003) |
| 2′-*C*-methyladenosine | **5** | | | | 0.17 | 6.6 | 38.8 | 1.9 | 84 | 44.2 | Migliaccio et al. (2003) |
| 2′-*C*-methylguanosine | **6** | | | | 1.4 | 13.4 | 9.6 | 0.1 | 20 | 200.0 | Migliaccio et al. (2003) |
| MK-0608 | **7** | | | | 0.3 | 10.1 | 33.7 | 0.07 | 25 | 357.1 | Olsen et al. (2004) |
| PSI-6130 | **11** | 0.82 | 0.42 | 0.5 | 0.31 | 0.75 | 2.4 | 0.023[a] | 0.044 | 1.9 | Ali et al. (2008) |
| β-D-2′-deoxy-2′-fluoro-2′-*C*-methyluridine | **12** | | | | | | | 0.14[a] | 2.97 | 21.2 | Ali et al. (2008) |
| Mericitabine | **13** | 0.58 | 3.2[b] | 5.5 | 0.32 | 1.08 | 3.4 | | | | Ali et al. (2008) |
| BMS-986094 | **15** | 0.006 | 0.006[b] | 1.0 | 0.006 | 0.074 | 12.3 | | | | Vernachio et al. (2011) |
| GS-9851 | **16** | 0.52[c] | 0.39[b] | 0.8 | 0.45 | 7.4 | 16.4 | | | | Lam et al. (2010) |
| Sofosbuvir | **17** | 0.23[c] | 0.11[b] | 0.5 | 0.42 | 7.8 | 18.6 | | | | Sofia et al. (2010) |
| PSI-353661 | **18** | 0.012[c] | 0.009[b] | 0.7 | 0.009 | 0.014 | 1.5 | | | | Furman et al. (2011) |
| PSI-352938 | **21** | 0.59[c] | 0.53[b] | 0.9 | 1.38 | 1.61 | 1.2 | | | | Lam et al. (2011b) |
| 4′-Azidocytidine | **23** | 1.96 | 8.2 | 4.2 | 1.9 | 0.53 | 0.3 | 0.32 | 0.29 | 0.9 | Le Pogam et al. (2006); Klumpp et al. (2006) |
| 2′-α-Deoxy-2′-β-hydroxy-4′-azidocytidine | **27** | 0.8 | 0.62 | 0.8 | 0.8 | 1.16 | 1.5 | | | | Klumpp et al. (2008) |
| 2′-α-Deoxy-2′-β-fluoro-4′-azidocytidine | **28** | 0.019 | 0.031 | 1.6 | 0.019 | 0.03 | 1.5 | | | | Klumpp et al. (2008) |

[a] Data represent $K_i$.
[b] S96T/N142T double mutant.
[c] Data represent $EC_{90}$.

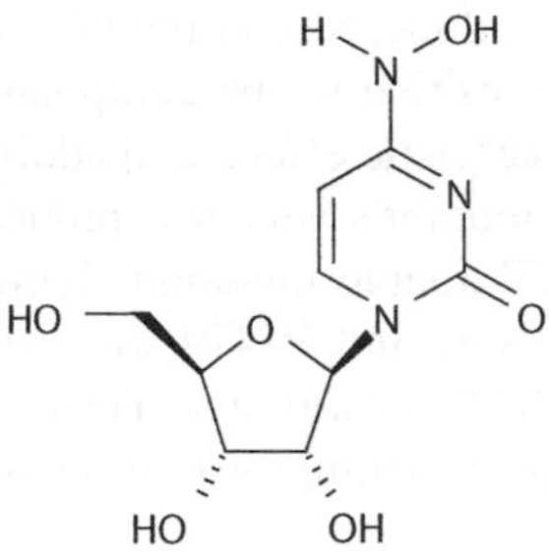

β-D-$N^4$-Hydroxycytidine
**(31)**

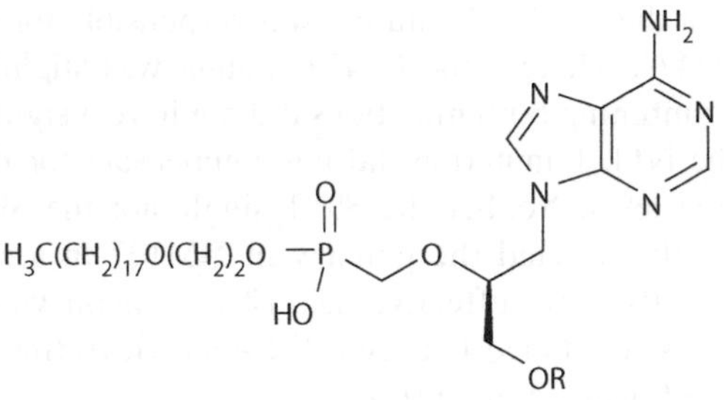

ODE-(*S*)-HPMPA
(R = H)
**(32)**

ODE-(*S*)-MPMPA
(R = $CH_3$)
**(33)**

**FIGURE 3.15** Chemical structures of compounds not belonging to one of the major classes of HCV inhibitors.

Figure 3.15), while its 3′-*O*-methyl derivative **33** was similarly active and showed improved selectivity. The mechanism of action of these compounds has not been elucidated (Wyles et al. 2009; Valiaeva et al. 2011).

### 3.3.4 A High Barrier to Emergence of HCV Resistant to 2′- and 4′-Modified Nucleosides

Resistance is always of concern in the development of antiviral drugs. Therefore, to test whether clones resistant to 2′-*C*-methyladenosine (**5**) would emerge, HCV replicon harboring Huh7 cells were cultured in the presence of increasing concentrations of this inhibitor. Resistant colonies could indeed be isolated, and sequence analysis revealed a serine to threonine mutation at amino acid position 282 (S282T) of the NS5B polymerase common to all resistant replicons. Preparation of replicons and recombinant HCV RdRp with an S282T substitution confirmed that this single mutation was sufficient to increase the $EC_{50}$ of 2′-*C*-methyladenosine almost 40-fold in a cell assay and the $IC_{50}$ more than 40-fold in an enzyme assay (Table 3.6). The same mutation also significantly reduced the potency of 2′-*C*-methylguanosine (**6**), whereas the $EC_{50}$ of 2′-*O*-methylcytidine (**4**) was hardly affected and its $IC_{50}$ was only moderately increased by about fourfold compared to the wild-type polymerase (Migliaccio et al. 2003). Later studies showed that the S282T substitution generally confers resistance on HCV RdRp against 2′-*C*-methyl modified nucleoside inhibitors, including MK-0608 (**7**) (Olsen et al. 2004) and NM107 (**2**) (Ali et al. 2008) (Table 3.6). Viruses harboring this mutation could also be isolated from some HCV-infected chimpanzees, which were treated with MK-0608 (Carroll et al. 2009).

Clones resistant to 4′-azidocytidine (R1479, **23**) evolved only after culturing replicon harboring cells over an extended period of time with exposure to increasing concentrations of the inhibitor. During subsequent analysis, an S96T/N142T double mutation in the NS5B sequence was identified in the vast majority of resistant replicons. Further investigations using engineered mutant replicons revealed that the S96T substitution was responsible for the observed increase in the $EC_{50}$ of R1479, whereas the N142T mutant was slightly more sensitive to the compound. Combining both mutations did not have a significant synergistic effect, and adding the N142T mutation did not compensate for the S96T mutant's very low replication level. Neither the S96T single nor the S96T/N142T double mutation significantly affected the potency of NM107 in a replicon assay, and R1479 was even slightly more effective against a replicon with the S282T substitution. Thus, no cross-resistance between 2′-*C*-methylcytidine and 4′-azidocytidine was observed (Le Pogam et al. 2006).

Selecting replicons resistant to 2′-deoxy-2′-fluoro-2′-*C*-methylcytidine (PSI-6130, **11**) proved very difficult; on the other hand, clones resistant to a nonnucleoside RdRp inhibitor (HCV-796) or an HCV protease inhibitor (VX-950) readily emerged (McCown et al. 2008). It took prolonged culturing (more than 20 passages) of HCV replicon harboring cells in the presence of increasing concentrations of **11** to finally isolate resistant clones. Sequencing revealed multiple mutations in the polymerase; among them, S282T was the only substitution recurring in all selected and analyzed replicons. In contrast to other 2′-*C*-methyl modified nucleosides, the potency of 2′-fluoro-2′-*C*-methylcytidine and its prodrug mericitabine (**13**) against the mutant NS5B was only moderately affected with a less than 3.5-fold increase in $EC_{50}$ in the replicon assay and a twofold increase in $K_i$ in an enzyme assay compared to wild type. Interestingly, 2′-deoxy-2′-fluoro-2′-*C*-methyluridine (**12**, **16**, sofosbuvir **17**), a metabolite of PSI-6130, was more susceptible to the S282T mutation in a replicon assay with a 19-fold higher $EC_{90}$ compared to wild-type replicons—a value similar to that observed for 2′-*C*-methylcytidine (Table 3.6). The S96T mutation, which reduced the efficacy of R1479, hardly affected the potency of PSI-6130, meaning that no cross-resistance existed between these two compounds. In fact, PSI-6130 and its uridine analog sofosbuvir were twice as potent against an S96T, or an S96T/N142T mutant, respectively, as against the wild type (Ali et al. 2008; Sofia et al. 2010).

Molecular modeling suggested that the S282T substitution in NS5B leads to a steric clash between the OH– and $CH_3$– group of threonine and the 2′-methyl group of NM107 (Dutartre et al. 2006). A 2′-methyl group is also present in 2′-deoxy-2′-fluoro-2′-*C*-methylcytidine (PSI-6130) and 2′-deoxy-2′-fluoro-2′-*C*-methyluridine (PSI-7977); however, the decrease in activity of PSI-6130 against the S282T mutant is smaller than that measured for the two related compounds, NM-107 and PSI-977, hinting that the triphosphate of PSI-6130 may bind in a unique way to the HCV RdRp active site.

The activity of the 2′-deoxy-2′-fluoro-2′-*C*-methylguanosine derivatives, PSI-353661 (**18**) and PSI-352938 (**21**), against a replicon bearing the S282T mutation was virtually the same as against wild type (Furman et al. 2011; Lam et al. 2011b), whereas the related 2′-*C*-methylguanosine (**6, 15, 20**) showed an almost tenfold reduced potency against the mutant replicon (Table 3.6)—reminiscent of the observations made with 2′-fluoro-2′-*C*-methylcytidine and 2′-*C*-methylcytidine. Resistance against **18** and **21** emerged only with prolonged culturing of cells containing a GT2a JFH-1 replicon in the presence of increasing inhibitor concentrations. Sequencing revealed various amino acid mutations, among them C223H, in the NS5B region that increased the $EC_{50}$ of **18** 2.2-fold and that of **21** 3.7-fold compared to wild type. A combination of multiple mutations was necessary to confer a significant level of resistance. The $EC_{50}$s of **18** and **21** for a replicon, bearing the three amino acid changes S15G/C223H/V321I, were 12 and 17-fold higher, respectively, than for wild type. The replication fitness of this triple mutant was not significantly reduced. None of the mutations identified in this study were selected by other inhibitors; therefore, no cross-resistance exists between 2′-deoxy-2′-fluoro-2′-*C*-methylguanosine and other 2′- or 4′-modified nucleoside inhibitors of HCV RdRp (Lam et al. 2011a).

The high barrier to the emergence of HCVs resistant to either R1479/balapiravir or PSI-6130/mericitabine was confirmed in clinical trials in which the drugs were administered as monotherapy for 14 days, and no resistant viruses could be detected (Reddy et al. 2007; Roberts et al. 2008). A possible explanation for the favorable resistance profile of the two inhibitors is their only moderately decreased potency against S96T or S282T NS5B bearing replicons in combination with these mutants' greatly impaired replication capacity of 4% and 15% of wild type, respectively. Drug resistance was not a problem, either in combination trial of mericitabine or sofosbuvir, on hundreds of patients treated for a full cycle of 12–48 weeks (Pawlotsky et al. 2012; Svarovskaia et al. 2012).

## 3.4 CONCLUSION AND PERSPECTIVES

Even though the recent introduction of new protease inhibitors provides a clear benefit for patients, HCV infection remains a global health problem for which less toxic, more potent, and more convenient treatment options are needed. One of the most promising targets for anti-HCV therapies is the HCV polymerase NS5B, a key enzyme that catalyzes the synthesis of viral RNA needed as a template for protein translation and amplification of the genomic material. The development of biochemical and cell-based assays has helped us to understand the molecular parameters that govern the regulation of NS5B. Moreover, it has become increasingly evident that inhibiting NS5B with nucleotide analogs used as chain terminators offers several advantages in terms of drug development. Because nucleotide analogs act as substrates for the polymerase and bind to conserved residues in its active site, they are considered to have a broad spectrum of activity and be capable of inhibiting equally all six genotypes of HCV. The second advantage of interacting with conserved residues is the high genetic barrier to resistance, which means that nucleoside analogs do not easily select resistance mutations over time. In addition, these mutations are often detrimental for virus fitness. In spite of these advantages, the development of nucleoside analogs has been slow and difficult because of suboptimal potency/efficacy, requiring high doses and exposing patients to undermining side effects. These limitations are best exemplified by valopicitabine, the first anti-NS5B nucleoside analog, which entered clinical trials almost ten years ago.

One of the major breakthroughs in the development of anti-HCV nucleosides came with the development of monophosphate prodrugs that are able to achieve significantly higher NTP levels inside liver cells, bypassing the first and rate-limiting phosphorylation step. The most advanced nucleotide prodrug, sofosbuvir, is currently in late-stage phase 3 clinical trials. At the time this chapter was written, Gilead Sciences had submitted a new drug application to the U.S. Food and Drug Administration (FDA) for approval of sofosbuvir. Sofosbuvir would be given once daily together with RBV as an all-oral therapy for patients with GT2 or GT3 HCV infection and in combination with RBV and pegylated interferon for treatment-naïve patients with GT1, -4, -5, or -6 HCV infection. It should also be mentioned that sofosbuvir is being codeveloped in parallel as a single formulation tablet together with the NS5A inhibitor ledipasvir, which underscores the potential for future DAA combination therapies that would not require the use of interferon and possibly RBV (for recent reviews on current and future DAAs, see De Clercq 2012; Pockros 2012; Poordad and Chee 2012). Therefore, it is very likely that superior combination therapy options relying on the use of DAAs, will emerge from such studies and that nucleoside analog inhibitors of NS5B will become the backbone for the next generation of HCV therapeutics.

## REFERENCES


Adachi T, Ago H, Habuka N, et al. (2002). The essential role of C-terminal residues in regulating the activity of hepatitis C virus RNA-dependent RNA polymerase. *Biochim Biophys Acta* 1601:38–48.

Ago H, Adachi T, Yoshida A, et al. (1999). Crystal structure of the RNA-dependent RNA polymerase of hepatitis C virus. *Structure* 7:1417–26.

Ali S, Leveque V, Le Pogam S, et al. (2008). Selected replicon variants with low-level in vitro resistance to the hepatitis C virus NS5B polymerase inhibitor PSI-6130 lack cross-resistance with R1479. *Antimicrob Agents Chemother* 52(12):4356–69.

Alter HJ, Seeff LB. (2000). Recovery, persistence, and sequelae in hepatitis C virus infection: A perspective on long-term outcome. *Semin Liver Dis* 20:17–35.

Berke JM, Vijgen L, Lachau-Durand S, et al. (2011). Antiviral activity and mode of action of TMC647078, a novel nucleoside inhibitor of the hepatitis C virus NS5B polymerase. *Antimicrob Agents Chemother* 55:3812–20.

Bichko V, Tausek M, Qu L, et al. (2005). Enhanced antiviral activity of NM107, alone or in combination with interferon a. *J Hepatol* 42:S154.

Boonstra A, van der Laan LJ, Vanwolleghem T, et al. (2009). Experimental models for hepatitis C viral infection. *Hepatology* 50:1646–55.

Bressanelli S, Tomei L, Rey FA, et al. (2002). Structural analysis of the hepatitis C virus RNA polymerase in complex with ribonucleotides. *J Virol* 76:3482–92.

Bressanelli S, Tomei L, Roussel A, et al. (1999). Crystal structure of the RNA-dependent RNA polymerase of hepatitis C virus. *Proc Natl Acad Sci U S A* 96:13034–9.

Bristol-Myers Squibb Company. (2012). Bristol-Myers Squibb discontinues development of BMS-986094, an investigational NS5B nucleotide for the treatment of Hepatitis C (Press release of August 23. Internet) Available from: http://news.bms.com/press-release/financial-news/bristol-myers-squibb-discontinues-development-bms-986094-investigationa (accessed June 14, 2013).

Buckwold VE, Beer BE, Donis RO. (2003). Bovine viral diarrhea virus as a surrogate model of hepatitis C virus for the evaluation of antiviral agents. *Antiviral Res* 60:1–15.

Bukh J. (2004). A critical role for the chimpanzee model in the study of hepatitis C. *Hepatology* 39:1469–75.

Caillet-Saguy C, Simister PC, Bressanelli S. (2011). An objective assessment of conformational variability in complexes of hepatitis C virus polymerase with non-nucleoside inhibitors. *J Mol Biol* 414:370–84.

Carroll SS, Koeplinger K, Vavrek M, et al. (2011). Antiviral efficacy upon administration of a HepDirect prodrug of 2′-C-methylcytidine to hepatitis C virus-infected chimpanzees. *Antimicrob Agents Chemother* 55:3854–60.

Carroll SS, LaFemina R, Hall DL, et al. (2002). Nucleoside derivatives as inhibitors of RNA-dependent RNA viral polymerase. Patent publication no. WO/2002/057425.

Carroll SS, Ludmerer S, Handt L, et al. (2009). Robust antiviral efficacy upon administration of a nucleoside analog to hepatitis C virus-infected chimpanzees. *Antimicrob Agents Chemother* 53:926–34.

Carroll SS, Tomassini JE, Bosserman M, et al. (2003). Inhibition of hepatitis C virus RNA replication by 2′-modified nucleoside analogs. *J Biol Chem* 278:11979–84.

Chang W, Bao D, Chun BK, et al. (2011). Discovery of PSI-353661, a novel purine nucleotide prodrug for the treatment of HCV infection. *ACS Med Chem Lett* 2:130–5.

Chen CM, He Y, Lu L, et al. (2007). Activity of a potent hepatitis C virus polymerase inhibitor in the chimpanzee model. *Antimicrob Agents Chemother* 51:4290–6.

Choo QL, Kuo G, Weiner AJ, et al. (1989). Isolation of a cDNA clone derived from a blood-borne non-A, non-B viral hepatitis genome. *Science* 244:359–62.

Clark JL, Mason JC, Hollecker L, et al. (2006). Synthesis and antiviral activity of 2′-deoxy-2′-fluoro-2′-C-methyl purine nucleosides as inhibitors of hepatitis C virus RNA replication. *Bioorg Med Chem Lett* 16:1712–15.

Coelmont L, Paeshuyse J, Windisch MP, et al. (2006). Ribavirin antagonizes the in vitro anti-hepatitis C virus activity of 2′-C-methylcytidine, the active component of valopicitabine. *Antimicrob Agents Chemother* 50:3444–6.

Cretton-Scott E, Perigaud C, Peyrottes S, et al. (2008). In vitro antiviral activity and pharmacology of IDX184, a novel and potent inhibitor of HCV replication. *J Hepatol* 48:S220.

De Clercq E. (2012). The race for interferon-free HCV therapies: A snapshot by the spring of 2012. *Rev Med Virol* 22:392–411.

Deval J, Powdrill MH, D'Abramo CM, et al. (2007). Pyrophosphorolytic excision of non-obligate chain terminators by hepatitis C virus NS5B polymerase. *Antimicrob Agents Chemother* 51:2920–8.

Di Marco S, Volpari C, Tomei L, et al. (2005). Interdomain communication in hepatitis C virus polymerase abolished by small molecule inhibitors bound to a novel allosteric site. *J Biol Chem* 280:29765–70.

Ding Y, Girardet JL, Hong Z, et al. (2005). Synthesis of 9-(2-beta-C-methyl-beta-d-ribofuranosyl)-6-substituted purine derivatives as inhibitors of HCVRNA replication. *Bioorg Med Chem Lett* 15(3):709–13.

Dorner M, Horwitz JA, Robbins JB, et al. (2011). A genetically humanized mouse model for hepatitis C virus infection. *Nature* 474:208–11.

Dutartre H, Bussetta C, Boretto J, et al. (2006). General catalytic deficiency of hepatitis C virus RNA polymerase with an S282T mutation and mutually exclusive resistance towards 2′-modified nucleotide analogues. *Antimicrob Agents Chemother* 50:4161–9.

Eldrup AB, Allerson CR, Bennett CF, et al. (2004a). Structure-activity relationship of purine ribonucleosides for inhibition of hepatitis C virus RNA-dependent RNA polymerase. *J Med Chem* 47:2283–95.

Eldrup AB, Prhavc M, Brooks J, et al. (2004b). Structure-activity relationship of heterobase-modified 2′-C-methyl ribonucleosides as inhibitors of hepatitis C virus RNA replication. *J Med Chem* 47:5284–97.

Erion MD, Reddy KR, Boyer SH, et al. (2004). Design, synthesis, and characterization of a series of cytochrome P(450) 3A-activated prodrugs (HepDirect prodrugs) useful for targeting phosph(on)ate-based drugs to the liver. *J Am Chem Soc* 126:5154–63.

Ferrari E, He Z, Palermo RE, et al. (2008). Hepatitis C virus NS5B polymerase exhibits distinct nucleotide requirements for initiation and elongation. *J Biol Chem* 283:33893–901.

Fogt J, Januszczyk P, Framski G, et al. (2008). Synthesis and antiviral activity of novel derivatives of 2′-beta-C-methylcytidine. *Nucleic Acids Symp Ser (Oxf)* 52:605–6.

Furman PA, Lam AM, Murakami E. (2009). Nucleoside analog inhibitors of hepatitis C viral replication: Recent advances, challenges and trends. *Future Med Chem* 1:1429–52.

Furman PA, Murakami E, Niu C, et al. (2011). Activity and the metabolic activation pathway of the potent and selective hepatitis C virus pronucleotide inhibitor PSI-353661. *Antiviral Res* 91:120–32.

Gane EJ, Rodriguez-Torres M, Nelson DR, et al. (2010). Sustained virologic response (SVR) following RG7128 1500mg BID/PEG-IFN/RBV for 28 days in HCV genotype 2/3 prior non-responders. *J Hepatol* 52:S16.

Gane EJ, Stedman CA, Hyland RH, et al. (2013). Nucleotide polymerase inhibitor sofosbuvir plus ribavirin for hepatitis C. *N Engl J Med* 368:34–44.

Golitsina NL, Danehy FT, Jr, Fellows R, et al. (2010). Evaluation of the role of three candidate human kinases in the conversion of the hepatitis C virus inhibitor 2′-C-methyl-cytidine to its 5′-monophosphate metabolite. *Antiviral Res* 85:470–81.

Gouklani H, Bull RA, Beyer C, et al. (2012). Hepatitis C virus nonstructural protein 5B is involved in virus morphogenesis. *J Virol* 86:5080–8.

Gunic E, Girardet JL, Ramasamy K, et al. (2007). Cyclic monophosphate prodrugs of base-modified 2′-C-methyl ribonucleosides as potent inhibitors of hepatitis C virus RNA replication. *Bioorg Med Chem Lett* 17:2452–55.

Hang JQ, Yang Y, Harris SF, et al. (2009). Slow binding inhibition and mechanism of resistance of non-nucleoside polymerase inhibitors of hepatitis C virus. *J Biol Chem* 284:15517–29.

Harrus D, Ahmed-El-Sayed N, Simister PC, et al. (2010). Further insights into the roles of GTP and the C terminus of the hepatitis C virus polymerase in the initiation of RNA synthesis. *J Biol Chem* 285:32906–18.

Herbst DA, Jr, Reddy KR. (2013). Sofosbuvir, a nucleotide polymerase inhibitor, for the treatment of chronic hepatitis C virus infection. *Expert Opin Investig Drugs* 22:527–36.

Hernandez-Santiago BI, Beltran T, Stuyver L, et al. (2004). Metabolism of the anti-hepatitis C virus nucleoside beta-d-N4-hydroxycytidine in different liver cells. *Antimicrob Agents Chemother* 48:4636–42.

Hong Z, Cameron CE, Walker MP, et al. (2001). A novel mechanism to ensure terminal initiation by hepatitis C virus NS5B polymerase. *Virology* 285:6–11.

Idenix Pharmaceuticals.(2013). Idenix Pharmaceuticals provides update on IDX184 and IDX19368 development programs (Press release of February 4, 2013. Internet) Available from: http://ir.idenix.com/releasedetail.cfm?ReleaseID=737733 (Accessed June 14, 2013).

Ilan E, Arazi J, Nussbaum O, et al. (2002). The hepatitis C virus (HCV)-trimera mouse: A model for evaluation of agents against HCV. *J Infect Dis* 185:153–61.

Jensen DM. (2011). A new era of hepatitis C therapy begins. *N Engl J Med* 364:1272–4.

Jin Z, Leveque V, Ma H, et al. (2012). Assembly, purification, and pre-steady-state kinetic analysis of active RNA-dependent RNA polymerase elongation complex. *J Biol Chem* 287:10674–83.

Klumpp K, Kalayanov G, Ma H, et al. (2008). 2′-deoxy-4′-azido nucleoside analogs are highly potent inhibitors of hepatitis C virus replication despite the lack of 2′-alpha-hydroxyl groups. *J Biol Chem* 283:2167–75.

Klumpp K, Leveque V, Le Pogam S, et al. (2006). The novel nucleoside analog r1479 (4′-azidocytidine) is a potent inhibitor of NS5B-dependent RNA synthesis and hepatitis C virus replication in cell culture. *J Biol Chem* 281:3793–9.

Koh YH, Shim JH, Wu JZ, et al. (2005). Design, synthesis, and antiviral activity of adenosine 5′-phosphonate analogues as chain terminators against hepatitis C virus. *J Med Chem* 48:2867–75.

Kowdley KV, Lawitz E, Crespo I, et al. (2013). Sofosbuvir with pegylated interferon alfa-2a and ribavirin for treatment-naive patients with hepatitis C genotype-1 infection (atomic): An open-label, randomised, multicentre phase 2 trial. *Lancet* 381: 2100–7.

Lalezari J, Asmuth D, Casiro A, et al. (2012). Short-term monotherapy with IDX184, a liver-targeted nucleotide polymerase inhibitor, in patients with chronic hepatitis C virus infection. *Antimicrob Agents Chemother* 56:6372–8.

Lalezari J, Poordad F, Mehra P, et al. (2010). Antiviral activity, pharmacokinetics and safety of IDX184 in combination with pegylated interferon (pegifn) and ribavirin (rbv) in treatment-naive HCV genotype 1-infected subjects. *J Hepatol* 52:S469.

Lam AM, Espiritu C, Bansal S, et al. (2011a). Hepatitis C virus nucleotide inhibitors PSI-352938 and PSI-353661 exhibit a novel mechanism of resistance requiring multiple mutations within replicon RNA. *J Virol* 85(23):12334–42.

Lam AM, Espiritu C, Murakami E, et al. (2011b). Inhibition of hepatitis C virus replicon RNA synthesis by PSI-352938, a cyclic phosphate prodrug of beta-d-2′-deoxy-2′-alpha-fluoro-2′-beta-C-methylguanosine. *Antimicrob Agents Chemother* 55:2566–75.

Lam AM, Murakami E, Espiritu C, et al. (2010). Psi-7851, a pronucleotide of {beta}-d-2′-deoxy-2′-fluoro-2′-C-methyluridine monophosphate, is a potent and pan-genotype inhibitor of hepatitis C virus replication. *Antimicrob Agents Chemother* 54:3187–96.

Lawitz E, Lalezari JP, Hassanein T, et al. (2011a). Once-daily PSI-7977 plus Peg/RBV in treatment naïve patients with HCV GT1: Robust end of treatment response rates are sustained post treatment. *Hepatology* 54:472A.

Lawitz E, Lalezari JP, Hassanein T, et al. (2013a). Sofosbuvir in combination with peginterferon alfa-2a and ribavirin for non-cirrhotic, treatment-naive patients with genotypes 1, 2, and 3 hepatitis C infection: A randomised, double-blind, phase 2 trial. *Lancet Infect Dis* 13:401–8.

Lawitz E, Lalezari JP, Freilich B, et al. (2012a). BMS-986094 (INX-08189) plus peginterferon alfa-2a and ribavirin results in increased rates of rapid virologic response (RVR) and eRVR in treatment-naïve HCV-genotype 2/3 patients compared to peginterferon alfa-2a and ribavirin: Week 12 results. *Hepatology* 56:566A.

Lawitz E, Box TD, Pruitt R, et al. (2012b). High rates of rapid virologic response (RVR) and complete early virologic response (cEVR) with IDX184, pegylated interferon and ribavirin in genotype 1 HCV infected subjects: Interim results. *Hepatology* 56:1006A.

Lawitz E, Lalezari JP, Rodriguez-Torres M, et al. (2010). High rapid virologic response (RVR) with PSI-7977 QD plus PEG-IFN/RBV in a 28-day phase 2A trial. *Hepatology* 52:706A.

Lawitz E, Rodriguez-Torres M, Denning J, et al. (2011b). Once daily dual-nucleotide combination of PSI-938 and PSI-7977 provides 94% HCVRNA< lod at day 14: First purine/pyrimidine clinical combination data (the nuclear study). *J Hepatol* 54:S543.

Lawitz E, Rodriguez-Torres M, Denning JM, et al. (2013b). Pharmacokinetics, pharmacodynamics, and tolerability of GS-9851, a nucleotide analog polymerase inhibitor, following multiple ascending doses in patients with chronic hepatitis C infection. *Antimicrob Agents Chemother* 57:1209–17.

Le Pogam S, Jiang WR, Leveque V, et al. (2006). In vitro selected Con1 subgenomic replicons resistant to 2′-C-methyl-cytidine or to R1479 show lack of cross resistance. *Virology* 351:349–59.

Lesburg CA, Cable MB, Ferrari E, et al. (1999). Crystal structure of the RNA-dependent RNA polymerase from hepatitis C virus reveals a fully encircled active site. *Nat Struct Biol* 6:937–43.

Leveque VJ, Wang QM. (2002). RNA-dependent RNA polymerase encoded by hepatitis C virus: Biomedical applications. *Cell Mol Life Sci* 59:909–19.

Li F, Maag H, Alfredson T. (2008). Prodrugs of nucleoside analogues for improved oral absorption and tissue targeting. *J Pharm Sci* 97:1109–34.

Lindenbach BD, Evans MJ, Syder AJ, et al. (2005). Complete replication of hepatitis C virus in cell culture. *Science* 309:623–6.

Lindenbach BD, Meuleman P, Ploss A, et al. (2006). Cell culture-grown hepatitis C virus is infectious in vivo and can be recultured in vitro. *Proc Natl Acad Sci U S A* 103:3805–9.

Lohmann V, Korner F, Herian U, et al. (1997). Biochemical properties of hepatitis C virus NS5BRNA-dependent RNA polymerase and identification of amino acid sequence motifs essential for enzymatic activity. *J Virol* 71:8416–28.

Lohmann V, Korner F, Koch J, et al. (1999a). Replication of subgenomic hepatitis C virus RNAs in a hepatoma cell line. *Science* 285:110–13.

Lohmann V, Overton H, Bartenschlager R. (1999b). Selective stimulation of hepatitis C virus and pestivirus NS5BRNA polymerase activity by GTP. *J Biol Chem* 274:10807–15.

Lohmann V, Roos A, Korner F, et al. (1998). Biochemical and kinetic analyses of NS5BRNA-dependent RNA polymerase of the hepatitis C virus. *Virology* 249:108–18.

Love RA, Parge HE, Yu X, et al. (2003). Crystallographic identification of a noncompetitive inhibitor binding site on the hepatitis C virus NS5BRNA polymerase enzyme. *J Virol* 77:7575–81.

Luo G, Hamatake RK, Mathis DM, et al. (2000). De novo initiation of RNA synthesis by the RNA-dependent RNA polymerase (NS5B) of hepatitis C virus. *J Virol* 74:851–63.

Ma H, Jiang WR, Robledo N, et al. (2007). Characterization of the metabolic activation of hepatitis C virus nucleoside inhibitor beta-D-2′-deoxy-2′-fluoro-2′-C-methylcytidine (PSI-6130) and identification of a novel active 5′-triphosphate species. *J Biol Chem* 282:29812–20.

Ma H, Leveque V, De Witte A, et al. (2005). Inhibition of native hepatitis C virus replicase by nucleotide and non-nucleoside inhibitors. *Virology* 332:8–15.

Maag D, Castro C, Hong Z, et al. (2001). Hepatitis C virus RNA-dependent RNA polymerase (NS5B) as a mediator of the antiviral activity of ribavirin. *J Biol Chem* 276:46094–8.

Maag H, Rydzewski RM, McRoberts MJ, et al. (1992). Synthesis and anti-HIV activity of 4′-azido- and 4′-methoxynucleosides. *J Med Chem* 35:1440–51.

McCarville JF, Dubuc G, Donovan E, et al. (2011). No resistance to IDX184 was detected in 3-day and 14-day clinical studies of IDX184 in genotype 1 infected HCV subjects. *J Hepatol* 54:S488–9.

Madela K, McGuigan C.(2012). Progress in the development of anti-hepatitis C virus nucleoside and nucleotide prodrugs. *Future Med Chem* 4:625–650.

McCown MF, Rajyaguru S, Le Pogam S, et al. (2008). The hepatitis C virus replicon presents a higher barrier to resistance to nucleoside analogs than to nonnucleoside polymerase or protease inhibitors. *Antimicrob Agents Chemother* 52:1604–12.

McGuigan C, Gilles A, Madela K, et al. (2010a). Phosphoramidate protides of 2′-C-methylguanosine as highly potent inhibitors of hepatitis C virus. Study of their in vitro and in vivo properties. *J Med Chem* 53:4949–57.

McGuigan C, Kelleher MR, Perrone P, et al. (2009). The application of phosphoramidate ProTide technology to the potent anti-HCV compound 4′-azidocytidine (R1479). *Bioorg Med Chem Lett* 19:4250–4.

McGuigan C, Madela K, Aljarah M, et al. (2010b). Design, synthesis and evaluation of a novel double pro-drug: INX-08189. A new clinical candidate for hepatitis C virus. *Bioorg Med Chem Lett* 20:4850–4.

McGuigan C, Madela K, Aljarah M, et al. (2011a). Phosphorodiamidates as a promising new phosphate prodrug motif for antiviral drug discovery: Application to anti-HCV agents. *J Med Chem* 54:8632–45.

McGuigan C, Madela K, Aljarah M, et al. (2011b). Dual pro-drugs of 2′-C-methyl guanosine monophosphate as potent and selective inhibitors of hepatitis C virus. *Bioorg Med Chem Lett* 21:6007–12.

Mehellou Y, Balzarini J, McGuigan C. (2009). Aryloxy phosphoramidate triesters: A technology for delivering monophosphorylated nucleosides and sugars into cells. *Chem Med Chem* 4:1779–91.

Mercer DF. (2011). Animal models for studying hepatitis C and alcohol effects on liver. *World J Gastroenterol* 17:2515–19.

Migliaccio G, Tomassini JE, Carroll SS, et al. (2003). Characterization of resistance to non-obligate chain-terminating ribonucleoside analogs that inhibit hepatitis C virus replication in vitro. *J Biol Chem* 278:49164–70.

Moradpour D, Penin F, Rice CM. (2007). Replication of hepatitis C virus. *Nat Rev Microbiol* 5:453–63.

Mosley RT, Edwards TE, Murakami E, et al. (2012). Structure of hepatitis C virus polymerase in complex with primer-template RNA. *J Virol* 86:6503–11.

Murakami E, Niu C, Bao H, et al. (2008). The mechanism of action of beta-d-2′-deoxy-2′-fluoro-2′-C-methylcytidine involves a second metabolic pathway leading to beta-d-2′-deoxy-2′-fluoro-2′-C-methyluridine 5′-triphosphate, a potent inhibitor of the hepatitis C virus RNA-dependent RNA polymerase. *Antimicrob Agents Chemother* 52:458–64.

Nelson DR, Zeuzem S, Andreone P, et al. (2012). Balapiravir plus peginterferon alfa-2a (40kd)/ribavirin in a randomized trial of hepatitis C genotype 1 patients. *Ann Hepatol* 11:15–31.

Nilsson M, Kalayanov G, Winqvist A, et al. (2012). Discovery of 4′-azido-2′-deoxy-2′-C-methyl cytidine and prodrugs thereof: A potent inhibitor of hepatitis C virus replication. *Bioorg Med Chem Lett* 22: 3265–8.

Nyanguile O, Devogelaere B, Vijgen L, et al. (2010). 1a/1b subtype profiling of nonnucleoside polymerase inhibitors of hepatitis C virus. *J Virol* 84:2923–34.

Ohara E, Hiraga N, Imamura M, et al. (2011). Elimination of hepatitis C virus by short term NS3-4A and NS5B inhibitor combination therapy in human hepatocyte chimeric mice. *J Hepatol* 54:872–8.

Olsen DB, Eldrup AB, Bartholomew L, et al. (2004). A 7-deaza-adenosine analog is a potent and selective inhibitor of hepatitis C virus replication with excellent pharmacokinetic properties. *Antimicrob Agents Chemother* 48:3944–53.

Palmer M. (2013). The future of hepatitis C beyond protease inhibitors-will hepatitis C be a disease of the past? *PractHepatol* 36:13–22.

Pawlotsky JM, Chevaliez S, McHutchison JG. (2007). The hepatitis C virus life cycle as a target for new antiviral therapies. *Gastroenterology* 132:1979–98.

Pawlotsky JM, Najera I, Jacobson I. (2012). Resistance to mericitabine, a nucleoside analogue inhibitor of HCVRNA-dependent RNA polymerase. *Antivir Ther* 17:411–23.

Perrone P, Daverio F, Valente R, et al. (2007a). First example of phosphoramidate approach applied to a 4′-substituted purine nucleoside (4′-azidoadenosine): Conversion of an inactive nucleoside to a submicromolar compound versus hepatitis C virus. *J Med Chem* 50:5463–70.

Perrone P, Luoni GM, Kelleher MR, et al. (2007b). Application of the phosphoramidate protide approach to 4′-azidouridine confers sub-micromolar potency versus hepatitis C virus on an inactive nucleoside. *J Med Chem* 50:1840–9.

Peyrottes S, Egron D, Lefebvre I, et al. (2004). Sate pronucleotide approaches: An overview. *Mini Rev Med Chem* 4:395–408.

Pierra C, Amador A, Benzaria S, et al. (2006). Synthesis and pharmacokinetics of valopicitabine (NM283), an efficient prodrug of the potent anti-HCV agent 2′-C-methylcytidine. *J Med Chem* 49:6614–20.

Pockros PJ. (2012). Interferon-free hepatitis C therapy: How close are we? *Drugs* 72:1825–31.

Pockros PJ, Jensen D, Tsai N, et al. (2013). JUMP-C: A randomized trial of mericitabine plus peginterferon alfa-2a/ribavirin for 24 weeks in treatment-naive HCV genotype 1/4 patients. *Hepatology* 58: 514–23.

Pockros PJ, Nelson D, Godofsky E, et al. (2008). R1626 plus peginterferon alfa-2a provides potent suppression of hepatitis C virus RNA and significant antiviral synergy in combination with ribavirin. *Hepatology* 48:385–97.

Poordad F, Chee GM. (2012). Interferon free hepatitis C treatment regimens: The beginning of another era. *Curr Gastroenterol Rep* 14:74–7.

Poordad F, Lawitz EJ, Gitlin N, et al. (2007). Efficacy and safety of valopicitabine in combination with pegylated interferon-alpha (pegifn) and ribavirin (RBV) in patients with chronic hepatitis C. *Hepatology* 46:866A.

Powdrill MH, Tchesnokov EP, Kozak RA, et al. (2011). Contribution of a mutational bias in hepatitis C virus replication to the genetic barrier in the development of drug resistance. *Proc Natl Acad Sci U S A* 108:20509–13.

Qin W, Luo H, Nomura T, et al. (2002). Oligomeric interaction of hepatitis C virus NS5B is critical for catalytic activity of RNA-dependent RNA polymerase. *J Biol Chem* 277:2132–7.

Quezada EM, Kane CM. (2013). The stimulatory mechanism of hepatitis C virus NS5A protein on the NS5B catalyzed replication reaction in vitro. *Open Biochem J* 7:11–14.

Ranjith-Kumar CT, Gutshall L, Sarisky RT, et al. (2003). Multiple interactions within the hepatitis C virus RNA polymerase repress primer-dependent RNA synthesis. *J Mol Biol* 330:675–85.

Reddy KR, Rodriguez-Torres M, Gane EJ, et al. (2007). Antiviral activity, pharmacokinetics, safety, and tolerability of R7128, a novel nucleoside HCVRNA polymerase inhibitor, following multiple, ascending, oral doses in patients with HCV genotype 1 infection who have failed prior interferon therapy. *Hepatology* 46:862A–3.

Reddy PG, Chun BK, Zhang HR, et al. (2011). Stereoselective synthesis of PSI-352938: A beta-d-2′-deoxy-2′-alpha-fluoro-2′-beta-C-methyl-3′,5′-cyclic phosphate nucleotide prodrug for the treatment of HCV. *J Org Chem* 76:3782–90.

Roberts SK, Cooksley G, Dore GJ, et al. (2008). Robust antiviral activity of R1626, a novel nucleoside analog: A randomized, placebo-controlled study in patients with chronic hepatitis C. *Hepatology* 48:398–406.

Rodriguez-Torres M, Lawitz E, Hazan L, et al. (2011). Antiviral activity and safety of INX-08189, a nucleotide polymerase inhibitor, following 7 days of oral therapy in naïvegenotype-1 chronic HCV patients. *Hepatology* 54:535A.

Rodriguez-Torres M, Lawitz E, Kowdley KV, et al. (2013). Sofosbuvir (GS-7977) plus peginterferon/ribavirin in treatment-naive patients with HCV genotype 1: A randomized, 28-day, dose-ranging trial. *J Hepatol* 58: 663–8.

Rondla R, Coats SJ, McBrayer TR, et al. (2009). Anti-hepatitis C virus activity of novel beta-d-2′-C-methyl-4′-azido pyrimidine nucleoside phosphoramidate prodrugs. *Antivir Chem Chemother* 20:99–106.

Scrima N, Caillet-Saguy C, Ventura M, et al. (2012). Two crucial early steps in RNA synthesis by the hepatitis C virus polymerase involve a dual role of residue 405. *J Virol* 86:7107–17.

Shi ST, Herlihy KJ, Graham JP, et al. (2009). Preclinical characterization of pf-00868554, a potent nonnucleoside inhibitor of the hepatitis C virus RNA-dependent RNA polymerase. *Antimicrob Agents Chemother* 53:2544–52.

Shimakami T, Hijikata M, Luo H, et al. (2004). Effect of interaction between hepatitis C virus NS5A and NS5B on hepatitis C virus RNA replication with the hepatitis C virus replicon. *J Virol* 78(6):2738–48.

Shirota Y, Luo H, Qin W, et al. (2002). Hepatitis C virus (HCV) NS5A binds RNA-dependent RNA polymerase (RdRP) NS5B and modulates RNA-dependent RNA polymerase activity. *J Biol Chem* 277:11149–55.

Smith DB, Kalayanov G, Sund C, et al. (2009a). The design, synthesis, and antiviral activity of monofluoro and difluoro analogues of 4′-azidocytidine against hepatitis C virus replication: The discovery of 4′-azido-2′-deoxy-2′-fluorocytidine and 4′-azido-2′-dideoxy-2′,2′-difluorocytidine. *J Med Chem* 52:2971–8.

Smith DB, Kalayanov G, Sund C, et al. (2009b). The design, synthesis, and antiviral activity of 4′-azidocytidine analogues against hepatitis C virus replication: The discovery of 4′-azidoarabinocytidine. *J Med Chem* 52:219–23.

Smith DB, Martin JA, Klumpp K, et al. (2007). Design, synthesis, and antiviral properties of 4′-substituted ribonucleosides as inhibitors of hepatitis C virus replication: The discovery of R1479. *Bioorg Med Chem Lett* 17:2570–6.

Smith KL, Lai VC, Prigaro BJ, et al. (2004). Synthesis of new 2′-beta-C-methyl related triciribine analogues as anti-HCV agents. *Bioorg Med Chem Lett* 14:3517–20.

Sofia MJ, Bao D, Chang W, et al. (2010). Discovery of a beta-d-2′-deoxy-2′-alpha-fluoro-2′-beta-C-methyluridine nucleotide prodrug (PSI-7977) for the treatment of hepatitis C virus. *J Med Chem* 53:7202–18.

Sommadossi JP, La Colla P. (2001a). Methods and compositions for treating flaviviruses and pestiviruses. Patent publication no. WO/2001/092282.

Sommadossi JP, La Colla P. (2001b). Methods and compositions for treating hepatitis C virus. Patent publication no. WO/2001/090121.

Sorbera LA, Castaner J, Leeson PA. (2006). Valopicitabine. *Drugs Fut* 31:320–4.

Standring DN, Lanford R, Wright T, et al. (2003). NM 283 has potent antiviral activity against genotype 1 chronic hepatitis C virus (HCV-1) infection in the chimpanzee. *J Hepatology* 38:3.

Stuyver LJ, McBrayer TR, Tharnish PM, et al. (2006). Inhibition of hepatitis C replicon RNA synthesis by beta-d-2′-deoxy-2′-fluoro-2′-C-methylcytidine: A specific inhibitor of hepatitis C virus replication. *Antivir Chem Chemother* 17:79–87.

Stuyver LJ, McBrayer TR, Whitaker T, et al. (2004). Inhibition of the subgenomic hepatitis C virus replicon in huh-7 cells by 2′-deoxy-2′-fluorocytidine. *Antimicrob Agents Chemother* 48:651–4.

Stuyver LJ, Whitaker T, McBrayer TR, et al. (2003). Ribonucleoside analogue that blocks replication of bovine viral diarrhea and hepatitis C viruses in culture. *Antimicrob Agents Chemother* 47:244–54.

Svarovskaia ES, Dvory-Sobol H, Gontcharova V, et al. (2012). Comprehensive resistance testing in patients who relapsed after treatment with sofosbuvir (GS-7977)-containing regimens in phase 2 studies. AASLD abstract 753. *Hepatology* 56:551A.

Tomassini JE, Boots E, Gan L, et al. (2003). An in vitro flaviviridae replicase system capable of authentic RNA replication. *Virology* 313:274–85.

Tomassini JE, Getty K, Stahlhut MW, et al. (2005). Inhibitory effect of 2′-substituted nucleosides on hepatitis C virus replication correlates with metabolic properties in replicon cells. *Antimicrob Agents Chemother* 49:2050–8.

Tomei L, Altamura S, Bartholomew L, et al. (2003). Mechanism of action and antiviral activity of benzimidazole-based allosteric inhibitors of the hepatitis C virus RNA-dependent RNA polymerase. *J Virol* 77:13225–31.

Tomei L, Vitale RL, Incitti I, et al. (2000). Biochemical characterization of a hepatitis C virus RNA-dependent RNA polymerase mutant lacking the C-terminal hydrophobic sequence. *J Gen Virol* 81:759–67.

Toniutto P, Fabris C, Bitetto D, et al. (2007). Valopicitabine dihydrochloride: A specific polymerase inhibitor of hepatitis C virus. *Curr Opin Investig Drugs* 8:150–8.

Valiaeva N, Wyles DL, Schooley RT, et al. (2011). Synthesis and antiviral evaluation of 9-(S)-[3-alkoxy-2-(phosphonomethoxy)propyl]nucleoside alkoxyalkyl esters: inhibitors of hepatitis C virus and HIV-1 replication. *Bioorg Med Chem* 19:4616–25.

Vernachio JH, Bleiman B, Bryant KD, et al. (2011). INX-08189, a phosphoramidate prodrug of 6-o-methyl-2′-C-methyl guanosine, is a potent inhibitor of hepatitis C virus replication with excellent pharmacokinetic and pharmacodynamic properties. *Antimicrob Agents Chemother* 55:1843–51.

Vo NV, Tuler JR, Lai MM. (2004). Enzymatic characterization of the full-length and C-terminally truncated hepatitis C virus RNA polymerases: Function of the last 21 amino acids of the C terminus in template binding and RNA synthesis. *Biochemistry* 43:10579–91.

Wakita T, Pietschmann T, Kato T, et al. (2005). Production of infectious hepatitis C virus in tissue culture from a cloned viral genome. *Nat Med* 11:791–6.

Wang QM, Hockman MA, Staschke K, et al. (2002). Oligomerization and cooperative RNA synthesis activity of hepatitis C virus RNA-dependent RNA polymerase. *J Virol* 76:3865–72.

Wedemeyer H, Jensen D, Herring R, Jr, et al. (2013). PROPEL: A randomized trial of mericitabine plus peginterferon alfa-2a/ribavirin therapy in treatment-naive HCV genotype 1/4 patients. *Hepatology* 58:524–37.

Wyles DL, Kaihara KA, Korba BE, et al. (2009). The octadecyloxyethyl ester of (S)-9-[3-hydroxy-2-(phosphonomethoxy) propyl]adenine is a potent and selective inhibitor of hepatitis C virus replication in genotype 1a, 1b, and 2a replicons. *Antimicrob Agents Chemother* 53:2660–2.

Yamashita T, Kaneko S, Shirota Y, et al. (1998). RNA-dependent RNA polymerase activity of the soluble recombinant hepatitis C virus NS5B protein truncated at the C-terminal region. *J Biol Chem* 273:15479–86.

Zamyatkin DF, Parra F, Alonso JM, et al. (2008). Structural insights into mechanisms of catalysis and inhibition in Norwalk virus polymerase. *J Biol Chem* 283:7705–12.

Zeuzem S, Andreone P, Pol S, et al. (2011). Telaprevir for retreatment of HCV infection. *N Engl J Med* 364:2417–28.

Zhong J, Gastaminza P, Chung J, et al. (2006). Persistent hepatitis C virus infection in vitro: Coevolution of virus and host. *J Virol* 80:11082–93.

Zhou XJ, Pietropaolo K, Chen J, et al. (2011). Safety and pharmacokinetics of IDX184, a liver-targeted nucleotide polymerase inhibitor of hepatitis C virus, in healthy subjects. *Antimicrob Agents Chemother* 55:76–81.

# 4 Mechanisms of Hepatitis C Virus Clearance by Interferon and Ribavirin Combination

## *Lessons Learned from In Vitro Cell Culture*

*Srikanta Dash, Partha K. Chandra, Kurt Ramazan, Robert F. Garry, and Luis A. Balart*

## 4.1 INTRODUCTION TO HEPATITIS C VIRUS

Hepatitis C virus (HCV) is an enveloped, positive-stranded RNA virus that belongs to the *Flaviviridae* family. This family also includes a variety of other important human pathogens such as yellow fever and dengue viruses (Simmonds et al. 2005). Cloning and sequencing of the HCV genome has increased our understanding of the HCV molecular virology and led to model systems to study viral replication and develop antiviral drugs (Choo et al. 1989; Moradpour et al. 2007). The HCV genome is organized into a highly conserved 5′-untranslated region (5′-UTR), a large open reading frame (ORF), and a 3′-untranslated region (3′-UTR) (Figure 4.1). The 5′-UTR binds to host ribosomes to translate the HCV polyprotein using an internal ribosome entry site (IRES) mechanism. A number of viruses (including HCV) utilize the IRES mechanism for gene expression, which is distinct from cap-dependent translation of cellular genes (Sachs et al. 1997; Honda et al. 1999). The 5′-UTR is highly conserved among different HCV genotypes and clinical strains (Bukh et al. 1992). Therefore, this region of HCV has been used for genotyping and quantification of viral RNA in clinical samples by real-time polymerase chain reaction (RT-PCR) (Furione et al. 1999; Hara et al. 2013). The antiviral mechanisms of interferon (IFN) and ribavirin (RBV) also target the highly conserved 5′-UTR (Dash et al. 2005; Chandra et al. 2011b; Panigrahi et al. 2013). The HCV genome contains a large ORF encoding for a 3011 amino acid-long polyprotein, which is proteolytically processed at the endoplasmic reticulum (ER) membrane into ten different mature viral proteins by cellular and viral proteases. The core protein and the two glycoproteins E1 and E2 are referred to as structural proteins because they

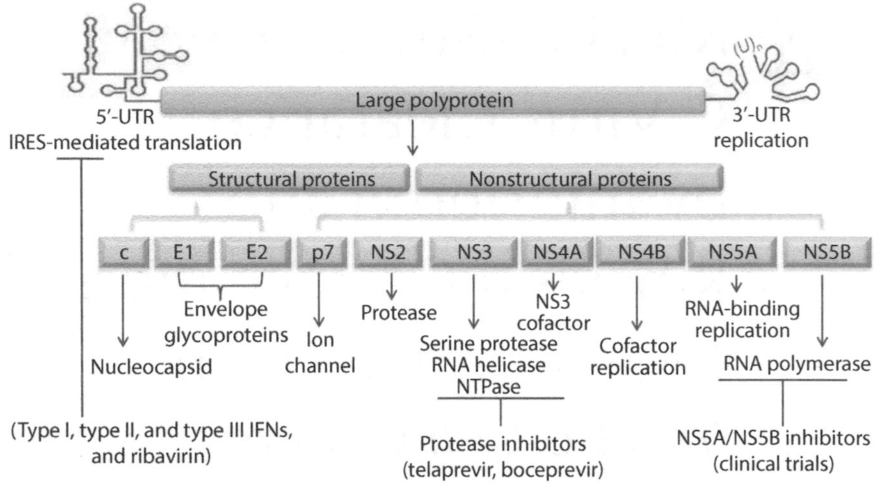


**FIGURE 4.1** Structure of HCV RNA genome and ten different mature proteins produced from the single large open reading frame. Type I, type II, and type III IFNs and ribavirin specifically inhibit the IRES (5′-UTR) function. Core, E1, and E2 are structural proteins and NS2–NS5 are known as nonstructural proteins. The nonstructural proteins (NS3/4A, NS5A, and NS5B) are the targets of antiviral drug discovery. The 3′-UTR is important for HCV genome replication.

are required for viral particle assembly and infection. E1 and E2 bind to a number of cellular receptors implicated in HCV infection of hepatocytes. HCV infection is initiated by the attachment and entry of virus particles into host cells by receptor-mediated endocytosis (Blanchard et al. 2006; Cocquel et al. 2006). The nonstructural proteins (p7, NS2, NS3, NS4A NS4B, NS5A, and NS5B) are required for replication of the viral genome. The nonstructural proteins (NS3 protease, NS4A, and NS5B polymerase) have been targets of intense research efforts to develop antiviral drugs against HCV. The highly conserved 3′-UTR present at the very end of the HCV genome is important for initiation of viral RNA replication (Friebe and Bartenschlager 2002).

Six genotypes of HCV have been sequenced throughout the world (Simmonds et al. 2005; Kuiken et al. 2006). There is up to 40% genetic heterogeneity in the nucleotide sequences among different HCV genotypes. The nucleotide sequences within HCV genotype subtypes (i.e., HCV genotype 1a, 1b) differ by about 20%–25%. Also, minor genomic variations of HCV can be found in the same infected individuals, called viral quasispecies. The sequence variation with HCV genotype is assumed to be due to the lack of a proofreading activity of the viral RNA-dependent RNA polymerase. The relative worldwide distribution of HCV genotypes varies considerably. Genotype-1 (subspecies 1a and 1b) viruses comprise the most common subtypes in the United States and worldwide (Kuiken et al. 2006). Genotype 1a is also common in Northern Europe. Genotypes 1b, 2a, and 2b have a worldwide distribution. Genotype 3 is the most common genotype in the Indian subcontinent. Genotype 4 is the most common genotype in Africa and the Middle East. Genotype 5 is found in South Africa. Genotype 6 is found in Hong Kong and Southeast Asia. In the United States, 75% of chronic hepatitis C cases belong to genotypes 1a and 1b, 13%–15% to genotypes 2a and 2b, and 6%–7% to genotype 3a. Most of the people exposed to HCV develop chronic infection, and only very few recover from virus infections naturally. It has been suggested that HCV overcomes the host's innate and adaptive immune responses to establish chronic persistent infection. The major component of the innate immune response to HCV infection is induction of the endogenous IFN system. The endogenously produced IFNs inhibit virus replication in infected cells via autocrine or paracrine mechanisms by inducing the expression of antiviral genes called IFN-stimulated genes (ISGs). Cell culture studies provide evidence that HCV has the unique capacity to inactivate essential components of pathways that lead to IFN production in hepatocytes (i.e., Toll/IL-1R-domain-containing adapter-inducing interferon-β (TRIF) and mitochondrial antiviral signaling (MAVS)) through the proteolytic activity of its nonstructural protein NS3/4A (Lemon 2010). Understanding innate antiviral mechanisms of HCV in an animal model system has been difficult because HCV infects naturally only humans and chimpanzees.

The World Health Organization (WHO) estimates that approximately 2% of people worldwide have been infected with HCV and that more than 180 million individuals are chronically infected with HCV (Lavanchy 2009). In the United States, approximately 4.1 million people—accounting for 2% of the general population—have been infected with HCV (Shepard et al. 2005). The infection rate in U.S. military veterans has been estimated to be 5.4%–20% and as high as 44% in homeless veterans (Dominitz et al. 2005).

## 4.2 ANTIVIRAL THERAPY REDUCES THE RISK OF HEPATOCELLULAR CARCINOMA RELATED TO CHRONIC HCV INFECTION

Hepatocellular carcinoma (HCC) is the most common malignant liver disease, accounting for approximately 90% of primary liver cancers (Bosch et al. 2004; Hoofnagle 2004; Altekruse et al. 2009; Yang and Robert 2010; El-Serag 2012). Approximately 80%–90% of HCC cases are associated with liver cirrhosis due to chronic hepatitis B virus (HBV) or HCV (Davila et al. 2004; El-Serag 2012). The risk of developing HCC is increased 10- to 100-fold in the case of HBV infection and 15- to 20-fold in patients who are chronically infected with HCV, providing very strong association of hepatitis virus infection with the development of liver cancer (Beasley et al. 1981; Nguyen et al. 2009; El-Serag 2012). There are many similarities and differences in the natural history, epidemiology, and mechanisms of HCC development among people infected with HBV or HCV (Table 4.1). Approximately 5% of the world's population (350 million people) is chronically infected with HBV. The highest incidences and prevalence of HBV infection are in Asian countries—except Japan (El-Serag 2012). In Africa and Asia, 60% of HCC is associated with HBV infection, whereas only 20% of HCC is related to HCV infection. On the other hand, about 2% of the world's population is infected with HCV, accounting for 180 million chronic HCV infections. HCV infection is highly prevalent in developed regions such as Europe, North America, and Japan. Among those areas, HCV infection is highest in Japan (80%–90%), followed by Italy (44%–66%), and then the United States

**TABLE 4.1**
**Similarities and Dissimilarities in Hepatitis B and Hepatitis C Viruses**

| Features | HBV | HCV |
|---|---|---|
| Virus family | Hepadnaviridae | Flaviviridae |
| Virus | 42 nm, enveloped | 40–70 nm, enveloped |
| Genome | dsDNA | ssRNA |
| Reverse transcription | Yes | No |
| Genome length | 3.2 kB | 9.6 kB |
| Transmission | Parenteral, sexual, and perinatal | Parenteral |
| Integration into the host genome | Yes | No |
| Global prevalence | ~350 million | ~200 million |
| Geographic distribution | Asian and African countries | USA, Europe, Egypt, and Japan |
| Chronic infection | 5% | 85% |
| Hepatocellular carcinoma | Yes | Yes |

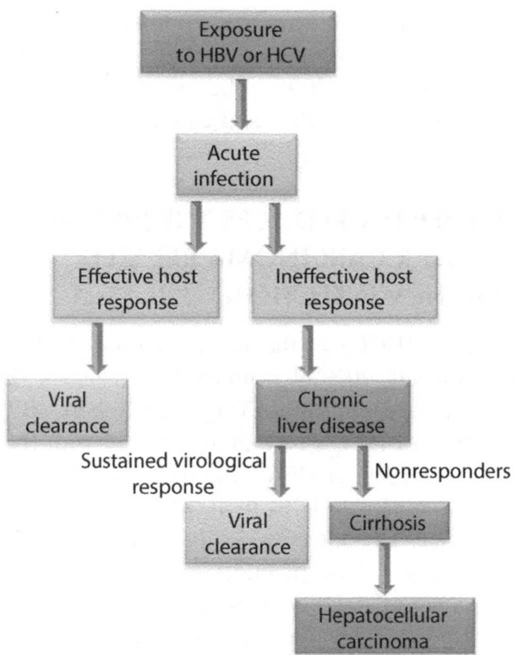


**FIGURE 4.2** Diagram showing the evolution of chronic HCV infection in humans and how antiviral therapy induces viral clearance and reduces the incidence of liver cirrhosis and hepatocellular carcinoma.

(30%–50%) (Tanaka et al. 2006; El-Serag and Rudolph 2007). The incidence of HCC is increasing, and hepatitis infection remains an important public health problem.

Chronic HBV and HCV infections share many common features that promote cancer development (Figure 4.2). The mechanism behind how chronic virus infections develop

HCC has been studied for a number of years and appears to be highly complex. There is no effective treatment for patients with liver cirrhosis and HCC. Patients with a small liver tumor are treated with surgical resection or liver transplantation, but relapse is common. Additionally, localized systemic chemotherapy has been used in many cases with limited success. The multikinase inhibitor sorafenib (approved by the U.S. Food and Drug Administration [FDA] for the treatment of HCC) has limited success because most HCC cases develop resistance to this drug. Thus, most of the currently available treatment for HCC is not effective; therefore, the mortality rate remains very high.

Antiviral therapy that results in the clearance of virus infection reduces the incidence of HCV-related liver diseases (Hino and Okita 2004; Arase et al. 2007). Chronic HCV patients who achieve sustained viral clearance (HBV or HCV) by standard IFN–RBV therapy have a significantly reduced incidence of liver cirrhosis and HCC (Shiratori et al. 2005; Hung et al. 2011). This hypothesis has been verified using prospective and retrospective studies in Japan, Europe, and the United States (Morgan et al. 2010; Bacus et al. 2011; Kimer et al. 2012; Qu et al. 2012; van der Meer et al. 2012). A large study using a population of Veteran's Administration (VA) patients from the United States examined the impact of sustained virological response (SVR) on all causes of mortality by HCV infection from January 2001 to June 2007. These investigators found that SVR reduces mortality among patients infected with HCV genotype 1, 2, and 3 viruses (Bacus et al. 2011). Another study, utilizing 530 patients from the Netherlands, reported that SVR was associated with improved survival in patients with chronic HCV infection and advanced hepatic fibrosis (van der Meer et al. 2012). A report from a large Hepatitis C Antiviral Long-term Treatment against Cirrhosis (HALT-C) study pointed out that patients with advanced cirrhosis, who achieved an SVR, showed marked reduction in overall mortality (Morgan et al. 2010). A study from Japan that included a total of 1505 patients from eight different randomized controlled trials also verified that IFN therapy efficiently reduced HCC development in HCV-related cirrhosis (Qu et al. 2012). Overall, studies from different parts of the world suggest that successful antiviral therapy has made a significant contribution in reducing the incidence of virally induced hepatocarcinogenesis.

Here, we provide a comprehensive review of new developments in the treatment of HCV infection. We also review the different hosts and viral-related factors that interfere with the mechanism of IFN and RBV treatment responses in chronic HCV infection in humans. The mechanisms of interferon-alpha (IFN-α), interferon-lambda (IFN-λ), and RBV combination treatment to inhibit HCV replication in cell culture model are discussed. Work performed in our and many other laboratories to understand HCV resistance mechanisms of IFN-α and RBV combination treatment, using in replicon and persistently infected in vitro cell cultures, are summarized. Finally, we present some new results indicating that IFN-λ, which is a type III IFN, is important in inducing HCV clearance in cell culture.

## 4.3 NEW STANDARD OF CARE FOR CHRONIC HCV INFECTION

The use of recombinant human IFN-α was approved by the FDA in 1992 for the treatment of chronic HCV infection. IFN-α given three times a week for 6–12 months resulted in an SVR in only 15%–25% of chronic HCV patients. RBV, a guanosine analog, has been used for the treatment of diseases caused by a number of RNA viruses including respiratory syncytial virus (RSV) and Lassa virus. IFN-α and RBV combination therapy received FDA approval in 1998 for the treatment of chronic HCV infection. This combination therapy has significantly improved the SVR rate by up to 40% (McHutchison et al. 1998). Clinical studies have shown that IFN-α and RBV combination therapy is more effective in the treatment of chronic HCV infection than the treatment with a single agent. Subsequently, another development took place, conjugating polyethyl glycol to recombinant IFN-α, which improved the stability (Harris and Chess 2003). This modification allowed IFN-α injection only once per week instead of three times a week. The pegylated IFN-α and RBV therapy resulted in increased SVR rates (Manns et al. 2001; Fried et al. 2002). This therapy, which has been used as a standard of care for a decade in patients with hepatitis C, leads to an SVR in 42%–52% of patients with HCV genotype 1; 65%–85% of patients with HCV genotype 4, 5, and 6; and 76%–82% of patients with HCV genotype 2 or 3 (Feld and Hoofnagle 2005). The clearance of HCV infection by IFN-α and RBV in patients infected with HCV genotype 1 remained low compared to infections with other HCV genotypes. For other HCV genotypes, IFN-α and RBV therapy is still used as a standard of care for patients with chronic HCV infection.

In 2011, two NS3/4A protease inhibitors (boceprevir and telaprevir) received FDA approval for the treatment of chronic HCV infection (Mathews and Lancaster 2012). The new standard of care for chronic HCV genotype 1 involves the combination of IFN-α, RBV, and one of the protease inhibitors, which resulted in 65%–80% clearance of HCV genotype 1 (Jacobson et al. 2011; Poordad et al. 2011; Zeuzem et al. 2011). The success of triple combination therapy as well as IFN-α and RBV combination therapy is determined by the measurement of serum HCV RNA levels by molecular techniques with a sensitivity of 10–15 IU/mL. It has been reported by a number of investigators that HCV RNA levels among individual patients during the antiviral response can vary significantly (Feld and Hoofnagle 2005; Asselah et al. 2010) (Figure 4.3). Some individuals clear the virus and become HCV RNA negative four weeks after treatment; this is defined as rapid virological response (RVR). Some individuals show an early reduction of HCV RNA levels of more than 2 logs and become HCV RNA negative at 12 weeks. This type of response is called an early virological response (EVR). Some individuals take a longer time, up to 24 weeks, to be clear of HCV in the serum. A sustained antiviral response (SVR) is defined as absence of HCV RNA up to 24 weeks after stopping the treatment. In selected individuals, HCV RNA reappears during the antiviral therapy; this is called breakthrough. In some patients, HCV RNA reappears when the therapy is stopped; this is called relapse. Patients who fail to clear HCV RNA in their serum after 24 weeks are termed nonresponders.

Although the introduction of the two protease inhibitors, boceprevir and telaprevir, has resulted in very high SVR rates, the new triple combination therapy has a number of limitations such as: (1) the failure rates in treatment of naïve patients are 20%–30% on triple therapy. In previously treated patients, the failure rates ranged as high as 50%–60%. The results of the REALIZE and RESPOND-2 trials suggested that about 40%–60% of previously identified null responders or partial

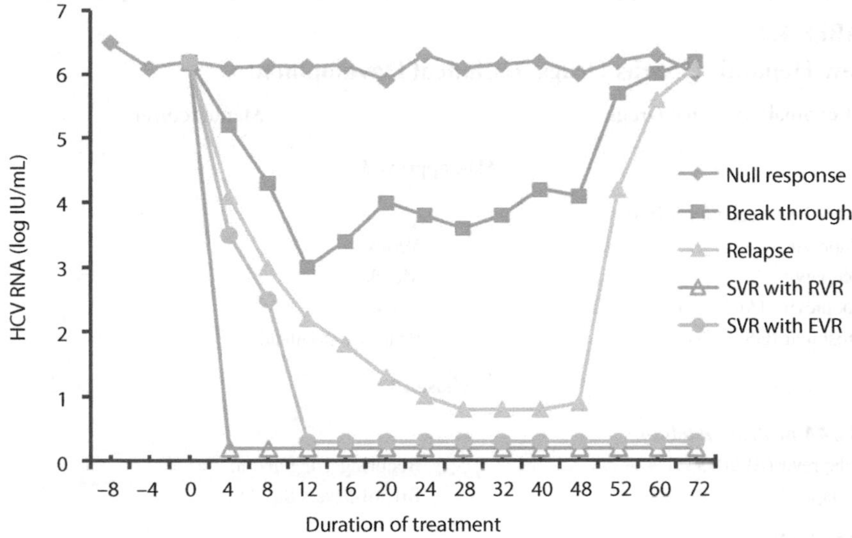


**FIGURE 4.3** Diagram illustrating how the triple therapy treatment response, which is determined on the basis of serum HCV levels, varies among patients. The success of HCV treatment is determined by the measurement of serum viral RNA levels by real-time RT-PCR over several weeks. (From Feld JJ, Hoofnagle JH, *Nature* 436, 967–972, 2005; Asselah T et al., *Liver Int* 30, 1259–1269, 2010.)

responders failed to clear the infection (Dabbouseh and Jensen 2013). (2) Protease inhibitors have more adverse side effects such as increased incidence of rash, anemia, dry skin, and fatigue than IFN-α and RBV alone. (3) The two protease inhibitors (telaprevir and boceprevir) are restricted to HCV genotype 1. (4) Long-term treatments with protease inhibitors lead to the development of viral variants carrying mutations in the protease domain—the evolution of drug-resistant variants due to repeated drug treatment may lead to viral breakthrough and disease progression. (5) The cost of protease-inhibitor-based treatment remains very high, which limits its widespread use, and so forth. Due to these limitations, new therapeutic strategies that ensure broader genotype coverage, better efficacy, better tolerance, and minimum resistant variants remain the long-term objectives of HCV antiviral drug development.

## 4.4 EMERGING NEW ANTIVIRALS FOR CHRONIC HCV INFECTION IN CLINICAL TRIALS

There are three classes of direct acting antivirals (DAAs) that have been evaluated in clinical trials (Table 4.2). Progress in antiviral drug development against HCV has been reviewed recently (Pawlotsky 2012; Dabbouseh and Jensen 2013). The new small-molecule drugs are directed against the nonstructural proteins that are critical for HCV replication, namely NS3/4A serine protease, NS5A protein, and NS5B RNA-dependent RNA polymerase.

## TABLE 4.2
## New Hepatitis C Virus Drugs in Clinical Development

| List of Small-Molecule Drugs | Manufacturer |
|---|---|
| | **FDA approved** |
| *NS3/4A protease inhibitors* | |
| Telaprevir | Vertex |
| Boceprevir | Merck |
| Simeprevir (TMC-435) | Janssen |
| Sofosbuvir (GS-7977) | Pharmasset/Gilead |
| | **Phase III** |
| *NS3/4A protease inhibitors* | |
| Faldaprevir (BI-201335) | Boehringer-Ingelheim |
| Asunaprevir | Bristol-Myers Squibb |
| *NS5A inhibitor* | |
| Daclatasvir (BMS-790052) | Bristol-Myers Squibb |
| *NS5B nucleoside/nucleotide inhibitor* | |
| *Cyclophillin inhibitors* | |
| Alisporivir | Novartis |
| | **Phase II** |
| *NS3/4A protease inhibitors* | |
| Vaniprevir | Merck |
| Danoprevir | Roche/Genentech |
| MK-5172 | Merck |
| GS-9451 | Gilead |
| GS-9256 | Gilead |
| ABT-450 | Abbott |
| Narlaprevir | Merck |
| NS5A inhibitors | |
| ABT-267 | Abbott |
| GS-5885 | Gilead |
| MK-8742 | Merck |
| NS5B nucleoside/nucleotide inhibitors | |
| IDX-184 | Idenix |
| Mericitabine | Roche/Genentech |
| ALS-2200; ALS-2158 | Vertex Pharmaceuticals Inc and Alios BioPharma Inc |
| NS5B non-nucleoside/nucleotide inhibitors | |
| Filibuvir | Pfizer |
| VX-222 | Vertex |
| *Cyclophillin inhibitors* | |
| SCY-465 | Scynexis |

### 4.4.1 NS3/4A Protease Inhibitors

NS3 protease is a bifunctional enzyme that contains serine protease in the N-terminal 180 amino acid residues. The search for small molecule inhibitors led to the identification of the first-generation protease inhibitors: telaprevir and boceprevir. Additional second generation NS3/4A protease inhibitors simeprevir (TMC435), faldaprevir, and danoprevir are in the clinical stage of development. Preliminary studies using them in naïve patients suggested that each of these second-generation protease inhibitors produces a better antiviral response than the first-generation protease inhibitors.

### 4.4.2 NS5A Inhibitors

The NS5A nonstructural protein is a phosphoprotein, but it has no enzyme activity. This protein is associated with the host cell membranes through its N-terminal amphipathic helix and is also associated with HCV replication complex. The NS5A protein has three domains: the first 213 amino acids in the N-terminal end is called domain 1, which contains a zinc-binding domain indicating that this protein is a zinc metalloprotein. Domains 2 and 3 are in the C-terminal portion. NS5A binds directly with NS5B polymerase, and it is part of the HCV replication complex. Small-molecule inhibitors targeted to NS5A protein have been identified, and some are in clinical studies. Declatasvir, or BMS 790052, is a potent inhibitor of HCV replication (Gao et al. 2010).

### 4.4.3 NS5B Inhibitors

NS5B is an RNA-dependent RNA polymerase that performs HCV genome replication and thus is an important target for antiviral drug development. Replication of HCV genome begins with the synthesis of a full-length negative strand. Negative-strand HCV RNA then serves as a template for the synthesis of a new positive-strand HCV genome. This process is critical for additional protein synthesis, continuing viral RNA replication and packaging of new virus particles. The synthesis of HCV RNA genome in the infected cells utilizes a single nucleotide as a primer. Inhibitors of NS5B include nucleosidic inhibitors and non nucleosidic inhibitors. Nucleosidic inhibitors are nucleotide analogs that cause chain termination when they are incorporated in the viral genome. Nucleoside inhibitors act across all genotypes because the NS5B polymerase is highly conserved among different HCV genotypes. Non nucleosidic inhibitors can bind to one of the several allosteric sites and induce a conformational change of the NS5B polymerase. Sofosbuvir (GS-7977) is one of several nucleoside agents showing very promising results in clinical trials. Initial trails with HCV patients treated with sofosbuvir resulted in rapid decline of viral load. This drug has activity against a broad range of HCV genotypes and seems to have 90% SVR in the treatment of naïve patients when given along with IFN-α and RBV (Smith 2013).

### 4.4.4 Development of Experimental Therapeutics for HCV

A number of DAAs are in the developmental stage and, hopefully, some of them will be approved soon to treat HCV infection with or without the use of IFN-α and RBV. One of the limitations of using DAAs to treat chronic HCV infection is the development of drug-resistant virus. Therefore, alternative therapeutic strategy may be required for the treatment of drug-resistant HCV. Here, we describe some of the novel small RNA–based experimental therapies that have been developed over the years and that can be utilized to treat drug-resistant HCV.

Two classes of small RNA–based antiviral approaches—called small interfering RNAs (siRNAs) and microRNAs (miRNAs)—have been developed. Some of the anti-HCV strategies using small RNAs (siRNA and miRNA) have shown promising results in the cell culture, animal models, and clinical trials. Small RNAs are the specificity components of a protein machinery called RNA-induced silencing complex (RISC) that uses the small RNAs to recognize complementary motifs in target nucleic acids and degrades the target RNA using a specific silencing mechanism. HCV is an RNA virus that replicates in the cytoplasm inhibition of HCV replication by intracellular delivery of siRNA or miRNA to offer an alternative intracellular therapeutic approach. These small RNAs can be used as powerful antivirals against a number of viruses that cannot be cleared by small-molecule drugs. The miRNA-122 has been shown to modulate HCV replication by binding to the 5′-UTR of HCV genome (Joplin et al. 2005). It has been shown that miRNA-122 antagonist reduces HCV titers in HCV-infected chimpanzees (Lanford et al. 2010). Recent clinical studies show that antisense oligonucleotide miravirsen, which binds to miRNA-122, has no adverse events and inhibits chronic HCV infection in humans; thus, it is the best candidate to be licensed to use in the clinic. Phase II clinical studies using miRNA-122 show a great promise as a molecular approach to hepatitis C treatment in humans (Janssen et al. 2013). Another proof of principle study performed in our laboratory shows that intracellular delivery of a combination of two siRNAs targeted to the HCV 5′-UTR minimizes escape mutant viruses and leads to rapid inhibition of HCV replication in cell culture and in an animal model (Chandra et al. 2012). These studies provide a hope for development of alternative therapies for drug-resistant HCV in the future.

## 4.5 MECHANISMS OF IFN-α AND RIBAVIRIN TREATMENT FAILURE IN CHRONIC HCV INFECTION

Several years of clinical research have provided evidence that a number of viral- and host-related factors interfere with IFN-α and RBV treatment induced virus clearance. These include virus genotype, viral load, human immunodeficiency virus (HIV) coinfection, gender, race, obesity, insulin resistance, fibrosis, interleukin-28B (IL-28B) genotype, pre activation of IFN-inducible genes, and others (Figure 4.4).

### 4.5.1 Virus Genotype

Clinical studies found that HCV genotypes 1 and 4 are more resistant to IFN-α and RBV therapy than HCV genotypes 2 and 3 (Manns et al. 2001; Fried et al. 2002; Hadziyannis et al. 2004; Roulot et al. 2007). The clearance of HCV is slow in

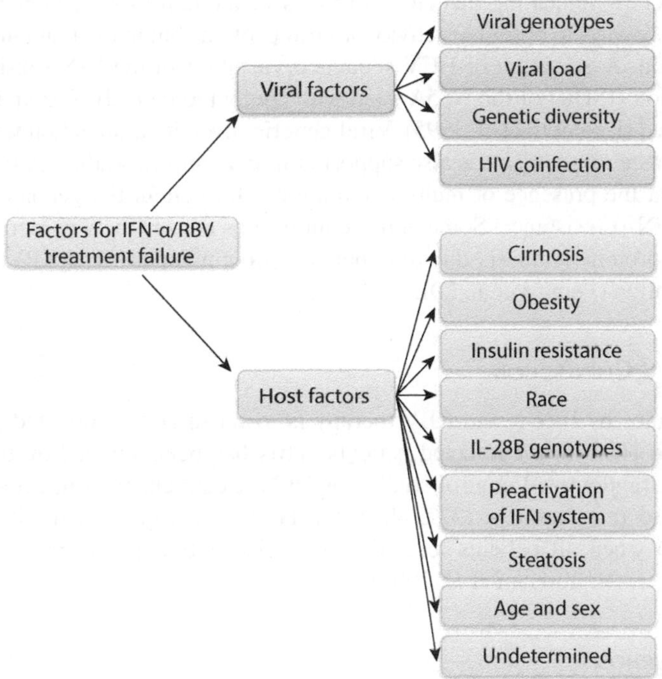


**FIGURE 4.4** Diagram illustrating how virus- and host-related factors influence the IFN and RBV treatment response in chronic HCV infection in humans.

patients infected with genotype 1a or 4 viruses as compared to genotypes 2 and 3. The mechanisms underlying these differences are unclear.

### 4.5.2 High Viral Load

Numerous studies have reported that patients with a very high baseline viral load are less likely to clear HCV by IFN-α and RBV therapy than those with a low baseline viral load (Marcellin et al. 1995; Martinot-Peignoux et al. 1995, 1998; Berg et al. 2003; von Wagner et al. 2005; Jensen et al. 2006; Zeuzem et al. 2006). The titer of HCV greater than 400,000–800,000 IU/mL is considered as a high viral load (Reddy et al. 2010).

### 4.5.3 Viral Genetic Diversity

The role of virus genetic diversity and amino acid covariance network involvement in IFN and RBV therapy in humans has been verified in the Virahep-C clinical study. The investigators sequenced the large ORF of HCV in 94 patients infected with genotype 1, and they found that genetic variability is significantly higher in responders to IFN-α and RBV therapy than in nonresponders (Aurora et al. 2009). The explanation of their findings is that higher HCV genetic diversity correlates with response therapy indicating that HCV survives in the nonresponders because there are only a few ways to optimize activity of viral proteins but many that interfere with their function. Association of HCV sequence variation in the IFN-sensitive determining region (ISDR) of the NS5A gene with the outcome of IFN-α and RBV has been reported (Enomoto et al. 1995). Viral genetic diversity association with IFN-α/RBV resistance mechanisms is also supported in cell culture studies. A recent study reported that the presence of multiple mutations throughout the genome is associated with IFN-α resistance. Some of these mutations are involved in increased shut-off of host protein synthesis due to enhanced protein kinase R (PKR) and eIF2α phosphorylation (Perales et al. 2013).

### 4.5.4 HIV Coinfection

HCV clearance by IFN-α and RBV therapy is lower in HIV-coinfected patients as compared to HCV mono infected patients. This has been verified by three large, multicenter, randomized control studies of HCV treatment in coinfected patients. The response rate was low (27%–40% for HCV genotype 1 and 14%–29% for genotype 4) when all patients received 48 weeks of therapy (Carrat et al. 2004; Chung et al. 2004; Torriani et al. 2004).

### 4.5.5 Cirrhosis

The SVR rate in patients with liver cirrhosis is low. A review article that collected information from published literature after 2011 has summarized the SVR response among 1133 cirrhotic patients all over the world, indicating that the overall SVR among cirrhotic patients was 37%, while the SVR rate was approximately 20% in patients with genotype 1 or 4 infection and 55% in patients with genotype 2 or 3 infection (Bota et al. 2013). Another retrospective cohort study in European and Canadian patients has reported that the SVR rate for chronic HCV patients who had biopsy-proved advanced fibrosis and cirrhosis was 29.6% (Veldt et al. 2007). Studies from the United States also support these results indicating that the overall SVR rate of combination therapy of peg IFN-α and RBV among patients with compensated cirrhosis is 30% (Wright 2002).

### 4.5.6 Obesity

Steatosis or nonalcoholic fatty liver disease is the most common cause of chronic liver diseases in North America and other parts of the world (Cheung and Sanyal 2008; Younossi and McCullough 2009; Hwang and Lee 2011; Alsio et al. 2012). Obesity has been identified as a risk factor for nonresponse to IFN and RBV combination therapy (Cheung and Sanyal 2008). Chronic HCV patients who are obese have 80% lower chance of achieving SVR as compared to nonobese patients (Bressler et al. 2003). The mechanism by which obesity affects the treatment response is not well understood. It is assumed that obese patients have increased accumulation of lipid droplets within the hepatocytes that interfere with the IFN-induced cellular Janus kinase/signal transducer and activator (JAK-STAT) signaling or that the presence of free fatty acids (FFAs) in the hepatocytes increases viral replication (Gunduz et al. 2012). The other possible explanation for this impaired response is that it might be caused by suboptimal serum concentration of antiviral drugs due to the poor lymphatic circulation in obese people (Cheung and Sanyal 2008).

### 4.5.7 Diabetes/Insulin Resistance

*Insulin resistance* is defined as an impaired clearance of blood glucose levels at a given level of insulin. Insulin resistance is associated with a number of metabolic syndromes including nonalcoholic fatty liver disease and diabetes. HCV infection induces insulin resistance and impairs clearance of glucose in hepatic cell culture models (Banerjee et al. 2008). Insulin resistance and type 2 diabetes mellitus have been shown to reduce the chances of achieving SVR by IFN and RBV (Romero-Gomez et al. 2005; Younossi and McCullough 2009). Treatment with the insulin sensitizing agent pioglitazone in combination with peginterferon and RBV has been shown to increase SVR and RVR rates during HCV treatment in some patients (Khattab et al. 2010). However, another group has reported that this treatment did not improve the SVR rate (Harison et al. 2012). Basic studies have provided evidence that HCV infection induces insulin resistance through down regulation of insulin receptors 1 and 2 (Kawaguchi et al. 2004; Shintani et al. 2004; Pazienza et al. 2007). This reason for impaired response to IFN-α in the condition of insulin resistance has been supported by basic studies indicating reciprocal interference between IFN-α and insulin signaling in liver cells (Franceschini et al. 2011).

### 4.5.8 African Americans

HCV infection is more common in African Americans than in members of other racial groups in the United States (Pearlman 2006; Burton et al. 2012). Many groups have reported that the response rate to peginterferon-α and RBV antiviral therapy is low among African Americans as compared to Caucasians (28% vs. 52%) (Muir et al. 2004; Conjeevaram et al. 2006). Triple combination therapy using peg IFN-α, RBV, and protease inhibitors is also less effective in African Americans (Jacobson et al. 2011; Poordad et al. 2011). The reason for this different response to antiviral therapy is not clear; however, two explanations have been proposed for this lower SVR rate in African Americans. First, the African American population has a higher prevalence of HCV genotype 1 infection (Alter 2007). Second, a higher prevalence of IL-28B T/T risk allele has been identified in African American patients with a lower SVR rate (Ge et al. 2009).

### 4.5.9 Interleukin-28B Genotype

Genome-wide association studies (GWAS) have opened a completely new direction to our understanding of the HCV clearance by IFN-α and RBV treatment (Donnelly et al. 2011; Urban et al. 2012). Five large studies (North America, Japan, Northern Europe, Australia, Switzerland) identified single nucleotide polymorphisms (SNPs) in the IL-28B region that are associated with response to IFN and RBV in chronic HCV infection (Rauch et al. 2010a). Two groups identified SNP rs12979860, located 3 kb upstream of IL-28B gene, to have the strongest correlation with treatment response. Individuals with the CC genotype had two- to six-fold greater SVR than those with the TT genotype (Ge et al. 2009; McCarthy et al. 2010). Three groups have shown that rs8099917 SNPs in the IL-28B gene have the strongest association with treatment failure (Suppiah et al. 2009; Tanaka et al. 2009; Rauch et al. 2010b). Individuals carrying one or two copies of the rs8099917 risk G-allele have been found to have higher risks of treatment failure as compared with individuals carrying the genotype TT. All SNPs have been found to be located close to IL-28B gene on chromosome 19. The IL-28B gene polymorphisms associated with HCV clearance by peg IFN-α and RBV have been confirmed by many clinical studies. Clinical tests for IL-28B genotype are used as a part of the standard of care for the treatment of chronic HCV1a infection.

The mechanism of how IL-28B contributes to IFN-α and RBV treatment response is not clear. There are two explanations that can account for the functional significance of IL-28B gene polymorphisms. The first explanation is supported by a few studies that indicated that the risk allele is associated with treatment failure and lower expression of IL-28A/B (Diegelmann et al. 2010; Langhans et al. 2011; Dolganiuc et al. 2012; Shi et al. 2012). A second explanation is supported by the observation that the risk allele associated with treatment failure in the presence of higher constitutive expression of ISGs in HCV-infected liver is due to increased IL-28A/B expression (Honda et al. 2010; Urban et al. 2010, 2012). Overall, the components of the molecular pathways linking the IL-28B genotype, IFN-λ expression, and nonresponse IFN and RBV therapy are unknown. The response to this triple combination therapy also depends on a patient's initial successful or unsuccessful response to peg IFN-α and RBV treatment. Some clinical studies have found that IL-28B genotype testing might also be predictive of triple therapy treatment response (Akuta et al. 2010; Jacobson et al. 2011). Understanding the complex interaction between the IFN-λ axis and HCV clearance should provide new insights for improving SVRs in chronic HCV infection.

## 4.6 IFN-α ANTIVIRAL ACTIVITY AGAINST HCV IS MEDIATED THROUGH JAK-STAT SIGNALING

IFNs are proteins produced by host cells in response to pathogens including viruses, bacteria, and parasites. IFNs are divided into three types based on the nature of signaling they induce after binding to the cell surface: type I IFN, type II IFN, and type III IFN (Figure 4.5). Type I IFNs bind to the IFN-α receptor (IFNAR) complex consisting of two major subunits: IFNAR1 and IFNAR2 (De Weerd et al. 2007). Type II IFNs are referred to as interferon gamma (IFN-γ) and are produced primarily by T lymphocytes and natural killer cells. IFN-γ binds to a distinct cell surface receptor called the interferon gamma receptor (IFNGR) consisting of two subunits, IFNGR1 and IFNGR2 (Hemmi et al. 1994). Type III IFNs are referred to as IFN-λ. They are also known as IL28A (IFN-λ2), IL28B (IFN-λ3), and IL29 (IFN-λ1), respectively. Type III IFNs signal through a distinct IFN-λ receptor (IFNLR) complex composed of the IFN-λR1 and IL-10R2 chain shared with the IL-10 receptor (Kotenko et al. 2003; Sheppard et al. 2003). The type III IFN genes are located on chromosome 19.

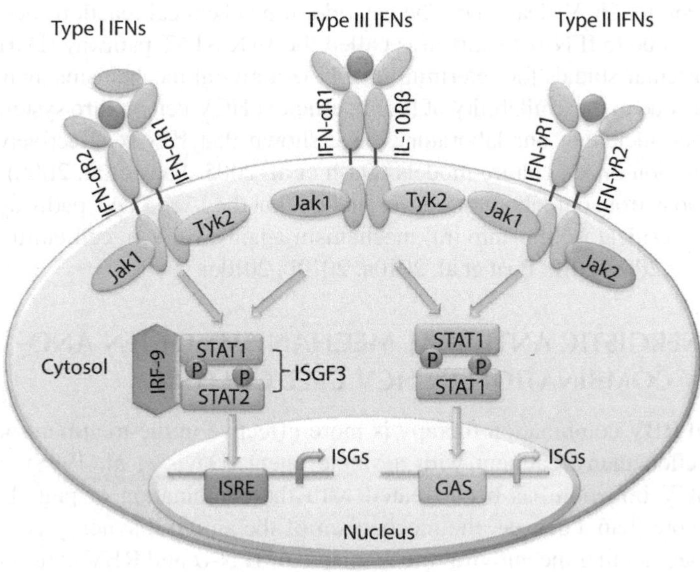


**FIGURE 4.5 (See color insert.)** Comparison of JAK-STAT signaling activation by different IFNs. Type I (IFN-α), type II (IFN-γ), and type III (IFN-λ) bind to separate cell surface receptors that activate two tyrosine kinases, Jak1 and Tyk2, which then phosphorylate the STAT. Phosphorylated STAT1 and STAT2 combine with IRF-9 to form ISGF3 complex, which activates the type I IFN genes. Type II IFN activates Jak1 and Jak2 tyrosine kinases and only phosphorylate STAT1 protein. The phosphorylated STAT1 protein forms homodimer and activate type II IFN genes. Type III IFN activates each pathway.

The coding region for each of these genes is divided into five exons. Type I and type III IFNs are induced by viral infections (Onoguchi et al. 2007). The promoters for the IL-28A and IL-28B genes are very similar and share several transcriptional regulatory elements indicating that they are regulated in a similar manner (Osterlund et al. 2007). Type III IFNs signal through JAK-STAT signaling and induce an antiviral state similar to type I IFN (Zhang et al. 2011). IFN-α first binds to the cell surface receptors (IFNAR1 and IFNAR2) that activate a cascade of signal transduction pathway mediated by two receptor-associated tyrosine kinases called Janus kinase 1 (Jak1) and tyrosine kinase 2 (Tyk2). IFNAR1 binds to tyrosine kinase 2 (Tyk2) and IFNAR2 to Jak1. Upon binding to the type I IFNs receptors, Tyk2 and Jak1 undergo mutual tyrosine phosphorylation. These kinases also phosphorylate IFNAR1 providing a docking site for the STAT2. Tyk2 or Jak1 also phosphorylates the STAT2 protein on its tyrosine residue 690. STAT1 subsequently is recruited to the membrane for phosphorylation and activation. The activated STAT1 and STAT2 monomers are released back to the cytoplasm where they assemble as heterodimers in a complex with IFN regulatory factor 9 (p48). This active transcription complex is called IFN-stimulated gene factor 3 (ISGF3). This complex is translocated to the nucleus where it binds to IFN-stimulated response element (ISRE) in the promoters of various ISGs to initiate antiviral gene transcription. The JAK-STAT pathways activate a large number of antiviral genes that are important for HCV clearance. The cascade of biochemical reactions occurring in normal cells due to IFN-α treatment is called the JAK-STAT pathway (Darnell et al. 1994). Molecular studies for determining IFN-α antiviral mechanisms against HCV are possible due to the availability of highly efficient HCV cell culture systems. Many investigators, including our laboratory, have shown that IFN-α effectively inhibits HCV replication in cell culture models (Dash et al. 2005; Jiang et al. 2008). A series of publications from our laboratory have verified that the JAK-STAT pathway induced by IFN-α is critical for the antiviral mechanism against HCV in cell culture models (Hazari et al. 2007, 2010; Poat et al. 2010a, 2010b, 2010c).

## 4.7 SYNERGISTIC ANTIVIRAL MECHANISMS OF IFN AND RBV COMBINATION IN HCV CELL CULTURE

IFN-α and RBV combination therapy is more effective in the treatment of chronic HCV infection than treatment with a single agent (Davis et al. 1998). Although chronic HCV infection has been treated with the combination of peg IFN-α and RBV for more than a decade, the mechanism of the antiviral synergy is not understood. Understanding the antiviral mechanisms of IFN-α and RBV action using the improved HCV cell culture system is important and may open newer interventions to improve the clinical response. Molecular studies of IFN-α and RBV action against HCV are also possible due to the availability of in vitro cell culture systems. The antiviral effects of IFN-α and RBV combination treatment were examined using a stable subgenomic cell line (S3-green fluorescent protein [GFP]) and infected cell culture system. The sustained antiviral effect in S3-GFP replicon cell culture was determined by three different methods: looking at the G-418 resistant cell colonies, measurement of GFP expression under a fluorescence microscope, and flow analysis. Because the subgenomic replicon system does not express HCV structural proteins or produce infectious virus, we verified the antiviral effect of IFN-α and RBV combination treatment using an infectious JFH1-Rluc chimera virus, the method described previously (Chandra et al. 2011a, 2012). The antiviral effect of IFN-α and RBV combination was measured by Rluc activity and core protein expression by immunostaining. The success of combination treatment was determined by measuring the percentage cells expressing HCV core protein. Results using replicon and infected cells indicate that IFN-α and RBV combination treatment inhibits HCV replication significantly more than treatment with IFN-α or RBV alone. The fact that viability of cells in these HCV cell culture systems was not significantly altered by the combination of IFN-α and RBV treatment indicated that the drugs do not have significant toxic effects (Panigrahi et al. 2013).

It was reported previously by our group that type I and type II IFN inhibit HCV replication by targeting the 5′-UTR of HCV RNA genome used for IRES-mediated translation (Dash et al. 2005). Therefore, we examined whether IFN-α and RBV combination treatment could also inhibit the HCV IRES–mediated translation. The mechanisms of IFN-α and RBV action on HCV translation were examined using HCV IRES-GFP or HCV IRES-Rluc-based subgenomic clones. HCV IRES–mediated translation of GFP was inhibited by IFN-α and RBV at increasing concentration of both the drugs as evidenced by fluorescence imaging and Western blot analysis. Cap-dependent translation was not inhibited by addition of these two drugs (Figure 4.6a–d). We quantified the relative antiviral activity of IFN-α, IFN-λ, and RBV at the level of HCV IRES translation using a HCV IRES *Renilla* luciferase plasmid. IFN-α, IFN-λ1, and RBV each inhibit HCV replication. Combination treatment of IFN-α and IFN-λ with RBV showed a stronger inhibitory effect on HCV IRES-Rluc expression. Analysis of synergistic, additive, and agonist effect of IFN-α and RBV combination treatment was determined according to the median effect principle using the CalcuSyn and MacSynergy II programs (Panigrahi et al. 2013). This analysis revealed that IFN-α and IFN-λ1 treatment show slight antagonism, whereas RBV treatment in combination with either IFN-α or IFN-λ shows synergistic interactions (Figure 4.6e).

Both IFN-α and RBV inhibit HCV IRES–mediated translation through prevention of polyribosome formation. Western blot analysis shows that IFN-α and RBV each induce PKR and eIF2α levels (Figure 4.7a). RBV also blocks polyribosome loading of HCV IRES-GFP mRNA through the inhibition of cellular inosine-5'-monophosphate dehydrogenase (IMPDH) activity and induction of PKR and eIF2α phosphorylation. Knockdown of PKR or IMPDH prevented RBV induces HCV IRES–mediated GFP translation (Figure 4.7a–d). IFN-α treatment induces levels of PKR and eIF2α

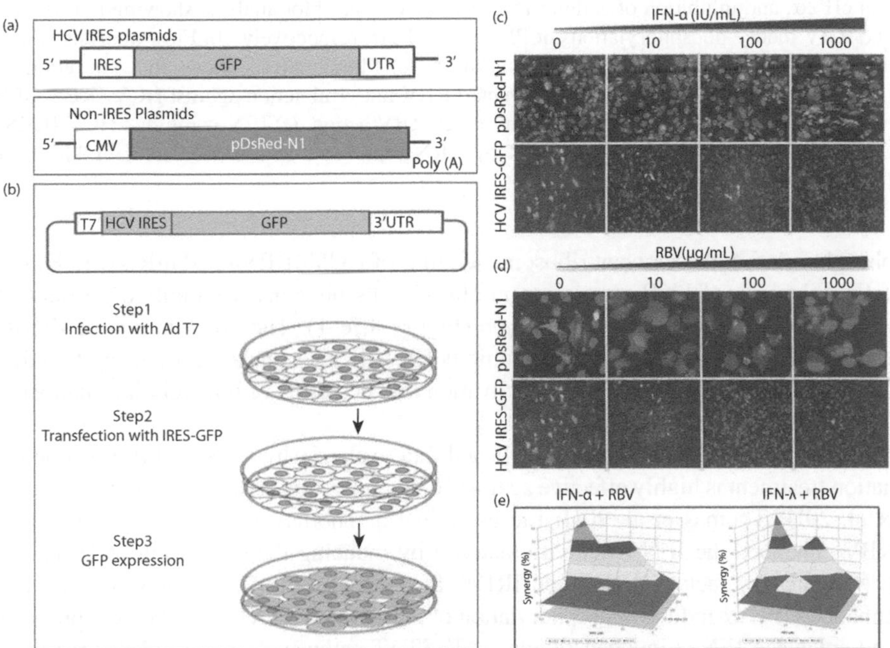


**FIGURE 4.6 (See color insert.)** IFN-α and RBV synergistically inhibit HCV IRES–mediated translation. (a) The structure of IRES and non-IRES plasmid clones used in our study. (b) The steps used to express HCV IRES-GFP or HCV IRES-luciferase using a recombinant adenovirus expressing T7 RNA polymerase. (c,d) Effect of IFN-α and RBV treatment on IRES (green fluorescence) and by non-IRES mechanism (red fluorescence) of translation. (e) Three-dimensional synergy inhibition plot for IFN and RBV combination treatment against HCV IRES.

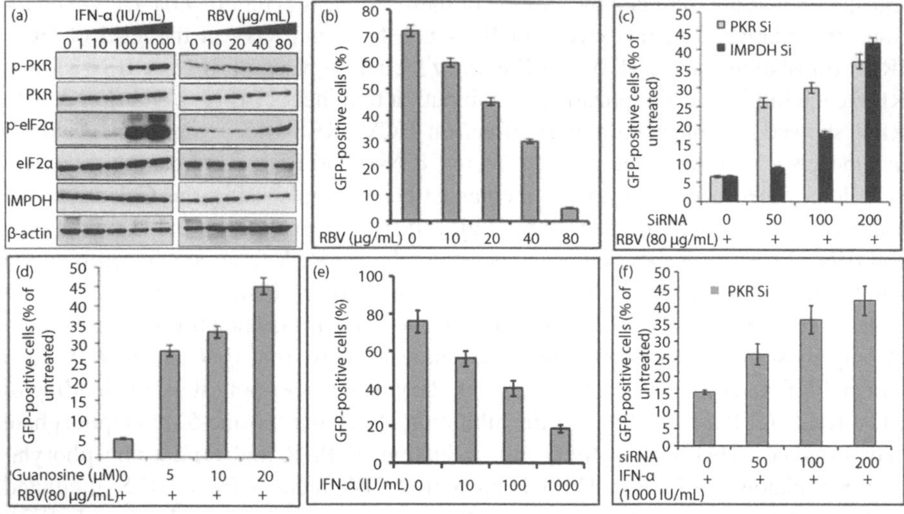


**FIGURE 4.7** IFN-α and RBV synergy antiviral mechanism involves the activation of PKR and eIF2α, and inhibition of cellular IMPDH. (a) Western blot analysis showing that IFN-α and RBV induce phosphorylations of PKR and eIF2α, respectively. (b) Flow analysis shows that RBV inhibits HCV IRES-GFP translation. (c) Flow analysis shows that inhibition of IMPDH and PKR levels by siRNA prevented RBV antiviral action against HCV IRES-GFP translation. (d) Guanosine treatment neutralizes RBV action. (e) IFN-α inhibits HCV IRES-GFP translation. (f) Inhibition of PKR by siRNA prevents IFN-α-mediated inhibition of HCV IRES-GFP translation.

phosphorylation that prevent ribosome loading of HCV IRES-GFP mRNA in Huh-7 cells. Silencing of PKR expression in Huh-7 cells prevents the inhibitory effect of IFN-α on HCV IRES-GFP expression (Figure 4.7e–f). The combination of IFN-α and RBV treatment synergistically inhibits HCV IRES translation by using two different mechanisms involving PKR activation and depletion of intracellular guanosine pool through inhibition of IMPDH.

Only a few studies have been published that explain why RBV and IFN-α combination treatment is highly effective against HCV replication (Liu et al. 2007; Stevenson et al. 2011; Thomas et al. 2011; Liu et al. 2012). Thomas et al. (2011) showed that RBV enhanced the IFN-α antiviral activity by inducing the expression of ISGs and IFN regulatory factors, IRF-7 and IRF-9. Stevenson et al. (2011) showed that RBV enhanced IFN-α-induced phosphorylation of STAT1 and STAT3 and MxA expression and enhanced IFN-α-induced cellular JAK-STAT pathway. Liu et al. (2012) reported that RBV enhances IFN-α signaling through activation of separate antiviral signaling by inducing the expression of cellular p53. A previous report by Liu et al. (2007) indicates that RBV enhances the IFN-α antiviral activity through the upregulation of PKR activity. However, none of these studies have shown the synergistic antiviral effect of IFN-α and RBV combination treatment using HCV cell culture.

Our study indicated that IFN-α and RBV combination treatment synergistically inhibits HCV replication in replicon and infected cell culture models. IFN-α directly inhibits HCV IRES translation by preventing polyribosome loading through PKR-mediated eIF2α phosphorylation. RBV also inhibits HCV IRES translation by preventing the polyribosome loading of HCV IRES mRNA. RBV-mediated blockage of polyribosome loading involves two mechanisms that involve PKR and IMPDH. First, RBV-mediated PKR and eIF2α phosphorylation inhibits the recycling of eIF2α and therefore inhibits HCV IRES translation. Second, RBV-mediated inhibition of IMPDH activity decreases the cellular GTP pool, which inhibits the HCV IRES translation by preventing polyribosome loading. This is supported by the observation that pretreatment with guanosine prevented RBV-mediated HCV IRES-GFP translation. Based on these observations, we propose a model explaining how RBV-mediated depletion of the cellular GTP pool and the activation of PKR by IFN-α and RBV combination treatment could be playing an important role in the synergistic antiviral mechanism (Figure 4.8).

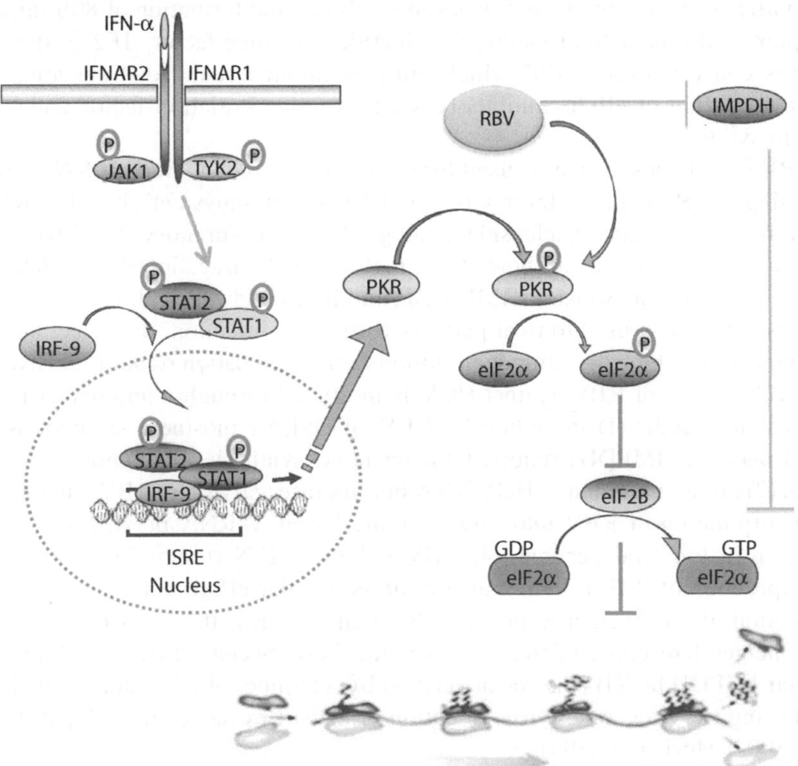


**FIGURE 4.8 (See color insert.)** Diagram summarizes the proposed IFN-α and RBV synergy antiviral mechanisms. IFN-α activates the JAK-STAT pathway leading to the activation of PKR. The activated PKR phosphorylates the eIF2α. Phosphorylation of eIF2α inhibits the recycling of initiation factors and inhibits HCV IRES translation. RBV activates the PKR and eIF2α phosphorylation that inhibits HCV IRES translation. RBV also inhibits HCV IRES by inhibiting IMPDH and GTP pool.

The mechanism of IFN-α antiviral activity through inhibition of HCV IRES–mediated translation is supported by a number of studies (Kato et al. 2002; Koev et al. 2002; Rivas-Estillas et al. 2002; Shimazaki et al. 2002; Wang et al. 2003; Dash et al. 2005). The newly discovered type III IFN, called IFN-λ, also inhibits IRES-mediated translation of HCV and hepatitis A (Kanda et al. 2012). There is an agreement among many studies indicating that type I, type II, and type III IFNs inhibit HCV replication by blocking HCV IRES–mediated translation that involves the PKR-induced phosphorylation of eIF2α. The eIF2α is a eukaryotic initiation factor required for protein translation (Dabo and Meurs 2012). This eIF2 protein exists as heterodimer consisting of eIFα, eIFβ, and eIFγ. The eIF2 protein complexes with GTP and initiator t-RNA to form the 43S preinitiation complex. The 43S preinitiation complex binds to an AUG codon on the target mRNA to initiate protein translation. The dissociation of the complex occurs when the eIF2's GTP is hydrolyzed by eIF5 (a GTPase-activating protein). The conversion causes eIF2-GDP to be released from the 48S complex, allowing translation to begin after recruitment of a 60S ribosome subunit and formation of 80S initiation complex. With the help of guanine nucleotide exchange factor eIF2-β, the eIF2-GDP is converted eIF2-GTP, which initiates another round of translation. The phosphorylation of eIF2α inhibits recycling of this initiation factor and blocks protein synthesis.

RBV is a guanosine analog used for the treatment of a number of RNA viruses including the RSV, Lassa fever virus, and HCV (Paeshuyse et al. 2011). RBV is a synthetic guanosine nucleoside analog (1-b-D-ribofuranosyl-1,2,4-triazole-3-carboxamide) that has been shown to be metabolized intracellularly into RBV and mono-, di-, and triphosphates (RMP, RDP, and RTP) (Sidwell et al. 1972). Although RBV is extensively used to treat patients with HCV infection, the direct antiviral mechanism by which the compound inhibits viral replication remains elusive. The antiviral activity of RBV against HCV is mediated through a number of mechanisms that include: (1) inhibition by RBV of cellular inosine 5′-monophosphate dehydrogenase (IMPDH) required for de novo synthesis of guanosine triphosphate; (2) direct inhibition of HCV RNA polymerase activity by RBV triphosphate; (3) incorporation of RBV into viral genome by HCV RNA polymerase causing mutation in the viral genome; (4) RBV enhances IFN-α signaling by inducing the expression of ISG; (5) RBV also inhibits cellular eIF4E activity required for translation of viral genome; and (6) RBV helps to clear the virus by stimulating the T helper 1 response of the host. Among these mechanisms, the inhibition of cellular IMPDH by RBV has been verified by a number of laboratories (including ours) using HCV and other virus infection models (Leyssen et al. 2005, 2006; Shu et al. 2008; Mori et al. 2011).

## 4.8 PERSISTENTLY INFECTED HCV CELL CULTURES SHOW IMPAIRED RESPONSE TO IFN-α AND RIBAVIRIN

Over the past several years, many studies have been performed to understand how the mechanisms of IFN-α antiviral action against HCV are impaired by cellular and viral proteins (Blindenbacher et al. 2003; Bode et al. 2003;

Duong et al. 2004, 2006, 2010; Lin et al. 2006; Randal et al. 2006; Christen et al. 2007; Kim et al. 2009; Sarasin-Filipowicz et al. 2009; Bellecave et al. 2010). These reports indicate that IFN-α signaling is controlled by a number of negative regulators such as suppressor of cytokine signaling (SOCS), ubiquitin-specific peptidase 18 (USP18), protein inhibitor of STAT1 (PIAS), and protein phosphatases. The SOCS family members, SOCS1 and SOCS3, prevent Stat phosphorylation by inhibiting the IFN-α receptor-associated Jak kinases (Kim et al. 2009). PIAS1 inhibits binding of STAT1 dimers to the response elements in the promoter region of the IFN-target genes. The upregulation of protein phosphatase 2A (PP2A) in HCV-infected cells inhibits IFN-α signaling. USP18 is a classical ISG that provides a strong negative feedback loop at the level of the receptor kinase complex. The mechanisms by which many patients develop resistance to RBV are not well understood. One group reported that reduced RBV uptake by HCV-infected cells contributed to an impaired antiviral response (Ibarra and Pfeiffer 2009; Ibarra et al. 2011). These investigators demonstrated that RBV uptake was reduced in the infected cells. No additional systematic study has been performed to understand how the IFN-α and RBV synergistic antiviral mechanism is impaired in HCV infection.

Studies conducted in our laboratory using stable subgenomic replicon cell lines and an infectious HCV cell culture system provide evidence that the mechanism of HCV resistance to exogenous IFN-α is related to expression of α chain of the IFNAR1 (Datta et al. 2011; Chandra et al. 2014). The replicon cells express NS3 to NS5B protein required for replication of HCV subgenomic RNA, but they lack structural proteins and do not produce infectious virus. We have isolated nine stable IFN-α-resistant Huh-7-based replicon cell lines (HCV1b) after long-term treatment with IFN-α. All nine IFN-α-resistant Huh-7 replicon cells showed reduced activation of pISRE-firefly luciferase promoter and impaired phosphorylation of Stat proteins (Datta et al. 2011). IFN-α-resistant phenotype in these cells is not attributed to the subgenomic HCV RNA replication because eliminating HCV from each replicon did not rescue the ISRE promoter activation, and a defect remained in the JAk-STAT signaling. The IFN-α resistance mechanism of the replicon cells was addressed by a complementary study that utilized the full-length plasmid clones of IFNAR1, IFNAR2, Jak1, Tyk2, STAT1, STAT2 and the ISRE-luciferase promoter plasmid. Expression of the full-length IFNAR1 clone alone restored the defective JAK-STAT signaling as well as STAT1, STAT2, and STAT3 phosphorylation, nuclear translocation, and antiviral response against HCV in all IFN-α-resistant cell lines (R-15, R-17, and R-24). Moreover, RT-PCR, Southern blotting, and DNA sequence analysis revealed that cells from R-15 and R-24 series of IFN-α-resistant cell lines have 58 amino acid deletions in the extracellular subdomain 1 (SD1) of IFNAR1. In addition, cells from the R-17 series have 50 amino acids deletion in the subdomain 4 (SD4) of IFNAR1 protein leading to impaired activation of Tyk2 kinase. These results are consistent with a number of studies from other laboratories indicating that impaired expression of IFNAR1 leads to the development of IFN-α resistance (Naka et al. 2005). Because the subgenomic replicon-based cell line does not produce infectious virus, we have verified our results using a persistently infected cell culture system.

The mechanism of HCV resistance to IFN-α and RBV combination treatment was examined in a persistently infected HCV cell culture model developed in our laboratory (Chandra et al. 2011, 2014). In the persistently infected Huh-7.5 cells, replication of HCV was measured by *Renilla* luciferase activity, core protein expression by Western blotting, immunostaining, and flow analysis. We found that approximately 70% of the cells were infected with HCV. The persistently infected cells secreted infectious virus particles as confirmed by a multicycle infectivity assay using three different liver-derived cell lines and one non-liver-derived cell line. IFN-α, RBV, and IFN-α plus RBV inhibited HCV replication ($P < 0.002$) but did not completely clear the virus. Interestingly, IFN-λ had marked antiviral activity and cleared the virus in this persistently infected Huh-7.5 cell culture model (Figure 4.9a). The impaired clearance of HCV by IFN-α was not related to a lack of biological activity because we demonstrated that the same IFN-α preparation used in a S3-GFP stable replicon cell line showed significant concentration-dependent antiviral activity. We have shown that persistently infected HCV cell culture impairs type I IFN-α receptor-1 expression but not the type III IFN receptors (Figure 4.9b and c). These results were consistent with the findings, using subgenomic replicon cell lines, that the mechanism

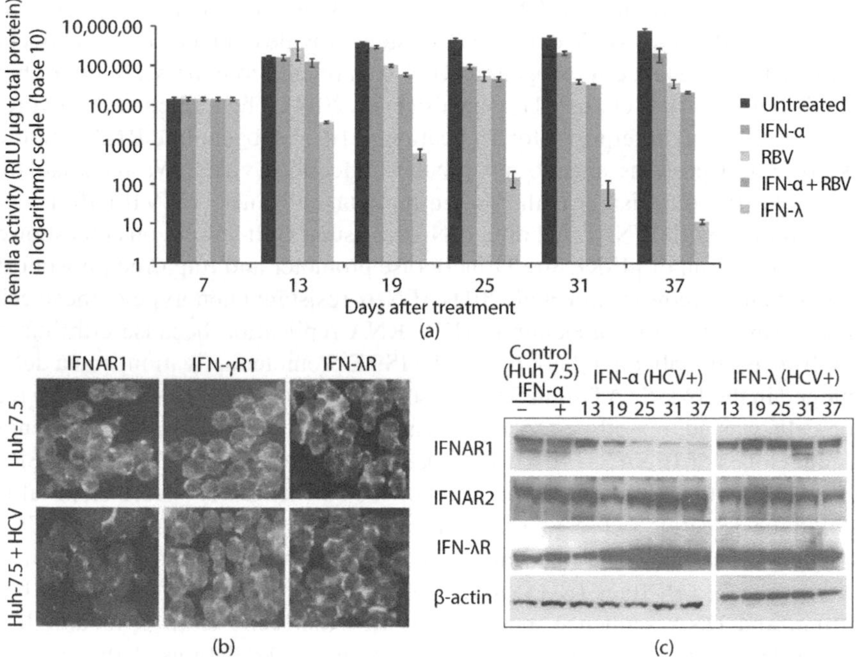


**FIGURE 4.9 (See color insert.)** Persistent HCV replication with high viral titer is partly resistant to IFN-α/RBV but not to IFN-λ treatment. (a) The antiviral action of IFN-α, RBV, and IFN-λ in persistently infected HCV culture measured by *Renilla* luciferase activity. (b) Immunofluorescence image showing the cell surface expression of IFN-α, IFNAR2, and IFN-λ receptor. (c) Western blot analysis showing selective downregulation of IFNAR1 in the persistently infected Huh-7.5 cells, whereas expression of IFN-γ or IFN-λ receptors did not change.

of HCV resistance to IFN-α is related to impaired expression of IFNAR1. Our observation is also consistent with another study showing that cells stably replicating HCV genomic replicon promote phosphorylation-dependent ubiquitination and degradation of IFNAR1 (Liu et al. 2009). Impaired expression of IFNAR1 is also seen in human liver biopsies and liver cirrhosis (Gunduz et al. 2010; Poat et al. 2010c).

The mechanism of HCV resistance to RBV treatment was addressed in HCV cell culture. Persistently infected cells showed reduced RBV uptake with respect to cytidine (Figure 4.10a). RBV is transported into cells via the nucleoside transporters ENT1 and CNT1 (Kong et al. 2004; Likura et al. 2012). The expression of the ENT1 and CNT1 nucleoside transporters in the infected and uninfected cells were examined by Western blotting. The expression of ENT1 and CNT1 were notably reduced in the IFN-α plus RBV resistant cells as compared to that in HCV-infected cells (Figure 4.10b). The expression level of ENT1 and CNT1 in uninfected Huh-7.5 cells did not change significantly when treated with IFN-α alone or with IFN-α plus RBV.

In order to explain why an HCV persistently infected cell culture is resistant to IFN-α and RBV combination treatment, we have examined activation of ER stress and autophagy in HCV cell culture (Chandra et al. 2014). ATF6 luciferase activity shows that persistently infected HCV culture also induces a chronic ER stress response. It is known that cellular autophagy response is activated secondary to ER stress as a cell survival defense mechanism. Induction of an autophagy response in the persistently infected HCV culture was verified by a number of autophagy assays including the processing of LC3-I into LC3-II by Western blotting and the induction of the key autophagy gene, Beclin1, in the persistently infected culture. The expression of p62 was reduced significantly in the infected cells in a time-dependent manner, which indicated that HCV replication induced an autophagy response. We also investigated whether persistent HCV replication could induce autophagosomes, which then progress to autophagolysosomes through fusion with acidic lysosomes detectable by using acridine

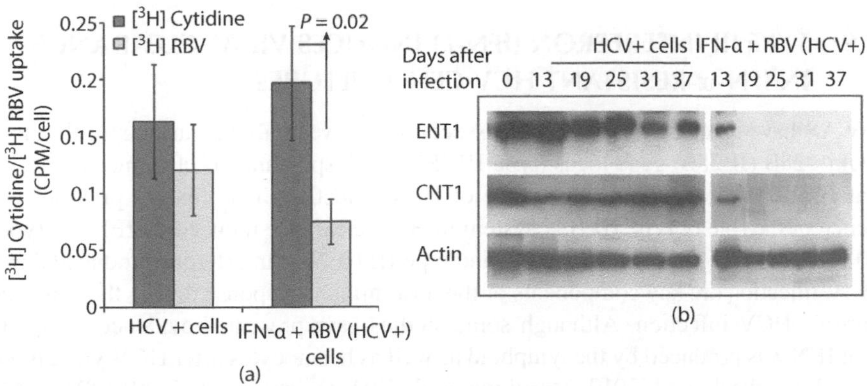


**FIGURE 4.10** Impaired RBV uptake and reduced expression of nucleoside transporters in HCV-infected cell culture. (a) Reduced uptake of RBV in HCV cell culture. (b) Western blot analysis of nucleoside transporters (ENT1 and CNT1) using equal amount of protein lysates from HCV-infected Huh-7.5 cells, and IFN-α plus RBV resistant persistently HCV-infected cells at different time points. ENT1 and CNT1 levels are decreased in persistently infected cells.

orange staining. Considerably higher numbers of membrane localized autophagic vacuoles were observed in the persistently infected cells as compared to uninfected cells. We further verified our hypothesis that the lack of cell surface expression of IFNAR1 is the cause of IFN-α resistance by analysis of HCV-positive and HCV-negative cell populations after flow sorting. These results showed that the IFN-α-resistant cells had a notably lower level of IFNAR1 than the IFN-α-sensitive cells.

We then investigated the role of ER stress and the induced autophagy response in creating defective JAK-STAT signaling and IFN-α resistance. HCV-infected Huh-7.5 cells were pretreated with either Phenylbutyric acid (PBA) (ER stress) or Hydroxychloroquine (HCQ) (autophagy) inhibitors for 24 hours, and then the antiviral activity of IFN-α was determined 72 hours posttreatment by measuring *Renilla* luciferase activity. Both PBA and HCQ significantly inhibited HCV replication in a dose-dependent manner. We examined whether treatment of the HCV-infected culture with PBA or HCQ in combination with IFN-α, or with IFN-α plus RBV, could clear HCV replication effectively. Results of these studies indicated that IFN-α significantly inhibits HCV replication in the presence of either PBA or HCQ. The concentrations of PBA and HCQ used in this study were not toxic to cells as determined by using an 3-(4,5-dimethylthiazol-2-yl)-2,5 diphenyltetrazolium bromide (MTT) assay. In a different approach, three cellular ER stress sensors (PERK, IRE1α, and ATF6) and one of the autophagy genes (ATG7) were silenced using siRNA. The siRNAs were transfected into the HCV cell culture and then, after three hours, cells were treated with IFN-α (250 IU/mL) and RBV (40 μg/mL). In our study (Chandra et al. 2014), we showed that the antiviral action of IFN-α and RBV was significantly improved after silencing these genes. These experiments verified the causal relationship of ER stress and the autophagy response in HCV resistance mechanisms to IFN-α and RBV. These study results indicate that inhibiting ER stress and the autophagy response could overcome IFN-α plus RBV resistance mechanisms associated with HCV infection.

## 4.9 TYPE III INTERFERON (IFN-λ) INDUCES VIRAL CLEARANCE IN IFN-α-RESISTANT HCV CELL CULTURES

GWAS have revealed a very strong association between allelic variants of the interleukin 28B (IL28B) gene locus (type III IFN) with spontaneous clearance as well as antiviral treatment-induced clearance of HCV infection. This finding provides evidence of the role of type III IFN system in HCV clearance induced by IFN-α. Type I (IFN-α and IFN-β), type II (IFN-λ), and type III (IFN-λ) interferons produced due to HCV infection are key components of the host immune response during the acute and chronic HCV infection. Although some studies have presented evidence indicating that IFN-λ is produced by the lymphoid as well as hepatocytes after HCV virus infection (Lauterbach et al. 2010; Marukian et al. 2011; Dolganiuc et al. 2012; Park et al. 2012; Zhang et al. 2013), it is not clear how the different SNPs of IL-28B control the HCV infection and antiviral clearance. Our interest in exploring the biology of IFN-λ started based on our preliminary results indicating that IFN-λ (IL-28A and IL-29) not only inhibits HCV replication but also clears HCV replication. We reported that type III IFN overcame IFN-α resistance mechanisms. In one study, we showed that

IFN-λ (IL-29) has a robust antiviral effect against IFN-α-resistant cell line R4-GFP developed in our laboratory. Moreover, IFN-λ 1 (IL-29) could induce HCV clearance in R4-GFP replicon cells that are resistant to IFN-α due to expression of truncated IFNAR1 (Hazari et al. 2011). In another study, we examined the clearance of HCV replication by type I, type II, and type III IFNs using a free fatty acid HCV coculture model. FFAs induce ER stress and block antiviral activity of IFN-α against HCV in cell culture. HCV-infected Huh 7.5 cells were cultured with or without a mixture of saturated (palmitate) and unsaturated (oleate) long-chain FFAs. Clearance of HCV replication in FFA cell culture after long-term therapy with IFN-α, IFN-λ, and IFN-γ was compared by the measurement of *Renilla* luciferase activity and HCV core by immunostaining (Figure 4.11). FFA-treatment-induced dose-dependent hepatocellular

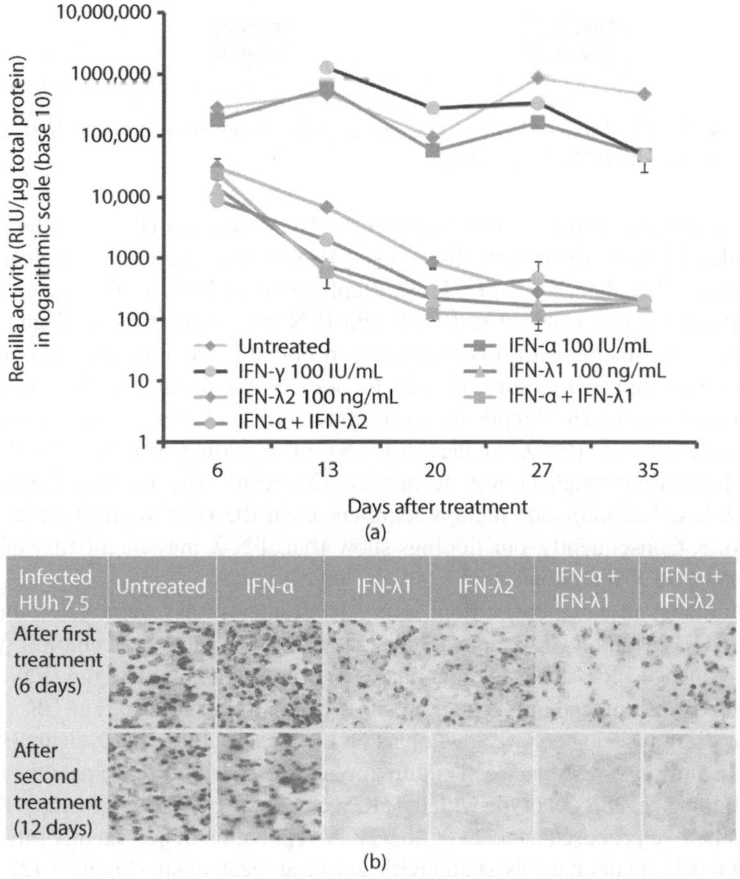


**FIGURE 4.11 (See color insert.)** IFN-λ1- or IFN-λ2-induced HCV clearance in IFN-α-resistant free fatty acid HCV cell cultures. (a) The luciferase activity indicating HCV replication in the presence and absence of different IFNs. IFN-λ1 (IL-29) and IFN-λ2 clears HCV replication. IFN-α and IFN-γ are not effective. (b) The core protein expression by immunostaining. IFN-λ-treated culture shows negative expression of HCV core protein, which was not seen in IFN-α and IFN-γ treatment.

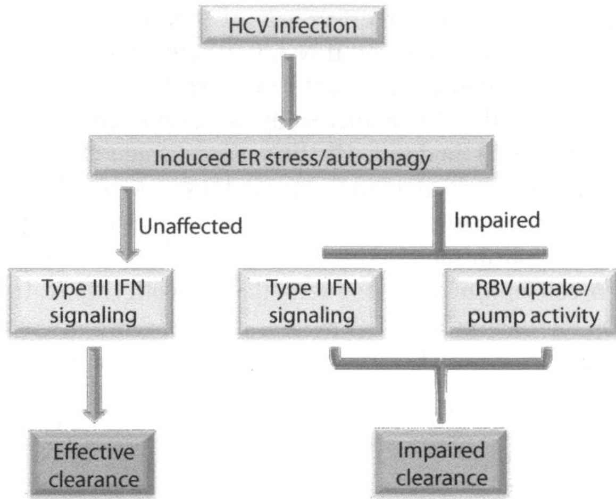


**FIGURE 4.12** HCV-induced ER stress and autophagy response block IFN-α and RBV signaling but not the IFN-λ1 signaling.

steatosis and lipid droplet accumulation in the HCV-infected Huh 7.5 cells (Gunduz et al. 2012). FFA treatment partially blocked IFN-α response and viral clearance by reducing the phosphorylation of STAT1. Replication of HCV in the presence of the FFA cell culture was resistant to IFN-α and IFN-γ treatments. IFN-λ1 and IFN-λ2 alone or in combination with IFN-α strongly inhibited HCV replication to completion in the FFA-treated cell culture. *Renilla* luciferase activity and HCV core immune staining confirm that HCV replication was undetectable after a second treatment with IFN-λ1 and IFN-λ2. IFN-λ2 induces JAK-STAT signaling in IFN-α-resistant FFA HCV cell culture through a novel mechanism. Our results suggested that IFN-λ alone or IFN-λ-based combination therapy regimens clear the HCV replication in the FFA cell culture. Consequently, our findings show that IFN-λ may be a better choice for treatment of HCV infection.

In summary, research carried out in our laboratory over several years found that IFN-α and RBV inhibit HCV replication in replicon and infected cell culture. We found that HCV-induced ER stress and autophagy response downregulate the IFN-α receptor 1 and RBV pump and uptake and impair the antiviral clearance. The ER stress and autophagy response do not have any impact on IFN-λ signaling. IFN-λ clears HCV replication in an IFN-α resistance cell culture model, which supports the results of clinical studies linking IL-28B genotype with the HCV clearance by antiviral therapy. Based on these results, we proposed a model of how HCV replication in persistently infected cell culture model impairs the IFN-α and RBV antiviral mechanism (Figure 4.12).

## 4.10 CONCLUSIONS

HCV infection is one of the risk factors for the development of hepatocellular carcinoma. Significant progress has been made in the antiviral drug development for HCV, which opens new hope for improved treatment. The risk of

HCV-related HCC can be minimized by a new standard of care for chronic HCV infection including combination of antiviral drugs along with IFN-α and RBV. However, the new therapies are expensive and are associated with an unsatisfactory safety profile. All of these antiviral molecules are now being used along with pegylated IFN-α and RBV. Our study showed that chronic HCV infection impairs the IFN-α signaling. Therefore, the use of IFN-λ is supposed to overcome these problems. There is a hope that antiviral therapy may be able to eradicate chronic HCV infection and reduce the incidence of human hepatocellular carcinomas.

## ACKNOWLEDGMENTS

This review was possible due to continuous financial support from National Institutes of Health (NIH) grants CA12748, CA089121, and AI103106 and funds received from the Louisiana Cancer Research Consortium (LCRC), Tulane University Health Sciences Center. The authors acknowledge important work of other investigators, which could not be cited due to space limitations.

## REFERENCES

Akuta N, Suzuki F, Hirakawa M, et al. (2010). Amino acid substitution in hepatitis C virus core region and genetic variation near the interleukin 28B gene predict viral response to telaprevir with peginterferon and ribavirin. *Hepatology* 52: 421–429.

Alsio A, Rembeck K, Askarieh G, et al. (2012). Impact of obesity on the bioavailability of peginterferon-alpha 2a and ribavirin and treatment outcome for chronic hepatitis C genotype 2 or 3. *PLos One* 7: 37521–37528.

Altekruse SF, McGlynn KA, Reichman ME. (2009). Hepatocellular carcinoma incidence, mortality, and survival trends in the United States from 1975 to 2005. *J Clin Oncol* 27: 1485–1491.

Alter MJ. (2007). Epidemiology of hepatitis C virus infection. *World J Gastroenterol* 13: 2436–2441.

Arase Y, Ikeda K, Suzuki F, et al. (2007). Interferon-induced prolonged biochemical response reduces hepatocarcinogenesis in hepatitis C virus infection. *J Med Virol* 79: 1485–1490.

Asselah T, Estrabaud E, Bieche I, et al. (2010). Hepatitis C: Viral and host factors associated with non-response to pegylated interferon plus ribavirin. *Liver Int* 30: 1259–1269.

Aurora R, Donlin MJ, Cannon NA, et al. (2009). Genome-wide hepatitis C virus amino acid covariance networks can predict response to antiviral therapy in humans. *J Clin Invest* 119: 225–236.

Bacus LI, Boothroyd DB, Phillips BR, et al. (2011). A sustained virologic response reduces risk of all-causes mortality in patients with hepatitis C. *Clin Gastroenterol Hepatol* 9: 509–516.

Banerjee S, Saito K, Ait-Goughoulte M, et al. (2008). Hepatitis C virus core protein upregulates serine phosphorylation of insulin receptor substrate 1 and impairs the downstream akt/protein kinase B signaling pathway for insulin resistance. *J Virol* 82: 2606–2612.

Beasley RP, Hwang LY, Lin CC, et al. (1981). Hepatocellular carcinoma and HBV. A prospective study of 22,707 men in Taiwan. *Lancet* 2: 1129–1133.

Bellecave P, Sarasin-Filipowicz M, Donze O, et al. (2010). Cleavage of mitochondria antiviral signaling protein in the liver of patients with chronic hepatitis C correlate with a reduced activation of the endogenous interferon system. *Hepatology* 51: 1127–1136.

Berg T, Sarrazin C, Herrmann E, et al. (2003). Prediction of treatment outcome in patients with chronic hepatitis C: Significance of baseline parameters and viral dynamics during therapy. *Hepatology* 37: 600–609.

Blanchard E, Belouzard S, Goueslain L, et al. (2006). Hepatitis C virus entry depends on clathrin-mediated endocytosis. *J Virol* 80: 6964–6972.

Blindenbacher A, Duong FHT, Hunziker L, et al. (2003). Expression of hepatitis C virus proteins inhibits interferon alpha signaling in the liver of transgenic mice. *Gastroenterology* 124: 1465–1475.

Bode JG, Ludwig S, Ehrhardt C, et al. (2003). Interferon alpha antagonistic activity of HCV core protein involves induction of suppressor cytokine signaling 3. *FASEB J* 17: 488–490.

Bosch FX, Ribes J, Diaz M, et al. (2004). Primary liver cancer: Worldwide incidence and trends. *Gastroenterology* 127(Suppl 1): S5–S16.

Bota S, Sporea I, Sirli R, et al. (2013). Severe adverse events during antiviral therapy in hepatitis C virus cirrhotic patients: A systematic review. *World J Hepatol* 5: 120–126.

Bressler BL, Guindi M, Tomlinson G, et al. (2003). High body mass index is an independent risk factor for nonresponse to antiviral treatment in chronic hepatitis C. *Hepatology* 38: 639–644.

Bukh J, Purcell RH, Miller RH. (1992). Sequence analysis of the 5′-nontranslated region of the hepatitis C virus. *Proc Natl Acad Sci U S A* 89: 4942–4946.

Burton MJ, Passarella MJ, McGuire BM. (2012). Telaprevir and boceprevir in African Americans with genotype 1 chronic hepatitis C: Implications for patients and providers. *South Med J* 105: 431–436.

Carrat F, Bani-Sadr F, Pol S, et al. (2004). Pegylated interferon alfa-2b vs. standard interferon alfa-2b, plus ribavirin, for chronic hepatitis C in HIV-infected patients: A randomized controlled trial. *JAMA* 292: 2839–2848.

Chandra PK, Bao L, Song K, et al. (2014). Persistently infected HCV cell culture impairs type I but not the type III IFN signaling. *Am J Pathol* 184(1): 214–29.

Chandra PK, Hazari S, Gunduz F, et al. (2011a). Hepatitis C virus infection induced ER-stress and autophagy response down regulates IFNAR1 leading to the defective JAK-STAT signaling and impaired IFN-alpha response. *Hepatology* 54: 795.

Chandra PK, Kundu AK, Hazari S, et al. (2012). Inhibition of hepatitis C virus replication by intracellular delivery of multiple siRNAs by nanosomes. *Mol Ther* 20: 1724–1736.

Chandra S, Hazari S, Chandra PK, et al. (2011b). Ribavirin and IFN-alpha each inhibits HCV IRES mediated translation and synergistically inhibits HCV replication in cells with a functional JAK-STAT pathway. *Hepatology* 54: 798–799.

Cheung O, Sanyal AJ. (2008). Hepatitis C infection and nonalcoholic fatty liver disease. *Clin Liver Dis* 12: 573–585.

Choo QL, Kuo G, Weiner AJ, et al. (1989). Isolation of a cDNA clone derived from a blood-borne non-A, non-B viral hepatitis genome. *Science* 244: 359–362.

Christen V, Treves S, Duong FH, et al. (2007). Activation of endoplasmic reticulum stress response by hepatitis C viruses up-regulates protein phosphatase 2A. *Hepatology* 46: 558–565.

Chung RT, Anderson J, Volberding P, et al. (2004). Peginterferon Alfa-2a plus ribavirin versus interfern-alpha-2a plus ribavirin for chronic hepatitis C in HIV-coinfected persons. *N Engl J Med* 351: 451–459.

Cocquel L, Voisset C, Dubuison J. (2006). Hepatitis C virus entry: Potential receptors and their biological functions. *J Gen Virol* 87: 1075–1084.

Conjeevaram HS, Fried MW, Jeffers LJ, et al. (2006). Peginterferon and ribavirin treatment of African Americans and Caucasian American Patients with hepatitis C genotype 1. *Gastroenterology* 131: 470–477.

Dabbouseh NM, Jensen DM. (2013). Future therapies for chronic hepatitis C. *Nat Rev Gastroenterol Hepatol* 10: 268–276.

Dabo S, Meurs EF. (2012). DsRNA-dependent protein kinase PKR and its role in stress, signaling and HCV infection. *Viruses* 4: 2598–2635.

Darnell JE, Jr, Kerr IM, Stark GR. (1994). JAK-STAT pathways and transcriptional activation in response to IFNs and other extracellular signaling proteins. *Science* 264: 1415–1421.

Dash S, Prabhu R, Hazari S, et al. (2005). Interferon alpha, beta, gamma each inhibits hepatitis C virus replication at the levels of internal ribosome entry site mediated translation. *Liver Int* 25: 580–594.

Datta S, Hazari S, Chandra PK, et al. (2011). Mechanisms of HCV's resistance to IFN-alpha in cell culture involves expression of functional IFN-alpha receptor 1. *Virol J* 8: 351–369.

Davila JA, Morgan RO, Shaib Y, et al. (2004). Hepatitis C infection and the increasing incidence of hepatocellular carcinoma: A population-based study. *Gastroenterology* 127: 1372–1380.

Davis GL, Esteban-Mur R, Rustgi V, et al. (1998). Interferon alfa-2b alone or in combination with ribavirin for the treatment of relapse of chronic hepatitis C. *N Engl J Med* 339: 1493–1499.

De Weerd NA, Samarajiwa SA, Hertzog PJ. (2007). Type I interferon receptors: Biochemistry and biological functions. *J Biol Chem* 282: 20053–20057.

Diegelmann J, Beigel F, Zitzmann K, et al. (2010). Comparative analysis of the lambda-interferons IL-28A and IL-29 regarding their transcriptome and their antiviral properties against hepatitis C virus. *PLos One* 5: 15200–15213.

Dolganiuc A, Kodys K, Marshall C, et al. (2012). Type III interferons, IL-28 and Il-29 are increased in chronic HCV infection and induce myeloid dendritic cell-mediated FoxP3+ regulatory T cells. *PLos One* 7: 44915–44925.

Dominitz JA, Boyko EJ, Koepsell TD, et al. (2005). Elevated prevalence of hepatitis C virus infection in users of the United States Veterans Medical Centers. *Hepatology* 41: 88–96.

Donnelly RP, Dickensheets H, O'Brien TR. (2011). Interferon-lambda and therapy for chronic hepatitis C virus infection. *Trends Immunol* 32: 443–450.

Duong FH, Christen V, Filipowicz M, et al. (2006). S-Adenosylmethionine and betaine correct hepatitis C virus induced inhibition of interferon signaling in vitro. *Hepatology* 43: 796–806.

Duong FH, Christen V, Lin S, et al. (2010). Hepatitis C virus induced up-regulation of protein phosphatase 2A inhibits histone modification and DNA damage repair. *Hepatology* 51: 741–751.

Duong FH, Filiowicz M, Tripodi M, et al. (2004). Hepatitis C virus inhibits interferon signaling through up-regulation of protein phosphatase 2A. *Gastroenterology* 126: 263–277.

El-Serag HB. (2012). Epidemiology of viral hepatitis and hepatocellular carcinoma. *Gastroenterology* 142: 1264–1273.

El-Serag HB, Rudolph KL. (2007). Hepatocellular carcinoma epidemiology and molecular carcinogenesis. *Gastroenterology* 132: 2557–2576.

Enomoto N, Sakuma I, Asahina Y, et al. (1995). Comparison of full-length sequences of interferon-sensitive and resistant hepatitis C virus 1b. Sensitivity to interferon is conferred by amino acid substitutions in the NS5A region. *J Clin Invest* 96: 224–230.

Feld JJ, Hoofnagle JH. (2005). Mechanism of action of interferon and ribavirin in treatment of hepatitis C. *Nature* 436: 967–972.

Franceschini L, Realdon S, Marcolongo M, et al. (2011). Reciprocal interference between insulin and interferon alpha signaling in hepatic cells: A vicious circle of clinical significance. *Hepatology* 54: 484–494.

Friebe P, Bartenschlager R. (2002). Genetic analysis of the sequences in the 3′ UTR of hepatitis C virus that are important for RNA replication. *J Virol* 76: 5326–5338.

Fried MW, Shiffman ML, Reddy KR, et al. (2002). Peginterferon alfa-2a plus ribavirin for chronic hepatitis C virus infection. *N Engl J Med* 347: 975–982.

Furione MS, Simoncini L, Gatti M, et al. (1999). HCV genotyping by three methods: Analysis of discordant results based on sequencing. *J Clin Virol* 13: 121–130.

Gao M, Nettles R, Belema M, et al. (2010). Chemical genetics strategy identifies an HCV NS5A inhibitor with potent clinical effect. *Nature* 465: 96–100.

Ge D, Fellay J, Thompson AJ, et al. (2009). Genetic variation in IL-28B predicts hepatitis C treatment-induced viral clearance. *Nature* 461: 399–401.

Gunduz F, Aboulnasr FM, Chandra PK, et al. (2012). Free fatty acids induce ER stress and block antiviral activity of interferon alpha against hepatitis C virus in cell culture. *Virol J* 9: 143–155.

Gunduz F, Mallikarjun C, Hazari S, et al. (2010). Interferon alpha induced intrahepatic pStat1 levels inversely correlate with serum HCV RNA levels in chronic hepatitis C virus infection. *Hepatology* 52(Suppl 4): 742.

Hadziyannis SJ, Sette H, Jr, Morgan TR, et al. (2004). Peginterferon-alpha 2a and ribavirin combination therapy in chronic hepatitis C: A randomized study of treatment duration and ribavirin dose. *Ann Intern Med* 140: 346–355.

Hara K, Rivera MM, Koh C, et al. (2013). Important factors in reliable determination of hepatitis C virus genotype by use of the 5′-untranslated region. *J Clin Microbiol* 51: 1485–1489.

Harison SA, Hamzeh FM, Han J, et al. (2012). Chronic HCV genotype 1 patients with insulin resistance with pioglitazone and peginterferon alpha-2a plus ribavirin. *Hepatology* 56: 464–473.

Harris JM, Chess RB. (2003). Effect of pegylation on pharmaceuticals. *Nat Rev Drug Discov* 2: 214–221.

Hazari S, Chandra PK, Poat B, et al. (2010). Impaired antiviral activity of interferon alpha against hepatitis C virus 2a in Huh-7 cells with a defective JAK-STAT pathway. *Virol J* 7: 36–52.

Hazari S, Nayak S, Chandra PK, et al. (2011). Robust antiviral activity of lambda interferon (IL-29) against HCV in interferon alpha resistant Huh-7 cells with a defective JAK-STAT signaling. *Hepatology* 54: 793.

Hazari S, Taylor L, Haque S, et al. (2007). Reduced expression of Jak-1 and Tyk-2 proteins leads to interferon resistance in hepatitis C virus replicon. *Virol J* 18: 89–102.

Hemmi S, Bohni R, Stark G, et al. (1994). A novel member of the interferon receptor family complements functionality of the murine interferon gamma receptor in human cells. *Cell* 76: 803–810.

Hino K, Okita K. (2004). Interferon therapy as chemoprevention of hepatocarcinogenesis in patients with chronic hepatitis C. *J Antimicrob Chemother* 53: 19–22.

Honda M, Beard MR, Ping LH, et al. (1999). A phylogenetically conserved stem-loop structure at the 5′ border of the internal ribosome entry site of hepatitis C virus is required for cap-independent viral translation. *J Virol* 73: 1165–1174.

Honda M, Sakai A, Yamashita T, et al. (2010). Hepatic ISG expression is associated with genetic variation in interleukin 28B and the outcome of IFN therapy for chronic hepatitis C. *Gastroenterology* 139: 499–509.

Hoofnagle JH. (2004). Hepatocellular carcinoma: Summary and recommendation. *Gastroenterology* 127: S319–S323.

Hung CH, Lu SN, Wang JH, et al. (2011). Sustained HCV clearance by interferon-based therapy reduces hepatocellular carcinoma in hepatitis B and C dually infected patients. *Antivir Ther* 16: 959–968.

Hwang SJ, Lee SD. (2011). Hepatic steatosis and hepatitis C: Still unhappy bedfellows. *J Gastroenterol* 26(Suppl 1): 96–101.

Ibarra KD, Pfeiffer JK. (2009). Reduced ribavirin antiviral efficacy via nucleoside transporter-mediated drug resistance. *J Virol* 83: 4538–4547.

Ibarra KD, Jain MK, Pfeiffer JK. (2011). Host-based ribavirin resistance influences hepatitis C virus replication and treatment response. *J Virol* 85: 7273–7283.

Jacobson IM, McHutchison JG, Dusheiko G, et al. (2011). Telaprevir for previously untreated chronic hepatitis C virus infection. *N Engl J Med* 364: 2405–2416.

Janssen HL, Lawitz EJ, Zeuzem S, et al. (2013). Treatment of HCV infection by targeting microRNA. *N Engl J Med* 368: 1685–1694.

Jensen DM, Morgan TR, Marcellin P, et al. (2006). Early identification of HCV genotype 1 patients responding to 24 weeks peginterferon alpha-2a (40 kd)/ribavirin therapy. *Hepatology* 43: 954–960.

Jiang D, Guo H, Xu C, et al. (2008). Identification of three interferon-inducible cellular enzymes that inhibit the replication of hepatitis C virus. *J Virol* 82: 1665–1678.

Joplin CL, Yi M, Lancaster AM, et al. (2005). Modulation of hepatitis C virus RNA abundance by a liver specific microRNA. *Science* 309: 1577–1581.

Kanda T, Wu S, Kiyohara T, et al. (2012). Interleukin-29 suppresses hepatitis A and hepatitis C viral internal ribosomal entrysite-mediated translation. *Viral Immunol* 25: 379–386.

Kato J, Kato N, Moriyama M, et al. (2002). Interferons specifically suppress the translation from the internal ribosome entry site of hepatitis C virus through a double-stranded RNA- activated protein kinase-independent pathway. *J Infect Dis* 186: 155–163.

Kawaguchi T, Yoshida T, Harada M, et al. (2004). Hepatitis C virus down-regulates insulin receptor substrates 1 and 2 through up-regulation of suppressor of cytokine signaling 3. *Am J Pathol* 165: 1499–1508.

Khattab M, Emad M, Abdelaleem A, et al. (2010). Pioglitazone improves virological response to peginterferon α-2b/ribavirin combination therapy in hepatitis C genotype 4 patients with insulin resistance. *Liver Int* 30: 447–454.

Kim KA, Lin W, Tai AW, et al. (2009). Hepatic SOCS3 expression is strongly associated with non-response to therapy and race in HCV and HCV/HIV infection. *J Hepatol* 50: 705–711.

Kimer N, Dahl EK, Gluud LL, et al. (2012). Antiviral therapy for prevention of hepatocellular carcinoma in chronic hepatitis C: Systemic review and meta-analysis of randomized controlled trials. *BMJ Open* 2: 1313–1320.

Koev G, Duncan RF, Lai MM. (2002). Hepatitis C virus IRES-dependent translation is insensitive to an eIF2alpha-independent mechanism of inhibition by interferon in hepatocyte cell lines. *Virology* 297: 195–202.

Kong W, Engel K, Wang J. (2004). Mammalian nucleoside transporters. *Curr Drug Metab* 5: 63–84.

Kotenko SV, Gallagher G, Baurin VV, et al. (2003). IFN-lambdas mediate antiviral protection through a distinct class II cytokine receptor complex. *Nat Immunol* 4: 69–77.

Kuiken C, Mizokani M, Deleage G, et al. (2006). Hepatitis C databases, principles and utility to researchers. *Hepatology* 43: 1157–1165.

Lanford RE, Hildebrandt-Eriksen ES, Petri A, et al. (2010). Therapeutic silencing of microRNA-122 in primates with chronic HCV infection. *Science* 327: 198–201.

Langhans B, Kupfer B, Braunschweiger I, et al. (2011). Interferon lambda serum levels in hepatitis C. *J Hepatol* 54: 859–865.

Lauterbach H, Bathke B, Gilles S, et al. (2010). Mouse CD8alpha+ DC and human BDCA3+ DC are major producers of IFN-lambda in response to poly IC. *J Exp Med* 207: 2703–2717.

Lavanchy D. (2009). The global burden of hepatitis C. *Liver Int* 29: 74–81.

Lemon SM. (2010). Induction and evasion of innate antiviral responses by hepatitis C virus. *J Biol Chem* 285: 22741–22747.

Leyssen P, Balzarini J, De Clercq E, et al. (2005). The predominant mechanism by which ribavirin exerts its antiviral activity in vitro against flaviviruses and paramyxoviruses is mediated by inhibition of IMP dehydrogenase. *J Virol* 79: 1943–1947.

Leyssen P, De Clercq E, Neyts J. (2006). The anti-yellow fever virus activity of ribavirin is independent of error-prone replication. *Mol Pharmacol* 69: 1461–1467.

Likura M, Furihata T, Mizuguchi M, et al. (2012). ENT1, a ribavirin transporter, plays a pivotal role in antiviral efficacy of ribavirin in hepatitis C virus replication cell system. *Antimicrob Agents Chemother* 56: 1407–1413.

Lin W, Kim SS, Yeung E, et al. (2006). Hepatitis C virus core protein blocks interferon signaling by interaction with the Stat1 SH2 domain. *J Virol* 80: 9226–9235.

Liu J, Huang Fu WC, Suresh Kumar KG, et al. (2009). Virus-induced unfolded protein response attenuates antiviral defenses via phosphorylation-dependent degradation of the type I interferon receptor. *Cell Host Microbe* 5: 72–83.

Liu WL, Su WC, Cheng CW, et al. (2007). RBV upregulates the activity of double stranded RNA-activated protein kinase and enhances the action of IFNα against HCV virus. *J Infect Dis* 196: 425–434.

Liu WL, Yang HC, Su WC, et al. (2012). RBV enhances the action of IFNα against hepatitis C virus by promoting the p53 activity through ERK1/2 pathway. *PLos One* 9: 43824–43837.

Manns MP, McHutchison JG, Gordon SC, et al. (2001). Peginterferon alfa-2b plus ribavirin compared with interferon alfa-2b plus ribavirin for initial treatment of chronic hepatitis C: A randomized trial. *Lancet* 358: 958–965.

Marcellin P, Pouteau M, Martinot-Peignoux M, et al. (1995). Lack of benefit of escalating dosage of interferon alfa in patients with chronic hepatitis C. *Gastroenterology* 109: 156–165.

Martinot-Peignoux M, Boyer N, Pouteau M, et al. (1998). Predictors of sustained response to alpha interferon therapy in chronic hepatitis C. *J Hepatol* 29: 214–223.

Martinot-Peignoux M, Marcellin P, Pouteau M, et al. (1995). Pretreatment serum hepatitis C virus RNA levels and hepatitis C virus genotype are the main and independent prognostic factors of sustained response to interferon alfa therapy in chronic hepatitis C. *Hepatology* 22: 1050–1056.

Marukian S, Andrus L, Sheahan TP. (2011). Hepatitis C virus induces interferon-lambda and interferon stimulated genes in primary liver cultures. *Hepatology* 54: 1913–1923.

Mathews SJ, Lancaster JW. (2012). New drug review: Telaprevir: A hepatitis C NS3/4 protease inhibitor. *Clin Ther* 34: 1857–1882.

McCarthy JJ, Li JH, Thompson A, et al. (2010). Replicated association between an IL-28B gene variant and sustained response to pegylated interferon and ribavirin. *Gastroenterology* 138: 2307–2314.

McHutchison JG, Gordon SC, Schiff ER, et al. (1998). Interferon alfa-2b alone or in combination with ribavirin as initial treatment for chronic hepatitis C. Hepatitis Interventional Therapy Group. *N Engl J Med* 339: 1485–1492.

Moradpour D, Penin F, Rice CM. (2007). Replication of hepatitis C virus. *Nat Rev Microbiol* 5: 453–563.

Morgan TR, Ghany MG, Kim HY, et al. (2010). Outcome of sustained virological responders with histologically advanced chronic hepatitis C. *Hepatology* 52: 833–844.

Mori K, Ikeda M, Ariumi Y, et al. (2011). Mechanism of action of ribavirin in a novel hepatitis C virus replication cell system. *Virus Res* 157: 61–70.

Muir AJ, Bornstein JD, Killenberg PG. (2004). Peginterferon and ribavirin for the treatment chronic hepatitis C in Blacks and non-Hispanic whites. *N Engl J Med* 350: 2265–2271.

Naka K, Takemoto K, Abe K, et al. (2005). Interferon resistance of hepatitis C virus replicon-harbouring cells is caused by functional disruption of type 1-interferon receptors. *J Gen Virol* 86: 2787–2792.

Nguyen VT, Law MG, Dore GJ. (2009). Hepatitis B related hepatocellular carcinoma: Epidemiological characteristics and disease burden. *J Viral Hepat* 16: 453–463.

Onoguchi K, Yoneyama M, Takemura A. (2007). Viral infections activate types I and III interferon genes through a common mechanism. *J Biol Chem* 282: 7576–7581.

Osterlund PI, Pietila TE, Veckman V. (2007). IFN regulatory factor family members differentially regulate the expression of type III IFN (IFN-lambda) genes. *J Immunol* 179: 3434–3442.

Paeshuyse J, Dallmeier K, Neyts J. (2011). Ribavirin for the treatment of chronic hepatitis C virus infection: A review of the proposed mechanisms of action. *Curr Opin Virol* 1: 590–598.

Panigrahi R, Hazari S, Chandra S, et al. (2013). Interferon and ribavirin combination treatment synergistically inhibit HCV internal ribosome entry site mediated translation at the level of polyribosome formation. *PLos One* 8: e72791.

Park H, Serti E, Eke O, et al. (2012). IL-29 is the dominant type-III interferon produced by hepatocytes during acute hepatitis C virus infection. *Hepatology* 56: 2060–2070.

Pawlotsky JM. (2012). New antiviral agents for hepatitis C. *F1000 Biol Rep* 4: 5–12.

Pazienza V, Clement S, Pugnale P, et al. (2007). The hepatitis C virus core protein of genotypes 3a and 1b downregulates insulin receptor substrate 1 through genotype-specific mechanisms. *Hepatology* 45: 1164–1171.

Pearlman BL. (2006). Hepatitis C virus infection in African Americans. *Clin Infect Dis* 42: 82–91.

Perales C, Beach NM, Gallego I, et al. (2013). Response of hepatitis C virus to long-term passage in the presence of interferon-alpha: Multiple mutations and a common phenotype. *J Virol* 87: 7593–7607.

Poat B, Hazari S, Chandra PK, et al. (2010a). SH2 modified STAT1 induces HLA-1 expression and improves IFN-γ signaling and inhibits HCV RNA replication. *PLos One* 5: 13117–13131.

Poat B, Hazari S, Chandra PK, et al. (2010b). Intracellular expression of IRF-9 Stat fusion protein overcome defective JAK-STAT signaling and inhibits HCV RNA replication. *Virol J* 7: 265–277.

Poat B, Hazari S, Gunduz F, et al. (2010c). Impaired JAK-STAT signaling of interferon alpha in chronically HCV infected liver tissue. *Hepatology* 52:A886: 746.

Poordad F, McCone J, Jr, Bacon BR, et al. (2011). Boceprevir for untreated chronic HCV genotype 1 infection 1 infection. *N Engl J Med* 364: 1195–1206.

Qu LS, Chen H, Kuai XL, et al. (2012). Effects of interferon therapy on development of hepatocellular carcinoma in patients with hepatitis C-related cirrhosis: A meta-analysis of randomized controlled trials. *Hepatol Res* 42: 782–789.

Randal G, Chen L, Panis M, et al. (2006). Silencing USP18 potentiates the antiviral activity of interferon against hepatitis C virus infection. *Gastroenterology* 131: 1584–1591.

Rauch A, Kutalik Z, Desombes P, et al. (2010a) Genetic variation in IL28B is associated with chronic hepatitis C and treatment failure: A genome-wide association study. *Gastroenterology* 138: 1338–1345.

Rauch A, Rohrbach J, Bochud PY. (2010b). The recent breakthroughs in the understanding of host genomics in hepatitis C. *Eur J Clin Invest* 40: 950–959.

Reddy KR, Shiffman ML, Rodriguez-Torres M, et al. (2010). Induction pegylated interferon alfa-2a and high dose ribavirin do not increase SVR in healthy patients with genotype 1a and high viral loads. *Gastroenterology* 139: 1972–1983.

Rivas-Estillas AM, Svitkin Y, Lastra ML, et al. (2002). PKR-dependent mechanisms of gene expression from a subgenomic hepatitis C virus clone. *J Virol* 76: 10637–10653.

Romero-Gomez M, Del Mar Viloria M, Andrade RJ, et al. (2005). Insulin resistance impairs sustained virological response rate to peginterferon plus ribavirin in chronic HCV patients. *Gastroenterology* 128: 636–641.

Roulot D, Bourcier V, Grando V, et al. (2007). Epidemiological characteristics and response to peginterferon plus ribavirin treatment of hepatitis C virus genotype 4 infection. *J Viral Hepat* 14: 460–467.

Sachs AB, Sarnow P, Hentze MW. (1997). Starting at the beginning, middle and end: Translation initiation in eukaryotes. *Cell* 89: 831–838.

Sarasin-Filipowicz M, Wang X, Yan M, et al. (2009). Alpha interferon induces long-lasting refractoriness of JAK-STAT signaling in the mouse liver through induction of USP18/UBP43. *Mol Cell Biol* 29: 4841–4851.

Shepard CW, Finelli L, Alter MJ. (2005). Global epidemiology of hepatitis C virus infection. *Lancet Infect Dis* 5: 558–567.

Sheppard P, Kindsvogel W, Xu W, et al. (2003). IL-28, IL-29 and their class II cytokine receptor IL-28R. *Nat Immunol* 4: 63–68.

Shi X, Pan Y, Wang M, et al. (2012). IL28B genetic variation is associated with spontaneous clearance of hepatitis C virus, treatment response, serum IL-28B levels in Chinese Population. *PLos One* 7: 37054–37062.

Shimazaki T, Honda M, Kaneko S, et al. (2002). Inhibition of internal ribosomal entry site-directed translation of HCV by recombinant IFN-alpha correlates with a reduced La protein. *Hepatology* 35: 199–208.

Shintani Y, Fujie H, Miyoshi H, et al. (2004). Hepatitis C virus infection and diabetes: Direct involvement of the virus in the development of insulin resistance. *Gastroenterology* 126: 840–848.

Shiratori Y, Ito Y, Yokosuka O, et al. (2005). Antiviral therapy for cirrhotic hepatitis C: Association with reduced hepatocellular carcinoma development and improved survival. *Ann Intern Med* 142: 105–114.

Shu QN, Nair V. (2008). Inosine monophosphate dehydrogenase (IMPDH) as a target in drug discovery. *Med Res Rev* 28: 219–232.

Sidwell RW, Huffman JH, Khare GP, et al. (1972). Broad-spectrum antiviral activity of virazole-1-beta-d-ribofuranosyl-1, 2,4-triazole-3-carboxamide. *Science* 177: 705–706.

Simmonds P, Bukh J, Combet C, et al. (2005). Consensus proposals for a unified system of nomenclature of hepatitis C virus genotypes. *Hepatology* 42: 962–973.

Smith K. (2013). Sofosbuvir: A new milestone in HCV treatment. *Nat Rev Gastroenterol Hepatol* 10: 258.

Stevenson NJ, Murphy AG, Bourke NM, et al. (2011). RBV enhances IFN-alpha signaling and MxA expression: A novel immune modulation mechanism during treatment of HCV. *PLos One* 6: 27866–27871.

Suppiah V, Moldovan M, Ahlenstiel G, et al. (2009). IL28B is associated with response to chronic hepatitis C interferon-alpha and ribavirin therapy. *Nat Genet* 41: 1100–1104.

Tanaka Y, Kurbanov F, Mano S, et al. (2006). Molecular tracing of the global hepatitis C virus epidemic predicts regional patterns of hepatocellular carcinoma mortality. *Gastroenterology* 130: 703–714.

Tanaka Y, Nishida N, Sugiyama M, et al. (2009). Genome-wide association of IL28B with response to pegylated interferon alpha and ribavirin therapy for chronic hepatitis C. *Nat Genet* 41: 1105–1109.

Thomas E, Feld JJ, Li Q, et al. (2011). RBV potentiates interferon action by augmenting interferon-stimulated gene induction in hepatitis C virus cell culture models. *Hepatology* 53: 32–41.

Torriani FJ, Rodriguez-Torres M, Rockstroh J, et al. (2004). Peginterferon-alpha-2a plus ribavirin for chronic hepatitis C virus infection in HIV-infected patients. *N Engl J Med* 351: 438–450.

Urban T, Charlton MR, Goldstein DB. (2012). Introduction to the genetics and biology of interleukin-28. *Hepatology* 56: 361–366.

Urban TJ, Thompson AJ, Bradrick SS, et al. (2010). IL28B genotype is associated with differential expression of intrahepatic interferon-stimulated genes in patients with chronic hepatitis C. *Hepatology* 52: 1888–1896.

Van der Meer AJ, Veldt BJ, Feld JJ, et al. (2012). Association between sustained virological response and all-cause mortality among patients with chronic hepatitis C and advanced fibrosis. *JAMA* 308: 2584–2593.

Veldt BJ, Heathcote EJ, Wedemeyer H, et al. (2007). Sustained virologic response and clinical outcomes in patients with chronic hepatitis C and advanced fibrosis. *Ann Intern Med* 147: 677–684.

von Wagner M, Huber M, Berg T, et al. (2005). Peginterferon-alpha-2a (40 KD) and ribavirin for 16 or 24 weeks in patients with genotype 2 or 3 chronic hepatitis C. *Gastroenterology* 129: 522–527.

Wang C, Pflugheber J, Sumpter R, Jr, et al. (2003). Alpha interferon induces distinct translational control programs to suppress hepatitis C virus RNA replication. *J Virol* 77: 3898–3912.

Wright TL. (2002). Treatment of patients with hepatitis C and cirrhosis. *Hepatology* 36: S185–S192.

Yang JD, Robert LR. (2010). Hepatocellular carcinoma: A global view. *Nat Rev Gastroenterol Hepatol* 7: 448–458.

Younossi ZM, McCullough AJ. (2009). Metabolic syndrome, non-alcoholic fatty liver disease and hepatitis C virus; impact on disease progression and treatment response. *Liver Int* 29: 3–12.

Zeuzem S, Andreone P, Pol S, et al. (2011). Telaprevir for retreatment of HCV infection. *N Engl J Med* 364: 2417–2428.

Zeuzem S, Buti M, Ferenci P, et al. (2006). Efficacy of 24 weeks treatment with peginterferon alfa-2b plus ribavirin in patients with chronic hepatitis C infected with genotype 1 and low pretreatment viremia. *J Hepatol* 44: 97–103.

Zhang L, Jilg N, Shao RX, et al. (2011). IL28B inhibits hepatitis C virus replication through the JAK-STAT pathway. *J Hepatol* 55: 289–298.

Zhang S, Kodys K, Li K, et al. (2013). Human type-2 myeloid dendritic cells produce interferon-lambda and amplify interferon-alpha in response to hepatitis C virus infection. *Gastroenterology* 144: 414–425.

# 5 Hepatitis B Virus Infection and Hepatocellular Carcinoma

*Rakhi Maiwall and Manoj Kumar Sharma*

## 5.1 INTRODUCTION

Hepatitis B virus (HBV) infection is a global health problem that affects almost one-third of the world's population on the basis of serologic evidence of past or present infection. Nearly 360–400 million people have chronic HBV infection (Gish and Gadano 2006). Of these, 15–40% develop cirrhosis, hepatic decompensation, and hepatocellular carcinoma (HCC) during their lifetime. HBV-related, end-stage liver disease or HCC is responsible for more than 0.5–1 million deaths per year and currently represent 5%–10% of cases of liver transplantation (Hall and Wild 2003). Worldwide, HCC ranks among the top ten cancers in prevalence and mortality of which the majority (nearly 85%) of HCCs occur in the developing world (Bosch et al. 2004). HBV is a leading risk factor for HCC, with more than 80% of HCC cases occurring in regions that are also endemic for hepatitis B (Michielsen and Ho 2011).

Upon exposure to HBV, individuals with a vigorous and broad immune response to the virus (>95% of adults) develop an acute self-limited infection that may result in acute hepatitis with clearance of hepatitis B surface antigen (HBsAg). Individuals who do not mount a broad and vigorous immune response do not clear the virus but develop persistent infection and become chronically infected with HBV. However, HBV persists in the body even after serological recovery from acute hepatitis B (Gupta et al. 1990).

## 5.2 RISK OF HCC IN CHRONIC HBV INFECTION

Overall, cohort studies show that the incidence of HCC was less than 0.01%–0.2% per year in so-called "inactive HBsAg carriers," less than 0.3%–0.9% per year in chronic hepatitis B (CHB), and 1.5%–6.0% in cirrhosis (Fattovich et al. 2008; Chu and Liaw 2006; Chen et al. 2007). Chronic HBV infection acquired in adults has a much lower risk of HCC development than infections resulting from early life transmission, possibly reflecting differences in duration of infection. The risk of HCC among white asymptomatic chronic HBV-infected subjects from Western developed countries (such as those in North America and Europe) is low, with an annual risk of 1–13 per 100,000 individuals (Villeneuve et al. 1994; Papatheodoridis et al. 2005). These figures are much lower than the annual risk of 240–1169 per 100,000 individuals reported from Asian countries such as Taiwan and Japan (Liaw et al. 1988; Sakuma et al. 1988). Among subjects with CHB, the overall annual incidence of HCC in seven studies involving 732 patients from Europe and the United States was calculated to be 300 per 100,000 individuals as compared to 600 per 100,000 individuals, calculated from nine studies involving 5661 patients from Taiwan, China, Korea, and Japan (Fattovich et al. 2008). Similarly, the calculated annual risk of HCC in cirrhotics based on studies from Europe and the United States was 2.2% as compared to 3.7% based on reports from Taiwan, Japan, and other parts of Asia (Fattovich et al. 2008).

Case–control studies have shown that estimates of odds ratio (OR) for HBsAg positivity range from 5–50 in Asian countries of high or intermediate prevalence of HCC (Tanaka et al. 1991; Tsai et al. 1994; Zhang 1995; Zhang et al. 1998; Kumar et al. 2007). There is also a significantly high risk or HBsAg negativity but antibody positivity (anti-HBe or IgG anti-HBc) among cirrhotic and noncirrhotic patients, in case–control studies (Kumar et al. 2007).

### 5.2.1 Factors Associated with Increased Risk of Developing HCC in Chronic HBV-Infected Subjects

Though HCC is more common in the setting of cirrhosis, with data showing that approximately 75% of HCC cases occur in patients with cirrhosis, it is not necessary that cirrhosis should precede HCC in HBV-infected people. Several potential factors have been identified to be associated with a higher risk of development of HCC. Complex interactions among viral factors, patient factors including genetic factors, and other environmental factors lead to HCC development in chronic HBV-infected patients.

#### 5.2.1.1 Viral Factors

##### *5.2.1.1.1 HBV DNA Levels*

Cross-sectional, case–control, and cohort studies have shown a significant dose–response association between serum HBV DNA levels at initial evaluation and subsequent risk of HCC (Chen et al. 2009b). Recent analysis of 3160 participants in the Risk Evaluation of Viral Load Elevation and Associated Liver Disease/Cancer–Hepatitis B Virus (REVEAL–HBV) study has shown that, regardless of baseline HBV DNA levels, participants with spontaneous viral load reduction to $<10^4$ copies/mL during follow-up had a risk of HCC similar to those with a baseline HBV DNA $<10^4$ copies/mL. The HCC risk was only slightly higher for participants whose follow-up levels of HBV DNA spontaneously decreased to <10,000 copies/mL compared to those with baseline levels of HBV DNA <10,000 copies/mL (control group; HR 2.25; 95% CI 0.68–7.37). Compared with the control group, the HRs (95% CI) for long-term levels of HBV DNA that persisted at 10,000–100,000 copies/mL decreased to/persisted at 100,000–1,000,000 copies/mL or decreased to/persisted at 1,000,000–10,000,000 copies/mL were 3.12 (1.09–8.89), 8.85 (3.85–20.35), and 16.78 (7.33–38.39), respectively. Thus, the group of long-term HBV DNA change was a strong independent risk predictor of HCC after taking age, gender, long-term alanine aminotransferase (ALT) pattern, HBV genotype, and habits of cigarette smoking and alcohol consumption into consideration. Thus, patients with similar HBV DNA levels at baseline have different risks of HCC depending on their HBV DNA levels during follow-up, with greater decreases in serum HBV DNA during the follow-up being associated with lower HCC risk. The HCC risk was primarily determined by HBV DNA levels at enrollment for the participants who had persistent HBV DNA levels during follow-up (Chen et al. 2011).

##### *5.2.1.1.2 HBV Genotypes*

HBV is classified into eight genotypes, A–H. These can be further segregated into subgenotypes based on a 4% (but 8%) difference in the entire nucleotide sequence, which revealed an intergroup divergence of >8%, and an intragroup divergence <5.6%. The genotypes have different geographic and ethnic distribution (Fang et al. 1998; Kao et al. 2000). Genotype A is prevalent in sub-Saharan Africa, North America, as well as Europe; B and C are common in Asia; D is ubiquitous and found worldwide; E is common in Africa; and F and H are localized to Central and South America. Genotype G is infrequent; hence, its epidemiology has not been determined. Genotypes are further subdivided into subgenotypes. Evidence for the influence of HBV genotypes/subgenotypes has been increasing with subgenotype Aa/A1 identified in HBV isolates from South Africa by phylogenetic analysis of preS2/S sequences (Chan et al. 2004) and two subgenotypes of genotype B, that is, (Bj/B1) indigenous to Japan and (Ba/B2), which predominates in Asia and has a recombination with genotype C over the precore region and core gene (Zhong et al. 2000). Subgenotypes have also been recognized in genotypes C and D. In Asia, genotype C is associated with an increased risk of HCC compared to genotype B (Chen et al. 2009a); however, these findings have not been substantiated by other studies (Sumi et al. 2003; Yuen et al. 2003). Also, genotype B was more commonly seen in HCC patients in the younger age group usually in the absence of cirrhosis (Kao et al. 2000). A study in India demonstrated a higher impact of HCC in subjects infected with HBV genotype D as compared to genotype A. Genotype D was more prevalent in HCC patients of less than 40 years of age, as compared to asymptomatic carriers (63% vs. 44%, $P = 0.06$) (Thakur et al. 2002). In a cohort of 1176 Alaska Native people with chronic HBV infection, including 47 patients with HCC, it has been shown that genotype F, which is endemic to North America, has a particularly severe phenotype, and genotype F was found in 68% of patients with HCC versus 18% of those without HCC ($P < 0.001$). For patients with genotype F, the median age at diagnosis of HCC was lower than that for patients with other genotypes (22.5 vs. 60 years, respectively; $P = 0.002$) (Livingston et al. 2007).

HBV subgenotypes have also demonstrated important effects on the development of liver disease, and it was shown that in patients with genotype C and subgenotype Ce, there was an increased the risk of HCC with CHB (Chan et al. 2008). Intriguingly, HBV genotype C is more prone to the T1762/A1764 mutation than genotype B, and the study from Taiwan showed a hazard ratio of 1.73 for patients who had these mutations (Yang et al. 2008a). Hence, it needs to be addressed if an increased association between genotype C and HCC risk is merely due to the high percentage of basal core promoter (BCP) mutations in patients with genotype C.

In a recent study, it was seen that genotype C, high viral load, and presence of mutations including A2962G, preS2, T105C, T1753V, and A1762T/G1764A are independently associated with increased risk of HCC (Liu et al. 2011).

##### *5.2.1.1.3 Pre-S2/S Mutation*

The truncated form of the *pre-S2/S* gene that is commonly seen in the 3′ terminus of *pre-S1* and the 5′ terminus of pre-S2 is a consequence of deleted integrated viral genomes (Tai et al. 2002) and is known to have transactivational abilities causing activation of cellular genes, including *c-myc, c-fos,* and *c-Haras* (Schlüter et al. 1994; Luber et al. 1996). The *pre-S* mutant can cause increased malignant transformation via inducing telomerase activation (Liu et al. 2007); activation of the c-Raf-1/MEK/extracellular signal-related kinase (ERK) signal transduction pathway; activation of the endoplasmic reticulum (ER) stress (Wang and Weinman 2006); and upregulation of *cyclooxygenase-2*, cyclin A [39], and vascular endothelial growth factor-A (VEGF-A). The presence of truncated *pre-S2/S* was investigated; its presence correlated with development of progressive liver diseases and HCC (Chen et al. 2006).

##### *5.2.1.1.4 Precore and Basal Core Promoter Mutations*

HBV is a partially double-stranded DNA virus and has four different promoters. The core promoter (CP) is composed of the upper regulatory region (nt. 1613–1742) and the BCP (nt. 1742–1849), and it acts to direct transcription of pregenomic mRNA and precore mRNA (Kramvis and Kew 1999). The common mutations in the BCP are an A to T mutation at 1762 and a G to A mutation at 1764. These T1762/A1764 double mutations have been reported in the development of HCC (Baptista et al. 1999; Yang et al. 2000; Kao et al. 2003; Kuang et al. 2004; Yuen et al. 2004, 2008; Liu et al. 2006; Tanaka et al. 2006). However, consistency in results could not be shown by other studies (Chan et al. 2000; Chun et al. 2000). Enh II is located in nt.1636–1741 and is composed of the boxes a (nt.1646–1668) and b (nt.1704–1715); it acts to stimulate the transcriptional activity of surface, core, and X gene promoters (Yuh and Ting 1990, 1993). In recent studies from Asia, T1766, A1768, and V1753 in BCP and T1653 in box a of Enh II were found to occur more frequently in HCC patients than in non-HCC patients (Takahashi et al. 1999; Guo et al. 2008; Yuen et al. 2008). These associations have also been confirmed by a recent meta-analysis (Liu et al. 2009).

#### 5.2.1.2 Environmental Factors

##### *5.2.1.2.1 Coexistent or Superadded Infections*

Coexistent or superadded infections with HCV, HIV, and HDV have also been shown to increase the fibrosis progression and development of HCC.

##### *5.2.1.2.2 Aflatoxin Exposure*

Among the important environmental factors, linkage of aflatoxin exposure to HCC in CHB patients was reported in a study done in villages near Shanghai. By causing P53 gene mutation, aflatoxin increases the risk of hepatocarcinogenesis (Yu and Yuan 2004; Kirk et al. 2006; Fattovich et al. 2008).

##### 5.2.1.2.3 Alcohol, Obesity, Diabetes, and Metabolic Syndrome

Alcohol consumption also increases the risk of HCC development as shown from studies done in Japan and Korea (Yang et al. 2000; Hassan et al. 2002). In the Japanese study, a fivefold increase in the relative risk was seen in HBV patients with alcohol consumption (>27 gm/d), although in the Korean study, a 50 gm/d alcohol consumption was associated with an increased risk of HCC. In another recent study by Chen et al (2007) in a cohort of 20,069 participants contributing to a total of 262,122 person-years, followed for a mean duration of 13.1 years, the annual incidence rates of HCC and liver-related death for inactive HBV carriers versus controls were 0.06% and 0.04% and 0.02% and 0.02%, respectively. The multivariate-adjusted hazard ratios for carriers of inactive HBV were 4.6 as compared to 2.1 for controls. Age and consumption of alcohol were identified as independent predictors for HBV carriers developing HCC (Chen et al. 2010). Obesity, diabetes, metabolic syndrome, nonalcoholic fatty liver disease, and inherited metabolic conditions such as hemochromatosis and alpha-1 antitrypsin deficiency are other known factors for risk of HCC development.

#### 5.2.1.3 Host Factors

The important host factors that increase the susceptibility for the development of this malignancy are explored in the following sections.

##### 5.2.1.3.1 Age and Gender

Among the HBV-infected patients, those who are above 40 have greater risk of development of HCC as compared to those who are below 40 (Nguyen et al. 2009). Further, males are at higher risk than females with an adjusted relative risk of 2.1–3.6 (Nguyen et al. 2009).

##### 5.2.1.3.2 Family History

Patients with a positive family history of HCC are at an increased risk, which was shown in a study from Taiwan where 553 CHB patients with HCC were compared to 4686 patients without HCC. It was seen that the first-degree relatives of the patients in the HCC group had a higher incidence of HCC. Similar results were found in an Italian study where a ratio of 70:1 was seen for patients who had CHB as well as a positive family history (Donato et al. 1999; Yu et al. 2000; Hassan et al. 2009).

##### 5.2.1.3.3 Biochemical and Histological Factors

An elevated serum ALT level signifies the presence of active disease and increases risk, particularly if the ALT is persistently or intermittently elevated over years. Persistently elevated alpha-fetoprotein (AFP) level is a reflection of enhanced regenerative state in the liver; the increased rate of cell division increases the risk of mutation, leading to increased risk of HCC. A low platelet count suggests the presence of cirrhosis, which itself increases the risk of HCC.

In a study of histological severity (i.e., presence of necro-inflammation and fibrosis in CHB) done on 796 biopsy-confirmed CHB patients as a determinant for the development of HCC, on multivariate analysis, age over 40 years, advanced fibrosis, and severe lobular activity at presentation were identified as independent risk factors for the development of HCC. Hence, presence of advanced fibrosis and severe lobular activity on histology at presentation of CHB are independent predisposing risk factors for the development of HCC (Lee et al. 2011). Other histologic risk factors revealed at biopsy include dysplasia, geographic morphologic changes that suggest clonal populations of cells, and a positive stain for proliferating cell nuclear antigen (Koskinas et al. 2005).

##### *5.2.1.3.4 Liver Stiffness*

Recently, liver stiffness measurement (LSM or FibroScan) has been an accurate noninvasive modality for the detection of liver fibrosis and cirrhosis and has provided reproducible and reliable results. In a study that aimed to investigate the usefulness of LSM as a predictor of HCC development in patients with CHB, it was found that patients with a higher LSM (>8 kPa) at baseline were at a significantly greater risk of HCC development. Hazard ratios were 3.07 (95% CI 1.01–9.31; $P = 0.047$) for LSM 8.1–13 kPa; 4.68 (95% CI 1.40–15.64; $P = 50.012$) for LSM 13.1–18 kPa; 5.55 (95% CI 1.53–20.04; $P = 0.009$) for LSM 18.1–23 kPa; and 6.60 (95% CI 1.83–23.84; $P = 0.004$) for LSM >23 kPa (Jung et al. 2011). Thus, LSM could be an important noninvasive modality that could prove useful in prediction of HCC development in patients with CHB.

#### 5.2.1.4 Genetic Factors

A study by Zhang et al. of Chinese patients provided the first evidence for a causative role of genetic susceptibility in HBV-associated HCC using a genome-wide association approach. A total of 440,794 single-nucleotide polymorphisms (SNPs) were genotyped in 715 patients with chronic HBV infection, 355 with HCC, and 360 controls without HCC. Only five SNPs showed a consistent significant association, of which only one *SNP (rs17401966)* was confirmed in combined analysis of the three independent case–control populations (1962 cases and 1430 controls) from different Chinese regions. Hence, a significant association of a reduced HCC risk was seen in the presence of the mutant [G] allele of the *rs17401966 SNP* (Zhang et al. 2010).

## 5.3 PREDICTIVE SCORES FOR HCC DEVELOPMENT IN CHRONIC HBV INFECTION

The prognosis of patients with advanced HCC is often grim; earlier detection of small tumors offers the potential for curative therapy. Regular HCC surveillance by ultrasound scan and/or AFP may identify early cancers and reduce mortality (Wong et al. 2008). HCC surveillance has been demonstrated to be cost effective in patients at high risk of developing cancer (Nouso et al. 2008). In addition, the use of antiviral therapy among high-risk patients could reduce the incidence of hepatic decompensation and possibly the development of HCC. Because the global burden of HBV infection is enormous, it is virtually impossible to screen all infected patients for HCC. Therefore, the identification of risk factors, target populations, and predictive models are needed to help focus screening on the infected population that is most at risk for the development of HCC. Accurate risk stratification for HCC development for patients with chronic HBV infection is very important. For patients assessed as having a high-risk score, management strategies including more intensive follow-up, more frequent HCC surveillance with use of sensitive imaging techniques such as computed tomography (CT) and magnetic resonance imaging (MRI), and the initiation of antiviral therapy may be considered.

The risk scores for development of HCC in chronic HBV-infected patients can also provide guidance about treatment, particularly in patients who do not meet existing treatment initiation recommendations according to current treatment guidelines. Risk calculation can lead to better communication of risk of HCC to patients, which could lead to better treatment acceptance and compliance. Risk stratification also can be useful from a public health perspective by helping in long-term resource planning and allocation (Sarin and Kumar 2012).

Various studies have evaluated the use of prediction models for HCC in patients with chronic HBV. Yuen et al. derived a prediction score from a cohort of 820 patients with chronic HBV observed for a mean duration of 76.8 months. Male gender ($P$ = 0.025, relative risk [RR] 2.98), increasing age ($P$ < 0.001, RR 1.07), higher HBV DNA levels ($P$ = 0.02, RR 1.28), CP mutations ($P$ = 0.007, RR 3.66), and presence of cirrhosis ($P$ < 0.001, RR 7.31) were independent risks for the development of HCC. A risk score was derived and validated with sensitivity >84% and specificity >76% to predict the five- and ten-year risks for the development of HCC. The Area under the curve (AUC) for the five- and ten-year predictions was 0.88 and 0.89, respectively. Cirrhosis was seen in 15.1% of the study patients, and 43.4% were hepatitis B envelope antigen (HBeAg) positive; 50% of the patients had ALT < 1 × upper limit of normal (ULN). Eighty-eight of 820 patients were treated. It was found that a formula including sex, age, HBV DNA, CP mutations, and cirrhosis had good accuracy in the prediction of HCC (Yuen et al. 2009). Another study from Hong Kong, China, evaluated 1005 patients and found that the following five factors independently predicted HCC development: age, albumin, bilirubin, HBV DNA, and cirrhosis. These variables were used to construct a prediction score ranging from 0 to 44.5. The score was validated in another prospective cohort of 424 patients. During a median follow-up of 10 years, 105 patients (10.4%) in the training cohort and 45 patients (10.6%) in the validation cohort developed HCC. Cutoff values of 5 and 20 best discriminated HCC risk. By applying the cutoff value of 5, the score excluded future HCC development with high accuracy (negative predictive value = 97.8% and 97.3% in the training and validation cohorts, respectively). In the validation cohort, the five-year, HCC-free survival rates were 98.3, 90.5, and 78.9% in the low-, medium-, and high-risk groups, respectively. The hazard ratios for HCC in the medium- and high-risk groups were 12.8 and 14.6, respectively; 38.1% in training cohort and 16.3% in validation cohort had cirrhosis by ultrasound; 14.2% in training cohort and 10.4% in validation cohort had ALT < 1 × ULN; 15.1% and 25% of patients received antiviral treatment in training and validation cohort, respectively (Wong et al. 2010). Another recent study from Taiwan used data from REVEAL. Two-thirds of the REVEAL-HBV study cohort was allocated for model derivation ($n$ = 2435), and the remaining third was allocated for model validation ($n$ = 1218) for developing a nomogram for risk of HCC in CHBV-infected patients. Independent risk predictors, included in three Cox proportional hazards regression models, were sex, age, family history of HCC, alcohol consumption habit, serum ALT level, HBeAg sero-status, serum HBV DNA level, and HBV genotype. All Area under the receiver operating curve's (AUROC) for risk prediction nomogram were ≥0.82 in both model derivation and validation sets. The correlation coefficients between the observed HCC risk and the nomogram-predicted risk were greater than 0.90 in all model derivation and validation sets. Most of the participants were noncirrhotics; 94% in training cohort and 94.2% in validation cohort had ALT < 1 × ULN; 85% and 83.7% in the training and validation cohort, respectively, were HBeAg negative. Antiviral therapy was not given during the follow-up (Yang et al. 2010).

Another study from Taiwan included sex, age, serum ALT concentration, HBeAg status, and serum HBV DNA level for calculating the risk score. The development cohort consisted of patients from the population-based prospective REVEAL-HBV database of 3584 HBsAg-positive and anti-HCV-negative patients, aged 30–65 years, who did not have cirrhosis and did not receive antiviral treatment during the median follow-up of 12.0 years (IQR 11.5–12.4) during which time 131 developed HCC. The validation cohort consisted of 1505 patients from three hospitals in Hong Kong and South Korea, with a mean follow-up of 7.3 years. No patients received antiviral therapy throughout the follow-up period and 111 patients developed HCC during follow-up. None of the patients in the development cohort had cirrhosis at baseline, whereas 82% of patients (1228 of 1505) in the validation cohort did not have cirrhosis at baseline; 84.4% and 61.6% of subjects in the development and validation cohorts, respectively, were HBeAg negative. An empirical 17-point risk score was developed, with HCC risk ranging from 0.0% to 23.6% at three years, 0.0% to 47.4% at five years, and 0.0% to 81.6% at ten years for patients with the lowest and highest HCC risk, respectively. To predict risk, AUROCs were 0.811 (95% CI 0.790–0.831) at three years, 0.796 (0.775–0.816) at five years, and 0.769 (0.747–0.790) at ten years in the validation cohort and, correspondingly, 0.902 (0.884–0.918), 0.783 (0.759–0.806), and 0.806 (0.783–0.828) after exclusion of 277 patients with cirrhosis in the validation cohort (Yang et al. 2011).

One important issue is what constitutes high risk of developing HCC in CHB. Studies need to be done to prospectively establish and test threshold scores that may trigger different clinical interventions including the need for therapy or enhanced surveillance for HCC. Treatment with antiviral agents and the resulting lowering of HBV DNA and ALT levels and HBeAg seroclearance and seroconversion can change the risk profile of an individual patient. Also because the responses are not durable in many patients, the risk profile may be dynamic. It will be important to establish whether lowering the risk score by successful anti-HBV treatment therapy eventually results in a reduction of future HCC risk. There is clearly a need for developing predictive scores for different phases of chronic HBV infection and stages of liver disease. It seems quite logical that different risk scores for different stages of chronic HBV infection would be more precise and helpful for clinical practice and policy decisions. Predictive scores also need to be developed for different ethnic groups and in geographical areas that might be different as regards to age at infection, genetic background, HBV genotype or other variants, and exposure to environmental factors. Although to be widely applicable, predictive scores should use readily available parameters with ready availability in future HBV genotypic variant analyses and genome-wide association studies (GWAS); the combination of genetic and nongenetic factors may promise a more personalized approach to predicting HCC in chronic HBV-infected patients, ultimately leading to improved patient care (Sarin and Kumar 2012).

## 5.4 CHRONIC HBV INFECTION AND LIVER ONCOGENESIS

### 5.4.1 HBV Virology

HBV is a partially double-stranded DNA virus of the hepadnavirus family with a relaxed circular DNA molecule of 3.2 kb. It shows a high degree of species specificity and infects only humans and higher primates the same as do other members of the hepadnavirus family (such as duck, woodchuck, and ground squirrel hepatitis viruses). The viral genome is organized in a compact manner with four partially overlapping open reading frames (ORF): S, C, P, and X (Figure 5.1).

The HBV genome is unique in that has a partially double-stranded DNA—a circular conformation of DNA; it depends on the enzyme reverse transcriptase (RT); and it persists in the infected cells as either integrated forms or an episomal form, that is, a covalently closed circular form of DNA (cccDNA) that is essential for its life cycle as the template for viral RNA transcription. HBV virions infect hepatocytes by binding to cellular surface receptors followed by membrane fusion, and the viral inner cores are transported into the nucleus where the formation of a cccDNA occurs,

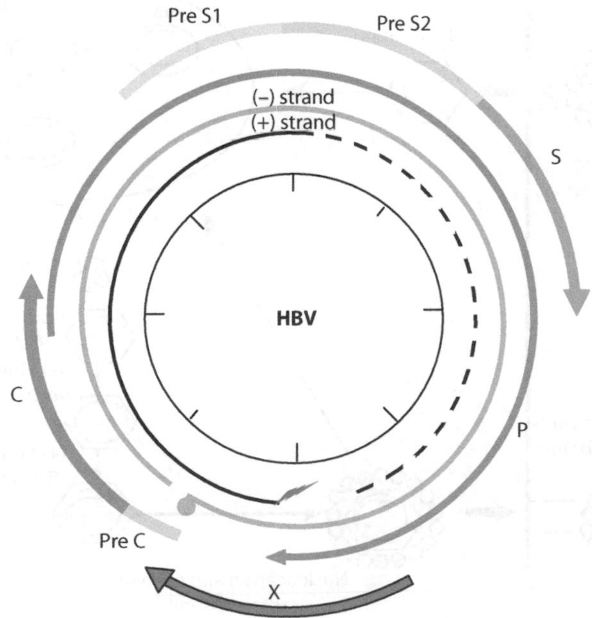


**FIGURE 5.1 (See color insert.)** A pictorial depiction of hepatitis B virus genome structure.

which is transcribed by the host RNA polymerase II into a series of viral mRNA. These transcripts are then transported to the cytoplasm, where translation into viral proteins occurs. Assembly of the viral core proteins occurs in the cytosol, where a single molecule of pregenomic RNA and a viral DNA polymerase are packed with core proteins forming particles that are coated with the viral lipoprotein envelopes by budding into ER; then these particles are exported from the cell as mature virions. The remaining core particles are transported back to the nucleus to maintain a stable pool of cccDNA (Figure 5.2).

HBV exerts its oncogenic potential through a multifactorial process that includes direct and indirect mechanisms that likely act synergistically (Hino and Kajino 1994). Liver cirrhosis itself, resulting from sustained inflammatory damage and hepatocyte regeneration, has been considered as a preneoplastic condition (Okuda 2007). HBV DNA sequences are frequently found to be integrated in the host genes, encoding for proteins related to the control of cell signaling, cellular proliferation, and viability. This leads to a cascade of interactive events that ultimately transforms normal hepatocytes into malignant cells (Brechot et al. 1980). The transactivating potential of several viral oncoproteins (such as HBx and the truncated *pre-S2/S*) on the regulatory cellular pathways is a further crucial oncogenic consequence of the integration process (Schlüter et al. 1994).

The HBV polymerase lacks proofreading ability and is, therefore, prone to generate mutations. Common mutations include the precore (G1896A) mutation, BCP

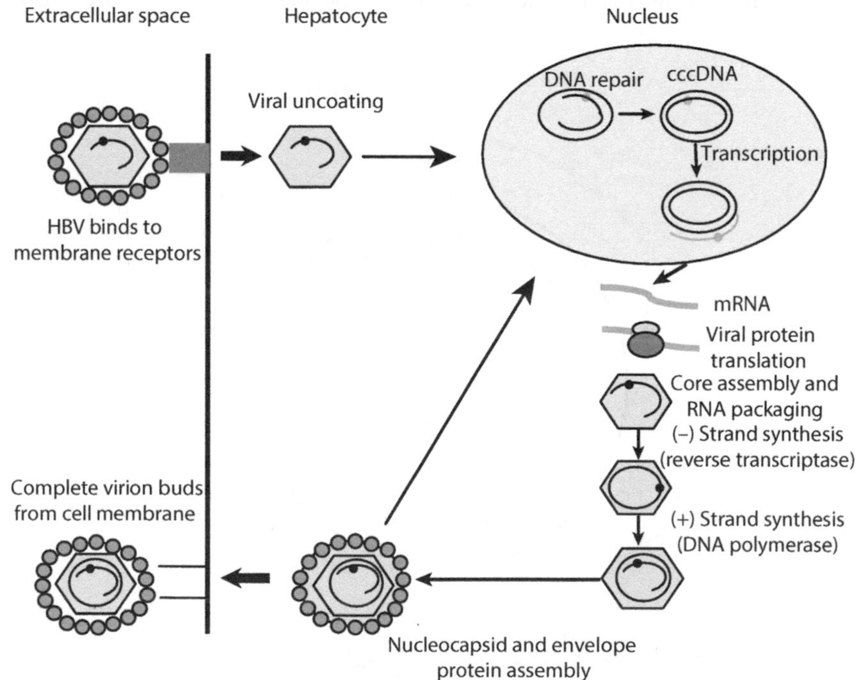


**FIGURE 5.2 (See color insert.)** A schematic representation of hepatitis B virus life cycle.

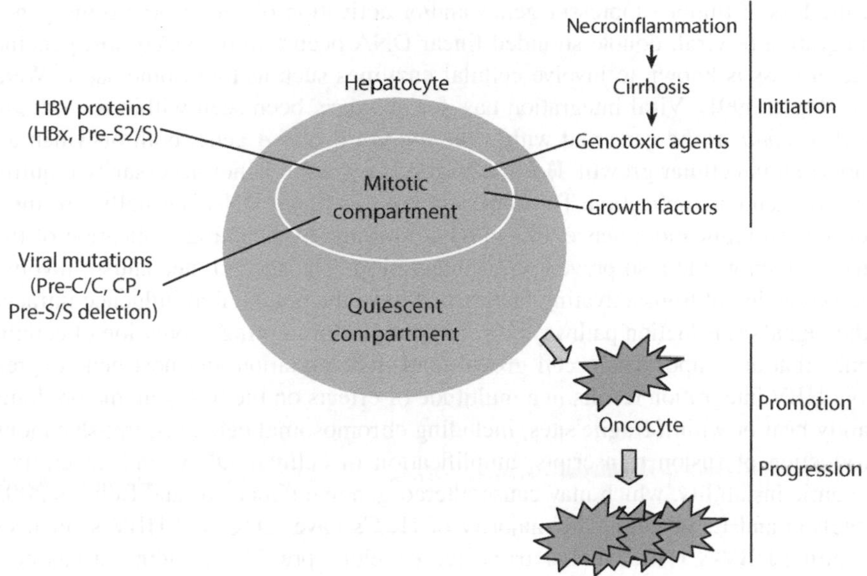


**FIGURE 5.3** A schematic representation of mechanisms of liver carcinogenesis.

mutations (A1762T/G1764A), and deletion mutations of *pre-S/S* genes (Locarnini et al. 2003). These mutations that arise during the course of HBV chronic infection have consequences at clinical and epidemiological levels. In fact, several HBV mutant strains including mutations in pre-C/C and CP and deletion in *pre-S/S* genes are involved in the pathogenesis of progressive liver disease and HCC development (Figure 5.3) (Kuang et al. 2004).

Although a majority of liver cancers develop in cirrhotic livers, a significant fraction of HBV-related HCCs occurs in a background of CHB in the absence of liver cirrhosis. The incidence of cirrhosis appears to be about twofold higher in HBeAg-negative compared to HBeAg-positive chronic hepatitis (Fattovich et al. 1995). Distinctive gene expression profiles have been detected in the nontumoral livers of chronic HBV carriers, such as activated expression of genes implicated in pro-apoptotic, inflammatory, and DNA repair responses, suggesting specific pathways triggered by CHB (Honda et al. 2001, 2006). Thus, besides the effects of host immune responses, HBV replication might trigger various signaling cascades in the infected hepatocyte.

### 5.4.2 Viral Integration

Viral genomic integration usually precedes the development of HCC and is seen in more than 85%–90% of HBV-related HCCs and also in nontumor tissue in patients with chronic HBV (Brechot et al. 1980; Shafritz et al. 1981). The process is accelerated by oxidative stress, chronic inflammation, and DNA damage (Dandri et al. 2002; Guerrero and Roberts 2005). Viral genome integration in the host genome results in the loss of tumor suppressor genes and/or activation of tumor-promoting genes. Integration of viral, double-stranded linear DNA occurs in damaged host genome. The process is known to involve cellular enzymes such as topoisomerase I (Wang and Rogler 1991). Viral integration has, for instance, been seen within the *retinoic acid receptor alpha* gene and within the *human cyclin A* gene, both of which are important in cellular growth. However, viral integration is not necessarily required for viral genome replication. The generation of this linear DNA is usually a result of faulty viral replication; hence, this can be inhibited by blocking replication of this virus, which would also prevent viral integration. The second mechanism involves the expression of transactivating factors that have the potential to influence intracellular signal transduction pathways by disrupting or promoting expression of cellular genes that are important in cell growth and differentiation and host gene expression. HBV integration results in a multitude of effects on the host genome predominantly near or within fragile sites, including chromosomal deletions, translocations, production of fusion transcripts, amplification of cellular DNA, and generalized genomic instability, which may cause altered genome (Guerrero and Roberts 2005; Feitelson and Lee 2007). The majority of HCCs have integrated HBV sequences, encoding HBV-X (HBx) and/or truncated envelope pre-S2/S proteins. It has been shown that HBV integration often targets gene families, such as human telomerase reverse transcriptase (HTERT), Platelet derived growth factor (PDGF) receptor, calcium signaling-related genes, and 60s ribosomal protein, which are involved in cell proliferation and hence can be considered the first hit in the tumorigenicity.

## 5.5 HBV PROTEINS

There are several kinds of proteins in HBV that are important for their life cycle and infection. Their structure and functions are described in the following sections.

### 5.5.1 HBx Protein

HBx gene is the smallest of the four partially overlapping open reading frames of HBV and comprises 452 nucleotides that together encode a 154-amino acid regulatory protein with a molecular mass of 17 kDa. Integration of the HBx sequence into host DNA is a common event in HCC. After integration, HBx promotes genetic instability through a variety of mechanisms (Figure 5.4).

#### 5.5.1.1 Effect on Signal Transduction Pathways

The HBx protein has indirect effects on hepatocyte proliferation by causing transcriptional activation of signal transduction pathways (Feitelson and Lee 2007), which include NF-kB activation in a Ras-dependent manner (Su and Schneider 1996), Jak-Stat pathway, Ras, Raf, stress-activated protein kinase (SAPK)/c-Jun, mitogen-activated protein kinase (MAPK), focal adhesion kinase (FAK), and protein kinase C pathways, as well as Src-dependent and phosphatiylinositol-3 kinase signaling cascades (Feitelson and Duan 1997; Murakami 2001), Wnt activation by binding to nuclear transcription factors, including cAMP response element binding (CREB),

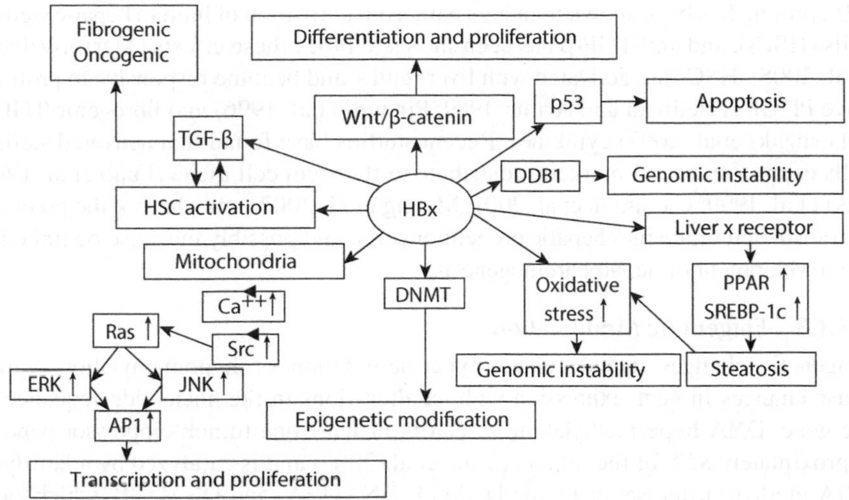


**FIGURE 5.4** A pictorial depiction of mechanisms of carcinogenesis by HBx.

activating transcription factor 2 (ATF-2), and Oct -1 and TATA box-binding protein (TBP) (Feitelson and Lee 2007; Feitelson et al. 2009). HBx expression also increases cytosolic calcium levels, causing activation of tyrosine kinase 2, which further causes activation of Ras and the Ras-Raf-ERK and Ras-MEKK-JNK signaling cascades and transcription of *c-fos* and *c-jun* (Benn and Schneider 1994; Bouchard et al. 2001; Oh et al. 2003). HBx also decreases proteasomal degradation of β-*catenin* and causes β-catenin accumulation in HCC tissue (Ding et al. 2005) as well as increases downstream targets of aberrant activation seen in around 20%–40% of HCCs, *c-myc*, and *cyclin D-1*. HCCs with β-catenin mutation are characterized by absence of steatosis and intrahepatic cholestasis, a differentiated morphology and low proliferative index (Audard et al. 2007), and hence a better prognosis. The presence of the *HBV-X* gene expression in HBsAg-HCC or occult HBV has been mainly attributed to the persistent viral transcription and replication in the liver tissue; hence, there is evidence that occult HBV is a risk factor for the development of HCC (Miyakawa and Mizokami 2000; Kuang et al. 2004).

### 5.5.1.2 Effect on Fibrogenesis via Transforming Growth Factor Beta and Hepatic Stellate

#### *5.5.1.2.1 Cell Activation*

Transforming growth factor beta (TGF-β) is an important mediator of fibrosis and signals through phosphorylation of Smad3 at its middle linker and/or C-terminal regions. It has complex actions because it is persistently induced during active hepatitis and causes extracellular matrix deposition leading to fibrogenesis and acceleration of cirrhosis. Overexpression or upregulation of TGF-β occurs in HCC tissue (Yoo et al. 1996) via HBx-mediated shift in TGF-β signaling from the tumor-suppressive pSmad3C pathway (i.e., Smad3 phosphorylated at the C-terminal region) to the oncogenic pSmad3L pathway (i.e., Smad3 phosphorylated at the linker region).

HBx protein has been shown to induce paracrine activation of human hepatic stellate cells (HSCs), and anti-TGF-β has been shown to block these effects (Martin-Vilchez et al. 2008). HSCs are activated with liver injury and become responsive to proliferative PDGF (Friedman and Arthur 1989; Pinzani et al. 1996) and fibrogenic TGF-β (Matsuzaki et al. 2007) cytokines. Recent studies have found that activated stellate cells express stem cell markers, contribute to the stem cell niche (Fujio et al. 1994, Niki et al. 1999; Cassiman et al. 2001; Myung et al. 2007), may harbor the potential to transdifferentiate into hepatic progenitor cells, and possibly may also be linked to the development of hepatocarcinogenesis.

#### 5.5.1.3 Epigenetic Modification

Epigenetic changes include either hypermethylation or hypomethylation, which cause changes in gene expression without alterations in the nucleotide sequence of the gene. DNA hypermethylation is seen in at least one tumor suppressor gene in approximately 82% of the tumors (Yang et al. 2003) and is catalyzed by a family of DNA-methytransferase, including DNMT1, DNMT3A, and DNMT3B, which catalyze addition of a methyl group using S-adenosylmethionine as the methyl donor to the 5′-CpG dinucleotide of the cytosine ring to form methylcytosine (Herman and Baylin 2003; Chen and Li 2004). The expression levels of the DNMTs are regulated by the tumor suppressor gene, *p53*. The *p16INK4A* gene is also frequently altered in HCC, and hypermethylation of *p16INK4A* gene correlates with HBx protein levels in the tumor tissue (Jung et al. 2007; Zhu et al. 2007). This gene encodes a negative regulator of cell cycle progression.

#### 5.5.1.4 Effects on Apoptosis

Apoptosis, or programmed cell death, is responsible for the removal of damaged or virally infected cells and is regulated by caspases and "death receptors," including tumor necrosis factor-alpha (TNF-α), Tumor necrosis factor (TNF)-related apoptosis-inducing ligand, and Fas, which act as initiators of apoptosis; members of the Bcl-2 proteins acting as inducers or suppressors; and transcription factors, such as *p53, Fos, Jun,* and *myc* (Ellis et al. 1991; White 1996; Arbuthnot et al. 2000). The effects of HBx protein on apoptosis are complex. It produces anti-apoptotic and pro-apoptotic effects. The possible explanation for these contrary effects is that high levels of HBx protein promote apoptosis, whereas low levels inhibit apoptosis (Ye et al. 2008).

Anti-apoptotic effects are predominantly via inhibition of p53-mediated apoptosis, which has been demonstrated in vivo and in vitro and by accumulation of the anti-apoptotic protein, survivin (Kuo and Chao 2010). HBx forms a complex with p53 in the cytoplasm, preventing its transcriptional transactivating properties by blocking its entry into the nucleus. HBx affects p-53-mediated apoptosis via indirect effects on phosphoinositide-3 kinase–protein kinase B–Bad (P3k-Akt-Bad) pathway activation, phosphatidyl inositol-3 kinase/AKt pathway activation, cyclooxygenase 2 (COX-2)/prostaglandin E2 (PGE2) pathway activation, p38/MAPK pathway activation, and by caspase 3 inactivation. HBx protein can also be pro-apoptotic and induce apoptosis by death-receptor-activating stimuli, which has been shown in HBx transgenic livers, in primary hepatocytes in culture, and in mice with a p53 null mutation (Kim et al. 1998; Terradillos et al. 1998).

#### 5.5.1.5 Effect on Telomerase Activity

Telomeres are composed of structural proteins, which stabilize and regulate the length of the telomeres, and are comprised of conserved homogeneous repeats (TTAGGG)n that extend up to 10–15 kb in length and are found at the ends of the eukaryotic chromosomes (Zakian 1989; Counter et al. 1992; Aviv et al. 2003). Normally the length of the telomeres reduces with each cell division, and cells fail to divide after telomeres reach a certain length. Malignant transformation occurs after chromosomal instability, when the cells continue to divide irrespective of the telomere length that can be maintained by telomerase. Telomerase is a ribonuclease enzyme and consists of two proteins, telomerase reverse transcriptase (TERT) and telomerase-associated protein (TEP1), and an RNA template (TERC). Activation of telomerase, which is seen in 80% of HCCs, results in an uncontrolled cell proliferation or malignant transformation because it synthesizes new telomerase de novo (Djojosubroto et al. 2003). Studies have shown that HBx protein increases TERT expression and telomerase activity in HBx-transfected cells, HBx protein-positive HCC tissues, and cultured HCC cells (Liu et al. 2004; Zhang et al. 2005).

#### 5.5.1.6 Effect on Nucleotide Excision Repair

The nucleotide excision repair (NER) helps in maintenance of genomic integrity and causes repair of damaged DNA that is achieved by a series of some 30 genes including *XPA to XPG* (Wood 1997; Taylor and Lehmann 1998). HBx compromises NER by inactivating the ultraviolet-damaged DNA-binding protein and inhibiting cell cycle checkpoint mechanisms inactivating p53. HBx also binds to the carboxylic acid (COOH) terminus, which is important for the gene's transactivational properties and for maintenance of cell viability, the deletion of which has also been proposed in HBx induced hepatocarcinogenesis (Poussin et al. 1999).

## 5.6 PRE-S2/S PROTEIN

The truncated form of the *pre-S2/S* gene is commonly found from deleted integrated viral genomes (Tai et al. 2002). The pre-S2/S product also has transactivational abilities. These proteins transactivate cellular genes, including c-myc, *c-fos*, and *c-Haras* (Schlüter et al. 1994). Pre-S2 may increase malignant transformation of human HCC cell lines by upregulating human TERT and inducing telomerase activation (Liu et al. 2007). The pre-S2 activator activates PKC and causes phosphorylation of the pre-S domain, and then activates the *c-Raf-1/MEK/ERK* signal transduction cascades (Hildt et al. 2002). The pre-S mutant large surface antigens can activate ER stress to induce oxidative DNA damage and genomic instability (Wang and Weinman 2006). The pre-S mutant also can upregulate *cyclooxygenase-2* and cyclin A to induce cell cycle progression and proliferation of hepatocytes. A recent study found that VEGF-A is upregulated by pre-S mutants and that pre-S mutant-expressed Huh-7 cells exhibited activation of Akt/mTOR (mammalian target of rapamycin) signaling and increased growth advantage, which could be inhibited by VEGF-A neutralization (Yang et al. 2009).

## 5.7 PATHOGENESIS OF HCC

Development of HCC involves a complex, multistep pathway where normal hepatocytes proliferate in the setting of chronic inflammation and alter intracellular signaling, developing into regenerative nodules with evolution to dysplastic nodules and cancer (Hui and Makuuchi 1999). The most important molecular event that occurs in cancer development is aberrant gene expression. Genetic and epigenetic changes may occur. In genetic changes, direct integration into the cellular DNA and "injury" with genetic mutations and translocations may occur, while in epigenetic changes, epigenetic modulations of genes such as DNA methylation—resulting in altered gene expression without any changes in the base sequence of DNA—can occur (Laurent-Puig and Zucman-Rossi 2006). Modifications of DNA sequences that do not directly encode mRNA and proteins can modify the generation of microRNA (miRNA); this can then result in dysregulation of primary gene signaling (Budhu et al. 2008). Cancer cells do not invent new pathways of signaling; instead, they use preexisting pathways that are dormant or less active in normal cells. The changes in genetics, epigenetics, or miRNA allow these pathways to become overactive in cancer cells, increasing cell proliferation and signaling and decreasing cell apoptosis, while escaping from the normal growth inhibition and regulation of growth from the extracellular matrix.

The pathogenesis of HCC is incredibly complex because the development of a malignant signature can occur with several different mechanisms. This process results in different subtypes of HCC that are similar on pathologic examination but different in manners of clinically behavior. Elevated expressions of TGF-α and insulin-like growth factor (IGF)-2 are responsible for accelerated hepatocyte proliferation (Coulouarn et al. 2008; Tovar et al. 2010). This increased expression reflects the expression of several genes that are upregulated by epigenetic mechanisms during the preneoplastic stage, resulting from the combined actions of cytokines produced by chronic inflammation, viral transactivation, and the regenerative response of the liver to cell loss. As hepatocytes repeatedly proliferate in response to these signals, monoclonal hepatocyte populations develop with progressive telomere shortening that predisposes to the development of dysplasia. A variety of genetic alterations and marked heterogeneity of gene expression profiles occur among HCC cases, suggesting that HCC might be one of the most complex and heterogeneous solid tumors in humans (de La Coste et al. 1998; Buendia 2000; Laurent-Puig and Zucman-Rossi 2006). Deregulation of various signaling pathways in HCC subsets may occur, such as *Wnt/β-catenin* signaling (de La Coste et al. 1998) and the p14ARF/p53 pathway (Hussain et al. 2007), TGF-β signaling, Ras/MAPK signaling, and the phosphatase and tensin homolog-protein kinase 3 and mammalian target of rapamycin (PTEN-Akt-Mtor) pathways (Laurent-Puig and Zucman-Rossi 2006; Villanueva et al. 2008). Additionally, altered expression of growth factors, such as hepatocyte growth factor (HGF), IGFs, and amphyregulin, as well as genes involved in angiogenesis may participate in the development and progression of HCC (Breuhahn et al. 2006). HBV-related HCCs display higher rates of chromosomal alterations than HCCs related to other risk factors (Marchio et al. 2000). Two major routes of hepatocarcinogenesis that differ by the extent of genetic instability have been identified, one of which being predominant in HBV-related tumors (Laurent-Puig et al. 2001). Genetically unstable tumors, poor differentiation, p53 mutations, and an unfavorable prognosis have been associated with HBV infection. HBV infection also displays activation of the mitotic cell cycle, deregulated expression of developmental and imprinted genes (such as *IGF2*), as well as activation of the AKT pathway (Boyault et al. 2007). Many studies on molecular classification of HCC based on gene expression profiles have unveiled the predominant importance of the tumor cell differentiation status for defining molecular subtypes and predicting disease outcome (Lee et al. 2006; Yamashita et al. 2008).

One meta-analysis, integrating transcriptome data from eight independent cohorts, has led to the identification of three major molecular subclasses of HCC associated with tumor differentiation and serum AFP levels (Hoshida et al. 2009). In this study, chronic HBV infection was found to be significantly associated with HCC clusters featured by moderately poor differentiation of tumor cells. Also, molecular classification of HBV-related HCCs has been used to predict early recurrence. The group with a high risk of recurrence showed an increased expression of genes involved in cell cycle progression, cell proliferation, migration and motility, and the notch signaling pathway (Woo et al. 2008). Future studies are needed to gain a comprehensive view of the signaling networks operating in liver cell transformation due to CHB.

## 5.8 GENETIC ALTERATIONS

Nearly all liver cancer cells contain many genes and genetic loci that have structural alterations, but specific genetic or allelic alterations have been found to rarely affect more than half of liver cancers studied. The amount of variation among these genetic alterations has prompted the discovery of genetic and molecular subclassifications of HCC. Many different genetic alterations have been discovered in HCC, including gains in chromosomes 1q, 6p, 8q, 17q, and 20q and losses in 4q, 8p, 13q, 16q, and 17p (Thorgeirsson and Grisham 2002). HBV-induced HCC is closely linked to p53 alterations. In tumors with a p53 mutation, loss of heterozygosity (LOH) of 17p was found in 95% of liver cancers (Oda et al. 1992). These tumors are more often poorly differentiated and at a more advanced stage of disease (Oda et al. 1992). Many tumor suppressor genes on several different chromosomes have been found to have LOH, including 17p (*p53*), 13q *(retinoblastoma 1, RB1)*, 16p (*axis inhibition protein 1, AXIN1*), 9p (*cyclin-dependent kinase inhibitor 2A, CDKN2A*), and 6q (*insulin-like growth factor 2 receptor, IGFR2R*) (Laurent-Puig and Zucman-Rossi 2006). These changes can result in further inactivation of tumor suppressor genes and signal the development of malignant cells.

Another tumor suppressor gene, *retinoblastoma gene (Rb)*, which is located on chromosome 13q14, also appears to be involved in HCC development, with 43% of tumors showing alteration in this gene (Zhang et al. 1994). Inactivation of Rb contributes to tumor progression and metastasis in many patients rather than tumor initiation and, therefore, tends to occur at more advanced stages of disease (Weinberg 1995). LOH on chromosome 16, corresponding to tumor suppressor gene *AXIN1,* occurs in more than 50% of the cases of HCC and occurs more frequently in large, poorly differentiated, and metastatic liver cancers but not early liver cancers (Laurent-Puig and Zucman-Rossi 2006). This LOH also plays a role in tumor aggressiveness rather than tumor initiation. DNA methylation is a key process in the epigenetic control of gene expression. Hypermethylation can lead to the inactivation of tumor suppressor genes and invasion suppressor genes and predispose for allelic loss (Herman et al. 1995; Yoshiura et al. 1995). Hypomethylation of DNA increases the expression of oncogenes (Cheah et al. 1984). The level of DNA methylation is controlled by DNA methyltransferase, which transfers a methyl group from S-adenosylmethionine to the 5-position of cytidine.

Increased expression of DNA methyltransferase is another step involved in human carcinogenesis (El-Deiry et al. 1991). DNA methyltransferase mRNA levels are significantly higher in liver tissue showing chronic hepatitis or cirrhosis than in normal liver tissue and are even higher in HCC cells (Kanai et al. 1999), suggesting that increased DNA methyltransferase expression is also a step responsible for hepatocarcinogenesis, especially in the early stages of disease. Changes in DNA methylation are reversible, and correcting this aberrant process seems to be more feasible than correcting genetic alterations, such as gene mutations. Reducing DNA methyltransferase activity has been shown in experimental models to inhibit tumor development; thus, it may be a potential target for the prevention and subsequent treatment of HCC (Ramchandani et al. 1997). Microsatellite instability also occurs in hepatocytes as a preneoplastic lesion. In the early preneoplastic stage, structural alterations in DNA occur slowly, but as dysplasia and neoplasia develop, the rate of development of structural alterations and amount of alterations increases sharply (Laurent-Puig et al. 2001). Individual liver cancers often contain multiple allelic deletions and chromosomal losses and gains concurrently, which emphasize the complexity of HCC development. Through the progression of HCC from small, well-differentiated tumors to large, poorly differentiated, and metastatic tumors, further chromosomal alterations occur, resulting in a tumor with areas of diverse mixtures of genomic aberrations (Salvucci et al. 1999).

## 5.9 GENETIC CLASSIFICATION OF HCC

Many genetic alterations accumulate during HCC development and may involve more than two dozen different genes, with at least three different signaling pathways affected.

1. The *Wnt* pathway, a term derived from the genetic mutation in *Drosophila melanogaster* (fruit fly), which produces a wingless phenotype. The wingless gene is a homolog to the *int-1* gene of vertebrates, which was originally identified near insertion sites for mammillary tumor virus in mice where virus integration causes amplification of *Wnt* genes. *Wnt* is a catenation of wingless and int.
2. The ERK pathway.
3. The mammalian target of rapamycin (mTOR) pathway.

Genetic classifications may be useful in clinical management and treatment of HCC. One study found that LOH was the most frequent genetic alteration, and half the tumors exhibited LOH on more than five chromosome arms (Pathway I). In the other half of HCCs, tumors were chromosome stable, and the β-catenin mutation was the one most frequently observed (Pathway II). The tumors in Pathway I were predominantly from HBV-infected patients and poorly differentiated, having a poor prognosis. Pathway II tumors were generally non-HBV infected, large, and well differentiated (Laurent-Puig et al. 2001).

Katoh et al. analyzed chromosomal alteration profiles, named cluster A and cluster B, that were associated with viral infection status, serum AFP levels, and presence of intrahepatic metastasis as well as patient survival. All liver cancers had chromosomal alterations to varying degrees, but more frequent and pronounced alterations were observed in cluster A, which had a worse prognosis and poorer survival than cluster B (Katoh et al. 2007). In addition, subclusters were noted that may respond well to different treatments, such as a subcluster A1 that was characterized by amplification on 1q and/or 6p, and that may respond to molecular inhibitors available for VEGF. A subcluster A3 showed a higher rate of 17q gain/amplification, and results of in vitro experiments showed that these cell lines respond well to rapamycin, an mTOR inhibitor that directly inhibits the activity of S6 kinases (Sabatini 2006). These data suggested that molecular stratification of individual liver cancers into genetically homogeneous subclasses could provide an opportunity for developing optimal tailor-made therapeutic agents.

Yet another genetic classification system found six main subgroups based on genetic alterations in the tumors. The primary clinical determinant of class membership in this scheme was HBV infection, with other main determinants being genetic and epigenetic alterations, including chromosomal instability, β-*catenin and p53* mutations, and parental imprinting (Boyault et al. 2007). Many other studies have attempted to develop HCC classification systems based on genetic alteration of the tumor cells (Cillo et al. 2004; Lee et al. 2004, 2006; Katoh et al. 2006; Thorgeirsson et al. 2006). The classification systems can identify tumors that are at high risk for being poorly differentiated histologically and for having vascular invasion and poor patient survival. Further studies are needed before these classification systems may be used to determine the targeted therapies that would be best suited to particular patients in each unique subclass of HCC.

## 5.10 NONTUMOR TISSUE AND PROGRESSION OF HCC

In addition to the genetic alterations in tumor tissue, when HCC is resected, the genetic alterations in the tissue adjacent to the tumor also seem to affect survival. In an analysis of 186 gene signatures, 113 were found to be associated with a good prognosis and 73 with a poor prognosis. Early recurrence of tumor is associated with clinical and histopathological factors of resected tumor, but late recurrence (more than two years after resection and likely to be new primary tumors rather than actual recurrences) is associated with the gene expression signature of the nontumor liver tissue. The poor prognosis signature contained gene sets that were associated with inflammation, such as interferon (INF) signaling, activation of nuclear factor kappa B (NF-kB), and signaling by TNF-α (Hoshida et al. 2008). The impact of inflammation on the risk of metastasis remains unclear. Another study investigating the impact of inflammatory changes on the risk of metastasis in HCC found that patients with metastatic disease had a significantly different microenvironment in nontumor tissue compared to patients without metastatic disease. Patients without metastasis seemed to have an increase in inflammatory cytokines compared to patients with metastatic disease. In fact, patients with metastatic disease had a global decrease in proinflammatory cytokines (Budhu et al. 2006). The use of nontumoral gene expression profiling may be used to identify patients at the highest risk of recurrence in order to target intensive follow-up or use of adjuvant therapies. However, further study is needed before clinical tests that can be applied to patient care can be embraced.

## 5.11 miRNAs

miRNAs are small, noncoding RNA gene products that play key regulatory roles in mRNA translation and degradation (Lagos-Quintana et al. 2001). In addition, each miRNA has the capability to regulate the expression of hundreds of coding genes and modulate multiple cellular pathways including those for proliferation, apoptosis, and stress response; miRNAs are frequently produced from "fragile" DNA sites, common break points, or regions of amplification or LOH. Research has clearly shown that miRNAs play a significant role in human hepatocarcinogenesis (Gregory and Shiekhattar 2005). Mature miRNAs are relatively stable; this characteristic may make miRNAs excellent molecular markers that may be used as a tool for further cancer diagnosis and prognosis.

One study analyzed a 20-miRNA signature that could significantly predict metastatic (M) from nonmetastatic (NM) status with an accuracy of 76%. In this cohort of patients with early-stage HCC, the patients with M status had a 500-day survival rate of 38% compared to 73% 500-day survival in patients with NM status. This outcome was related to the presence of metastatic disease and higher recurrence rate in patients with miRNA profiles consistent with M status (Ji et al. 2009). Other important miRNAs are miRNA-26, miR-122, and miR-221, which are associated with a poor prognosis as well (Gramantieri et al. 2007; Coulouarn et al. 2009). Many other miRNAs are being studied to elucidate their roles in liver cancer. The ability to understand the alterations in miRNA expression in liver cancer may be used in the future to accurately predict prognosis and potentially offer adjuvant therapy.

## 5.12 FIBROSIS, CIRRHOSIS, AND HCC

There is convincing evidence that the incidence rate of HCC is about fivefold higher among infected patients with cirrhosis than in HBV asymptomatic carriers, suggesting that cirrhosis is a preneoplastic condition per se (Kirosawa 2002). The strong association between cirrhosis and HCC suggests a hepatocarcinogenic process that is largely mediated by inflammation, leading to repeated cycles of cell death and regeneration that increase hepatocyte proliferation turnover (Schirmacher et al. 1993).

The sustained stimulation of liver cell to progress toward the cell cycle can ultimately overcome DNA repair mechanisms in the presence of mutational events. The accumulation of critical variants in the host genome may heavily contribute to transformation of hepatocytes into malignant clones, and the cells designed for the elimination through the apoptosis program or immune response will become fully transformed (Chisari 2000). Concurrently, liver fibrosis disrupts the architecture of hepatic structure. As a consequence, cell-to-cell interactions are modified, and this ultimately leads to loss of control over cell growth. Thus, the persistent inflammatory changes, caused by chronic infection, promote liver cancer development through an integrated multistep process.

## 5.13 OCCULT HBV INFECTION AND HCC

A peculiar aspect of chronic HBV infection is represented by the persistence of HBV genomes in the absence of circulating HBs antigen. This so-called "occult" infection can occur not only in individuals with anti-HBs and/or anti-HBc antibodies but also in those who are negative to HBV markers (Brechot et al. 2001). Extensive studies have demonstrated that occult HBV infection represents an entity with clinical relevance as risk of transmission through organ transplantation, blood transfusion, perinatal transmission, acute exacerbation, or even fulminant hepatitis after immunosuppression or chemotherapy (Mulrooney-Cousins and Michalak 2007). Occult HBV infection has a worldwide diffusion, and its distribution is related to the prevalence of HBV infection in different geographical areas. Its prevalence is also high in certain patient populations, such as those who are chronically infected by HCV and those affected by cryptogenic liver disease and HCC (Chemin and Trepo 2005; Pollicino et al. 2007).

The awareness of occult HBV infection emerged following the development of molecular sensitive technology that was able to detect very low levels (<103 viral genomes per milliliter) of HBV DNA in the serum samples and/or in the liver. However, the detection of HBV genomes in the liver tissue remains the most truthful way to identify the occult infection, and this may strongly limit the impact of viral persistence on real incidence, viral viability, and pathogenic consequences (Koike et al. 1998). Despite a low replication rate, silent HBV detection can be associated with increased cytonecrotic activity and advanced liver diseases. The mechanisms leading to occult HBV infection remain poorly understood. Occult infection results from a multifactorial process, likely involving host and viral factors. For instance, the host immune response may play a role in maintaining low levels of intrahepatic HBV replication and transcription (Mulrooney-Cousins and Michalak 2007). Among the viral factors, mutations that may alter HBsAg production and negatively influence viral DNA multiplication, coinfection with HCV and presence of circulating deleted viral particles might affect the replication rate of HBV. The HBV DNA variants could act by modifying the antigenicity of viral proteins; for example, rearrangements in the *pre-S1* and *S genes* have been associated with reduced HBsAg expression, changes in the *X gene* might affect viral replication fitness, mutations in the overlapping CP region may influence the low replicative potential, and variants in precore/core sequences may reduce HBV replication efficiency through the epsilon signal functional structure that is essential for pregenomic encapsidation and starting HBV DNA synthesis (Poussin et al. 1999; Weinberger et al. 2000; Muroyama et al. 2006).

It is widely accepted that occult HBV persistence is an important risk factor for liver cell clonal expansion and HCC development. This association was suggested by epidemiological and molecular studies and supported by animal models. The HBV genome has been detected in tumor tissue of HBsAg-negative patients with HCC in a prevalence ranging from 30% to 80% (Paterlini et al. 1993; Pollicino et al. 2004). Studies on the rodent models showed that the HBV infection had an increased risk of developing HCC, despite the apparent clearance of the virus by serological tests (Michalak et al. 1999). Moreover, a high proportion of HCV-related HCC cases showed occult HBV infection suggesting that an interplay underlying HCV/HBV occult coinfection might contribute to HCC development because it occurs in clear HCV/HBV coinfection (Paterlini et al. 1990). In vitro experiments revealed that HCV core proteins suppress HBV expression in cell cultures, thus potentially favoring the occult status (Shih et al. 1993). Occult HBV strain populations harbor a genetic heterogeneity in viral regions (pre-S/S, precore/core, X, polymerase) and regulatory elements (CP, Enhancer I and II) potentially involved in viral replication and/or gene expression. However, point variations or deletions, a particular genotype or a pattern of changes able to predict oncogenic transformation, remain to be identified.

At first, the mechanism by which occult HBV infection may promote HCC development would seem to depend on the viral DNA integration into the hepatocyte genome. However, the intrahepatic persistence of HBV cccDNA replicative intermediate suggests that the occult status has to be referred to a low replication rate rather than integration capacity (Squadrito et al. 2002). On the other hand, in a recent study carried out in Taiwan, multiple genetic variants in the pre-S2 (M1I and Q2K) and Enhancer II (G1721A) domains of viral genome have been found in HCC carriers of HBV occult infection compared to HCC with clear HBV infection. This pattern of mutations, distinctive of the occult status, was proposed as a viral marker for HCC in occult HBV carriers that may aid in the identification of the cases with HBsAg, progressive chronic hepatitis, and high risk of HCC development (Chen et al. 2009a).

## 5.14 PREVENTION OF HCC

### 5.14.1 Prevention before Development of HBV Infection by Universal Vaccination

In developing countries, HBV is primarily acquired at an early age, most often through vertical transmission from mother to child and horizontal transmission from child to child. However, sexual transmission accounts for a majority of the transmission in adult life. In 1991, the Global Advisory Group of the Expanded Program on Immunization (EPI) of World Health Organization (WHO) recommended that hepatitis B vaccine be integrated into national immunization programs in all countries with a hepatitis B carrier prevalence of 8% or greater; the WHO, Western Pacific Region, has been committed to reducing chronic HBV infection in children under five years of age to less than 2% by 2012 (Clements et al. 2006). As a consequence of this, hepatitis B vaccination is currently a part of the national infant immunization schedule in 162 countries (Kao and Chen 2002). Several cohort studies in HBV endemic countries and regions have shown a noticeable reduction in HBsAg prevalence in school children, national servicemen, and antenatal women after introduction of universal hepatitis B vaccination (Goh 1997; Chang et al. 2000; Tsebe et al. 2001; Ng et al. 2005). The impact on adult HBsAg carriers has been reduced in countries such as Korea, China, Thailand, Singapore, and Taiwan, which adopted infant HBsAg immunization very early. Taiwan was the first to demonstrate a convincing decrease in the annual incidence of childhood HCC from 0.52 to 0.13 per 100,000 children (Chang et al. 1997; Lin et al. 2003).

## 5.14.2 Prevention of HCC in Patients Infected with Hepatitis B

For preventing HCC in patients who are infected with hepatitis B, the aim should first be to target the high-risk groups that include patients with cirrhosis and those who had a locoregional or surgical resection of HBV-related tumor. Second, it is also important to cause adequate suppression of serum HBV DNA below the limits of detection and/or *HBeAg* seroconversion by use of appropriate antiviral therapy.

### 5.14.2.1 Nucleoside/Nucleotide Analogs

With nucleos(t)ide analogs (NAs), a consistent suppression of HBV DNA, histological reversal of cirrhosis, has been demonstrated after five years of treatment (Chang et al. 2010). However, whether treatment with NAs also causes a reduction in the incidence of HCC remains to be seen. A prospective randomized, placebo-controlled trial of lamivudine (Cirrhosis and Lamivudine Monotherapy [CALM] study) was made by Liaw et al. (2004) in patients with chronic HBV infection and advanced fibrosis or cirrhosis. The trial was terminated early after a median of 32 months' follow-up when the primary end point was achieved with $P < 0.001$ because the cumulative incidence of HCC was 3.9% in the lamivudine-treated group as compared to 7.4% in the placebo-treated controls. The role of HBV DNA was confirmed when patients who developed viral resistance during the study had the advantages of treatment blunted. In other case–control study, 142 HBeAg-positive patients were compared to 124 controls matched for age, HBeAg status, and absence of cirrhosis. Cases were treated with lamivudine. Controls were seen to develop HCC and/or cirrhosis at a significantly higher rate ($P = 0.03$). Similar results were seen in case–control studies from Japan and Korea. A meta-analysis addressing the impact of nucleoside analogs (mainly lamivudine) showed a 78% (RR 0.22; 95% CI 0.10–0.50) reduction in risk of HCC after treatment with nucleoside analogs compared to controls (Matsumoto et al. 2005; Yuen et al. 2007; Eun et al. 2010).

### 5.14.2.2 Interferon

In order to examine long-term outcomes of IFN-treated HBeAg-positive hepatitis B patients, Lin et al. (2007) performed a large case–control study in Taiwan with 233 patients and controls matched for age, sex, baseline ALT, and HBV DNA level with CHB, which exhibited a significant benefit of interferon in terms of HCC development (2.7% vs. 13%, $P = 0.011$). A multivariate analysis showed that IFN therapy has better long-term outcomes against HBeAg seroconversion and genotype B HBV infection as independent risk factors. Another small cohort study from Japan with a follow-up of 15 years of IFN-treated HBV-related cirrhosis patients showed HCC development in 34% of patients without HBV DNA clearance compared to 8.0% with HBV DNA clearance ($P = 0.026$). The only randomized control study included patients treated with IFN-α with or without prednisolone priming or placebo (Lin et al. 2004). In a study on patients from two randomized control trials treated by lymphoblastoid IFN for 12 weeks as part of prednisolone priming, initially at an escalating dose then maintained at 6 MIU/m2 in the first study, while IFN-α 2a was given as either 9 MIU three times weekly for 16 weeks or an identical regimen followed by IFN-α 2a, 3 MIU three times weekly for an additional 12 weeks in the second study, a reduction in the cumulative incidence of HCC was found to be significantly different: 20% in controls compared to <7% in IFN-α treated groups after 72 months of posttreatment follow-up. However, the study suffered from some flaws, such as a significant difference in the follow-up in both groups and the exact statistical significance of reduction in the incidence of HCC was not clear (Cammà et al. 2001). In a recent meta-analysis (Sung et al. 2008), 12 studies were included that compared IFN treatment, in which there were a single randomized controlled trial, a case–control study, and the remaining ten were cohort studies. IFN treatment was seen to reduce the risk of HCC by 34% (RR 0.66, 95% CI 0.48–0.89), more so in patients with cirrhosis. Thus, IFN has been shown to reduce the risk of HCC even though the quality of the studies assessing this has been quite variable, with differences in types of patients included and response rates and wide variation in the interferon regimen administered. In most patients, the response was secondary to viral suppression and more often seen in patients with cirrhosis.

#### 5.14.2.3 Surgical Resection or Curative Locoregional Therapy

This constitutes one of the most important high-risk groups for development of HCC. In a cohort study of 157 patients (89 nonviremic and 68 viremic) who underwent resection for HCC, there was a statistically significant difference in the five-year cumulative recurrence rate of the viremic group (73%) as compared to the nonviremic group (55%) ($P = 0.043$). Similar results were seen in another cohort study where patients with HBV DNA of more than 2000 IU/mL had a significantly higher risk of HCC recurrence after resection, with an OR of 22.3 (95% CI 3.3–151, $P = 0.001$) as compared with patients who were treated with lamivudine ($n = 10$), none of which developed recurrence (Kuzuya et al. 2007; Hung et al. 2008; Kim et al. 2008).

CHB is a major risk factor for HCC in Asia, and the most important strategy is prevention of the infection by adoption of a universal hepatitis B vaccination program. Weak evidence supports benefit with interferon therapy and first-generation oral antiviral (lamivudine), both of which act via suppression of HBV DNA. Long-term outcome data with newer antivirals, which have a higher genetic barrier to drug resistance and are more potent, are needed.

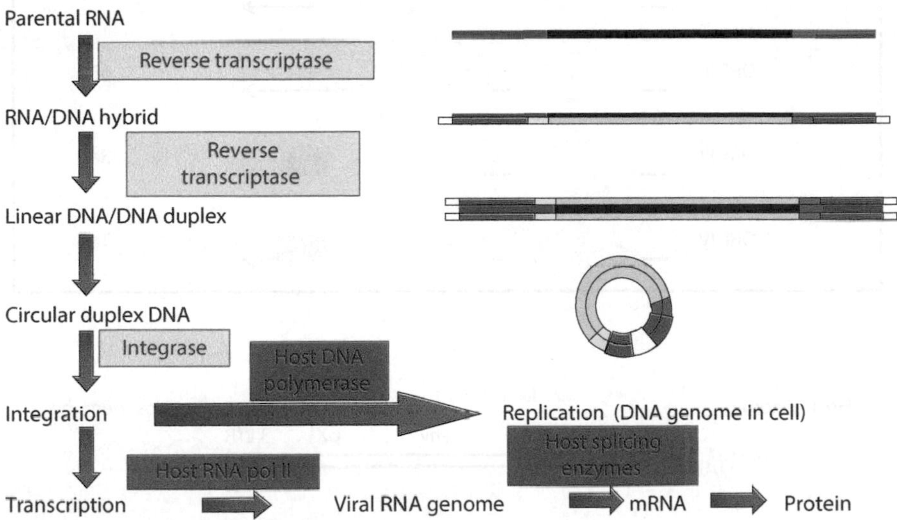


**FIGURE 1.1** The pathway of RNA virus infection.

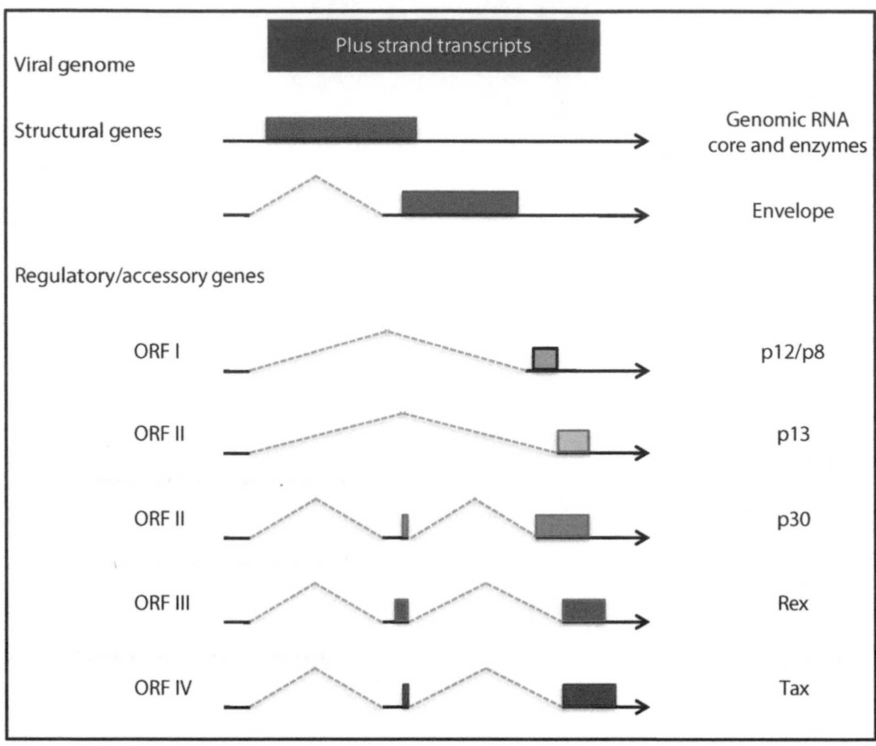


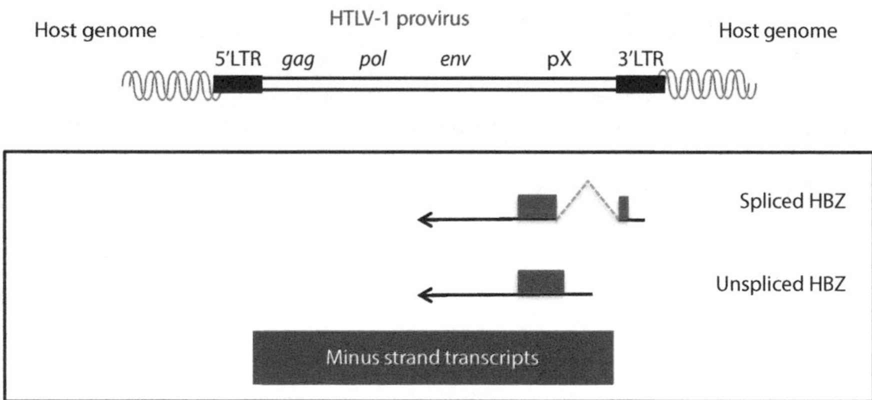


**FIGURE 2.1** HTLV-1-encoded genes. The HTLV-1 proviral genome consists of *gag*, *pol,* and *env* genes flanked by long terminal repeat (LTR) sequences at both ends. The pX region encodes several regulatory/accessory proteins in four open reading frames (ORFs) and the minus strand.

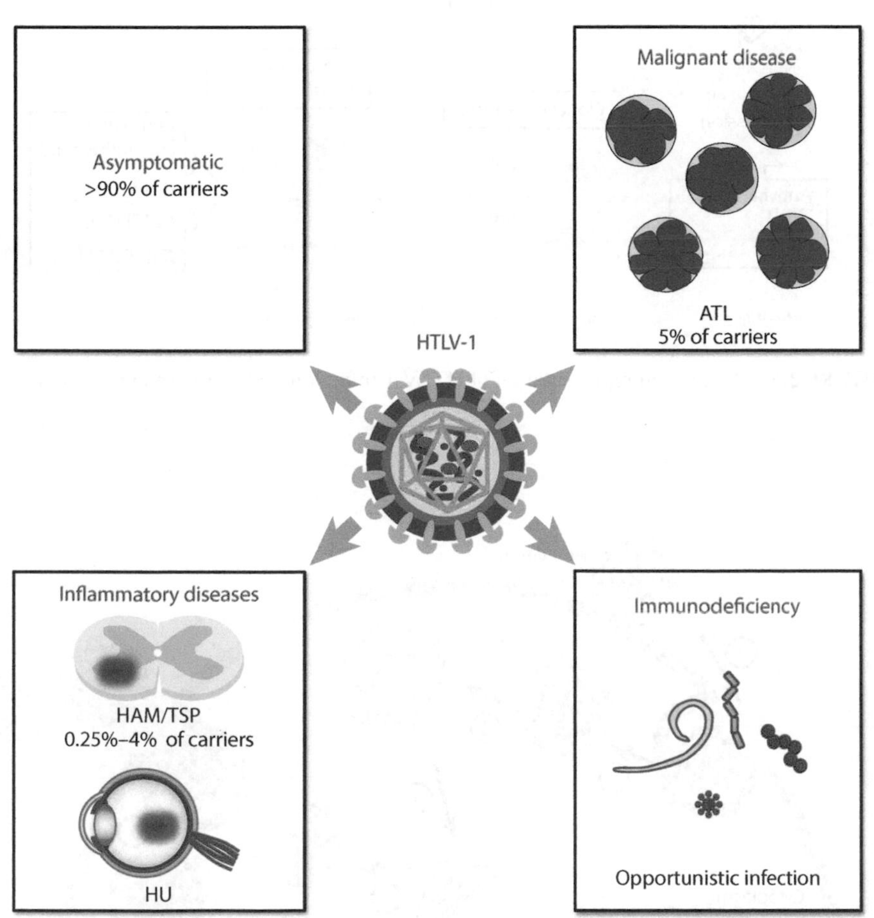


**FIGURE 2.2** A pictorial representation of HTLV-1-associated diseases.

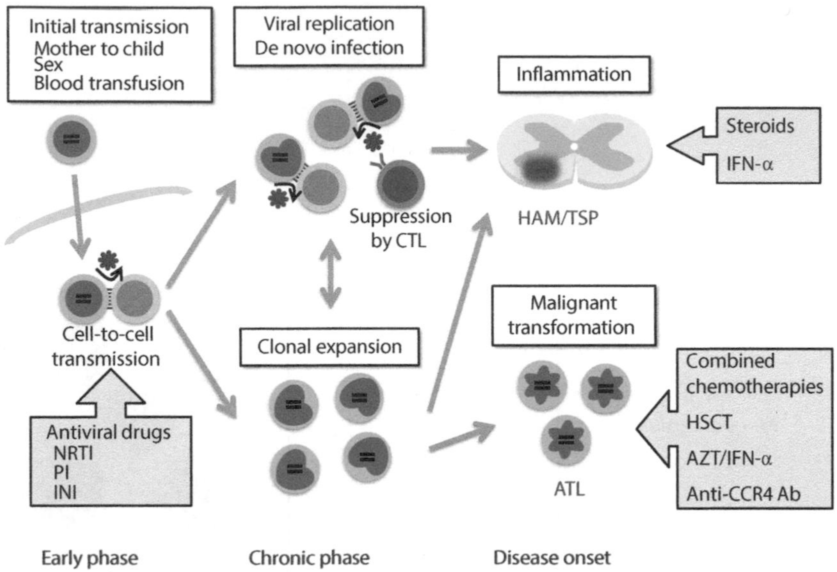


**FIGURE 2.3** A pictorial representation of HTLV-1 infection and its treatment strategies.

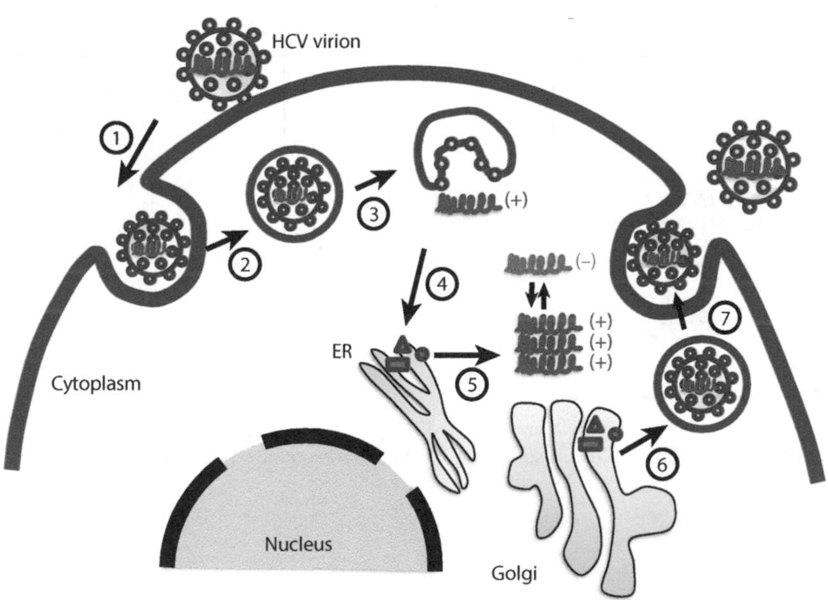


**FIGURE 3.2** Replication cycle of hepatitis C virus: (1) attachment, (2) endocytosis, (3) uncoating, (4) translation, (5) RNA replication, (6) virion assembly and maturation, and (7) exocytosis.

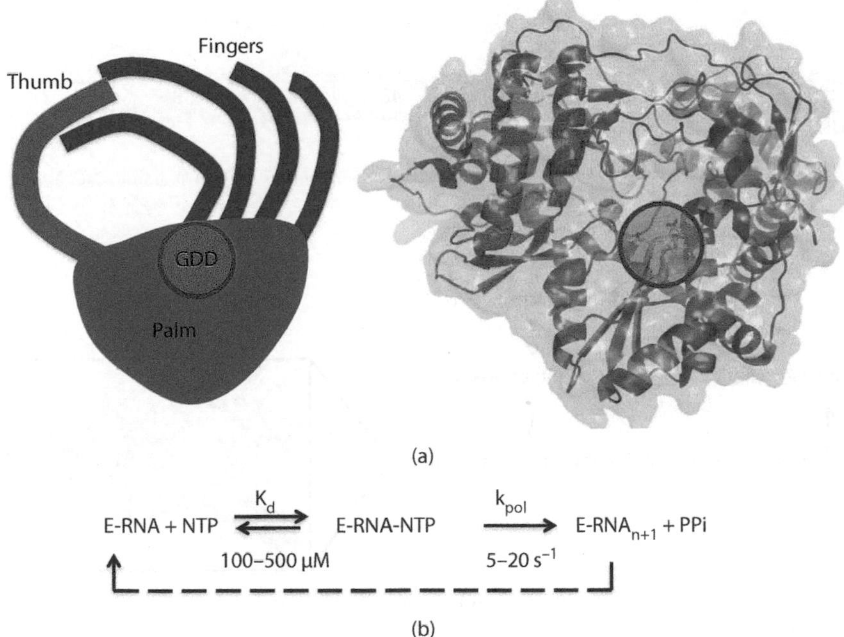


**FIGURE 3.4** Three-dimensional structure of NS5B. (a) The polymerase of HCV adopts the shape of a right hand in a closed-state, where fingers (blue) and thumb (green) subdomains form extensive interaction. The catalytic site containing the GDD motif (yellow circle) is at the base of the palm subdomain (magenta); (b) simplified scheme of the nucleotide incorporation reaction. Under the elongation mode, NS5B incorporates nucleotides at a rate of 5–20 bases per second. The enzyme is processive, which means it does not dissociate from the RNA between each cycle of the catalytic reaction.

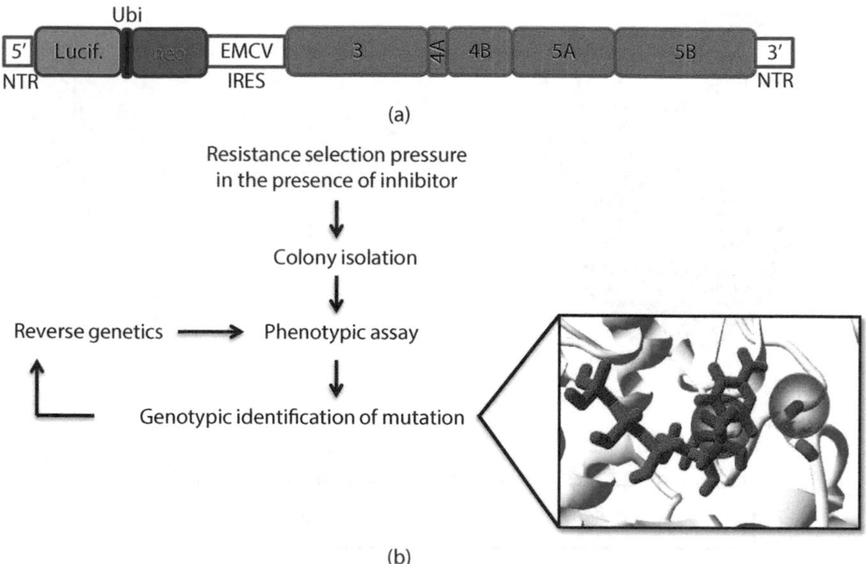


**FIGURE 3.6** Use of subgenomic replicon in cell-based assays. (a) Example of a subgenomic HCV replicon construct. In this case, the bis-cistronic replicon is first composed of the 5′-NTR, followed by the reporter gene (i.e., luciferase), and the selectable marker gene neomycin joined in the first cistron by a cleavable ubiquitin sequence. The internal ribosome entry site (IRES) enables the translation of the HCV nonstructural genes NS3 to NS5B and the construct ends with a 3′-NTR. (b) Principle of the characterization of resistance mutations.

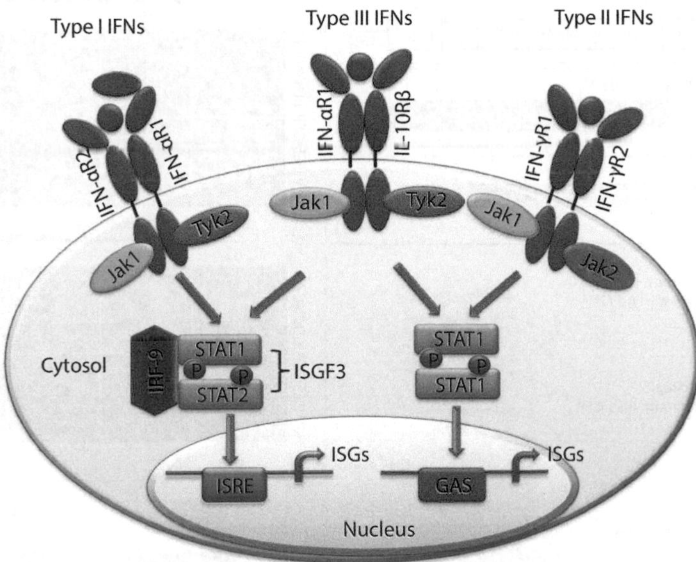


**FIGURE 4.5** Comparison of JAK-STAT signaling activation by different IFNs. Type I (IFN-α), type II (IFN-γ), and type III (IFN-λ) bind to separate cell surface receptors that activate two tyrosine kinases, Jak1 and Tyk2, which then phosphorylate the STAT. Phosphorylated STAT1 and STAT2 combine with IRF-9 to form ISGF3 complex, which activates the type I IFN genes. Type II IFN activates Jak1 and Jak2 tyrosine kinases and only phosphorylate STAT1 protein. The phosphorylated STAT1 protein forms homodimer and activate type II IFN genes. Type III IFN activates each pathway.

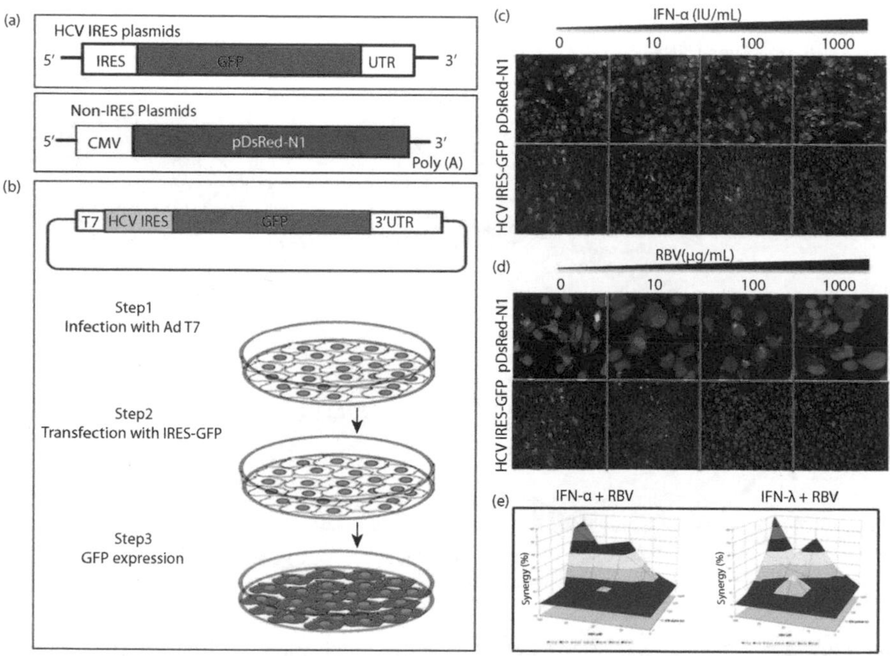


**FIGURE 4.6** IFN-α and RBV synergistically inhibit HCV IRES–mediated translation. (a) The structure of IRES and non-IRES plasmid clones used in our study. (b) The steps used to express HCV IRES-GFP or HCV IRES-luciferase using a recombinant adenovirus expressing T7 RNA polymerase. (c,d) Effect of IFN-α and RBV treatment on IRES (green fluorescence) and by non-IRES mechanism (red fluorescence) of translation. (e) Three-dimensional synergy inhibition plot for IFN and RBV combination treatment against HCV IRES.

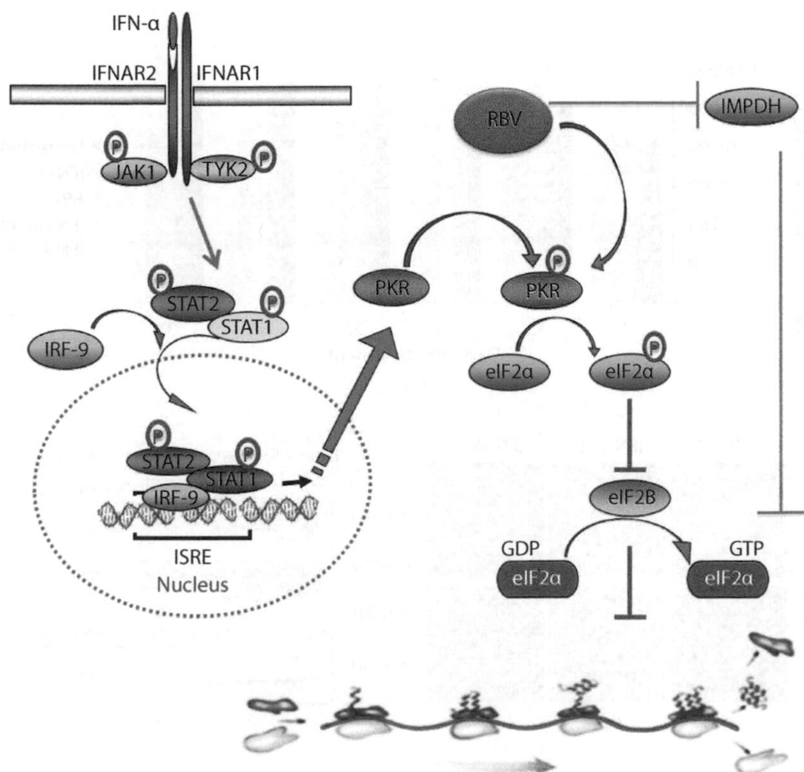


**FIGURE 4.8** Diagram summarizes the proposed IFN-α and RBV synergy antiviral mechanisms. IFN-α activates the JAK-STAT pathway leading to the activation of PKR. The activated PKR phosphorylates the eIF2α. Phosphorylation of eIF2α inhibits the recycling of initiation factors and inhibits HCV IRES translation. RBV activates the PKR and eIF2α phosphorylation that inhibits HCV IRES translation. RBV also inhibits HCV IRES by inhibiting IMPDH and GTP pool.

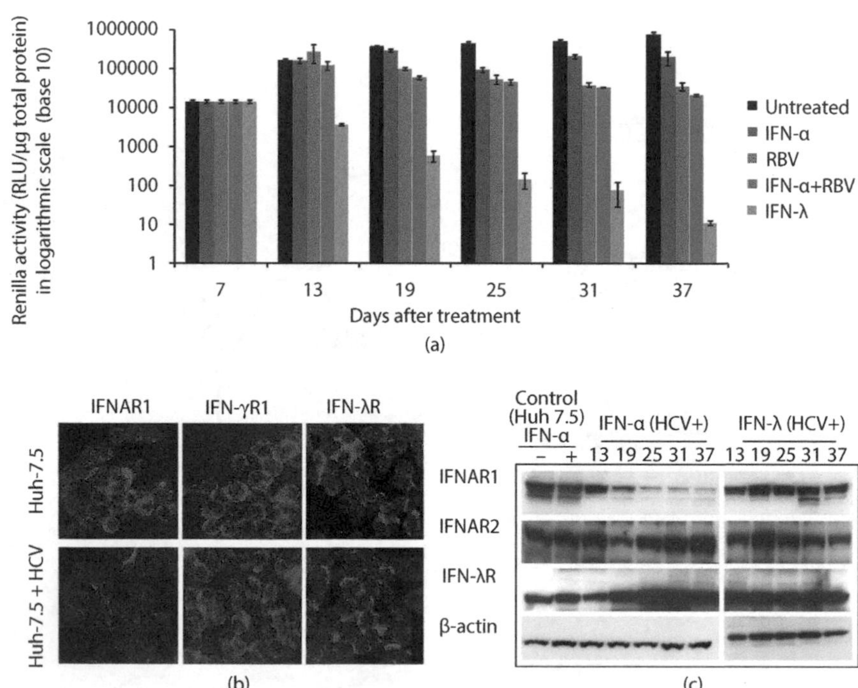


**FIGURE 4.9** Persistent HCV replication with high viral titer is partly resistant to IFN-α/RBV but not to IFN-λ treatment. (a) The antiviral action of IFN-α, RBV, and IFN-λ in persistently infected HCV culture measured by *Renilla* luciferase activity. (b) Immunofluorescence image showing the cell surface expression of IFN-α, IFNAR2, and IFN-λ receptor. (c) Western blot analysis showing selective downregulation of IFNAR1 in the persistently infected Huh-7.5 cells, whereas expression of IFN-γ or IFN-λ receptors did not change.

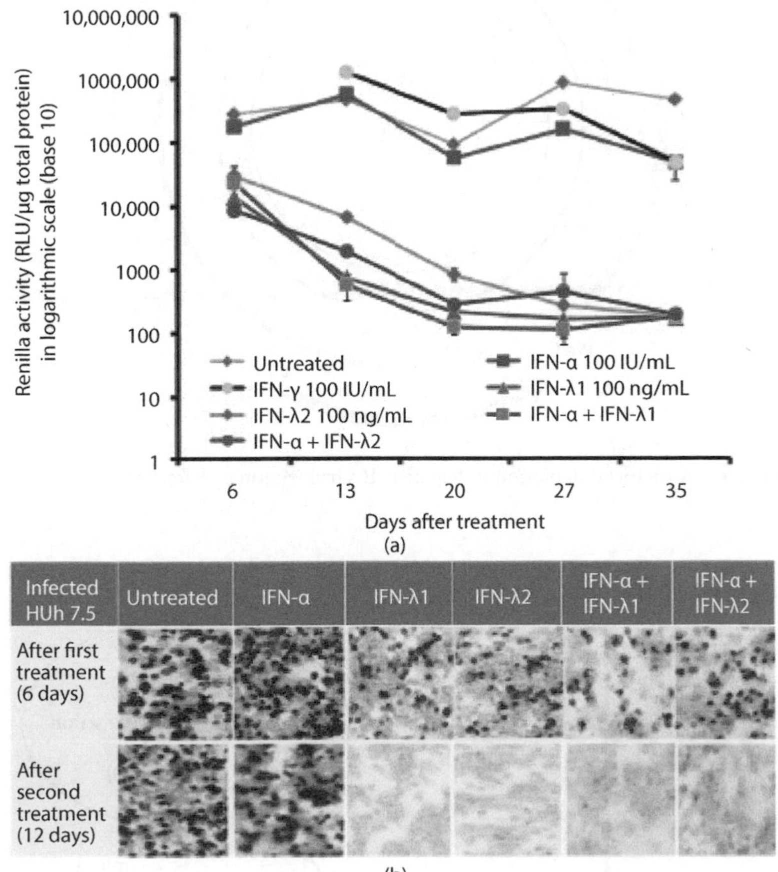


**FIGURE 4.11** IFN-λ1- or IFN-λ2-induced HCV clearance in IFN-α-resistant free fatty acid HCV cell cultures. (a) The luciferase activity indicating HCV replication in the presence and absence of different IFNs. IFN-λ1 (IL-29) and IFN-λ2 clears HCV replication. IFN-α and IFN-γ are not effective. (b) The core protein expression by immunostaining. IFN-λ-treated culture shows negative expression of HCV core protein, which was not seen in IFN-α and IFN-γ treatment.

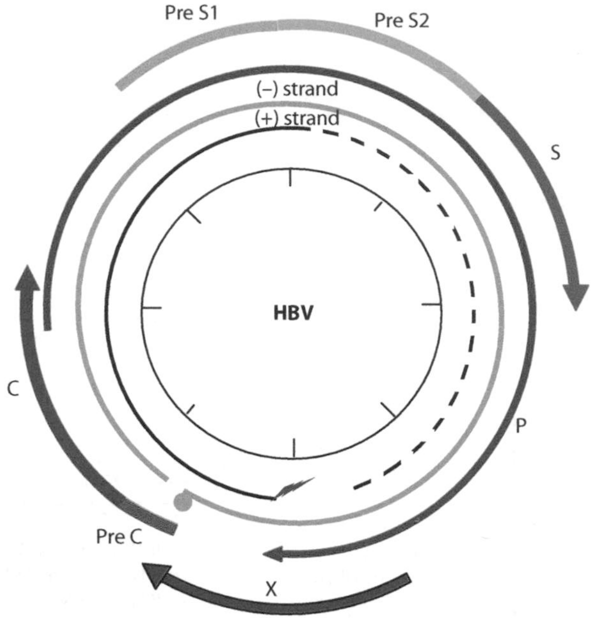


**FIGURE 5.1** A pictorial depiction of hepatitis B virus genome structure.

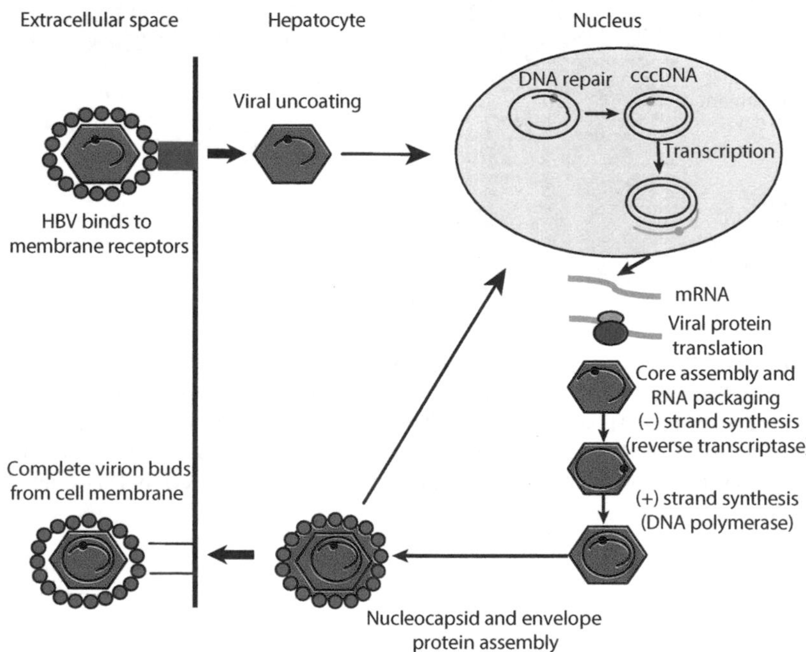


**FIGURE 5.2** A schematic representation of hepatitis B virus life cycle.

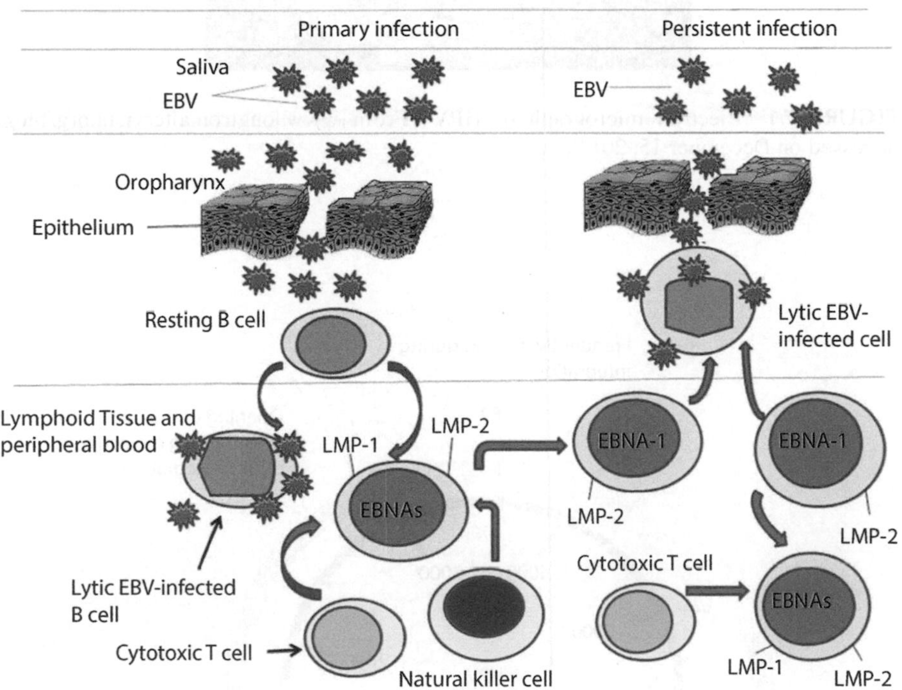


**FIGURE 6.4** Schematic representation of primary and persistent infection of EBV.

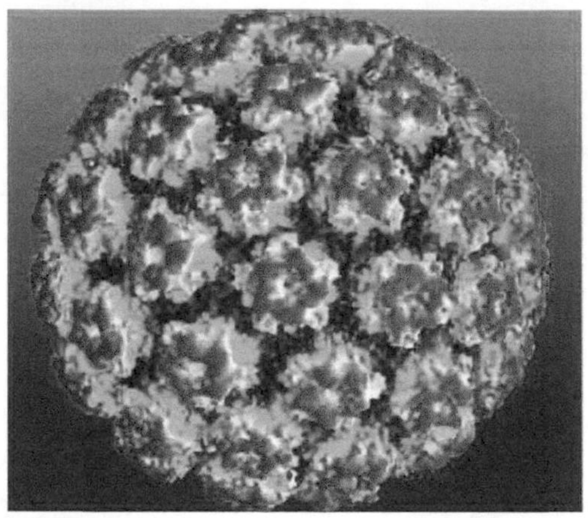

**FIGURE 7.1** Electron micrograph of HPV (From www.longtron.altervista.org/blog, accessed on December 15, 2013).

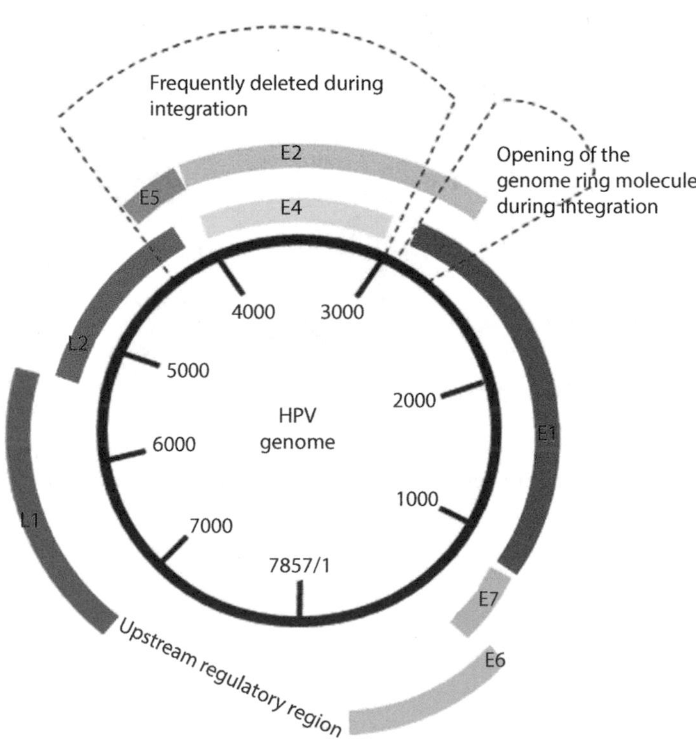


**FIGURE 7.2** The organization of the HPV genome with the early genes (E1–E7), the capsid genes (L1 and L2), and the upstream regulatory region, with its integration into the host genome.

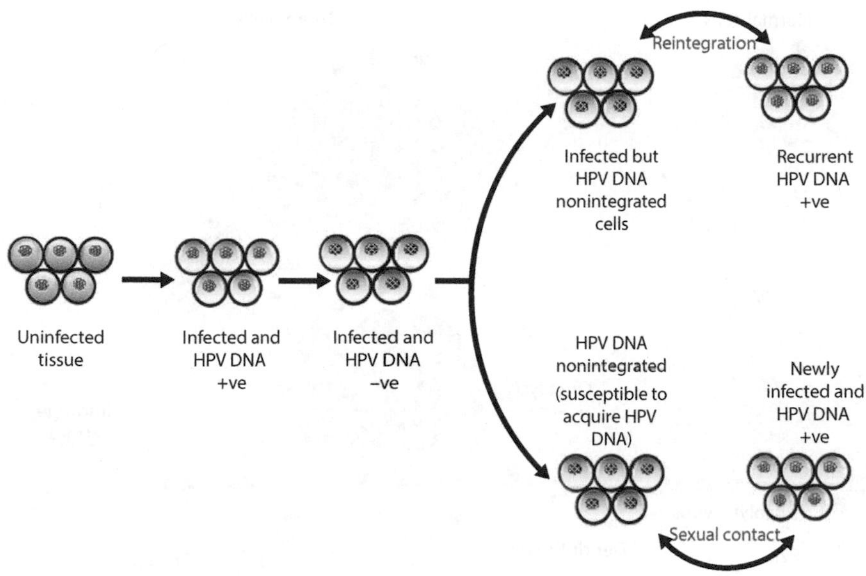


**FIGURE 7.5** Integration of HPV DNA in infected individuals.

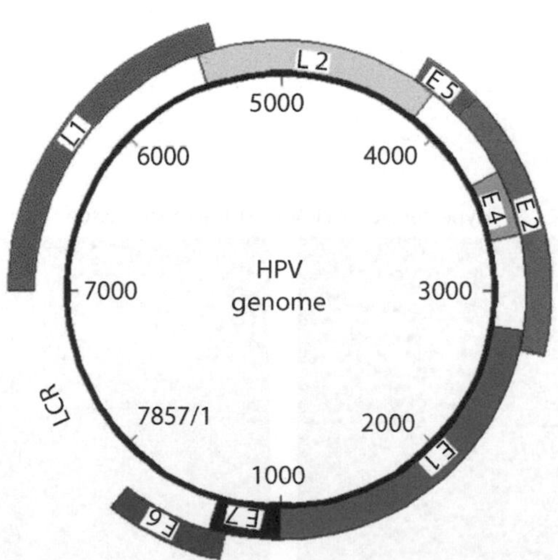


**FIGURE 8.1** Schema of HPV genome. The genome can be divided into three segments: ~4000 bp section (E) that encodes the proteins primarily involved in viral DNA replication and oncogenic task in the infected cell, ~3000 bp section (L) that encodes the structural proteins of the virions, and an ~1000 bp noncoding section (LCR) that contains the origin of viral DNA replication, and transcriptional regulatory elements.

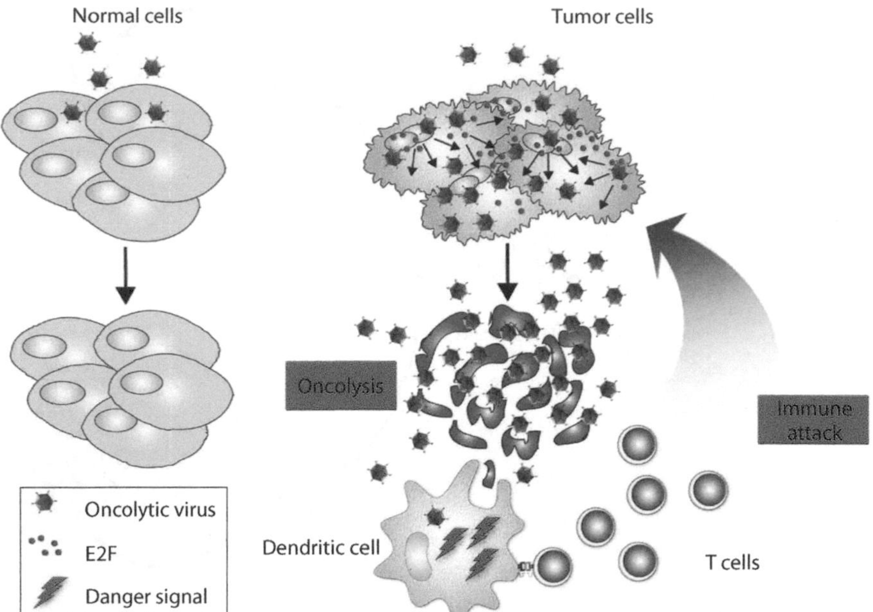


**FIGURE 13.1** Schematic representation of oncolytic virus mechanisms. Oncolytic adenovirus can only replicate in tumor cells due to abundance of eukaryotic initiation factor 2F. In normal cells, tumor suppressor Rb binds and inactivates E2F. Recruited to the site of inflammation, antigen-presenting cells (dendritic cells) phagocytize tumor material, and virus-triggered danger signals also drive adaptive T-cell immune responses against tumor antigens.

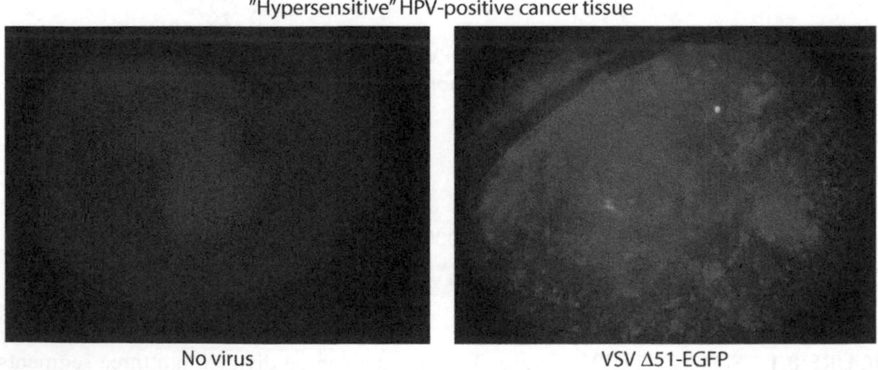


**FIGURE 13.2** Targeting of HPV by oncolytic vesicular stomatitis virus. Fresh surgical cervical cancer tissue was infected ex vivo with IFN-sensitive oncolytic VSV. Infection was established only in IFN-unresponsive tissue.

## 5.15 CONCLUSION

Several potential factors have been identified as associated with a higher risk of development of HCC. Complex interactions between viral factors and patient factors, including genetic factors and other environmental factors, lead to HCC development in chronic HBV-infected patients. HBV-related carcinogenesis is a multistep process, encompassing the combination of different, not mutually exclusive, effects such as the induction of chronic liver inflammation and regeneration, its integration into the hepatocyte genome, and the transactivating and transforming activity of several viral proteins (HBx and truncated Pre-S2/S) that may stimulate cellular oncogenes or suppress growth-regulating genes. Data related to the influence of different HBV genotypes and the emergence of selective variants as biomarkers of HCC development still remain controversial. At last, recent studies on occult HBV infection, as defined by serologically undetectable hepatitis B surface antigen (HBsAg–), despite the presence of circulating HBV DNA, suggest that the occult viral strains, maintaining the transcriptional activity and the pro-oncogenetic assets of the clear HBV infection (HBsAg+), may harbor a potential risk for liver cancer development. It is now possible to prevent HBV-related HCC. The most effective method is hepatitis B vaccination, which prevents chronic HBV infection and resulting chronic liver disease. Interferon therapy appears to confer benefit, but the evidence is weaker. First-generation oral antiviral (lamivudine) reduces HCC risk, particularly in cirrhotics. Long-term outcome data with newer, more potent HBV antivirals that have a higher genetic barrier to drug resistance are eagerly awaited.

## REFERENCES

Arbuthnot P, Capovilla A, Kew MC. (2000). The putative role of hepatitis B virus X protein in hepatocarcinogenesis: Effects on apoptosis, DNA repair, and MAP kinase and JAK/STAT pathways. *J Gastroenterol Hepatol* 15: 357–368.

Audard V, Grimber G, Elie C, et al. (2007). Cholestasis is a marker for hepatocellular carcinomas displaying beta-catenin mutations. *J Pathol* 212: 345–352.

Aviv A, Levy D, Mangel M. (2003). Growth, telomere dynamics and successful and unsuccessful human aging. *Mech Ageing Dev* 124: 829–837.

Baptista M, Kramvis A, Kew MC. (1999). High prevalence of 1762(T) 1764(A) mutations in the basic core promoter of hepatitis B virus isolated from black Africans with hepatocellular carcinoma compared with asymptomatic carriers. *Hepatology* 29: 946–953.

Benn J, Schneider RJ. (1994). Hepatitis B virus HBx protein activates Ras-GTP complex formation and establishes a Ras, Raf, MAP kinase signaling cascade. *Proc Natl Acad Sci U S A* 91: 10350–10354.

Bosch FX, Ribes J, Díaz M. (2004). Primary liver cancer: Worldwide incidence and trends. *Gastroenterology* 127: S5–S16.

Bouchard MJ, Wang LH, Schneider RJ. (2001). Calcium signaling by HBx protein in hepatitis B virus DNA replication. *Science* 294: 2376–2378.

Boyault S, Richman DS, de Reynies A. (2007). Transcriptome classification of HCC is related to gene alterations and to new therapeutic targets. *Hepatology* 45: 42–52.

Brechot C, Pourcel C, Louise A, et al. (1980). Presence of integrated hepatitis B virus DNA sequences in cellular DNA of human hepatocellular carcinoma. *Nature* 286: 533–535.

Brechot C, Thiers V, Kremsdorf D, et al. (2001). Persistent hepatitis B virus infection in subjects without hepatitis B surface antigen: Clinically significant or purely "occult"? *Hepatology* 34: 194–211.

Breuhahn K, Longerich T, Schirmacher P. (2006). Dysregulation of growth factor signaling in human hepatocellular carcinoma. *Oncogene* 25: 3787–3800.

Budhu A, Forgues M, Ye Q-H, et al. (2006). Prediction of venous metastases, recurrence, and prognosis in hepatocellular carcinoma based on a unique immune response signature of the liver microenvironment. *Cancer Cell* 10: 99–111.

Budhu A, Jia HL, Forgues M, et al. (2008). Identification of metastasis-related micro-RNAs in hepatocellular carcinoma. *Hepatology* 47: 897–907.

Buendia MA. (2000). Genetics of hepatocellular carcinoma. *Semin Cancer Biol* 10: 185–200.

Cammà C, Giunta M, Andreone P, et al. (2001). Interferon and prevention of hepatocellular carcinoma in viral cirrhosis: An evidence-based approach. *J Hepatol* 34: 593–602.

Cassiman D, Denef C, Desmet VJ, et al. (2001). Human and rat hepatic stellate cells express neurotrophins and neurotrophin receptors. *Hepatology* 33: 148–158.

Chan HL, Leung NW, Hussain M, et al. (2000). Hepatitis B e antigen-negative chronic hepatitis B in Hong Kong. *Hepatology* 31: 763–768.

Chan HL, Tse CH, Mo F, et al. (2008). High viral load and hepatitis B virus subgenotype Ce are associated with increased risk of hepatocellular carcinoma. *J Clin Oncol* 26: 177–182.

Chan HL-Y, Hui AY, Wong ML, et al. (2004). Genotype C hepatitis B virus infection is associated with an increased risk of hepatocellular carcinoma. *Gut* 53: 1494–1498.

Chang MH, Chen CJ, Lai MS, et al. (1997). Universal hepatitis B vaccination in Taiwan and the incidence of hepatocellular carcinoma in children. Taiwan Childhood Hepatoma Study Group. *N Engl J Med* 336: 1855–1859.

Chang MH, Shau WY, Chen CJ, et al. (2000). Hepatitis B vaccination and hepatocellular carcinoma rates in boys and girls. *JAMA* 284: 3040–3042.

Chang TT, Liaw YF, Wu SS, et al. (2010). Long-term entecavir therapy results in the reversal of fibrosis/cirrhosis and continued histological improvement in patients with chronic hepatitis B. *Hepatology* 52: 886–893.

Cheah MS, Wallace CD, Hoffman RM, et al. (1984). Hypomethylation of DNA in human cancer cells: A site-specific change in the c-myc oncogene. *J Natl Cancer Inst* 73: 1057–1065.

Chemin I, Trepo C. (2005). Clinical impact of occult HBV infections. *J Clin Virol* 43: S15–S21.

Chen BF, Liu CJ, Jow GM, et al. (2006). High prevalence and mapping of pre-S deletion in hepatitis B virus carriers with progressive liver diseases. *Gastroenterology* 130: 1153–1168.

Chen CF, Lee WC, Yang HI, et al. (2011). Risk evaluation of viral load elevation and associated liver disease/cancer in HBV (reveal-HBV) Study Group. Changes in serum levels of HBV DNA and alanine aminotransferase determine risk for hepatocellular carcinoma. *Gastroenterology* 141(4): 1240–1248.

Chen C-H, Changchien C-S, Lee CM, et al. (2009a). A study on sequence variations in pre-S/surface, X and enhancer II/core promoter/precore regions of occult hepatitis B virus in non-B, non-C hepatocellular carcinoma patients in Taiwan. *Int J Cancer* 125: 621–629.

Chen CJ, Yang HI, Iloeje UH, et al. (2009b). REVEAL-HBV Study Group. Hepatitis B virus DNA levels and outcomes in chronic Hepatitis B. *Hepatology* 49: S72–S84.

Chen JD, Yang HI, Iloeje UH, et al. (2010). Carriers of inactive hepatitis B virus are still at risk for hepatocellular carcinoma and liver-related death. *Gastroenterology* 138(5): 1747–1754.

Chen T, Li E. (2004). Structure and function of eukaryotic DNA methyltransferases. *Curr Top Dev Biol* 60: 55–89.

Chen YC, Chu CM, Yeh CT, et al. (2007). Natural course following the onset of cirrhosis in patients with chronic hepatitis B: A long-term follow-up study. *Hepatol Int* 1(1): 267–273.

Chisari FV. (2000). Rous-Whipple award lecture. Viruses, immunity, and cancer: Lessons from hepatitis B. *Am J Pathol* 156: 1117–1132.

Chu CM, Liaw YF. (2006). Hepatitis B virus-related cirrhosis: Natural history and treatment. *Semin Liver Dis* 26(2): 142–152.

Chun YK, Kim JY, Woo HJ, et al. (2000). No significant correlation exists between core promoter mutations, viral replication, and liver damage in chronic hepatitis B infection. *Hepatology* 32: 1154–1162.

Cillo U, Bassanello M, Vitale A, et al. (2004). The critical issue of hepatocellular carcinoma prognostic classification: Which is the best tool available? *J Hepatol* 40: 124–131.

Clements CJ, Baoping Y, Crouch A, et al. (2006). Progress in the control of hepatitis B infection in the Western Pacific Region. *Vaccine* 24: 1975–1982.

Coulouarn C, Factor VM, Anderon JB, et al. (2009). Loss of miRNA-122 expression in liver cancer correlates with suppression of the hepatic phenotype and gain of metastatic properties. *Oncogene* 28(40): 3526–3536.

Coulouarn C, Factor VM, Thorgeirsson SS, et al. (2008). Transforming growth factor-beta gene expression signature in mouse hepatocytes predicts clinical outcome in human cancer. *Hepatology* 47: 2059–2067.

Counter CM, Avilion AA, Le Feuvre CF, et al. (1992). Telomere shortening associated with chromosomal instability is arrested in immortal cells which express telomerase activity. *EMBO J* 11: 1921–1929.

Dandri M, Burda MR, Bürkle A, et al. (2002). Increase in de novo HBV DNA integrations in response to oxidative DNA damage or inhibition of poly(ADP-ribosyl)ation. *Hepatology* 35: 217–223.

de La Coste A, Romagnolo B, Billuart P, et al. (1998). Somatic mutations of the beta-catenin gene are frequent in mouse and human hepatocellular carcinomas. *Proc Natl Acad Sci U S A* 95: 8847–8851.

Ding Q, Xia W, Liu JC, et al. (2005). Erk associates with and primes GSK-3beta for its inactivation resulting in upregulation of beta-catenin. *Mol Cell* 19: 159–170.

Djojosubroto MW, Coi YS, Lee HW, et al. (2003). Telomeres and telomerase in aging, regeneration, and cancer. *Mol Cells* 15: 164–175.

Donato F, Gelatti U, Chiesa R, et al. (1999). A case-control study on family history of liver cancer as a risk factor for hepatocellular carcinoma in North Italy. Brescia HCC Study. *Cancer Causes Control* 10(5): 417–421.

El-Deiry WS, Nelkin BD, Celano P, et al. (1991). High expression of the DNA methyltransferase gene characterizes human neoplastic cells and progression stages of colon cancer. *Proc Natl Acad Sci U S A* 88: 3470–3474.

Ellis RE, Yuan J, Horvitz HR, et al. (1991). Mechanisms and functions of cell death. *Annu Rev Cell Biol* 7: 663–698.

Eun JR, Lee HJ, Kim TN, et al. (2010). Risk assessment for the development of hepatocellular carcinoma: According to on-treatment viral response during long-term lamivudine therapy in hepatitis B virus-related liver disease. *J Hepatol* 53(1): 118–125.

Fang ZL, Ling R, Wang SS, et al. (1998). HBV core promoter mutations prevail in patients with hepatocellular carcinoma from Guangxi, China. *J Med Virol* 56: 18–24.

Fattovich G, Bortolotti F, Donato F. (2008). Natural history of chronic hepatitis B: Special emphasis on disease progression and prognostic factors. *J Hepatol* 48: 335–352.

Fattovich G, Giustina G, Schalm SW, et al. (1995). Occurrence of hepatocellular carcinoma and decompensation in western European patients with cirrhosis type B. The EUROHEP Study Group on hepatitis B virus and cirrhosis. *Hepatology* 21: 77–82.

Feitelson MA, Duan LX. (1997). Hepatitis B virus X antigen in the pathogenesis of chronic infections and the development of hepatocellular carcinoma. *Oncogene* 50: 1141–1157.

Feitelson MA, Lee J. (2007). Hepatitis B virus integration, fragile sites and hepatocarcinogenesis. *Cancer Lett* 252: 157–170.

Feitelson MA, Reis HM, Tufan NL, et al. (2009). Putative roles of hepatitis B x antigen in the pathogenesis of chronic liver disease. *Cancer Lett* 286: 69–79.

Friedman SL, Arthur MJ. (1989). Activation of cultured rat hepatic lipocytes by Kupffer cell conditioned medium. Direct enhancement of matrix synthesis and stimulation of cell proliferation via induction of platelet-derived growth factor receptors. *J Clin Invest* 84: 1780–1785.

Fujio K, Evarts RP, Hu Z, et al. (1994). Expression of stem cell factor and its receptor, c-kit, during liver regeneration from putative stem cells in adult rat. *Lab Invest* 70: 511–516.

Gish RG, Gadano AC. (2006). *World Health Organization: Fact Sheet No. 204. Hepatitis B, Revised August 2008.* Geneva: World Health Organization.

Goh KT. (1997). Prevention and control of hepatitis B virus infection in Singapore. *Ann Acad Med Singapore* 26: 671–681.

Gramantieri L, Ferracin M, Fornari F, et al. (2007). Cyclin G1 is a target of miRNA-122a, a microRNA frequently down-regulated in human hepatocellular carcinoma. *Cancer Res* 67(13): 6092–6099.

Gregory RJ, Shiekhattar R. (2005). MicroRNA biogenesis and cancer. *Cancer Res* 65: 3509–3512.

Guerrero RB, Roberts LR. (2005). The role of hepatitis B virus integrations in the pathogenesis of human hepatocellular carcinoma. *J Hepatol* 42: 760–777.

Guo X, Jin Y, Qian G, et al. (2008). Sequential accumulation of the mutations in core promoter of hepatitis B virus is associated with the development of hepatocellular carcinoma in Qidong, China. *J Hepatol* 49: 718–725.

Gupta S, Govindarajan S, Fong TL, et al. (1990). Spontaneous reactivation in chronic hepatitis B: Patterns and natural history. *J Clin Gastroenterol* 12: 562.

Hall AJ, Wild CP. (2003). Liver cancer in low and middle income countries. *BMJ* 326: 994–995.

Hassan MM, Hwang LY, Hatten CJ, et al. (2002). Risk factors for hepatocellular carcinoma: Synergism of alcohol with viral hepatitis and diabetes mellitus. *Hepatology* 36: 1206–1213.

Hassan MM, Spitz MR, Thomas MB, et al. (2009). The association of family history of liver cancer with hepatocellular carcinoma: A case control study in US. *J Hepatol* 50: 334–341.

Herman JG, Baylin SP. (2003). Gene silencing in cancer in association with promoter hypermethylation. *N Engl J Med* 349: 2042–2054.

Herman JG, Merlo A, Mao L, et al. (1995). Inactivation of the CDKN2/p16/MTS1 gene is frequently associated with aberrant DNA methylation in all common human cancers. *Cancer Res* 55: 4525–4530.

Hildt E, Munz B, Saher G, et al. (2002). The Pre S2 activator MHBs(t) of hepatitis B virus activates c-raf-1/Erk2 signaling in transgenic mice. *EMBO J* 21: 525–535.

Hino O, Kajino K. (1994). Hepatitis virus-related hepatocarcinogenesis. *Intervirology* 37: 133–135.

Honda M, Kaneko S, Kawai H, et al. (2001). Differential gene expression between chronic hepatitis B and C hepatic lesion. *Gastroenterology* 120: 955–966.

Honda M, Yamashita T, Ueda T, et al. (2006). Different signaling pathways in the livers of patients with chronic hepatitis B or chronic hepatitis C. *Hepatology* 44: 1122–1138.

Hoshida Y, Nijman SM, Kobayashi M, et al. (2009). Integrative transcriptome analysis reveals common molecular subclasses of human hepatocellular carcinoma. *Cancer Res* 69: 7385–7392.

Hoshida Y, Villanueva A, Kobayashi M, et al. (2008). Gene expression in fixed tissues and outcome in hepatocellular carcinoma. *N Engl J Med* 359(19): 1995–2004.

Hui AM, Makuuchi M. (1999). Molecular basis of multistep hepatocarcinogenesis: Genetic and epigenetic events. *Scand J Gastroenterol* 8: 737–742.

Hung IF, Poon RT, Lai CL, et al. (2008). Recurrence of hepatitis B-related hepatocellular carcinoma is associated with high viral load at the time of resection. *Am J Gastroenterol* 103: 1663–1673.

Hussain SP, Schwank J, Staib F, et al. (2007). TP53 mutations and hepatocellular carcinoma: Insights into the etiology and pathogenesis of liver cancer. *Oncogene* 26: 2166–2176.

Ji J, Shi J, Budhu A, et al. (2009). MicroRNA expression, survival, and response to interferon in liver cancer. *N Engl J Med* 361(15): 1437–1447.

Jung JK, Arora P, Pagano JS, et al. (2007). Expression of DNA methyltransferase 1 is activated by hepatitis B virus X protein via a regulatory circuit involving the p16$^{INK4A}$-cyclin D1-CDK4/6-p Rb-E2F1 pathway. *Cancer Res* 67: 5771–5778.

Jung KS, Kim SU, Ahn SH, et al. (2011). Risk assessment of hepatitis B virus-related hepatocellular carcinoma development using liver stiffness measurement (FibroScan). *Hepatology* 53(3): 885–894.

Kanai Y, Hui AM, Sun L, et al. (1999). DNA hypermethylation at the D17S5 locus and reduced HIC-1 m RNA expression are associated with hepatocarcinogenesis. *Hepatology* 1999; 29: 703–709.

Kao JH, Chen DS. (2002). Global control of hepatitis B virus infection. *Lancet Infect Dis* 2: 395–403.

Kao JH, Chen PJ, Lai MY, et al. (2000). Hepatitis B genotypes correlate with clinical outcomes in patients with chronic hepatitis B. *Gastroenterology* 118: 554–559.

Kao JH, Chen PJ, Lai MY, et al. (2003). Basal core promoter mutations of hepatitis B virus increase the risk of hepatocellular carcinoma in hepatitis B carriers. *Gastroenterology* 124: 327–334.

Katoh H, Ojima H, Kokubu A, et al. (2007). Genetically distinct and clinically relevant classification of hepatocellular carcinoma: Putative therapeutic targets. *Gastroenterology* 133: 1475–1486.

Katoh H, Shibata T, Kokubu A, et al. (2006). Epigenetic instability and chromosomal instability in hepatocellular carcinoma. *Am J Pathol* 168: 1375–1384.

Kim BK, Park JY, Kim do Y, et al. (2008). Persistent hepatitis B viral replication affects recurrence of hepatocellular carcinoma after curative resection. *Liver Int* 28: 393–401.

Kim H, Lee H, Yun Y, et al. (1998). X gene product of hepatitis B virus induces apoptosis in liver cells. *J Biol Chem* 273: 381–385.

Kirk GD, Bah E, Montesano R, et al. (2006). Molecular epidemiology of human liver cancer: Insights into etiology, pathogenesis and prevention from The Gambia, West Africa. *Carcinogenesis* 27(10): 2070–2082.

Kirosawa K. (2002). Trend of liver cirrhosis as precancerous lesions. *Hepatol Res* 24: 40–45.

Koike K, Kobayashi M, Gondo M, et al. (1998). Hepatitis B virus DNA is frequently found in liver biopsy samples from hepatitis C virus-infected chronic hepatitis patients. *J Med Virol* 54: 249–255.

Koskinas J, Petraki K, Kavantzas N, et al. (2005). Hepatic expression of the proliferative marker Ki-67 and p53 protein in HBV or HCV cirrhosis in relation to dysplastic liver cell changes and hepatocellular carcinoma. *J Viral Hepatitis* 12: 635–641.

Kramvis A, Kew MC. (1999). The core promoter of hepatitis B virus. *J Viral Hepat* 6: 415–427.

Kuang SY, Jackson PE, Wang JB, et al. (2004). Specific mutations of hepatitis B virus in plasma predict liver cancer development. *Proc Natl Acad Sci U S A* 101: 3575–3580.

Kumar M, Kumar R, Hissar SS, et al. (2007). Risk factor analysis for hepatocellular carcinoma in HCC patients with and without cirrhosis: A case-control study of 213 HCC patients from India. *J Gastroenterol Hepatol* 22(7): 1104–1111.

Kuo TC, Chao CC. (2010). Hepatitis B virus X protein prevents apoptosis of hepatocellular carcinoma cells by upregulating SATB1 and HURP expression. *Biochem Pharmacol* 80: 1093–1102.

Kuzuya T, Katano Y, Kumada T, et al. (2007). Efficacy of antiviral therapy with lamivudine after initial treatment for hepatitis B virus-related hepatocellular carcinoma. *J Gastroenterol Hepatol* 22: 1929–1935.

Lagos-Quintana M, Rauhut R, Lendeckel W, et al. (2001). Identification of novel genes encoding for small expressed RNAs. *Science* 294: 853–858.

Laurent-Puig P, Legoix P, Bluteau O, et al. (2001). Genetic alterations associated with hepatocellular carcinomas define distinct pathways of hepatocarcinogenesis. *Gastroenterology* 120: 1763–1773.

Laurent-Puig P, Zucman-Rossi J. (2006). Genetics of hepatocellular tumors. *Oncogene* 25: 3778–3786.

Lee JS, Chu IS, Heo J, et al. (2004). Classification and prediction of survival in hepatocellular carcinoma by gene expression profiling. *Hepatology* 40: 667–676.

Lee JS, Heo J, Libbrecht L, et al. (2006). A novel prognostic subtype of human hepatocellular carcinoma derived from hepatic progenitor cells. *Nat Med* 12: 410–416.

Lee SH, Chung YH, Kim JA, et al. (2011). Histological characteristics predisposing to development of hepatocellular carcinoma in patients with chronic hepatitis B. *J Clin Pathol* 64(7): 599–604.

Liaw YF, Sung JJ, Chow WC, et al. (2004). Lamivudine for patients with chronic hepatitis B and advanced liver disease. *N Engl J Med* 351: 1521–1531.

Liaw YF, Tai DI, Chu CM, et al. (1988). The development of cirrhosis in patients with chronic type B hepatitis: A prospective study. *Hepatology* 8(3): 493–496.

Lin HH, Wang LY, Hu CT, et al. (2003). Decline of hepatitis B carrier rate in vaccinated and unvaccinated subjects: Sixteen years after newborn vaccination program in Taiwan. *J Med Virol* 69: 471–474.

Lin SM, Tai DI, Chien RN, et al. (2004). Comparison of long-term effects of lymphoblastoid interferon alpha and recombinant interferon alpha-2a therapy in patients with chronic hepatitis B. *J Viral Hepat* 11: 349–357.

Lin SM, Yu ML, Lee CM, et al. (2007). Interferon therapy in HBeAg positive chronic hepatitis reduces progression to cirrhosis and hepatocellular carcinoma. *J Hepatol* 46: 45–52.

Liu CJ, Chen BF, Chen PJ, et al. (2006). Role of hepatitis B virus precore/core promoter mutations and serum viral load on noncirrhotic hepatocellular carcinoma: A case-control study. *J Infect Dis* 194: 594–599.

Liu H, Luan F, Ju Y, et al. (2007). In vitro transfection of the hepatitis B virus PreS2 gene into the human hepatocarcinoma cell line Hep G2 induces upregulation of human telomerase reverse transcriptase. *Biochem Biophys Res Commun* 355: 379–384.

Liu S, Zhang H, Gu C, et al. (2009). Associations between hepatitis B virus mutations and the risk of hepatocellular carcinoma: A meta-analysis. *J Natl Cancer Inst* 101: 1066–1082.

Liu SJ, Xie JX, Yin JH, et al. (2011). A matched case-control study of hepatitis B virus mutations in the preS and core promoter regions associated independently with hepatocellular carcinoma. *J Med Virol* 83: 45–53.

Liu YC, Chen CVJ, Wu HS, et al. (2004). Telomerase and c-myc expression in hepatocellular carcinomas. *Eur J Surg Oncol* 30: 380–390.

Livingston SE, Simonetti JP, Mc Mahon BJ, et al. (2007). Hepatitis B virus genotypes in Alaska Native people with hepatocellular carcinoma: Preponderance of genotype F. *J Infect Dis* 195: 5–11.

Locarnini S, Mc Millan J, Bartholomeusz A. (2003). The hepatitis B virus and common mutants. *Sem Liver Disease* 23: 5–20.

Luber B, Arnold N, Stürzl M, et al. (1996). Hepatoma-derived integrated HBV DNA causes multistage transformation in vitro. *Oncogene* 12: 1597–1608.

Marchio A, Pineau P, Meddeb M, et al. (2000). Distinct chromosomal abnormality pattern in primary liver cancer of non-B, non-C patients. *Oncogene* 19: 3733–3738.

Martin-Vilchez S, Sanz-Cameno P, Rodríguez-Muñoz Y, et al. (2008). The hepatitis B virus X protein induces paracrine activation of human hepatic stellate cells. *Hepatology* 47: 1872–1883.

Matsumoto A, Tanaka E, Rokuhara A, et al. (2005). Efficacy of lamivudine for preventing hepatocellular carcinoma in chronic hepatitis B: A multicenter retrospective study of 2795 patients. *Hepatol Res* 32: 173–184.

Matsuzaki K, Murata M, Yoshida K, et al. (2007). Chronic inflammation associated with hepatitis C virus infection perturbs hepatic transforming growth factor beta signaling, promoting cirrhosis and hepatocellular carcinoma. *Hepatology* 46: 48–57.

Michalak TI, Pardoe IU, Coffin CS, et al. (1999). Occult lifelong persistence of infectious hepadnavirus and residual liver inflammation in woodchucks convalescent from acute viral hepatitis. *Hepatology* 29: 928–938.

Michielsen P, Ho E. (2011). Viral hepatitis B and hepatocellular carcinoma. *Acta Gastroenterol Belg* 74(1): 4–8.

Miyakawa Y, Mizokami M. (2000). Classifying hepatitis B virus genotypes. *Intervirology* 46: 329–338.

Mulrooney-Cousins PM, Michalak TI. (2007). Persistent occult hepatitis B virus infection: Experimental findings and clinical implications. *World J Gastroenterol* 13: 5682–5686.

Murakami S. (2001). Hepatitis B virus X protein: A multifunctional viral regulator. *J Gastroenterol* 36: 651–604.

Muroyama R, Kato N, Yoshida H, et al. (2006). Nucleotide change of codon 38 in the X gene of hepatitis B virus genotype C is associated with an increased risk of hepatocellular carcinoma. *J Hepatol* 45: 805–812.

Myung SJ, Yoon JH, Gwak GY, et al. (2007). Wnt signaling enhances the activation and survival of human hepatic stellate cells. *FEBS Lett* 581: 2954–2958.

Ng KP, Saw TL, Baki A. (2005). Impact of the expanded program of immunization against hepatitis B infection in school children in Malaysia. *Med Microbiol Immunol* 194: 163–168.

Nguyen VT, Law MG, Dore GJ. (2009). Hepatitis B-related hepatocellular carcinoma: Epidemiological characteristics and disease burden. *J Viral Hepat* 16(7): 453–463.

Niki T, Pekny M, Hellemans K, et al. (1999). Class VI intermediate filament protein Nestin is induced during activation of rat hepatic stellate cells. *Hepatology* 29: 520–527.

Nouso K, Tanaka H, Uematsu S, et al. (2008). Cost effectiveness of the surveillance program of hepatocellular carcinoma depends on the medical circumstances. *J Gastroenterol Hepatol* 23: 437–444.

Oda T, Tsuda H, Scarpa A, et al. (1992). p53 gene mutation spectrum in hepatocellular carcinoma. *Cancer Res* 52(22): 6358–6364.

Oh JC, Jeong DL, Kim IK, et al. (2003). Activation of calcium signaling by hepatitis B virus-X protein in liver cells. *Exp Mol Med* 35: 301–309.

Okuda H. (2007). Hepatocellular carcinoma development in cirrhosis. *Best Pract Res Clin Gastroenterol* 21: 161–173.

Papatheodoridis GV, Dimou E, Dimakopoulos K, et al. (2005). Outcome of hepatitis B e antigen-negative chronic hepatitis B on long-term nucleos(t)ide analog therapy starting with lamivudine. *Hepatology* 42(1): 121–129.

Paterlini P, Driss F, Nalpas B, et al. (1993). Persistence of hepatitis B and hepatitis C viral genomes in primary liver cancers from HBs Ag-negative patients: A study of a low-endemic area. *Hepatology* 17: 20–29.

Paterlini P, Gerken G, Nakajima E, et al. (1990). Polymerase chain reaction to detect hepatitis B virus DNA and RNA sequences in primary liver cancers from patients negative for hepatitis B surface antigen. *N Engl J Med* 323: 80–85.

Pinzani M, Milani S, Herbst H, et al. (1996). Expression of platelet derived growth factor and its receptor in normal human liver and during active hepatic fibrogenesis. *Am J Pathol* 148: 785–800.

Pollicino T, Raffa G, Costantino L, et al. (2007). Molecular and functional analysis of occult hepatitis B virus isolates from patients with hepatocellular carcinoma. *Hepatology* 45: 277–285.

Pollicino T, Squadrito G, Cerenzia G, et al. (2004). Hepatitis B virus maintains its pro-oncogenic properties in the case of occult HBV infection. *Gastroenterology* 126: 102–110.

Poussin K, Dienes H, Sirma H, et al. (1999). Expression of mutated hepatitis B virus X genes in human hepatocellular carcinomas. *Int J Cancer* 192: 111–118.

Ramchandani S, Mac Leod AR, Pinard M, et al. (1997). Inhibition of tumorigenesis by a cytosine DNA, methyltransferase, antisense oligodeoxynucleotide. *Proc Natl Acad Sci U S A* 94: 684–689.

Sabatini DM. (2006). mTOR and cancer: Insights into a complex relationship. *Nat Rev Cancer* 6: 729–734.

Sakuma K, Saitoh N, Kasai M, et al. (1988). Relative risks of death due to liver disease among Japanese male adults having various statuses for hepatitis B s and e antigen/antibody in serum: A prospective study. *Hepatology* 8(6): 1642–1646.

Salvucci M, Lemoine A, Saffroy R, et al. (1999). Microsatellite instability in European hepatocellular carcinoma. *Oncogene* 18(1): 181–187.

Sarin SK, Kumar M. (2012). Predictive scores for hepatocellular carcinoma development in chronic hepatitis B virus infection: "Does one size fit all?" *Gastroenterology* 142(4): 1038–1040.

Schirmacher P, Rogler CE, Dienes HP. (1993). Current pathogenetic and molecular concepts in viral liver carcinogenesis. *Virchows Arch B Cell Pathol Incl Mol Pathol* 63: 71–89.

Schlüter V, Meyer M, Hofschneider PH, et al. (1994). Integrated hepatitis B virus X and 30 truncated preS/S sequences derived from human hepatomas encode functionally active transactivators. *Oncogene* 9: 3335–3344.

Shafritz DA, Shouval D, Sherman HI, et al. (1981). Integration of hepatitis B virus DNA into the genome of liver cells in chronic liver disease and hepatocellular carcinoma studies in percutaneous liver biopsies and post-mortem tissue specimens. *N Engl J Med* 305: 1067–1073.

Shih CM, Lo SJ, Miyamura T, et al. (1993). Suppression of hepatitis B virus expression and replication by hepatitis C virus core protein in Hu H-7 cells. *J Virol* 67: 5823–5832.

Squadrito G, Orlando ME, Pollicino T, et al. (2002). Virological profiles in patients with chronic hepatitis C and overt or occult HBV infection. *Am J Gastroenterol* 97: 1518–1523.

Su F, Schneider RJ. (1996). HBx protein activates transcription factor NF-kappaB by acting on multiple cytoplasmic inhibitors of related proteins. *J Virol* 70: 4558–4566.

Sumi H, Yokosuka O, Seki N, et al. (2003). Influence of hepatitis B virus genotypes on the progression of chronic type B liver disease. *Hepatology* 37: 19–26.

Sung JJ, Tsoi KK, Wong VW, et al. (2008). Meta-analysis: Treatment of hepatitis B infection reduces risk of hepatocellular carcinoma. *Aliment Pharmacol Ther* 28: 1067–1077.

Tai PC, Suk FM, Gerlich WH, et al. (2002). Hypermodification and immune escape of an internally deleted middle-envelope (M) protein of frequent and predominant hepatitis B virus variants. *Virology* 292: 44–58.

Takahashi K, Ohta Y, Kanai K, et al. (1999). Clinical implications of mutations C-to-T1653 and T-to-C/A/G1753 of hepatitis B virus genotype C genome in chronic liver disease. *Arch Virol* 144: 1299–1308.

Tanaka K, Hirohata T, Koga S, et al. (1991). Hepatitis C and hepatitis B in the etiology of hepatocellular carcinoma in the Japanese population. *Cancer Res* 51: 2842–2847.

Tanaka Y, Mukaide M, Orito E, et al. (2006). Specific mutations in enhancer II/core promoter of hepatitis B virus subgenotypes C1/C2 increase the risk of hepatocellular carcinoma. *J Hepatol* 45: 646–653.

Taylor EM, Lehmann AR. (1998). Conservation of eukaryotic DNA repair mechanisms. *Int J Radiat Biol* 74: 277–286.

Terradillos O, Pollicino T, Lecouer H, et al. (1998). p53-independent pro-apoptotic effects of the hepatitis B virus HBx protein in vivo and in vitro. *Oncogene* 17: 2115–2123.

Thakur V, Guptan RC, Kazim SN, et al. (2002). Profile, spectrum and significance of HBV genotypes in chronic liver disease patients in the Indian subcontinent. *J Gastroenterol Hepatol* 17: 165–170.

Thorgeirsson SS, Grisham JW. (2002). Molecular pathogenesis of human hepatocellular carcinoma. *Nat Genet* 31: 339–346.

Thorgeirsson SS, Lee J-S, Grisham JW. (2006). Functional genomics of hepatocellular carcinoma. *Hepatology* 43: S145–S150.

Tovar V, Alsinet C, Villanueva A, et al. (2010). IGF activation in a molecular subclass of hepatocellular carcinoma and pre-clinical efficacy of IGF-1R blockage. *J Hepatol* 52: 550–559.

Tsai JF, Chang WY, Jeng JE, et al. (1994). Hepatitis B and C virus infection as risk factors for liver cirrhosis and cirrhotic hepatocellular carcinoma: A case–control study. *Liver* 14: 98–102.

Tsebe KV, Burnett RJ, Hlungwani NP, et al. (2001). The first five years of universal hepatitis B vaccination in South Africa: Evidence for elimination of HBsAg carriage in under 5-year-olds. *Vaccine* 19: 3919–3926.

Villanueva A, Chiang DY, Newell P, et al. (2008). Pivotal role of m TOR signaling in hepatocellular carcinoma. *Gastroenterology* 135: 1972–1983.

Villeneuve JP, Desrochers M, Infante-Rivard C, et al. (1994). A long-term follow-up study of asymptomatic hepatitis B surface antigen-positive carriers in Montreal. *Gastroenterology* 106(4): 1000–1005.

Wang HP, Rogler CE. (1991). Topoisomerase I-mediated integration of hepadnavirus DNA in vitro. *J Virol* 65: 2381–2392.

Wang T, Weinman SA. (2006). Causes and consequences of mitochondrial reactive oxygen species generation in hepatitis C. *J Gastroenterol Hepatol* 21: S34–S37.

Weinberg RA. (1995). The retinoblastoma protein and cell cycle control. *Cell* 81: 323–330.

Weinberger KM, Bauer T, Böhm S, et al. (2000). High genetic variability of the group-specific a-determinant of hepatitis B virus surface antigen (HBsAg) and the corresponding fragment of the viral polymerase in chronic virus carriers lacking detectable HBsAg in serum. *J Gen Virol* 81: 1165–1174.

White E. (1996). Life, death, and the pursuit of apoptosis. *Genes Dev* 10: 1–15.

Wong GL, Wong VW, Tan GM, et al. (2008). Surveillance programme for hepatocellular carcinoma improves the survival of patients with chronic viral hepatitis. *Liver Int* 28: 79–87.

Wong VW, Chan SL, Mo F, et al. (2010). Clinical scoring system to predict hepatocellular carcinoma in chronic hepatitis B carriers. *J Clin Oncol* 28(10): 1660–1665.

Woo HG, Park ES, Cheon JH, et al. (2008). Gene expression-based recurrence prediction of hepatitis B virus-related human hepatocellular carcinoma. *Clin Cancer Res* 14: 2056–2064.

Wood RD. (1997). Nucleotide excision repair in mammalian cells. *J Biol Chem* 72: 23465–23468.

Yamashita T, Forgues M, Wang W, et al. (2008). EpCAM and alpha-fetoprotein expression defines novel prognostic subtypes of hepatocellular carcinoma. *Cancer Res* 68: 1451–1461.

Yang B, Guo M, Herman JG, et al. (2003). Aberrant promoter methylation profiles of tumor suppressor genes in hepatocellular carcinoma. *Am J Pathol* 163: 1101–1107.

Yang HI, Lu SN, Liaw YF, et al. (2000). Hepatitis B e antigen and the risk of hepatocellular carcinoma. *N Engl J Med* 347(3): 168–174.

Yang HI, Sherman M, Su J, et al. (2010). Nomograms for risk of hepatocellular carcinoma in patients with chronic hepatitis B virus infection. *J Clin Oncol* 28(14): 2437–2444.

Yang HI, Yeh SH, Chen PJ, et al. (2008a). Associations between hepatitis B virus genotype and mutants and the risk of hepatocellular carcinoma. *J Natl Cancer Inst* 100: 1134–1143.

Yang HI, Yuen MF, Chan HL, et al. (2011). Risk estimation for hepatocellular carcinoma in chronic hepatitis B (REACH-B): Development and validation of a predictive score. *Lancet Oncol* 12(6): 568–574.

Yang JC, Teng CF, Wu HC, et al. (2009). Enhanced expression of vascular endothelial growth factor-A in ground glass hepatocytes and its implication in hepatitis B virus hepatocarcinogenesis. *Hepatology* 49: 1962–1971.

Yang L, Wang Y, Mao H, et al. (2008b). Sonic hedgehog is an autocrine viability factor for myofibroblastic hepatic stellate cells. *J Hepatol* 48: 98–106.

Ye L, Dong N, Wang Q, et al. (2008). Progressive changes in hepatoma cells stably transfected with hepatitis B virus X gene. *Intervirology* 51: 50–58.

Yoo YD, Ueda H, Park K, et al. (1996). Regulation of transforming growth factor-beta 1 expression by the hepatitis B virus (HBV) X transactivator. Role in HBV pathogenesis. *J Clin Invest* 97: 388–395.

Yoshiura K, Kanai Y, Ochiai A, et al. (1995). Silencing of the E-cadherin invasion-suppressor gene by Cp G methylation in human carcinomas. *Proc Natl Acad Sci U S A* 92: 7416–7419.

Yu MC, Yuan JM. (2004). Environmental factors and risk for hepatocellular carcinoma. *Gastroenterology* 127(Suppl 1): S72–78.

Yu MW, Chang HC, Liaw YF, et al. (2000). Familial risk of hepatocellular carcinoma among chronic hepatitis B carriers and their relatives. *J Natl Cancer Inst* 92(14): 1159–1164.

Yuen MF, Sablon E, Yuan HJ, et al. (2003). Significance of hepatitis B genotype in acute exacerbation, HBeAg seroconversion, cirrhosis-related complications, and hepatocellular carcinoma. *Hepatology* 37: 562–567.

Yuen MF, Seto WK, Chow DH, et al. (2007). Long-term lamivudine therapy reduces the risk of long-term complications of chronic hepatitis B infection even in patients without advanced disease. *Antivir Ther* 12: 1295–1303.

Yuen MF, Tanaka Y, Fong DY, et al. (2009). Independent risk factors and predictive score for the development of hepatocellular carcinoma in chronic hepatitis B. *J Hepatol* 50: 80–88.

Yuen MF, Tanaka Y, Mizokami M, et al. (2004). Role of hepatitis B virus genotypes Ba and C, core promoter and precore mutations on hepatocellular carcinoma: A case control study. *Carcinogenesis* 25: 1593–1598.

Yuen MF, Tanaka Y, Shinkai N, et al. (2008). Risk for hepatocellular carcinoma with respect to hepatitis B virus genotypes B/C, specific mutations of enhancer II/core promoter/precore regions and HBV DNA levels. *Gut* 57: 98–102.

Yuh CH, Ting LP. (1990). The genome of hepatitis B virus contains a second enhancer: Cooperation of two elements within this enhancer is required for its function. *J Virol* 64: 4281–4287.

Yuh CH, Ting LP. (1993). Differentiated liver cell specificity of the second enhancer of hepatitis B virus. *J Virol* 67: 142–149.

Zakian VA. (1989). Structure and function of telomeres. *Annu Rev Genet* 23: 579–604.

Zhang H, Zhai Y, Hu Z, et al. (2010). Genome-wide association study identifies 1p36.22 as a new susceptibility locus for hepatocellular carcinoma in chronic hepatitis B virus carriers. *Nat Genet* 42: 755–758.

Zhang JY. (1995). The research progress of the relationship between hepatitis B, C infection and hepatocellular carcinoma. In: Zhang JY, Zheng PY, Zhang Y (eds.), *Research and Practice in Viral Liver Diseases*. Beijing: Chinese Science and Technology Press, pp. 148–167.

Zhang JY, Dai M, Wang X, et al. (1998). A case–control study of hepatitis B and C virus infection as risk factors for hepatocellular carcinoma in Henan, China. *Int J Epidemiol* 27: 574–578.

Zhang X, Dong N, Zhang H, et al. (2005). Effects of hepatitis B virus X protein on human telomerase reverse transcriptase expression and activity in hepatoma cells. *J Lab Clin Med* 145: 96–104.

Zhang X, Xu HJ, Murakami Y, et al. (1994). Deletions of chromosome 13q, mutations in Retinoblastoma 1, and retinoblastoma protein state in human hepatocellular carcinoma. *Cancer Res* 54: 4177–4182.

Zhong S, Chan JY, Yeo W, et al. (2000). Frequent integration of precore/core mutants of hepatitis B virus in human hepatocellular carcinoma tissues. *J Viral Hepat* 7: 115–123.

Zhu R, Li BZ, Li H, et al. (2007). Association of p16INK4 hypermethylation with hepatitis B virus X protein expression in the early stage of HBV-associated hepatocarcinogenesis. *Pathol Int* 57: 326–323.

# 6 Epstein–Barr Virus and Treatment of Its Infection

*Tarun Jha, Amit Kumar Halder, and Nilanjan Adhikari*

## 6.1 INTRODUCTION

Tumors occur as a result of genetic and epigenetic modulations that transform normal cells into immortalized highly proliferating malignant cells. Several factors contribute to malignant transformation and clonal expansions of malignant cells. Viruses serve as an important factor for diverse types of human cancers (Butel 2000). Viral oncogenes were first found as a part of viral genome in oncogenic animal viruses (Truyen and Lochelt 2006). Different viruses follow different strategies for cancer development. However, in all cases, oncogenic virus products function as tumor antigens and elicit specific T-cell responses that are required for protection against tumor development (Klein 2002). Oncogenic viruses consist of DNA and RNA viruses. The DNA oncogenic viruses encode viral proteins essential for viral replication. Examples of the DNA viruses causing human tumors include Epstein–Barr virus (EPV), human papillomavirus (HPV), polyomavirus, simian virus 40 (SV40), and so on. In most of these DNA-virus-induced tumors, virus-encoded protein antigens are found in the nucleus, cytoplasm, or plasma membrane of tumor cells. On the other hand, RNA viruses carry changed variants of normal host cell genes. Retroviral oncogene products possess similar malignant potential as cellular oncogenes (Judson et al. 1994).

EBV is a ubiquitous pathogen that has adopted a unique and effective strategy for infection, persistence, and spread in its only biological host—human beings. It infects more than 90% of the human population worldwide and is associated with a diverse array of proliferative diseases in different parts of the world (Kutok and Wang 2006). In the United States, the most common EBV-associated malignancies are Hodgkin and non-Hodgkin lymphoma. In China, nasopharyngeal carcinoma (NPC) is one of the most frequently diagnosed EBV-associated cancers. In Africa, especially in equatorial Africa, Burkitt's lymphoma (BL) is the most common childhood tumor caused by EBV. In Eastern Asia, Europe, and South America, approximately one million new cases of gastric carcinoma are reported every year, and EBV-associated cases comprise about 10% of all (GCs) worldwide. In addition to these, EBV was reported in 21% cases of breast cancers. Taken together, these data show the formidable health risks involved with this virus (Cohen et al. 2011). The available options for the prevention against this pathogen are still very limited. Although chemotherapy is the only effective strategy to treat EBV-associated malignancies, 50% of the patients die from secondary malignancies, disease relapse, or chemotherapy-associated complications. The current chapter describes the techniques available to prevent EBV infection and to treat EBV-associated malignancies. In addition to this, we also discuss the recent outcomes in the fields of synthetic and natural medicinal chemistry research that were solely aimed to fight against EBV infection as well as EBV-associated malignancies.

## 6.2 HISTORICAL BACKGROUND

EBV was named after its discoverers, Michael Epstein and Yvonne Barr. They documented the existence of this pathogen for the first time in 1964 (Epstein et al. 1964). However, the plausible role of this virus in malignancy was insinuated in 1958 by a surgeon, Denis Burkitt, who described a common cancer (rapidly growing B-cell lymphoma, which is developed in the jaw) affecting children in Uganda in Africa (Burkitt 1958). Later, this particular cancer was named BL. Denis Burkitt carefully examined the climatic and geographical distribution of this lymphoma, and he believed that a virus was involved in the manifestation of BL. In 1964, by using an electron microscope, Michael Epstein and his group observed a virus-like particle (Michael Epstein initially noted it was a "virus, like herpes") in the suspension culture derived from a BL biopsy specimen. The serological markers of this cell line were later developed by Werner Henle and his wife, Gertrude Szpingier (Henle), of Philadelphia's Children's Hospital (Henle and Henle 1966). The development of an immune fluorescent test for the detection of EBV antibodies revealed that BL patients contain highly elevated antibody titers against EBV antigens. Subsequently, this test also showed that EBV is responsible for NPCs (de Schryver et al. 1969). In 1968, Henle and coworkers unexpectedly identified EBV as a causative agent for infectious mononucleosis (IM) (the most common clinical manifestation of EBV infection) when one of their laboratory technicians (who had initially been diagnosed as seronegative to EBV antibodies) suddenly fell ill with IM and was diagnosed as seropositive to EPV antibodies immediate after infection (Henle et al. 1968). The subsequent detection of EBV DNA in BL cells and the experimental production of lymphomas in cotton-top marmosets and owl monkeys established EBV as the first virus clearly implicated in the development of human cancer.

## 6.3 MOLECULAR STRUCTURE AND BIOLOGY OF EBV

EPV is a member of the herpesvirus family and was formally designated as herpesvirus 4. It belongs to the Lymphocryptoviridae, the gamma 1 subtype of the subfamily Gammaherpesvirinae. It is one of the eight known herpesviruses and is one of the most common viruses in humans (Middeldorp et al. 2003). Mature virions are approximately 120 to 180 nm in diameter. It has a double-stranded, linear 172-kbp-long DNA genome that encodes more than 85 genes. The DNA is wrapped in a toroid-shaped protein core formed by a nucleocapsid with 162 capsomeres and an outer envelop with external glycoprotein spikes. A protein tegument lies between capsid and envelope. The envelope is embedded with several glycoproteins that are essential for host selection, receptor recognition, and viral entry to the cell. The most abundant structural glycoprotein is gp350/220, which helps the virions to bind cluster of differentiation 21 (CD21) of the host B cells.

The B95–8 strain (derived from IM) of EBV was the first genome that was completely cloned and sequenced from any herpesvirus (Baer et al. 1984). Later, complete sequences of two more strains of EBV were determined. These are GD1 (derived from a Chinese NPC patient) and AG876 (derived from a Ghanaian BL patient). The genome of EBV has a higher percentage of G + C (58%) than that of humans (42%). Like other gammaherpesviruses, the EBV genome has 0.5 kb terminal repeats (TR) at the end of its linear genome. The EBV genome contains an unusually large tandemly repeated DNA sequence within the genome, known as major internal repeats (IR1), which contains 5–10 copies of a sequence of 3072 bp. IR1 contains Wp promoter for the EBV nuclear antigens (EBNAs) that are important for viral infection and malignancy. Apart from promoter Wp, two other promoters (Cp and Qp) are required for the EBNA transcription. IR1 divides the EBV genome into long and short unique sequences ($U_L$ and $U_S$, respectively) filled with closely packed genes (Figure 6.1).

Although genes of other herpesviruses are named according to the positions in $U_L$ or $U_S$ region, EBV genes are systematically named according to *Bam*HI restriction fragments (because the prototype B95–8 genome was sequenced from an EBV DNA *Bam*HI fragment cloned library). The largest fragment is *Bam*H1-A and *Bam*H1-B is the second largest. A sequence within the *Bam*H1-C fragment is called origin of plasmid replication (oriP) (Figure 6.2); oriP consists of two cis-acting elements, a motif of 20 × 30 bp tandem repeats of cognate sequence (family of repeats, or FR) located 1000 bp from an area of dyad symmetry (DS) that contains four similar repeat units. The motif of repeats functions in vitro as a transcriptional enhancer. DNA replication starts in DS and proceeds bi-directionally and asymmetrically, and it stops in the motif of repeats (Figure 6.2b). The trans-acting viral replication protein EBV nuclear antigen-1 (EBNA-1) binds to the oriP sequence and helps to replicate viral episome as well as to transactivate other viral genes. Another origin of replication is ori-Lyt (Figure 6.2), which is used for the production of new virus progeny during the lytic phase (Farrell 2005).

On the basis of the gene sequence, two predominant EBV-strains were identified. These are type A (or type 1) and type B (or type 2) EBV. These two strains differ in the sequences of viral genes expressed during latent phase infection and in their ability to transform B lymphocytes. The type 2 EBV transforms B cells less efficiently than type 1 EBV in vitro. From a genetic context, these two strains differ from each other at the EBNA loci EBNA2 and EBNA3A-C. The type 1 EBV is mainly found in the Western Hemisphere and Southeast Asia, whereas type 2 virus infection is more prevalent in African regions (Sample et al. 1990).

## 6.4 PRIMARY EBV INFECTION

EBV is an orally transmitted human herpesvirus that passes from one host to another via saliva. EPV initially infects B lymphocytes because these circulate in close proximity to the respiratory (oropharyngeal) epithelium. EBV transforms normal B

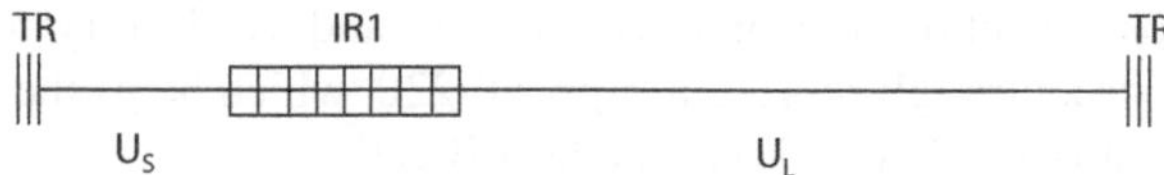


**FIGURE 6.1** Simplified diagram of linear EBV genome structure (TR = terminal repeats, IR1 = internal repeat 1, $U_L$ = unique long, $U_S$ = unique short).

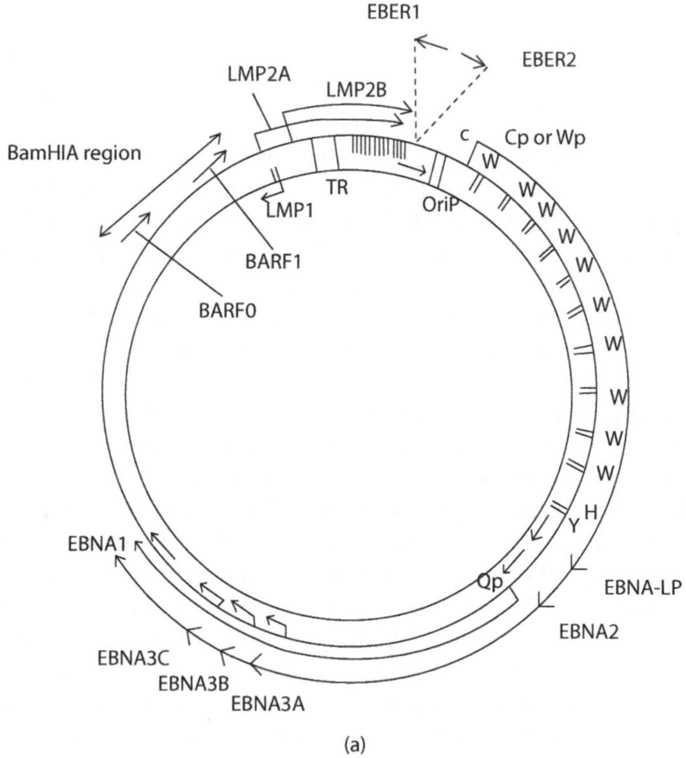

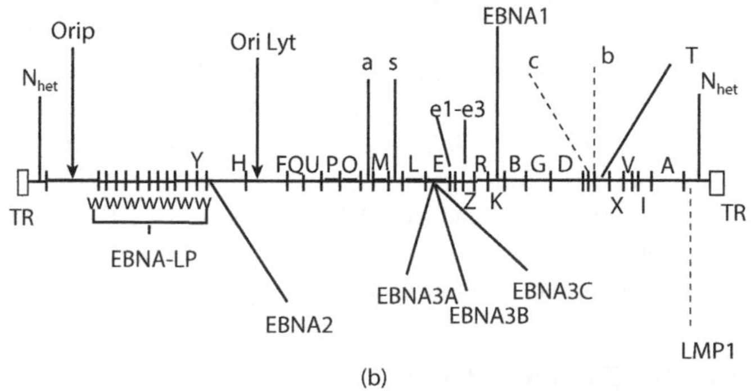

**FIGURE 6.2** (a) Location and transcription of EBV latent genes on the double-stranded viral DNA episome. (b) Location of open reading frames for EBV latent proteins on the *Bam*HI restriction-endonuclease map of the prototype B95.8 genome.

lymphocytes to latently infected lymphoblastoid cell lines (LCLs) that could grow long term in cell culture (Pope 1967). The B-cell infection begins with the attachment of the gp350/220 viral membrane glycoprotein to the CD21 present in the plasma membrane of the lymphocytes. The CD21 and human leukocyte antigen (HLA) class II molecules serve as the viral entry receptor and coreceptor, respectively. The virus may also infect epithelial cells and may initially replicate in the oropharynx before infecting B cells. Because the epithelial cells lack CD21, EBV protein BMRF-2 interacts with β1 integrins. EBV gH/gL envelope protein subsequently triggers fusion via interaction with αvβ6/8 integrins (Thompson and Kurzrock 2004). Immediately after infection, an EBV-specific cytotoxic T-cell lymphocyte (CTL) response reduces the level of primary infection to a considerable extent. Three antibodies (IgM, IgG, and IgA) are produced against EBV viral capsid antigen (early antigen D and R) (Linde 1996). The attachment of viral glycoprotein to CD21 leads to several events. Initially, CD21 receptors are cross-linked. This is followed by several events that prepare the host cell for infection within 0.5 h. These involve immediate activation of lck tyrosine kinase-induced calcium mobilization, increase in mRNA synthesis, blast transformation, cell adhesion, enlargement of nucleus, and cytoplasmic volume expansion. Moreover, CD23 is overexpressed and interleukin 6 (IL-6) production is also increased (Thompson and Kurzrock 2004). Infection in epithelial cells results in replication of the virus followed by lysis of the cell and the release of virions from the cell. Also, infection of B lymphocytes generally results in persistent infection without replication or release of virus. Therefore, two types of EBV-mediated cellular infection may occur. These are "lytic" infection and "latent" infection. In lytic infection, syntheses of viral DNA, RNA, and proteins are followed by assembling of viral particles and lysis of the infected cells. In latent infection, viral DNA is maintained in the host cell as episome that, depending on suitable circumstances, may be reactivated later to initiate lytic phase (Tsurumi et al. 2005).

## 6.5 LYTIC PHASE

In lytic phase, the nucleocapsid is dissolved, along with endosomal membrane, and the viral genome is transported into the host nucleus. It is replicated with the help of viral DNA polymerase. There are three temporal classes of viral lytic gene products; these are immediate early (IE), early (E), and late early (L). The IE products act as transactivators of other gene products and also help in immune escape very early in the lytic phase. The E gene products have a wide range of functions, such as replication, metabolism, and inhibition of immune response. They are used for the biosyntheses of viral DNA polymerase and thymidine kinase, which are essential for viral DNA replication during lytic phase. The L gene products tend to code some major envelope glycoproteins such as viral capsid antigens (VCA), gp350, and so on. Some other L gene products are also used for host immune evasion (Tsurumi et al. 2005).

### 6.5.1 Immediate Early Gene Products

Immediate early gene products are considered important because these are responsible for the translation from latent to lytic phase (Kieff and Rickinson 2007).

These gene products also transactivate E and L gene products. The best known IE gene product is BZLF1 or Zebra. It serves as a checkpoint for initiation of the replicative cycle. BZLF1 is a DNA-binding transactivator protein involved in triggering expression of the lytic genes and downregulation of latent genes. These lead to cell death and release of infectious virions. BZLF1 protein consists of three domains: (a) one transactivation domain, (b) a basic domain with high homology to a conserved region of the c-junc/c-fos family of transcription factor, and (c) a p53-interacting domain. Overexpression of p53 protein as well as gamma irradiation induces the expression of Zebra leading to initiation of lytic cycle. The Zebra expression was also found in some EBV-associated tumor cells (Middeldorp et al. 2003). It also helps in escaping immune response by interfering with interferon γ (IFNγ) signaling. The Zebra protein reduces the expression of IFNγ and thus results in inhibition of IFNγ-induced STAT1 tyrosine phosphorylation and subsequent disruption of MHC-II upregulation (Morrison et al. 2001). Other known IE products are BRLF1, BRRF1, and BI'LF4 transactivator genes. Transcription of IE genes occurs even in the presence of protein synthesis inhibitors.

### 6.5.2 Early Gene Products

Early gene products are normally induced by IE gene products. However, these may also be induced by irradiation and membrane receptors (Kieff and Rickinson 2007). The best known E gene product is BHRF1. Although early lytic genes are not significantly expressed in EBV-associated malignancies, BHRF1 is found to play some important roles in the manifestation of tumors. BHRF1 mRNAs are expressed from a promoter in *Bam*HI-H during early phase of lytic infection (Pearson et al. 1987). This protein is overexpressed in EBV-associated B-cell lymphoma, Hodgkin's disease, NPC, oral hairy leucopenia, and gastric carcinoma. The E genes lead to the production of different enzymes that are associated with replication of viral genes. For EBV replication, six viral proteins are required. These are: (a) BALF5—a viral DNA polymerase, (b) BALF2—a single-stranded DNA-binding protein analog, (c) BMRF1—a DNA polymerase processivity factor, (d) BSLF1—a primase analog, (e) BBLF4—a primase analog, and (f) BBLF2/3—a third component of primase/helicase complex. EBV lytic phase replicating enzymes are potential targets for many antiviral agents (Kenney 2007). BALF5 is a 110 kDa protein that requires a stabilizing effect from BMRF1 for the activity. Therefore, compounds that mimic the protein BMRF1 could be used for blocking EBV DNA replication. In addition to this, E genes also contribute in the inhibition of CTL-mediated apoptosis of the infected cells, resulting in the prolongation of virus production time. Moreover, BHRF1 has limited structural homology to human apoptotic inhibitor Bcl-2 (approximately 25% amino acid sequence identity and 42% amino acid sequence similarity). Like Bcl-2, BHRF1 is also located in the membranes of mitochondria, endoplasmic reticulum, and nucleus. The carbon terminus of BHRF1 contains a mitochondrial-targeting sequence that affects cell physiology through interfering with mitochondrial function. It may increase cell survival time by inhibiting apoptosis and may allow accumulation of oncogenic mutations and production of the maximum number of virions during the lytic phase (Oudejans et al. 1995).

### 6.5.3 Late Gene Products

Late gene products encode for viral structural proteins. Some important L products are the small capsid protein, the scaffold protein VCA p18 (BFRF3), the small capsid protein VCA-p40 (BdRF1), nuclear membrane protein gp125 (BALF4), envelope glycoprotein gp350/220 (BLLF1), and so on. The first three proteins are highly immunogenic in nature and are used as targets for serodiagnosis of EBV infection (van Grunsven et al. 1994). BCRF1 is one of the most important EBV L gene products. It shares 84% sequence homology with human IL-10, and it also produces viral IL10 (Moore et al. 1990). Human IL-10 inhibits activation and effector functions of immunological agents such as T cells, monocytes, macrophages, and natural killer (NK) cells. It also serves as a growth factor for B lymphocytes. Therefore, by mimicking these functions, viral IL-10 helps in manipulating normal immunogenic responses and inhibits the function of macrophages and NK cells. It also inhibits the production of IFNγ.

## 6.6 LATENT INFECTION

Latency is a continuous viral infection in the host system without active production of new viral progeny. In the latent phase, EBV mainly exists in the memory B cell and epithelial cells (Thorley-Lawson and Gross 2004). Approximately one in a million B cells of the host contains latent EBV genome. When EBV infects B lymphocytes, its genome circularizes to form episome in the nucleus of the host cell. The latently infected B cells express a limited number of viral encoded genes (approximately 10 out of 80) and corresponding gene products. Among these gene products, Epstein–Barr virus nuclear antigens (EBNA1, 2, 3A, 3B, 3C, and LPs) and latent membrane proteins (LMPs) (LMP1 and 2) are mainly expressed to maintain latency (Joseph et al. 2000). Depending on transcription strategies and gene expression patterns, EBV latency is classified into three types. In type I, EBV genomes replicate in dividing memory cells, and only EBNA1 is expressed with restricted expression of Epstein–Barr virus-encoded RNAs (EBERs) and BamHI-A rightward transcripts (BARTs). Type I is observed in BL. The type II latency is found mainly in NPC and Hodgkin lymphoma (HL) where viral genome replication is observed in normal B cells. EBNA1, EBERs, BARTs, and three LMPs are expressed in type II. Naïve B cells are activated during type III latency where nearly all latency gene products are fully expressed in acute IM as well as in immune-compromised infections (Odumade et al. 2011). The major latent gene products are discussed as follows.

### 6.6.1 Epstein–Barr Virus Nuclear Antigen 1

EBNA1 is expressed in all latency types, and it is indispensable for viral DNA replication and episomal maintenance. It is a DNA-binding phosphoprotein, the coding sequence of which lies in the BamHI fragment K rightward reading frame 1 (BKRF1) open reading frame (ORF). Because no EBNA1-specific antibody is expressed by CD8+ T cells, it constructs the main strategy for immune evasion

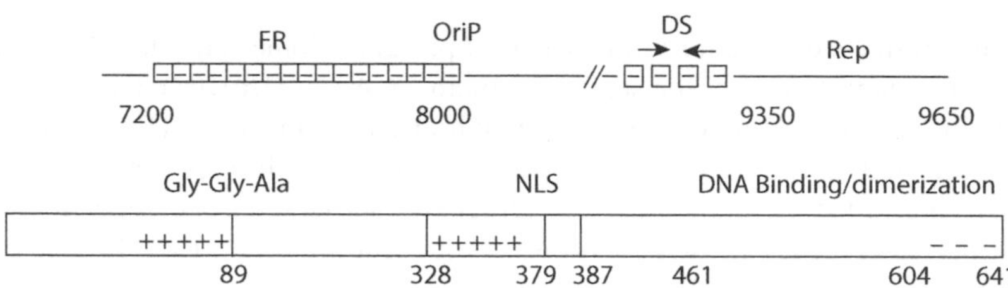


**FIGURE 6.3** Expanded representation of OriP.

during latency. Although EBNA1 is not recognized by CD8+ T cells, CD4+ T cells and antibodies reactive with EBNA1 are readily detected in most healthy virus carriers. However, EBNA1 is a poor diagnostic marker for the detection of EBV infection. The EBNA1 protein consists of a positively charged amino terminus, a glycine-alanine repeat segment, another short positively charged domain followed by nuclear localization sequence as well as DNA binding, and a dimerization segment (Middeldorp et al. 2003). The Gly-Ala repeat has an inhibitory effect on the proteosomal processing of CTLs; therefore, these are responsible for immune evasion.

EBNA1 binds three regions of episomes. Two of these are high-affinity sites and these are: (a) a family of 20 tandem repeats of the cognate sequence (or FR) and (b) DS (Leight and Sugden 2000). The third region is promoter Qp, which is discussed later. FR and DS form the origin of plasmid (oriP) replication (Figure 6.3). Both these elements contain 18-bp EBNA1-binding sites and are separated by 1 kb intervening sequence. After binding to oriP, EBNA1 uses host enzymes for regulating all necessary steps for viral episomal replication. EBNA1 is also attached to the host chromosomes during mitosis, and it helps in differentiating the episomes into host progeny nuclei. EBNA1 bound to the region of DS directs the assembly of the cellular DNA replication machinery at oriP and the initiation of DNA synthesis. FR-bound EBNA-1 helps in segregating EBV genomes to daughter cells by tethering its viral DNA cargo to host metaphase chromosomes by its association with a cellular chromosome-associated protein or direct interaction with the host DNA. Apart from oriP, EBNA1 binds at +10 and +34 nucleotide downstream of promoter Qp (third binding site). EBNA1 mainly suppresses the activity of Qp. Apart from being an important macromolecule in EPV episome replication and maintenance, the role of EBNA-1 in genetic alteration related oncogenicity was noticed (Yoshioka et al. 2008).

### 6.6.2 Epstein–Barr Virus Nuclear Antigen 2

EBNA2 and EBNA-LP are latent proteins that are also detected after EBV infection. EBNA2 is encoded in the primary transcripts started from Wp and Cp. EBNA2 serves mainly as a transcriptional coactivator protein that directs viral gene expressions. It also transactivates several viral and cellular genes that effect cellular immortalization, such as CD23, CD21, c-*myc*, c-*fgr*, LMP1, LMP2, and EBNA-C

(Middeldorp et al. 2003). CD23 and CD21 are the surface markers for infected B cells. The c-*myc* and c-*fgr* are the cellular proto-oncogenes. LMPs are viral genes that promote various infection-related events. Instead of binding the DNA directly, EBNA2 interacts with Cp-binding factor 1 and other transcription factors such as LMP promoters. Moreover, the interaction of EBNA2 with the Cp-binding factors depends on the association with the DNA-binding protein, recombinational signal-binding protein j-κ (RBP j-κ) (Henkel et al. 1994). RBP j-κ is normally a repressor protein, but its association with EBNA2 inhibits its repressor activity. RBP j-κ is also a component of the notch signaling pathway (Zimber-Strobl and Stroble 2001). The notch proteins are transmembrane proteins that undergo proteolytic cleavages by specific activators; the intracellular parts move to nucleus and transactivate cellular genes. It has been observed that EBNA-2 along with notch proteins transactivates several genes. EBNA2 and EBNA-LP cooperatively participate in the G0-G1 transition of the cell cycle.

### 6.6.3 Epstein–Barr Virus Nuclear Antigen-Latent Protein

EBNA-LP is also known as EBNA5. It is encoded in the leader of Wp-driven EBNA mRNAs and is co-expressed along with EBNA2 in freshly encoded B cells. EBNA2 and EBNA-LP interact with each other and drive resting B-cell lymphocytes into the G1 phase by binding and inactivating p53 and retinoblastoma protein (pRb) tumor suppressor gene products. This transformation is an important step in the early stages of EBV infection. Like EBNA2 (Szekely et al. 1993), EBNA-LP also interacts with some transcriptional factors that activate the notch signaling pathway. EBNA-LP causes redistribution of EBNA-3C in the nucleus (Radkov et al. 1997).

### 6.6.4 Epstein–Barr Virus Nuclear Antigen 3A, 3B, and 3C

The EBNA3 family members are encoded by three genes that lie adjacent to each other on the viral genome. EBNA3A, EBNA3B, and EBNA3C share 72–84% homology with each other. EBNA3A and EBNA3C are essential proteins for in vivo transformation and immortalization of B cells. However, EBNA-3B is dispensable for cell growth transformation. Its expression has a correlation with upregulation of CD40 and downregulation of CD77 in vitro. EBNA3C promotes the expression of LMP1, which facilitates cell growth and inhibits apoptosis. All EBNA3 proteins interact with Cp-binding factor 1 (Thompson and Kurzrock 2004). All EBNA3 genes control the expressions of certain cellular genes and bind to a variety of host proteins including RBP j-κ. Therefore, these are also involved in the upregulation of cellular and viral promoters.

### 6.6.5 Latent Membrane Protein 1

LMP1 is considered as a major EBV-transforming gene. It has growth-transforming effects in rodent fibroblasts and induces many changes associated with primary EBV infection and transformation of B cells. LMP1 inhibits apoptosis in B cells by inducing anti-apoptotic bcl-2 and A20 proteins and also by upregulating IL-10, which stimulates B-cell proliferation and inhibits local immune response. LMP1 mimics the CD40 ligand-mediated growth-promoting activities (Zimber-Strobl et al. 1996). LMP1 mimics CD40 by associating with tumor necrosis factor receptor (TNFR), TNFR-associated factors (TRAFs), and TNFR-associated death domain protein (TRADD). Interactions with TRAF1 and TRAF3 lead to NF-κβ activation resulting in the increased expressions of B-cell activation markers (CD23, CD39, CD40, CD44, MHC class II, etc.) and cellular adhesion molecules (LFA-1 and ICAM-1). LMP1 induces enhanced signaling through the Jak-Stat pathway by interacting with Jak3 kinase and Erk mitogen-activated protein kinase (MAPK) signaling. LMP1 also regulates expressions of numerous cytokines (e.g., lymphotoxin) that act as autocrine growth factors for EBV-infected cells (Thompson and Kurzrock 2004).

### 6.6.6 Latent Membrane Protein 2A and 2B

The LMP2 gene encodes two proteins—LMP2A and LMP2B. Both are integral membrane proteins that differ in their $NH_2$ terminal domains. LMP2A carries an extra 118-residue domain that is encoded in exon 1, whereas LMP2B is noncoding. The NH2-terminal domain of LMP2A is cytoplasmic and contains an immune receptor tyrosine-based activation motif. LMP2s are co-expressed with EBNA1 and are important for long-term persistence of EBV in the B cells (Babcock and Thorley-Lawson 2000).

### 6.6.7 EBV-Encoded RNAs 1 and 2

The Epstein–Barr virus–encoded RNAs 1 and 2 (EBER1 and 2) are small, nonpolyadenylated, uncapped, and noncoding RNAs that have a secondary structure with extensive intramolecular base pairing. EBER1 and EBER2 have 167 and 172 nucleotides, respectively (Swaminathan et al. 1991). In the nucleus, these are associated with cellular La antigen and EBER-associated protein. These are normally involved in splicing of other viral transcripts, including EBNAs and LMPs. Moreover, these may inhibit the transcription block induced by eukaryotic initiation factor 2 (eIF-2). These EBER-associated functions help in preventing apoptosis and immune evasion carried out by IFNα and IFNγ. EBER1 and 2 interact with an RNA-activated protein kinase (referred to as PKR), which is a vital mediator of interferon-induced antiviral effects. The detection of IL-10 expression in EBV-positive BL tumors and the observation that the EBERs can induce IL-10 expression in BL cell lines indicate that EBERs may have an important role in the pathogenesis of EBV-positive BL. However, the role of these proteins in the immortalization of infected B cells is controversial (Kitagawa et al. 2000).

### 6.6.8 BamHI-A Rightward Transcripts

Rightward transcripts of the BamHI-A region of the viral genome (BARTs) are detected in all types of EBV latency in several malignancies associated with EBV infection, especially NPC (Chen et al. 1992). BARTs encode a number of potential ORFs, including BARF0, RK-BARF0, A73, and RPMS1. It is not certain whether protein products of BARTs are expressed in malignancy or not.

However, anti-BARF0 antibodies and CTL responses were reported in some tumors suggesting that syntheses of these proteins do occur. In vitro studies suggested that these proteins may regulate EBNA2 and kinase signaling and may also be involved in notch activity (Smith et al. 2000).

### 6.6.9 BamHI-A Rightward Frame

A rightward frame from the BamHI-A region (BARF1) is expressed by its own promoter. It is a useful carcinoma-specific marker that was originally recognized as an early gene product. However, recent studies revealed that it is expressed in type 2 (e.g., nasopharyngeal and gastric carcinoma) and type I (EBV-immortalized epithelial cells) latency. It was suggested that though BARF1 is expressed during latent infection, its expression is increased several fold during the lytic phase. BARF1 protein has transforming properties in B cells and epithelial cells. BARF1 produces a protein (called p29) that shows homology to intercellular adhesion molecule-1 (ICAM-1) and human colony–stimulating factor 1 (CSF1). The p29 protein requires glycosylation, myristylation, and phosphorylation as important steps for posttranslational modification as well as activation. It forms a complex with CSF1 and modulates host immune response by inhibiting macrophage activation, cytokine release, and INFα secretion. BARF1 protein showed strong mitogenic activity. It is massively secreted in the serum of NPC patients. Therefore, BARF1 is responsible for both oncogenicity and immune evasion during EBV infection (Fiorini and Ooka 2008).

## 6.7 EPIDEMIOLOGY

EBV is present in all populations, infecting more than 95% of all individuals within the first four decades of life. In developing countries, infection occurs during childhood. Nearly the entire population of a developing country becomes infected before adolescence. Young children acquire primary EBV infection mainly by oral secretions. Primary infection is usually asymptomatic or produces an acute illness that is often unrecognized. In developed countries, infection is usually delayed until adolescence. In adolescents and young adults, EBV infection appears as a clinical syndrome, named infectious mononucleosis (IM). Three major syndromes of IM are fever, pharyngitis, and adenopathy. Infection during adolescence is also acquired mainly by oral transmission. Salivary tissues are the repositories of EBV, and periodic shedding from such tissue is a recognizable feature of EBV infection. Shedding is continued for a few months after primary infection and then starts falling gradually. For immune-compromised individuals, shedding is increased up to 80%. By the age of 25 years, most people become seropositive to EBV antigens and are not susceptible to reinfection. Apart from oral transmission, transplantation of hematopoietic cells and solid organs may be responsible for viral transmission. Transmission of EBV virus during sexual intercourse has been suggested but still remains controversial. The frequency of early EBV infection has a strong correlation with socioeconomic background because people living in highly populated countries (with low standards of hygiene) acquire primary infection much earlier than those living in less populated countries (with high standards of hygiene).

## 6.8 IMMUNE RESPONSE TO EBV

Like any other microbial infection, EBV infection also develops two types of immune responses; these are (a) innate immune responses and (b) adaptive immune responses.

### 6.8.1 Innate Immune Response

The innate immune system acts as a first line defense against all microbial infection, including viral infection. Viruses normally elicit type I interferon (IFN) response immediately after infection. EBV DNA and proteins are recognized by pattern recognition receptors such as Toll-like receptors (TLRs), which induce an IFN response that facilitates the activation of NK cells. In IM, inflammatory cytokines such as TNF, IL-2, IL-6, IL-1β, and IFN-γ are elevated by CTLs and NK cells. IFN-γ is likely to be responsible for different symptoms (fever, fatigue, headache, etc.) that are observed during IM. Apart from inflammatory cytokines, the immunosuppressive cytokines such as IL-10 and tumor growth factor-beta (TGF-β) are also increased during EBV infection. The NK cells may also play some important roles in the inhibition of EBV-induced transformation of resting B cells as well as in the control of lytic infections. This is substantiated by the fact that NK cell numbers are inversely correlated with virus load. Surprisingly, EBV does not exhibit any NK immune evasion strategy, in contrast to multiple escape routes that are noticed in cytomegaloviruses.

### 6.8.2 Adaptive Immune Response

The first humoral response detected in EBV infection is the IgM class antibody, which is directed against the VCA. Most of the patients develop anti-VCA IgG antibodies that show a peak during the first two to four months and then persist throughout the life of the host. On the other hand, EBNA1-specific IgG appears after three months of primary infection and persists permanently in the host's body. Other antibodies are anti-gp350, anti-early antigen diffuse (EA-D) antibodies. Although anti-gp350 is noticed in 60–80% of infected patients, these are not useful as diagnostic markers. It has been suggested that antibodies allow EBVs to be maintained in tonsillar epithelium when their initial reservoir (i.e., B cells) is depleted by CTLs. Apart from these, CD4 and CD8 T cells for latent and lytic antigens are also observed during primary infection. In IM, a robust response is observed for CD8 T cells specific for lytic antigens. The lymphocytosis in the blood that characterizes IM consists largely of CD8 T cells specific for lytic antigens. The CD4 and CD8 T cells for latent EBV antigens do not show a sudden upsurge in their levels after infection. However, these persist throughout the life of the host. CD8 T cells are crucial for the control of EBV as evidenced by the occurrence of EBV lymphomagenesis and lymphoproliferation in immunocompromised patients.

## 6.9 EBV-ASSOCIATED CANCERS

To become oncogenic, EBV must follow some biological strategies, and these include (a) maintenance of viral genome in the cell, (b) avoiding destruction of the host cell, and (c) protection from normal immune surveillance of the host

(Thompson and Kurzrock 2004). To maintain viral episome in the host cells, EBV establishes latent infection in B lymphocytes after primary infection and ensures transmission to cell progeny during replication of B lymphocytes. EBV latent genes induce an activated phenotype in the infected host cells. These cells are not transformed and remain as activated phenotypes. In the case of normal individuals, expansion of the activated cells is prevented by CTL cells (Figure 6.4). This is supported by the fact that infected B cells are able to grow in vitro only when they are cultured without T cells. If activated B-lymphocyte cells proceed unchecked by CTLs or acquire oncogenic mutations, these may become neoplastic. Gradually, these activated B cells enter the resting B-cells' memory compartments where only EBNA1 is expressed. Such restricted expression of oncogenic gene products renders these infected memory B-cells nonpathogenic. These resting EBV cells may intermittently enter the lytic cycle during which viral replication occurs. Viral replication is followed by the lysis of the host cells and the release of virions that infect a greater number of cells. The oncogenicity of EBV mainly stems from immunosuppression where infected cells increase in numbers and produce virally expressed genes that stimulate multiple interconnecting oncogenic pathways. Therefore, years after primary infection, different EBV-associated malignancies may emerge (Ocheni et al. 2010). Some of the EBV-mediated malignancies and their characteristics are shown in Table 6.1.

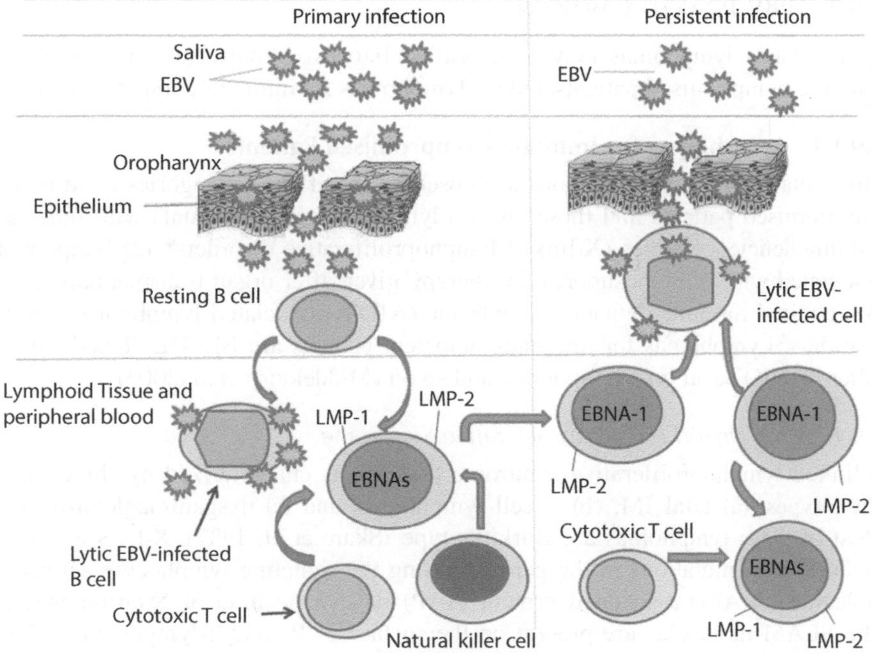


**FIGURE 6.4 (See color insert.)** Schematic representation of primary and persistent infection of EBV.

**TABLE 6.1**
**Characteristics of Some Epstein–Barr Virus-Mediated Malignancies**

| EBV-associated carcinomas | Role of EBV | Characteristics |
|---|---|---|
| Burkitt's lymphoma | 100% in endemic variant, 20% in sporadic variants, and 30–40% in immunodeficiency-related variant | Only EBNA-1 is detected together with EBERs and micro RNAs |
| Posttransplant, lymphoproliferative disorders | Overall, 60–80% PTLD patients, 100% in early-onset type | Latency type III pattern is observed, occasionally type II and III patterns are also noticed |
| Hodgkin lymphoma | 40% of classical Hodgkin's lymphoma (especially mixed-cellularity type) | Latency type II is observed |
| T/NK nasal type lymphoma | Approximately 100% | T-cell antigens and T-cells receptor gene arrangement are absent; NK cell marker CD56 is expressed |
| Nasopharyngeal carcinoma | Nearly 100% nonkeratinizing tumors | LMP-1 expression is noticed |
| Gastric carcinoma | 10% of all cases worldwide | Restricted latency type I is observed |

### 6.9.1 EBV-Induced Lymphomas

EBV-induced lymphomas may be classified into two groups: (a) lymphomas for immune-compromised patients and (b) lymphomas for immune-competent patients.

#### 6.9.1.1 Lymphomas for Immune-Compromised Patients

Three main types of lymphomas are observed in different categories of immune-compromised patients and these are: (1) lymphomas in individuals with inherited immunodeficiency states (X-linked lymphoproliferative disorders), (2) lymphomas associated with immunosuppressant therapy given after organ transplantation, and (3) acquired immune deficiency syndrome (AIDS)–associated lymphoproliferative disorders. Lymphomas for immune-competent patients are BL, HL, T cell/natural killer (T/NK) nasal type lymphoma, and so on (Middeldorp et al. 2003).

##### *6.9.1.1.1 X-Linked Lymphoproliferative Syndrome*

X-linked lymphoproliferative syndrome (X-LPS) is characterized by three major phenotypes: (a) fatal IM, (b) B-cell lymphomas, and (c) dysgammaglobulinemia. Most of these lymphomas are Burkitt's type (Skare et al. 1987). X-LPS is caused by hereditary mutations in the gene encoding the signaling lymphocyte activation molecule (SLAM)-associated protein (SAP) on q25 position of X-chromosomes. The SLAM molecules are present on the surface of B- and T-lymphocytes; these initiate immune system related signal transduction pathways. SAP is present only on T-lymphocytes, and it acts as a negative regulator of SLAM molecules. Abnormality in SAP protein therefore produces extensive overreaction of T-lymphocytes during primary EBV infection that leads to fatal IM followed by virus-associated hematophagocytic syndrome (Latour et al. 2001). This syndrome destroys bone marrow and liver. Seventy percent of patients generally die within a few weeks; whereas 100% of patients die by the age of 40 due to hepatic necrosis and bone marrow failure. The only curative treatment for this syndrome is allogeneic bone marrow transplantation.

###### *6.9.1.1.2 Posttransplant Lymphoproliferative Disorders*

There is a high risk of lymphoproliferative disorders after solid-organ and stem cell transplantations. The World Health Organization (WHO) classified posttransplant lymphoproliferative disorders (PTLDs) into two groups: (a) polyclonal early-onset PTLDs and (b) monoclonal late-onset PTLDs (Borish et al. 2001). A strong correlation exists between early-onset PTLDs and EBV infection but such relation is not observed with late-onset PTLDs. The late-onset PTLDs are further classified into three types and these include (a) Burkitt's or Burkitt's-like lymphoma, (b) diffuse large B-cell lymphoma, and (c) HL. The occurrence of PTLDs, however, depends on several factors such as the type of organ being transplanted, the status of EBV infection, the immunotherapy received, and the condition of the transplant recipient and the donor. Overall, the chances of developing PTLDs range from 0.5 to 30%, and it mainly varies depending on these factors. This disorder occurs most frequently in combined liver–kidney transplants, followed by cardiac, liver, lung, and kidney transplants. The chance of PTLD is more common for solid-organ transplant compared to hematopoietic stem cell transplantation. This may be due to the fact that the immunosuppressant therapy is more prolonged in the case of solid-organ transplant. In general, a balance exists between EBV load and the immune system of the transplant recipient. Immunosuppressant-induced loss of T cells may ultimately lead to the accumulation of EBV-infected cells in the body with enhanced viral replication (Ocheni et al. 2010). Tumor cells mainly express latency III pattern of EBV gene expression (Riddler et al. 1994). Occasionally, latency type I and type II are found. The expression of the latent antigens, especially LMP-1 and EBNA 2, was observed to have potential effects on the infected cell. A strong correlation was found between PTLD and polyclonal B-cell hyperplasia. Because the occurrence of PTLDs is directly correlated with immunosuppressant doses, withdrawal of therapy may help in increasing the numbers of $CD8^+$ CTLs that may in turn prevent PTLD. An alternate strategy involves the application of the monoclonal antibody rituximab that may effectively eliminate B cells from the host and thus may prevent PTLD. Prevaccination of high-risk patients with gp350 or peptide-based vaccine may reduce the problem of PTLD (Middeldorp et al. 2003).

###### *6.9.1.1.3 AIDS-Related Lymphomas*

Lymphomas associated with human immunodeficiency virus (HIV)/AIDS consist of a group of diverse lymphoproliferative disorders that arise mainly in the terminal phases of HIV infection. These include central nervous system lymphomas, HL, BL, primary effusion lymphoma, large B-cell lymphoma, and others. The AIDS-related lymphomas (ARLs) are mostly of B-cell origin and contain the patient's intrinsic EBV. Most ARLs contain monoclonal EBV genome. Both type I and type II latency gene products are expressed. In these cases, EBV infection and immunosuppression may increase the t(8;14) or t(8;2) or t(8;22) translocation of *c-myc* gene. It involves translocation of *c-myc* gene from chromosome 8 to the Ig locus on chromosome 14 (heavy chain), 2 or 22 (j or k light chain). It results in the overexpression of proto-oncogenic *c-myc* gene (Diebold et al. 1997). The *c-myc* plays a critical role in regulating cell proliferation, differentiation, and apoptosis. It is a sequence-specific transcriptional activator responsible for the expression of a large number of target genes. The *c-myc* expression is tightly regulated under normal condition. However after translocation, it is deregulated and gives rise to several oncogenic events.

#### 6.9.1.2 Lymphomas in Immune-Competent Patients

Burkitt's, Hodgkin's, and T/NK nasal type lymphomas are the most important EBV-mediated lymphomas that may occur in immune-competent patients (who do not have any history of immense acute or chronic immunosuppression).

##### *6.9.1.2.1 Burkitt's Lymphoma*

BL is an aggressive B-lymphocyte cell tumor that exhibits a very rapid cell proliferation rate (doubling time of tumor is 24–48 h). It was originally found in Africa where it accounts for approximately 50% of all childhood cancers. BL is classified into three types: (a) the endemic type, (b) the sporadic type, and (c) the immune-deficient type. In all types of BL, translocation of *c-myc* gene is observed. In 80% of cases, t(8;14) translocation is observed. In rest of the cases, either t(8;2) or t(8;22) translocation is involved. Irrespective of types, BL is morphologically composed of monomorphic medium-sized cells that are often scattered with macrophages, called "starry sky macrophages." The BL tumor cells usually express IgM, B-cell markers (e.g. CD19, CD20, CD22, etc.), markers of germinal center centroblasts (e.g., CD10, BCL64, etc.), and the human germinal center-associated lymphoma (HGAL) protein. The involvement of germinal center is indispensable for the pathogenesis of all types of BL. The endemic type of BL is ascribed to the climate-dependent distribution of the disease. An astonishingly high chance of BL has been found in regions where malaria prevails (e.g., the lymphoma belt of equatorial Africa). Although the exact relation between malaria and BL is not properly understood, it is hypothesized that malaria-mediated immunosuppression (by lowering the expression of T-cell responses including those directed against EBV) results in B-cell activation (by a mechanism that involves cysteine-rich interdomain region 1$\alpha$ protein or enhanced expression of IL-10) (Brady et al. 2007). In endemic tumors, the latent gene expression pattern of EBV is restricted. Only EBNA1 is expressed along with EBERs (latency type I). LMPs are rarely detected. This raises a question as to how these two proteins may give rise to tumor by *c-myc* translocation. It was noticed that EBNA1 inhibition in the cell increases the rate of apoptosis before any change is noticed in the EBV genome. This suggests that EBNA1 expression is somehow related to inhibition of apoptosis of the B cells. A group of investigators suggested that EBNA1-induced recombinase-activating gene (RAG-1 and -2) expressions may be responsible for translocation of the gene (Srinivas and Sixbey 1995). Also, the role of EBERs is also suspected because these are responsible for the production of B-cell growth factor IL-10, and EBV-positive cell lines are also found to be abundant with this growth factor. Sporadic-type BL is found outside the African region. Yet, it is morphologically similar to the endemic type, and it is found in all age groups (mainly affecting the abdominal cavities of the patients). Immune-deficient type BL primarily includes HIV-associated BL (described earlier). There is little difference between endemic and nonendemic BL. In endemic BL, bone marrow is less frequently involved, and patients are more responsive to chemotherapy (Ocheni et al. 2010).

###### *6.9.1.2.2 Hodgkin Lymphoma*

HL (previously known as Hodgkin's disease) is characterized by the presence of clonal, multinucleated Reed–Sternberg malignant cells that are derived from B cells. Five types of HL are known so far: (a) mixed-cellularity, (b) nodular-sclerosis, (c) lymphocyte-rich, (d) lymphocyte-depleted, and (e) lymphocyte predominant. The first four types are known as classical HL (cHL). All cHLs are related to EBV infection. The relation between HL and EBV is mainly justified by the fact that monoclonal EBV episomes are found in the Reed–Sternberg cells. In addition to this, people having a past history of IM have a fourfold higher risk of developing HL. Moreover, the antibody titers against EBV capsid antigen are also increased several fold during HL. EBV infection is associated with some 45% of mixed-cellularity cHL cases. EBV-positive cHL is more common in males and shows bimodal age distribution (<10 years and >50 years). EBV negative cHL is more common in adolescents, and it does not show any gender predisposition.

Reed–Sternberg cells show a type II latency pattern with higher expressions of EBNA-1, EBERs, LMP-1, -2A, and -2B. Like BL, malignant HL cells may also acquire serious Ig mutation. It is postulated that LMP-1 activates NF-κB, which in turn prevents apoptosis and contributes to the immortalization of B cells. However, the exact function of LMPs in the pathogenesis of HL is not properly understood. Like BL, EBV-induced HL also demonstrates geographical and socioeconomic predominance. It has been noticed that people having a low socioeconomic standard are more prone to develop this disease. Association between EBV and HL is less in developed countries. In North America and Europe, 20%–50% HL is associated with EBV. For developing countries, this association is little bit higher. For example, China has 60–65% HL cases liked to EBV. However, for underdeveloped countries, this association may approach even 90%–100%. This astonishingly higher percentage of EBV-induced HL in underdeveloped countries may be due to immunosuppression similar to that observed in African BL in a malaria-infected population.

###### *6.9.1.2.3 T cell/NK Nasal Type Lymphoma*

In addition to B cells, EBV may also infect other cells. Several types of non-B cell, non-HL are related to EBV. Among non-HLs, two important diseases are discussed here: these are T/NK cell lymphoma and angioimmunoblastic lymphadenopathy. T/NK cell lymphomas occur in the nasal or upper aerodigestive area and show several unique genotypic and phenotypic features. This lymphoma is characterized by the absence of T-cell antigens, expression of NK cell marker CD56, and the absence of T-cells receptor gene arrangement. Angioimmunoblastic lymphadenopathy is a typical B-cell lymphoma where B-cell clones are present behind T-cell clones. EBV was found in the neoplastic and non-neoplastic T cells as well.

### 6.9.2 Epithelial Malignancies

EBV is also associated with many epithelial malignancies. Among these, NPC is the most important. Apart from this, EBV is also a causative factor for gastric carcinoma, breast cancer, and leiomyosarcoma.

#### 6.9.2.1 Nasopharyngeal Carcinoma

NPC is the most common cancer originating in the nasopharynx. It is the uppermost region of the pharynx (throat), behind the nose—where the nasal passages and auditory tubes join the remainder of the upper respiratory tract. Although undifferential NPC (where cancer cells do not seem different from normal cells) is directly associated with EBV-mediated carcinogenesis, such association has remained controversial for differential NPCs (Niedobitek 2000). Undifferentiated NPC is mainly common in males in their mid-40s. NPCs have been reported in almost all parts of the world. However, there is an exceptionally high prevalence of this cancer in the Chinese population, especially in Cantonese males irrespective of whichever part of the world they live. This malignancy is primarily found in North African and Inuit populations. The association between NPC and EBV infection is well known. However, the exact pathogenesis is still obscure. This is because of the fact that mature NPC cells do not show EBV infection after transformation. However, before transformation, these cells exhibit EBV infection. Therefore, it is believed that EBV infection is actually affected by other environmental factors. This hypothesis is strengthened by the fact that some dietary carcinogens (i.e., nitroso derivatives such as *N*-nitrosopyrrolidine, *N*-nitrosodimethylamine, and *N*-nitrosopiperidine, present in salted fish) possess a high risk for developing NPCs. Exposure to smoke or chemical pollutants such as nickel and other trace elements was also reported to be associated with NPCs. EBV latency gene expression in NPC was found to be intermediate compared to that seen in latency type I and type II (Heussinger et al. 2004). Latent viral EBNA1 and EBER genes are expressed along with LMPs. Although LMP1 is found in every NPC and the growth-promoting effect of LMP-2A is apparent, this later protein is found only in 50% of NPC cases. In addition, LMP-1 mRNA is found in 35% of cases. Early detection of NPC is frequently the result of detecting IgA antibody titers against VCA. Among all EBV-associated proteins, the role of LMP1 and E-cadherin (CDH-1) is prominent for the epigenetic alterations observed in NPCs. The CDH-1 promoter methylation was found to be highly correlated with EBV infection and subsequent rapid metastases of NPCs.

#### 6.9.2.2 Gastric Carcinoma

There are two different types of GCs: (a) intestinal and (b) diffuse. EBV is responsible for approximately 10% of both types of GCs. These EBV-mediated GCs demonstrate a characteristic diffuse "lace pattern." Unlike BL and NPCs, EBV-mediated GC is a nonendemic disease. The exact role of EBV in GC is unclear because several other factors, such as *Halicobacter pylori* infection, food indigestion, or mental stress, may also contribute to the formation of GCs. Several lines of evidence indicate that EBV infection of epithelial cells is due to "cell-to-cell" contact that may be the result of accidental EBV infection of gastric epithelial cells by EBV-infected activated lymphocytes. However, the exact mechanism of action for this transfer is still unclear. EBV shows a unique expression pattern in GCs, which is defined as restricted latency type I. The expressed EBV genes are EBNA-1, EBER-1, EBER-2, and BARF0. Expressions of LMPs (LMP-1, -2A) are rarely noticed (Ocheni et al. 2010). EBER-1 and -2 are expressed in all cases. However, their role in this malignancy is uncertain. Although LMP2A is expressed in small amounts in some cases of EBV-mediated GCs, some mechanisms of this protein have been proposed. LMP-2A leads to the induction of surviving protein that inhibits apoptosis of infected malignant cells via NF-κB signaling. LMP-2A is also found to activate the Ras-PI3k-Akt signaling pathway that inhibits TGFβ-induced apoptosis. Growth promotion by EBV infection is also observed through the secretion of insulin-like growth factor-1 (IGF-1), which is an important autocrine growth factor (Lizasa et al. 2012).

## 6.10 DIAGNOSIS OF EBV-MEDIATED MALIGNANCIES

Currently available techniques for the diagnosis of EBV-mediated malignancies are as follows:

### 6.10.1 In Situ Hybridization

In this method, EBER transcripts are detected by an in situ hybridization (ISH) procedure. This is the most common diagnostic process to date.

### 6.10.2 Immunohistochemistry and Immunocytochemistry

In some malignancies (e.g., PTLD, HL, and IM), immunochemistry of LMP1 is applied as an alternative procedure of ISH. However, it is less reliable for EBV-induced non-HL. Apart from LMP1, other targets for immunochemical estimation include EBNA1, EBNA2, and LMP2A.

### 6.10.3 Sequence-Based Viral Nucleic Acid Amplification

Sequence-based viral nucleic acid amplification is a rapid as well as sensitive technique for the estimation of viral load. Here, EBV nucleic acids are detected by sequence-based amplification of viral nucleic acids—with the help of polymerase chain reaction (PCR). The major limitation is that it cannot differentiate between latency and malignancy. However, it may be used for the detection of undifferentiated carcinomas as well as for monitoring PTLD and NPC.

### 6.10.4 Serological Approach

This method involves EBV-specific Enzyme-linked immunosorbent assay (ELISA) or immunofluorescent assays of early antigens and immunoglobulins. However, the technique is only useful for immunocompetent hosts. In NPC, IgA antibodies are elevated and these may be detected by this approach.

### 6.10.5 Miscellaneous Approaches

These include Southern blot analysis of EBV DNA, in vitro culture of EBV and EBV-infected lymphocytes, electron microscopy analyses of morphological changes in EBV infection, and others.

## 6.11 TREATMENT OF EBV INFECTIONS

Despite growing understanding of the role of EBV in the pathogenesis of malignancy, the optimal management of EBV-associated tumors remains unsatisfactory. However, recent applications of antiviral agents, immune-based therapies, and specific monoclonal antibodies demonstrated some encouraging outcomes. In the posttransplant setting, EBV-related lymphomas may also be managed by reducing the level of immune suppression, although this strategy may threaten the integrity of the transplant.

### 6.11.1 Antivirals

There are several antiviral compounds that entered the clinical setting and showed potential anti-EBV activities. However, it has been difficult to demonstrate reproducible antitumor effects, and reports of tumor regression remain inconsistent to date. The majority of these drugs are broad-spectrum, antiherpesvirus and anticytomegalovirus agents that vary in their efficacies against EBV. These include acyclic nucleoside analogs acyclovir (Colby et al. 1980, 1981), ganciclovir (GCV) (Jacobson et al. 1987; Purifoy et al. 1993), penciclovir (Boyd et al. 1987, 1988; Harnden et al. 1987), and the prodrugs valciclovir (Purifoy et al. 1993), valganciclovir (Cvetkovic and Wellington 2005), famcyclovir (Vere Hodge et al. 1989), valaciclovir (a prodrug of acyclovir), foscarnet (Wagstaff and Bryson 1994), cidofovir (Plosker and Noble 1999), and adefovir (De Clercq et al. 1989) (Figure 6.5).

Acyclovir and ganciclovir are not drugs of choice because these require activation by viral thymidine kinase. These pharmacological agents are nucleoside analogs, which are converted to their monophosphate form by viral thymidine kinase and subsequently to active triphosphates by host cellular enzymes. The activated triphosphate then inhibits viral DNA polymerase. In EBV-associated lymphoid disorders (in contrast to the situation in EBV lytic disease), the virus rarely exhibits lytic replication (only a small number of cells are in lytic phase and these also die within a few days) (Hong et al. 2005). Therefore, viral thymidine kinase enzyme is not expressed.

To circumvent this problem, arginine butyrate, which selectively activates EBV thymidine kinase genes in lymphoma cells, has been administered along with gancyclovir; this combination demonstrated some efficacy in patients with EBV-associated lymphoproliferations after solid-organ transplantation, especially in patients who are resistant to conventional radiation and/or chemotherapy. This combination produced effective response in a clinical trial because pathologic examination showed complete remission of the EBV lymphoma (with no residual disease) in more than 50% of patients in a single three-week course of this combination therapy. Possible side effects of this therapy include nausea and reversible lethargy but only at the highest doses (Mentzer et al. 2001).

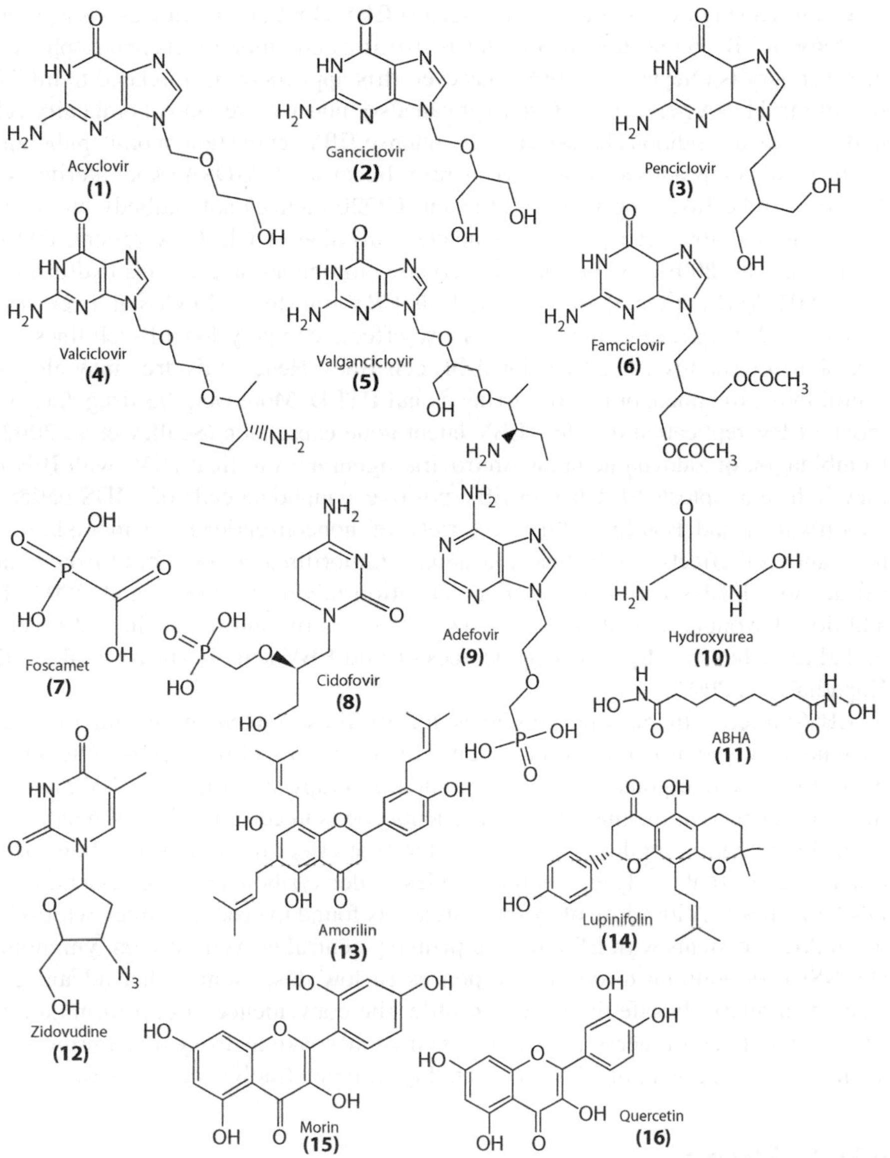

**FIGURE 6.5** Antiviral drugs active against EBV.

Another antiviral agent (foscarnet) does not require activation by viral thymidine kinase, and it acts directly against the viral DNA polymerase. Treatment with foscarnet demonstrates continuous as well as complete remission in patients with EBV-associated lymphoproliferations. These clinical experiences may justify the efficacy of antiviral treatment in EBV-associated lymphoproliferations (Oertel and Riess 2002).

Cidofovir (Figure 6.5) is also active against EBV DNA polymerase and is a potent inhibitor of EBV replication in vitro. It has striking antitumor effects in nasopharyngeal xenografts (Neyts et al. 1998); however, this appears to be unrelated to inhibition of viral DNA polymerase. Cidofovir can also induce regression of oral hairy cell leukoplakia (a condition characterized by intense EBV replication in oral epithelium of immune-compromised patients (generally because of AIDS)) (Abdulkarim and Bourhis 2001). Taken together with the anti-CD20 monoclonal antibody known as rituximab, cidofovir can produce complete remission of CD20-expressing PTLD (Hanel et al. 2001). A histone deacetylase inhibitor—azelaic bis-hydroxamic acid (ABHA) (Figure 6.5)—effectively kills EBV-transformed LCLs at a very low dose ($IC_{50}$ 2–5 μg/mL) within 48 h. It is also effective in polyclonal B-cell lines and was shown to be toxic to seven clonal BL cell lines. Hence, this drug may also be useful in the treatment of late-occurring clonal PTLD. Moreover, this drug does not induce EBV replication or affect EBV latent gene expression (Sculley et al. 2002). Combination of zidovudine (an antiretroviral agent used to treat HIV) with IFN-α may induce apoptosis (in vitro) in EBV-positive lymphoma cells of AIDS patients (Abdulkarim and Bourhis 2001). A variety of nonconventional compounds may have antiviral effects. For instance, lavanones (amorilin and lupinifolin) from plant extracts block EBV early-antigen (EA) activation in vitro (Itoigawa et al. 2002). In addition, flavonoid derivatives synthesized from morin and quercetin and several herbal remedies may have varying degrees of anti-EBV activity (Iwase et al. 2001; Kapadia et al. 2002).

EBV-targeted therapeutic strategies for viral-associated malignant diseases have not received primary consideration. This is because of the fact that latent herpesviruses are irresponsive to antiviral chemotherapy, and the role EBV plays in the maintenance of the malignant cell phenotype is uncertain. The antineoplastic agent hydroxyurea produces an antiproliferative effect by inhibiting ribonucleotide reductase that catalyzes ribonucleotides to deoxyribonucleotides essential for DNA synthesis. A low dose of hydroxyurea was found to produce direct cytotoxicity in AIDS patients with EBV-related primary central nervous system lymphoma (PCNSL). In addition to durable responses in low dose, some other advantages (such as a relatively safe therapeutic profile, the convenience of oral formulation, and excellent central nervous system penetration) make hydroxyurea a promising agent for the treatment of EBV-associated lymphoma (Slobod et al. 2000).

### 6.11.2 Vaccines

Prophylactic vaccines against some pathogenic viruses are of excellent public health significance in terms of safety and effectiveness. There is a great demand for effective vaccines against EBV. In spite of several attempts to formulate EBV vaccine, no candidate vaccine has yet been proven sufficiently effective for commercial purpose (Cohen et al. 2011). EBV vaccines should be able to either block primary EBV infection or significantly reduce the EBV load during primary EBV infection. Because immunization with whole viral proteins did not elicit an efficient CTL response, focus was directed to developing peptide vaccines based on defined epitope sequence. Two broad approaches have been taken into consideration for designing effective vaccines for controlling EBV-associated diseases. One is EBV structural antigens as target antigens and another is latent antigens as potential vaccine candidates.

#### 6.11.2.1 Epstein–Barr Virus Structural Antigens as Target Antigens

EBV is enveloped by a membrane composed of four major virus-specific proteins (i.e., gp350, gp220, gp85, and p140.40). Most strategies for developing EBV vaccines have focused on the virus membrane antigen, which consists of at least three glycoproteins. Prophylactic vaccines are known to function primarily via the induction of virus-neutralizing antibodies; gp350 contains the main neutralization epitope and is the primary target of the virus-neutralizing antibody response. In the past decades, several efforts were made to develop subunit preparation of gp350 (recombinant and affinity purified). Recombinant formulations of gp350 showed significant protection against EBV-induced B-cell lymphomas in cotton-top tamarins. The recombinant gp350 vaccine was found to produce neutralizing antibodies in a phase I/II trial (Moutschen et al. 2007). This formulation showed a good safety profile and was well tolerated. Although the vaccine was proven effective in preventing the development of EBV-induced IM, it showed no efficacy in preventing asymptomatic EBV infection (Sokal et al. 2007). Highly purified gp350 induced high levels of neutralizing antibodies and also inhibited tumor formation in cotton-top tamarins when administered subcutaneously with an adjuvant such as muramyl dipeptide or immune-stimulation complexes (Morgan et al. 1988a). A number of recombinant vectors, including vaccinia-gp350 and adenovirus 5-gp350, have also been successfully used in these animals to block tumor outgrowth (Morgan et al. 1988b). Nevertheless, the development of neutralizing antibody titers in vaccinated animals did not show any overall correlation with protection from EBV infection (Wilson et al. 1996). Very low levels of neutralizing anti-gp350 antibodies are present in the saliva of healthy EBV-immune donors. This finding may suggest that such antibodies are unlikely to be the basis of long-term immunity in healthy seropositive individuals (Yao et al. 1991). Overall, a vaccine solely based on gp350 did not completely prevent the infection of every single B lymphocyte or epithelial cell.

It has been suggested that polyantigens containing several antigenic determinants of gp220 and gp350 may elicit a more promising response because these proteins are able to modulate the immune responses of patients suffering from diseases such as NPC, IM, or EBV-related BL (Wolf et al. 1997). The B-cell activation and immunoglobulin secretion by costimulation of the receptor may also be enhanced for antigen gp350/220 (Mond and Lees 2002); gp85 is another potential target for vaccine design (Khanna et al. 1999). Khanna et al. (1999) identified CTL epitopes within the EBV structural antigen gp85. Using ex vivo primary effectors, strong reactivity to gp85 peptides was observed. An animal model system further revealed that gp85 epitopes are capable of generating structural antigen-specific CTL responses and reducing infections with the virus expressing gp85.

#### 6.11.2.2 Latent Antigens as Potential Vaccine Candidates

The EBV structural antigens are not expressed in latently infected B lymphocytes. Hence, therapeutic EBV vaccine efforts have been focused on latency antigens expressed in EBV-associated diseases. EBNA1 has been identified as a vaccine antigen. In a specific embodiment, a purified protein corresponding to EBNA1 elicited a strong CD4+ T-cell response. LMP1 and LMP2 are target antigens available for expanding CTL responses in patients with HD and NPC. Duraiswamy et al. (2003) developed a recombinant poxvirus vaccine that encodes a polyepitope protein derived from LMP1. Human cells infected with this vaccine were efficiently recognized by LMP1-specific CTLs from HLA A2 healthy individuals. The outgrowth of LMP1-expressing tumors in HLAA2/Kb mice was also reversed by this vaccine. Because EBNA1 is the only viral protein expressed in all EBV-positive proliferating cells in healthy EBV carriers and in all EBV-associated malignancies (Villegas et al. 2010), a possible vaccine should include EBNA1 added to another latent or lytic gene. Taylor et al. (2004) developed such a modified vaccine with CD4+ epitope-rich C-terminal domain of EBNA1 fused to full-length LMP2. In this vaccine, LMP2 was the source of subdominant CD8+ T-cell epitopes. This vaccine showed immunogenicity to CD4+ and CD8+ T cells.

#### 6.11.2.3 Problems in EBV Vaccine Development

The development of EBV vaccine has been delayed due to several reasons. Some of these are as follows: (a) Although gp350 is the most promising agent for vaccine development, it is unknown whether its combination with other EBV antigens might be more effective in prophylactic treatment; (b) there is no good surrogate marker for EBV-mediated tumor development, and it poses great difficulty in performing clinical trials; (c) the lack of knowledge about exact immune response against EBV infection is still a problem for effective vaccine development; (d) there are limitations in animal models for studying protection against EBV infection and disease; and (e) some additional information is required regarding the socioeconomic dependence of IM and accessing cost-effective prophylactic vaccines (Cohen et al. 2005).

### 6.11.3 Immunotherapy

The major drawback of antivirals is that these have no influence on the underlying immune suppression that favors EBV-driven tumorigenesis. Adoptive immunotherapy using EBV-specific CTLs, though time consuming and work intensive, may overcome this disadvantage (Gottschalk et al. 2002). The adoptive transfer of antigen-specific cytotoxic T lymphocytes offers a safe and effective therapy for certain viral infections and could prove useful in the eradication of tumor cells. Heslop et al. (1996) reported the long-term detection of gene-marked EBV-specific CTLs in immune-compromised patients at risk for development of EBV lymphoproliferative disease. Infusions of these cell lines not only restored cellular immune responses against EBV but also established populations of CTL precursors that could effectively respond to viral infection for as long as 18 months. The adoptive transfer of EBV CTLs was successfully applied in the treatment of PTLD.

In 2010, Heslop et al. (2010) tried to address the long-term efficacy, safety, and practicality of EBV-specific CTL immunotherapy. They studied 114 patients who received infusions of EBV-specific CTLs to prevent or treat PTLD. None of the 101 patients developed EBV-positive PTLD. A gene-marking component was used to demonstrate the persistence of functional CTLs for up to nine years. The conclusion is that CTL lines provide a safe and effective prophylaxis or treatment for PTLD. However, Subklewe et al. (2005) compared dendritic cells (DCs) with LCLs for T-cell stimulation against dominant and subdominant EBV antigens. The DCs expand tenfold more EBNA3A- and LMP2-specific T cells than LCLs and expand EBV-specific T-cell responses more efficiently than LCLs. Kuzushima et al. (2009) have introduced EBNA1 and LMP1 mRNAs into APCs. These modified cells can induce EBV-specific CTLs, inhibit the outgrowth of EBV-infected B lymphocytes, and then lyse EBV-infected NK lymphomas and NK cells.

### 6.11.4 Gene Therapy

Gene therapy strategies for introducing novel compounds or cytotoxic gene products (e.g., HSV1-TK gene into EBV-infected tumor cells followed by GCV therapy) have also gained attention in the past few decades. Such strategies involve the inhibition of EBV oncoproteins or cellular genes that are critical for virus-associated oncogenesis. Chia et al. (2003) administered a nucleic acid molecule that could limit tumor cell growth and/or cause tumor cell death. The molecule comprises an EBNA1 responsive promoter region operatively linked to a gene necessary for viral replication. This method may be used to treat EBV-associated tumors. Franken et al. (1996) introduced a suicide gene regulated by the expression of EBNA2 into latent EBV-infected cells. Cells expressing EBNA2 are demonstrated to be more selectively sensitive to GCV. There is also a complete macroscopic regression of established B-cell lymphomas in severe combined immunodeficiency (SCID) mice. However, gene therapy suffers from a common problem of accurate delivery to the appropriate disease sites.

### 6.11.5 SiRNA Therapy

Therapies using drugs targeted at latent proteins mainly expressing in tumors such as LMP1, LMP2A, or EBNA1 are promising. These proteins are critical to the immortalization and proliferation of cells and also for evading immune responses. The efficacy of small interfering RNA (siRNA) was manifested by Mei et al. (2006) who constructed a plasmid encoding a 21-nt siRNA that specifically and efficiently interfered with LMP1; siRNA can induce apoptosis in EBV-positive lymphoma cells.

### 6.11.6 Monoclonal Antibodies

The anti-CD20 monoclonal antibody designated rituximab has shown significant success in the treatment of a variety of CD20-expressing lymphomas. It is also an effective agent in the management of EBV-related lymphoproliferative disorders. A response rate of 69% (mostly complete responses) was reported in a group of transplant recipients (either solid-organ or hematopoietic stem cell transplant) (Haddad et al. 2001). Because these lymphomas use IL-6 as a growth factor, anti-IL-6 monoclonal antibodies have also been tried. The reported response rate was found to be 67% (8 out of 12 patients treated) (Hong et al. 2005).

## 6.12 STRATEGIES FOR NOVEL SYNTHETIC NUCLEOSIDE ANTIVIRALS

Nucleoside antivirals have been used in the clinical treatment of EBV-associated diseases since the late 1970s. In spite of the impressive efficiency of these nucleoside analogs in the treatment of EBV infection, all these compounds suffer from the same drawbacks, including toxic side effects, poor bioavailability, and potential mutagenesis. Nearly all clinically effective nucleoside analogs target the same active sites on viral DNA polymerase molecules (Coen and Schaffer 2003). Because of undesirable adverse effects as well as resistance, the search for more effective novel compounds is still in progress. In the past few decades, several authors reported their work on the syntheses and biological evaluation of novel nucleoside analogs active against EBV infection and malignancies. In most cases, African American BL (Daudi cell line) or H-1 (EBV-immortalized B cells) were used for the determination of antiviral potencies of these compounds. Qiu et al. (1998a) synthesized (*R*)-(-)- and (*S*)-(+)-synadenol. Both these compounds were active against EBV, but (*R*)-(-)-synadenol (compound **17**) (Figure 6.6) was found to be more potent ($EC_{50}$ = 0.09 μM) than S-enantiomer as well as the standard drug GCV. This group of authors (Qiu et al. 1998b) synthesized some methylenecyclopropane nucleoside analogs. Four compounds were found to be more active than ganciclovir. Among these, compound **18** (Figure 6.6) was the most potent against H-1 ($IC_{50}$ = 0.2 μM) and Daudi ($IC_{50}$ = 3.2 μM) cells. Guan et al. (2000) synthesized some spirocyclic analogs of 2′-deoxyadenosine and 2′-deoxyguanosine. These compounds were active against Daudi cell lines. Phosphoralaninate (compound **19**) (Figure 6.6) was the most active against EBV, with $EC_{50}$ = 2.8 μM.

Some *Z*- and E-methylenecyclopropane adenosine analogs were synthesized and tested antiviral activities against HCMV, HSV-I, HSV-2, EBV, VZV, HBV, and HIV-1 (Qiu et al. 2000a). Compound **20** (Figure 6.6) exhibited moderate activity against H-1 cells ($EC_{50}$ = 36 μM, $CC_{50}$>50 μM), but it was inactive against Daudi cells. Qiu et al. (2000b) synthesized some *R* and *S* enantiomers of 2-aminopurine methylenecyclopropane analogs of nucleosides. One of the *S* enantiomers, aminopurine methylcyclopropane analog (compound **21**, Figure 6.6) showed the most potent activity ($EC_{50}$ = 0.44 μM against H-1 cell and 1.0 μM against Daudi cells). Wang et al. (2002a) reported compound **22** (Figure 6.6) as having activity against EBV.

In another communication, Wang et al. (2002b) synthesized some methylenecyclobutane analogs of 2′-deoxyadenosine, 2′-deoxyguanosine, and 2′-deoxycytidine, and the corresponding phosphoralaninate pronucleotides as potential antiviral agents. Phosphoralaninate of adenine methylenecyclobutane (**23**, Figure 6.6) was reported as a potent inhibitor of EBV replication in Daudi cell culture. Kushner et al. (2003) tested some purine analogs, where compound **24** (Figure 6.6) showed the best activity against EBV ($EC_{50}$ = 0.6 μM). Zhou et al. (2004a) synthesized some fluoromethylenecyclopropane analogs where compound **25** (Figure 6.6) was found to be the most potent (activity <0.32 μM against Daudi cells). In another report, Zhou et al. (2004b) reported some second generation of methylenecyclopropane nucleoside analogs, in which compound **26** (Figure 6.6) was found to be the most active ($EC_{50}$ = 0.30 μM). Kern et al. (2005) reported some second-generation 2, 2-bis-hydroxymethyl derivatives and tested them against EBV to find that compound **27**

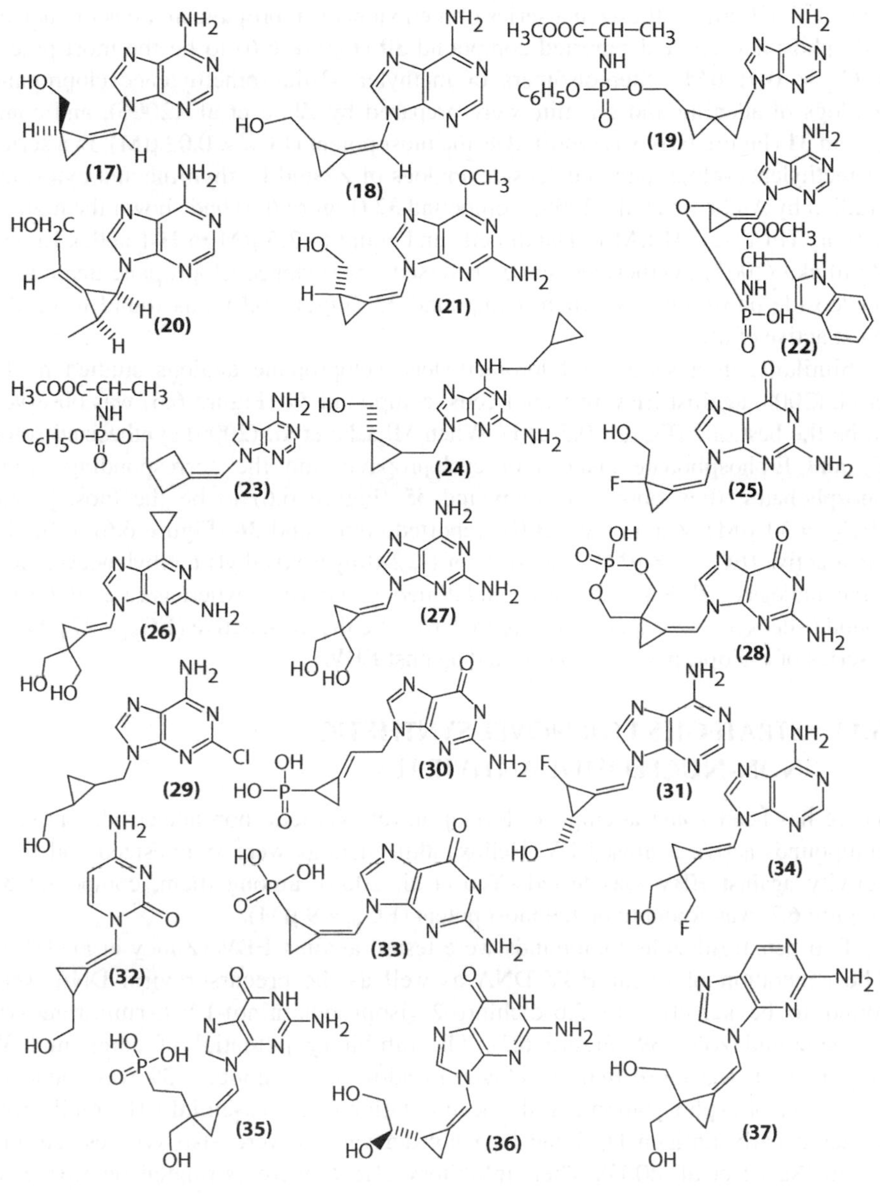

**FIGURE 6.6** Synthetic nucleoside antivirals (**17–37**) active against EBV.

(Figure 6.6) was the most active ($EC_{50}$ < 0.30 μM). In a series of phenylmethylphosphor-L-alaninate pronucleotide analogs studied by Yan et al. (2005), compound **28** (Figure 6.6) was found to be the most active against Daudi cell ($EC_{50}$ 0.96 μM).

Similarly, in a series of methylenecyclopropane nucleotides reported by Qin et al. (2006), Z and E isomers of 2-chloropurine methylenecyclopropane (**29**, Figure 6.6) were found to be noncytotoxic as well as effective against EBV ($EC_{50}$ < 0.08 μM).

Yan et al. (2006) synthesized a series of methylenecyclopropane analogs of nucleoside phosphonates and reported compound **30** (Figure 6.6) to be the most potent ($EC_{50}$ < 0.03 μM). Stereoisomers of methylene-3-fluoromethylenecyclopropane analogs of adenine and guanine were prepared by Zhou et al. (2006), and compound **31** (Figure 6.6) was reported as the most potent ($EC_{50}$ < 0.03 μM). In a series of methylenecyclopropane nucleoside analogs of Z- and E -thymine and cytosines studied by Ambrose et al. (2005), compound **32** (Figure 6.6) had shown the highest activity ($EC_{50}$ < 0.41 μM in Daudi cells and equal to 2.5 μM in H-1 cells). Li and Zemlicka (2007) synthesized some "reversed" methylenecyclopropane analogs of nucleoside phosphonates, where compound **33** (Figure 6.6) was found to be the most active of all.

Similarly, in a series of fluoromethylenecyclopropane analogs studied by Li et al. (2008) against EBV in Daudi cells, compound **34** (Figure 6.6) was observed to be the best one ($EC_{50}$ = 0.5 μM). When Mhaske et al. (2009) synthesized some Z- and E-phosphonate analogs of cyclopropavir and the corresponding cyclic phosphonates, they observed compound **35** (Figure 6.6) to be the most potent ($EC_{50}$ = 3.1 μM). Zhou et al. (2009) reported compound **36** (Figure 6.6) to be the most active ($EC_{50}$ = 8 μM) in a series of (1, 2-dihydroxyethyl) methylenecyclopropane analogs of 2′-deoxyadenosine and 2′-deoxyguanosine; whereas Li et al. (2012) found 6-deoxycyclopropavir (**37**, Figure 6.6) to be the most active ($EC_{50}$ = 27 μM) in a series of cyclopropavir analogs tested against EBV.

## 6.13 STRATEGIES FOR NOVEL SYNTHETIC NON-NUCLEOSIDE ANTIVIRALS

There has been some attempt to develop novel synthetic non-nucleoside antiviral compounds active against EBV. Helioxanthin analogs were synthesized and their activity against EBV was tested (Yeo et al. 2005); among them, compound **38** (Figure 6.7) was found to be the most potent ($EC_{50}$ = 9 μM).

Two benzimidazole compounds were tested against EBV (Zancy et al. 1999). The generation of linear EBV DNA as well as the precursor viral DNA were found to be sensitive to 5,6-dichloro-2 (isopropylamino)-1-b-L-ribofuranosyl-1H-benzimidazole (**39**, Figure 6.7). The inhibitory potential of compound **39** (Figure 6.7) was more than acyclovir. In addition, compound **39** was found to inhibit the phosphorylation and the accumulation of the essential EBV replicative cofactor, early antigen D. Some lapachol derivatives were also synthesized and tested (Sacau et al. 2003). Their inhibitory effects were estimated on EBV-EA, which was induced by 12-O-tetradecanoylphorbol-13-acetate (TPA). Compound **40** (Figure 6.7) was found to be the most potent compound of all. When a series of benzimidazole derivatives was synthesized and evaluated against the EBV-EA activation induced by TPA, compound **41** (Figure 6.7) was found to be the best (Ramla et al. 2007).

Benzylated pyrimidines were synthesized and found to be effective against EBV in Akata cell culture (Novikov et al. 2010). In this series, compound **42** (Figure 6.7) had exhibited the highest potency ($EC_{50}$ = 2.3 μM). Similarly, compound **43** (Figure 6.7) was reported (Abdel-Mohsen et al. 2010) as an active compound against EBV-EA.

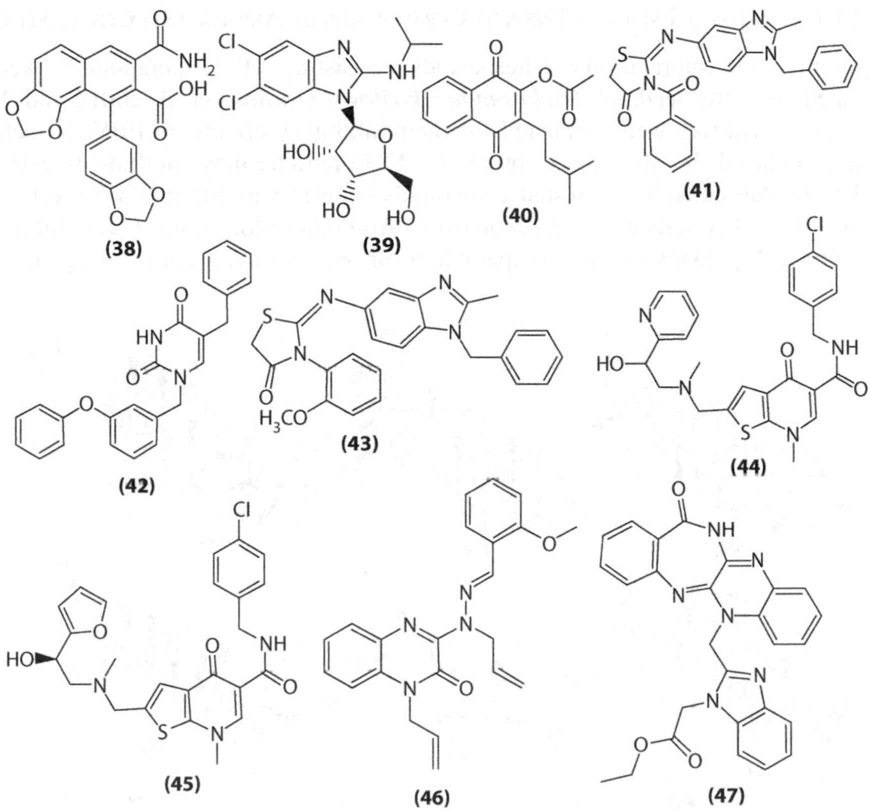

**FIGURE 6.7** Synthetic non-nucleoside antivirals (**38–47**) active against EBV.

When Schnute et al. (2007) synthesized some carboxamide derivatives and tested them against EBV, compound **44** (Figure 6.7) was found to be the most active ($IC_{50}$ = 70 nM). The same group of authors (Schnute et al. 2008) reported the synthesis and biological assay of carboxylate esters, in which compound **45** (Figure 6.7) was found to be the most active ($IC_{50}$ = 400 nM). Quinoxaline derivatives were also found to be active against EBV (Galal et al. 2011) and, when evaluated against EBV-EA activation induced by TPA, compound **46** (Figure 6.7) was found to be the best. Similarly, in a series of benzimidazoles studied against EBV-EA activation induced by TPA, Omar et al. (2012) reported compound **47** (Figure 6.7) as most active.

## 6.14 STRATEGIES FOR NOVEL NATURAL PRODUCT DERIVED NON-NUCLEOSIDE ANTIVIRALS

Natural compounds are valuable sources for obtaining molecules that may be effective against EBV-EA inactivation. In the past few decades, several natural compounds that showed potential activity against EBV-induced carcinomas were reported by different, but limited, groups of authors.

### 6.14.1 Natural Product Derived Compounds by Akihisa and Coworkers

Forty-nine multiflorane-type triterpenoids consisting of 11 compounds were isolated from the seeds of *Trichosanthes kirilowii* (Akihisa et al. 2001), and 38 of their derivatives were evaluated for their inhibitory effects on EBV-EA activation induced by the tumor promoter 12-O-tetradecanoylphorbol-13-acetate (TPA) in Raji cells. These tested compounds showed variable inhibitory effects against EBV-EA activation, and 5-dehydrokarounidiol (**48**, Figure 6.8) exhibited activity with potencies either comparable to or stronger than that of glycyrrhetic

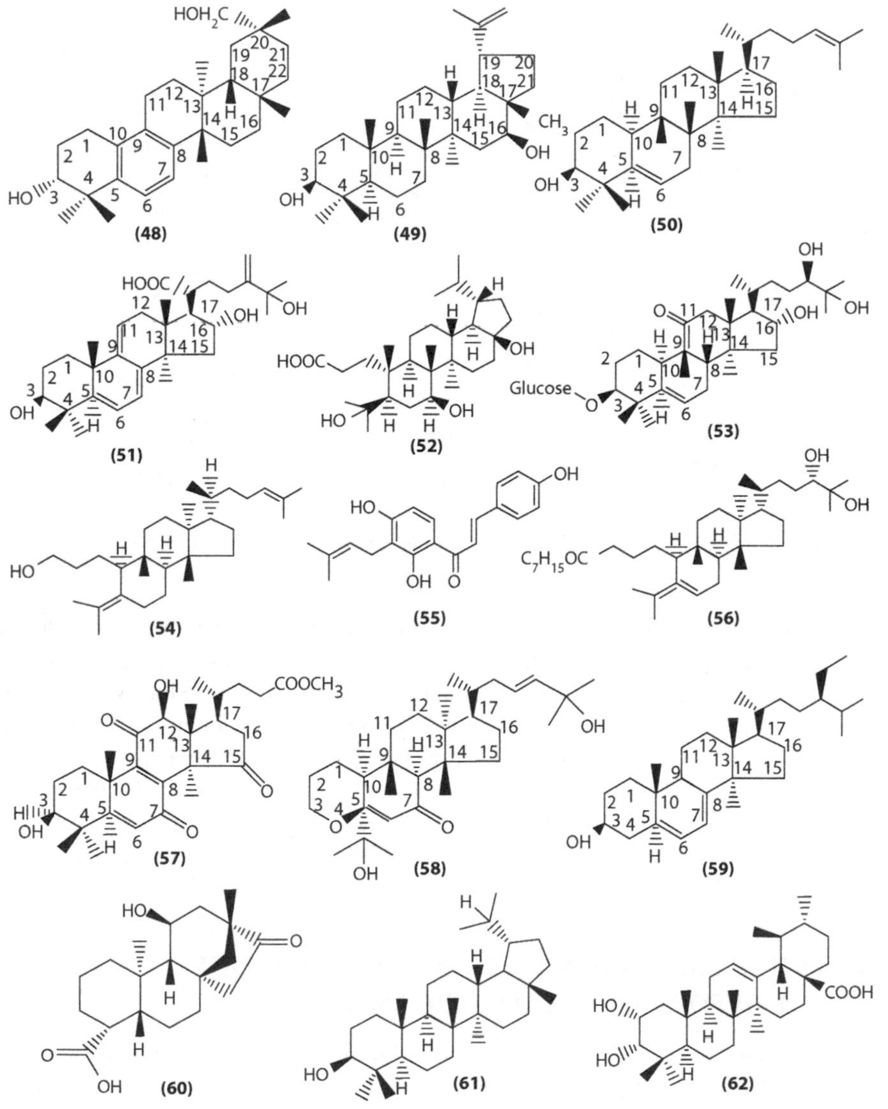

**FIGURE 6.8** Natural compounds (**48–62**) found to be effective against EBV.

acid (a well-known natural antitumor promoter). The structure–activity relationship (SAR) study suggested that C-3 hydroxyl group (Figure 6.8) is essential for activity because acetylation or benzoylation reduced the potency of the compound. Moreover, the 3-oxo derivatives also showed lower activity. Unsaturation in the phenyl ring system was found to be conducive. The presence of the oxo group at both or either of C-7 and C-11 as well as the $\Delta_8$ compound reduced the activity. Furthermore, esterification with a bulky benzoyl group of the hydroxymethylene moiety greatly reduced the potency.

In another attempt to find novel natural products, Ukiya et al. isolated 15 pentacyclic triterpene diols and triols from the nonsaponifiable lipid fraction of the edible flower extract of *Chrysanthemum morifolium* (Ukiya et al. 2002a). All these compounds showed inhibitory effects against EBV-EA activation. Among these, seven compounds showed remarkably high inhibitory effects at higher concentration (100% inhibition at $1 \times 10^3$ mol ratio). Calenduladiol (**49**, Figure 6.8) was found to be the most potent compound among these.

Some triterpenoid compounds were isolated from the fruit of *Momordica grosvenori*, belonging to the family of Cucurbitaceae (Ukiya et al. 2002b). Again, all tested compounds showed potent inhibitory effects on EBV-EA induction (70–100% inhibition at $1 \times 10^3$ mol ratio/TPA); 10α-cucurbitadienol (**50**, Figure 6.8) was found as the most active compound. Ukiya et al. (2002c) tested some triterpene compounds from the sclerotium of *Poria cocos* (family: Polyporaceae). Compound **51** (Figure 6.8) was found to be the most active. In another communication, Akihisa et al. (2002a) reported another six triterpenes that were tested for their inhibitory effects against EBV-EA activation. All these triterpene derivatives demonstrated potent inhibitory effects. However, compound **52** (Figure 6.8) appeared to be the most active among all.

Seven triterpene glycosides (bryoniosides A-G) and cabenoside D and bryoamaride were isolated from the methanol extract of the roots of *Bryonia dioica* (family: Cucurbitaceae) (Ukiya et al. 2002d). All these compounds showed potent inhibitory effects on EBV-EA induction (100% inhibition at $1 \times 10^3$ mol ratio/TPA), but compound **53** (Figure 6.8) was found to be the most active among them. Akihisa et al. (2002b) reported some triterpene alcohols isolated from the latex of *Euphorbia antiquorum* (Euphorbiaceae) and tested their inhibitory effects on EBV-EA activation induced by the tumor promoter TPA. Compound **54** (Figure 6.8) was found to be the most active among all.

Akihisa et al. (2003) also isolated a group of 17 compounds from an ethyl acetate-soluble fraction of the exudate obtained from stems of *Angelica keiskei* belonging to the family of Umbelliferae. These compounds were evaluated against TPA induced EBV-EA in Raji cells. With the exception of three compounds, all other tested compounds showed potent inhibitory effects on EBV-EA induction in higher concentration (92–100% inhibition at $1 \times 10^3$ mol ratio/TPA). Isobavachalcone (**55**, Figure 6.8) was the most active compound. Thirty terpenoid compounds were isolated from the diethyl ether extract of the pollen grains of *Helianthus annuus* belonging to the family Compositae (Ukiya et al. 2003a). Twenty-one compounds possessing a di- or a polycyclic ring system showed potent inhibitory effects on EBV-EA induction (91–100% inhibition at $1 \times 10^3$ mol ratio/TPA). Compound **56** (Figure 6.8) was found to be the most potent one.

Iwatsuki et al. (2003a) have reported triterpenoids fruiting bodies of *Ganoderma lucidum* belonging to the family of Polyporaceae. All these compounds showed potent inhibitory effects (96–100% inhibition at $1 \times 10^3$ mol ratio/TPA) and compound **57** (Figure 6.8) was found to be the most active. Six triterpenoids were isolated from diethyl ether extract of the pollen grains of a sunflower (*Helianthus annuus*) belonging to the family of Compositae (Ukiya et al. 2003b). All these tested compounds showed 97–100% inhibition (at $1 \times 10^3$ mol ratio/TPA), and compound **58** (Figure 6.8) was found to be the best. Some sterol compounds were also found to be effective against TPA-induced EBV-EA activation in Raji cells (Iwatsuki et al. 2003b). The most active compound was Schottenol (**59**, Figure 6.8). Akihisa et al. (2004a) reported some isosteviol analogs, and compound **60** (Figure 6.8) was the best among these. All the tested diterpenes showed potent inhibitory effects along with the five metabolites exhibited more potency.

Seven triterpenoids were isolated from the nonsaponifiable lipid of the seed oil of *Camellia japonica* (family: Theaceae) (Akihisa et al. 2004c). All seven triterpenoids showed inhibitory effects against EBV-EA activation induced by TPA in Raji cells. Three compounds showed potent inhibitory effects against EBV-EA induction ($IC_{50}$ values of 277–420 mol ratio/32 pmol TPA). Lupane-3β, 20-diol (compound **61**, Figure 6.8) was the most potent among all. Nine triterpene acids were isolated from the ethanolic extract of *Perilla frutescens* (Banno et al. 2004). The inhibitory activity of these compounds was tested against EBV-EA activation by TPA in Raji cells; of these, five compounds showed potent inhibitory activity, with 3-epicorosolic acid (**62**, Figure 6.8) having the highest potency.

Another study conducted by the same group (Akihisa et al. 2004b) reported the inhibitory effects of 22 fatty acids; of these, three n-3 polyunsaturated acids—namely eicosapentaenoic acid (EPA), docosapentaenoic acid (DPA), and docosahexaenoic acid (DHA)—were found to exhibit potent inhibitory effects on EBV-EA activation. Of these three, EPA (**63**, Figure 6.9) was the best. Six azaphilones, two furanoisophthalides, and two amino acids were isolated from the extracts of *Monascus pilosus* (Eurotiaceae family) (Akihisa et al. 2005b). These compounds were evaluated for their inhibitory effects on TPA-induced EBV-EA activation in Raji cells. Among these compounds, six exhibited potent inhibitory effects on EBV-EA activation. Rubropunctatin (**64**, Figure 6.9) was found to be the most active compound. Sixteen triterpene acids were isolated from the ethyl acetate-soluble fraction of the methanol extract of the leaves of *Eriobotrya japonica* (Rosaceae family) (Banno et al. 2005). All these compounds showed potent activity against the TPA-induced EBV-EA activation, and 1β-hydroxyeuscaphic acid (**65**, Figure 6.9) was the most potent of them. Seven sterols and eight polyisoprenepolyols were isolated from the nonsaponifiable lipid fraction of the dichloromethane extract of an edible mushroom, *Hypsizigus marmoreus* (Buna-shimeji) (Akihisa et al. 2005a). These compounds were evaluated for their inhibitory effects on EBV-EA activation induced by TPA in Raji cells, and compound **66** ([22E, 24R]-5α, 6α -Epoxyergosta-8, 22-diene-3β, 7β –diol) (Figure 6.9) was found to be the most active.

Ten oleanane-type triterpene glycosides along with five flavonol glycosides were isolated from the flowers of a marigold (*Calendula officinalis*) belonging to the Asteraceae family (Ukiya et al. 2006). Most of the compounds showed moderate inhibitory activity against EBV-EA activation induced by TPA, among which

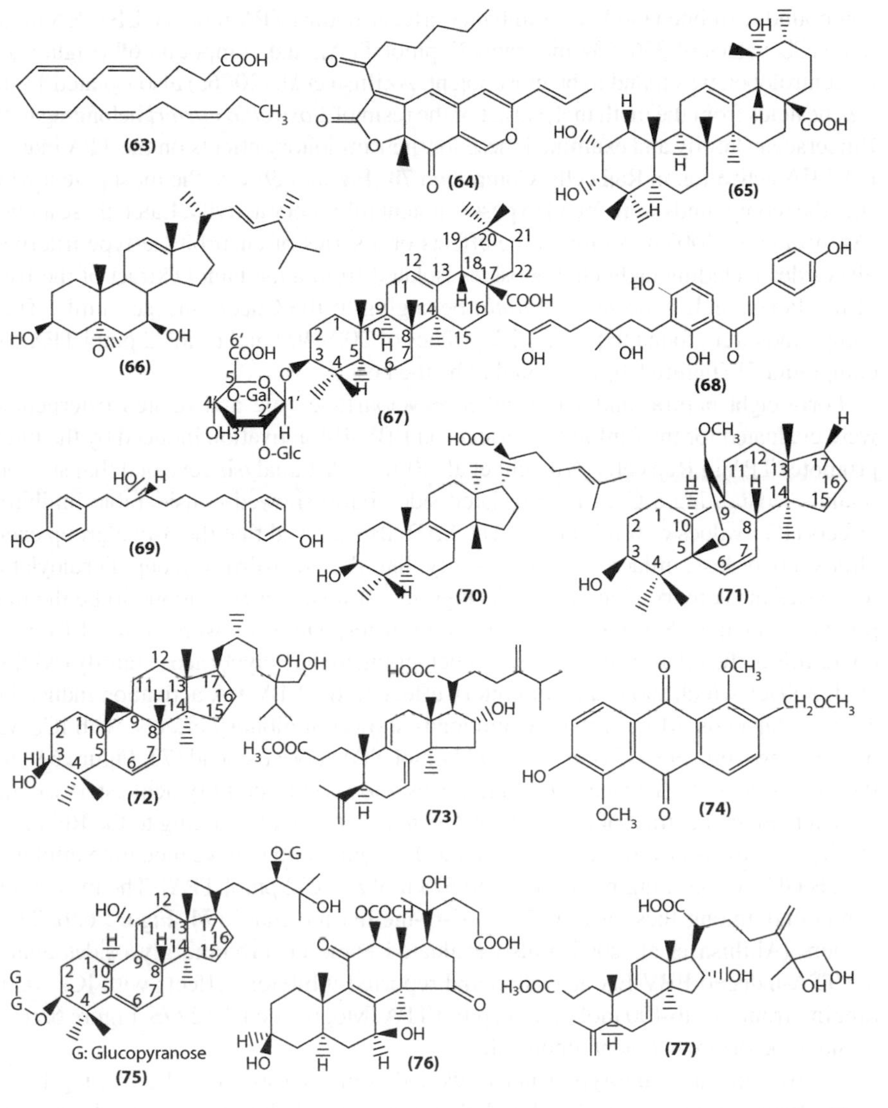

**FIGURE 6.9** Structures of natural compounds (**63–77**) found to be effective against EBV.

calendula gycoside B (**67**, Figure 6.9) was found to be the most active. Three new chalcones, nine aromatic compounds, and a diacetylene were isolated from an ethyl acetate-soluble fraction of exudates of stems of *Angelica keiskei* belonging to the family Umbelliferae (Akihisa et al. 2006c). Some chalcones and some aromatic compounds were found to have inhibitory effects against TPA-induced EBV-EA activation in Raji cells with xanthoangelol J (**68**, Figure 6.9) having the best activity.

A cyclic diarylheptanoid (i.e., acerogenin M), nine known diarylheptanoids, and two known phenolic compounds were isolated from a methanol extract of the stem bark of *Acer nikoense* belonging to the Aceraceae family (Akihisa et al. 2006b). All of these compounds exhibited moderate inhibitory effects against TPA-induced EBV-EA induction ($IC_{50}$ values of 356–534 mol ratio/32 pmol TPA), and compound **69** (Figure 6.9), (-)-centrolobol, was found to be most potent. Akihisa et al. (2006a) also isolated 15 triterpene acids from the methanol extract of the resin of *Boswellia carteri* belonging to the Burseraceae family and examined these for their inhibitory effects on the TPA induced EBV-EA activation in Raji cells. Compound **70** (Figure 6.9) was the most potent while six other compounds were found to posses potent inhibitory activity. Later, these authors (Akihisa et al. 2007a) examined the effects of a series of cucurbitane-type triterpene glycosides including eight charantosides isolated from a methanol extract of the fruits of the Japanese *Momordica charantia* belonging to the Cucurbitaceae family. These compounds were found to possess $IC_{50}$ values of 200–409 mol ratio/32 pmol TPA, and compound **71** (Figure 6.9) was found to be the best.

Forty-eight natural and semisynthetic cycloartane-type and related triterpenoids were evaluated for their inhibitory effects on EBV-EA activation induced by the tumor promoter TPA in Raji cells (Kikuchi et al. 2007). SAR analysis revealed that six compounds (containing a C-24 hydroxylated side chain) showed considerable inhibitory effects ($IC_{50}$ values of 6.1–7.4 μM). The SAR also showed that the 3-oxo group exerts almost no, or less, influence on the activity than the 3β-hydroxy group. Feruloylation decreases the activity. Compound **72** (Figure 6.9) was, however, found to be the most potent compound. Seventeen lanostane-type triterpene acids were isolated from the epidermis of the sclerotia of *Poria cocos* belonging to the Polyporaceae family (Akihisa et al. 2007e), which showed an inhibitory effect on the EBV-EA activation induced by TPA in Raji cells. Most of these compounds showed inhibitory effects, with $IC_{50}$ values ranging from 195 to 340 mol ratio/32 pmol TPA; poricoic acid (**73**, Figure 6.9) had the highest activity. Thirteen anthraquinones and saccharide fatty acid esters isolated from a methanol extract of the fruits of *Morinda citrifolia* belonging to the Rubiaceae family were examined by Akihisa et al. (2007c) and found to have moderate inhibitory effects ($IC_{50}$ values ranging from 386 to 578 mol ratio/32 pmol TPA). The most potent compound among these was 1, 5, 15-tri-O-methyl morindol (**74**, Figure 6.9). These authors (Akihisa et al. 2007b) also evaluated some cucurbitane glycosides against the TPA-induced EBV-EA activation and reported inhibitory effects with $IC_{50}$ values ranging from 346 to 400 mol ratio/32 pmol TPA. Mogroside IIIA2 (**75**, Figure 6.9) was found to be the most potent among all.

A series of lanostane-type triterpenes and some sterols from the fruiting bodies of the fungus *Ganoderma lucidum* belonging to the Polyporaceae family was also isolated and evaluated against the induction of TPA-induced EBV-EA activation in Raji cells by Akihisa et al. (2007d). Most of the compounds showed potent inhibitory effects on EBV-EA induction ($IC_{50}$ values of 235–370 mol ratio/32 pmol TPA), with 20-hydroxylucidenic acid P (**76**, Figure 6.9) having the highest potency. Eighteen lanostane-type triterpene acids and a known diterpene acid were isolated from the epidermis of the sclerotia of *Poria cocos* belonging to the Polyporaceae family (Akihisa et al. 2009b). All these compounds were found to exhibit inhibitory effects against the EBV-EA activation induced by TPA in Raji cells ($IC_{50}$ values ranging from 187 to 348 mol ratio/32 pmol TPA), with compound **77** (Figure 6.9) being the most potent.

A series of nortriterpenoids, including 28 limonoids, three degraded limonoids, and one diterpenoid were isolated from the seed extract of the neem plant *Azadirachta indica* (Akihisa et al. 2009a). All compounds exhibited moderate to potent inhibitory effects ($IC_{50}$ values of 230–501 mol ratio/32 pmol TPA) against the EBV-EA activation induced by TPA among which compound **78** (Figure 6.10) had the highest activity. Four triterpene acetates and four triterpene cinnamates were isolated from the shea tree *Vitellaria paradoxa* belonging to the Sapotaceae family (Akihisa et al. 2010). All these compounds exhibited moderate inhibitory effects against the EBV-EA activation induced by TPA. Among these, butyrospermol cinnamate (**79**, Figure 6.10) was found to be the most active. A series of triterpenoids including a sesquiterpenoid were isolated from dammar resin obtained from

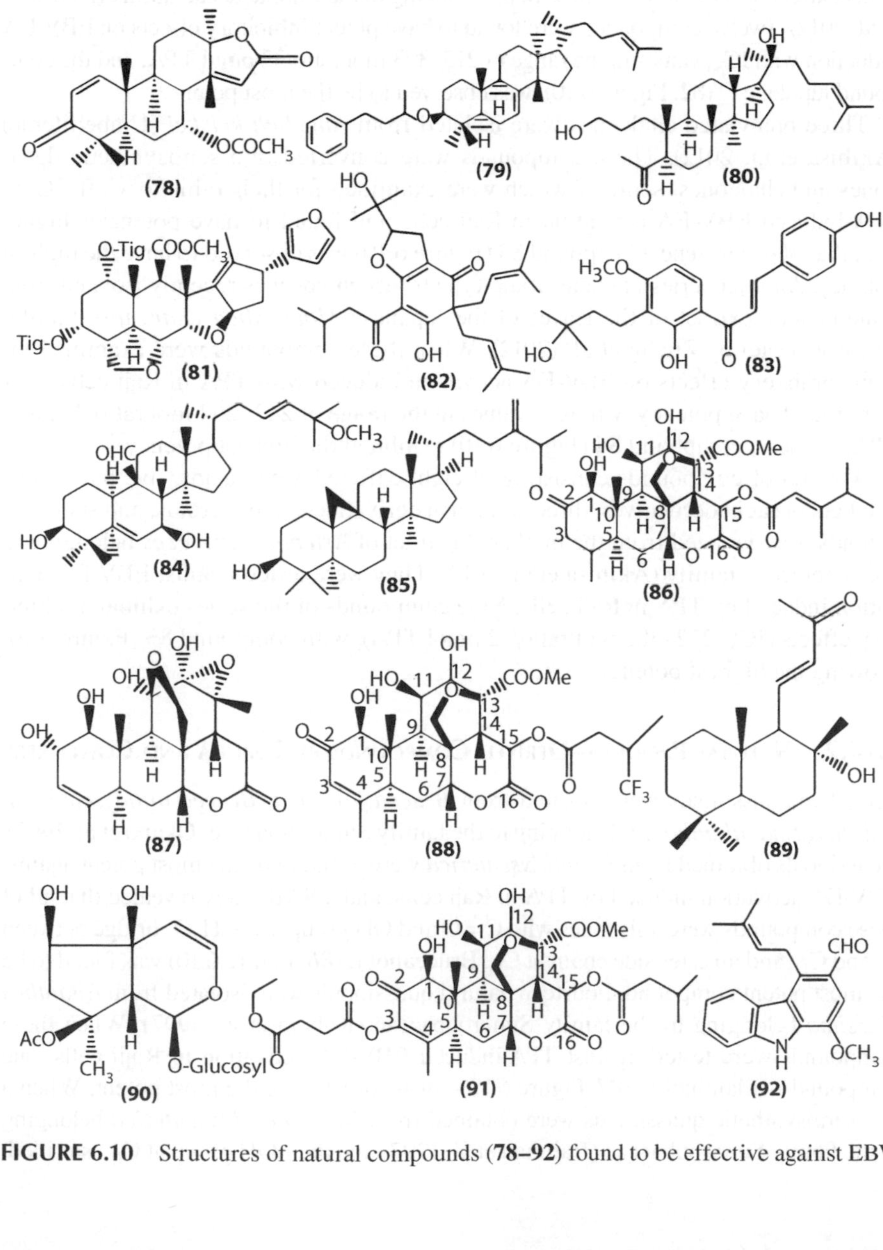

**FIGURE 6.10** Structures of natural compounds (**78–92**) found to be effective against EBV.

*Shorea javanica* K & V (family: Dipterocarpaceae) (Ukiya et al. 2010), and one of the acidic triterpenoids—dammarenolic acid—was converted to the other 14 semisynthestic derivatives. All these compounds were found to be quite potent. The highest activity was found with compound **80** (Figure 6.10). A series of limonoids (tetranortriterpenoids) was isolated by Akihisa et al. (2011) from the n-hexane extract of *Azadirachta indica* (neem) seeds. Some of these compounds were found to exert moderate inhibitory effects ($IC_{50}$ values of 410–471 mol ratio/32 pmol TPA); salannin (**81**, Figure 6.10) exhibited the highest activity. Some phloroglucinol derivatives, some chalcones, four flavanones, two flavonol glycosides, and five triterpenoids were isolated from the female inflorescence pellet extracts of *Humulus lupulus* (hop) belonging to the Cannabaceae family (Akazawa et al. 2012). Twelve compounds were found to have potent inhibitory effects on EBV-EA induction with $IC_{50}$ values in the range of 215–393 mol ratio/32 pmol TPA, and the compound lupulone C (**82**, Figure 6.10), was observed to be the most potent.

Three prenylated chalcones were isolated from *Angelica keiskei* (Umbelliferae) (Akihisa et al. 2012). These compounds were converted into semisynthetic flavanones and chalcones, some of which were examined for their inhibitory effects on TPA-induced EBV-EA activation in Raji cells and found to have potencies higher than that of β-carotene. Compound **83** (Figure 6.10) was observed to have the highest potency. Another series of cucurbitane-type triterpenoids was recently isolated from a methanolic extract of the leaves of the Japanese *Momordica charantia* (family: Cucurbitaceae) by Zhang et al. (2012). When these compounds were examined for their inhibitory effects on EBV-EA activation induced with TPA in Raji cells, four were found have potency with $IC_{50}$ values in the range of 242–264 mol ratio/32 pmol TPA; of these, compound **84** (Figure 6.10) exhibited the highest potency.

A series of compounds consisting of eight caffeoylquinic acids, five flavonoids, two benzoic acid derivatives, three coumarin derivatives, four steroids, and six triterpenoids were isolated from the methanol extract of *Artemisia princeps* belonging to the Asteraceae family (Akihisa et al. 2013). They were tested against EBV-EA activation induced by TPA in Raji cells. Nine compounds of the series exhibited inhibitory effects ($IC_{50}$ 272–382 mol ratio/32 pmol TPA), with compound **85** (Figure 6.10) showing the highest potency.

### 6.14.2 Natural Product–Derived Compounds by Tokuda and Coworkers

Forty-five quassinoids were isolated from *Brucea javanica*, *Brucea antidysenterica*, and *Picrasma ailanthoides* belonging to the family Simaroubaceae (Okano et al. 1995). Quassinoids obtained from *B. antidysenterica* were found to be the most potent against EBV-EA activation induced by TPA in Raji cells, and a SAR study revealed that all of these compounds were aglycones with C = O and OH groups, a -$CH_2O$- bridge between $C_8$ and $C_{13}$, and an ester side chain at $C_{15}$. Bruceanol E (**86**, Figure 6.10) was found to be the most potent compound. Fourteen natural quassinoids were isolated from *Ailanthus altissima* belonging to the family Simaroubaceae (Kubota et al. 1997). When these compounds were tested against TPA-induced EBV-EA activation in Rajii cells, one compound—ailantinol-C (**87**, Figure 6.10)—was found to be the most potent. When a few semisynthetic quassinoids were obtained from *Brucea antidysenterica* belonging to the family Simaroubaceae (Rahman et al. 1997) and their OH group at $C_{15}$ esterified, an assay of these compounds against TPA induced EBV-EA activation revealed compound **88** (Figure 6.10), that had a fluorinated aliphatic side chain, to be the most potent compound. Fluorinated aromatic side chain compounds were also active.

Some biflavonoids were isolated from an ethanolic extract of the stem bark of *Calophyllum panciflorum* belonging to the family Guttiferae and found to exhibit significant inhibitory activity against TPA induced EBV-EA activation in Raji cells (Ito et al. 1999). Among them, talbotaflavone (**89**, Figure 6.10) was the most potent compound. Some compounds isolated from the flowering whole plant of *Ajuga decumbens* belonging to the Labiatae family were tested against TPA-induced EBV-EA activation (Takasaki et al. 1999). Among these, compound **90** (Figure 6.10) was found to be the most potent. Tamura et al. (2000) introduced a lipophilic senecioyl group into shinjulactones B and C as well as esterified the diosphenol moiety in brusatol and brucein A and found that the resulting compounds were active against EBV-EA activation. Compound **91** (Figure 6.10) showed the highest inhibitory activity. It showed that esterification of disophenol group of quassinoids is important for higher activity. Some diterpenoid compounds were isolated from the stem bark of *Picea glehni* (family: Pinaceae) and all of these were found to have potent inhibitory effects against TPA-induced EBV-EA activation (Kinouchi et al. 2000) with compound **92** (Figure 6.10) having the highest potency of all (Kinouchi et al. 2000). Likewise, some phenol glycosides isolated from the dried leaves of *Eucalyptus cypellocarpa* (family: Myrtaceae) by Ito et al. (2000c) were found to have antitumor-promoting activity against EBV-EA; compound **93** (Figure 6.11) was found to be the most potent among these.

Some clauslactones were isolated from the leaves of *Clausena excavate* belonging to the family Rutaceae by Ito et al. (2000a). Among these, clauslactone C (**94**, Figure 6.11) was observed to be the most potent compound. Carbazole alkaloids were also isolated by Ito et al. (2000b) from *Clausena anisata* belonging to the family Rutaceae. One compound—ekeberginine (**95**, Figure 6.11)—was found to be the most potent while others also showed potent inhibitory activity against EBV-EA activation. Compound **96** (Figure 6.11) was found to be the most potent depsidone from *Garcinia assigu* (family: Guttiferae) (Ito et al. 2001). Tamura et al. (2002) synthesized seven shinjulactone C (quassinoid isolated from *Ailanthus altissima*) derivatives and evaluated their antitumor-promoting effects against EBV-EA. Among these synthetic derivatives, compound **97** (Figure 6.11), having a 3′, 3′-dimethylsuccinate moiety, showed the highest inhibition. A SAR study suggested that succinate derivatives have better activity than glutarates. Some diterpene compounds were isolated from the cones of *Pinus luchuensis* (family: Pinaceae) by Minami et al. (2002), and 15-nor-14-oxolabda-8(17)-12E-dien-19-oic acid (**98**, Figure 6.11) exhibited the highest activity. Some sarcophine analogs were also tested against EBV-EA by Katsuyama et al. (2002) where some compounds showed higher chemopreventive activity; the highest activity was associated with **99** (Figure 6.11). Some xanthones were isolated from the stem bark of *Calophyllum brasilienses* (family: Guttiferae) by Ito et al. (2002). Some of these compounds showed promising inhibitory activity against TPA-induced EBV-EA activation in Raji cells, and brasixanthone B (**100**, Figure 6.11) was found to be the most potent.

Ito et al. (2003a) isolated five new cinnamylphenols from plants of the *Dalbergia cultrate* belonging to the family Leguminosae and evaluated them

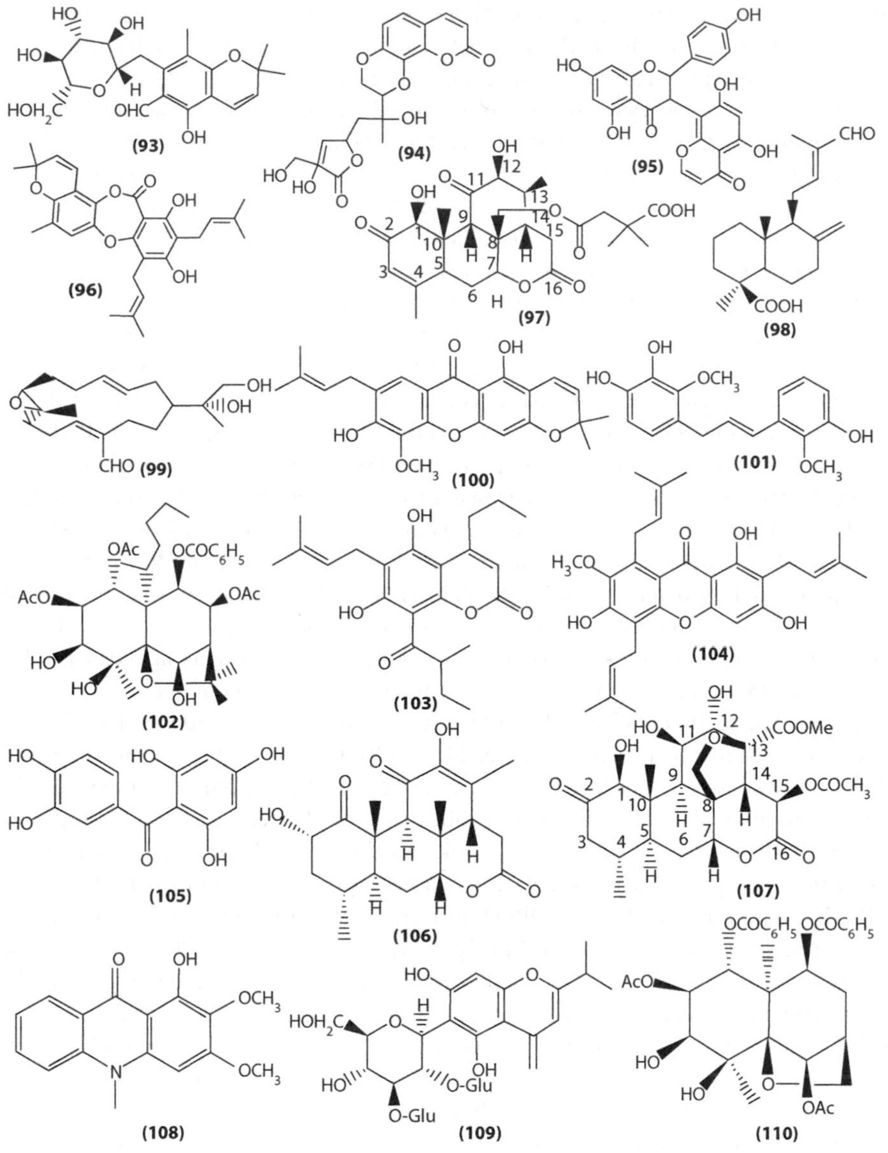

**FIGURE 6.11** Natural compounds (**93–110**) found to be effective against EBV.

against EBV-EA activation induced by TPA in Raji cells. All these compounds were found to show remarkably potent activities, suggesting that some of these might be valuable as potential cancer chemopreventive agents. Among these, dalberatin E (**101**, Figure 6.11) was the most potent. Some sesquiterpenoids isolated from *Crossopetalum tonduzii* (belonging to the Celastraceae family) by Jiménez et al. (2003) possessed strong antitumor-promoting effects on EBV-EA activation, with compound **102** (Figure 6.11) having the highest activity. Similarly, among four substituted coumarins isolated by Ito et al. (2003b) from the stem bark of *Calophyllum brasiliense* (family: Guttiferae), the compound mammea B/BB (**103**, Figure 6.11) was found to be the most potent. Ito et al. (2003c) also isolated some xanthone derivatives from the stem bark of *Garcinia fusca* belonging to the same family (Guttiferae). All these xanthones consisted of a terpenoid (prenyl and/or geranyl) side chain(s) in their molecules and showed potent inhibitory effects. One compound, 7-O-methylgarcinone (**104**, Figure 6.11), was found to be the most active among these. Some benzophenone analogs were isolated by Ito et al. (2003d) from another member of Guttiferae family, *Garcinia assigu*. The compound macurin (**105**, Figure 6.11) was found to be the most potent among those tested. When Tamura et al. (2003) isolated three new quassinoids (ailantinol E, F, and G) from *Ailanthus altissima* (family: Simaroubaceae), ailantinol F (**106**, Figure 6.11) was found to be the most active against EBV-EA activation induced by TPA.

Murakami et al. (2004) reported six cytotoxic quassinoids and four canthin alkaloids; compound **107** (Figure 6.11) was found to be the most active. A SAR study suggested that quassinoids that had an α-oriented methyl group at $C_4$ exhibited stronger inhibitory effects than those that had β-oriented groups. Three new carbazole alkaloids (glybomines A, B, and C) along with monomeric alkaloids belonging to the carbazole, quinazoline, furoquinoline, quinolone, and acridone classes were isolated from stems of the plant *Glycosmis arborea* (family: Rutaceae) by Ito et al. (2004a). These alkaloids were tested for their inhibitory effects on TPA-induced EBV-EA activation, and arborinine (**108,** Figure 6.11) had the highest activity. Some flavonol glycosides were isolated from the leaf extract of *Kunzea ambigua* (family: Myrtaceae) by Ito et al. (2004b); among these, compound **109** (Figure 6.11) was found to exhibit potent inhibitory effects on TPA-induced EBV-EA activation in Raji cells. Similarly, in a series of sesquiterpenes isolated by Perestelo et al. (2010) from the leaves of *Maytenus jelskii* (family: Celastraceae), and tested against EBV-EA activation induced by TPA, compound **110** (Figure 6.11) was found to be an effective antitumor-promoting agent.

## 6.15 CONCLUSION

EBV infection is associated with the pathogenesis of several malignancies. Although work on EBV and EBV-mediated cancers has been conducted for almost 50 years, therapy for EBV-mediated tumors remains largely in the nascent stage. The current chapter focuses on the biology of EBV and current available therapies for the treatment of EBV-associated malignancies. The recent data show that there are diverse approaches to fight against this deadly pathogen; these include the developments of antivirals, vaccines, monoclonal antibodies, siRNA, and so on. It is believed that a combination of immunotherapy and antivirals may be highly effective. Several studies are aimed at the development of new natural or synthetic antivirals effective against EBV. On the other hand, the development of an EBV vaccine has been slow. More resources should be devoted to the development of an effective vaccine that could eliminate the chance of EBV infection in the first place. Recent studies also showed that adoptive immunotherapy with EBV-specific CTLs or the administration of the targeted antibodies hold considerable promise for the treatment of EBV-mediated malignancies. However, the impetus for this work is being fuelled by the encouraging developments of new antivirals and immunological approaches. Some recent studies have showed interesting results. Further work should be aimed at optimizing these different treatment protocols.

## REFERENCES


Abdel-Mohsen HT, Ragab FAF, Ramla MM, et al. (2010). Novel benzimidazole–pyrimidine conjugates as potent antitumor agents. *Eur J Med Chem* 45: 2336–2344.

Abdulkarim B, Bourhis J. (2001). Antiviral approaches for cancers related to Epstein–Barr virus and human papillomavirus. *Lancet Oncol* 2: 622–630.

Akazawa H, Kohno H, Tokuda H, et al. (2012). Anti-inflammatory and anti-tumor-promoting effects of 5-deprenyllupulonol C and other compounds from Hop (*Humulus lupulus* L.). *Chem Biodivers* 9: 1045–1054.

Akihisa T, Franzblau SG, Tokuda H, et al. (2005a). Antitubercular activity and inhibitory effect on Epstein–Barr virus activation of sterols and polyisoprenepolyols from an edible mushroom, *Hypsizigus marmoreus*. *Biol Pharm Bull* 28: 1117–1119.

Akihisa T, Hamasaki Y, Tokuda H, et al. (2004a). Microbial transformation of isosteviol and Inhibitory effects on Epstein–Barr virus activation of the transformation products. *J Nat Prod* 67: 407–410.

Akihisa T, Hayakawa Y, Tokuda H, et al. (2007b). Cucurbitane glycosides from the fruits of *Siraitia grosvenorii* and their inhibitory effects on Epstein–Barr virus activation. *J Nat Prod* 70: 783–788.

Akihisa T, Higo N, Tokuda H, et al. (2007a). Cucurbitane-type triterpenoids from the fruits of *Momordica charantia* and their cancer chemopreventive effects. *J Nat Prod* 70: 1233–1239.

Akihisa T, Kawashima K, Orido M. (2013). Antioxidant and melanogenesis-inhibitory activities of caffeoylquinic acids and other compounds from moxa. *Chem Biodivers* 10: 313–327.

Akihisa T, Kojima N, Kikuchi T, et al. (2010). Anti-inflammatory and chemopreventive effects of triterpene cinnamates and acetates from shea fat. *J Oleo Sci* 59(6): 273–280.

Akihisa T, Matsumoto K, Tokuda H, et al. (2007c). Anti-inflammatory and potential cancer chemo-preventive constituents of the fruits of *Morinda citrifolia* (Noni). *J Nat Prod* 7: 754–757.

Akihisa T, Motoi T, Seki A, et al. (2012). Cytotoxic activities and anti-tumor-promoting effects of microbial transformation products of prenylated chalcones from *Angelica keiskei*. *Chem Biodivers* 9: 318–330.

Akihisa T, Nakamura Y, Tagata M, et al. (2007d). Anti-inflammatory and anti-tumor-promoting effects of triterpene acids and sterols from the fungus *Ganoderma lucidum*. *Chem Biodivers* 4: 224–231.

Akihisa T, Nakamura Y, Tokuda H, et al. (2007e). Triterpene acids from *Poria cocos* and their anti-tumor-promoting effects. *J Nat Prod* 70: 948–953.

Akihisa T, Noto T, Takahashi A, et al. (2009a). Melanogenesis inhibitory, anti-inflammatory, and chemopreventive effects of limonoids from the seeds of *Azadirachta indicia* A. Juss. (neem). *J Oleo Sci* 58: 581–594.

Akihisa T, Tabata K, Banno N, et al. (2006a). Cancer chemopreventive effects and cytotoxic activities of the triterpene acids from the resin of *Boswellia carteri*. *Biol Pharm Bull* 29: 1976–1979.

Akihisa T, Taguchi Y, Yasukawa K, et al. (2006b). Acerogenin M, a cyclic diarylheptanoid, and other phenolic compounds from acer nikoense and their anti-inflammatory and anti-tumor-promoting effects. *Chem Pharm Bull* 54: 735–739.

Akihisa T, Takahashi A, Kikuchi T, et al. (2011). The melanogenesis-inhibitory, anti-inflammatory, and chemopreventive effects of limonoids in n-hexane extract of *Azadirachta indica* A. Juss. (neem) seeds. *J Oleo Sci* 60: 53–59.

Akihisa T, Takamine Y, Yoshizumi K, et al. (2002a). Microbial transformations of two lupane-type triterpenes and anti-tumor-promoting effects of the transformation products. *J Nat Prod* 65: 278–282.

Akihisa T, Tokuda H, Hasegawa D, et al. (2006c). Chalcones and other compounds from the exudates of *Angelica keiskei* and their cancer chemopreventive effects. *J Nat Prod* 69: 38–42.

Akihisa T, Tokuda H, Ichiishi E, et al. (2001). Anti-tumor promoting effects of multiflorane-type triterpenoids and cytotoxic activity of karounidiol against human cancer cell lines. *Cancer Lett* 173: 9–14.

Akihisa T, Tokuda H, Ogata M, et al. (2004b). Cancer chemopreventive effects of polyunsaturated fatty acids. *Cancer Lett* 205: 9–13.

Akihisa T, Tokuda H, Ukiya M, et al. (2003). Chalcones, coumarins, and flavanones from the exudates of *Angelica keiskei* and their chemopreventive effects. *Cancer Lett* 201: 133–137.

Akihisa T, Tokuda H, Ukiya M, et al. (2004c). 3-Epicabraleahydroxylactone and other triterpenoids from camellia oil and their inhibitory effects on Epstein–Barr virus activation. *Chem Pharm Bull* 52: 153–156.

Akihisa T, Tokuda H, Yasukawa K, et al. (2005b). Azaphilones, furanoisophthalides, and amino acids from the extracts of *Monascus pilosus*-fermented rice (red-mold rice) and their chemopreventive effects. *J Agric Food Chem* 53: 562–565.

Akihisa T, Uchiyama E, Kikuchi T, et al. (2009b). Anti-tumor-promoting effects of 25-methoxyporicoic acid A and other triterpene acids from *Poria cocos*. *J Nat Prod* 72: 1786–1792.

Akihisa T, Wijeratne EMK, Tokuda H, et al. (2002b). Eupha-7,9(11), 24-trien-3α-ol ("Antiquol C") and other triterpenes from *Euphorbia antiquorum* latex and their inhibitory effects on Epstein–Barr virus activation. *J Nat Prod* 65: 158–162.

Ambrose A, Zemlicka J, Kern ER, et al. (2005). Phosphoralaninate pronucleotides of pyrimidine methylcyclopropane analogs of nucleosides: Synthesis and antiviral activity. *Nucleo Nucleo Nucleic Acid* 24: 1763–1774.

Babcock GJ, Thorley-Lawson DA. (2000). Tonsillar memory B-cells, latently infected with Epstein–Barr virus, express the restricted pattern of latent genes previously found only in Epstein–Barr virus–associated tumors. *Proc Natl Acad Sci U S A* 97: 12250–12255.

Baer R, Bankier AT, Biggin MD, et al. (1984). DNA sequence expression of the B95.8 Epstein–Barr virus genome. *Nature* 310: 207–211.

Banno N, Akihisa T, Tokuda H, et al. (2004). Triterpene acids from the leaves of *Perilla frutescens* and their anti-inflammatory and antitumor-promoting effects. *Biosci Biotechnol Biochem* 68: 85–90.

Banno N, Akihisa T, Tokuda H, et al. (2005). Anti-inflammatory and antitumor-promoting effects of the triterpene acids from the leaves of *Eriobotrya japonica*. *Biol Pharm Bull* 28: 1995–1999.

Borish B, Raphael M, Swerdlow SH. (2001). Lymphoproliferative diseases associated with primary immune disorders. In: Jaffe ES, Harris NL, Stein H (eds.), *World Health Organization Classification of Tumors, Pathology and Genetics of Tumors of Haematopoietic and Lymphoid Tissues*. Lyon, France: IACR Press, pp. 257–259.

Boyd MR, Bacon TH, Sutton D. (1988). Antiherpesvirus activity of 9-(4-hydroxy-3-hydroxy-methylbut-1-yl) guanine (BRL 39123) in animals. *Antimicrob Agents Chemother* 32: 358–363.

Boyd MR, Bacon TH, Sutton D, et al. (1987). Antiherpesvirus activity of 9-(4-hydroxy-3-hydroxy-methylbut-1-yl) guanine (BRL 39123) in cell culture. *Antimicrob Agents Chemother* 31: 1238–1242.

Brady G, MacAurthur GJ, Farell PJ. (2007). Epstein–Barr virus and Burkitt's lymphoma. *J Clin Pathol* 60: 1397–1402.

Burkitt D. (1958). A sarcoma involving the jaws of African children. *Br J Surg* 46: 218–223.

Butel JS. (2000). Viral carcinogenesis: Revelation of molecular mechanisms and etiology of human disease. *Carcinogenesis* 21: 405–426.

Chen HL, Lung MM, Sham JS, Choy DT, Griffin BE, Ng MH. (1992). Transcription of the BamHI A region of the EBV genome in NPC tissues and B-cells. *Virology* 191: 193–201.

Chia M, Klamut HJ, Li JH. (2003). Nucleic acids and methods for treating EBV-positive cancers thereof. Patent WO2003047634A2.

Coen DM, Schaffer PA. (2003). Antiherpesvirus drugs: A promising spectrum of new drugsand drug targets. *Nat Rev Drug Discov* 2: 278–288.

Cohen JI, Fauci AS, Varmuc H, et al. (2011). Epstein–Barr virus: An important vaccine target for cancer prevention. *Sci Transl Med* 3: 107fs7.

Cohen JI, Mocarski ES, Raab-Traub N, et al. (2005). The need and challenges for development of Epstein–Barr virus vaccine. *Vaccine* 79: 13993–14003.

Colby BM, Furman PA, Shaw JE, et al. (1981). Phosphorylation of acyclovir [9-(2-hydroxyethoxymethyl) guanine] in Epstein–Barr virus-infected lymphoblastoid cell lines. *J Virol* 38: 606–611.

Colby BM, Shaw JE, Elion GB, et al. (1980). Effect of acyclovir [9-(2-hydroxyethoxymethyl) guanine] on Epstein–Barr virus DNA replication. *J Virol* 34: 560–568.

Cvetkovic RS, Wellington K. (2005). Valganciclovir: A review of its use in the management of CMV infection and disease in immunocompromised patients. *Drugs* 65: 859–878.

De Clercq E, Holy A, Rosenberg I. (1989). Efficacy of phosphonylmethoxyalkyl derivatives of adenine in experimental herpes simplex virus and vaccinia virus infections in vivo. *Antimicrob Agents Chemother* 33: 185–191.

de Schryver A, Friberg S, Jr, Klein G, et al. (1969). Epstein–Barr virus associated antibody pattern in carcinoma of the post nasal space. *Clin Exp Immunol* 5: 443–459.

Diebold J, Raphael M, Prevot S, et al. (1997). Lymphomas associated with HIV infection. *Cancer Rev* 30: 263–293.

Duraiswamy J, Sherritt M, Thomson S, et al. (2003). Therapeutic LMP1 polyepitope vaccine for EBV-associated Hodgkin disease and nasopharyngeal carcinoma. *Blood* 101: 3150–3156.

Epstein MA, Achong BG, Barr YM. (1964). Virus particles in cultured lymphoblasts from Burkitt's lymphoma. *Lancet* 1: 702–703.

Farrell PJ. (2005). Epstein–Barr virus genome. In: Robertson ES (ed.), *Epstein–Barr Virus*. Norfolk, England: Caister Academic Press, pp. 263–287.

Fiorini S, Ooka T. (2008). Secretion of Epstein–Barr virus encoded BARF1 oncoprotein from latently infected B cells. *Virology J* 5: 70–73.

Franken M, Estabrooks A, Cavacini L, et al. (1996). Epstein–Barr virus-driven gene therapy for EBV-related lymphomas. *Nat Med* 2: 1379–1382.

Galal SA, Abdelsamie AS, Tokuda H, et al. (2011). Part I: Synthesis, cancer chemopreventive activity and molecular docking study of novel quinoxaline derivatives. *Eur J Med Chem* 46: 327–340.

Gottschalk S, Heslop H, Roon C. (2002). Treatment of Epstein–Barr virus-associated malignancies with specific T cells. *Adv Cancer Res* 84: 175–201.

Guan HP, Ksebati MB, Cheng YC, et al. (2000). Spiropentane mimics of nucleosides: Analogs of 2'-deoxyadenosine and 2'-deoxyguanosine. Synthesis of all stereoisomers, isomeric assignment, and biological activity. *J Org Chem* 65: 1280–1290.

Haddad E, Paczesny S, Leblond V, et al. (2001). Treatment of B-lymphoproliferative disorder with a monoclonal anti-interleukin-6 antibody in 12 patients: A multicenter phase I–II clinical trial. *Blood* 97: 1590–1597.

Hanel M, Fiedler F, Thorns C. (2001). Anti-CD20 monoclonal antibody (Rituximab) and Cidofovir as successful treatment of an EBV associated lymphoma with CNS involvement. *Onkologie* 24: 491–494.

Harnden MR, JarvestRL, Bacon TH, et al. (1987). Synthesis and antiviral activity of 9-[4-hydroxy-3-(hydroxymethyl)-but-1-yl] purines. *J Med Chem* 30: 1636–1642.

Henkel T, Ling PD, Hayward SD, et al. (1994). Mediation of Epstein–Barr virus EBNA2 transactivation by recombination signal binding protein J-κ. *Science* 265: 92–95.

Henle G, Henle W. (1966). Immunofluorescence in cells derived from Burkitt's lymphoma. *J Bacteriol* 91: 1248–1256.

Henle G, Henle W, Diehl V. (1968). Relation of Burkitt's tumor-associated herpes-type virus to infectious mononucleosis. *Proc Natl Acad Sci U S A* 59: 94–101.

Heslop HE, Ng CYC, Li C, et al. (1996). Long-term restoration of immunity against Epstein–Barr virus infection by adoptive transfer of gene-modified virus-specific T lymphocytes. *Nat Med* 2: 551–555.

Heslop HE, Slobod KS, Pule MA, et al. (2010). Long-term outcome of EBV-specific T-cell infusions to prevent or treat EBV-related lymphoproliferative disease in transplant recipients. *Blood* 115: 925–935.

Heussinger N, Burtner M, Ott G, et al. (2004). Expression of the Epstein–Barr virus (EBV)-encoded latent membrane protein 2A (LMP2A) in EBV associated nasopharyngeal carcinoma. *J Pathol* 203: 696–699.

Hong GK, Gulley ML, Feng WH, et al. (2005). Epstein–Barr virus lyric infection contributes to lymphoproliferative disease in a SCID mouse model. *J Virol* 79: 13993–14003.

Ito C, Itoigawa M, Kanematsu T, et al. (2003a). New cinnamylphenols from *Dalbergia* species with cancer chemopreventive activity. *J Nat Prod* 66: 1574–1577.

Ito C, Itoigawa M, Katsuno S, et al. (2000a). Chemical constituents of *Clausena excavata*: Isolation and structure elucidation of novel furanone-coumarins with inhibitory effects for tumor-promotion. *J Nat Prod* 63: 1218–1224.

Ito C, Itoigawa M, Mishina Y, et al. (2001). Cancer chemopreventive agents. New depsidones from *Garcinia* plants. *J Nat Prod* 64: 147–150.

Ito C, Itoigawa M, Mishina Y, et al. (2002). Chemical constituents of *Calophyllum brasiliensis*: Structure elucidation of seven new xanthones and their cancer chemopreventive activity. *J Nat Prod* 65: 267–272.

Ito C, Itoigawa M, Mishina Y, et al. (2003b). Chemical constituents of *Calophyllum brasiliense*. 2. Structure of three new coumarins and cancer chemopreventive activity of 4-substituted coumarins. *J Nat Prod* 66: 368–371.

Ito C, Itoigawa M, Miyamoto Y, et al. (1999). A new biflavonoid from *Calophyllum panciflorum* with antitumor-promoting activity. *J Nat Prod* 62: 1668–1671.

Ito C, Itoigawa M, Miyamoto Y, et al. (2003d). Polyprenylated benzophenones from *Garcinia assigu* and their potential cancer chemopreventive activities. *J Nat Prod* 66: 206–209.

Ito C, Itoigawa M, Sato A, et al. (2004a). Chemical constituents of *Glycosmis arborea*: Three new carbazole alkaloids and their biological activity. *J Nat Prod* 67: 1488–1491.

Ito C, Itoigawa M, Takakura T, et al. (2003c). Chemical constituents of *Garcinia fusca*: Structure elucidation of eight new xanthones and their cancer chemopreventive activity. *J Nat Prod* 66: 200–205.

Ito C, Katsuno S, Itoigawa M, et al. (2000b). New carbazole alkaloids from *Clausena anisata* with antitumor promoting activity. *J Nat Prod* 63: 125–128.

Ito H, Kasajima N, Tokuda H, et al. (2004b). Dimeric flavonol glycoside and galloylated C-glucosylchromones from *Kunzea ambigua*. *J Nat Prod* 67: 411–415.

Ito H, Koreishi M, Tokuda H, et al. (2000c). Cypellocarpins A-C, phenol glycosides esterified with oleuropeic acid, from *Eucalyptus cypellocarpa*. *J Nat Prod* 63: 1253–1257.

Itoigawa M, Ito C, Ju-ichi M, et al. (2002). Cancer chemopreventive activity of flavanones on Epstein–Barr virus activation and two-stage mouse skin carcinogenesis. *Cancer Lett* 176: 25–29.

Iwase Y, Takemura Y, Ju-ichi M, et al. (2001). Inhibitory effect of flavonoid derivatives on Epstein–Barr virus activation and two-stage carcinogenesis of skin tumors. *Cancer Lett* 173: 105–109.

Iwatsuki K, Akihisa T, Tokuda H, et al. (2003a). Lucidenic acids P and Q, methyl lucidenate P, and other triterpenoids from the fungus *Ganoderma lucidum* and their inhibitory effects on Epstein–Barr virus activation. *J Nat Prod* 66: 1582–1585.

Iwatsuki K, Akihisa T, Tokuda H, et al. (2003b). Sterol ferulates, sterols, and 5-alk(en)ylresorcinols from wheat, rye, and corn bran oils and their inhibitory effects on Epstein–Barr virus activation. *J Agric Food Chem* 51: 6683–6688.

Jacobson MA, de Miranda P, Cederberg DM, et al. (1987). Human pharmacokinetics and tolerance of oral ganciclovir. *J Antimicrob Agents Chemother* 31: 1251–1254.

Jiménez IA, Bazzocchi IL, Núñez MJ, et al. (2003). Absolute configuration of sesquiterpenes from *Crossopetalum tonduzii* and their inhibitory effects on Epstein–Barr virus early antigen activation in Raji cells. *J Nat Prod* 66: 1047–1050.

Joseph AM, Babcock GJ, Thorley-Lawson DA. (2000). EBV persistence involves strict election of latently infected B cells. *J Immunol* 165: 2975–2981.

Judson HF, Lewin B, Stent GS, et al. (1994). Basic genetic mechanisms. In: Alberts B, Bray D, Lewis J, et al. (eds.), *Molecular Biology of the Cell*. 3rd edn. New York: Garland Science, pp. 273–287.

Kapadia G, Azuine M, Tokuda H, et al. (2002). Inhibitory effect of herbal remedies on 12-O-tetradecanoylphorbol-13-acetate-promoted Epstein–Barr virus early antigen activation. *Pharmacol Res* 45: 213–220.

Katsuyama I, Fahmy H, Zjawiony JK, et al. (2002). Semisynthesis of new sarcophine derivatives with chemopreventive activity. *J Nat Prod* 65: 1809–1814.

Kenney SC. (2007). Reactivation and lytic replication of EBV. In: Arvin A, Campadelli-Fiume G, Mocarski E, et al. (eds.), *Human Herepsviruses: Biology, Therapy, and Immunoprophylaxis*. Cambridge: Cambridge University Press, pp. 403–433.

Kern ER, Kushner NL, Hartline CB, et al. (2005). In vitro activity and mechanism of action of methylenecyclopropane analogs of nucleosides against herpesvirus replication. *Antimicrob Agent Chemo* 49: 1039–1045.

Khanna R, Sherritt M, Burrows SR. (1999). EBV structural antigens, gp350 and gp85, as targets for ex vivo virus-specific CTL during acute infectious mononucleosis: Potential use of gp350/gp85 CTL epitopes for vaccine design. *J Immunol* 162: 3063–3069.

Kieff E, Rickinson AB. (2007). Epstein–Barr virus and its replication. In: Fields BN, Knipe DM, Howley PM (eds.), *Fields Virology*. Philadelphia, PA: Lippincott-Williams & Wilkins, pp. 2603–2654.

Kikuchi T, Akihisa T, Tokuda H, et al. (2007). Cancer chemopreventive effects of cycloartane-type and related triterpenoids in in vitro and in vivo models. *J Nat Prod* 70: 918–922.

Kinouchi Y, Ohtsu H, Tokuda H, et al. (2000). Potential antitumor-promoting diterpenoids from the stem bark of *Picea glehni*. *J Nat Prod* 63: 817–820.

Kitagawa N, Goto M, Kurozumi K, et al. (2000). Epstein–Barr virus-encoded poly(A)(-) RNA supports Burkitt's lymphoma growth through interleukin-10 induction. *EMBO J* 19: 6742–6750.

Klein G. (2002). Perspectives in studies of human tumor viruses. *Front Biosci* 7: 268–274.

Kubota K, Fukamiya N, Tokuda H, et al. (1997). Quassinoids as inhibitors of Epstein–Barr virus early antigen activation. *Can Lett* 113: 165–168.

Kushner NL, Williams SL, Hartline CB, et al. (2003). Efficacy of methylenecyclopropane analogs of nucleosides against herpesvirus replication in vitro. *Nucleo Nucleo Nucleic Acid* 22: 2105–2119.

Kutok JL, Wang FW. (2006). Spectrum of Epstein–Barr virus-associated diseases. *Anuu Rev Pathol Mech Dis* 1: 375–404.

Kuzushima KA, Yoshinori I, Okamura A. (2009). Cytotoxic-cell epitope peptides that specifically attack Epstein–Barr virus-infected cells and uses thereof. US Patent 2009035324.

Latour S, Gish G, Helgason CD, et al. (2001). Regulation of SLAM-mediated signal transduction by SAP, the X-linked lymphoproliferative gene product. *Nat Immunol* 2: 681–690.

Leight ER, Sugden B. (2000). EBNA-1: A protein pivotal to latent infection by Epstein–Barr virus. *Rev Med Virol* 10: 83–100.

Li C, Prichard MN, Korba BE, et al. (2008). Fluorinated methylenecyclopropane analogs of nucleosides. Synthesis and antiviral activity of (Z)- and (E)-9-{[(2-fluoromethyl-2-hydroxymethyl)-cyclopropylidene]methyl}adenine and guanine. *Bioorg Med Chem* 16: 2148–2155.

Li C, Quenelle DC, Prichard MN, et al. (2012). Synthesis and antiviral activity of 6 deoxycyclopropavir, a new prodrug of cyclopropavir. *Bioorg Med Chem* 20: 2669–2674.

Li C, Zemlicka J. (2007). Synthesis of "reversed" methylenecyclopropane analogs of antiviral phosphonate analogs of antiviral phosphonates. *Nucleo Nucleo Nucleic Acid* 26: 111–120.

Linde A. (1996). Diagnosis of Epstein–Barr virus related diseases. *Scand J Infect Dis* 28(Suppl 100): 83–88.

Lizasa H, Nanbo A, Nishikawa J, et al. (2012). Epstein–Barr virus (EBV)-associated gastric carcinoma. *Viruses* 4: 3420–3439.

Mei YP, Zhu XF, Zhou JM, et al. (2006). siRNA targeting LMP1-induced apoptosis in EBV-positive lymphoma cells is associated with inhibition of telomerase activity and expression. *Cancer Lett* 232: 189–198.

Mentzer S, Perrine S, Faller D. (2001). Epstein–Barr virus posttransplant lymphoproliferative disease and virus-specific therapy: Pharmacological re-activation of viral target genes with arginine butyrate. *Transpl Infect Dis* 3: 177–185.

Mhaske SB, Ksebati B, Prichard MN, et al. (2009). Phosphonate analogs of cyclopropavir phosphates and their E-isomers. Synthesis and antiviral activity. *Bioorg Med Chem* 17: 3892–3899.

Middeldorp JM, Brink AATP, van den Brule AJC, et al. (2003). Pathogenic roles for Epstein–Barr virus (EBV) gene products in EBV-associated proliferative disorders. *Crit Rev Oncol Hepatol* 45: 1–36.

Minami T, Wada S, Tokuda H, et al. (2002). Potential antitumor-promoting diterpenes from the cones of Pinus luchuensis. *J Nat Prod* 65: 1921–1923.

Mond, JJ, Lees A. (2002). Enhancement of B cell activation and immunoglobulin secretion by co-stimulation of receptors for antigen and EBV Gp350/220. US Patent 6432679.

Moore KW, Vieira P, Fiorentino DF, et al. (1990). Homology of cytokine synthesis inhibitory factor (IL-10) to the Epstein–Barr virus gene BCRF1. *Science* 248: 1230–1234.

Morgan AJ, Finerty S, Lovgren K, et al. (1988a). Prevention of Epstein–Barr (EB) virus-induced lymphoma in cotton-top tamarins by vaccination with the EB virus envelope glycoprotein gp340 incorporated into immune-stimulating complexes. *J Gen Virol* 69: 2093–2096.

Morgan AJ, Mackett M, Finerty S, et al. (1988b). Recombinant vaccinia virus expressing Epstein–Barr virus glycoprotein gp340 protects cotton top tamarins against EB virus-induced malignant lymphomas. *J Med Virol* 25: 189–195.

Morrison TE, Mauser A, Wong A, et al. (2001). Inhibition of IFNγ signaling by an Epstein–Barr virus immediate early protein. *Immunity* 15: 787–799.

Moutschen M, Leonard P, Sokal EM, et al. (2007). Phase I/II studies to evaluate safety and immunogenicity of a recombinant gp350 Epstein–Barr virus vaccine in healthy adults. *Vaccine* 25: 4697–4705.

Murakami C, Fukamiya N, Tamura S, et al. (2004). Multidrug-resistant cancer cell susceptibility to cytotoxic quassinoids, and cancer chemopreventive effects of quassinoids and canthin alkaloids. *Bioorg Med Chem* 12: 4963–4968.

Neyts J, Sadler R, De Clercg E, et al. (1998). The antiviral agent cidofovir has pronounced activity against nasopharyngeal carcinoma grown in nude mice. *Cancer Res* 58: 384–388.

Niedobitek G. (2000). Epstein–Barr virus infection in the pathogenesis of nasopharyngeal carcinoma. *J Clin Pathol Mol Pathol* 53: 248–254.

Novikov MS, Buckheit RW, Jr, Temburnikar K, et al. (2010). 1-Benzyl derivatives of 5-(arylamino)uracils as anti-HIV-1 and anti-EBV agents. *Bioorg Med Chem* 18: 8310–8314.

Ocheni S, Olusina DB, Oyekunle AA, et al. (2010). EBV-associated malignancies. *Open Infect Dis J* 4: 101–112.

Odumade OA, Hogquist KA, Balfour HH, Jr. (2011). Progress and problems in understanding and managing primary Epstein–Barr virus infections. *Clin Microbiol Rev* 24: 193–209.

Oertel SH, Riess H. (2002). Antiviral treatment of Epstein–Barr virus-associated lymphoproliferations. Recent results. *Cancer Res* 159: 89–95.

Okano M, Fukamiya N, Tagahara K, et al. (1995). Inhibitory effects of quassinoids on Epstein–Barr virus activation. *Can Lett* 94: 139–146.

Omar MA, Shaker YM, Galal SA, et al. (2012). Synthesis and docking studies of novel antitumor benzimidazoles. *Bioorg Med Chem* 20: 6989–7001.

Oudejans JJ, van de Brule AJC, Jiwa NM, et al. (1995). BHRF1, the Epstein–Barr virus homologue of the bcl-2 (proto-) oncogene, is transcribed in EBV associated B-cell lymphomas and in reactive lymphocytes. *Blood* 86: 1891–1902.

Pearson GR, Luka J, Petti L, et al. (1987). Identification of an Epstein–Barr virus early gene encoding a second component of the restricted early antigen complex. *Virology* 160: 151–161.

Perestelo NR, Jiménez IA, Tokud H, et al. (2010). Sesquiterpenes from *Maytenus jelskii* as potential cancer chemopreventive agents. *J Nat Prod* 73: 127–132.

Plosker GL, Noble S. (1999). Cidofovir: A review of its use in cytomegalovirus retinitis in patients with AIDS. *Drugs* 58: 325–345.

Pope J. (1967). Establishment of cell lines from peripheral leukocytes in infectious mononucleosis. *Nature* 216: 810–811.

Purifoy DJ, Beauchamp LM, de Miranda P, et al. (1993). Review of research leading to new anti-herpes virus agents in clinical development: Valaciclovir hydrochloride (256 U, the L-valyl ester of acyclovir) and 882C, a specific agent for varicella zoster virus. *J Medic Virol* 1: 139–145.

Qin X, Chen X, Wang K, et al. (2006). Synthesis, antiviral, and antitumor activity of 2-substituted purine methylenecyclopropane analogs of nucleosides. *Bioorg Med Chem* 14: 1247–1254.

Qiu YL, Geiser F, Kira T, et al. (2000b). Synthesis and enantioselectivity of the antiviral effects of (R,Z)-,(S,Z)-methylenecyclopropane analogs of purine nucleosides and phosphoralaninate prodrugs: Influence of heterocyclic base, type of virus and host cells. *Antivir Chem Chem* 11: 191–202.

Qiu YL, Hempel A, Camerman N, et al. (1998a). (R)-(-)- and (S)-(+)-Synadenol: Synthesis, absolute configuration, and enantioselectivity of antiviral effect. *J Med Chem* 41: 5257–5264.

Qiu YL, Ksebati MB, Ptak RG, et al. (1998b). (Z)-and (E)-2-((hydroxymethyl)-cyclopropylidene) methyladenine and -guanine. New nucleoside analogs with a broad-spectrum antiviral activity. *J Med Chem* 41: 10–23.

Qiu YL, Ksebati MB, Zemlicka J. (2000a). Synthesis of (Z)-and (E)-9[(2-((hydroxyethylene)-cyclopropyl]adenine—new methylenecyclopropane analogs of adenosine and their substrate activity for adenosine deaminase. *Nucleo Nucleo Nucleic Acid* 19: 31–37.

Radkov S, Bain M, Farrell P, et al. (1997). Epstein–Barr virus EBNA3C represses Cp, the major promoter for EBNA expression, but has no effect on the promoter of the cell gene CD21. *J Virol* 71: 8552–8562.

Rahman S, Fukamiya N, Ohno N, et al. (1997). Inhibitory effects of quassinoid derivatives on Epstein–Barr virus early antigen activation. *Chem Pharm Bull* 45: 675–677.

Ramla MM, Omar MA, Tokudab H, et al. (2007). Synthesis and inhibitory activity of new benzimidazole derivatives against Burkitt's lymphoma promotion. *Bioorg Med Chem* 15: 6489–6496.

Riddler SA, Breinig MC, MvKnight JLC. (1994). Increased levels of circulating Epstein–Barr virus-infected lymphocytes and decreased Epstein–Barr virus nuclear antigen antibody responses are associated with the development of posttransplant lymphoproliferative disease in solid organ transplant recipients. *Blood* 84: 972–984.

Sacau EP, Estévez-Braun A, Ravelo AG, et al. (2003). Inhibitory effects of lapachol derivatives on Epstein–Barr virus activation. *Bioorg Med Chem* 11: 483–488.

Sample J, Young L, Martin B, et al. (1990). Epstein–Barr virus type 1 and type 2 differ in the EBNA-3A, EBNA-3B and EBNA-3C genes. *J Virol* 64: 4084–4092.

Schnute ME, Anderson DJ, Brideau RJ, et al. (2007). 2-Aryl-2-hydroxyethylamine substituted 4-oxo-4, 7-dihydrothieno- [2,3-b]pyridines as broad-spectrum inhibitors of human herpesvirus polymerases. *Bioorg Med Chem Lett* 17: 3349–3353.

Schnute ME, Brideau RJ, Sarah A, et al. (2008). Synthesis of 4-oxo-4,7-dihydrofuro[2,3-b] pyridine-5-carboxamides with broad-spectrum human herpesvirus polymerase inhibition. *Bioorg Med Chem Lett* 18: 3856–3859.

Sculley T, Buck M, Gabrielli B, et al. (2002). A histone deacetylase inhibitor, azelaicbishydroxamic acid, shows cytotoxicity on Epstein–Barr virus transformed B-cell lines: A potential therapy for post-transplant lymphoproliferative disease. *Transplantation* 73: 271–279.

Skare JC, Milunsky A, Byron KS, et al. (1987). Mapping the X-linked lymphoproliferative syndrome. *Proc Natl Acad Sci U S A* 84: 2015–2018.

Slobod K, Taylor G, Sandlund J, et al. (2000). Epstein–Barr virus-targeted therapy for AIDS-related primary lymphoma of the central nervous system. *Lancet* 356: 1493–1494.

Smith PR, Jesus O, Turner D, et al. (2000). Structure and coding content of CST (BART) family RNAs of Epstein–Barr virus. *J Virol* 74: 3082–3092.

Sokal EM, Hoppenbrouwers K, Vandermeulen C, et al. (2007). Recombinant gp350 vaccine for infectious mononucleosis: A phase 2, randomized, double-blind, placebo-controlled trial to evaluate the safety, immunogenicity, and efficacy of an Epstein–Barr virus vaccine in healthy young adults. *J Infect Dis* 196: 1749–1753.

Srinivas S, Sixbey JW. (1995). Epstein–Barr virus induction of recombinase activating genes RAG1 and 2. *J Virol* 85: 8155–8158.

Subklewe M, Sebelin K, Block A, et al. (2005). Dendritic cells expand Epstein–Barr virus specific CD8+ T cell responses more efficiently than EBV transformed B cells. *Hum Immunol* 66: 938–949.

Swaminathan S, Tomkinson B, Kieff E. (1991). Recombinant Epstein–Barr virus with small RNA (EBER) genes deleted transforms lymphocytes and replicates in vivo. *Proc Natl Acad Sci U S A* 88: 1546–1550.

Szekely L, Selivanova G, Magnusson KP, et al. (1993). EBNA-5, an Epstein–Barr virus-encoded nuclear antigen, binds to the retinoblastoma and p53 proteins. *Proc Natl Acad Sci U S A* 90: 5455–5459.

Takasaki M, Tokuda H, Nishino H, et al. (1999). Cancer chemopreventive agents (antitumor-promoters) from Ajuga decumbens. *J Nat Prod* 62: 972–975.

Tamura S, Fukamiya N, Mou XY, et al. (2000). Conversion of quassinoids for enhancement of inhibitory effect against Epstein–Barr virus early antigen activation. Introduction of lipophilic side chain and esterification of diosphenol. *Chem Pharm Bull* 48: 876–878.

Tamura S, Fukamiya N, Okano M, et al. (2002). Cancer chemopreventive effect of quassinoid derivatives. Introduction of side chain to shinjulactone C for enhancement of inhibitory effect on Epstein–Barr virus activation. *Can Lett* 185: 47–51.

Tamura S, Fukamiya N, Okano M, et al. (2003). Three new quassinoids, ailantinol E, F, and G, from *Ailanthus altissima. Chem Pharm Bull* 51: 385–389.

Taylor GS, Haigh TA, Gudgeon NH, et al. (2004). Dual stimulation of Epstein–Barr Virus (EBV)-specific CD4+- and CD8+-T-cell responses by a chimeric antigen construct: Potential therapeutic vaccine for EBV-positive nasopharyngeal carcinoma. *J Virol* 78: 768–778.

Thompson MP, Kurzrock R. (2004). Epstein–Barr virus and cancer. *Clin Cancer Res* 10: 803–821.

Thorley-Lawson DA, Gross A. (2004). Persistence of the Epstein–Barr virus and origins of associated lymphomas. *N Eng J Med* 350: 1328–1337.

Truyen U, Lochelt M. (2006). Relevant oncogenic viruses in veterinary medicine: Original pathogens and animal models for human disease. *Contrib Microbiol* 13: 101–117.

Tsurumi T, Fujita M, Kudoh A. (2005). Latent and lytic Epstein–Barr virus replication strategies. *Rev Med Virol* 15: 3–15.

Ukiya M, Akihisa T, Tokuda H, et al. (2002a). Constituents of Compositae plants III. Anti-tumor promoting effects and cytotoxic activity against human cancer cell lines of triterpene diols and triols from edible chrysanthemum flowers. *Cancer Lett* 177: 7–12.

Ukiya M, Akihisa T, Tokuda H, et al. (2002b). Inhibitory effects of cucurbitane glycosides and other triterpenoids from the fruit of *Momordica grosvenori* on Epstein–Barr virus early antigen induced by tumor promoter 12-O-tetradecanoylphorbol-13-acetate. *J Agric Food Chem* 50: 6710–6715.

Ukiya M, Akihisa T, Tokuda H, et al. (2002c). Inhibition of tumor-promoting effects by poricoic acids G and H and other lanostane-type triterpenes and cytotoxic activity of poricoic acids A and G from Poria cocos. *J Nat Prod* 65: 462–465.

Ukiya M, Akihisa T, Tokuda H, et al. (2003a). Isolation, structural elucidation, and inhibitory effects of terpenoid and lipid constituents from sunflower pollen on Epstein–Barr virus early antigen induced by tumor promoter, TPA. *J Agric Food Chem* 51: 2949–2957.

Ukiya M, Akihisa T, Tokuda H, et al. (2003b). Sunpollenol and five other rearranged 3,4-seco-tirucallane-type triterpenoids from sunflower pollen and their inhibitory effects on Epstein–Barr virus activation. *J Nat Prod* 66: 1476–1479.

Ukiya M, Akihisa T, Yasukawa K, et al. (2002d). Anti-inflammatory and anti-tumor promoting effects of cucurbitane glycosides from the roots of Bryonia dioica. *J Nat Prod* 65: 179–183.

Ukiya M, Akihisa T, Yasukawa K, et al. (2006). Anti-inflammatory, anti-tumor-promoting, and cytotoxic activities of constituents of marigold (*Calendula officinalis*) flowers. *J Nat Prod* 69: 1692–1696.

Ukiya M, Kikuchi T, Tokuda H, et al. (2010). Antitumor-promoting effects and cytotoxic activities of Dammar resin triterpenoids and their derivatives. *Chem Biodivers* 7: 1871–1884.

van Grunsven WM, Spaan WJ, Middeldorp JM. (1994). Localization and diagnostic application of immunodominant domains of the BFRF3-encoded Epstein–Barr virus capsid protein. *J Infect Dis* 170: 13–19.

Vere Hodge RA, Sutton D, Boyd MR, et al. (1989). Selection of an oral prodrug (BRL 42810; famciclovir) for the antiherpesvirus agent BRL 39123 [9-(4-hydroxy-3-hydroxymethyl-but-l-yl) guanine; penciclovir]. *Antimicrob Agents Chemother* 33: 1765–1773.

Villegas E, Santiago O, Sorlozano A, et al. (2010). New strategies and patent therapeutics in EBV-associated diseases. *Mini Rev Med Chem* 10: 914–927.

Wagstaff AJ, Bryson HM. (1994). Foscarnet. A reappraisal of its antiviral activity, pharmacokinetic properties and therapeutic use in immunocompromised patients with viral infections. *Drugs* 48: 199–226.

Wang R, Corbett TH, Cheng YC, et al. (2002a). Tryptophanyl phosphoramidates as prodrugs of synadenol and its E-isomer: Synthesis and biological activity. *Bioorg Med Chem Lett* 12: 2467–2470.

Wang R, Kern ER, Zemlicka J. (2002b). Synthesis of methylenecyclobutane analogs of nucleosides with axial chirality and their phosphoralaninates: A new pronucleotide effective against Epstein–Barr virus. *Antivir Chem Chem* 13: 251–262.

Wilson AD, Shooshstari M, Finerty S, et al. (1996). Virus-specific cytotoxic T cell responses are associated with immunity of the cotton top tamarin to Epstein–Barr virus (EBV). *Clin Exp Immunol* 103: 199–205.

Wolf HJ, Jagerhuber J. (1997). DNA sequences of the EBV genome, recombinant DNA molecules, processes for preparing EBV-related antigens, diagnostic compositions and pharmaceutical compositions containing said antigens. US Patent 5679774.

Yan Z, Kern ER, Gullen E, et al. (2005). Nucleotides and pronucleotides of 2,2-bis(hydroxymethyl)methylenecyclopropane analogs of purine nucleosides: Synthesis and antiviral activity. *J Med Chem* 48: 91–99.

Yan Z, Zhou S, Kern ER, et al. (2006). Synthesis of methylenecyclopropane analogs of antiviral nucleoside phosphonates. *Tetrahedron* 62: 2608–2615.

Yao QY, Rowe M, Morgan AJ, et al. (1991). Salivary and serum IgA antibodies to the Epstein–Barr virus glycoprotein gp340: Incidence and potential for virus neutralization. *Int J Cancer* 48: 45–50.

Yeo H, Li Y, Fu L, et al. (2005). Synthesis and antiviral activity of helioxanthin analogs. *J Med Chem* 48: 534–546.

Yoshioka M, Crum MM, Sample JT. (2008). Autoexpression of Epstein–Barr virus antigen 1 expression by inhibition of pre-mRNA processing. *J Virol* 82: 1679–1687.

Zancy VL, Gershburg E, Davis MG, et al. (1999). Inhibition of Epstein–Barr virus replication by a benzimidazole L-riboside: Novel antiviral mechanism of 5,6-dichloro-2-(isopropylamino)-1-b-L-ribofuranosyl-1H-benzimidazole. *J Virol* 73: 7271–7277.

Zhang J, Huang Y, Kikuchi T, et al. (2012). Cucurbitane triterpenoids from the leaves of *Momordica charantia*, and their cancer chemopreventive effects and cytotoxicities. *Chem Biodivers* 9: 428–440.

Zhou S, Breitenbach JM, Borysko KZ, et al. (2004b). Synthesis and antiviral activity of (Z)- and (E)-2,2-[bis(hydroxymethyl)cyclopropylidene]methylpurines and -pyrimidines: Second-generation methylenecyclopropane analogs of nucleosides. *J Med Chem* 47: 566–575.

Zhou S, Drach JC, Prichard MN, et al. (2009). (Z)- and (E)-2-(1,2-Dihydroxyethyl)methylenecyclopropane analogs of 2′-deoxyadenosine and 2′-deoxyguanosine. Synthesis of all stereoisomers, absolute configuration, and antiviral activity. *J Med Chem* 52: 3397–3407.

Zhou S, Kern ER, Gullen E, et al. (2004a). (Z)- and (E)-[2-Fluoro-2-(hydroxymethyl)cyclopropylidene]methylpurines and -pyrimidines, a new class of methylenecyclopropane analogs of nucleosides: Synthesis and antiviral activity. *J Med Chem* 47: 6964–6972.

Zhou S, Kern ER, Gullen E, et al. (2006). 9-[3-Fluoro-2-(hydroxymethyl) cyclopropylidene] methyl-adenines and -guanines. Synthesis and antiviral activity of all stereoisomers. *J Med Chem* 49: 6120–6128.

Zimber-Strobl U, Kempkes B, Marschall G, et al. (1996). Epstein–Barr virus latent membrane protein (LMP1) is not sufficient to maintain proliferation of B cells but both it and activated CD40 can prolong their survival. *EMBO J* 15: 7070–7078.

Zimber-Stroble U, Stroble LJ. (2001). EBNA2 and notch signaling in Epstein–Barr virus mediated immortalization of B lymphocytes. *Semin Cancer Biol* 11: 423–434.

# 7 Landscape of Papillomavirus in Human Cancers

## *Prevention and Therapeutic Avenues*

*Susri Ray Chaudhuri (Guha),* Anirban Roy,* Indranil Chatterjee, Rahul Roy Chowdhury, and Snehasikta Swarnakar*

### 7.1 INTRODUCTION

Human papillomavirus (HPV) is a member of the Papillomaviridae family of viruses, which are so far implicated for causing epithelial warts, mostly benign in nature. However, this particular viral entity has shot into prominence due to its increasingly distinct association with severe human maladies such as cervical and oropharyngeal squamous cell carcinomas (OSCCs). Interestingly enough, there is a silver lining in the form of explorable therapeutic options that may pave the path for fruitful research and subsequent deployment in clinical medicine. Various aspects of these options will be elucidated in the ensuing paragraphs.

#### 7.1.1 Structure and Properties of HPV

The first visualization of papillomavirus particles in human warts by electron microscopy was reported in 1949 (zur Hausen 1996). In 1978, Stefania Ginsburg-Jabłońska, a Polish physician and dermatologist at the Pasteur Institute, was the first scientist to discover the association between HPV and skin cancer (Orth et al. 1978b). Two years earlier, in 1976, Harald zur Hausen (who later received a Nobel Prize in medicine) published his legendary hypothesis on HPV that explored the involvement of HPV with cervical cancer. In 1983–1984, he and his collaborators identified two strains of HPV, HPV-16 and HPV-18, which are almost 100% associated with cervical cancer (Castellsagué 2008; Denny 2009).

Papillomaviruses are essentially species specific with a predilection for the site of infection (Stanley 2012a). There are more than 200 subtypes already isolated.

* Both of these authors contributed equally.

HPVs infect birds and mammals alike, though they preferentially infect either human skin or mucosa (Bernard 2005; Schäfer et al. 2011).

The papillomavirus family represents an extremely heterogeneous group of viruses, and the complete genome structure of which was first elucidated by Crawford and Crawford in 1963 (zur Hausen 1996). This is a large family of small, nonenveloped, circular, double-stranded DNA viruses that produce benign epithelial warts (Liao 2006; Zheng et al. 2006; Chow et al. 2010). The HPV DNA genome consists of approximately 8000 base pairs, packed into round particles approximately 55 nm in diameter, with icosahedral symmetry (Münger et al. 2004; Trottier 2012). The capsid comprises of 72 capsomeres in a skew arrangement (as depicted in Figure 7.1), and each capsomere has two capsid proteins—major L1 and minor L2 (Belnap 1996; Finnen et al. 2003). The genome has a guanine plus cytosine content of 40%–50%, with two coding regions and one noncoding regulatory region (Finley and Luo 1998). They are divided into late (L) and early (E) genes, depending on their expression during the course of an infection. The early regions encode eight open reading frames (ORFs), responsible for regulatory functions such as integration of the virus into the host genome, viral replication, and activation of the lytic cycle (De Villiers et al. 2004; Kanodia et al. 2007). The late gene encodes the viral capsid proteins (Doorbar 2005). Integration generally occurs in the E1/E2 regions of the viral genome by disrupting the E2 viral transcriptional regulatory circuit (as depicted in Figure 7.2) (Howley 1991; Raybould et al. 2011). The virus infects the keratinocytes of the basal layer of stratified squamous epithelium via minor abrasions in the skin (Burd 2003; Stanley 2012a) and delivers its DNA to the nucleus. The infected basal cells divide, and their progeny take HPV DNA with them (Longworth and Laimins 2004; Pyeon et al. 2009). During the early phases of infection, the viral genome exists as extrachromosomal DNA circle (copy number varies within 50–100) that replicates as the host cell chromosomes replicate (Kurg 2011). The first genes expressed are E6 and E7, which are involved in cell transformation and cause precancerous changes in the host cells (Burd 2003).

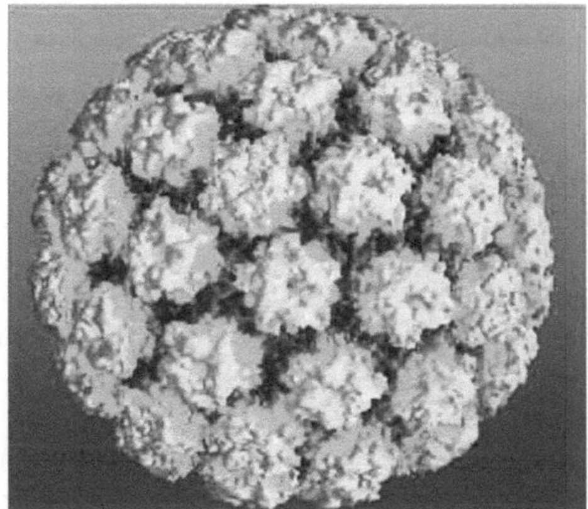

**FIGURE 7.1 (See color insert.)** Electron micrograph of HPV (From www.longtron.altervista.org/blog, accessed on December 15, 2013).

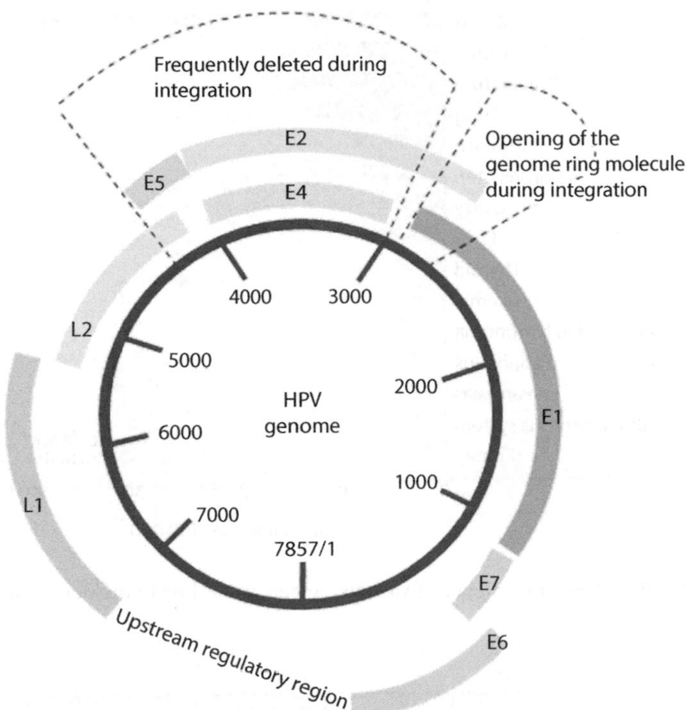


**FIGURE 7.2 (See color insert.)** The organization of the HPV genome with the early genes (El–E7), the capsid genes (Ll and L2), and the upstream regulatory region, with its integration into the host genome.

### 7.1.2 Epidemiology of HPV Infection

The worldwide incidence of cervical cancer is approximately 510,000 new cases annually, with some 288,000 deaths worldwide. In India, the current estimates indicate approximately 132,000 new cases diagnosed and 74,000 deaths annually (Figure 7.3), accounting for nearly one-third of cervical cancer deaths worldwide (Kaarthigeyan 2012). Every year in the United States, about 12,000 women are diagnosed with cervical cancer, and about 4000 women die from this disease (NIH 1996).

At any given time, about 6.6% of women in the general population are estimated to harbor cervical HPV infection. HPV-16 and -18 account for nearly 76.7% of cervical cancer worldwide (Kaarthigeyan 2012), while high-risk HPV types are detected in 99% of cervical cancers. Overall, it is estimated that 5.2% of all cancers are attributable to HPV (Figure 7.4) (Stanley 2012b). The abundance of non-HPV-16 and -18 related cervical cancers varies with geographical regions. In 1986, the HPV DNA was first identified in an invasive head and neck squamous cell carcinoma (HNSCC) by Southern blot hybridization (Gillison et al. 2000). Since 1986, HPV sequences have been detected repeatedly in a variable proportion of HNSCC, from less than

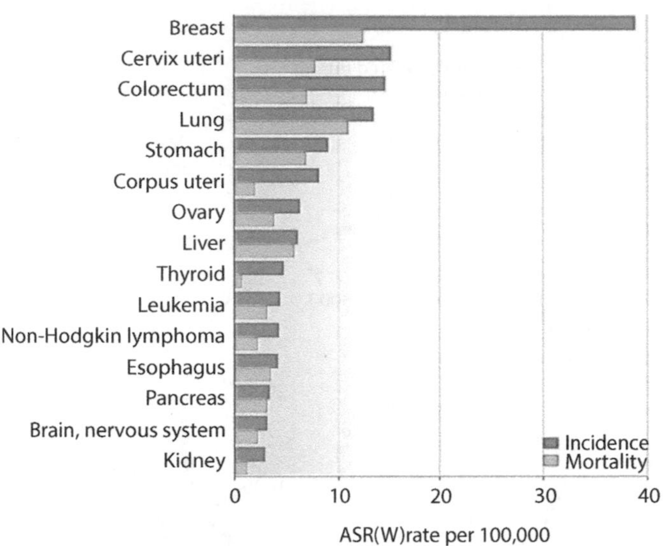


**FIGURE 7.3** Worldwide prevalence of various cancers according to incidence and mortality.

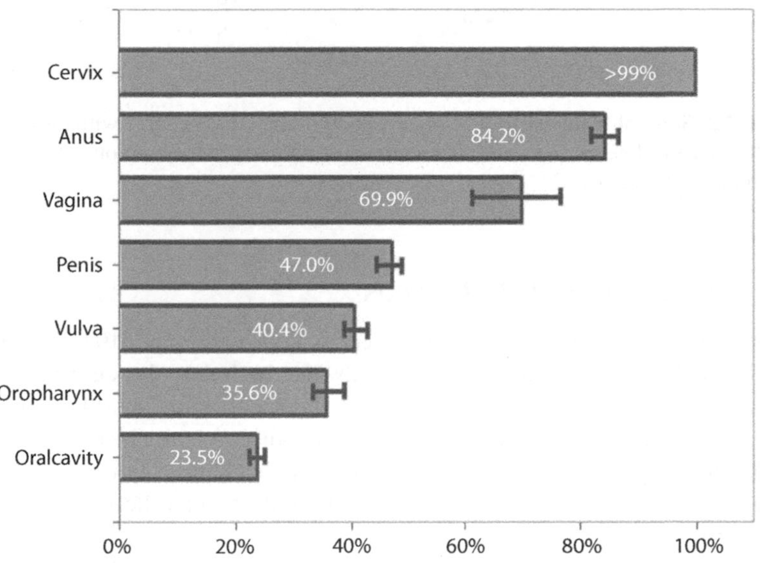


**FIGURE 7.4** Estimated HPV contribution in various cancers.

10% to up to 90%, irrespective of anatomic sites of tumor and HPV signature. Recent studies showed that the incidence of HPV-associated OSCC is increasing. A study of all OSCC in Sweden conducted by Näsman (2009) showed that the proportion of HPV-associated tonsillar cancers increased dramatically from 23% oropharyngeal cancers in 1970, to 28% in the 1980s, and further to 57% in the 1990s. In addition to oropharyngeal cancers, HPV also has been detected in a subset of laryngeal (24%) and oral cavity (23%) cancers (Kreimer et al. 2005). Cohort studies from the 1990s suggested that approximately 50% of oropharyngeal cancers were attributable to HPV, while recent reports documented 70%–80% of HPV-related malignancies in North America and Europe (Marur et al. 2010).

### 7.1.3 Risk Stratification of HPV Infection

HPVs have been designated as high-risk and low-risk, based on whether the HPV-mediated lesions are linked to risk for malignancy. HPV types found in cervical intraepithelial neoplasia (CIN) and other anogenital cancers are considered as high-risk types (HPV-16 and -18) that increase the risk of malignancy (Hawes and Kiviat 2002). The low-risk types (HPV-6 and -11) are primarily associated with venereal warts or condyloma acuminata and these rarely progress to malignancy (Todd et al. 2002). In 1995, the International Agency for Research on Cancer (IARC) endorsed the view that HPV types 16 and 18 are carcinogenic in humans. Together, these two HPV types are responsible for approximately 70% of cervical cancer cases, and other high-risk HPV (e.g., HPV-42, -43, -44, -45, -51, -52, -56, -58, -59, and -68) types account for the remaining number of cases (Burd 2003). HPV viral genes modify the properties of infected epithelial cells extensively, and the high-risk type viruses possess remarkable oncogenic properties (Zheng and Baker 2006). In vitro assay on human keratinocytes (which resemble the normal target cell of the virus) transfected with low-risk and high-risk HPV types showed that viruses from the high-risk group have the ability to immortalize primary human keratinocytes, while the low-risk types are unable to do so (Burd 2003). In addition, the low-risk viruses perform poorly in experiments concerned with the malignant transformation of rodent cells in comparison to the high-risk HPV types (Howley 1991; zur Hausen 1999).

### 7.1.4 HPV and Human Carcinogenesis

Malignancy of the uterine cervix is the most widely accepted case associated with HPV infection. In addition, high-risk HPVs are associated with other anogenital carcinomas, including vulvar, anal, penile, and a proportion of head and neck cancers. In the cervix, they cause cervical dysplasia, which—if identified early—can be successfully treated, securing a complete cure. Similarly, HPV may also cause anal intraepithelial neoplasms (AIN), vaginal intraepithelial neoplasms (VaIN), vulval intraepithelial neoplasms (VIN), and so on (Alba et al. 2009). Recent studies have demonstrated that high-risk HPVs might be involved in the pathogenesis of breast cancer as well (Burd 2003). HPV DNA detection in extragenital cancers is reported, although the etiological involvement of HPV in such malignancies is debatable.

HPV causes condyloma (warts) and dysplasia (precancer), which are the two types of pathophysiological conditions. The most common clinically significant manifestation of persistent genital HPV infection is CIN. Within a few years of infection, low-grade CIN (i.e., CIN 1) may develop, which may spontaneously resolve as the infection clears out (Wright 2006). Persistent HPV infections progress directly to high-grade CIN (i.e., CIN 2 or CIN 3). High-risk HPV types (including types 16, 18, 31, 33, 35, 39, 45, 51, 52, 56, 58, 59, 68, 69, 73, and 82) can cause high-grade as well as low-grade cellular abnormalities that are precursors to cervical and anogenital cancers.

Epidemiological and molecular pathological evidences strongly suggest that HPV-16 has a causal association with a subset of oropharyngeal SCC (Gillison et al. 2000). It is unclear why the oropharynx is more susceptible to HPV transformation, although its similarities to the uterine cervix in terms of easy access for infection and the same embryonic development from endoderm have been suggested (Shaib et al. 2012). The tonsils also contain deep invagination of the mucosal surface (crypts), thought to favor the capture and processing of viral antigens, which may aid the entrance of HPV to basal cells. Patients with OSCC typically have a sore throat, ear pain, odynophagia, or hoarseness, and often lymph node involvement in the neck. Incidence trends show a significant increase in HPV-related OSCC among white males aged 40–59 years, with a decrease among all races/sexes older than 40 years (Haddad 2007). There are reports of cutaneous malignancies associated with HPV-5 and -8 (Tomlins and Storey 2010). A childhood disease characterized by reddish, flat warty lesions, epidermodysplasia verruciformis (EV), evolves into carcinoma in a majority of cases (Jovanoviæ and Karadagliæ 2000). Molecular evidence indicates the involvement of transcriptionally active HPV, which is further established (by ultraviolet radiation) as a cocarcinogen in such lesions.

The male counterpart of CIN is penile intraepithelial neoplasm (PIN). The occurrences of PIN are much lower in comparison to CIN, but they are highly associated with high-risk HPV infections and can lead to penile carcinoma (Morris et al. 2011). In a Spanish study, it was found that out of 49 penile carcinoma patients, 38 were HPV positive. Among them, 32 were HPV-16 positive (Pascual et al. 2007). A Danish study suggested that 65% of penile cancer cases were associated with HPV; out of those, 92% were HPV-16 positive (Madsen et al. 2008).

## 7.2 GENOTYPES OF HPVs AND THEIR ASSOCIATION IN CERVICAL CANCER

The use of recombinant DNA technology and molecular cloning has made it clear that there are multiple HPV types and that the warts on different tissue locations are caused by different HPV types with tropisms for mucosal or cutaneous squamous surfaces (Orth et al. 1978b). It is apparent that some members of the HPV family, particularly a subset infecting the anogenital tract, are true human carcinogens that cause of carcinoma of the cervix, the second most common cancer in women worldwide (zur Hausen et al. 1981).

HPVs are not classified into serotypes but into genotypes on the basis of DNA sequence. In vitro growth of HPV is difficult, and HPV infection is determined by the detection of HPV DNA in biopsy specimens, swabs, or scrapes from mucosal or cutaneous surfaces, using sensitive molecular hybridization methods. HPVs have a predilection for either cutaneous or mucosal epithelial surfaces or fall into two groups: low-risk types that predominantly cause benign warts and high-risk types that may result in malignant disease as a rare consequence of infection. This risk profile is shown clearly in the genital tract, where 30–40 HPVs regularly or sporadically infect the mucosal epithelium in men and women. The two most common low-risk mucosal HPVs are HPV-6 and -11, which together cause about 90% of genital warts and almost all recurrent respiratory papillomas (RRP) (Lacey et al. 2006) as well as a proportion of low-grade CIN 1, VIN 1, VaIN 1, and AIN 1 (Moscicki et al. 2006). High-risk HPV is associated with CIN 2 and 3, VIN 2 and 3, and AIN 2 and 3 (Arbyn et al. 2011).

Meta-analysis studies, published in 1995–2009, detecting HPV in women with normal cytology revealed a 11.7% global prevalence of HPV, with 24% in sub-Saharan Africa, 21% in Eastern Europe, and 16.1% in Latin America; a total of 1,016,719 women were studied (Bruni et al. 2010). Among the combined sample of more than 15,000 women aged 15–74 years from 11 countries, the crude prevalence of any HPV was 9%. The age-standardized HPV prevalence of all types varied more than 20-fold, with the lowest prevalence in Spain and the highest in Nigeria. Grouping by region showed that HPV prevalence was five times higher in sub-Saharan Africa than in Europe, with intermediate rates in Asia and South America (Clifford et al. 2005). HPV-16 was twice as frequent as any other high-risk type in all regions except sub-Saharan Africa, where HPV-35 was equally common. The proportion of HPV-16 infection was 26% in Europe and 12%, 18%, and 21% in Africa, Asia, and South America, respectively. The next most common HPV types (worldwide) among cytologically normal women were HPV-42, -58, -31, and -18, with significant differences among regions. The second most common type was HPV-33 in Asia, HPV-58 in South America, and HPV-31 in Europe (Castellsagué et al. 2006). In India, HPV types 16 and 18 are the two dominant types with the highest attribution to cervical cancer (Basu et al. 2009).

### 7.2.1 How Does Infection Occur?

Genital HPV infection is primarily a sexually transmitted infection. This virus may be transmitted by skin to skin contact during hand to genital touch or by contact of other mucosal surfaces other than penetrative vaginal or anal sexual intercourse (Moscicki et al. 2006). It is estimated that 50%–80% of sexually active men and women will acquire a genital HPV (high- and low-risk) infection in their lifetime. The peak period of infection is soon after the start of sexual activity (Peto et al. 2004), and the risk increases with the number of sexual partners (Koutsky 1997), though HPV infection has been also noted in women with single partners. HPV is highly infectious, with an incubation period ranging from few weeks to months or even years, and the duration of this latent period probably relates to the dose of virus received. Eventually, viral DNA is detected and the infection cycle commences. This phase of active replication also persists over a variable length of time, but eventually the vast majority of infected individuals mount an effective immune response with subsequent clinical remission from disease (Stanley 2006). Cell-mediated immune (CMI) response to the early proteins, principally E2 and E6 (Woo et al. 2010), is necessary for lesion regression followed by antibody to the major capsid protein L1. About 10%–20% of infected individuals do not effectively clear the virus and remain

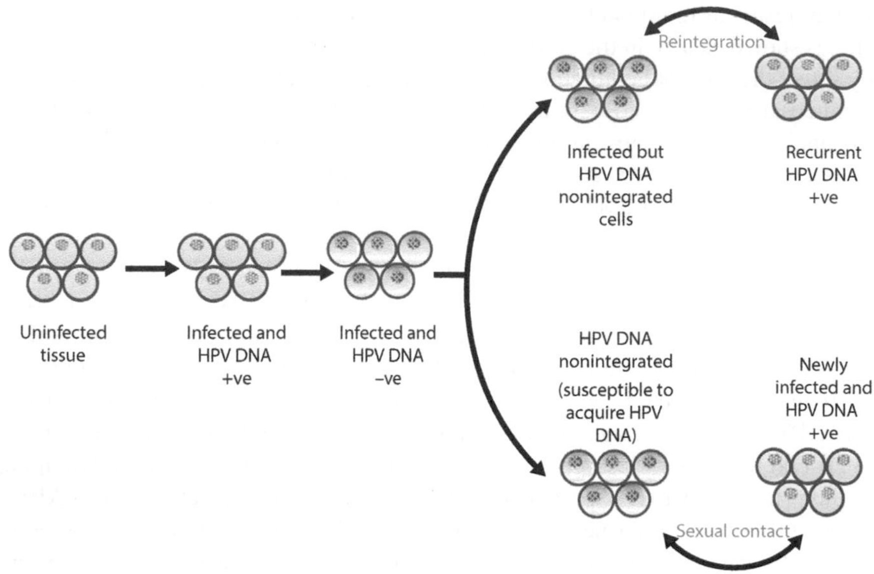


**FIGURE 7.5 (See color insert.)** Integration of HPV DNA in infected individuals.

DNA positive with a persistent active viral infection. These individuals are at risk for progression to high-grade precancers in CIN 2 and 3 and then invasive cancer (Figure 7.5).

HPV infections are exclusively intraepithelial. The virus infects the basal epithelial cells (probably in low copy number) and amplifies in a first round of DNA amplification to 100 nuclear episomes per cell. Maximal viral gene expression and viral assembly occur only in differentiating keratinocytes in the stratum spinosum and granulosum of squamous epithelium (Doorbar 2005). HPV exists as a nuclear episome in productive viral infection, but the integration of high-risk HPV DNA into the host genome is an important step in neoplastic progression in the cervix (Pett et al. 2004; Wentzensen et al. 2004). A variable segment of the HPV genome including the E6 and E7 oncogenes and the upstream regulatory region are retained in spite of deletion or disruption of the virus regulatory E2 gene during integration of the virus into the host genome (Alazawi et al. 2002). HPV-16 integration causes disruption or deletion of E2 and allows increased expression of the viral oncogenes (Pett et al. 2006).

### 7.2.2 Immune Response to HPV

It is clear from the natural history of HPV (Moscicki et al. 2006) and the progress of HPV infection in immunocompromised individuals (Benton and Arends 1996; Ahdieh et al. 2001) that there is an immune response that clears HPV infection (Figure 7.6). Immunological studies of regressing genital warts have provided detailed insight into this process of immune response (Coleman et al. 1994).

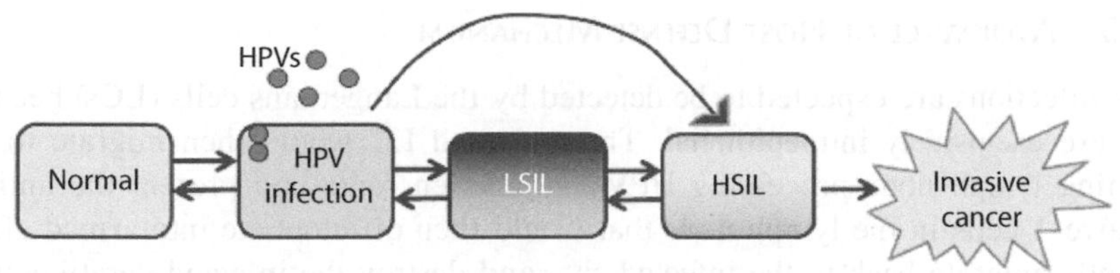


**FIGURE 7.6** The conversions and interconversions from a normal tissue to a malignant tissue due to HPV infection.

Nonregressing genital warts are characterized by a lack of immune cells; the few intraepithelial lymphocytes are CD8+ T cells and mononuclear cells that are present mainly in the stroma. Wart regression is characterized by a massive mononuclear cell infiltrate in stroma and epithelium. The lymphocytes are predominantly CD4+, but many CD8+ cells are present that are concentrated in the epithelium. The wart keratinocytes have upregulated major histocompatibility complex (MHC) I and II, adhesion molecules, and local release of cytokines such as interferon gamma and interleukin 12 (IL-12). These are characteristic of a thymus (Th1)-based lymphocyte response. Mucosal papillomavirus infections in animal models have shown the immunological events of the entire wart cycle from infection to regression (Rouse and Suvas 2004).

Genital HPV infection in sexually active young women is extremely common with a cumulative prevalence of 60%–80%. Mostly these infections will clear (i.e., DNA for that specific HPV type can no longer be detected), and most individuals will seroconvert. On an average, 8–16 months for clearance for the high-risk HPVs and about 4–8 months time for clearance of low-risk HPVs are reported (Giuliano et al. 2002; Brown et al. 2005). A persistent infection (i.e., the production of an infectious virus) is established when the CMI response fails to clear or control the infection. These types of individuals have the increased probability of progression to CIN 3 and invasive carcinoma (Muñoza et al. 2006).

In genital HPV infections, T-cell responses to E2 and E6 are important. Longitudinal studies following women with biopsy proven CIN 1 show that lesions that regress are associated with E2-specific systemic CD4+ T-cell response with progressive lesions. Regressing lesions are characterized as cell lesions with cytotoxic capability. E2- and E6-specific T helper (Th) 1 and Th2 type CD4+ T-cell responses (de Jong et al. 2002) and E6- and E7- specific cytotoxic T lymphocyte (CTL) (Nakagawa et al. 1997) responses are frequently detected in peripheral blood mononuclear cell (PBMC) cultures of healthy individuals. HPV-16 with specific CD4+ T and CD8+ T cells is able to migrate from the circulation to the epithelium upon antigenic challenge in healthy subjects, whereas failure to mount such responses has been seen in patients with CIN 2 and 3 or invasive cervical cancer. Therefore, CD8+ T cells fail to migrate into the lesions, and with induction of HPV-16-specific T regulatory cells ($T_r$), there is an influx of these cells into the lesion (van der Burg et al. 2004). Thus, the data suggest that effective immune control of HPV-16 infection in the cervix causes generation of interferon gamma (IFN-$\gamma$) secreting CD4+ T cells that are specific for E2 and E6.

### 7.2.3 Avoidance of Host Defense Mechanism

HPV infections are expected to be detected by the Langerhans cells (LCs) because they are exclusively intraepithelial. The activated LC would then migrate to the draining lymph node processing HPV antigens en route and present the antigen to naïve T cells in the lymph node that would then differentiate into armed effector cells, migrate back to the infected site, and destroy the infected keratinocytes. This protective cycle is hampered in a number of ways. The HPV infectious cycle is in itself an immune evasion mechanism that inhibits host detection of the virus. Inflammation and cell death do not occur due to HPV replication and release because the differentiating keratinocyte is already programmed to die, and any death due to natural causes does not act as a danger signal in the infected site. During most of the duration of the HPV infectious cycle, there is little or no release into the local milieu of the proinflammatory cytokines. This is important for dendritic cell activation and migration; thus, the central signals to kick start the immune responses are absent (Stanley 2006).

HPV-infected keratinocytes should activate the powerful antiviral defense system, type 1 IFN secretion—even in the absence of viral-induced cytolysis, and cell death. The antiviral, antiproliferative, antiangiogenic, and immunostimulatory properties of the type 1 IFNs, IFN-$\alpha$ and IFN-$\beta$, act as a bridge between innate and adaptive immunity and activate immature dendritic cells (Le Bon and Tough 2002; Theofilopoulos et al. 2005). Papillomaviruses have evolved mechanisms for inhibiting IFN synthesis and signaling. High-risk HPVs downregulate IFN-$\alpha$-inducible gene expression (Chang and Laimins 2000; Nees et al. 2001), and E6 and E7 oncoproteins of HPV-16 interact directly with components of the interferon signaling pathways (Ronco et al., 1998; Barnard and McMillan 1999) followed by uptake of HPV capsids (Fausch et al. 2005). This phenomenon would inhibit LC migration and maturation and the priming of the immune response against the capsid proteins. The HPV efficiently escapes the innate immune response by delaying the activation of the adaptive immune response. Local immune nonresponsiveness may be established in the infected mucosa as the host dendritic cells are exposed to low levels of viral proteins in a noninflammatory milieu for protracted time periods (Kobayashi et al. 2004). In this operationally HPV antigen-tolerant milieu, host defenses become irrevocably compromised. Also, HPV antigen-specific effector cells are not recruited to the infected area or their activity is downregulated, or both. Therefore, if during a persistent HPV infection there is deregulation of high-risk HPV E6 and E7 with increased protein expression that does not result in an armed effector CMI response, HPV-mediated progression to high-grade squamous intraepithelial lesions and invasive carcinoma is unimpeded.

### 7.2.4 Invasion of the Cell: Integration into the Cell Cycle

Development of effective therapies for HPV infection has been slow, largely due to unknown pathogenesis of the unique and complex replication cycle of the virus. HPVs are exclusively intraepithelial pathogens with a replication cycle. High-level expression of viral proteins and viral assembly occur only in differentiating keratinocytes in the stratum spinosum and granulosum of squamous epithelium (Doorbar 2005). Throughout the infectious cycle, viral genes are differentially expressed, temporally and spatially. The number of gene products is much larger due to the complex use of sites within the genome (Schwartz 2008). Latency is the further complication of these infections. HPV DNA can remain latent within cells even though others have entered the productive cycle. Spontaneous wart regression is immune mediated, but this does not result in virus clearance. Viral genomes can be detected in apparently normal epithelium (Abramson et al. 2004) many months and years after wart regression (Moore et al. 2002). The strong, local, cell-mediated immunity that facilitates regression of HPV-infected lesions can control latent infection and, in healthy immunocompetent individuals, recurrence of disease is rare. High levels of genital HPV infection and neoplasia are seen in immunosuppressed organ transplant recipients and those with human immunodeficiency virus (HIV) infection (Palefsky et al. 2006). Antiviral chemotherapies are essential for such patients. HPV infection of the anogenital skin and mucosae results in lesions with two morphologies: anogenital warts (condyloma acuminata) and squamous intraepithelial lesions. Condylomata, which are polypoid growths, are associated mostly with infection by low-risk types such as HPV-6 and HPV-11. They have a low to negligible risk of malignant progression. Squamous intraepithelial lesions are classified histologically and form a distinct spectrum of histological atypia. In the cervix, they are graded on the degree of loss of cytoplasmic maturation and exhibition of cytological atypia. The majority of lesions maintains the virus as an episome and supports a complete virus replication cycle; the viral gene expression is tightly regulated. Virus particles are generated due to late gene expression. A complete viral infectious cycle is not supported because of the defects in cellular differentiation that characterize these lesions. Late gene expression is either lost or significantly reduced, the viral DNA sequences may be integrated into the host genome, or E6–E7 oncogene expression is deregulated. The high-grade lesions' genetic instability and the changes in viral gene expression, such as the loss of E2 and El expression, occur in association with the same therapeutic strategies as low-grade lesions.

There is a massive upregulation of viral gene expression and viral DNA replication, with amplification of the viral copy number to many thousands of copies per cell, abundant expression of the E6 and E7 early genes, and expression of late genes from the late promoter with invasion of the infected keratinocyte in the differentiating compartment (Doorbar 2007; Chow et al. 2010). High E7 expression in cervical cancer having episomal (pure episome or concomitant) viral genomes with intact E2 is attributable to loss of E2 repressor activity due to E2BS-I/II methylation in HPV-16-related cervical cancer pathogenesis, a finding that will help to classify the status of the viral genomes within the host cells and their impact on disease prognosis (Das Ghosh et al. 2012).

The infectious cycle of HPVs is tailored to the differentiation program of the target cell, the keratinocyte, which raises several important issues with respect to immune recognition. Infection and vegetative growth are completely dependent upon the program of keratinocyte differentiation. The time from infection to virus release takes about three weeks because this is the time taken for the basal keratinocyte to move up through the epithelium, undergo complete differentiation, and desquamate. The period between infection and the appearance of lesions is highly variable and can range from weeks to months; this suggests that the virus does effectively evade host defenses. In addition, there is no cytolysis or cytopathic death as a consequence of virus replication and assembly. Minimal amounts of virus are exposed to immune defenses because HPV is an exclusively intraepithelial pathogen, where there is no blood-borne phase of the life cycle (Stanley 2006).

## 7.3 HPV AND ITS ASSOCIATION WITH HEAD AND NECK CANCER

### 7.3.1 Short Overview on Head and Neck Cancer

"Head and neck cancer" by definition implies squamous cell carcinoma of several anatomic sites—namely lip, oral cavity, nose and paranasal sinuses, nasopharynx, oropharynx, hypopharynx, and larynx. With a steady rise in incidence over the last three decades, HNSCC is the fifth leading cause of cancer by incidence and the sixth leading cause of cancer mortality worldwide (Parkin and Bray 2006; Deshpande and Wong 2008). Although the singular and synergistic entente of tobacco and alcohol has gained acceptance as the prime culprits triggering HNSCC, recent data attribute human papillomaviral etiology with being the trigger, especially to the oropharyngeal and oral subsets of HNSCC (Goon et al. 2009). The scenario of HNSCC has changed significantly over the past two decades. The well-documented association between HPV infection and cervical cancer is common knowledge, but the association of high-risk HPVs in the development of head and neck carcinoma in some tobacco and alcohol negative subjects is also backed by epidemiological, molecular, and clinical evidence (Psyrri and DiMiao 2008). HPV-positive HNSCC patients are younger than HPV-negative HNSCC patients by three to five years (Chaturvedi et al. 2008).

In Finland, Stina Syrjänen et al. (1983) provided the first evidence of the association between HPV and HNSCC. They reported that 40% of the cancers in their study contained histological and morphological resemblance to HPV-associated lesions (Goon et al. 2009). Since then, growing epidemiological evidence strongly points to the role of HPV in a subset of HNSCC. HPV is detected in approximately 25% of all HNSCCs (Kreimer et al. 2005), and the majority of these HPV-associated HNSCCs are oropharyngeal (tonsillar and base of the tongue) squamous cell cancers (Lajer and Buchwald 2011). HPV has also been detected in a smaller subset of laryngeal (24%) and oral (23%) cancers (Kreimer et al. 2005; Hobbs et al. 2006; Hocking et al. 2011).

### 7.3.2 Infection of HPV in Head and Neck Cancer

The risk of acquiring HPV infection is linked primarily to sexual behavior, including having multiple sexual partners. D'Souza and colleagues recently showed in a case-control study (D'Souza and Dempsey 2011) that a high number (26 or more) of lifetime vaginal-sex partners and six or more lifetime oral-sex partners were associated with an increased risk of HNSCC (Goon et al. 2009). Oral-vaginal contact, direct mouth-to-mouth contact, and mouth-to-penis contact are modes of HPV transmission in decreasing order of significance. Individuals may be more likely to acquire an oral HPV infection when performing oral sex on a woman than on a man. An increased risk of HPV-associated HNSCC has been reported in husbands of women with in situ carcinoma and invasive cervical cancer (Ragin et al. 2007; Psyrri and DiMiao 2008). This suggests that the transmission of HPV infection can occur by an orogenital route. HPV transmission during vaginal sex is more common from an infected cervix to the penis than vice versa. Penile infection targets the glans more frequently than the shaft. The anus is also an additional source site other than the cervix (Hernandez et al. 2008). HPV cannot be transmitted through blood or breast feeding because it lacks a viremic phase. In one study, the correlation between high-risk types of HPV in the cervix of women and the semen of their sex partners also hints at the role of sperm as a possible vector for HPV (Bosch et al. 2006). Current smoking and HIV infection are equally associated with drastically increased oral HPV prevalence, suggesting that tobacco-related and HIV-related immunosuppression may propagate oral HPV natural history. The risk of acquiring HPV-associated HNSCC by HIV-infected persons is approximately two- to sixfold higher (Engels et al. 2008; Gillison 2009).

HPV-16, the most prevalent genotype in cervical carcinoma, is also the most frequently detected HPV type in HNSCC and is found in up to 90% of HPV-positive cases (De Villiers et al. 1985). Since 1986, when HPV-16 DNA was first detected in an invasive SCC of the HNSCC by Southern blot hybridization (Weinberger et al. 2006), HPV sequences have been repeatedly detected in a variable proportion of HNSCC, from as few as 10% to 100% (Steinberg and DiLorenzo 1996). HPV-associated OSCC tends to be poorly differentiated, often basaloid in histology, and frequently present at an advanced stage (Psyrri and DiMiao 2008). The viral load of HPV in head and neck cancers appears to vary considerably. Available data suggest that oral cavity HPV viral load measured by DNA is much lower than in the cervix, penis, or anus (Herrero et al. 2003). Tonsillar cancers appear to show a wide variation in HPV copy numbers.

### 7.3.3 Differences between Tobacco-Induced OSCC and HPV-Induced OSCC

Several lines of clinical evidence suggest that HPV-associated OSCC could be biologically different from traditional OSCC (Psyrri et al. 2009). Tobacco- and alcohol-associated OSCCs have a male predilection, while HPV-associated OSCCs can affect both sexes equally. In addition, patients with HPV-associated OSCC are often nonsmokers, nondrinkers, and on an average are five years younger than their tobacco-use-associated counterparts (Psyrri and DiMiao 2008). From a molecular perspective, tobacco- and alcohol-induced carcinomas are characterized by down-regulation of tumor suppressor protein p16 and mutation of p53 as well as the Rb gene (Brennan et al. 1995; Sartor et al. 1999). The p16 protein acts as a tumor suppressor by binding to cyclin D1-CDK4/CDK6 complex and inhibits Rb protein phosphorylation. In contrast, overexpression of p16 has been reported repeatedly in HPV-related cancers. HPV also inactivates p53 and Rb proteins by binding with its own proteins E6 and E7, respectively (Chaturvedi et al. 2008). Thus, both the tumor suppressor pathways are silenced. Removal of HPV E6 and E7 expression leads to restoration of the apoptotic pathways, rendering the tumor more sensitive to chemoradiation therapy (Goon et al. 2009). Rampias et al. (2009) showed that E6 and E7 oncogene repression transcripts through shRNA treatment restore p53 and retinoblastoma (Rb) tumor suppressor pathways, thereby triggering apoptosis in HPV-16-positive oropharyngeal cancer cell lines—a ray of hope for the patients of HPV-positive OSCC.

### 7.3.4 HPV-Induced OSCC

Several studies suggested that p16 overexpression is a surrogate marker for HPV association with oropharyngeal cancers (Marur et al. 2010; El-Naggar and Westra 2012). But almost 50% of OSCCs are not associated with HPV. Considering this fact, Weinberger and his colleagues designed a study where the cases were grouped into the three categories, as shown in Table 7.1 (Weinberger et al. 2006).

The negative control for this study was tobacco-induced OSCC, which causes a loss of function of p16. The entire study was of a cohort of 107 OSCC patients, and p16 expression levels were determined by quantitative in situ methods of protein analysis or immunohistochemistry. Weinberger et al. found that the cases assigned to HPV-16 DNA-positive and p16 overexpressors (class III in Table 7.1) were less likely to be recurrent in comparison to the cases in other groups. Moreover, patients in class II had much better disease-free survival (DFS) rates (75%), overall survival (OS) rates (79%), and five-year local recurrence rates (only 14%) compared to other groups. They showed that the HPV 16 DNA-positive and p16-overexpressive group (class I) tumors were 4.8 times less likely to recur compared to the HPV-negative cohort.

## 7.4 DISEASE PATHOGENESIS

### 7.4.1 Genetic and Immunological Causes

The challenge for the virus is to exploit the cellular DNA polymerases and other replication factors that are produced only in mitotically active cells. To solve this problem, HPVs encode proteins that act in the context of the viral life cycle, initiate cellular DNA synthesis in noncycling cells, and inhibit the apoptosis of infected keratinocyte, thus creating an environment suitable for viral DNA replication.

**TABLE 7.1**
**HPV-16 DNA Expression and p16 Expression Status according to Weinberger's Study**

| | Class I | Class II | Class III |
|---|---|---|---|
| HPV-16 DNA presence and expression level of p16 | HPV-16 DNA −ve p16 low | HPV-16 DNA +ve p16 low | HPV-16 DNA +ve p16 high |
| Overall survival | 5–20% | 5–18% | 79% |
| Disease-free survival | 15% | 13% | 75% |
| Local recurrence rate in five years | 45% | 74% | 14% |

The E6 and E7 genes are central to these functions of HPVs. The E7 gene of the high-risk HPVs binds with the unphosphorylated form of the retinoblastoma (Rb) protein, overriding the G1/S checkpoint of the cell cycle, and the E6 gene of the high-risk HPVs binds with p53 and targets it for degradation. Viral DNA replication in non-cycling cells occurs due to cumulative actions of E6 and E7 proteins. Deregulation of growth control in the infected cells occurs with development of carcinoma (Moody and Laimins 2010). Keratinocytes constitutively secrete or can be induced to secrete several cytokines, including interleukins, IL-6, IL-10, and IL-18, and tumor necrosis factors (TNFs) (Arend et al. 2008). IL-1 is a key keratinocyte cytokine with a broad range of pleiotropic effects, including activation of T helper (Th) cells and dendritic cells, and the promotion of B cell maturation and clonal expansion. Keratinocytes synthesize pro-IL-1α and pro-IL-1β under normal conditions but cannot process and secrete them in their activated form. After inflammasome activation, processing and secretion of IL-1 α (as the activated cytokine) occur. The secretion of proinflammatory cytokines by keratinocytes is central to the activation of tissue-resident immune cells, such as LCs and macrophages, and the recruitment of effector T cells. Together they play major roles in adaptive immune responses to the local injury or infection. HPV infection dampens these crucial responses almost from the start of the infectious cycle where IL-1 and IL-6 are central regulators in this immune response. The host interferon response for HPV infection (Black et al. 2007) is actively suppressed by high-risk HPV proteins E6 and E7. In general, HPV evades the immune response, thus delaying the activation of adaptive immunity.

Eukaryotic cells express germ line-encoded receptors of the innate immune system, pathogen recognition receptors (PRRs) that recognize invariant molecular motifs known as pathogen-associated molecular patterns (PAMPs) (Medzhitov and Janeway 1997). Genital tract keratinocytes express several toll-like receptors (TLRs) located either on the cell surface (TLR1, TLR2, TLR4, TLR5, and TLR6) or in the endosome (TLR3 and TLR9) (Nasu and Narahara 2010). TLR7 expression is induced on keratinocytes by triggering TLR3 with double-stranded RNA, a feature of viral infections, thus activating IFN-responsive genes (Kalali et al. 2008). Type I IFNs that elicit predominantly Th1-type cytotoxic responses are produced by activation of TLRs on keratinocytes (Miller and Modlin 2007).

### 7.4.2 Molecular Aspects of HPV-Related Pathogenesis

E1 and E2 are the most highly conserved papillomavirus proteins and are essential for viral DNA replication (Yang et al. 1991; Chiang et al. 1992). E1 is the only viral enzyme with adenosine triphosphatase and helicase activity. The protein is divided into an N9 domain (Amin et al. 2000), a sequence-specific DNA-binding domain (DBD) (Titolo et al. 2003), and a C9 helicase domain (Seo et al. 1993; Titolo et al. 2000). E2 is a sequence-specific DNA-binding protein having roles in DNA replication, transcription (Hegde 2002), and partitioning of viral genomes to daughter cells during mitosis (Ilves et al. 1999). The protein is organized into two functional domains: an N9 transactivation domain involved in transcriptional regulation via direct association with E1 and a C9 DBD/dimerization domain (Hegde 2002). Both domains are separated by a hinge region, the function of which remains poorly characterized.

El monomers are specifically recruited to the viral ori through interaction with the El DBD as well as protein–protein interaction with the E2 transactivation domain (TAD) via the E2 DBD (Sanders and Stenlund 1998). The El DBD dimerizes on the ori, additional El monomers are recruited, and a double hexamer forms and in the process melts the duplex DNA, resulting in the assembly of a hexamer around each strand of the ori DNA (Fouts et al. 1999). ATP plays an important role in these processes. ATP binding enhances the El–E2 interaction with the viral ori (Titolo et al. 1999) but also changes the conformation of the El–E2 protein interaction (White et al. 2001). As the El double hexamers are formed, E2 dissociates and DNA pol-A is recruited to the viral ori (Masterson et al. 1998) via p70 and El (Lusky et al. 1994), the full replication complex assembles for viral DNA replication to proceed. In theory, any one of these interactions could be targeted by small-molecule inhibitors.

### 7.4.3 Pathogenesis in Head and Neck Cancer

The dominance of HPV-16 in HNSCC (85%–95%) is almost similar to that seen in cervical carcinoma. This probably reflects a similarity in life cycles of high-risk HPV subtypes in different mucosal sites, with an allied difference in mucosal immune responses. Meta-analyses have shown that the HPV subtypes associated with HNSCC are broadly similar (but not identical) with those seen in cervical carcinoma. HPV is generally detected in the tumor of the oropharyngeal cancers, where it is localized to the cell nuclei, transcriptionally active, clonal, and not found in the surrounding benign tissues (Gillison et al. 2000). These effects are predominantly due to the E6 and E7 oncogenes, which enhance degradation of p53 and Rb tumor suppressor genes, respectively, leading to cell proliferation, impaired apoptosis, and—ultimately—chromosomal instability. Moreover, HPV-6 and -11 have also been implicated in malignancies such as Ackerman's tumor (verrucous carcinoma of the oral cavity).

Integration of HPV DNA into host genomic DNA usually causes disruption and/or deletion of HPV El or E2 ORFs, which is imperative for viral replication and transcription. E2 acts also as a repressor of E6 and E7; thus, increased E6 and E7 expression maintains the immortalized phenotype. However, despite the dominance of the integrated HPV genome in terms of cervical carcinogenesis, 15%–30% of cervical cancers contain HPV only in the episomal form. In some cases, investigators have found deletions in the YY1-binding sites of the LCR (long control region) of HPV-16 episomal DNA, which may allow elevated activity of the E6/E7 promoter. In the case of HNSCC, HPV-16 is the dominant subtype and is found in 84% of HPV-positive cancers. It is reported that of the HPV-16-positive samples, 48% are integrated, 35% are episomal, and 17% are of mixed episomal and integrated forms as analyzed by Real time-PCR. Tonsillar carcinomas have been reported to have the highest prevalence rate of HPV DNA within cancerous cells (51%) of all the forms of head and neck cancers. Mellin et al. (2002) reported that all 11 cases of HPV-positive tonsillar carcinomas in their series contain HPV DNA in episomal form. In 1992, an earlier study reported that two HPV-16-positive tonsillar carcinomas had episomal HPV DNA. Moreover, two HPV-33-positive tonsillar carcinomas were found—one integrated and the other in mixed forms.

E6 binds to and inhibits p53, which is active in repressing the cell cycle in the event of DNA damage and also in triggering apoptosis if the damage is too severe to be repaired by the cell (Taylor and Stark 2001). E6 also activates cellular telomerase, the enzyme that synthesizes the telomere repeat sequences at the ends of eukaryotic chromosomes. This allows the cells to replicate indefinitely (Newbold 2002). E7 promotes cell division by binding to Rb, a tumor suppressor protein that usually binds to and inactivates a transcription factor known as E2F. E2F unbound from Rb causes transcription of genes involved in DNA replication and cell division. This provided experimental evidence as to the association of OSCC and HPV. The sites of incidence are almost always the Waldeyer's oropharyngeal ring (tonsils and base of tongue).

## 7.5 PREVENTION AND THERAPEUTICS

### 7.5.1 Prophylactic HPV Vaccine

HPV vaccines offer a promising approach to the prevention of HPV and associated diseases. However, they do not replace other prevention strategies such as regular cervical cancer screening using the Pap test because the vaccines will not prevent all HPV types. There are two licensed HPV L1 virus-like particle (VLP) prophylactic vaccines: Cervarix®, a bivalent HPV-16 and -18 vaccine from GlaxoSmithKline Biologicals, Rixensart, Belgium, and Gardasil®, a quadrivalent HPV-16, -18, -6, and -11 vaccine from Merck and Co., Inc., West Point, PA. Both the vaccines have been shown in randomized control trials to be highly efficacious against HPV-16 and -18 mediated CIN 2 and 3 in 15- to 26-year olds under a three-shot immunization schedule (0, 1 or 2, and 6 months) (Kjaer et al. 2009; Paavonen et al. 2009). Additional trial endpoints were evaluated for the quadrivalent vaccine with high efficacy (> 96%) against HPV-6/-11/-16/-18 mediated VIN and external genital warts (Dillner et al. 2010). In 16- to 23-year-old heterosexual men, the quadrivalent vaccine has been shown to achieve more than 90% efficacy against HPV-6, -11, -16, and -18 mediated external genital warts and more than 73% efficacy against AIN in homosexual men. The vaccines do not contain thimerosal or mercury as a preservative. The quadrivalent vaccine uses alum and the bivalent vaccine uses AS04 (500 μg aluminum hydroxide, 50 μg 3-O-deacyl-4′-monophosphoryl lipid A) as adjuvants. Both the vaccines should be delivered through a series of three intramuscular injections over a six-month period. The second and third doses should be given one (bivalent) or two (quadrivalent) and six months, respectively, after the first dose. Syncope can occur after vaccination and has been observed among adolescents and young adults. To avoid serious injury related to syncopal episodes, vaccine providers should consider observing patients for 15 minutes after they are vaccinated.

Ideally, patients should be vaccinated before the onset of sexual activity. Patients who have been infected with one or more HPV types still get protection from the HPV types they have not acquired. These vaccines can be given to lactating women and patients with minor or acute illnesses (such as diarrhea), mild upper respiratory tract infections, with or without fever, having an equivocal or abnormal Pap test, or having genital warts. The vaccine should also be given to immune-compromised patients.

However, the immune response to vaccination and vaccine efficacy might be less in immune-compromised people. HPV vaccines are contraindicated for persons with a history of immediate hypersensitivity to any vaccine component because the quadrivalent vaccine is produced in *Saccharomyces cerevisiae* and is contraindicated for persons having hypersensitivity to yeast. Prefilled syringes of bivalent HPV have latex in the rubber stopper and should not be used in persons with an anaphylactic latex allergy; HPV-2 single-dose vials contain no latex.

HPV vaccines should not be given to patients (a) with a history of immediate hypersensitivity to any vaccine component (quadrivalent HPV vaccine is contraindicated for persons with a history of immediate hypersensitivity to yeast; bivalent HPV vaccine in prefilled syringes is contraindicated for persons with anaphylactic latex allergy); (b) patients with moderate or severe acute illnesses (patients should wait until the illness improves before getting vaccinated); and (c) pregnant women. Although the vaccine has not been causally associated with adverse pregnancy outcomes or adverse events to the developing fetus, data on vaccination in pregnancy are limited.

HPV L1 VLP vaccines are highly immunogenic after the third immunization, and the antibody levels are 1–4 logs higher than those in natural infections (Villa et al. 2006). It is reported that serum-neutralizing antibody persists with geometric mean titers about 1 log greater than natural infection for the seven- to nine-year duration. Mathematical modeling predicts slow decay of the antibody over a 30- to 50-year period and, therefore, protection over that time (David et al. 2009). VLP vaccines generate type-specific and cross-neutralizing antibodies. Concentrations of cross-neutralizing species are on average 1–2 logs lower than type specific (Smith et al. 2007). The enhanced immunogenicity of VLP immunization in contrast to natural infection probably relates to the route of immunization. VLPs are surface bound to antigen-presenting cells (APCs) or local immunocytes and are transformed into the lymph node encountering and priming naïve B cells in the follicle. This event will result in protective immunity (Lenz et al. 2005).

Preclinical studies have demonstrated that bovine or rabbit immunizations with L2 polypeptides protect against papillomavirus type 4 at mucosal sites in the bovine and at cutaneous sites in the cottontail rabbit papillomavirus (Lin et al. 1992; Campo et al. 1993; Embers et al. 2002). This suggests that immunization against the minor capsid protein 2 might work as a pan-HPV vaccine against various genotypes of papillomaviruses. Inoculation of amino-terminal L2 polypeptides also induced protection against heterologous papillomavirus types. Vaccination with HPV-16 L2 (amino acids 11–200) protects against cottontail rabbit papillomavirus (CRPV) and rabbit oral papillomavirus, both evolutionarily divergent from HPV-16 (Gambhira et al. 2007). Vaccination with BPV-1 L2 (amino acids 1–88) peptides produces sera with cross-neutralizing activity against different HPVs. Human volunteers vaccinated with the candidate prophylactic/therapeutic vaccine HPV-16 L2E6E7 fusion protein induced L2-specific antibodies that also neutralized a divergent type of HPV (Gambhira et al. 2006; WHO 2006). The monovalent L2 immunogens generate neutralizing titers that are greater for the homologous type of virus than for a heterologous type papillomavirus. The lower immune response to heterologous HPVs could severely limit the breadth and duration of protection of an L2-based vaccine.

To address this issue and provide broader immunity, the L2-neutralizing epitope was inserted on the surfaces of VLPs increasing the titers of neutralizing antibodies nearly tenfold (Slupetzky et al. 2007). A synthetic L2 lipopeptide in which the cross neutralizing L2 peptide is linked to both a T-helper epitope and a ligand for TLR2. Tandem repetition of the same peptide displayed on bacterial thioredoxin, multi-type L2 fusion proteins from different papillomavirus types have been utilized in inducing cross-neutralizing antibodies against several clinically relevant HPV types (Alphs et al. 2008; Jagu et al. 2009).

HPV vaccines were studied in thousands of people in many countries around the world including the United States. More than 46 million doses of HPV vaccine had been distributed in the United States as of June 2012. During these studies, mild adverse effects including pain where the injection was given, fever, dizziness, and nausea were noticed. Merck conducted six phase 1 and phase 2 clinical studies between 1997 and 2004. Four smaller phase 1 or early phase 2 studies evaluated monovalent HPV VLP vaccines (serotypes 11, 16, or 18) in order to characterize safety and immune responses among different doses. Two larger phase 2 studies were conducted between 2000 and 2004 that included clinical endpoints in addition to the safety and immune-response endpoints. The studies suggested an acceptable safety profile for further clinical development. The larger phase 2 studies provided supportive evidence of vaccine activity (Jemal et al. 2006).

Two randomized, double-blinded, placebo-controlled phase 3 studies, namely, FUTURE I and FUTURE II, evaluated the clinical efficacy and safety of the quadrivalent compound. The studies evaluated the clinical endpoints of CIN 2/3 and external genital lesions due to HPV. In the FUTURE II study, a total of 10,585 subjects, or 87% of all enrolled subjects, met the criteria for the per protocol population for the primary HPV-16/18 efficacy analysis. The results demonstrated a vaccine efficacy of 100% for HPV-16/18 (Wright et al. 2003).

Results from a large clinical trial of the HPV vaccine Cervarix confirm its protection against cervical precancer and cancer and also provide evidence of cross-protection against other high-risk types of HPV in the Papilloma Trial against Cancer in Young Adults (PATRICIA). The study was conducted in 14 countries, where more than 18,000 young women between the ages of 15 and 25 were enrolled. Initial results indicated that Cervarix substantially reduced the risk of cervical precancer and cancer. Among women with no HPV exposure at the start of the study, Cervarix provided complete protection against HPV-16 or HPV-18 related CIN 3 or worse. Cervarix provided less protection to women who may have had prior exposure to HPV. In addition, bivalent vaccine provides cross-protection against four other HPV strains other than HPV-16 and HPV-18, namely HPV-33, -31, -45, and -51 (Wheeler et al. 2012).

### 7.5.2 Screening in Cervical Cancer

Although vaccination provides the primary prevention, screening for precancer and cancer provides secondary prevention for cervical cancer. However, colposcopy is used to detect early abnormalities of the cervix that, if untreated, could lead to cervical cancer. Over the course of many years, the cells lining the surface of the cervix undergo a series of changes. In rare cases, these changed cells can become cancerous. However, cell changes in the cervix can be detected at a very early stage; thus, the risk of cervical cancer development may be arrested.

The Papanicolaou (Pap) stain test is used to screen for cervical cancer in women. About one in 20 women was found to have an abnormal result in a Pap test and required further treatment. Most of these changes will not lead to cervical cancer. Treatment can be given to prevent cancer from developing in women with abnormal cells. The incidence of cervical cancer is low in individuals under the age of 25, and the prevalence of transient HPV infection after coitarche is high. Almost one in six cervical cytology samples taken from this age group are abnormal. If screening starts at an earlier age, low-grade CIN or cervical precancer resolve spontaneously. Screening may thus lead to unnecessary colposcopy, with the possibility of increased anxiety, overtreatment, and potentially negative consequences in women of the reproductive age group. Cervical screening is less efficient in detecting CIN 3 in older women. More cervical cytology samples are required to detect a case of CIN 3 after the age of 50, but it is more efficient at preventing invasive cancer.

In countries such as the United Kingdom, where cervical screening is mandatory, it is recommended that women who are between 25 and 49 years of age are screened every three years, and women between 50 and 64 years are screened every five years. This is based on an audit of screening histories by Sasieni et al. that concludes that five-yearly screening offers considerable protection (83%) against cancer at ages 55–69 and even annual screening offers only modest additional protection (87%). Three-yearly screening offers additional protection (84%) over five-yearly screening (73%) for cancers at ages 40–54 but is almost as good as annual screening (88%). As with all screening tests, cervical screening is not 100% accurate. Regular screening can prevent up to 75% of cancers from developing but not in every case.

Cervical sampling, using liquid-based cytology (LBC), was implemented throughout the United Kingdom by October 2008 and is the current standard method of screening there. In 2003, the National Institute for Clinical Excellence (NICE) recommended that LBC be used as the primary means of processing samples in the cervical screening program in England and Wales. Meta-analysis published in the Cochrane Database suggests that the introduction of LBC does not increase detection of cervical precancer.

In low-resource countries, VIA has been used as a simple screening to detect precancer and early cervical cancers. In India, VIA is undertaken by trained health workers who apply 4%–5% acetic acid onto the cervix and observe the aceto-white changes after one minute. Biopsy from the aceto-white areas is then obtained by using a colposcope, and treatment is undertaken by biopsy-positive women. In a large randomized trial involving 151,000 women in the slums of Mumbai, India, use of VIA reduced cervical cancer mortality from 16.2 to 11.2 deaths per 100,000 women—a 31% reduction compared to the unscreened group. Primary screening with high-risk HPV DNA testing generally detects more than 90% of all CIN 2, CIN 3, or cancer cases, and it is approximately 25% more sensitive than cytology at a cutoff of borderline dyskaryosis or worse. However, it is about 6% less specific. The reduced specificity of HPV DNA testing may be improved by adding HPV typing, for example, or by detecting the presence of biomarkers such as p16 and mRNA coding for viral E6 and E7 proteins.

HPV testing using Hybrid Capture© (HC2) is the most sensitive test of cure and is superior to either cytology or colposcopy. The addition of cytology to HPV testing improves the sensitivity of sampling. The optimum time for the double test (cytology and HC2) is probably 18–24 months, but testing at 12–18 months represents a pragmatic fit with the current timing of follow-up cytology. The high negative predictive value of HPV testing combined with cytology is useful. There is some evidence to suggest that all women who have had treatment for CIN should have an HPV HC2 test after six months as well as a combined cervical cytology and HPV test after 12–18 months. If all three of these tests are negative, the patient can be returned to routine three-yearly (or, for those over 49, five-yearly) testing.

A landmark study undertaken by India's International Agency for Research on Cancer (IARC) in conjunction with Tata Memorial Centre (TMC) and the Nargis Dutt Memorial Cancer Hospital (NDMCH) to evaluate the effectiveness of HPV testing in Indian women found that, in a low-resource setting, a single round of HPV testing was associated with a significant reduction in the numbers of advanced cervical cancers and deaths from cervical cancer. The randomized trial was conducted among a total of 131,746 healthy women between the ages of 30 and 59 years (Sankaranarayanan et al. 2009).

### 7.5.3 Therapeutic Options: Treatment of HPV Infection

Cytotoxic agents are widely used in the treatment of genital warts (Viera et al. 2010). They are topical preparations that kill cells on contact, irrespective of HPV status, by antiproliferative or chemodestructive modes of action. Podophyllotoxin, a cream (Europe) or a gel (United States) that can be self-applied by the patient, is a treatment for genital warts (Lacey 2005), achieving 50% clearance but with recurrence rates of 25–30% (Maw 2004). The binding of podophyllotoxin to microtubule proteins may arrest cell cycle at metaphase and thus prevent genital warts.

Trichloracetic acid (TCA) is a clinic-based topical therapy that has a local caustic action effectively generating a chemical burn of the wart. It is as effective as podophyllotoxin but can result in ulceration, dermal scarring, and secondary infections, if inappropriately applied. It has, however, no systemic toxicity and can be used during pregnancy.

IFNs are the only antiviral drugs approved for the therapy of benign HPV-related lesions. Although IFN-α, IFN-β, and IFN-γ have all been tested against condyloma acuminate, the most information is available on IFN-α. IFN-α appears efficacious via a number of routes of administration, schedules, and dosages with an acceptable safety profile. Success with IFN-α therapy, in terms of reduced recurrence rates of condylomas, was reported from studies in which all visible lesions were surgically removed with subsequent administration of subcutaneous local IFN-37 (Cirelli and Tyring 1994).

Imiquimod is a pharmacological agent that can modulate innate immune responses. It is an agonist for TLR 7, ligation of which activates dendritic cells, macrophages, and keratinocytes to release type I IFNs and other proinflammatory cytokines (Stanley 2002). Randomized clinical trials with imiquimod 5% cream applied topically to genital warts have shown efficacy and safety and a reduced recurrence rate (12%) compared with placebo (30%) (Beutner et al. 1998). Although imiquimod has a therapeutic effect on intraepithelial disease and small trials on VIN (Terlou et al. 2011) and AIN (Fox et al. 2010) have demonstrated efficacy, the drug is not licensed for these diseases. The inflammatory side effects of imiquimod (erythema, edema, itching, and pain) have limited its use on mucosal surfaces.

Polyphenon E, which is a standardized extract of green tea leaves, *Camellia sinensis,* contains epigallocatechin gallate (ECGC), a molecule that impacts on multiple signaling pathways inducing cell cycle arrest or apoptosis. This has shown efficacy against genital warts when used in the form of a topical ointment. Randomized control trials show that it exhibits greater clearance of warts (54%) and lower recurrence rates when compared to rates of those who were given a placebo (35%) (Tatti et al. 2010).

Cidofovir is a monophosphate nucleotide analog of deoxycytidine (dCTP). After undergoing cellular phosphorylation, it competitively inhibits the incorporation of dCTP into viral DNA by viral DNA polymerase. Incorporation of the drug disrupts further chain elongation (Van Pachterbeke et al. 2009).

Curcumin is a potent antioxidative agent and an active compound of the perennial herb that also exhibits anti-inflammatory and antitumor activity (Aggarwal et al. 2003). Current studies using curcumin on HPV-positive human cervical cancer cell lines have shown that curcumin downregulates the AP-1 and NF-κB expression in a dose- and time-dependent manner (Prusty and Das 2005). The AP-1 plays a crucial role during the development of cervical cancer by binding to its potential cognate binding sites within HPV upstream regulatory region (URR) and is absolutely indispensable for efficient HPV oncogene expression. AP-1 has been shown to be regulating the transcription of almost all HPV types investigated so far. Curcumin treatment of HPV-18-positive cervical cancer cells selectively suppresses the HPV-18 transcription (Divya and Pillai 2006). Based on these leads, a polyherbal cream, Basant, has been formulated using curcumin and other herbal components. Basant is found to inhibit HPV entry into the cervical cells in vitro as well as to inhibit *Neisseria gonorrhoeae*, various species of candida, and HIV-1 in cultures (Talwar et al. 2008).

ADXS-HPV is an immunotherapy that is designed to target cells expressing the HPV gene E7. Expression of the E7 gene from high-risk HPV variants is responsible for the transformation of infected cells into dysplastic and malignant tissues. Dysplasia or malignancy can be eliminated by eliminating the cells expressing E7. ADXS-HPV is designed to infect APCs and direct them to generate a powerful, cellular immune response to HPV E7. The resulting cytotoxic T-cells infiltrate and attack the tumors while specifically inhibiting tumor Tregs, the tumors that are protecting it (Jones et al. 1995).

Small-molecule inhibitors targeting the ATPase activity of HPV-6 E1 were identified by high-throughput screening of a large collection of in-house compounds by scientists working for Boehringer–Ingelheim in Canada (White et al. 2011). The lead molecules from this screen were biphenylsulfonacetic acid analogs, the mode of action of which probably affected ATP binding by an allosteric mechanism (White et al. 2005). Unfortunately, this class of inhibitors was not active in cell-based assays.

Therapeutic vaccines are the bridge for the temporal deficit by attacking already persistent HPV infections and treating cervical cancer in women. Vaccination with VLPs such as Gardasil and Cervarix has demonstrated efficacy in HPV prophylaxis but these vaccines lack therapeutic potential. Considering the vital role of HPV-16 in carcinogenesis, most of the efforts for developing therapeutic vaccines have been directed toward development of vaccines against HPV-16 viral oncogenes E6 and E7. Based upon the nature of immunogen, the therapeutic vaccines can be divided into two major classes: (1) proteins/peptide-based vaccines and (2) DNA-based vaccines. The majority of the studies conducted as phase 1 clinical trials both in healthy individuals and in patients having high-grade lesions or frank malignancies have used whole protein or peptides derived from HPV-16 or HPV-18 E6 and E7 mainly because of their oncogenic potential and they are invariably retained and expressed throughout HPV-related disease progression and carcinogenesis. Several animal studies have shown promising results indicating that therapeutic HPV vaccine may regress disease progression (Gomez-Gutierrez et al. 2007). Phase 1 trials that recorded these parameters indicated that despite effective anti-HPV immunity, lesion regression as well as HPV clearance was suboptimal and was only observed in a subset of patients. Perhaps this could be the potential reason why these therapeutics could not enter in phase 2 and phase 3 studies. Less-expensive, therapeutic DNA-based vaccines are being prepared against HPV-16 E7. These efforts are also directed toward generation of effective immune response against HPV by intradermal administration of DNA vaccines. This represents a feasible strategy for delivering DNA directly into the professional APCs of the skin, but the efficacy of this strategy for regression or downstaging of the lesion is yet to be proven. Apart from therapeutic HPV vaccines, there have been attempts to pulse dendritic cells with tumor lysate expressing HPV-16 antigens (Adams et al. 2003).

## 7.6 CLINICAL ASPECTS OF HPV-INDUCED CANCERS

### 7.6.1 Cervical Cancer

Cervical cancer is a worldwide disease where detection happens at late stages. The cervix is the exposed part of the uterus, which makes it easily accessible to viral infection through sexual intercourse; 7.7% of women worldwide have HPV infection without any clinical lesion. Majority of the infected women are able to get rid of HPV through their immune defense mechanism, while those who are unable to do so develop cancer. The incubation period from infection to the various stages of precancer and then on to cancer can take up to 10–15 years. The precancerous stages include CIN, where the abnormality is confined within the basement membrane and is graded as I, II, and III, depending on whether the abnormality extends up to one-third, two-thirds, or up to the complete epithelial cell layer. CIN 1 is also called a low-grade lesion (or LSIL); CIN 2 and CIN 3 are graded as high-grade lesions or (HSILs). With an effective screening strategy, CIN 1, CIN 2, and CIN 3 are detectable in cervical smears using PAP stain. Here they are graded as mild, moderate, or severe dyskaryosis on a cervical smear test. Detection of a precancerous lesion may also be undertaken by colposcopy when the cervix is bathed in 5% acetic acid and that abnormality is detectable as a white patch on the cervical epithelium.

In India, cervical cancer is staged according to The International Federation of Gynecology and Obstetrics (FIGO) classification, which was revised in 2009.

Stage IA and IB1 are considered early stages of disease, while IB2 and the stages thereafter are advanced stages of the disease. Essentially, early stages of the disease can be treated by surgery, while the advanced stages are treated by radiotherapy. Among the advanced stages, IB2 and IIA may sometimes be amenable to surgery but may also require radiotherapy as an adjuvant modality of treatment.

The surgery primarily consists of the removal of the uterus and the cervix along with the parametrium, and a 1–2 cm cuff of vagina and pelvic lymph nodes. The surgery, known as a radical hysterectomy with pelvic lymph node dissection (or Werthiem's hysterectomy), is performed in the early stage of the disease. The ovaries may or may not be removed. The recent guidelines from the National Comprehensive Cancer Network (NCCN) suggest that along with pelvic lymphadenectomy, para-aortic Lymph node (LN) sampling should be undertaken to identify disease beyond the pelvis, which has an impact on treatment and prognosis of cancer. If pelvic lymphnode parametrium or lymphovascular spaces are positive for cancer cells in histological sections after surgery, then patients require radiotherapy as an adjuvant treatment.

Radiotherapy for advanced disease or residual disease after surgery consists of 23–25 fractions of external beam radiotherapy delivering a total dose of 45–50 Gy. Thereafter, two to three further doses of radiation are given as brachytherapy where the fractions of radiation are given per vaginum at a dose of 7 Gy each, one week apart. If there is a tumor residual (defined as a tumor present within six months of treatment), then chemotherapy may become necessary. Tumor recurrence within three years of treatment (defined as the presence of a tumor six months after treatment) requires chemotherapy; localized tumor growth after three years may be subjected to repeat radiotherapy. Chemotherapy has been tried in cervical cancer in the neoadjuvant setting and is the subject of various ongoing trials. In addition, chemotherapy in a low dose is also used in concurrence with radiotherapy to enhance the effect of radiation.

### 7.6.2 HPV-Induced HNSCC: A Bane Turns Boon

HPV is not sufficient to induce malignant transformation to infected epithelial cells, though initiation of carcinogenesis occurs by its infection. The number of HNSCCs with HPV DNA positive far exceeds the number that actually expresses HPV oncoprotein E6 and E7 (Mannarini et al. 2009). Studies have also shown that not all HPV DNA positive HNSCCs contain transcriptionally active HPV. Braakhuis et al. (2004) showed that 17% of tumors contained HPV-16 DNA but only 8% of them expressed E6/E7 mRNAs. The favorable outcome of HPV-induced oropharyngeal cancers might be attributable to the absence of field cancerization or to enhanced radiation sensitivity (Psyrri et al. 2008). The term "field cancerization" is used to describe the presence of carcinogen-induced early genetic changes in the epithelium from which multiple independent lesions arise, leading to the development of multifocal tumors (Dakubo et al. 2007). Despite the poorly differentiated histology of tumors, however, patients with p16-expressing oropharyngeal tumors seem to have a favorable prognosis. In addition, patients having lower expression of p53 and Rb due to lower alcohol and tobacco addiction may escape aggressive therapies (Weinberger et al. 2006).

Moreover, unlike tobacco-associated oropharyngeal cancers that cause mutation to the p53 gene, the apoptotic response of tumors to radiation and chemotherapy might be restored. It might be important to distinguish HPV-associated HNSCCs with better prognosis from tobacco/alcohol-associated HNSCCs with poor prognosis (Psyrri and DiMiao 2008). Careful clinical trials are required to determine the optimum management of HPV-dependent and HPV-independent HNSCCs. The diagnosis of HPV-associated OSCCs should be considered in all OSCCs, especially those arising from the tonsils in patients addicted to tobacco or alcohol. Immunocompromised patients and those with basaloid or poorly differentiated tumors need proper diagnosis. Survival of HPV-positive HNSCC patients is notably better than that of HPV-negative HNSCC patients (three-year survival of 84% vs. 57%, respectively) (D'Souza and Dempsey 2011). The worldwide five-year relative survival rate for oral cancer due to HPV is generally less than 50%, although females tend to have a higher relative survival rate than males (Ragin et al. 2007).

## 7.7 FUTURE AVENUES

The load of high-risk HPV DNA in a cervical sample has been suggested as a parameter to distinguish HPV infections of clinical relevance (Burd 2003). However, the initial optimism regarding its clinical value has been questioned by inconsistencies in the relation among viral load, duration of infection, HPV clearance, subsequent risk of acquisition, and progression of disease. Most studies using quantitative HPV PCR methods showed a substantial overlap of viral load values among women with and without high-grade CIN, especially in the range of high viral loads, limiting the clinical applicability of viral load analysis (Boulet et al. 2008). A recent report documented the association of high HPV-16 load with prevalence of cervical cancer. The population-based model generates an estimate of avertable burden and potential cost of HPV-16 and HPV-18 vaccinations in young adolescent girls prior to the age of 12 (herein referred to as pre-adolescent vaccination) where cervical cancer screening is not considered (Goldie et al. 2008; Kim et al. 2008). Vaccination strategies are distinguished by age of vaccination coverage level (defined as completion of a three-dose course) and vaccine efficacy.

Detection of RNA transcripts of HPV-infected tissues enables differentiation between asymptomatic and high-risk lesion and could therefore be considered as a better risk factor (Kanodia et al. 2007; Lazarczyk et al. 2009). RNA, instead of DNA, is prescribed as a potential tool for routine clinical diagnostics for better reproducibility and specificity (Espy et al. 2006). Currently, an RNA-based HPV assay is commercially available as the PreTect HPV Proofer (Ratnam et al. 2010), which incorporates nucleic acid sequence-based amplification (NASBA) of E6/E7 mRNA transcripts before type-specific detection for HPV-16, -18, -31, -33, and -45 (Boulet et al. 2010; Ratnam et al. 2010). It has been suggested for triaging HPV DNA-positive women and women with equivocal cytology. A broad spectrum mRNA test (15 types), Aptima (GenProbe), is currently under development (Boulet et al. 2008; Trope et al. 2009). Preliminary evaluation of the prototype assay showed that high-risk HPV E6/E7 mRNA detection improved the specificity of cervical precancer and cancer (Castle et al. 2007; Cuschieri and Wentzensen 2008). Quantitative reverse transcriptase PCR can be used to evaluate E6/E7 mRNA expression levels in biopsies (Bellone et al. 2009) that may lead to a prognostic tool to identify women at increased risk of developing cervical cancer.

Recent studies showed that host cellular micro-RNAs (miRNAs) modulate expression of several viral genes resulting in perturbation of the host–pathogen interaction network. Likewise, viruses also encode miRNAs that interfere with the host miRNA pathway and that help them to protect themselves from host immune response (Reshmi and Pillai 2008). In contrast to general miRNAs, aberrant miRNAs oppose the pathophysiologic outcome in accordance with their targeted miRNAs. Disruption of miRNA expression correlates with hematopoietic malignancies (Zhao et al. 2013) as well as breast (Iorio et al. 2005), ovarian (Iorio et al. 2007), cervical (Lee et al. 2008), prostate (Tong et al. 2009), colorectal (Motoyama et al. 2009), gastric (Katada et al. 2009), nasopharyngeal (Chen et al. 2009), lung (Ortholan et al. 2009), and renal cancers (Petillo et al. 2009); miRNA expression profiles in cervical cancer cell lines showed that miR-24, miR-27a, and miR-205 were most abundant in frank cervical cancer. Downregulation of miR-34a, to which p53 binds and triggers cell proliferation, is a signature of HPV-induced cervical carcinogenesis (Wang et al. 2009). HPV-16 DNA-positive cervical cancer tissues showed downregulation of miR-23b that regulates urokinase-type plasminogen activator expression. In addition, downregulation of miR-218 and miR-203 are also noticed in HPV-16 DNA-integrated cervical cancer cell lines (Martinez et al. 2008). Experimental results on changes in miRNA expression during chemotherapy and radiotherapy raise hope of using miRNA as a therapeutic agent for HPV-related disorders. On the other hand, miRNA signature might be a potential prognostic marker for cervical cancer (Lee et al. 2008). Further experiments on miRNA transgenic and knock-out models are needed regarding safety and efficacy of miRNA therapy for cervical or head and neck cancers. In this line, use of locked nucleic acid (LAN)-modified oligonucleotides, anti miRNA oligonucleotides (AMOs), is in the infancy stages of evaluation for their therapeutic potency in cancers other than cervical cancer (Wu et al. 2006). Nonetheless, miRNA-based prognosis and therapeutics are welcomed for successful treatment of HPV-related cancers.

## REFERENCES


Abramson AL, Nouri M, Mullooly V, Fisch G, Steinberg BM. (2004). Latent human papillomavirus infection is comparable in the larynx and trachea. *J Med Virol* 72: 473–7.

Adams M, Navabi H, Jasani B, Man S, Fiander A, Evans AS, et al. (2003). Dendritic cell (DC) based therapy for cervical cancer: Use of DC pulsed with tumor lysate and matured with a novel synthetic clinically non-toxic double stranded RNA analogue poly [I]:poly[C(12)U] (Ampligen R). *Vaccine* 21: 787–90.

Aggarwal BB, Kumar A, Bharti AC. (2003). Anticancer potential of curcumin: Preclinical and clinical studies. *Anticancer Res* 23: 363–98.

Ahdieh L, Klein RS, Burk R, Cu-Uvin S, Schuman P, Duerr A, et al. (2001). Prevalence, incidence, and typespecific persistence of human papillomavirus in human immunodeficiency virus (HIV)–positive and HIV-negative women. *J Infect Dis* 184: 682–90.

Alazawi W, Pett M, Arch B, Scott L, Freeman T, Stanley MA, et al. (2002). Changes in cervical keratinocyte gene expression associated with integration of human papillomavirus 16. *Cancer Res* 62: 6959–65.

Alba A, Cararach M, Rodríguez-Cerdeira C. (2009). The Human Papillomavirus (HPV) in human pathology: Description, pathogenesis, oncogenic role, epidemiology and detection techniques. *Open Dermatol J* 3: 90–102.

Alphs HH, Gambhira R, Karanam B, Roberts J, Jagu S, Schiller J, et al. (2008). Protection against heterologous human papillomavirus challenge by a synthetic lipopeptide vaccine containing a broadly cross-neutralizing epitope of L2. *Proc Natl Acad Sci U S A* 105: 5850–5.

Amin AA, Titolo S, Pelletier A, Fink D, Cordingley MG, Archambault J. (2000). Identification of domains of the HPV11 E1protein required for DNA replication in vitro. *Virology* 272: 137–50.

Arbyn M, Castellsagué X, de Sanjosé S, Bruni L, Saraiya M, Bray F, et al. (2011). Worldwide burden of cervical cancer in 2008. *Ann Oncol* 22: 2675–86.

Arend WP, Palmer G, Gabay C. (2008). IL-1, IL-18, and IL-33 families of cytokines. *Immunol Rev* 223: 20–38.

Barnard P, McMillan NA. (1999). The human papillomavirus E7 oncoprotein abrogates signaling mediated by interferon-α. *Virology* 259: 305–13.

Basu P, Roychowdhury S, Das Bafna U, Chaudhury S, Kothari S, Sekhon R, et al. (2009). Human papillomavirus genotype distribution in cervical cancer in India: Results from a multi-center study. *Asian Pac J Cancer Prev* 10: 27–34.

Bellone S, El-Sahwi K, Cocco E, Casagrande F, Cargnelutti M, Palmieri M, et al. (2009). Human papillomavirus type 16 (HPV-16) virus-like particle L1-specific $CD8^+$ cytotoxic T lymphocytes (CTLs) are equally effective as E7-specific $CD8^+$ CTLs in killing autologous HPV-16-positive tumor cells in cervical cancer patients: Implications for L1 dendritic cell-based therapeutic vaccines. *J Virol* 83: 6779–89.

Belnap DM, Olson NH, Cladel NM, Newcomb WW, Brown JC, Kreider JW, et al. (1996). Conserved features in papillomavirus and polyomavirus capsids. *J Mol Biol* 259: 249–63.

Benton EC, Arends MJ. (1996). Human papillomavirus in the immunosuppressed. In: Lacey C (ed.), *Papillomavirus Reviews: Current Research on Papillomaviruses*. Leeds: Leeds University Press, pp. 271–9.

Bernard HU. (2005). The clinical importance of the nomenclature, evolution and taxonomy of human papillomaviruses. *J Clin Virol* 32: S1–6.

Beutner KR, Tyring SK, Trofatter KFJ, Jr, Douglas JMJ, Jr, Spruance S, Owens ML, et al. (1998). Imiquimod, a patient-applied immune-response modifier for treatment of external genital warts. *Antimicrob Agents Chemother* 42: 789–94.

Black AP, Ardern-Jones MR, Kasprowicz V, Bowness P, Jones L, Bailey AS, et al. (2007). Human keratinocyte induction of rapid effector function in antigen-specific memory $CD4^+$ and $CD8^+$ T cells. *Eur J Immunol* 37: 1485–93.

Bosch FX, Qiao YL, Castellsagué X. (2006). The epidemiology of human papillomavirus infection and its association with cervical cancer. *Int J Gynecol Obst* 94: S8–S21.

Boulet GAV, Horvath CAJ, Berghmans S, Bogers J. (2008). Human papillomavirus in cervical cancer screening: Important role as biomarker. *Cancer Epidemiol Biomarkers Prev* 17: 810–17.

Boulet GAV, Micalessi IM, Horvath CAJ, Benoy IH, Depuydt CE, Bogers JJ. (2010). Nucleic acid sequence-based amplification assay for human papillomavirus mRNA detection and typing: Evidence for DNA amplification. *J Clin Microbiol* 48: 2524–9.

Braakhuis BJM, Snijders PJF, Keune WH, Meijer CJLM, Ruijter-Schippers HJ, Leemans CR, et al. (2004). Genetic patterns in head and neck cancers that contain or lack transcriptionally active human papillomavirus. *J Natl Cancer Inst* 96: 998–1006.

Brennan JA, Boyle JO, Koch WM, Goodman SN, Hruban RH, Eby YJ. (1995). Association between cigarette smoking and mutation of the p53 gene in squamous-cell carcinoma of the head and neck. *N Engl J Med* 332: 712–17.

Brown DR, Shew ML, Qadadri B, Neptune N, Vargas M, Tu W, et al. (2005). A longitudinal study of genital human papillomavirus infection in a cohort of closely followed adolescent women. *J Infect Dis* 191: 182–92.

Bruni L, Diaz M, Castellsagué X, Ferrer E, Bosch FX, de Sanjosé S. (2010). Cervical human papillomavirus prevalence in 5 continents: Meta-analysis of 1 million women with normal cytological findings. *J Infect Dis* 202: 1789–99.

Burd EM. (2003). Human papillomavirus and cervical cancer. *Clin Microbiol Rev* 1: 1–17.

Campo MS, Grindlay GJ, O'Neil BW, Chandrachud LM, McGarvie GM, Jarrett WF. (1993). Prophylactic and therapeutic vaccination against a mucosal papillomavirus. *J Gen Virol* 74: 945–53.

Castellsagué X. (2008). Natural history and epidemiology of HPV infection and cervical cancer. *Gynecol Oncol* 110: S4–S7.

Castellsagué X, Diaz M, de Sanjose S, Munõz N, Herrero R, Franceschi S, et al. (2006). Worldwide human papillomavirus etiology of cervical adenocarcinoma and its cofactors: Implications for screening and prevention. *J Natl Cancer Inst* 98: 303–15.

Castle PE, Dockter J, Giachetti C, Garcia FAR, McCormick MK, Mitchell AL, et al. (2007). A cross-sectional study of a prototype carcinogenic human papillomavirus E6/E7 messenger RNA assay for detection of cervical precancer and cancer. *Clin Cancer Res* 13(9): 2599–605.

Chang YE, Laimins LA. (2000). Microarray analysis identifies interferon-inducible genes and Stat-1 as major transcriptional targets of human papillomavirus type 31. *J Virol* 74: 4174–82.

Chaturvedi AK, Engels EA, Anderson WF, Gillison ML. (2008). Incidence trends for human papillomavirus-related and -unrelated oral squamous cell carcinomas in the United States. *J Clin Oncol* 26: 612–19.

Chen CH, Chen GH, Chen SJ. (2009). MicroRNA deregulation and pathway alterations in nasopharyngeal carcinoma. *Br J Cancer* 100: 1002–11.

Chiang CM, Ustav M, Stenlund A, Ho TF, Broker TR, Chow LT. (1992). Viral E1 and E2 proteins support replication of homologous and heterologous papillomaviral origins. *Proc Natl Acad Sci U S A* 89: 5799–803.

Chow LT, Broker TR, Steinberg BM. (2010). The natural history of human papillomavirus infections of the mucosal epithelia. *Acta Pathol Microbiol Immunol Scandinavica* 118: 422–49.

Cirelli R, Tyring SK. (1994). Interferons in human papillomavirus infections. *Antiviral Res* 24: 191–204.

Clifford GM, Gallus S, Herrero R, Munõz N, Snijders PJ, Vaccarella S, et al. (2005). Worldwide distribution of human papillomavirus types in cytologically normal women in the International Agency for Research on Cancer HPV prevalence surveys: A pooled analysis. *Lancet* 366: 991–8.

Coleman N, Birley HDL, Renton AM, Hanna NF, Ryait BK, Byrne M, et al. (1994). Immunological events in regressing genital warts. *Am J Clin Pathol* 102: 768–74.

Cuschieri K, Wentzensen N. (2008). Human papillomavirus mRNA and p16 detection as biomarkers for the improved diagnosis of cervical neoplasia. *Cancer Epidemiol Biomarkers Prev* 17: 2536–45.

Dakubo GD, Jakupciak JP, Birch-Machin MA, Parr RL. (2007). Clinical implications and utility of field cancerization. *Cancer Cell Int* 7: 1–12.

Das Ghosh D, Bhattacharjee B, Sen S, Premi L, Mukhopadhyay I, Chowdhury RR, et al. (2012). Some novel insights on HPV16 related cervical cancer pathogenesis based on analyses of LCR methylation, viral load, E7 and E2/E4 expressions. *PLoS One* 7: e44678.

David MP, Van Herck K, Hardt K, Tibaldi F, Dubin G, Descamps D, et al. (2009). Long-term persistence of anti-HPV-16 and -18 antibodies induced by vaccination with the AS04-adjuvanted cervical cancer vaccine: Modeling of sustained antibody responses. *Gynecol Oncol* 115: S1–6.

de Jong A, O'Neill T, Khan AY, Kwappenberg KM, Chisholm SE, Whittle NR, et al. (2002). Enhancement of human papillomavirus (HPV) type 16 E6 and E7-specific T-cell immunity in healthy volunteers through vaccination with TA-CIN, an HPV-16 L2E7E6 fusion protein vaccine. *Vaccine* 20: 3456–64.

Denny L. (2009). Human papillomavirus infections: Epidemiology, clinical aspects and vaccines. *Open Infect Dis J* 3: 135–42.

Deshpande AM, Wong DT. (2008). Molecular mechanisms of head and neck cancer. *Expert Rev Anticancer Ther* 8: 799–809.

De Villiers EM, Fauquet C, Broker TR, Bernard HU, zur Hausen H. (2004). Classification of papillomaviruses. *Virology* 324: 17–27.

De Villiers EM, Weidauer H, Otto H, Hausen H. (1985). Papillomavirus DNA in human tongue carcinomas. *Int J Cancer* 36: 575–8.

Dillner J, Kjaer SK, Wheeler CM, Sigurdsson K, Iversen OE, Hernandez-Avila M, et al. (2010). Four year efficacy of prophylactic human papillomavirus quadrivalent vaccine against low grade cervical, vulvar, and vaginal intraepithelial neoplasia and anogenital warts: Randomised controlled trial. *BMJ* 341: c3493.

Divya CS, Pillai MR. (2006). Antitumor action of curcumin in human papillomavirus associated cells involves downregulation of viral oncogenes, prevention of NFκB and AP-1 translocation, and modulation of apoptosis. *Mol Carcinog* 45: 320–32.

Doorbar J. (2005). The papillomavirus life cycle. *J Clin Virol* 32(Suppl 1): S7–15.

Doorbar J. (2007). Papillomavirus life cycle organization and biomarker selection. *Dis Markers* 23: 297–313.

D'Souza G, Dempsey A. (2011). The role of HPV in head and neck cancer and review of the HPV vaccine. *Prev Med* 53: S5–11.

El-Naggar AK, Westra WH. (2012). p16 expression as a surrogate marker for HPV-related oropharyngeal carcinoma: A guide for interpretative relevance and consistency. *Head Neck* 34: 459–61.

Embers ME, Budgeon LR, Pickel M, Christensen ND. (2002). Protective immunity to rabbit oral and cutaneous papillomaviruses by immunization with short peptides of L2, the minor capsid protein. *J Virol* 76: 9798–805.

Engels EA, Biggar RJ, Hall HI, Cross H, Crutchfield A, Finch JL, et al. (2008). Cancer risk in people infected with human immunodeficiency virus in the United States. *Int J Cancer* 123: 187–94.

Espy MJ, Uhl JR, Sloan LM, Buckwalter SP, Jones MF, Vetter EA. (2006). Real-time PCR in clinical microbiology: Applications for routine laboratory testing. *Clin Microbiol Rev* 19: 165–256.

Fausch SC, Da Silva DM, Kast WM. (2005). Heterologous papillomavirus virus-like particles and human papillomavirus virus-like particle immune complexes activate human Langerhans cells. *Vaccine* 23: 1720–9.

Finley JB, Luo M. (1998). X-ray crystal structures of half the human papilloma virus E2 binding site: d(GACCGCGGTC). *Nucleic Acids Res* 26: 5719–27.

Finnen RL, Erickson KD, Chen XS, Garcea RL. (2003). Interactions between papillomavirus L1 and L2 capsid proteins. *J Virol* 77: 4818–26.

Fouts ET, Yu X, Egelman EH, Botchan MR. (1999). Biochemical and electron microscopic image analysis of the hexameric E1 helicase. *J Biol Chem* 274: 4447–58.

Fox PA, Nathan M, Francis N, Singh N, Weir J, Dixon G, et al. (2010). A double-blind, randomized controlled trial of the use of imiquimod cream for the treatment of anal canal high-grade anal intraepithelial neoplasia in HIV-positive MSM on HAART, with long-term follow-up data including the use of open-label imiquimod. *AIDS* 24: 2331–5.

Gambhira R, Gravitt PE, Bossis I, Stern P, Viscidi RP, Roden R. (2006). Vaccination of healthy volunteers with human papillomavirus type 16 L2E7E6 fusion protein induces serum antibody that neutralizes across papillomavirus species. *Cancer Res* 66: 11120–4.

Gambhira R, Jagu S, Karanam B, Gravitt PE, Culp TD, Christensen ND, et al. (2007). Protection of rabbits against challenge with rabbit papillomaviruses by immunization with the N terminus of human papillomavirus type 16 minor capsid antigen L2. *J Virol* 81: 11585–92.

Gillison ML. (2009). Oropharyngeal cancer: A potential consequence of concomitant HPV and HIV infection. *Curr Opin Oncol* 21: 439–44.

Gillison ML, Koch WM, Capone RB, Spafford M, Westra WH, Wu L, et al. (2000). Evidence for a causal association between human papillomavirus and a subset of head and neck cancers. *J Natl Cancer Inst* 92: 709–20.

Giuliano AR, Harris R, Sedjo RL, Baldwin S, Roe D, Papenfuss MR, et al. (2002). Incidence, prevalence, and clearance of type-specific human papillomavirus infections: The Young Women's Health Study. *J Infect Dis* 186: 462–9.

Goldie SJ, O'Shea M, Campos NG, Diaz M, Sweet S, Kim SY. (2008). Health and economic outcomes of HPV 16,18 vaccination in 72 GAVI-eligible countries. *Vaccine* 26: 4080–93.

Gomez-Gutierrez JG, Elpek KG, Montes de Oca-Luna R, Shirwan H, Sam Zhou H, McMasters KM. (2007). Vaccination with an adenoviral vector expressing calreticulin-human papillomavirus 16 E7 fusion protein eradicates E7 expressing established tumors in mice. *Cancer Immunol Immunother* 56: 997–1007.

Goon PKC, Stanley MA, Ebmeye J, Steinsträsser L, Upile T, Jerjes W, et al. (2009). HPV & head and neck cancer: A descriptive update. *Head Neck Oncol* 1: 36.

Haddad RI. (2007). Human papillomavirus infection and oropharyngeal cancer. Available from: http://www.oralcancerfoundation.org/facts/pdf/hpv_Infection

Hawes SE, Kiviat NB. (2002). Are genital infections and inflammation cofactors in the pathogenesis of invasive cervical cancer? *J Natl Cancer Inst* 94: 1592–3.

Hegde RS. (2002). The papillomavirus E2 proteins: Structure, function, and biology. *Ann Rev Biophys Biomol Struct* 31: 343–60.

Hernandez BY, Wilkens LR, Zhu X, Thompson P, McDuffie K, Shvetsov YB, et al. (2008). Transmission of human papillomavirus in heterosexual couples. *Emerg Infect Dis* 14: 888–94.

Herrero R, Castellsagué X, Pawlita M, Lissowska J, Kee F, Balaram P, et al. (2003). Human papillomavirus and oral cancer: The International Agency for Research on Cancer Multicenter Study. *J Natl Cancer Inst* 95: 1772–83.

Hobbs CGL, Sterne JAC, Bailey M, Heyderman RS, Birchall MA, Thomas SJ. (2006). Human papillomavirus and head and neck cancer: A systemic review and meta-analysis. *Clin Otolaryngol* 31: 259–66.

Hocking JS, Stein A, Conway EL, Regan D, Grulich A, Law M, et al. (2011). Head and neck cancer in Australia between 1982 and 2005 show increasing incidence of potentially HPV-associated oropharyngeal cancers. *Br J Cancer* 104: 886–91.

Howley PM. (1991). Role of the human papillomaviruses in human cancer. *Cancer Res* 51: 5019s–22s.

Ilves I, Kivi S, Ustav M. (1999). Long-term episomal maintenance of bovine papillomavirus type 1 plasmids is determined by attachment to host chromosomes, which is mediated by the viral E2 protein and its binding sites. *J Virol* 73: 4404–12.

Iorio MV, Ferracin M, Liu CG, Veronese A, Spizzo R, Sabbioni S, et al. (2005). MicroRNA gene expression deregulation in human breast cancer. *Cancer Res* 65: 7065–70.

Iorio MV, Visone R, Di Leva G, Donati V, Petrocca F, Casalini P, et al. (2007). MicroRNA signatures in human ovarian cancer. *Cancer Res* 67: 8699–707.

Jagu S, Karanam B, Gambhira R, Chivukula S, Chaganti R, Lowy D, et al. (2009). Concatenated multitype L2 fusion proteins as candidate prophylactic pan-human papillomavirus vaccines. *J Natl Cancer Inst* 101: 782–92.

Jemal A, Siegel R, Ward E, Murray T, Xu J, Thun MJ. (2006). Cancer statistics, 2006. *CA Cancer J Clin* 56: 106–30.

Sideri M, Jones RW, Wilkinson EJ, Preti M, Heller DS, Scurry J, et al. (2005). Squamous vulvar intraepithelial neoplasia: 2004 modified terminology, ISSVD Vulvar Oncology Subcommittee. *J Reprod Med* 50:807–10.

Jovanoviæ MA, Karadagliæ Ð. (2000). Pathogenetic mechanisms in epidermodysplasia verruciformis. *Archive Oncol* 8: 119–26.

Kaarthigeyan K. (2012). Cervical cancer in India and HPV vaccination. *Indian J Med Paed Oncol* 33: 7–12.

Kalali BN, Köllisch G, Mages J, Müller T, Bauer S, Wagner H, et al. (2008). Double-stranded RNA induces an antiviral defense status in epidermal keratinocytes through TLR3-, PKR-, and MDA5/RIGI-mediated differential signaling. *J Immunol* 181: 2694–704.

Kanodia S, Fahey LM, Kast WM. (2007). Mechanisms used by human papillomaviruses to escape the host immune response. *Curr Cancer Drug Targets* 7: 79–89.

Katada T, Ishiguro H, Kuwabara Y, Kimura M, Mitui A, Mori Y, et al. (2009). microRNA expression profile in undifferentiated gastric cancer. *Int J Oncol* 34: 537–42.

Kim JJ, Brisson M, Edmunds WJ, Goldie SJ. (2008). Modeling cervical cancer prevention in developed countries. *Vaccine* 26: K76–86.

Kjaer SK, Sigurdsson K, Iversen OE, Hernandez-Avila M, Wheeler CM, Perez G, et al. (2009). A pooled analysis of continued prophylactic efficacy of quadrivalent human papillomavirus (types 6/11/16/18) vaccine against high-grade cervical and external genital lesions. *Cancer Prev Res (Phila)* 2: 868–78.

Kobayashi A, Greenblatt RM, Anastos K, Minkoff H, Massad LS, Young M, et al. (2004). Functional attributes of mucosal immunity in cervical intraepithelial neoplasia and effects of HIV infection. *Cancer Res* 64: 6766–74.

Koutsky L. (1997). Epidemiology of genital human papillomavirus infection. *Am J Med* 102: 3–8.

Kreimer AR, Clifford GM, Boyle P, Franceschi S. (2005). Human papillomavirus types in head and neck squamous cell carcinomas worldwide: A systematic review. *Cancer Epidemiol Biomarkers Prev* 14: 467–75.

Kurg R. (2011). The role of E2 proteins in papillomavirus DNA replication. In: Seligmann H (ed.), *DNA Replication – Current Advances*. InTech, pp. 613–38.

Lacey CJ. (2005). Therapy for genital human papillomavirus-related disease. *J Clin Virol* 32: S82–90.

Lacey CJ, Lowndes CM, Shah KV. (2006). Chapter 4: Burden and management of non-cancerous HPV-related conditions: HPV-6/11 disease. *Vaccine* 24: S35–S41.

Lajer CB, Buchwald CV. (2011). The role of human papillomavirus in head and neck cancer. *J Am Dent Assoc* 142: 905–14.

Lazarczyk M, Cassonnet P, Pons C, Jacob Y, Favre M. (2009). The EVER proteins as a natural barrier against papillomaviruses: A new insight into the pathogenesis of human papillomavirus infections. *Microbiol Mol Biol Rev* 73: 348–70.

Le Bon A, Tough DF. (2002). Links between innate and adaptive immunity via type I interferon. *Curr Opin Immunol* 14: 432–6.

Lee JW, Choi CH, Choi JJ, Park YA, Kim SJ, Hwang SY, et al. (2008). Altered microRNA expression in cervical carcinomas. *Cancer Res* 14: 2535–42.

Lenz P, Lowy DR, Schiller JT. (2005). Papillomavirus virus-like particles induce cytokines characteristic of innate immune responses in plasmacytoid dendritic cells. *Eur J Immunol* 35: 1548–56.

Liao JB. (2006). Viruses and human cancer. *Yale J Biol Med* 79: 115–22.

Lin YL, Borenstein LA, Selvakumar R, Ahmed R, Wettstein FO. (1992). Effective vaccination against papilloma development by immunization with L1 or L2 structural protein of cottontail rabbit papillomavirus. *Virology* 187: 612–19.

Longworth MS, Laimins LA. (2004). Pathogenesis of human papillomaviruses in differentiating epithelia. *Microbiol Mol Biol Rev* 68: 362–72.

Lusky M, Hurwitz J, Seo YS. (1994). The bovine papillomavirus E2 protein modulates the assembly of but is not stably maintained in a replication-competent multimeric E1-replication origin complex. *Proc Natl Acad Sci U S A* 91: 8895–9.

Madsen BS, van den Brule AJC, Jensen HL, Wohlfahrt J, Frisch M. (2008). Risk factors for squamous cell carcinoma of the penis—Population-based case-control study in Denmark. *Cancer Epidemiol Biomarkers Prev* 17: 2683–91.

Mannarini L, Kratochvil V, Calabrese L, Silva GL, Morbini P, Betka J, et al. (2009). Human papilloma virus (HPV) in head and neck region: Review of literature. *Acta Otorhinolaryngol Ital* 29: 119–26.

Martinez I, Gardiner AS, Board KF, Monzon FA, Edwards RP, Khan SA. (2008). Human papillomavirus type 16 reduces the expression of microRNA-218 in cervical carcinoma cells. *Oncogene* 27: 2575–82.

Marur S, D'Souza G, Westra WH, Forastiere AA. (2010). HPV-associated head and neck cancer: A virus-related cancer epidemic. *Lancet Oncol* 11: 781–9.

Masterson PJ, Stanley MA, Lewis AP, Romanos MA. (1998). A C-terminal helicase domain of the human papillomavirus E1 protein binds E2 and the DNA polymerase alpha-primase p68 subunit. *J Virol* 72: 7407–19.

Maw R. (2004). Critical appraisal of commonly used treatment for genital warts. *Int J STD AIDS* 15: 357–64.

Medzhitov R, Janeway CA, Jr. (1997). Innate immunity: The virtues of a non-clonal system of recognition. *Cell* 91: 295–8.

Miller LS, Modlin RL. (2007). Human keratinocyte toll-like receptors promote distinct immune responses. *J Invest Dermatol* 127: 262–3.

Mellin H, Dahlgren L, Munck-Wikland E, Lindholm J, Rabbani H, Kalantari M, et al. (2002). Human papillomavirus type 16 is episomal and a high viral load may be correlated to better prognosis in tonsillar cancer. *Int J Cancer* 102:152–8.

Moody CA, Laimins LA. (2010). Human papillomavirus oncoproteins: Pathways to transformation. *Nat Rev Cancer* 10: 550–60.

Moore RA, Nicholls PK, Santos EB, Gough GW, Stanley MA. (2002). Absence of canine oral papillomavirus DNA following prophylactic L1 particle-mediated immunotherapeutic delivery vaccination. *J Gen Virol* 83: 2299–301.

Morris BJ, Gray RH, Castellsague X, Bosch FX, Halperin DT, Waskett JH, et al. (2011). The strong protective effect of circumcision against cancer of the penis. *Adv Urol* 812368: 1–21.

Moscicki AB, Schiffman M, Kjaer S, Villa LL. (2006). Chapter 5: Updating the natural history of HPV and anogenital cancer. *Vaccine* 24: S42–S51.

Motoyama K, Inoue H, Takatsuno Y, Tanaka F, Mimori K, Uetake H, et al. (2009). Over- and under-expressed microRNAs in human colorectal cancer. *Int J Oncol* 34: 1069–75.

Münger K, Baldwin A, Edwards KM, Hayakawa H, Nguyen CL, Owens M, et al. (2004). Mechanisms of human papillomavirus-induced oncogenesis. *J Virol* 78: 11451–60.

Muñoza N, Castellsagué X, Berrington de Gonzálezc A, Gissmann L. (2006). Chapter 1: HPV in the etiology of human cancer. *Vaccine* 24: S1–10.

Nakagawa M, Stites DP, Farhat S, Sisler JR, Moss B, Kong F, et al. (1997). Cytotoxic T lymphocyte responses to E6 and E7 proteins of HPV type 16: Relationship to cervical intraepithelial neoplasia. *J Infect Dis* 175: 927–31.

Näsman A. (2009). *Studies on the Influence of Human Papillomaviruses (HPV) and Other Biomarkers on the Prevalence of Oropharyngeal Cancer and Clinical Outcome.* Karolinska Institute, pp. 1–84.

Nasu K, Narahara H. (2010). Pattern recognition via the toll-like receptor system in the human female genital tract. *Mediators Inflam* 1–13.

National Institutes of Health. (1996). Cervical cancer. *NIH Consensus Statement* 14: 1–38.

Nees M, Geoghegan JM, Hyman T, Frank S, Miller L, Woodworth CD. (2001). Papillomavirus type 16 oncogenes downregulate expression of interferon-responsive genes and upregulate proliferation-associated and NF-κB-responsive genes in cervical keratinocytes. *J Virol* 75: 4283–96.

Newbold RF. (2002). The significance of telomerase activation and cellular immortalization in human cancer. *Mutagenesis* 17: 539–50.

Orth G, Jablonska S, Breitburd F, Favre M, Croissant O. (1978a). The human papillomaviruses. *Bull Cancer* 65: 151–64.

Orth G, Jablonska S, Favre M, Croissant O, Jarzabek-Chorzelska M, Rzesa G. (1978b). Characterization of two types of human papillomaviruses in lesions of epidermodysplasia verruciformis. *Proc Natl Acad Sci U S A* 75: 1537–41.

Ortholan C, Puissegur MP, Ilie M, Barbry P, Mari B, Hofman P. (2009). MicroRNAs and lung cancer: New oncogenes and tumor suppressors, new prognostic factors and potential therapeutic targets. *Curr Med Chem* 16: 1047–61.

Paavonen J, Naud P, Salmerón J, Wheeler CM, Chow S, Apter D, et al. (2009). Efficacy of human papillomavirus (HPV)-16/18 AS04-adjuvanted vaccine against cervical infection and precancer caused by oncogenic HPV types (PATRICIA): Final analysis of a double-blind, randomised study in young women. *Lancet* 374: 301–14.

Palefsky JM, Gillison ML, Strickler HD. (2006). Chapter 16: HPV vaccines in immunocompromised women and men. *Vaccine* 24: S140–6.

Parkin DM, Bray F. (2006). Chapter 2: The burden of HPV-related cancers. *Vaccine* 24: 11–25.

Pascual A, Pariente M, Godínez JM, Sánchez-Prieto R, Atienzar M, Segura M. (2007). High prevalence of human papillomavirus 16 in penile carcinoma. *Histol Histopathol* 22: 177–83.

Petillo D, Kort EJ, Anema J, Furge KA, Yang XJ, Teh BT. (2009). MicroRNA profiling of human kidney cancer subtypes. *Int J Oncol* 35: 109–14.

Peto J, Gilham C, Deacon J, Taylor C, Evans C, Binns W, et al. (2004). Cervical HPV infection and neoplasia in a large population-based prospective study: The Manchester cohort. *Br J Cancer* 91: 942–53.

Pett MR, Alazawi WOF, Roberts I, Dowen S, Smith DI, Stanley MA, et al. (2004). Acquisition of high-level chromosomal instability is associated with integration of human papillomavirus type 16 in cervical keratinocytes. *Cancer Res* 64: 1359–68.

Pett MR, Herdman MR, Palmer RD, Yeo GSH, Shivji MK, Stanley MA, et al. (2006). Selection of cervical keratinocytes containing integrated HPV16 associates with episome loss and an endogenous antiviral response. *Proc Natl Acad Sci U S A* 103: 3822–7.

Prusty BK, Das BC. (2005). Constitutive activation of transcription factor AP-1 in cervical cancer and suppression of human papillomavirus (HPV) transcription and AP-1 activity in HeLa cells by curcumin. *Int J Cancer* 113: 951–60.

Psyrri A, DiMiao D. (2008). Human papillomavirus in cervical and head-and-neck cancer. *Nat Clin Pract Oncol* 5: 24–31.

Psyrri A, Gouveris P, Vermorken JB. (2009). Human papillomavirus-related head and neck tumors: Clinical and research implication. *Curr Opin Oncol* 21: 201–5.

Pyeon D, Pearce SM, Lank SM, Ahlquist P, Lambert PF. (2009). Establishment of human papillomavirus infection requires cell cycle progression. *Plos Pathogens* 5: e1000318, 1–9.

Ragin CCR, Modugno F, Gollin SM. (2007). The epidemiology and risk factors of head and neck cancer: A focus on human papillomavirus. *J Dent Res* 86: 104–14.

Rampias T, Sasaki C, Weinberger P, Psyrri A. (2009). E6 and E7 gene silencing and transformed phenotype of human papillomavirus 16–positive oropharyngeal cancer cells. *J Natl Cancer Inst* 101: 412–23.

Ratnam S, Coutlee F, Fontaine D, Bentley J, Escott N, Ghatage P. (2010). Clinical performance of the PreTect HPV-Proofer E6/E7 mRNA assay in comparison with that of the hybrid capture 2 test for identification of women at risk of cervical cancer. *J Clin Microbiol* 48: 2779–85.

Raybould R, Fiander A, Hibbitts S. (2011). Human papillomavirus integration and its role in cervical malignant progression. *Open Clin Cancer J* 5: 1–7.

Reshmi G, Pillai MR. (2008). Beyond HPV: Oncomirs as new players in cervical cancer. *FEBS Lett* 582: 4113–16.

Ronco LV, Karpova AY, Vidal M, Howley PM. (1998). Human papillomavirus 16 E6 oncoprotein binds to interferon regulatory factor-3 and inhibits its transcriptional activity. *Genes Dev* 12: 2061–72.

Rouse BT, Suvas S. (2004). Regulatory cells and infectious agents: Detentes cordiale and contraire. *J Immunol* 173: 2211–15.

Sanders CM, Stenlund A. (1998). Recruitment and loading of the E1 initiator protein: An ATP-dependent process catalysed by a transcription factor. *EMBO J* 17: 7044–55.

Sankaranarayanan R, Nene BM, Shastri SS, Jayant K, Muwonge R, Budukh AM, et al. (2009). HPV screening for cervical cancer in rural India. *N Engl J Med* 360: 1385–94.

Sartor M, Steingrimsdottir H, Elamin F, Gäken J, Warnakulasuriya S, Partridge M, et al. (1999). Role of p16/MTS1, cyclin D1 and RB in primary oral cancer and oral cancer cell lines. *Br J Cancer* 80: 79–86.

Schäfer K, Neumann J, Waterboer T, Rös F. (2011). Serological markers for papillomavirus infection and skin tumor development in the rodent model *Mastomys coucha*. *J Gen Virol* 92: 383–94.

Schwartz S. (2008). HPV-16 RNA processing. *Front Biosci* 13: 5880–91.

Seo YS, Müller F, Lusky M, Gibbs E, Kim HY, Phillips B, et al. (1993). Bovine papilloma virus (BPV)-encoded E2 protein enhances binding of E1 protein to the BPV replication origin. *Proc Natl Acad Sci U S A* 90: 2865–9.

Shaib W, Kono S, Saba N. (2012). Antiepidermal growth factor receptor therapy in squamous cell carcinoma of the head and neck. *J Oncol* 2012: 521215.

Slupetzky K, Gambhira R, Culp TD, Shafti-Keramat S, Schellenbacher C, Christensen ND, et al. (2007). A papillomavirus-like particle (VLP) vaccine displaying HPV16 L2 epitopes induces cross-neutralizing antibodies to HPV11. *Vaccine* 25: 2001–10.

Smith JF, Brownlow M, Brown M, Kowalski R, Esser MT, Ruiz W, et al. (2007). Antibodies from women immunized with gardasil cross-neutralize HPV 45 pseudovirions. *Hum Vaccines* 3: 109–15.

Stanley M. (2006). Immune responses to human papillomavirus. *Vaccine* 24: S16–S22.

Stanley MA. (2002). Imiquimod and the imidazoquinolones: Mechanism of action and therapeutic potential. *Clin Exp Dermatol* 27: 571–7.

Stanley MA. (2012a). Epithelial cell responses to infection with human papillomavirus. *Clin Microbiol Rev* 25: 215–22.

Stanley MA. (2012b). Genital human papillomavirus infections: Current and prospective therapies. *J Gen Virol* 93: 681–91.

Steinberg BM, DiLorenzo TP. (1996). A possible role for human papillomaviruses in head and neck cancer. *Cancer Metastasis Rev* 15: 91–112.

Syrjänen K, Lamberg M, Pyrhönen S, et al. (1983). Morphological and immunohistochemical evidence suggesting human papillomavirus (HPV) involvement in oral squamous cell carcinogenesis. *Int J Oral* Surg 12:418–24.

Talwar GP, Dar SA, Rai MK, Reddy KV, Mitra D, Kulkarni SV, et al. (2008). A novel polyherbal microbicide with inhibitory effect on bacterial, fungal and viral genital pathogens. *Int J Antimicrob Agents* 32: 180–5.

Tatti S, Stockfleth E, Beutner KR, Tawfik H, Elsasser U, Weyrauch P, et al. (2010). Polyphenon E: A new treatment for external anogenital warts. *Br J Dermatol* 162: 176–84.

Taylor WR, Stark GR. (2001). Regulation of the G2/M transition by p53. *Oncogene* 20: 1803–15.

Terlou A, van Seters M, Ewing PC, Aaronson NK, Gundy CM, Heijmans-Antonissen C, et al. (2011). Treatment of vulvar intraepithelial neoplasia with topical imiquimod: Seven years median follow-up of a randomized clinical trial. *Gynecol Oncol* 121: 157–62.

Theofilopoulos AN, Baccala R, Beutler B, Kono DH. (2005). Type I interferons (α/β) in immunity and autoimmunity. *Ann Rev Immunol* 23: 307–36.

Titolo S, Brault K, Majewski J, White PW, Archambault J. (2003). Characterization of the minimal DNA binding domain of the human papillomavirus E1 helicase: Fluorescence anisotropy studies and characterization of a dimerization-defective mutant protein. *J Virol* 77: 5178–91.

Titolo S, Pelletier A, Pulichino AM, Brault K, Wardrop E, White PW, et al. (2000). Identification of domains of the human papillomavirus type 11 E1 helicase involved in oligomerization and binding to the viral origin. *J Virol* 74: 7349–61.

Titolo S, Pelletier A, Sauvé F, Brault K, Wardrop E, White PW, et al. (1999). Role of the ATP-binding domain of the human papillomavirus type 11 E1 helicase in E2-dependent binding to the origin. *J Virol* 73: 5282–93.

Todd R, Hinds PW, Munger K, Rustgi AK, Opitz OG, Suliman Y, et al. (2002). Cell cycle dysregulation in oral cancer. *Crit Rev Oral Biol Med* 13: 51–61.

Tomlins C, Storey A. (2010). Cutaneous HPV5 E6 causes increased expression of Osteoprotegerin and interleukin 6 which contribute to evasion of UV-induced apoptosis. *Carcinogenesis* 31: 2155–64.

Tong AW, Fulgham P, Jay C, Chen P, Khalil I, Liu S, et al. (2009). MicroRNA profile analysis of human prostate cancers. *Cancer Gene Ther* 16: 206–16.

Trope A, Sjøborg K, Eskild A, Cuschieri K, Eriksen T, Thoresen S, et al. (2009). Performance of human papillomavirus DNA and mRNA testing strategies for women with and without cervical neoplasia. *J Clin Microbiol* 47: 2458–64.

Trottier H. (2012). Epidemiology of mucosal human papillomavirus (HPV) infections among adult and children. In: Broeck DV (eds.), *Human Papillomavirus and Related Diseases—From Bench to Bedside—Research Aspects*. Croatia: InTech, pp. 3–18.

van der Burg SH, Piersma SJ, de Jong A, van der Hulst JM, Kwappenberg KM, van den Hende M, et al. (2004). Association of cervical cancer with the presence of CD4+ regulatory T cells specific for human papillomavirus antigens. *Proc Natl Acad Sci U S A* 104: 12087–92.

Van Pachterbeke C, Bucella D, Rozenberg S, Manigart Y, Gilles C, Larsimont D, et al. (2009). Topical treatment of CIN 2+ by cidofovir: Results of a phase II, double-blind, prospective, placebo-controlled study. *Gynecol Oncol* 115: 69–74.

Viera MH, Amini S, Huo R, Konda S, Block S, Berman B. (2010). Herpes simplex virus and human papillomavirus genital infections: New and investigational therapeutic options. *Int J Dermatol* 49: 733–49.

Villa LL, Ault KA, Giuliano AR, Costa RL, Petta CA, Andrade RP, et al. (2006). Immunologic responses following administration of a vaccine targeting human papillomavirus types 6, 11, 16, and 18. *Vaccine* 24: 5571–83.

Wang X, Wang HK, McCoy JP, Banerjee NS, Rader JS, Broker TR. (2009). Oncogenic HPV infection interrupts the expression of tumor-suppressive miR-34a through viral oncoprotein E6. *RNA* 15: 637–47.

Weinberger PM, Yu Z, Haffty BG, Kowalski D, Harigopal M, Brandsma J, et al. (2006). Molecular classification identifies a subset of human papillomavirus-associated orophayngeal cancers with favorable prognosis. *J Clin Oncol* 24: 736–46.

Wentzensen N, Vinokurova S, von Knebel Doeberitz M. (2004). Systematic review of genomic integration sites of human papillomavirus genomes in epithelial dysplasia and invasive cancer of the female lower genital tract. *Cancer Res* 64: 3878–84.

Wheeler CM, Castellsague X, Garland SM, Szarewski A, Paavonen J, Naud P, et al. (2012). Cross-protective efficacy of HPV-16/18 AS04-adjuvanted vaccine against cervical infection and precancer caused by non-vaccine oncogenic HPV types: End-of-study analysis of the randomised, double-blind PATRICIA trial. *Lancet Oncol* 13: 100–10.

White PW, Faucher AM, Goudreau N. (2011). Small molecule inhibitors of the human papillomavirus E1-E2 interaction. *Curr Top Microbiol Immunol* 348: 61–88.

White PW, Faucher AM, Massariol MJ, Welchner E, Rancourt J, Cartier M, et al. (2005). Biphenylsulfonacetic acid inhibitors of the human papillomavirus type 6 E1 helicase inhibit ATP hydrolysis by an allosteric mechanism involving tyrosine 486. *Antimicrob Agents Chemother* 49: 4834–42.

White PW, Pelletier A, Brault K, Titolo S, Welchner E, Thauvette L, et al. (2001). Characterization of recombinant HPV6 and 11 E1 helicases: Effect of ATP on the interaction of E1 with E2 and mapping of a minimal helicase domain. *J Biol Chem* 276: 22426–38.

WHO. (2006). Guidelines to assure the quality, safety and efficacy of recombinant HPV virus-like particle vaccines. Available from: http://screening.iarc.fr/doc/WHO_vaccine_guidelines_2006.pdf

Woo YL, van den Hende M, Sterling JC, Coleman N, Crawford RA, Kwappenberg KM, et al. (2010). A prospective study on the natural course of low-grade squamous intraepithelial lesions and the presence of HPV16 E2-, E6- and E7-specific T-cell responses. *Int J Cancer* 126: 133–41.

Wright TC, Jr. (2006). Pathology of HPV infection at the cytologic and histologic levels: Basis for a 2-tiered morphologic classification system. *Int J Gynecol Obst* 94: S22–S31.

Wright TC, Cox JT, Massad LS, Carlson J, Twiggs LB, Wilkinson EJ, et al. (2003). Consensus guidelines for the management of women with cervical intraepithelial neoplasia. *Am J Obs Gynecol* 189: 295–304.

Wu W, Sun M, Zou GM, Chen J. (2006). MicroRNA and cancer: Current status and prospective. *Int J Cancer* 120: 953–60.

Yang L, Li R, Mohr IJ, Clark R, Botchan MR. (1991). Activation of BPV-1 replication in vitro by the transcription factor E2. *Nature* 353: 628–32.

Zhao JL, Rao DS, O'Connell RM, Garcia-Flores Y, Baltimore D. (2013). MicroRNA-146a acts as a guardian of the quality and longevity of hematopoietic stem cells in mice. *Dev Biol Stem Cells Immunol* 190: 1–24.

Zheng ZM, Baker CC. (2006). Papillomavirus genome structure, expression, and post-transcriptional regulation. *Front Biosci* 11: 2286–302.

zur Hausen H. (1996). Papillomavirus infections—A major cause of human cancers. *Biochim Biophys Acta* 1288: F55–78.

zur Hausen H. (1999). Papillomaviruses in human cancers. *Proc Assoc Am Physicians* 111:581–7.

zur Hausen H, de Villiers EM, Gissmann L. (1981). Papillomavirus infections and human genital cancer. *Gynecol Oncol* 12: S124–8.

# 8 Role of High-Risk Human Papillomaviruses in Breast Carcinogenesis

*Ala-Eddin Al Moustafa*

## 8.1 INTRODUCTION

Breast cancer is the most frequently diagnosed malignancy in women worldwide with approximately 1.2 million new cases and 226,000 deaths each year as estimated by the World Health Organization (WHO). The majority of breast cancer related deaths are due to complications from metastatic disease. In view of the limited success of available treatment modalities for breast metastatic cancer, preventive strategies and new targeted therapies need to be developed (Weigelt et al. 2005).

It is well established that high-risk human papillomavirus (HPV) infections are associated with the development of anogenital cancers including those of the cervix (Castellsagué et al. 2006; Smith et al. 2007), while infections by the low-risk HPVs induce only benign genital warts (de Villiers et al. 2004; Bernard 2005). HPV is mainly a sexually transmitted infection (Rylander et al. 1994). The lifetime risk of HPV infection is greater than 50% for sexually active men and women (Sawaya and Smith-McCune 2007), and approximately 5%–30% of those infected carry multiple subtypes (Revzina and Diclemente 2005). In parallel, it has been reported that high-risk HPV infection alone, produced by only one type of these viruses, is not sufficient to induce neoplastic transformation of human normal cervical epithelial cells; the high-risk HPV-infected cells must undergo additional genetic changes. High-risk HPVs are also important risk factors for other human cancers such as colorectal and head and neck (HN) cancer; roughly 80% and 30% of these cancers are positive for high-risk HPVs, respectively (Daling et al. 2004; Venuti et al. 2004; Ragin and Taioli 2007). Moreover, it was observed that the presence of high-risk HPVs serves as a prognostic factor in early-stage cervical, colorectal, and HN cancers and is associated with vascular invasion, lymph node metastases, and tumor size (Begum et al. 2003; Graflund et al. 2004; Zuna et al. 2004; Umudum et al. 2005; Varnai et al. 2006). Nevertheless, during high-risk HPVs infection, E6/E7 oncoproteins are expressed and, as a result, the restraint on cell cycle progression is abolished and normal terminal differentiation is retarded (Sherman et al. 2007). Meanwhile, it was recently reported that E5 oncoprotein deregulates cell cycle progression and therefore cell differentiation through its interaction with epidermal growth factor-receptor 1 (EGF-R1) signaling as well as with other genes, such as Bub1 and Mad2 (Pedroza-Saavedra et al. 2010; Liao et al. 2013).

Several earlier studies, including ours, reported that high-risk HPVs are present in human breast cancers (Liu et al. 2001; de Villiers et al. 2005; Kan et al. 2005; Akil et al. 2008; Antonsson et al. 2011; Glenn et al. 2012); however, controversially, a few studies revealed that HPVs could not be detected in breast cancer and normal mammary tissues (Gopalkrishna et al. 1996; Lindel et al. 2007; Hachana et al. 2010). Moreover, studies that found HPV-positive breast cancer tissues revealed that certain types of high-risk HPV infections are linked to specific geographic locations. Recently, it was pointed out that the E6/E7 oncoproteins of high-risk HPVs convert non-invasive and non-metastatic breast cancer cells to invasive and metastatic forms (Yasmeen et al. 2007b). This chapter discusses the presence and the distribution of HPVs in human breast cancers in several regions of the world and presents a review of the role of E5, E6, and E7 of high-risk HPVs in breast carcinogenesis and metastasis.

## 8.2 HUMAN PAPILLOMAVIRUSES

Human papillomaviruses (HPVs) are small, double-stranded DNA viruses that generally infect cutaneous and mucosal epithelial tissues of the anogenital tract. To date, more than 120 different viral types have been identified, and about one-third of these infect epithelial cells in the genital tract. HPVs are classified as either high-risk or low-risk viruses, with the former type being associated with cancer formation and the latter type not. For example, HPV types 6 and 11 are classified as low-risk types; infection with these types results in the proliferation of epithelial cells and manifests as warts or papillomas on the skin (de Villiers et al. 2004; Bernard 2005). Infections with low-risk types are generally self-limiting and do not lead to malignancy. On the other hand, infections with high-risk HPVs (type 16, 18, 31, 33, 35, 39, 45, 51, 52, 55, 56, 58, 59, 68, 73, 82, and 83) are associated with the development of cervical cancers; more than 96% of these cancers are positive for high-risk HPVs (Castellsagué et al. 2006; Smith et al. 2007).

The HPV genome is about 7.9 kb in size and encodes early (E) and late (L) proteins and noncoding region (NCR) (Figure 8.1). Early are designated as E1, E2, E4, E5, E6, and E7, and late proteins as L1 and L2. In a host cell, the viral DNA is transcribed as a polycistronic mRNA that is cleaved to yield the different viral proteins. The E1 and E2 genes are expressed first upon viral entry into the host cell, and they encode viral DNA replication proteins (Motoyama et al. 2004). The E5 protein along with E1, E2, and E4 are replication proteins that allow the viral DNA (Figure 8.1) to be replicated as an episome in low copy number (Longworth and Laimins 2004a; Doorbar 2005). During the process of differentiation of epithelial cells, the p670 promoter on viral DNA causes increased expression of E1, E2, E4, and E5 proteins, resulting in increased viral DNA amplification. Therefore, E5 is a viral replication protein that helps in replicating the viral episomal DNA (Moody and Laimins 2009).

Early genes are responsible for modulating epithelial cell function so as to favor virus production. For example, their gene products modulate keratinocyte differentiation, promote viral DNA replication and segregation, and inhibit viral clearance by the immune system. All early HPV proteins perform multiple functions and interact with a variety of cellular partners (Tungteakkhun et al. 2008; Moody

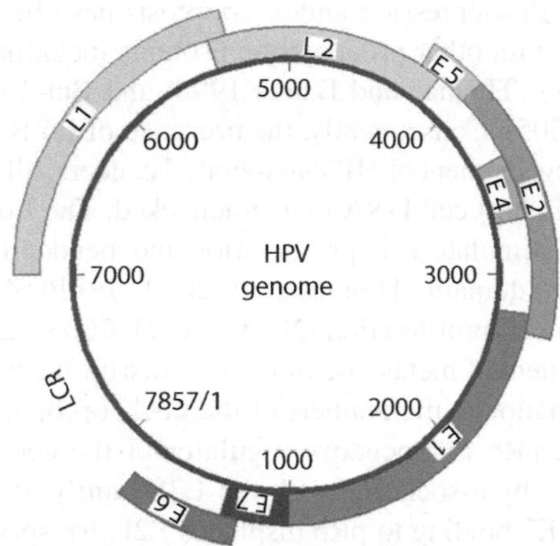


**FIGURE 8.1 (See color insert.)** Schema of HPV genome. The genome can be divided into three segments: ~4000 bp section (E) that encodes the proteins primarily involved in viral DNA replication and oncogenic task in the infected cell, ~3000 bp section (L) that encodes the structural proteins of the virions, and an ~1000 bp noncoding section (LCR) that contains the origin of viral DNA replication, and transcriptional regulatory elements.

and Laimins 2009). These interactions in turn modify the flow through numerous cellular pathways, including apoptosis. The core proteins of the late genes, L1 and L2, together with HPV DNA, participate in the assembly of virus particles. During the normal viral life cycle, the HPV genome exists in host cells as an episome. In rare cases, however, integration of the viral genome into that of the host can occur. Interestingly, this integration is not required for the HPV life cycle and is, in fact, detrimental to the virus because integrated viral sequences do not reproduce. However, integration is closely tied to the development of cancer because most cases of HPV-induced cervical cancer feature an integrated form of the HPV genome (Moody and Laimins 2009; Doorbar 2013). Such integration typically leads to an increase in the expression of HPV early proteins, including the E5, E6, and E7 oncoproteins, and a consequential increase in cellular alteration and the probability of HPV-induced carcinogenesis (Grm et al. 2005; Yuan et al. 2012). Several recent studies demonstrated that E5 oncoprotein can play an important role in cell transformation and consequently carcinogenesis through its interaction with EGF-R1 signaling pathways (MAP kinase and Phosphatidylinositol-3-kinase-Akt) and proapoptotic proteins (Kim et al. 2006; Suprynowicz et al. 2008; Oh et al. 2010).

E6 and E7 are thought to work together in lesions caused by high-risk HPV types such as HPV-6. These two proteins are expressed from bicistronic micro-RNA (mRNA) (Stacey et al. 2000) initiated from the viral early promoter (p97). Both E6 and E7 have functions that stimulate cell cycle progression, and both can associate with regulators of the cell cycle (Münger et al. 2001).

The viral E6 protein complements the role of E7 and is thought to prevent the induction of apoptosis in response to unscheduled S-phase entry mediated by E7 (Ghittoni et al. 2010). Although the association of E6 with p53 and the inactivation of p53-mediated growth suppression and/or apoptosis have been well documented, E6 can also associate with other proapoptotic proteins including Bcl-2 homologous antagonist/killer (Bak) (Thomas and Banks 1998) and Bcl-2-assoicated X protein (Bax) (Magal et al. 2005). Consequently, the presence of E6 is considered a predisposing factor in the development of HPV-associated cancers, allowing the accumulation of chance errors in host cell DNA to go unchecked. The E6 protein of high-risk HPV types can also stimulate cell proliferation independently of E7 through its C-terminal PDZ-ligand domain (Thomas et al. 2002). E6-PDZ binding is sufficient to mediate suprabasal cell proliferation (Nguyen et al. 2003a, 2003b) and may contribute to the development of metastatic tumors by disrupting normal cell adhesion.

The E7 viral association with members of the pocket protein family such as pRb is well characterized; pRb is a negative regulator of the cell cycle that normally prevents S-phase entry by associating with the E2F family of transcription factors (Münger et al. 2001). E7 binding to pRb displaces E2F, irrespective of the presence of external growth factors, and leads to the expression of proteins necessary for DNA replication (Doorbar 2005). E7 can also associate with other proteins involved in cell proliferation, including histone deacetylases (Longworth and Laimins 2004b), components of the AP-1 transcription complex (Antinore et al. 1996), and the cyclin-dependent kinase inhibitors p21 and p27 (Funk et al. 1997). Despite the ability of E7 to stimulate cell proliferation, during productive infection, only a subset of cells in the parabasal layers is mitotically active. The expression of cyclin E is necessary for S-phase entry and is expressed during natural infection as a result of E7 expression and disruption of the E2F/pRb complex. In differentiating epithelial cells, however, the high levels of cyclin-dependent kinase inhibitors (p21cip1 and p27kip1) can lead to the formation of inactive complexes that contain E7, cyclin E/cdk2, and either p21 or p27 (Noya et al. 2001). It appears that, during natural infection, the ability of E7 to stimulate S-phase progression is limited to the subset of differentiated cells that have low levels of p21/p27 or that express high enough levels of E7 to overcome the block to S-phase entry (Feller et al. 2010). These studies demonstrate that E5, E6, and E7 oncoproteins deregulate numerous genes in high-risk-HPV-infected cells that could lead to full cell transformation in cooperation with other oncogenes.

Finally, and to address the role of E6/E7 genes in HPV-associated carcinogenesis in vivo, several transgenic mice have been developed expressing E6/E7 of HPV type 16. For example, Herber et al. (1996) and Song et al. (1999) generated transgenic mice carrying E6 or E7 of HPV type 16 individually and E6/E7 together, in which expression of these genes is directed from the human K14 promoter. These transgenic mice developed skin tumors, in general, and cervical cancer with chronic estrogen administration (Song et al. 1999; Riley et al. 2003). The K14-E6 and K14-E7 mice developed low-grade cervical dysplasia and high-grade cervical carcinomas, respectively. However, the cervical cancers were larger and more extensive and invasive in K14-E6/E7 transgenic mice (Riley et al. 2003). On the other hand and to examine the oncogenic properties of E5 in vivo, K14-E5 transgenic mice were created in which expression of E5 was directed to the basal layer of the stratified squamous epithelia. These mice display epidermal hyperplasia, aberrant differentiation of the epithelium, and are susceptible to spontaneous skin tumors (Genther Williams et al. 2005). E5 was shown to contribute to the promotion and progression stages in skin carcinogenesis similar to what was seen previously in E6 transgenic mice (Song et al. 2000; Maufort et al. 2007). Recently, it was reported that E5 transgenic mice developed severe neoplastic cervical disease after six months of treatment with estrogen (Maufort et al. 2010). These studies clearly show that high-risk HPVs—through their E5, E6, and E7 oncoproteins—play an important role in cervical, oral, as well as skin cancer initiation and progression.

## 8.3 HPVs IN BREAST CANCERS

Earlier studies have demonstrated that approximately 50% of human breast cancers are positive for high-risk HPV subtypes, especially 16, 18, and 33 (Yu et al. 2000; Liu et al. 2001; de Villiers et al. 2005; Kan et al. 2005; Akil et al. 2008; Antonsson et al. 2011; Glenn et al. 2012). This section details some of the important investigations that demonstrated the presence of HPVs in human breast cancers (Table 8.1).

Di Lonardo et al. (1992) was the first group to detect the presence of high-risk HPV type 16 in 29.4% of 40 breast cancer specimens from Italian women and in 17.1% of the lymph nodes containing breast cancer metastases using polymerase chain reaction (PCR). In Norwegian women, Hennig et al. (1999) found that HPV type 16 is present in 19 out of 41 (46%) women who had a history of high-grade cervical intraepithelial neoplasia (CIN III) and breast carcinoma using PCR. Yu et al. (1999, 2000) found high-risk HPV type 33 in 14 cases (43.8%) of invasive breast ductal carcinoma in Chinese and Japanese women, but high-risk HPV type 16 and HPV type 18 DNA were not detected in this investigation. We should note that this was the first report detecting high-risk HPV type 33 in human breast carcinomas. Damin et al. (2004) detected HPVs in 25 (24.75%) breast cancer specimens from Brazilian women, but none of the benign breast tissues were found to be positive in this study.

**TABLE 8.1**
**Representative Studies Confirming HPVs Presence in Human Breast Cancers**

| Country | Sample | Cases | % HPVs+ | References |
|---|---|---|---|---|
| Italy | Paraffin | 40 | 29.4 | Di Lonardo et al. 1992 |
| Norway | Paraffin | 41 | 46 | Hennig et al. 1999 |
| China and Japan | Paraffin | 53 | 34 | Yu et al. 1999, 2000 |
| Brazil | Paraffin | 101 | 24.75 | Damin et al. 2004 |
| Germany | Paraffin | 29 | 86 | de Villiers et al. 2005 |
| Australia | Frozen | 50 | 48 | Kan et al. 2005 |
| Turkey | Frozen | 50 | 74 | Gumus et al. 2006 |
| Canada | Paraffin | 27 | 85.2 | Yasmeen et al. 2007b |
| Syria | Paraffin | 113 | 61.06 | Akil et al. 2008 |
| Mexico | Paraffin | 51 | 29 | de León et al. 2009 |
| Australia | Frozen | 54 | 50 | Antonsson et al. 2011 |
| Iran | Paraffin | 79 | 25.9 | Sigaroodi et al. 2012 |
| Australia | Frozen | 50 | 50 | Glenn et al. 2012 |
| Argentina | Paraffin | 61 | 26 | Pereira Suarez et al. 2013 |

De Villiers et al. (2005) examined the presence of HPVs in areolar and tumor tissues from German female patients with breast cancer. HPVs were present in 25 of 29 breast cancer specimens and in 20 of 29 samples from the corresponding mammary tissues. Multiple HPV types were found in seven cancer tissues and in ten nipple tissue samples. These results reveal the occurrence of HPV in nipple and areolar tissue in patients with breast carcinomas and suggest that HPV may infect breast tissue through the nipple. Kan et al. (2005) examined 50 unselected invasive ductal breast cancer specimens from Australian women using PCR for HPV types 16, 18, or 33. Overall, 24 (48%) of the 50 samples were positive for HPV type 18.

In Greece, Kroupis et al. (2006) used four PCR methods to verify the presence of HPVs in frozen breast cancer tissues; they found that 17 samples out of 107 were positive (15.9%). Fourteen samples were revealed positive for HPV type 16 (67% of all detected HPV types), while three samples tested positive for HPV type 59, two for HPV type 58, and one each for HPV type 73 and HPV type 82. Gumus et al. (2006) tested 50 breast cancer and normal tissue samples from Turkish women for the presence of low-risk HPV type 11 and high-risk HPV types 16, 18, and 33. Thirty-seven malignant breast tissues (74%) revealed positives for HPVs. In addition, 16 normal breast tissue samples (32%) were also shown to be positive. HPV type 18 was detected in 20 of the HPV-positive malignant tissue samples (54.4%) and in nine of the HPV-positive normal tissue samples (56.3%). HPV type 33 was found in 35 (94.6%) of the HPV-positive cancer tissue samples and in 14 (87.5%) of the HPV-positive normal tissues.

Using DNA microarrays, Choi et al. (2007) examined DNA from 154 Korean women patients, including 123 breast cancer samples, 31 intraductal papillomas, and nipple tissue from 27 patients with cancer. HPVs were present in eight breast carcinomas (6.5%) but not in any intraductal papilloma tissues samples. All of the detected HPVs were high risk. There was a slightly increased prevalence of HPV in the papillary carcinomas (11.5%) and invasive ductal carcinomas with adjacent intraductal papillomas (11.8%) compared to the other histological subtypes (3.2–4.3%). Khan et al. (2008) examined samples from 124 Japanese female patients with breast carcinoma using PCR with the SPF10 primers that target the E6 region of HPV types 16, 18, and 33. HPVs were found in 26 (21%) breast carcinomas. The most frequently detected HPV types were 16, 6, 18, and 33, respectively.

In order to determine the role of high-risk HPV infections in human breast cancer cases in Canada, we investigated the presence of high-risk HPV types 16, 18, 31, 33, and 35 in 27 cases of invasive and metastatic and ten cases of in situ breast cancer as well as 18 normal mammary tissues. We found that all of our samples were negative for HPV types 18, 31, 33, and 35. However, HPV type 16 was present in all examined (17 samples) invasive and metastatic breast cancer tissues and in 6 of 10 samples from in situ carcinomas, while being present in only 2 of 18 samples of normal mammary human tissues. We assumed that those in situ breast carcinomas will ultimately progress into invasive carcinomas under the effect of HPV type 16 because they were already of intermediate to high nuclear grade (Yasmeen et al. 2007b). Afterward and to identify the presence of high-risk HPV in human breast cancer in Middle Eastern women, we investigated the incidence of high-risk HPV types 16, 18, 31, 33, and 35 in a cohort of 113 breast cancer samples from Syrian women by PCR analysis using specific primers for their E6 and/or E7 genes and tissue microarray analysis. Our study revealed that 69 (61.06%) of the 113 samples were HPV positive and 24 (34.78%) of these specimens were coinfected with more than one HPV type. Our study revealed that HPV types 16, 18, and 31 were present in only 10, 11, and 8 cancer tissues, respectively. In contrast, 63 and 42 cancer tissues were positive for HPV types 33 and 35, respectively (Akil et al. 2008). Therefore, we concluded that the most frequent high-risk HPVs in breast cancer in Syrian women are HPV types 33 and 35.

Ong et al. (2009) examined the presence of low- and high-risk HPVs in 92 breast cancer samples of Singaporean women by PCR analysis and using a commercially available HPV DNA genotyping chip for detection. They found that 32 of 92 (35%) were positive for HPVs. De León et al. (2009) investigated the presence of HPVs in paraffin-embedded specimens selected from 51 Mexican patients with breast cancer using PCR and sequencing techniques. These findings were compared with 43 cases of nonmalignant breast lesions matched according to patient age and tumor size. They found that 15 (29.4%) of the samples from patients with breast cancer were positive for HPVs; among these positive cases, 10 were positive for HPV type 16, three were positive for HPV type 18, and two were positive for both types. In contrast, all of the samples from the benign tissues were negative for HPVs. Heng et al. (2009) used in situ hybridization (ISH) and standard PCR to assess the presence of high-risk HPV in the cells of breast cancer tissues from Australia and in cell lines and found that the oncogenic characteristics of HPV-associated breast cancer are similar to HPV-associated cervical cancer, such as HPV types 16 and 18.

Aguayo et al. (2011) analyzed 55 breast cancer samples from Chile using conventional PCR for a 65 bp fragment of L1 region. Their study revealed that only four of 46 (8.7%) were positive for high-risk HPVs. Antonsson et al. (2011) examined the presence of high-risk HPVs in 54 fresh frozen breast cancer samples from Australian women using PCR and ISH and reported that 27 of 54 (50%) cases were positive for high-risk HPVs with sequence analysis, indicating all cases were positive for HPV type 18.

Sigaroodi et al. (2012) explored the presence of HPVs in a cohort of 130 breast tissue samples (79 cancers and 51 noncancer tissues) from Iranian women by PCR analysis. They found that 15 of 79 (25.9%) cancer samples were positive for HPVs, especially high-risk types 16 and 18. In contrast, their study revealed that only one case of 51 (2.4%) of noncancer samples was positive for HPV type 124. Glenn et al. (2012) examined the presence of HPVs by PCR in 50 fresh, frozen, unselected, invasive breast cancer samples from Australian women using primers for E6 and E7 of HPV types 16, 18, and 33. They found that only HPV type 18 was present in 25 of 50 (50%) of the cases.

Finally, a very recent study by Pereira Suarez et al. (2013) revealed that HPVs were present in 16 of 61 (26%) breast cancer samples from Argentina. This study revealed that the most frequent HPVs found in Argentinean women were low-risk HPV type 11 followed by high-risk HPV type 16. In this investigation, the authors used PCR and RT-PCR with MY09/MY11 primers as well as E6 and E7 primers, respectively. By using different methods of HPV detection, these studies provided evidence that high-risk HPVs are present in human breast cancers. Meanwhile, all these studies

**TABLE 8.2**
**Representative Studies Not Confirming HPV Presence in Human Breast Cancers**

| Country | Sample | Cases | References |
|---|---|---|---|
| USA | Paraffin | 43 | Bratthauer et al. 1992 |
| UK | Paraffin | 80 | Wrede et al. 1992 |
| India | Paraffin | 30 | Gopalkrishna et al. 1996 |
| Switzerland | Paraffin | 81 | Lindel et al. 2007 |
| France | Fresh | 50 | de Cremoux et al. 2008 |
| Tunisia | Paraffin | 123 | Hachana et al. 2010 |
| India | Biopsies | 228 | Hedau et al. 2011 |

show clearly that specific types of high-risk HPV infection in breast tissues are related to specific geographic locations (Table 8.1). However, as mentioned earlier, some studies did not support the theory of HPV presence in breast cancer and normal tissues (Table 8.2).

This idea has been also endorsed by several other investigations. Bratthauer et al. (1992) examined the presence of low-risk HPV types 6 and 11 and high-risk HPV types 16 and 18 by Southern blot in 15 papillomas, 15 papillary carcinomas, and 13 infiltrating ductal breast carcinomas in U.S. patients but could not detect any presence of these viruses in the samples examined. Also, in a series of 80 breast carcinomas from the United Kingdom, Wrede et al. (1992) were unable to detect the presence of any HPVs. Gopalkrishna et al. (1996) analyzed fine-needle aspirate cell samples from 26 patients with breast cancer and four breast cancer biopsies from India for the presence of HPV types 16 and 18 using PCR and Southern blot, but no positive results were obtained.

Lindel et al. (2007) investigated the presence of HPVs in 81 breast cancer tissues from Switzerland using PCR with the SPF1/2 primers that targeted a conserved domain in L1 and covered approximately 40 different low-risk and high-risk types. They found that all of the samples were negative for HPVs. de Cremoux et al. (2008) also did not find any HPVs in 50 breast carcinoma samples from French women. Using primers for L1, E1, E6, and E7, Hachana et al. (2010) examined a cohort of 123 breast cancer tissues from Tunisian women by PCR for the presence of HPVs but failed to detect high- or low-risk HPVs in any sample. Finally, Hedau et al. (2011) investigated the presence of HPVs, and especially HPV types 16 and 18, by PCR in 228 biopsies from Indian breast cancer patients but also failed to find HPVs in any sample investigated.

However, the above studies notwithstanding, many of the investigations (including ours) did confirm the presence of HPVs (especially the high-risk group) in human breast cancer compared with normal mammary tissues (Table 8.1). Two recent meta-analysis studies confirmed this fact (Li et al. 2011; Simões et al. 2012); thus, we believe that high-risk HPVs are present and can have important roles in the initiation and progression of human breast cancers through E5, E6, and E7 oncoproteins of these viruses.

## 8.4 ROLE OF E5, E6, AND E7 ONCOPROTEINS IN BREAST CARCINOGENESIS AND METASTASIS

As described here, E5 of HPV type 16 manipulates multiple cellular signaling pathways in order to create a transformed phenotype, which forms the basis for further progression of human cervical cancer into a malignant and metastatic form (Suprynowicz et al. 2008; Oh et al. 2010). Thus, targeting E5 oncoprotein could help to control tumorigenesis at an early stage because most of the functions of E5 occur during early stages of the viral life cycle. However, there are no studies regarding the role of the E5 oncoprotein of high-risk HPVs in human breast cancer (Figure 8.2).

On the other hand, there are only few studies that explore the role of E6/E7 oncoproteins in breast carcinogenesis and metastasis; these were principally performed by our group. Meanwhile, it is important to mention that the presence of high-risk HPV in human breast cancer (especially types 16 and 18) is correlated with invasive carcinomas (Kroupis et al. 2006; Yasmeen et al. 2007b; Glenn et al. 2012). To assess the role of high-risk HPVs in human breast cancer metastasis, we examined the effect of E6/E7 of HPV type 16 on cell invasive and metastatic abilities in two noninvasive breast cancer cell lines, MCF7 and BT20. We transduced MCF7 and BT20 cells with E6/E7 of HPV type 16 using a recombinant retroviral system (Halbert et al. 1991). In our study, we utilized polyclonal populations of MCF7-E6/E7 and BT20-E6/E7 cells; first, we confirmed that these cell lines express E6/E7 in

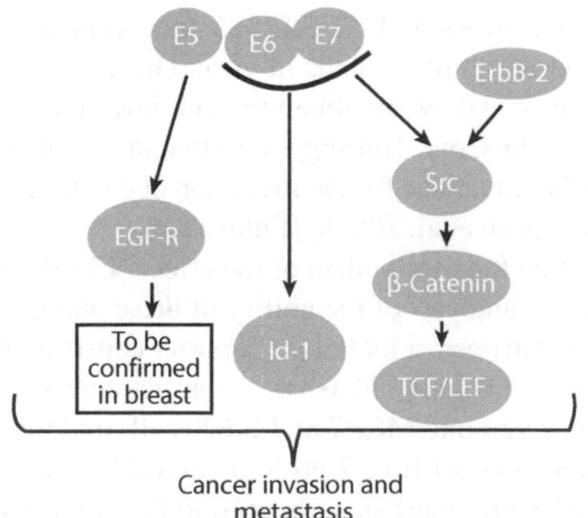


**FIGURE 8.2** E5/E6/E7 of high-risk HPV/oncogene interactions in human breast cancer cells. E5, E6, and E7 oncoproteins can enhance breast cancer progression through three possible pathways: (1) the activation of EGF-R by E5 oncoprotein, as it was demonstrated in cervical cancer (Kim et al. 2006); however, this pathway needs to be confirmed in breast cancer, (2) the upregulation of Id-1 via the activation of his promoter by E6/E7 oncoproteins, and (3) the cooperation between E6/E7 oncoproteins and ErbB-2 or other oncogenes which can phosphorylate β-catenin via Src activation and subsequently provokes a dissociation of E-cadherin/catenin complex. Afterwards, the free β-catenin enters the cell nucleus and activates Tcf/Lef transcription factors which can deregulate several oncogenes necessary to incite cancer progression.

comparison with their wild-type counterparts by RT-PCR analysis. Furthermore, we investigated the invasive and metastatic abilities of MCF7-E6/E7 and BT20-E6/E7 cells and their wild-type cell lines in vitro and in vivo, respectively. We demonstrated that E6/E7 of HPV type 16 induces cell invasive and metastatic abilities of MCF7 and BT20 in comparison with the wild-type cells. Consequently, MCF7-E6/E7 and BT20-E6/E7 cells exhibited a significant lung metastatic activity compared to the parental population cell lines (Yasmeen et al. 2007b).

On the other hand, it was reported that inhibitor of DNA binding-1 (Id-1) regulates cell invasion and metastasis of human breast cancer cells (Desprez et al. 1998; Lin et al. 2000; Fong et al. 2003). To determine if Id-1 is a target of cell invasion and metastasis induced by E6/E7 of HPV type 16 in human breast cancer cells, we examined the expression of Id-1 in MCF7, MCF7-E6/E7, BT20, and BT20-E6/E7 cell lines. We found that Id-1 is upregulated in MCF7 and BT20 cells expressing E6/E7 by comparison with their wild-type counterparts (Yasmeen et al. 2007b). Minn et al. (2005) identified a subset of genes that mediate lung metastasis of human breast cancer; Id-1 was revealed as one of these important genes. In our study, we noted that human breast cancer cells expressing E6/E7 display a major lung metastatic activity compared with their wild-type cells in vivo (Yasmeen et al. 2007b). Moreover, we have reported that E6/E7 of HPV type 16 induces cellular transformation of mouse normal embryonic fibroblast (NEF) cells (Al Moustafa et al. 2004b), which is accompanied by an overexpression of Id-1 in NEF-E6/E7-transformed cells. To identify the exact role of Id-1 in the invasive and metastatic abilities induced by E6/E7 of HPV type 16, we examined the effect of E6/E7 in mouse NEF and knockout Id-1 (Id-1$^{-/-}$) cells (Lyden et al. 1999). We found that E6/E7 provokes cell invasion of NEF-E6/E7-transformed cells similar to MCF7 and BT20 cell lines but not in Id-1$^{-/-}$ cells in vitro. Furthermore, using Id-1 antisense retroviruses (Fong et al. 2003), we inhibited the cell invasion ability of MCF7-E6/E7 and BT20-E6/E7 cells. Thus, our data suggested that Id-1 is the downstream target of E6/E7 of high-risk HPV infections for the induction of cell invasion and metastasis in breast cancer cells (Yasmeen et al. 2007b) (Figure 8.2).

Next, we reasoned that the mechanism of the effect of E6/E7 on Id-1 upregulation may be deduced from an analysis of the ability of these oncoproteins to activate the Id-1 promoter. For this purpose, a luciferase reporter construct bearing a full-length 2.2 kb long Id-1 promoter region (Id-1sbsluc) and its empty vector control (Singh et al. 2002) were transfected into MCF7 and BT20 cells and those expressing E6/E7. We found that the expression of E6/E7 proteins in MCF7 and BT20 cells markedly activates full-length Id-1 promoter as compared to the control vector. Furthermore, we used deleted mutants such as sbsΔ-3 (deletion of -272 to -145, which contains boxes 1, 2, and 3) and 5′del-6-δ2 (consisting of -272 to -145 region with deletion of box 2, which contains SP1 and NF1 sites) of an Id-1 promoter region previously shown to be crucial in the Id-1 promoter regulation in invasive cells (Singh et al. 2002). Our data revealed that the luciferase activity of sbsΔ-3 mutant is completely abolished in comparison with the full-length promoter in both cell lines, MCF7 and BT20. Moreover, the activity of the 5′del-6-δ2 mutant is reduced dramatically in BT20-E6/E7 cells and, to a lesser extent, in MCF7-E6/E7 cells as compared to 5′del-6 mutant (region -272 to -145 containing three boxes). These results indicated for the first time that SP1 and NF1 sites of Id-1 promoter play an important role in the regulation of Id-1 gene by E6/E7 of HPV type 16. In addition, these data provided a new basis for understanding the mechanisms of high-risk HPV infections and their relation to human breast cancer progression (Yasmeen et al. 2007b).

Alternatively, and in order to study the cooperation effect of E6/E7 oncoproteins of high-risk HPVs with other proto-oncogenes in breast tumorigenesis (Figure 8.2), we generated double transgenic mice carrying E6/E7 oncoproteins of high-risk HPV type 16 and activated *Neu* (rat ErbB-2) under human keratin 14 (K14) and mouse mammary tumor virus (MMTV) promoters, respectively. First, we created E6/E7 transgenic mice by crossbreeding K14-E6 mice with K14-E7 mice (Song et al. 1999; Riley et al. 2003), and then we crossed these mice with MMTV/activated *Neu* (rat ErbB-2) (Guy et al. 1992). The MMTV/ErbB-2 female mice developed in situ mammary cancers usually within one year, which could subsequently evolve into an invasive form. On the contrary, female K14-E6/E7 mice did not form any breast tumors within the same time period (Yasmeen et al. 2007c). Furthermore, the E6 or E7 of HPV type 16 individually and E6/E7 transgenic mice developed skin tumors and cervical cancer following chronic estrogen administration (Riley et al. 2003). However, the E6/E7/ErbB-2 double transgenic mice developed large and extensive breast tumors within a six-month period without any treatment. The histological analysis of E6/E7/ErbB-2 transgenic mouse tumors revealed that they are invasive high-grade breast cancers, while the breast tissue from E6/E7 or ErbB-2 transgenic mice showed normal mammary phenotype and in situ breast cancer, respectively (Yasmeen et al. 2007c).

To assess the outcome of E6/E7/ErbB-2 cooperation in human breast carcinogenesis, we investigated the effect of E6/E7 of HPV type 16 and ErbB-2 on the BT20 breast cancer cell line. We found that E6/E7/ErbB-2 cooperates in the BT20 cell line to induce large colony formation and cell migration in comparison to E6/E7, ErbB-2, and wild-type cells. Thus, we demonstrated, for the first time, that E6/E7 oncoproteins of high-risk HPV type 16 cooperate with ErbB-2 overexpression in breast carcinogenesis in vivo and in vitro. On the other hand, we reported that D-type cyclins are downstream targets of the nuclear β-catenin provoked by E6/E7/ErbB-2 cooperation in human normal oral epithelial-transformed cells (Al Moustafa et al. 2004a; Yasmeen et al. 2007a). Moreover, we noted that crossbreeding of E6/E7/ErbB-2 double transgenic mice with cyclin D1 knockout mice dramatically reduced breast tumor progression (unpublished data). Therefore, we reasoned that the mechanism of E6/E7/ErbB-2 cooperation in breast tumorigenesis may be deduced from the analysis of β-catenin expression patterns; accordingly, we examined the E6/E7/ErbB-2 double transgenic tumors for β-catenin expression patterns using immunohistochemistry. It was found that β-catenin is translocated from the undercoat membrane to the nucleus in E6/E7/ErbB-2 mouse tumors and BT20-E6/E7/ErbB-2 cells in comparison with their respective controls. Regarding the mechanism of β-catenin translocation, it was hypothesized that E6/E7/ErbB-2 cooperation provokes a dissociation of E-cadherin/catenin complex by tyrosine phosphorylation of β-catenin through pp60(c-Src) kinase phosphorylation. Subsequently, the free β-catenin enters the nucleus and modulates cell transcription via its association with Tcf/Lef transcription factors.

Using SKI-606, an Src family kinase inhibitor, it has been demonstrated that β-catenin is physically associated to activated pp60 (c-Src) kinase and constitutively phosphorylated on the tyrosine residue in human colorectal cancer cells (Coluccia et al. 2006). To assess this possibility in E6/E7/ErbB-2 cooperation, we examined the effect of SKI-606 on β-catenin regulation patterns in BT20 breast cancer cells and those expressing E6/E7, ErbB-2, and E6/E7/ErbB-2. We found that SKI-606 inhibits β-catenin phosphorylation and consequently its translocation to the nucleus in these cells (relocalizing it to undercoat membrane); thus, SKI-606 blocks cell migration through the conversion of β-catenin's role from transcriptional regulator to a cell–cell adhesion function in E6/E7/ErbB-2 breast cancer cells. This finding suggested that β-catenin phosphorylation and translocation induced by E6/E7/ErbB-2 cooperation occur via c-Src activation (Figure 8.2). Alternatively, Woods Ignatoski et al. (2005) revealed that E6/E7 oncoproteins of HPV type 16 can cooperate with ErbB-2 to induce cellular transformation of the human mammary MCF-10A immortalized cell line, which is acquired through EGF-R activation. However, we were able to immortalize human normal mammary epithelial cells by E6/E7 of HPV type 16 but unable to transform them under the effect of E6/E7/ErbB-2 cooperation (unpublished data). Finally, Shai et al. (2008) reported that E6/E7 of HPV type 16 cooperates with estrogen to induce invasive mammary tumors in p53-deficient mice.

Based on these few studies, it is evident that the presence of E5, E6, and E7 oncoproteins of high-risk HPVs could enhance the invasion and metastatic ability of breast cancer through its interaction with other oncogenes on the promoter's level, such as Id-1 in our cell model. Additionally, E5, E6, and E7 oncoproteins could also cooperate with other genes, such as ErbB-2, to initiate breast cancer progression via the β-catenin/Tcf/Lef transcription factor pathway. Finally, high-risk HPVs could initiate and enhance breast cancer development and progression through their association with estrogen in the absence of p53.

## 8.5 CONCLUSION

There has been substantial evidence presented in this chapter that suggests that high-risk HPVs are present and play an important role in breast carcinogenesis and metastasis through the interaction of E5, E6, and E7 oncoproteins with other genes such as EGF-R, Id-1, and Erb-2 as well as the absence of p53. However, we believe that further studies are required to elucidate the exact role of high-risk HPV in human breast carcinogenesis. In parallel, the development of new in vitro and in vivo models, such as cell lines and transgenic animal models, is necessary to identify the function of E5, E7, and the E7 oncoprotein in the initiation and progression of human breast cancers, which can lead to generate new targets for treating HPV-positive cancers including breast.

Alternatively and with regard to HPV infection prevention, we assume that currently available HPV vaccines, which is against only two high-risk HPV types (16 and 18) (Brown et al. 2009; Paavonen et al. 2009), might reduce the development of breast cancer and its metastasis in women (Al Moustafa et al. 2012), especially when supported by long-term follow-up of women who receive this vaccine at a young age. Nevertheless, we firmly believe that prevention of high-risk HPV infection by the new generation of HPV vaccine, which works against the seven most frequent high-risk HPVs worldwide (Merck & Co. Inc., NCT00543543), is likely to prevent many HPV-associated cancers (including breast) and their progression to the invasive form that is responsible for the majority of cancer-related deaths.

## ACKNOWLEDGMENTS

The author thanks Mrs. A. Kassab for reading this chapter and gratefully acknowledges the support received for his laboratory from the Canadian Institutes for Health Research (CIHR), the Cancer Research Society of Canada, the National Colorectal Cancer Campaign, and the Fonds de la Recherche en Santé du Québec (FRSQ-Réseau du Cancer).

## REFERENCES

Aguayo F, Khan N, Koriyama C, et al. (2011). Human papillomavirus and Epstein-Barr virus infections in breast cancer from Chile. *Infect Agent Cancer* 6: 7.

Akil N, Yasmeen A, Kassab A, et al. (2008). High-risk human papillomavirus infections in breast cancer in Syrian women and their association with Id-1 expression: At issue micro array study. *Br J Cancer* 99: 404–7.

Al Moustafa AE, Foulkes WD, Benlimame N, et al. (2004a). E6/E7 proteins of HPV type 16 and ErbB-2 cooperate to induce neoplastic transformation of primary normal oral epithelial cells. *Oncogene* 23: 350–8.

Al Moustafa AE, Foulkes WD, Wong A, et al. (2004b). Cyclin D1 is essential for neoplastic transformation induced by both E6/E7 and E6/E7/ErbB-2 cooperation in normal cells. *Oncogene* 23: 5252–6.

Al Moustafa AE, Yasmeen A, Ghabreau L, Akil N. (2012). Does the Syrian population have to wait for the new generation of human papillomaviruses vaccine? Hum Vaccin *Immunother* 8: 1867–8.

Antinore MJ, Birrer MJ, Patel D, et al. (1996). The human papillomavirus type 16 E7 gene product interacts with and trans-activates the AP1 family of transcription factors. *EMBO* 15: 1950–60.

Antonsson A, Spurr TP, Chen AC, et al. (2011). High prevalence of human papillomaviruses in fresh frozen breast cancer samples. *J Med Virol* 83: 2157–63.

Begum S, Gillison ML, Ansari-Lari MA, et al. (2003). Detection of human papillomavirus in cervical lymph nodes: A highly effective strategy for localizing site of tumor origin. *Clin Cancer Res* 9: 6469–75.

Bernard HU. (2005). The clinical importance of the nomenclature, evolution and taxonomy of human papillomaviruses. *J Clin Virol* 32 Suppl 1: S1–6.

Bratthauer GL, Tavassoli FA, O'Leary TJ. (1992). Etiology of breast carcinoma: No apparent role for papillomavirus types 6/11/16/18. *Pathol Res Pract* 188: 384–6.

Brown DR, Kjaer SK, Sigurdsson K, et al. (2009). The impact of quadrivalent human papillomavirus (HPV; types 6, 11, 16, and 18) L1 virus-like particle vaccine on infection and disease due to oncogenic nonvaccine HPV types in generally HPV-naive women aged 16–26 years. *J Infect Dis* 199: 926–35.

Castellsagué X, Díaz M, de Sanjosé S, et al., International Agency for Research on Cancer Multicenter Cervical Cancer Study Group. (2006). Worldwide human papillomavirus etiology of cervical adenocarcinoma and its cofactors: Implications for screening and prevention. *J Natl Cancer Inst* 98: 303–15.

Choi YL, Cho EY, Kim JH, et al. (2007). Detection of human papillomavirus DNA by DNA chip in breast carcinomas of Korean women. *Tumour Biol* 28: 327–32.

Coluccia AM, Benati D, Dekhil H, et al. (2006). SKI-606 decreases growth and motility of colorectal cancer cells by preventing pp60(c-Src)-dependent tyrosine phosphorylation of beta-catenin and its nuclear signaling. *Cancer Res* 66: 2279–86.

Daling JR, Madeleine MM, Johnson LG, et al. (2004). Human papillomavirus, smoking, and sexual practices in the etiology of anal cancer. *Cancer* 101: 270–80.

Damin AP, Karam R, Zettler CG, et al. (2004). Evidence for an association of human papillomavirus and breast carcinomas. *Breast Cancer Res Treat* 84: 131–7.

de Cremoux P, Thioux M, Lebigot I, et al. (2008). No evidence of human papillomavirus DNA sequences in invasive breast carcinoma. *Breast Cancer Res Treat* 109: 55–8.

de León DC, Montiel DP, Nemcova J, et al. (2009). Human papillomavirus (HPV) in breast tumors: Prevalence in a group of Mexican patients. *BMC Cancer* 9: 26.

de Villiers EM, Fauquet C, Broker TR, et al. (2004). Classification of papillomaviruses. *Virology* 324: 17–27.

de Villiers EM, Sandstrom RE, zur Hausen H, Buck CE. (2005). Presence of papillomavirus sequences in condylomatous lesions of the mamillae and in invasive carcinoma of the breast. *Breast Cancer Res* 7: R1–11.

Desprez PY, Lin CQ, Thomasset N, Sympson CJ, et al. (1998). A novel pathway for mammary epithelial cell invasion induced by the helix-loop-helix protein Id-1. *Mol Cell Biol* 18: 4577–88.

Di Lonardo A, Venuti A, Marcante ML. (1992). Human papillomavirus in breast cancer. *Breast Cancer Res Treat* 21: 95–100.

Doorbar J. (2005). The papillomavirus lifecycle. *J Clin Virol* 32: S7–15.

Doorbar J. (2013). Latent papillomavirus infections and their regulation. *Curr Opin Virol* 3: 416–21.

Feller L, Wood NH, Khammissa RA, Lemmer J. (2010). Human papillomavirus-mediated carcinogenesis and HPV-associated oral and oropharyngeal squamous cell carcinoma. Part1: Human papillomavirus-mediated carcinogenesis. *Head Face Med* 6: 14.

Fong S, Itahana Y, Sumida T, et al. (2003). Id-1 as a molecular target in therapy for breast cancer cell invasion and metastasis. *Proc Natl Acad Sci U S A* 100: 13543–8.

Funk JO, Waga S, Harry JB, et al. (1997). Inhibition of CDK activity and PCNA-dependent DNA replication by p21 is blocked by interaction with the HPV-16 E7 oncoprotein. *Genes Dev* 11: 2090–100.

Genther Williams SM, Disbrow GL, Schlegel R, et al. (2005). Requirement of epidermal growth factor receptor for hyperplasia induced by E5, a high-risk human papillomavirus oncogene. *Cancer Res* 65: 6534–42.

Ghittoni R, Accardi R, Hasan U, et al. (2010). The biological properties of E6 and E7 oncoproteins from human papillomaviruses. *Virus Genes* 40: 1–13.

Glenn WK, Heng B, Delprado W, et al. (2012). Epstein-Barr virus, human papillomavirus and mouse mammary tumour virus as multiple viruses in breast cancer. *PLoS One* 7: e48788.

Gopalkrishna V, Singh UR, Sodhani P, et al. (1996). Absence of human papillomavirus DNA in breast cancer as revealed by polymerase chain reaction. *Breast Cancer Res Treat* 39: 197–202.

Graflund M, Sorbe B, Sigurdardóttir S, Karlsson M. (2004). HPV-DNA, vascular space invasion, and their impact on the clinical outcome in early-stage cervical carcinomas. *Int J Gynecol Cancer* 14: 896–902.

Grm HS, Massimi P, Gammoh N, Banks L. (2005). Crosstalk between the human papillomavirus E2 transcriptional activator and the E6 oncoprotein. *Oncogene* 24: 5149–64.

Gumus M, Yumuk PF, Salepci T, et al. (2006). HPV DNA frequency and subset analysis in human breast cancer patients' normal and tumoral tissue samples. *J Exp Clin Cancer Res* 25: 515–21.

Guy CT, Webster MA, Schaller M, et al. (1992). Expression of the neu protooncogene in the mammary epithelium of transgenic mice induces metastatic disease. *Proc Natl Acad Sci U S A* 89: 10578–82.

Hachana M, Ziadi S, Amara K, et al. (2010). No evidence of human papillomavirus DNA in breast carcinoma in Tunisian patients. *Breast* 19: 541–4.

Halbert CL, Demers GW, Galloway DA. (1991). The E7 gene of human papillomavirus type 16 is sufficient for immortalization of human epithelial cells. *J Virol* 65: 473–8.

Hedau S, Kumar U, Hussain S, et al. (2011). Breast cancer and human papillomavirus infection: No evidence of HPV etiology of breast cancer in Indian women. *BMC Cancer* 11: 27.

Heng B, Glenn WK, Ye Y, et al. (2009). Human papillomavirus is associated with breast cancer. *Br J Cancer* 101: 1345–50.

Hennig EM, Suo Z, Thoresen S, et al. (1999). Human papillomavirus 16 in breast cancer of women treated for high grade cervical intraepithelial neoplasia (CIN III). *Breast Cancer Res Treat* 53: 121–35.

Herber R, Liem A, Pitot H, Lambert PF. (1996). Squamous epithelial hyperplasia and carcinoma in mice transgenic for the human papillomavirus type 16 E7 oncogene. *J Virol* 70: 1873–81.

Kan CY, Iacopetta BJ, Lawson JS, Whitaker NJ. (2005). Identification of human papillomavirus DNA gene sequences in human breast cancer. *Br J Cancer* 93: 946–8.

Khan NA, Castillo A, Koriyama C, et al. (2008). Human papillomavirus detected in female breast carcinomas in Japan. *Br J Cancer* 99: 408–14.

Kim SH, Juhnn YS, Kang S, et al. (2006). Human papillomavirus 16 E5 up-regulates the expression of vascular endothelial growth factor through the activation of epidermal growth factor receptor, MEK/ERK1,2 and PI3K/Akt. *Cell Mol Life Sci* 63: 930–8.

Kroupis C, Markou A, Vourlidis N, et al. (2006). Presence of high-risk human papillomavirus sequences in breast cancer tissues and association with histopathological characteristics. *Clin Biochem* 39: 727–31.

Li N, Bi X, Zhang Y, et al. (2011). Human papillomavirus infection and sporadic breast carcinoma risk: A meta-analysis. *Breast Cancer Res Treat* 126: 515–20.

Liao S, Deng D, Hu X, et al. (2013). HPV16/18E5, a promising candidate for cervical cancer vaccines, affects SCPs, cell proliferation and cell cycle, and forms a potential network with E6 and E7. *Int J Mol Med* 31: 120–8.

Lin CQ, Singh J, Murata K, et al. (2000). A role for Id-1 in the aggressive phenotype and steroid hormone response of human breast cancer cells. *Cancer Res* 60: 1332–40.

Lindel K, Forster A, Altermatt HJ, et al. (2007). Breast cancer and human papillomavirus (HPV) infection: No evidence of a viral etiology in a group of Swiss women. *Breast* 16: 172–7.

Liu Y, Klimberg VS, Andrews NR, et al. (2001). Human papillomavirus DNA is present in a subset of unselected breast cancers. *J Hum Virol* 4: 329–34.

Longworth MS, Laimins LA. (2004a). Pathogenes is of human papillomaviruses in differentiating epithelia. *Microbiol Mol Biol Rev* 68(2): 362–72.

Longworth MS, Laimins LA. (2004b). The binding of histone deacetylases and the integrity of zinc finger-like motifs of the E7 protein are essential for the life cycle of human papillomavirus type 31. *J Virol* 78: 3533–41.

Lyden D, Young AZ, Zagzag D, et al. (1999). Id1 and Id3 are required for neurogenesis, angiogenesis and vascularization of tumour xenografts. *Nature* 401: 670–7.

Magal SS, Jackman A, Ish-Shalom S, et al. (2005). Downregulation of Bax mRNA expression and protein stability by the E6 protein of human papillomavirus 16. *J Gen Virol* 86: 611–21.

Maufort JP, Shai A, Pitot HC, Lambert PF. (2010). A role for HPV 16 E5 in cervical carcinogenesis. *Cancer Res* 70: 2924–31.

Maufort JP, Williams SM, Pitot HC, Lambert PF. (2007). Human papillomavirus 16 E5 oncogene contributes to two stages of skin carcinogenesis. *Cancer Res* 67: 6106–12.

Minn AJ, Gupta GP, Siegel PM, et al. (2005). Genes that mediate breast cancer metastasis to lung. *Nature* 436: 518–24.

Moody CA, Laimins LA. (2009). Human papillomaviruses activate the ATMDNA damage pathway for viral genome amplification upon differentiation. *PLoS Pathog* 5: e1000605.

Motoyama S, Ladines-Llave CA, Luis Villanueva S, Maruo T. (2004). The role of human papilloma virus in the molecular biology of cervical carcinogenesis. *Kobe J Med Sci* 50: 9–19.

Münger K, Basile JR, Duensing S, et al. (2001). Biological activities and molecular targets of the human papillomavirus E7 oncoprotein. *Oncogene* 20: 7888–98.

Nguyen ML, Nguyen MM, Lee D, et al. (2003a). The PDZ ligand domain of the human papillomavirus type 16 E6 protein is required for E6's induction of epithelial hyperplasia in vivo. *J Virol* 77: 6957–64.

Nguyen MM, Nguyen ML, Caruana G, et al. (2003b). Requirement of PDZ-containing proteins for cell cycle regulation and differentiation in the mouse lens epithelium. *Mol Cell Biol* 23: 8970–81.

Noya F, Chien WM, Broker TR, Chow LT. (2001). p21cip1 degradation in differentiated keratinocytes is abrogated by costabilization with cyclin E induced by human papillomavirus E7. *J Virol* 75: 6121–34.

Oh JM, Kim SH, Cho EA, et al. (2010). Human papillomavirus type 16 E5 protein inhibits hydrogen-peroxide-induced apoptosis by stimulating ubiquitin-proteasome-mediated degradation of Bax in human cervical cancer cells. *Carcinogenesis* 31: 402–10.

Ong K, Koay ES, Putti TC. (2009). Detection of cutaneous HPV types 4 and 24 DNA sequences in breast carcinoma in Singaporean women of Asian ancestry. *Pathology* 41: 436–42.

Paavonen J, Naud P, Salmerón J, et al. (2009). Efficacy of human papillomavirus (HPV)-16/18 AS04-adjuvanted vaccine against cervical infection and precancer caused by oncogenic HPV types (PATRICIA): Final analysis of a double-blind, randomised study in young women. *Lancet* 374: 301–14.

Pedroza-Saavedra A, Lam EW, Esquivel-Guadarrama F, Gutierrez-Xicotencatl L. (2010). The human papillomavirus type 16 E5 oncoprotein synergizes with EGF-receptor signaling to enhance cell cycle progression and the down-regulation of p27(Kip1). *Virology* 400: 44–52.

Pereira Suarez AL, Lorenzetti MA, Gonzalez Lucano R, et al. (2013). Presence of human papilloma virus in a series of breast carcinoma from Argentina. *PLoS One* 8: e61613.

Ragin CC, Taioli E. (2007). Survival of squamous cell carcinoma of the head and neck in relation to human papillomavirus infection: Review and meta-analysis. *Int J Cancer* 121: 1813–20.

Revzina NV, Diclemente RJ. (2005). Prevalence and incidence of human papillomavirus infection in women in the USA: A systematic review. *Int J STD AIDS* 16: 528–37.

Riley RR, Duensing S, Brake T, et al. (2003). Dissection of human papillomavirus E6 and E7 function in transgenic mouse models of cervical carcinogenesis. *Cancer Res* 63: 4862–71.

Rylander E, Ruusuvaara L, Almströmer MW, et al. (1994). The absence of vaginal human papillomavirus 16 DNA in women who have not experienced sexual intercourse. *Obstet Gynecol* 83: 735–7.

Sawaya GF, Smith-McCune K. (2007). HPV vaccination—more answers, more questions. *N Engl J Med* 356: 1991–3.

Shai A, Pitot HC, Lambert PF. (2008). p53 Loss synergizes with estrogen and papillomaviral oncogenes to induce cervical and breast cancers. *Cancer Res* 68: 2622–31.

Sherman ME, Schiffman MH, Lorincz AT, et al. (2007). Cervical specimens collected in liquid buffer are suitable for both cytologic screening and ancillary human papillomavirus testing. *Cancer* 81: 89–97.

Sigaroodi A, Nadji SA, Naghshvar F, et al. (2012). Human papillomavirus is associated with breast cancer in the north part of Iran. *ScientificWorldJournal* 2012: 837191.

Simões PW, Medeiros LR, Simões Pires PD, et al. (2012). Prevalence of human papillomavirus in breast cancer: A systematic review. *Int J Gynecol Cancer* 22: 343–7.

Singh J, Murata K, Itahana Y, Desprez PY. (2002). Constitutive expression of the Id-1 promoter in human metastatic breast cancer cells is linked with the loss of NF-1/Rb/HDAC-1 transcription repressor complex. *Oncogene* 21: 1812–22.

Smith JS, Lindsay L, Hoots B, et al. (2007). Human papillomavirus type distribution in invasive cervical cancer and high-grade cervical lesions: A meta-analysis update. *Int J Cancer* 121: 621–32.

Song S, Liem A, Miller JA, Lambert PF. (2000). Human papillomavirus types 16 E6 and E7 contribute differently to carcinogenesis. *Virology* 267: 141–50.

Song S, Pitot HC, Lambert PF. (1999). The human papillomavirus type 16 E6 gene alone is sufficient to induce carcinomas in transgenic animals. *J Virol* 73: 5887–93.

Stacey SN, Jordan D, Williamson AJ, et al. (2000). Leaky scanning is the predominant mechanism for translation of human papillomavirus type 16 E7 oncoprotein from E6/E7 bicistronic mRNA. *J Virol* 74: 7284–97.

Suprynowicz FA, Disbrow GL, Krawczyk E, et al. (2008). HPV-16 E5 oncoprotein upregulates lipid raft components caveolin-1 and ganglioside GM1 at the plasma membrane of cervical cells. *Oncogene* 27: 1071–8.

Thomas M, Banks L. (1998). Inhibition of bak-induced apoptosis by HPV-18 E6. *Oncogene* 17: 2943–54.

Thomas M, Laura R, Hepner K, et al. (2002). Oncogenic human papillomavirus E6 proteins target the MAGI-2 and MAGI-3 proteins for degradation. *Oncogene* 21: 5088–96.

Tungteakkhun SS, Filippova M, Neidigh JW, et al. (2008). The interaction between human papillomavirus type 16 and FADD is mediated by a novel E6 binding domain. *J Virol* 82: 9600–14.

Umudum H, Rezanko T, Dag F, Dogruluk T. (2005). Human papillomavirus genome detection by in situ hybridization in fine-needle aspirates of metastatic lesions from head and neck squamous cell carcinomas. *Cancer* 105: 71–7.

Varnai AD, Bollmann M, Griefingholt H. (2006). HPV in anal squamous cell carcinoma and anal intraepithelial neoplasia (AIN). Impact of HPV analysis of anal lesions on diagnosis and prognosis. *Int J Colorectal Dis* 21: 135–42.

Venuti A, Badaracco G, Rizzo C, et al. (2004). Presence of HPV in head and neck tumours: High prevalence in tonsillar localization. *J Exp Clin Cancer Res* 23: 561–6.

Weigelt B, Peterse JL, van't Veer LJ. (2005). Breast cancer metastasis: Markers and models. *Nat Rev Cancer* 5: 591–602.

Woods Ignatoski KM, Dziubinski ML, Ammerman C, Ethier SP. (2005). Cooperative interactions of HER-2 and HPV-16 oncoproteins in the malignant transformation of human mammary epithelial cells. *Neoplasia* 7: 788–98.

Wrede D, Luqmani YA, Coombes RC, et al. (1992). Absence of HPV 16 and 18 DNA in breast cancer. *Br J Cancer* 65: 891–4.

Yasmeen A, Bismar TA, Dekhil H, et al. (2007c). ErbB-2 receptor cooperates with E6/E7 oncoproteins of HPV type 16 in breast tumorigenesis. *Cell Cycle* 6: 2939–43.

Yasmeen A, Bismar TA, Kandouz M, et al. (2007b). E6/E7 of HPV type 16 promotes cell invasion and metastasis of human breast cancer cells. *Cell Cycle* 6: 2038–42.

Yasmeen A, Hosein AN, Yu Q, Al Moustafa AE. (2007a). Critical role for D-type cyclins in cellular transformation induced by E6/E7 of human papillomavirus type 16 and E6/E7/ErbB-2 cooperation. *Cancer Sci* 98: 973–7.

Yu Y, Morimoto T, Sasa M, et al. (1999). HPV33 DNA in premalignant and malignant breast lesions in Chinese and Japanese populations. *Anticancer Res* 19: 5057–61.

Yu Y, Morimoto T, Sasa M, et al. (2000). Human papillomavirus type 33 DNA in breast cancer in Chinese. *Breast Cancer* 7: 33–6.

Yuan CH, Filippova M, Duerksen-Hughes P. (2012). Modulation of apoptotic pathways by human papillomaviruses (HPV): Mechanisms and implications for therapy. *Viruses* 4: 3831–50.

Zuna RE, Allen RA, Moore WE, et al. (2004). Comparison of human papillomavirus genotypes in high-grade squamous intraepithelial lesions and invasive cervical carcinoma: Evidence for differences in biologic potential of precursor lesions. *Mod Pathol* 17: 1314–22.

# 9 Prospective Therapies of Genital Human Papillomavirus Infections

*Ricardo Ney Oliveira Cobucci, Janaina Cristiana de Oliveira Crispim, and Ana Katherine da Silveira Gonçalves*

## 9.1 INTRODUCTION

Genital infection by human papillomavirus (HPV) is the most common sexually transmitted disease in the world. Its global prevalence among women without cervical abnormalities is 12%, with higher rates in Africa (24%), Eastern Europe (21%), and Latin America (16%). However, in the presence of high-grade genital lesions and cancer, prevalence increases to 90%. The serotypes most commonly found are HPV-16 (3.2%) and HPV-18 (1.4%) (Forman et al. 2012).

HPV infection is related to six types of carcinomas: cervix, penis, vulva, vagina, anus, and oropharynx. The World Health Organization (WHO) estimated that of the 12.7 million new cancer cases occurring in 2008, 610,000 were related to HPV, with 530,000 cases of cervical carcinoma. Furthermore, projections indicate that due to a lack of effective prevention programs, especially in less developed countries, by 2030 there will be an increase of 2% in the overall incidence of this cancer (Forman et al. 2012).

The persistence of high-risk HPV, especially HPV-16 or HPV-18, is a prerequisite for the development of cervical intraepithelial neoplasia and genital cancer. However, the natural history of the disease shows that in the vast majority of cases spontaneous resolution of HPV infection occurs and that only when the virus remains for more than a year will there be a greater risk of developing carcinoma (Bharti et al. 2009).

The HPV life cycle consists of four important and necessary steps for the pathological consequences to occur, and each are considered as potential targets for all anti-HPV therapeutic strategies. The virus entry is the first step in the pathogenesis of HPV, followed by persistence in epithelial cells. These two steps, establishing the viral genome into the host cell, have been the subject of therapeutic interventions through vaccination and prophylactic therapy. The third and most important step in the life cycle is HPV integration into the human genome, which is often found in high-grade or greater lesions. The integration of the HPV genome is the key point in which the progression to malignancy is facilitated. The fourth major step in the viral life cycle is the expression/positive regulation of transcription of viral oncogene, especially E6 and E7, during differentiation of the epithelium. All of these four steps represent potential targets for the development of new therapies against genital HPV infection (Bharti et al. 2009).

Finally, persistent infections with certain high-risk human papillomavirus (HPV) types, such as 16 and 18, can result in the development of cervical cancer. Neither of the two prophylactic vaccines against HPV-16 and -18 that are in current use have any therapeutic efficacy for prevalent HPV infections. Ablative therapy is widely used for the treatment of HPV cervical dysplasia; however, disease recurrence is a widely recognized problem. Thus, there is a continuing need for therapeutic approaches for the treatment of HPV infections. Future therapies will be directly or indirectly antiviral, targeting HPV protein functions or enhancing the ability of the immune system to resolve infection or inducing apoptosis indirectly in HPV-infected cells.

## 9.2 CHARACTERISTICS OF HPV

HPV is a small, nonenveloped DNA virus that is approximately 55 nm in diameter. The HPV genome consists of approximately 8000 base pairs in a double-stranded DNA molecule enclosed in an icosahedral protein capsid composed of 72 capsomers (de Villiers et al. 2004). The HPV genome contains eight open reading frames (ORFs) that can be divided into three functional regions: (1) noncoding regulatory region, termed as the upstream regulatory region (URR) or the long control region (LCR), which modulates viral DNA replication and gene transcription; (2) an early (E) region, which harbors the early genes (E1, E2, E4–7) that code for proteins involved in viral genome persistence and replication, and viral transcription and regulation of cell proliferation; and (3) a late (L) region that is composed of two genes, L1 and L2, that code for the major and minor capsid proteins (McMurray et al. 2001).

The taxonomy of 120 papillomaviruses has been described using sequence comparisons of the L1 ORF. There are 16 genotypes of papillomaviruses categorized. The alpha and beta genotypes contain the largest number of associated papillomaviruses. All HPVs have a special affinity for epithelial cells, and infection persists in the dividing basal cells of the cutaneous or mucosal epithelium. The small, circular, double-stranded DNA virus naturally exists episomally and has the ability to cause a wide range of lesions in humans. Papillomaviruses are characteristically epitheliotropic and cause proliferative lesions in infected epidermal or mucosal epithelia. They are commonly designated wart viruses, although many members of the group induce only discrete lesions that differ histologically from common warts. Certain types may cause benign and certain types may cause malignant tumors (Berkhout et al. 1995).

### 9.2.1 HPV Classification and Diseases Caused by HPV Infection

Papillomaviruses are strictly host-specific. To date, 120 different HPV types have been classified (Bernard et al. 2010). They are classified as genotypes, and each type is numbered in order of their discovery. HPVs belong to the Papillomaviridae family, which is divided into genera, species, types, subtypes, and variants. The taxonomy of HPV is based on comparison of the nucleotide sequences and homology of the L1 ORF. If the DNA sequences of the L1 genes differ more than 10% from the closest known HPV type, it is recognized as a new type. A subtype is defined with a 2%–10% difference in the DNA sequences. Less than 2% is defined as an intratype variant (de Villiers et al. 2004).

HPVs are grouped according to the type of epithelia they infect. At present, there are about 40 HPVs infecting the mucosal sites of the body, including the anogenital tract of both genders. All HPV genotypes infecting the genital tract belong to the alphapapillomavirus genus, which includes 15 species and 58 HPV genotypes. All genotypes are also classified according to their clinical behavior (i.e., association with malignancy) into high-risk (HR) types, low-risk (LR) types and probable HR types.

The LR HPVs include types 6, 11, 40, 42, 43, 44, 53, 54, 61, 72, and 73. Types 6 and 11 are the most common and are associated with 90% of genital warts and laryngeal papilloma (Gale et al. 1994; Brown et al. 1999; Gale 2005; Gale and Zidar 2006; Potocnik et al. 2007). Most HPV infections are cleared rapidly by the immune system and do not progress to cancer. Papilloma can form in weeks to years after contact with an HPV-infected individual. Hosts that are asymptomatic may become unknown reservoirs of infection.

More than 40 types of HPV infect the mucosal epithelial lining of the anogenital tract and other mucosal areas of the body. Mucosal epithelial infection is associated with a range of disease from benign anogenital warts to malignant carcinomas of the genital, oral, or conjunctival mucosa. Of the types that infect the anogenital tract, two groups can be established based on the virus's ability to cause malignancy. All of these types belong to the alpha genus. Sexually transmitted HR HPVs include types 16, 18, 31, 33, 35, 39, 45, 51, 52, 56, 58, 59, and 68 (Muñoz et al. 2003). They may lead to the development of cervical intraepithelial neoplasia (CIN), vulva intraepithelial neoplasia (VIN), penile intraepithelial neoplasia (PIN), and/or anal intraepithelial neoplasia (AIN). It is now widely accepted that HPV-16 and -18 are the causative agent of around 70% cervical cancer cases (Muñoz et al. 2004).

### 9.2.2 Life Cycle of Human Papillomavirus

HPV gains access to host cells through microabrasions in the epithelium. The target is the basal cells of the epithelium (Egawa 2003; Schmitt et al. 2006). The basal layer consists of basal epithelial cells including the stem cells of the epithelium. The replication of HPV is dependent upon complete keratinocyte differentiation. Following the access of viral particles to the basal layer keratinocytes, high-level expression of viral proteins and viral assembly occur only in the upper layers of the squamous epithelia (Doorbar 2005). After infecting the cells at low copy number, viral DNA replication amplifies the viral copy number approximately 50–100 copies/cell. These infected cells leave the basal layer and enter into the proliferation compartment of the epithelium. When the keratinocyte reaches the superficial layer and dies, viral genomes are repackaged into capsids and shed from the cell (Doorbar 2005; Stanley 2006).

It has been estimated that the time from infection to virus release takes at least three weeks. This is the time required for the keratinocyte to undergo complete differentiation and desquamation. In humans, the time from infection to appearance of HPV-induced lesions can vary from weeks to months (Doorbar 2007). This HPV infectious cycle effectively evades the immune system because there is no retention of HPV antigens until the infected cell reaches the epithelial surface (Stanley 2006; Wang 2007).

## 9.3 CLINICAL MANIFESTATIONS

### 9.3.1 Clinical Infection

An HPV infection produced by the LR types (most commonly HPV types 6 and 11) induces proliferation of the squamous epithelia leading to benign tumors such as warts, papilloma, and condyloma. These clinical lesions are a result of a productive HPV infection (with expression of all its proteins) in the maturing epithelial cells. This leads to morphological changes in the infected epithelium, including cellular proliferation (epithelial acanthosis) and degenerative changes in the nuclei and cytoplasm (koilocytosis). This productive infection is usually followed by spontaneous regression and virus clearance or maintenance of the viral genome as latent episomes in the basal cells (Doorbar 2007).

### 9.3.2 Subclinical Infection

Subclinical HPV infections are defined as lesions that are only visible under a colposcope; in histological specimens, these are those that demonstrate only minor epithelial changes that are not consistent with characteristic clinical HPV lesions (Syrjänen and Syrjänen 2000).

### 9.3.3 Latent Infection

HPV infection is considered to be latent when the virus can only be detected by sensitive molecular methods in an otherwise normal epithelium without any cytological, morphological, or colposcopic alterations (Syrjänen and Syrjänen 2000).

### 9.3.4 Malignant Transformation

HPV infections by the HR types are associated with premalignant lesions and cancer, in which the most frequent genotypes represent species 7 and 9 (Bosch et al. 2008). In addition to cancer and its precursors, these HR types are detected in women with no or only mild cytological abnormalities (Clifford et al. 2005). In most cases, however, HPV infections will regress within two years (Wang et al. 2009); but in some cases, the infection remains persistent for years and even decades, which eventually leads to the development of cervical cancer (CC).

The mechanisms of progression towards CC are not fully understood, but the crucial event is probably the uncontrolled expression of viral transforming proteins E6 and

E7 that occurs following integration of the viral genome into the host cell chromosome. Integrated HPV DNA is found in 100% and 80% of HPV-18- and HPV-16-positive CCs, respectively (Cheung et al. 2008; Saunier et al. 2008). Integration of HPV DNA into the host genome is a critical event in carcinogenesis, but controversy exists whether it is an early or a late event (Arias-Pulido et al. 2006; Kulmala et al. 2006).

Although HPV integration is a crucial event in malignant transformation, some cases of CC contain HPV as an episomal form, which suggests that mechanisms other than viral integration are present, such as promoter methylation or direct mutation of E2 (Turan et al. 2006, 2007).

## 9.4 DETECTION OF HUMAN PAPILLOMAVIRUS INFECTION

HPV cannot be cultured, and the detection methods of HPV involve morphological methods, HPV DNA detection, HPV RNA detection, and serology, which are described here.

### 9.4.1 Morphological Methods

#### 9.4.1.1 Visual Examination

At physical examination, the use of either acetic acid (VIA) or Lugol's iodine (VILI) makes cervical lesions visible to the "naked eye."

#### 9.4.1.2 Colposcopy

The colposcope provides a magnified visual impression of the labia, vagina, cervix (vagina), and transformation zone (TZ). Application of 5% acetic acid solution results in acetowhite staining of the abnormal areas in the epithelium. Women are referred for colposcopy after detection of an abnormal Pap test, usually Atypical squamous cells of undetermined significance (ASCUS) or dyskaryosis. Colposcopy is a descriptive diagnostic tool suggesting an abnormality, and directed punch biopsies are necessary to confirm the findings using light microscopy.

#### 9.4.1.3 Pap Smear Cytology

Cervicovaginal cytology is the time-honored diagnostic method used in screening for CC precursor lesions. This diagnostic tool is known as the Papanicolaou (Pap) test or Pap smear. Exfoliated cells from the vagina and uterine cervix are collected with a wooden spatula and a small brush (cytobrush), followed by fixation of the smear onto a glass slide. To classify the abnormalities in the Pap smear, different systems are in use. The 2001 Bethesda System (TBS 2001) is currently the most widely used classification system (Solomon et al. 2001).

The widespread use of Pap smear cytology for screening has reduced the incidence and mortality of CC in many countries, albeit the rates still vary depending on the level of implementation (Sankila et al. 2001; Anttila et al. 2004; Peto et al. 2004). In countries where organized screening programs have been active for a long time (e.g., in Finland, Sweden, British Columbia, and Canada), the incidence of CC has decreased as much as 70%–80% (Nieminen et al. 1999). Failure to reduce CC incidence and mortality, especially in developing countries, has been ascribed to nonavailability or low quality of screening, the low sensitivity of the conventional Pap smear, or low quality of colposcopy; treatment failures; and lack of follow-up practices. The key to all successful screening is high coverage and attendance rates among the total female population (Anttila et al. 1999).

#### 9.4.1.4 Liquid-Based Cytology

Liquid-based cytology (LBC) is a modification of Pap smear cytology, where the sample is collected from the cervix in the same way as with Pap smear cytology, but only plastic sampling devices may be used (Karnon et al. 2004). LBC has been widely accepted as the primary tool in CC screening. The cervical sample in this method involves making a suspension of the cells, which is then used to produce a thin layer of cells on the cytological slide (Arbyn et al. 2008).

#### 9.4.1.5 Histopathology

The histopathological examination is the gold standard in the diagnosis of CIN lesions and CC. CIN lesions are classified into three grades according to their severity. The basic histological criteria include epithelial differentiation indicated as loss of polarity, nuclear atypia, and abnormal mitotic figures.

### 9.4.2 HPV DNA Detection

#### 9.4.2.1 Nucleic Acid Hybridization

Following rapid technological development, several types of hybridization methods have become available for HPV testing since the early 1980s. All nucleic acid hybridization methods are based on HPV DNA or RNA detection, in which a probe sequence is bound to a complementary sequence in the sample. The most common methods used in HPV testing include Southern transfer hybridization (STH), dot blot hybridization (DB), and in situ hybridization (ISH). In routine HPV testing, all these have been mostly replaced by polymerase chain reaction (PCR)-based techniques.

#### 9.4.2.2 Polymerase Chain Reaction

PCR-based methods are commonly used, highly sensitive, and specific methods for HPV detection. PCR is a selective target amplification assay capable of exponential and reproducible increase of the HPV sequences present in biological specimens (Garland and Tabrizi 2006). It can theoretically produce one billion copies from a single-stranded DNA molecule after 30 cycles of amplification. When performing PCR, care must be taken to avoid false-positive results, which may be derived from cross-contaminating specimens or reagents with the PCR products of previous rounds. Several procedures are available to avoid this problem while using the PCR protocols for HPV detection (Iftner and Villa 2003). Therefore, the sensitivity and specificity of PCR techniques can vary depending, for instance, on the primer set, size of the PCR product, reaction conditions, performance of the DNA polymerase used in the reaction, as well as the spectrum of HPV types amplified and the ability to detect multiple types (Brink et al. 2007).

#### 9.4.2.3 Multiplex HPV Genotyping

Multiplex HPV genotyping (MPG) is a recent, simple, bead-based, high-throughput hybridization method based on Luminex® suspension array technology (Schmitt et al. 2006), which allows simultaneous detection and genotyping of up to 100 HPV types. MPG is based on the amplification of HPV DNA by the consensus primers GP5+/6+ and the subsequent detection of the products with type-specific oligonucleotide probes coupled to fluorescence-labeled polystyrene beads, which create a suspension array with unique absorption spectra. This allows up to 100 different targets to be measured simultaneously in a single reaction.

### 9.4.3 HPV RNA Detection

There are commercially available HPV RNA tests to detect HPV messenger RNA (mRNA) transcripts coding for E6/E7 and thereby the presence of oncogene activity (Castle et al. 2007; Tropé et al. 2009).

### 9.4.4 Serology

Until now, serology has played no role in the diagnosis of HPV infections. Serological techniques measure specific natural antibodies against different HPV types (Carter et al. 1996). Interest in serology has increased considerably during recent years when prophylactic HPV vaccines have become available worldwide. HPV antibodies can be detected only in half of those exposed to HPV. Therefore, antibody testing is unreliable for the diagnosis of current or past HPV infection of individual persons. Serology is considered to measure a past HPV exposure, and there are a variety of different technical modifications available. The most commonly used method is enzyme linked immunosorbent assay (ELISA) using virus-like particles (VLP) as antigens.

## 9.5 HPV PROPHYLAXIS

Currently, there are two prophylactic vaccines available: one bivalent is highly effective in preventing lesions caused by HPV-16 and -18 and another has high effectiveness for quadrivalent HPV-6, -11, -16, and -18. Despite avoiding 98% of the lesions caused by HPV, vaccination does not allow for a large reduction in the incidence of CC. This is due to the coverage being very low in most countries, which maintains a high number of new cases of anogenital injuries related to the virus. Thus, the development of more specific and effective antiviral treatments becomes a priority (Dillner et al. 2011).

## 9.6 CURRENT THERAPIES

Even after the establishment of the casual relationship between HPV and CC, the presence or absence of HPV currently has no impact on the decision of the treatment. Strategies are mainly anticancer and not antiviral. Several therapies are available for treatment of HPV-associated diseases, including cryotherapy, trichloroacetic acid, and high-frequency surgery. These treatment strategies are ablative in nature and aim to remove the lesion; they do not specifically target HPV infection.

These treatments, though effective, are applicable only when the lesion is visible and do not necessarily eliminate the HPV infection. Some of these procedures are also associated with significant morbidity such as bleeding, cervical stenosis (narrowing), and—in some cases—pelvic infection and cervical incompetence. Although these treatment modalities are effective, they are all associated with considerable recurrence ranging up to 10% of cases (Bharti et al. 2009; Stanley 2012).

## 9.7 FUTURE CHALLENGES

Progress in developing effective therapies for HPV infection has been slow, mainly due to difficulties in studying the biology and pathogenesis of these viruses, which have a unique and complex replication cycle. HPVs are exclusively intraepithelial pathogenic with a replication cycle that is both time and differentiation dependent (Stanley 2012).

Another complication of these infections is the phenomenon of latency. After infection, HPV DNA can remain latent within the cells, while others were used in the production cycle. Spontaneous regression of the lesion mediated by immune response is common in immunocompetent women, but this does not result in clearance of the virus; viral genomes can be detected in apparently normal epithelium many years after the lesion regression (Stanley 2012).

Due to the low coverage and high recurrence of infection following the treatments available today, it is essential to explore alternative methods for the control of HPV infection and for effective treatment for cancer of the cervix and other HPV-associated premalignant and malignant lesions (Bharti et al. 2009).

## 9.8 NEW THERAPIES

Various therapeutic strategies are being efficiently exploited and can reach several stages of viral infection, such as viral entry, viral latency, viral replication, and oncogenic expression. Because low- and high-grade lesions differ in respect to antiviral activity, treatment approaches are also different and are specific to the degree of the lesion (Bharti et al. 2009). Future therapies can be classified as therapeutic vaccines, immunotherapies, therapies based on RNA interference, antivirals, and natural derivatives/herbs. Many of these therapeutic agents are in advanced stages of clinical evaluation.

### 9.8.1 Therapeutic Vaccines

Vaccinations with VLP, such as Gardasil® and Cervarix®, have shown effectiveness in HPV prophylaxis, but these vaccines do not have therapeutic potential. Considering the HPV-16's vital role in carcinogenesis, efforts to develop therapeutic vaccines against HPV have been directed at the development of vaccines against viral oncogenes E6 and E7 of HPV-16. Based on the nature of the immunogenicity, therapeutic vaccines can be divided into two main classes: (1) protein/peptide-based vaccines and (2) DNA-based vaccines. Most of these clinical trials are at phase I. These studies, in healthy subjects and in patients with high-grade lesions or neoplasia, used vaccines with whole protein or peptides derived from HPV-16 E6 and E7 or HPV-18, especially because of the oncogenic potential of these viruses (Kaufmann et al. 2007; Santin et al. 2008; Welters et al. 2008).

These vaccines have proven to be highly effective in the generation of humoral immune response as well as cytotoxic T-cell response in most patients in the study, but their effect on lesion regression or HPV positive was determined in very few studies. Several animal studies have shown promising results, indicating that the vaccine therapy can prevent the progression of the disease (Kaufmann et al. 2007; Santin et al. 2008; Welters et al. 2008).

Phase I clinical trials indicate that even though the anti-HPV was effective in immunity, the regression of the lesion, as well as the elimination of HPV, was suboptimal and observed only in a subset of patients. Perhaps this is the reason why the studies with the vaccine have not yet reached phases II and III. Besides these, there are several other restrictions that must be overcome, such as the cost of vaccines. To overcome some of these limitations, less expensive second-generation therapeutic vaccines, based on DNA and E7 from HPV-16, are being tested (Muderspach et al. 2000; de Jong et al. 2002; Frazer et al. 2004; Peng et al. 2008).

DNA immunization is quite an inventive vaccination strategy that involves the direct introduction of plasmid DNA, encoding the desired antigen into the host. DNA vaccines expand strong protective responses against tumors and provide several important advantages over current vaccines, which are as follows: (1) DNA vaccines mimic the effects of live attenuated vaccines in their ability to induce major histocompatibility complex (MHC) class I restricted CD8+ T-cell responses, which may be advantageous compared with conventional protein-based vaccines, while mitigating some of the safety concerns associated with live vaccines; (2) DNA vaccines can be manufactured in a relatively cost-effective manner and stored with relative ease; and (3) DNA vaccines provide prolonged antigen expression, leading to the amplification of immune response and induce memory responses against infectious agents. In mice, results reveal that the target DNA vaccine can influence an E7-specific cytotoxic T-lymphocyte (CTL) response, which is imperative in the lysis of infected tumor cells, compared to negative control ($P < 0.05$). Additionally, treatment of tumor-bearing mice with pcDNA/E7 + heat-shock protein 70 (HSP70) plasmid generates stronger immune responses and significantly decreased tumor sizes. Coadministration of pcDNA/E7 + HSP70 plasmid was immunologically more effective than pcDNA/E7 alone and could be an efficient approach to induce dramatically E7-specific immune responses acting as a future CC vaccine (Farzanehpour et al. 2013).

### 9.8.2 Immunotherapies

Therapeutic interventions to induce effective immune response and to allow the control of established infection are possible attractive strategies. Such therapies have the potential to treat latent infection or clinically apparent lesions. Doing so would cause immune response mediated by Th1 cells. Treatments that induce this response, especially against the E6 and E7 oncogenes, are likely to be effective against high-grade lesions and carcinomas associated with HPV (Stanley 2012; Pang and Thierry 2013).

Immunity against HPV might be important for elimination of the virus. The innate immune responses involving macrophages, natural killer cells, and natural killer T cells may play a role in the first line of defense against HPV infection. In the second line of defense, adaptive immunity via CTLs targeting HPV-16 E2 and E6 proteins appears to eliminate cells infected with HPV-16. However, HPV can evade host immune responses. First, HPV does not kill host cells during viral replication and therefore neither presents viral antigen nor induces inflammation. HPV-16 E6 and E7 proteins downregulate the expression of type 1 interferon (IFN) in host cells. The lack of costimulatory signals by inflammatory cytokines, including IFNs, during antigen recognition may induce immune tolerance rather than the appropriate responses. Moreover, HPV-16 E5 protein downregulates the expression of HLA class 1, and it facilitates evasion of CTL attack. These mechanisms of immune evasion may eventually support the establishment of persistent HPV infection, leading to the induction of CC. Considering such immunological events, prophylactic HPV-16 and 18 vaccine appears to be the best way to prevent CC in women who are immunized in adolescence (Stanley 2012).

Although current studies have targeted proteins E6 and E7, there may be other options. A recombinant vaccine expressing E2 gene has been investigated in clinical trials with promising results (Nieto et al. 2010; Pang and Thierry 2013). Moreover, the deliberate activation of toll-like receptors (TLRs) by administering an agonist of TLR7/8 increases the production of CD8+ T cells in HPV-associated lesions, suggesting that therapy with TLR agonists may represent a viable treatment. Imiquimod, a potent agonist TRL7 stimulator of type 1 interferon and other cytokines, has been observed to demonstrate efficacy in humans against HPV and genital warts. The usefulness of Imiquimod as immunotherapy for premalignancy associated with persistent HPV infection was recently confirmed in a clinical study with TRL7 agonist for the topical treatment of VIN (Terlou et al. 2010; Bhat et al. 2011).

Experimental studies in humans with tumors induced by HPV support the idea that the infections and tumor-related virus environment are anti-inflammatory. Therefore, both the innate immune response as adaptive immune response may be affected in genital infections. Accordingly, the most effective immunotherapy is the induction of agents that combine the adaptive immune response and T-cell-mediated innate immunity causing local inflammation (Bhat et al. 2011).

### 9.8.3 Therapies Based on RNA Interference

After identification of target genes involved in neoplastic transformation and tumor growth in HPV-associated genital infections, specific therapeutic approaches are being extensively studied. In the last decade, some studies have allowed a major breakthrough in the technology of RNA interference specifically targeting oncogenes E6 and E7 of HPV without any damage to normal cellular RNA, which is often lacking in conventional therapies (Bharti et al. 2009).

The expression of E6 and E7 oncogenes is detected in both the upper layers of the epithelium, as well as during the initial phase in the advanced stage of the disease. The control of the expression of these proteins will eventually lead to treatment of active lesions. Although there are some therapies for RNA interference based on various oncogenes in phases I–III clinical trials, they are still being clinically tested for HPV-associated lesions. Several studies in vitro and in vivo show efficacy against HR HPV using antiviral ribozymes, antisense molecules (oligonucleotides), and short-interfering RNA (siRNA) (Bharti et al. 2009).

Despite being a totally rational antiviral strategy, the RNA interference operates only at the post-transcriptional level by suppressing the gene expression of viral oncogenes and has no impact on viral latency. Furthermore, due to the requirement of large amounts of therapeutic principle, the target cells with oncogenic RNA tend to lose their proliferative capacity (as desired). Therefore, in view of these defects, in principal, RNA interference may not be useful as therapy alone. Thus, some more rational approaches have been developed through a combination of siRNA against E6 and E7 with current forms of treatment for cancer of the cervix. As with other technologies, in vitro siRNA against CC has shown promising results (Yoshinouchi et al. 2003; Yamato et al. 2008).

RNA interference (RNAi) has great promise in providing highly specific therapies for all HPV diseases, yet this promise has not been realized. Furthermore, although RNAi has the ability to act as a direct antiviral via the specific targeting of HPV sequences, it is also useful as a tool to rapidly identify novel antiviral drug targets via large-scale screens, and this could contribute to potential therapies (Honegger et al. 2013).

### 9.8.4 Antivirals

Currently, there are no specific therapies for papillomavirus, but they are necessary for several reasons. Antiviral therapy has the potential to treat unapparent HPV infection as well as overt clinical disease. There is a substantial part of the population infected by HPV, some of these being immunocompromised individuals who at present cannot be treated with immunotherapy, and these drugs are the only option. Multifocal lesions such as VIN are not amenable to ablation and may not respond to immunotherapy but may be the target of antivirals. In addition, antiviral agents, unlike immunotherapies, may not be limited to HPV type in their more comprehensive effectiveness (Stanley 2012).

The papillomavirus genes are expressed in a different way, temporally and spatially, across the replication cycle. The treatment of active lesions can affect any protein. However, if the aim is to treat the infection and injury, the latently infected cells have to be eliminated. The problem is that the viral latency is poorly understood, but all evidence indicates that only the El and E2 genes are transcribed and thus are the only targets. El is a helicase-dependent adenosine triphosphate and acts as a primer for replication of the HPV ligand in a complex with E2 at the origin of replication. Disruption of these interactions could theoretically be achieved by small molecule inhibitors, and because El is the most conserved protein of HPV, it is possible that an inhibitor would have broad reactivity against many types of HPV E2, which plays a central role in the control of replication and transcription of HPV, is expressed throughout the life cycle of the virus, and is an attractive target for the development of an antiviral agent (Stanley 2012). An antiviral agent has shown efficacy in genital HPV-induced lesions. Cidofovir is a cytosine analog that inhibits viral DNA polymerase, but that seems to induce apoptosis in these lesions (Snoeck et al. 2001).

### 9.8.5 Natural Derivatives/Herbs

Because the productive life cycle of HPV is closely related to the differentiation of infected epithelial cells, several interventions have been tested for their activity on different stages of the life cycle of HPV and the host cell, including binding and entry of virus, the presence of episomal viral DNA in shape and its replication in host cells, and expression of host cell-dependent viral oncogenes after integration of its DNA into host cells. Because these cellular transcription factors serve as an important link between the oncogenic transcription host cell and also the expression of the viral gene, transcription factors provide a single target for the development of anti-HPV treatments. Theoretically, preventing the binding of transcription factors will eventually lead to suppression of expression of genes of oncogenic HPV and stop viral replication (Longworth and Laimins 2004).

In recent years, various herbal antioxidants, derivatives, and plant polyphenol compounds have been used for the selective suppression of transcription factors into host cells by HPV-infected cells. The removal and selective change in the composition of these factors have been demonstrated and are associated with negative regulation of HPV gene expression and the induction of apoptosis in infected cells (Divya and Pillai 2006). Some of the potent natural derivatives that are under development to be used as therapeutic anti-HPV are as follows:

- *Curcumin:* It is a potent antioxidant (diferuloylmethane) and an active compound of the perennial herb, which also has anti-inflammatory and antitumor activity. Studies using curcumin in cervical carcinoma cells infected with HPV-18 showed that it was able to selectively suppress the transcription of the virus (Prusty and Das 2005).
- *Epigallocatechin gallate (EGCG):* EGCG, a component of green tea, is known to possess anticancer and antiproliferative properties. Several studies conducted in recent years with EGCG showed inhibition of cervical adenocarcinoma cells, associated with pKi-67 suppression and the induction of apoptosis (Ahn et al. 2003; Yokoyama et al. 2004, 2008; Noguchi et al. 2006). EGCG has also been tested in a clinical trial, with polyphenols of green tea in 51 patients with cervical lesions, the result being a 69% response rate (35/51) in the treated group compared with 10% (4/39) in the control group. Taken together, these observations suggest that the polyphenols in green tea are effective against HPV infection and for treatment of cervical lesions and can be a potential drug to be used as a therapeutic anti-HPV agent (Ahn et al. 2003).
- *Indole-3-carbinol (I3C):* It is a chemical that is found in high concentrations in cruciferous vegetables. Because the phytochemical I3C is an anti estrogenic compound as well as an inducer of apoptosis, it is logical that the relative amounts of E2 and I3C would affect apoptosis in cervical cells. Interestingly, perhaps because of its anti estrogenic activities,

> I3C has become a treatment strategy for laryngeal papillomatosis (benign tumors with an HPV etiology) and, more recently, for cervical dysplasia. In vitro high concentrations of I3C need to be used because this compound is slowly converted into active condensation products (e.g., diindolylmethane), whereas in vivo benefits for cervical disease occur at concentrations obtainable from eating cruciferous vegetables (Bell et al. 2000).

It has been demonstrated that I3C has chemopreventive effects in preclinical studies. The effects of oral administration of I3C as a therapy for CIN were tested in a placebo-controlled study. A total of 30 patients with biopsies showing CIN were randomly assigned to receive placebo or I3C (200 and 400 mg/day, respectively) orally for a period of 12 weeks. None of the placebo group showed regression of CIN, although four of the eight 200 mg/day arm and four of the nine 400 mg/day arm showed regression. This small trial together with preclinical data in animals indicates that this may be an effective therapy in the future (Bell et al. 2000).

## 9.9 CONCLUSIONS

The limitations of current prophylactic HPV vaccines demonstrate a pressing need for novel approaches to the eradication of HPV-related neoplasia and suggest that the development of therapeutic vaccines for the treatment of HPV-associated lesions will remain an important goal even if worldwide prophylactic vaccine programs are successfully implemented. Therapeutic vaccines for cancer are an attractive alternative to conventional therapies because the latter result in serious adverse effects and in most cases are not effective against advanced disease. Vaccines targeting oncogenic viral proteins such as HPV-16 E6 and HPV-16 E7 are ideal candidates to elicit strong immune responses without generating autoimmunity because: (1) these products are not expressed in normal cells and (2) their expression is required to maintain the malignant phenotype.

The fact that there are many therapies available for the treatment of HPV-associated lesions is a reflection of the reality that, as a whole, these are not satisfactory. Surgical removal or ablations of the lesions by various local treatments (laser, electrodiathermy, cryotherapy, etc.) are still the most reliable methods for treating papilloma and intraepithelial neoplasia. The currently available therapies are not antivirals because they only cause physical removal of injury or inflammation and induce immune response to make the virus ineffective. A successful immunotherapy might therefore be a preferred mode of treatment because it can target all HPV-associated lesions irrespective of their location. Ideally, it would also induce long-lasting immunity, thus preventing recurrence.

Systemic and intralesional administration of interferon has been attempted in laryngeal papillomatosis and in recurrent condylomata acuminata. So far, the evidence for success is not fully convincing. Retinoic acid may have some therapeutic potential because this compound has been shown to suppress HPV transcriptional activity.

Recent progress in the identification and characterization of small-molecule antagonists of proteins E1, E2, E6, and E7, as well as therapies that alter the interactions between HPV and their cellular targets, indicates that infection control will improve.

Further studies are needed to develop clinically applicable combination therapies based on RNAi. Above all, to establish the proof of concept of RNAi-based combination therapeutics, the mechanism of synergy should be investigated between two treatments through in vivo experiments, demonstrating that the combination therapy is significantly superior to either modality alone.

Moreover, the success of new treatments depends on a greater understanding of immunology in the microenvironment of lesions/tumors induced by HPV and on the advancement of ongoing clinical trials with vaccines and therapies to elicit strong systemic immune response against viral antigens specifically linked with progression to malignancy, such as proteins E6 and E7, effective in the latent phase of infection, and thus allowing the elimination of the virus. With more discoveries of new drugs, development of new adjuvants, and the better understanding of tumor biology, we will have more opportunities to develop improved therapeutics against HPV-associated cancers.

## REFERENCES

Ahn WS, Huh SW, Bae SM, et al. (2003). A major constituent of green tea, EGCG, inhibits the growth of a human cervical cancer cell line, CaSki cells, through apoptosis, G(1) arrest, and regulation of gene expression. *DNA Cell Biol* 22: 217–24.

Anttila A, Pukkala E, Soderman B, et al. (1999). Effect of organized screening on cervical cancer incidence and mortality in Finland, 1963–1995: A recent increase in the incidence. *Int J Cancer* 83: 59–63.

Anttila A, Ronco G, Clifford G, et al. (2004). Cervical cancer screening programmes and policies in European countries. *Br J Cancer* 91: 935–41.

Arbyn M, Bergeron C, Klinkhamer P, et al. (2008). Liquid compared with conventional cervical cytology: A systematic review and meta-analysis. *Obstet Gynecol* 111: 167–77.

Arias-Pulido H, Peyton CL, Joste NE, et al. (2006). Human papillomavirus type 16 integration in cervical carcinoma in situ and in invasive cervical cancer. *J Clin Microbiol* 44: 1755–62.

Bell MC, Crowley-Nowick P, Bradlow HL, et al. (2000). Placebo-controlled trial of indole-3-carbinol in the treatment of CIN. *Gynecol Oncol* 78: 123–9.

Berkhout RJM, Tieben LM, Smits HL, et al. (1995). Nested PCR approach for detection and typing of epidermodysplasia verruciformis-associated human papillomavirus types in cutaneous cancers from renal transplant recipients. *J Clin Microbiol* 33: 690–5.

Bernard HU, Burk RD, Chen Z, et al. (2010). Classification of papillomaviruses (PVs) based on 189 PV types and proposal of taxonomic amendments. *Virology* 401: 70–9.

Bharti AC, Shukla S, Mahata S, et al. (2009). Anti-human papillomavirus therapeutics: Facts & future. *Indian J Med Res* 130: 296–310.

Bhat P, Mattarollo SR, Gosmann C, et al. (2011). Regulation of immune responses to HPV infection and during HPV-directed immunotherapy. *Immunol Rev* 239: 85–98.

Bosch FX, Burchell AN, Schiffman M, et al. (2008). Epidemiology and natural history of human papillomavirus infections and type-specific implications in cervical neoplasia. *Vaccine* 26: K1–16.

Brink AA, Snijders PJ, Meijer CJ. (2007). HPV detection methods. *Dis Markers* 23: 273–81.

Brown DR, Schroeder JM, Bryan JT, et al. (1999). Detection of multiple human papillomavirus types in condylomata acuminata lesions from otherwise healthy and immunosuppressed patients. *J Clin Microbiol* 37: 3316–22.

Carter JJ, Koutsky LA, Wipf GC, et al. (1996). The natural history of human papillomavirus type 16 capsid antibodies among a cohort of university women. *J Infect Dis* 174: 927–36.

Castle PE, Dockter J, Giachetti C, et al. (2007). A cross-sectional study of a prototype carcinogenic human papillomavirus E6/E7 messenger RNA assay for detection of cervical precancer and cancer. *Clin Cancer Res* 13: 2599–605.

Cheung JL, Cheung TH, Tang JW, et al. (2008). Increase of integration events and infection loads of human papillomavirus type 52 with lesion severity from low-grade cervical lesion to invasive cancer. *J Clin Microbiol* 46: 1356–62.

Clifford GM, Gallus S, Herrero R, et al. (2005). Worldwide distribution of human papillomavirus types in cytologically normal women in the International Agency for Research on Cancer HPV prevalence surveys: A pooled analysis. *Lancet* 366: 991–8.

de Jong A, O'Neill T, Khan AY, et al. (2002). Enhancement of human papillomavirus (HPV) type 16 E6 and E7-specific T-cell immunity in healthy volunteers through vaccination with TA-CIN, an HPV16 L2E7E6 fusion protein vaccine. *Vaccine* 20: 3456–64.

de Villiers EM, Fauquet C, Broker TR, et al. (2004). Classification of papillomaviruses. *Virology* 324: 17–27.

Dillner J, Arbyn M, Unger E, et al. (2011). Monitoring of human papillomavirus vaccination. *Clin Exp Immunol* 163: 17–25.

Divya CS, Pillai MR. (2006). Antitumor action of curcumin in human papillomavirus associated cells involves downregulation of viral oncogenes, prevention of NFkB and AP-1 translocation, and modulation of apoptosis. *Mol Carcinog* 45: 320–32.

Doorbar J. (2005). The papillomavirus life cycle. *J Clin Virol* 32: S7–15.

Doorbar J. (2007). Papillomavirus life cycle organization and biomarker selection. *Dis Markers* 23: 297–313.

Egawa K. (2003). Do human papillomaviruses target epidermal stem cells? *Dermatology* 207: 251–4.

Farzanehpour M, Soleimanjahi H, Hassan ZM, et al. (2013). HSP70 modified response against HPV based tumor. *Eur Rev Med Pharmacol Sci* 17: 228–34.

Forman D, Martel C, Lacey C, et al. (2012). Global burden of human papillomavirus and related diseases. *Vaccine* 30S: F12–23.

Frazer IH, Quinn M, Nicklin JL, et al. (2004). Phase 1 study of HPV16-specific immunotherapy with E6E7 fusion protein and ISCOMATRIX adjuvant in women with cervical intraepithelial neoplasia. *Vaccine* 23: 172–81.

Gale N. (2005). Papilloma/papillomatosis. In: Barnes L, Eveson JW, Reichart P, Sidransky D (eds.), *Pathology and Genetics of Head and Neck Tumours*. Lyon: IARC Press, pp. 144–5.

Gale N, Poljak M, Kambic V, et al. (1994). Laryngeal papillomatosis: Molecular, histopathologic, and clinical evaluation. *Virchows Arc* 425: 291–5.

Gale N, Zidar N. (2006). Benign and potentially malignant lesions of the squamous epithelium and squamous cell carcinoma. In: Cardesa A, Slootweg PJ (eds.), *Pathology of the Head and Neck*. Heidelberg: Springer, pp. 1–38.

Garland S, Tabrizi S. (2006). Methods for HPV detection: Polymerase chain reaction assays. In: Monsonego J (ed.), *Emerging Issues on HPV Infections: From Science to Practice*. Basel: Karger, pp. 63–72.

Honegger A, Leitz J, Bulkescher J, et al. (2013). Silencing of human papillomavirus (HPV) E6/E7 oncogene expression affects both the contents and the amounts of extracellular microvesicles released from HPV-positive cancer cells. *Int J Cancer* 133:1631–42.

Iftner T, Villa LL. (2003). Chapter 12: Human papillomavirus technologies. *J Natl Cancer Inst Monogr* 31: 80–8.

Karnon J, Peters J, Platt J, et al. (2004). Liquid-based cytology in cervical screening: An updated rapid and systematic review and economic analysis. *Health Technol Assess* 8: 1–78.

Kaufmann AM, Nieland JD, Jochmus I, et al. (2007). Vaccination trial with HPV16 L1E7 chimeric virus-like particles in women suffering from high grade cervical intraepithelial neoplasia (CIN 2/3). *Int J Cancer* 121: 2794–800.

Kulmala SM, Syrjanen UB, Gyllensten I, et al. (2006). Early integration of high copy HPV16 detectable in women with normal and low grade cervical cytology and histology. *J Clin Pathol* 59: 513–17.

Longworth MS, Laimins LA. (2004). Pathogenesis of human papillomaviruses in differentiating epithelia. *Microbiol Mol Biol Rev* 68: 362–72.

McMurray HR, Nguyen D, Westbrook TF, et al. (2001). Biology of human papillomaviruses. *Int J Exp Pathol* 82: 15–33.

Muderspach L, Wilczynski S, Roman L, et al. (2000). A phase I trial of a human papillomavirus (HPV) peptide vaccine for women with high-grade cervical and vulvar intraepithelial neoplasia who are HPV 16 positive. *Clin Cancer Res* 6: 3406–16.

Muñoz N, Bosch FX, de Sanjosé S, et al., International Agency for Research on Cancer Multicenter Cervical Cancer Study Group. (2003). Epidemiologic classification of human papillomavirus types associated with cervical cancer. *N Engl J Med* 348: 518–27.

Muñoz N, Méndez F, Posso H, et al., HPV Study Group. (2004). Incidence, duration, and determinants of cervical human papillomavirus infection in a cohort of Colombian women with normal cytological results. *J Infect Dis* 190: 2077–87.

Nieminen P, Kallio M, Anttila A, et al. (1999). Organized vs. spontaneous Pap-smear screening for cervical cancer: A case control study. *Int J Cancer* 83: 55–8.

Nieto K, Gissmann L, Shadilich L. (2010). Human papillomavirus-specific immune therapy: Failure and hope. *Antivir Ther* 15: 951–7.

Noguchi M, Yokoyama M, Watanabe S, et al. (2006). Inhibitory effect of the tea polyphenol, (-)-epigallocatechin gallate, on growth of cervical adenocarcinoma cell lines. *Cancer Lett* 234: 135–42.

Pang CL, Thierry F. (2013). Human papillomavirus proteins as prospective therapeutic targets. *Microb Pathog* 58: 55–65.

Peng S, Trimble C, Alvarez RD, et al. (2008). Cluster intradermal DNA vaccination rapidly induces E7-specific CD8+ T-cell immune responses leading to therapeutic antitumor effects. *Gene Ther* 15: 1156–66.

Peto J, Gilham C, Fletcher O, et al. (2004). The cervical cancer epidemic that screening has prevented in the UK. *Lancet* 364: 249–56.

Potocnik M, Kocjan BJ, Seme K, et al. (2007). Distribution of human papillomavirus (HPV) genotypes in genital warts from males in Slovenia. *Acta Dermatovenerol Alp Panonica Adriat* 16: 91–8.

Prusty BK, Das BC. (2005). Constitutive activation of transcription factor AP-1 in cervical cancer and suppression of human papillomavirus (HPV) transcription and AP-1 activity in HeLa cells by curcumin. *Int J Cancer* 113: 951–60.

Sankila R, Demaret E, Hakama M, et al. (2001). *Evaluation and Monitoring of Screening Programmes.* Brussels: European Commission, Europe Against Cancer Programme.

Santin AD, Bellone S, Palmieri M, et al. (2008). Human papillomavirus type 16 and 18 E7-pulsed dendritic cell vaccination of stage IB or IIA cervical cancer patients: A phase I escalating-dose trial. *J Virol* 82: 1968–79.

Saunier M, Monnier-Benoit S, Mauny F, et al. (2008). Analysis of human papillomavirus type 16 (HPV16) DNA load and physical state for identification of HPV16-infected women with high-grade lesions or cervical carcinoma. *Clin Microbiol* 46: 3678–85.

Schmitt M, Bravo IG, Snijders PJ, et al. (2006). Bead-based multiplex genotyping of human papillomaviruses. *J Clin Microbiol* 44: 504–12.

Snoeck R, Bossens M, Parent D, et al. (2001). Phase II double-blind, placebo-controlled study of the safety and efficacy of cidofovir topical gel for the treatment of patients with human papillomavirus infection. *Clin Infect Dis* 33: 597–602.

Solomon D, Davey D, Kurman R, et al., (2001). The 2001 Bethesda System: Terminology for reporting results of cervical cytology. *JAMA* 287: 2114–9.

Stanley M. (2006). Immune responses to human papillomavirus. *Vaccine* 24: S16–22.

Stanley MA. (2012). Genital human papillomavirus infections: Current and prospective therapies. *J Gen Virol* 93: 681–91.

Syrjänen K, Syrjänen S. (2000). *Papillomavirus Infections in Human Pathology*. London: Wiley, pp. 1–615.

Terlou A, vanSeters M, Kleinjan A, et al. (2010). Imiquimod induced clearance of HPV is associated with normalization of immune cell counts in usual type vulvar intraepithelial neoplasia. *Int J Cancer* 127: 2831–40.

Tropé A, Sjøborg K, Eskild A, et al. (2009). Performance of human papillomavirus DNA and mRNA testing strategies for women with and without cervical neoplasia. *J Clin Microbiol* 47: 2458–64.

Turan T, Kalantari M, Calleja-Macias IE, et al. (2006). Methylation of the human papillomavirus-18 L1 gene: A biomarker of neoplastic progression? *Virology* 349: 175–83.

Turan T, Kalantari M, Cuschieri K, et al. (2007). High-throughput detection of human papillomavirus-18 L1 gene methylation, a candidate biomarker for the progression of cervical neoplasia. *Virology* 361: 185–93.

Wang KL. (2007). Human papillomavirus and vaccination in cervical cancer. *Taiwan J Obstet Gynecol* 46: 352–62.

Wang S, Lang JH, Cheng XM. (2009). Cytologic regression in women with atypical squamous cells of unknown significance and negative human papillomavirus test. *Am J Obstet Gynecol* 201: 569. e1–6.

Welters MJ, Kenter GG, Piersma SJ, et al. (2008). Induction of tumor-specific CD4+ and CD8+ T-cell immunity in cervical cancer patients by a human papillomavirus type 16 E6 and E7 long peptides vaccine. *Clin Cancer Res* 14: 178–87.

Yamato K, Yamada T, Kizaki M, et al. (2008). New highly potent and specific E6 and E7 siRNAs for treatment of HPV 16 positive cervical cancer. *Cancer Gene Ther* 15: 140–53.

Yokoyama M, Noguchi M, Nakao Y, et al. (2004). The tea polyphenol, (-)-epigallocatechin gallate effects on growth, apoptosis, and telomerase activity in cervical cell lines. *Gynecol Oncol* 92: 197–204.

Yokoyama M, Noguchi M, Nakao Y, et al. (2008). Antiproliferative effects of the major tea polyphenol, (-)-epigallocatechin gallate and retinoic acid in cervical adenocarcinoma. *Gynecol Oncol* 108: 326–31.

Yoshinouchi M, Yamada T, Kizaki M, et al. (2003). In vitro and in vivo growth suppression of human papillomavirus 16-positive cervical cancer cells by E6 siRNA. *Mol Ther* 8: 762–8.

# 10 Human Herpesvirus Type 8/Kaposi Sarcoma Herpesvirus

## *An Overview of Viral Pathogenesis and Treatment of Associated Diseases*

*Hiba El Hajj, Raghida Abou Merhi, and Ali Bazarbachi*

## 10.1 HERPESVIRIDAE

Herpesviruses represent a very large and clearly defined group of viruses of considerable medical importance and uniqueness. Herpesviruses are responsible for—or involved in—many discomforts and diseases, among them are cold sores, skin eruptions and rashes, shingles, mononucleosis, eye infections, birth defects, and cancers.

Of the more than 100 known herpesviruses, eight routinely infect only humans. These are the human herpesviruses (HHVs), and they are divided into three subfamilies: Alphaherpesvirinae, infecting a variable host range and having a short replicative cycle (HHV-1, HHV-2, and HHV-3); Betaherpesvirinae, infecting a restricted host range and having a long replicative cycle (HHV-5, HHV-6, and HHV-7); and Gammaherpesvirinae (HHV-4 and HHV-8) (Table 10.1).

## 10.2 HUMAN HERPESVIRUS TYPE 8/KAPOSI SARCOMA-ASSOCIATED HERPESVIRUS

Of the viruses that infect humans, human herpesvirus 8 (HHV-8) or Kaposi's sarcoma-associated herpesvirus (KSHV) was described as a novel gamma-2 herpesvirus closely related to the human gamma-1 herpesvirus, the Epstein–Barr virus (EBV) (Viejo-Borbolla and Schulz 2003). However, HHV-8 shares a greater

**TABLE 10.1**
**Herpesviruses, Common Names, Modes of Transmission, and Life Cycle Characteristics**

| Subfamily | Common Name | Mode of Transmission | Characteristics of Life Cycle |
|---|---|---|---|
| | | **Alphaherpesvirinae** | |
| HHV-1 | Herpes simplex type 1 | Close contact | Short replicative cycle |
| HHV-2 | Herpes simplex type 2 | Sexually transmitted disease | Short replicative cycle |
| HHV-3 | Varicella-Zoster virus | Respiratory and close contact | Short replicative cycle |
| | | **Betaherpesvirinae** | |
| HHV-5 | Cytomegalovirus | Transfusion, transplants, congenital, and close contact | Long replicative cycle |
| HHV-6 | Herpes lymphotropic virus | Respiratory and close contact | Long replicative cycle |
| HHV-7 | - | Saliva? | Long replicative cycle |
| | | **Gammaherpesvirinae** | |
| HHV-4 | Epstein–Barr virus | Saliva | Long replicative cycle |
| HHV-8 | Kaposi's sarcoma herpesvirus | Sexual, saliva? | Long replicative cycle |

degree of homology with viruses of the genus *Rhadinovirus*, which includes a simian herpesvirus (the genus prototype herpesvirus saimiri, or HVS) (Moore et al. 1996b), the rhesus rhadinovirus (RRV) (Alexander et al. 2000), and the murine gammaherpesvirus 68 (Virgin et al. 1997). It was identified first from the Kaposi sarcoma (KS) of an acquired immunodeficiency syndrome (AIDS) patient (Chang et al. 1994) and belongs now to the human oncogenic herpesviruses (Verma et al. 2007). These are viral oncogenes that modulate the host cell machinery by constitutively activating growth-signaling pathways (Boshoff and Weiss 2002; Hengge et al. 2002).

### 10.2.1 HHV-8 Life Cycle

Like most herpesvirus infections, HHV-8 infection also exhibits two different cycles: a latent cycle and a lytic cycle, each with completely different viral gene expression profiles. During its latent cycle, the HHV-8 genomic DNA persists within the host cell with restricted viral gene expression (Zhong et al. 1996; Sarid et al. 1998; Jenner et al. 2001; Paulose-Murphy et al. 2001; Fakhari and Dittmer 2002; Nakamura et al. 2003). Indeed, there is no viral progeny produced, and the HHV-8 genome is maintained as multiple episomes that circularize from the viral linear DNA to form 50–150 copies per cell after infection (Moore et al. 1996b; Renne et al. 1996a). However, certain conditions, including environmental stress and chemical stimuli, induce the activation of the viral genome—the passage from the latent to the lytic cycle. This process leads to the release of the progeny viral particles from the host cell. Although, in most cases, HHV-8 maintains a stringent latent infection, 1–5% of virally infected cells undergo spontaneous lytic reactivation (Cesarman et al. 1995a,

1995b; Renne et al. 1996; Staskus et al. 1997; Cannon et al. 2000). Thus, HHV-8 lytic replication and the continued infection of new cells are likely to be essential for the maintenance of the population of infected cells and for the development of viral pathogenesis (Grundhoff and Ganem 2004; Wang et al. 2004a, 2004b). HHV-8 viral DNA is replicated once per cell cycle and partitioned into daughter cells along with the host cell chromosomes (Hu et al. 2002; Verma et al. 2007). Lytic replication is marked by an increase in gene expression and the production of infectious virus progeny. Although lytic replication of HHV-8 may help create a microenvironment that enhances the growth of latently infected cells, and some lytic genes may activate intracellular pathways regulating lymphocyte proliferation and survival, it is thought that the oncopathology of HHV-8 is mainly due to the viral products produced in latency; therefore, these genes are believed to be most directly involved in malignant transformation (Carbone and Gaidano 1997).

### 10.2.2 HHV-8 Transmission, Prevalence, and Epidemiology

HHV-8 virus was first thought to exclusively be a sexually transmitted disease (STD) in men with homosexual behavior. However, HHV-8 is transmitted through horizontal and vertical (mother to child and sexual) as well as parenteral (blood transfusion) routes (Edelman 2005). Finally, it can also be transmitted during organ transplantation (Edelman 2005). The virus is shed frequently from the oropharynx of immunocompetent and immunocompromised carriers in endemic areas. The development of serologic assays allowed seroepidemiologic studies and confirmed that HHV-8 prevalence varies widely. Among the general population, HHV-8 seropositivity is less than 10% in northern Europe, 1–3% of blood donors in North America and Asia, 10–30% in the Mediterranean region, and more than 50% in most of sub-Saharan Africa (Edelman 2005). The homosexual population exhibits higher positivity (8–25%) than the general population (Fukumoto et al. 2011).

### 10.2.3 HHV-8 Genome

The HHV-8 genome is housed in an icosahedral capsid approximately 1,200 angstroms in diameter (Trus et al. 2001) with a typical herpesvirus envelope and tegument. The HHV-8 genome was fully cloned and sequenced (Moore et al. 1996; Russo et al. 1996; Zhong et al. 1996). It consists of around 170 kb, double-stranded, linear, viral, genomic DNA flanked by multiple guanine cytosine (GC)-rich terminal repeats (Renne et al. 1996; Russo et al. 1996) and may exist in a circular episomal form during latency (Lagunoff and Ganem 1997). More than 80 open reading frames, prefaced with *orf*, were numbered consecutively from the left-hand side of the viral genome (Russo et al. 1996). Some frames were unique and specific to this virus and were designated as K1, K2, K3, … and K15 (Russo et al. 1996) (Figure 10.1).

The viral gene expression profile differs between the latent and lytic cell cycles (Figure 10.2). During latency, HHV-8 expresses mainly genes from the circular viral episomes that include the latency-associated nuclear antigens (LANA-1 and -2, encoded by *orf* 73 and 10.5, respectively) (Nador et al. 1996; Ballestas et al. 1999; Friborg et al. 1999; Simonelli et al. 2003), viral cyclin D (*orf*72), and viral Fas-ligand

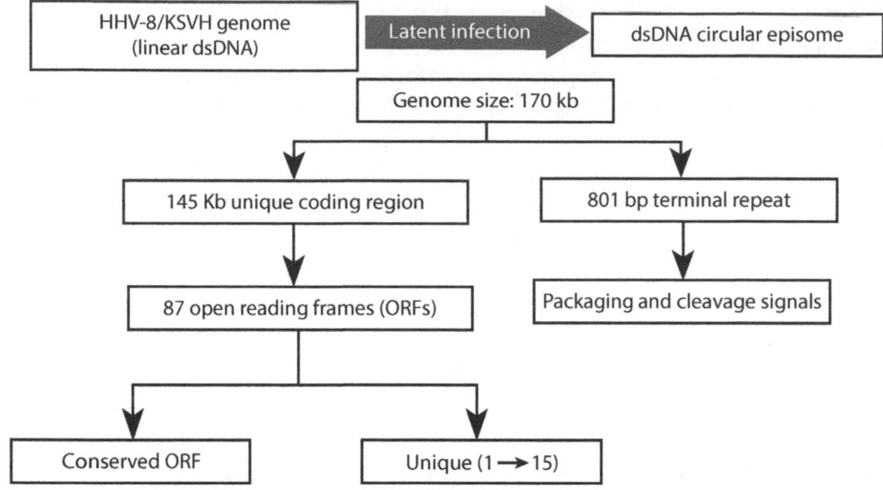


**FIGURE 10.1** HHV-8 genome structure and size.

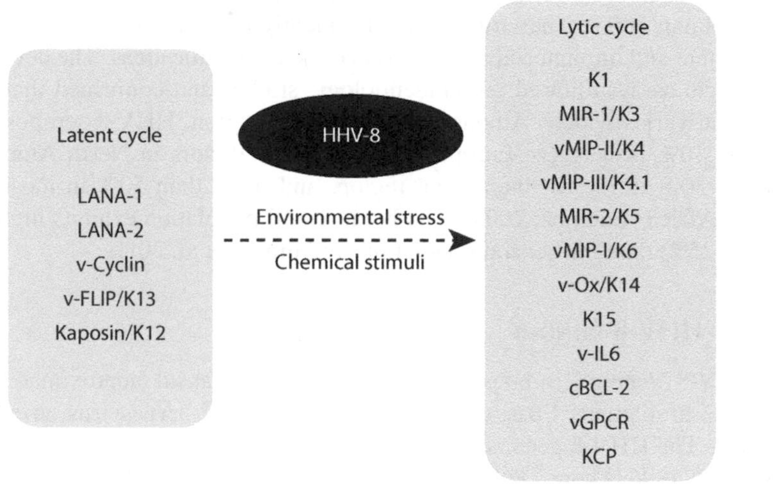


**FIGURE 10.2** HHV-8 protein expression profile during lytic and latent life cycles.

interleukin-1B-converting enzyme inhibitory protein (vFLIP encoded by K13) (Irmler et al. 1997; Swanton et al. 1997; Djerbi et al.1999). LANA-1, v-cyclin, and vFLIP are all expressed from the same locus and are transcribed from a common promoter as part of a multicistronic messenger RNA (mRNA) in the latently infected cells (Cesarman et al. 1995; Dittmer et al. 1998; Grundhoff and Ganem 2004). Importantly, transgenic mice that express LANA1, LANA2, vFLIP, or v-cyclin develop lymphoid malignancies with low frequency after a long latency (Verschuren et al. 2004; Chugh et al. 2005; Fakhari et al. 2006).

#### 10.2.3.1 Latency viral-expressed Genes

##### *10.2.3.1.1 LANA-1*

The most predominantly expressed HHV-8 antigen is LANA-1 (Komatsu et al. 2002). Its main function is to tether the viral episome (Ballestas et al. 1999) to the host chromosome during mitosis, ensuring right partitioning of the viral genome into progeny cells (Chadburn et al. 2004). Thus, it is essential for the equal segregation and maintenance of viral DNA during replication (Ballestas et al. 1999). During latency, LANA-1 ensures and maintains latency through autoactivating its own promoter, presumably to maintain the expression of latent proteins while suppressing other viral genes (Jeong et al. 2001, 2004; Renne et al. 2001). It has been shown that it causes impaired apoptosis due to p53 inhibition (Friborg et al. 1999; Si and Robertson 2006) and retinoblastoma (Rb) binding (Radkov et al. 2000). The consequences of this process include cell cycle progression, impaired apoptosis, and increased hypoxia-inducible factor-1$\alpha$ (HIF-1$\alpha$) levels, which lead to activation of genes involved in angiogenesis, cell proliferation, and survival (Cai et al. 2006) (Table 10.2). LANA-1 is a 1,162 amino acid peptide with a calculated molecular mass of 135 kDa (Gao et al. 1996; Kellam et al. 1997). It contains three distinct protein domains: an N-terminal basic domain of 337 aa, a middle 585-aa hydrophilic region, and a C-terminal basic 240-aa domain (Russo et al. 1996). LANA-1 is localized to the nucleus of HHV-8 latently infected cells; a nuclear localization sequence (NLS) between aa 24 and 30 was identified and is homologous to the NLS for EBV (Schwam et al. 2000). LANA-1 can repress or activate transcription (Groves et al. 2001; Hyun et al. 2001; Lim et al. 2001; Renne et al. 2001; Jeong et al. 2004).

##### *10.2.3.1.2 LANA-2*

Another latent protein, LANA-2, also known as viral interferon regulatory factor 3 (vIRF3), is expressed in the nuclei of virtually all HHV-8-infected lymphoid cells in primary effusion lymphoma (PEL) and in the majority of HHV-8-infected cells in multicentric Castleman's disease (MCD) but not in KS (Cunningham et al. 2003). LANA-2 is expressed during latency and antagonizes p53-mediated apoptosis in vitro (Wies et al. 2008). It stimulates c-Myc function (Lubyova et al. 2007) and stabilizes HIF-1$\alpha$ and pro-apoptotic cellular interferon regulatory factor-5 (IRF-5) (Shin et al. 2008). It was shown that it stimulates growth inhibition via G2/M cell cycle arrest (Table 10.2) (Bi et al. 2011). LANA-2 was also shown to directly interact with cellular IRF-3, IRF-7, and the transcriptional co-activator CBP/p300, and to stimulate their transcriptional activity, leading to enhanced expression of type I interferon and chemokine genes (Lubyova et al. 2004).

##### *10.2.3.1.3 v-Cyclin*

v-Cyclin is a viral homolog of cellular cyclin D that binds, as its cellular counterparts, to human cyclin-dependent kinase 6 (CDK6). CDK6 is known to phosphorylate Rb and histone H1 in vitro (Li et al. 1997), thus resulting in resistance to CDK inhibitors, progression through the cell cycle, and uncontrolled cell division (Godden et al. 1997; Adams 2001). v-Cyclin also stably associates with p27-KIP1 in PEL cells in vivo, promoting malignant cells' survival (Järviluoma et al. 2004). However, v-cyclin may also induce genomic instability through centrosomal abnormalities promoting malignant transformation (Table 10.2) (Koopal et al. 2007). It induces lymphoma in

**TABLE 10.2**
**Human Herpesvirus-8 Protein Functions**

| Function | Viral Proteins Latent Cycle | Viral Proteins Lytic Cycle |
|---|---|---|
| Proliferation | LANA-1 | vIL-6 |
| | LANA-2 | vGCPR |
| | v-Cyclin | K1 |
| | vFLIP | K15 |
| | Kaposin | |
| Apoptosis | LANA-1 | vBCL-2 |
| | LANA-2 | vIAP |
| | vFLIP | |
| Genetic instability | LANA-1 | |
| | v-Cyclin | |
| Immune modulation | v-Cyclin | vIRF |
| | | vMIP-I |
| | | vMIP-II |
| | | vMIP-III |
| | | MIR-1 |
| | | MIR-2 |
| | | KCP |
| | | v-Ox |

the absence of p53 in transgenic mice (Verschuren et al. 2002, 2004). In light of these findings, v-cyclin and LANA-1 seem to be synergistic in their tumorigenic activities.

#### *10.2.3.1.4 vFLIP*

vFLIP is a homolog of cellular FLIP, which functions as an inhibitor of FAS-mediated apoptosis and an activator of the transcription factor nuclear factor B (NF-κB) (Guasparri et al. 2004; Golks et al. 2006; Boulanger et al. 2008). vFLIP has been associated with cell survival, morphologic change, and inflammatory activation (Ye et al. 2008). It also inhibits apoptosis and ensures survival of malignant cells (Table 10.2).

#### *10.2.3.1.5 Kaposin*

Along with the proteins of the latency locus, kaposin (also known as K12) is expressed in most HHV-8-infected spindle tumor cells (Kliche et al. 2001). The most common mRNA transcript encoded by the K12 locus is actually 1.5–2.3 kb in size (Kliche et al. 2001), and the translation program for K12 is complex, using canonical and alternative translation initiation codons to encode three proteins (kaposin A, kaposin B, and kaposin C) (Sadler et al. 1999). Although functional characterization of kaposin A and B has been investigated and carried out, no function has been ascribed to kaposin C to date. Kaposin A is a highly hydrophobic type II transmembrane protein and is expressed on the cell surface (Kliche et al. 2001). Kaposin A might activate the extracellular signal regulated kinase (ERK)-1/2 MAPK activated protein Kinase 2 pathway (Table 10.2) and downstream AP-1235 through its interaction with cytohesin-1, a guanine nucleotide exchange factor that regulates the function of β-integrins (Muralidhar et al. 2000). Kaposin B is a 38 kDa protein (Tomkowicz et al. 2002) that upregulates cytokine expression by blocking the degradation of their mRNA transcripts and binds to and activates mitogen activated protein kinase (MK2) resulting in the inhibition of adenylate-uridylate (AU)-rich elements (AREs)-mediated mRNA degradation (McCormick and Ganem 2005).

#### 10.2.3.2 Lytic Viral-Expressed Genes/Viral Cellular Homologs

Under certain environmental factors and with exposure to certain chemicals, HHV-8 switches from its latent to its lytic life cycle. The protein expression profile differs completely in terms of function, and the viral-expressed genes during the lytic cycle are mainly cellular homologs (Figure 10.2). These proteins play several roles, including cell cycle promotion, tumor growth, immune system evasion, and clinical viral pathogenesis (Wang et al. 2004). They include the following.

##### *10.2.3.2.1 K1 Protein*

This is the most heterogeneous HHV-8 type 1 membrane glycoprotein (Lee et al. 2005) that is significantly induced during viral lytic replication. It modulates B cell signaling pathways through activation of NF-κB (Table 10.2) (Samaniego et al. 2001). K1 promotes cell growth and causes immortalization of primary endothelial cells (Wang et al. 2006). K1 transgenic mice develop sarcomas and plasmablastic lymphomas and display constitutive activation of NF-κB, Oct-2, and Lyn (Prakash et al. 2005). In addition, K1 can induce matrix metalloproteinase-9 and vascular endothelial growth factor (VEGF) in endothelial cells (Wang et al. 2004), the two factors implicated in angiogenesis and cell invasion.

##### *10.2.3.2.2 K15 Protein*

K15 has a complex expression profile. The body cavity based lymphoma (BCBL)-1 cell line, derived from HHV-8-associated PEL, shows differentially spliced protein products ranging from 281 to 489 aa in length and 23 to 55 kDa in size that localize to the endoplasmic reticulum, mitochondria, and plasma membrane surface (Choi et al. 2000). K15 is weakly expressed during latency but is strongly upregulated in the viral lytic cycle (Wong and Damania 2006). K15 structural characteristics include a highly hydrophobic N-terminus, a variable number of transmembrane domains, and a C-terminus containing motifs with high sequence identity to putative SH2 and SH3 sarama family kinases (Src) kinase binding domains (Choi et al. 2000; Brinkmann et al. 2003). K15 inhibits B cell receptor (BCR) activation (Choi et al. 2000) and localizes in the lipid rafts that are common carriers of signal transducers; its C-terminus binds to Tumor necrosis factor receptor associated factor (TRAF)-1, TRAF-2, and TRAF-3 (Brinkmann et al. 2003). K15 interaction with TRAF-2 strongly activates the NF-κB pathway and the AP-1 transcription factor through the Jun Nuclear Kinase (JNK) and ERK MAP kinase (Table 10.2) (Brinkmann et al. 2003).

##### *10.2.3.2.3 vIAP*

This viral protein is a homolog of the cellular protein survivin and is not found in any other herpesvirus. Both the viral and cellular survivins contain a conserved

BH2 domain and a partial IAP repeat domain (Wang et al. 2006); viral inhibition of apoptosis (VIAP) is a 19–21 kDa glycoprotein that localizes in the host cell mitochondrial membrane and inhibits apoptosis, induced by Fas, TRAIL, Bax, TNF-α, cycloheximide, staurosporine, and ceramide (Table 10.2) (Wang et al. 2002, 2006). It protects HHV-8-infected cells from apoptotic death by preventing mitochondrial damage (Feng et al. 2002) or by suppressing caspase 3 function (Wang et al. 2006).

#### *10.2.3.2.4 The Viral Analog of B Cell Leukemia-2*

This is a viral lytic protein encoded by ORF-16 and expressed in both spindle cells and monocytes in KS lesions. It only displays 15–20% amino acid identity to cellular B cell leukemia (Bcl-2) but contains the critical BH1 and BH2 domains required to heterodimerize with cellular Bcl-2 and potently suppresses caspase 3 death effector functions (Sarid et al. 1997). It plays a role in inhibiting apoptosis of virally infected cells (Table 10.2) (Friborg et al. 1998). Indeed, the viral analog of B cell leukemia-2 (vBcl-2) prevents apoptosis induced by v-cyclin.

#### *10.2.3.2.5 The Viral G-Protein-Coupled Receptor*

The viral G-protein-coupled receptor (vGPCR) is encoded by *orf*74 and represents a 7-transmembrane, IL-8 receptor homolog. It constitutively signals several pathways downstream of multiple G protein subunits in a phospholipase C- and phosphatidylinositol 3-kinase (PI-3K)-dependent manner (Cannon and Cesarman 2004). This protein is expressed during lytic viral cycle and induces endothelial cell survival and growth factor and cytokine production through activation of the protein kinase pathways PKC, PKB, Akt, NF-κB, and MAPK (Jenner and Boshoff 2002; Dourmishev et al. 2003; Greene et al. 2007) (Table 10.2). These signaling pathways could directly modulate the pathogenesis of the virus because mice transfected with vGPCR develop tumors with a histology similar to that of KS (Yang et al. 2000), while cells transfected with vGPCR secrete VEGF (Bais et al. 1998), which is found in excess in KS tumor tissue (Bais et al. 1998); vGPCR increases expression of a number of autocrine and paracrine cytokines and growth factors such as interleukin 1 β (IL-1β), IL-6, IL-8, granulocyte macrophage-colony stimulating factor (GM-CSF), tumor necrosis factor alpha (TNF-α), basic fibroblast growth factor (BFGF), growth-regulated protein alpha (Gro-α), VEGF, and monocyte chemotactic protein 1 (MCP-1), which might be essential for early KS cell proliferation and tumor angiogenesis and inflammation (reviewed in Greene et al. 2007). Expression of vGPCR in endothelial cells leads to constitutive activation of VEGF receptor and cell immortalization (Table 10.2) (Bais et al. 2003; Grisotto et al. 2006); transgenic mice expressing vGPCR develop highly vascular "KS-like" lesions (Yang et al. 2000). Interestingly, vGPCR may also regulate the expression of latent and lytic genes (Staudt and Dittmer 2006).

#### *10.2.3.2.6 The Viral Interleukin-6*

Viral interleukin 6 (vIL-6) represents another human cytokine analog that is directly produced by HHV-8 during its replication. Although its level of homology is less than the v-BCL2 and represents only 25% of homology with the cellular IL-6, it can stimulate multiple cellular pathways to promote cellular proliferation and survival

(Moore et al. 1996; Osborne et al. 1999). vIL-6 also plays an important role in HHV-8 immune modulation, protecting PEL cells from IFN-α-induced antiviral response (Chatterjee et al. 2002). In addition, vIL-6 induces the secretion of cellular IL-6 and VEGF to support cell growth of IL-6-dependent cells and contributes to hematopoiesis, tumorigenesis, and angiogenesis (Table 10.2) (Aoki et al. 1999).

##### *10.2.3.2.7 The Viral Macrophage Inflammatory Proteins*

The viral macrophage inflammatory proteins (vMIPs) are three HHV-8 lytic proteins, vMIP-I, vMIP-II and vMIP-III, also known as viral chemokine/cytokine ligands (vCCLs), vCCL-1 (K6), vCCL-2 (K4), and vCCL-3 (K4.1), respectively. They share 25–40% homology with human macrophage inflammatory protein1α (MIP-1α) (Moore et al. 1996). They all are encoded by the unique viral genes K4 and K6, playing a key role as proinflammatory and angiogenic factors. In addition, vMIP-II plays a role in immune system evasion (Table 10.2), possibly by restricting the recruitment of Th1 lymphocytes to HHV-8-infected cells (Weber et al. 2001).

##### *10.2.3.2.8 MIR-1/K3 and MIR-2/K5*

HHV-8 encodes two really interesting new gene (RING) finger E3 ubiquitin ligases (MIR1 and MIR2) that mediate ubiquitination and degradation of cellular proteins important for the establishment of an efficient antiviral immune response. MIR1 and MIR2 share 30% sequence identity, but their substrate preferences are different. MIR1 primarily downregulates major histocompatibility complex class I (MHC-I), whereas MIR2 can downregulate a wide range of cell surface proteins (Coscoy and Ganem 2000, 2001). Both MIR-1 and MIR-2 are expressed during lytic replication and are localized at the endoplasmic reticulum. They negatively modulate the cytotoxic T lymphocyte (CTL)-dependent immune response by triggering endocytosis of the surface MHC class I molecules for degradation (Table 10.2) (Coscoy and Ganem 2000; Ishido et al. 2000).

##### *10.2.3.2.9 The Viral Interferon Regulatory Factors*

HHV-8 encodes three homologs of human IFN regulatory factors, including viral interferon regulatory factor 1 (vIRF-1) and vIRF-2 encoded by K11.1 and the previously discussed vIRF-3 or LANA-2 (Gao et al. 1997; Burysek et al. 1999); vIRF-1 and vIRF-2 are produced by the virus as early lytic genes expressed in KS lesions (Wang et al. 2001; Dittmer 2003). With vMIP-II, they evade the immune system (Coscoy and Ganem 2000). This immune system evasion can also happen through other mechanisms such as repression of interferon transcription, inhibition of MHC class 2-mediated T-cell activation, prevention of activation by HHV-8 antigens through K1 (Lee et al. 2000), and reduction in number of class 1 MHC molecules on the cell surface via endocytosis induced by K3 and K5 (Table 10.2) (Coscoy and Ganem 2000).

##### *10.2.3.2.10 Complement-Control Protein*

The ORF4-encoded complement-control protein (known as KCP) inhibits the activation of the complement cascade (Spiller et al. 2003a, 2003b, 2006). It represents a complement, bridging innate and adaptive immune responses and humoral and cell-mediated immunity. KCP is also associated with the envelope of purified KSHV virions where it potentially protects them from complement-mediated immune response (Table 10.2) (Spiller et al. 2003b).

*10.2.3.2.11 K14 (vOX2)*

Viral OX2 (vOX2) is a homolog of the cellular protein OX2, a glycosylated cell surface protein and member of the immunoglobulin superfamily that restricts cytokine production in a paracrine fashion. It has a structure similar to cellular OX2. However, in contrast to cellular OX2, vOX2 potently activates inflammatory cytokine production (Chung et al. 2002). Indeed, vOX2 induces the production of IL-1$\beta$, TNF-$\alpha$, and IL-6 from various cell types, including monocytes/macrophages and dendritic cells, and cooperates with IFN-$\gamma$ in a paracrine manner to induce cytokine production in a B cell line (Table 10.2) (Chung et al. 2002).

## 10.3 HUMAN HERPESVIRUS-8-ASSOCIATED DISEASES

HHV-8 is associated with a wide spectrum of malignant disorders, including PEL, solid-extracavitary lymphoma, and KS, as well as a nonmalignant disorder—MCD (Cesarman et al. 1995; Soulier et al. 1995; Otsuki et al. 1996; DePond et al. 1997) (Table 10.3).

### 10.3.1 Multicentric Castleman's Disease

MCD was first described by Dr. Benjamin Castleman in 1954 in a patient who had a solitary hyperplastic mediastinal lymph node with a regressive germinal center (Castleman and Towne 1954). The most important feature of this disease is that it is a polyclonal and not a malignant disorder, and so it is called a lymphoproliferative disorder instead of a malignancy. It is part of a diverse group of lesions clinically

**TABLE 10.3**
**Comparison of MCD, PEL, and KS**

| | MCD | PEL | KS |
|---|---|---|---|
| Clonality | Polyclonal | Monoclonal | Monoclonal |
| Disease | Lymphoproliferative disorder | Malignant disorder | Malignant disorder |
| Involved organs | Lymph nodes and spleen | Body cavity (pleural, peritoneal, or pericardial), extranodal sites | Cutaneous or mucosal lesions; visceral involvement; hematologic manifestations |
| Cell type | Plasmablasts | Immunoblasts | Spindle-shaped KS cells |
| Latency/Lytic gene status | HHV-8 mostly negative but very few positive cells expressing lytic genes | HHV-8 positive; expressing mostly latency genes | HHV-8 positive; expressing mostly latency genes |
| EBV status | Negative | Positive in majority | Negative |

classified as either localized/unicentric disease (UCD) or MCD. UCD is a single, usually mediastinal, angiofollicular lymph node hyperplasia that resolves after surgical resection. MCD is associated with multisystem involvement and a generalized lymphadenopathy. HHV-8 infection is found in 100% of human immunodeficiency virus (HIV)-positive MCD cases but only in 40–50% of HIV-negative MCD cases (Soulier et al. 1995; Luppi et al. 1996; Chadburn et al. 1997; Gessain 1997). However, exacerbation of symptoms related to MCD correlates with an increased HHV-8 viral load in peripheral blood mononuclear cells, supporting its role in the pathogenesis of this disorder (Grandadam et al. 1997). The clinical manifestations of MCD include fever, weakness, generalized lymphadenopathy, and hypergammaglobulinemia due in part to raised serum concentrations of interleukin 6 (Oksenhendler et al. 1996; Grandadam et al. 1997). Morphologically, the cells harboring HHV-8 in MCD resemble immunoblasts with prominent central or marginal nuclei (Judde et al. 2000; Parravicini et al. 2000). These cells occur as isolated cells in the mantle zone of B cell follicles but may eventually form monoclonal microlymphomas (Du et al. 2001). In HHV-8-positive MCD, there are medium-sized cells with a moderate amount of amphophilic cytoplasm and a large vesicular nucleus containing up to three prominent nucleoli. These are called “plasmablasts.” As immunoblasts, these cells can be isolated in the mantle zone of B cell follicles but coalesce to form microlymphoma or frank plasmablastic lymphoma in some cases (Dupin et al. 2000; Chadburn et al. 2008). Most KSHV-positive plasmablasts are positive for the proliferation marker Ki67. Phenotypically, KSHV-positive plasmablasts resemble mature B cells (Dupin et al. 2000; Du et al. 2001; Li et al. 2006) (Table 3).

### 10.3.2 Primary Effusion Lymphoma

Shortly after the discovery of HHV-8 in KS, the presence of the viral genome was detected in a subset of lymphomas in AIDS patients. These patients presented lymphomatous effusions in body cavities that were known as body cavity-based non-Hodgkin lymphoma (Cesarman et al. 1995). Later, it was shown that these lymphomatous effusions present distinct clinical, morphological, immunophenotypic, and molecular characteristics and were called PELs (Nador et al. 1997).

A PEL is a rare aggressive lymphoma, accounting for approximately 3% of AIDS-related lymphomas (Carbone et al. 1996). It occurs mainly, but not exclusively, in HIV-positive patients, often middle-aged homosexual males. Typically, patients with PEL present with effusions in the pleural, pericardial, or abdominal cavities, usually in the absence of an obvious tumor mass, lymphadenopathy, or hepatosplenomegaly (Nador et al. 1997; Banks and Warnke 2001). The neoplastic cells are pleomorphic and show a range of cytomorphological appearances from features of large immunoblastic or plasmablastic cells to those of anaplastic cells (Nador et al. 1997; Banks and Warnke 2001). Immunophenotypically, expression of CD45 antigen (>90% of cases) confirms the hematolymphoid derivation of PEL cells, which exhibit an indeterminate immunophenotype because they usually lack classic B cell markers (CD19, CD20) and T-cell markers (CD2, CD3, CD5, CD7) but express CD45 and post-germinal center B cell/plasma cell-associated antigens (CD38, CD138) (Carbone and Gloghini 2008). On the contrary, immunogenotypic studies revealed that PEL is consistently represented by a monoclonal B cell population. Taken together, the immunophenotypic and immunogenotypic characteristics of PEL suggest that mostly this lymphoma represents the malignant counterpart of a B cell that has reached a mature stage of development and is shifting toward terminal plasma cell differentiation (Carbone et al. 2001).

In addition, LANA-1 expression is crucial for distinguishing PEL from a range of lymphoma subtypes such as Hodgkin lymphoma, Burkitt's lymphoma, T-lymphoblastic lymphoma, and anaplastic large cell lymphoma. LANA-1 expression can be detected by immunocytochemistry in the nuclei of most PEL cells and is considered the most valuable marker for diagnosis and differential diagnosis of PEL. In the majority of PEL, the neoplastic cells are coinfected by EBV but exhibit a restricted latency pattern with a lack of detectable expression of latent membrane proteins (LMPs) (Nador et al. 1997; Banks and Warnke 2001). Finally, the gene expression profile of PEL shares features of immunoblasts and plasma cells but is clearly distinct from both (Table 10.3) (Jenner et al. 2003; Klein et al. 2003).

### 10.3.3 Extracavitary PEL

Although the vast majority of PELs occur exclusively as lymphomatous effusions, some PEL patients subsequently develop secondary tissue masses while others present initially solid masses followed (or not) by lymphomatous effusion (Otsuki et al. 1996; Depond et al. 1997; Huang et al. 2004). In both cases, the lymphomatous effusion and the solid tissue cells show similar morphology and immunophenotype as well as identical immunoglobulin rearranged genes (Otsuki et al. 1996; Depond et al. 1997; Huang et al. 2004). PEL and extracavitary PEL are indistinguishable in their morphological and immunophenotypic presentations. These lymphomas do not express B cell antigens, are usually positive for EBV, and harbor mutated Ig genes. Later on, Chadburn et al. (2004) extensively investigated HHV-8-associated solid lymphoma in comparison with PEL and proposed that HHV-8-positive solid lymphomas be designated as extracavitary PELs.

### 10.3.4 Kaposi Sarcoma

KS was first described by the Viennese dermatologist Moritz Kaposi (1837–1902) more than a century ago (Kaposi 1872). Its association with AIDS was discovered in the early 1980s. In addition, HHV-8 was identified in KS lesions as compared to EBV, which was negative (Chang et al. 1994). KS is a low-grade vascular tumor that may involve the skin, mucosa, and viscera (Antman and Chang 2000). KS is the most prevalent malignancy among patients with AIDS as well as in patients with drug-related or transplant-associated immunosuppression. There are four recognized epidemiologic–clinical forms: classic, African (endemic), AIDS-associated (epidemic), and iatrogenic (or transplant-associated) KS (reviewed by Radu and Pantanowitz 2013).

African KS follows a more aggressive course than classic KS, especially the lymphadenopathy form that affects young individuals. AIDS-associated KS usually arises in HIV-positive patients with low CD4 T-cell counts (Sullivan et al. 2008). It is a very aggressive disease that typically manifests with disseminated lesions and visceral involvement. Iatrogenic KS is associated with immunosuppression due to drugs or after transplantation. KS occurs mainly in renal transplant recipients and, infrequently, after other solid organ or bone marrow transplants. Post-transplant KS may result from reactivation of latent HHV-8 infection in recipients or from tumor cells contributed from organ donors (reviewed by Radu and Pantanowitz 2013).

In all of the aforementioned forms of KS, KS lesions evolve from early (patch stage) macules into plaques (plaque stage) that may subsequently develop into larger nodules (tumor stage). These tumors may ulcerate, cause marked lymphedema, present as exophytic growths (e.g., cutaneous horns), or invade subjacent tissues (e.g., underlying bone). Different stages can coexist in the same individual at the same time. Similar stages of KS growth apply to cutaneous and mucosal lesions. Although KS commonly presents at mucocutaneous sites, tumors may involve lymph nodes as well as visceral organs, most notably the respiratory and gastrointestinal tracts (Table 10.3) (reviewed in Radu and Pantanowitz 2013).

## 10.4 TREATMENT STRATEGIES AGAINST HHV-8-ASSOCIATED DISEASES

Despite significant progress over the past decade in understanding the pathophysiology of HHV-8-associated diseases, treatment today remains toxic and incompletely efficacious. In addition to highly active antiretroviral therapy (HAART) in HIV-infected individuals, several treatment strategies are currently used against HHV-8-associated diseases. These include vascular endothelial growth factor inhibitors (VEGFIs), tyrosine kinase inhibitors (TKI), monoclonal antibody-based therapies, matrix metalloproteinase inhibitors (MMPIs), local therapy, and chemotherapy.

### 10.4.1 Treatment of Castleman's Disease

Complete surgical excision of the affected lymph node is the method of choice in UCD patients and has led so far to high cure rates (Bowne et al. 1999). Although the complete resection is technically difficult, radiotherapy yields good cure rates (Chronowski et al. 2001). MCD patients are more complicated in terms of diagnosis, prognosis, and treatment strategies. Several therapeutic options have been used so far. However, the gold standard is yet to be reached. The current adopted strategies include chemotherapy, antiviral therapy, immunomodulators, and targeted therapy (Table 10.4).

#### 10.4.1.1 Chemotherapy

Various single-agent drugs have been used to treat MCD, including chlorambucil, corticosteroids, cyclophosphamide, 2-chlorodeoxyadenosine, carmustine, vincristine, and bleomycin. Single-agent liposomal doxorubicin, oral etoposide, and vinblastine were reported to produce durable remission predominantly in HIV-positive patients (Oksenhendler et al. 1996; Scott et al. 2001; Loi et al. 2004). In addition to these single agents, and due to the complexity of the MCD and its poor response, multiple combinations have been tried. These include chlorambucil/prednisone;

## TABLE 10.4
## Summary of Treatment Strategies Used for Castleman's Disease

| Treatment Strategy | Unicentric Castleman's Disease (UCD) | Multicentric Castleman's disease (MCD) |
|---|---|---|
| Surgical resection | + | – |
| Radiotherapy | + | – |
| Chemotherapy | – | **Single agents**:<br>Chlorambucil, corticosteroids, cyclophosphamide, 2-chlorodeoxyadenosine, carmustine, vincristine, bleomycin, doxorubicin, oral etoposide, vinblastine<br>**Combinations:**<br>Chlorambucil/prednisone<br>Cyclophosphamide/prednisone, cyclophosphamide/procarbazine, cyclophosphamide/vincristine/prednisone, cyclophosphamide/doxorubicin/vincristine/ prednisone (CHOP)<br>Cyclophosphamide/vincristine/doxorubicin/ dexamethasone<br>High-dose melphalan followed by autologous stem cell transplantation |
| Antiviral therapy | | |
| Anti-HHV-8 therapy | – | Valganciclovir/ganciclovir |
| HAART | | **Nucleoside or nucleotide reverse transcriptase inhibitors:**<br>Efavirenz, tenofovir, emtricitabine, zidovudine, lamivudine, abacavir, stavudine or tenofovir disoproxil fumarate<br>**Non nucleoside reverse transcriptase inhibitors:**<br>Cidofovir, efavirenz, nevirapine, or etravirine<br>**Protease inhibitors:**<br>Atazanavir, saquinavir, ritonavir, indinavir, nelfinavir, fosamprenavir, lopinavir, darunavir, or tipranavir |
| Immunomodulators | – | + |
| IFN-alpha | – | + |
| Thalidomide | – | + |
| Targeted therapy | + | + |
| Monoclonal antibodies | Rituximab | Rituximab<br>Il-6 and Il-6-R antibodies (rhPM-1….)<br>IL-1 or IL1R (Anakinra) |
| Proteasome inhibitors | – | Bortezomib<br>Bortezomib/dexamethasone |

cyclophosphamide/prednisone; cyclophosphamide/procarbazine; cyclophosphamide/vincristine/prednisone;cyclophosphamide/doxorubicin/vincristine/prednisone(CHOP); cyclophosphamide/vincristine/doxorubicin/dexamethasone; and high-dose melphalan followed by autologous stem cell transplantation (Ogita et al. 2007; El-Osta and Kurzrock 2011) (Table 10.4).

#### 10.4.1.2 Antiviral Therapy

##### *10.4.1.2.1 Anti-HHV-8 Therapy*

The high degree of HHV-8 lytic replication in MCD made anti-HHV-8 viral therapy an attractive approach. The general antiherpes medications, ganciclovir, foscarnet, and cidofovir, were shown to have in vitro activity against HHV-8 (Neyts and De Clercq 1997; Friedrichs et al. 2004; Casper 2008). In HIV-associated MCD patients, treatment with valganciclovir/ganciclovir showed durable remission in two out of three patients (Table 10.4) (Casper et al. 2004).

##### *10.4.1.2.2 Highly Active Antiretroviral Therapy*

HAART is defined as treatment with at least three active antiretroviral medications typically using two nucleoside or nucleotide reverse transcriptase inhibitors (efavirenz, tenofovir, emtricitabine, zidovudine, lamivudine, abacavir, stavudine, or tenofovir disoproxil fumarate) plus a non nucleoside reverse transcriptase inhibitor (cidofovir, efavirenz, nevirapine, or etravirine) or a protease inhibitor (atazanavir, saquinavir, ritonavir, indinavir, nelfinavir, fosamprenavir, lopinavir, darunavir, or tipranavir).

The implementation of HAART to treat HIV/AIDS patients has dramatically boosted survival. In patients with HIV-associated MCD, the initiation of HAART led to MCD exacerbation (probably because of the immune reconstitution syndrome) (Zietz et al. 1999). Nevertheless, the overall survival rate was remarkably higher, and the progression rate to lymphoma was 20 times lower in HIV-related MCD (Table 10.4) (Aaron et al. 2002; Oksenhendler et al. 2002; Oksenhendler 2009).

#### 10.4.1.3 Immunomodulators

##### *10.4.1.3.1 Interferon Alpha*

Single-agent interferon-α was used first in HIV-associated MCD and showed significant clinical benefits (El-Osta and Kurzrock 2011). Furthermore, a long-term complete remission was observed in MCD patients off treatment after a four-year period (Table 10.4) (Tamayo et al. 1995).

##### *10.4.1.3.2 Thalidomide*

Thalidomide is widely used in plasma cell malignancies such as multiple myeloma in which IL-6 plays an important role in disease progression. Because IL-6 is highly involved in MCD, thalidomide was tried and showed a beneficial effect (Table 10.4) (El-Osta and Kurzrock 2011). Its therapeutic value in MCD patients might therefore result from disrupting IL-6 production. In a patient with MCD, associated with pemphigus vulgaris, the IL-6 level decreased significantly concordantly with clinical improvement (Miltenyi et al. 2009). Other reports demonstrated a decline in the level of c-reactive protein (CRP), a surrogate of IL-6 activity (Stary et al. 2008).

#### 10.4.1.4 Targeted Therapy

*10.4.1.4.1 Monoclonal Antibodies*

*10.4.1.4.1.1 Rituximab* Rituximab is a monoclonal chimeric antibody that targets CD20 on B cells, leading to their depletion via activating complement-dependent cytotoxicity and antibody-dependent cell-mediated cytotoxicity. Rituximab was first introduced in HIV-positive HHV-8-positive MCD patients (Marcelin et al. 2003). However its efficiency was detected in HIV-positive and HIV-negative patients (El-Osta and Kurzrock 2011). It was later combined with chemotherapy where 22 out of 24 (92%) HIV-positive patients achieved response, and 17 (71%) had a sustained remission at one year (Table 10.4) (Marcelin et al. 2003). In the same study, Marcelin et al. investigated the therapeutic effect of rituximab in five HIV-positive and HHV-8-positive patients. Of these, two patients died quickly after initiation of treatment, and three had complete remissions after a follow-up period of 4–14 months. Clinical response was associated with a decrease in viral load (Marcelin et al. 2003). Another study included 21 HIV-positive patients with plasmablastic type MCD and showed that 14 patients (67%) achieved a partial response to rituximab, whereas 29% had stable disease. The two-year overall survival rate was 95%; the progression-free survival rate was 92% at one year and 79% at two years. Interestingly, in that study, rituximab therapy was associated with a decline in the HHV-8 viral load as well as in IL-10 and IL-6 levels after treatment completion (Bower et al. 2007). Rituximab has also been used successfully in other circumstances, such as in UCD (Table 10.4) (Estephan et al. 2005).

*10.4.1.4.1.2 Targeting IL-6 or IL-6R* Many of the signs and symptoms of MCD may be mediated by the production of vIL-6, which has been pirated by HHV-8. A mouse model showed that the expression of IL-6 produces a disease that histologically and clinically resembles MCD (Brandt et al. 1990). Hence, targeting IL-6 or IL-6R represents an important goal in the treatment of MCD. Beck et al. reported a case of MCD associated with elevated IL-6 levels and used a murine anti-IL-6 antibody to treat it (Beck et al. 1994). However, this murine monoclonal antibody was shortly neutralized by human antibodies; hence, the response was transient, and the disease relapsed after termination of treatment (Beck et al. 1994). Subsequently, a humanized antibody to IL-6 receptor was developed and tested on 28 Japanese $HIV^-MCD^+$ patients, two of whom were HHV-8-positive (Nishimoto et al. 2005). Twelve of 23 (52%) participants who had lymphadenopathy saw a significant reduction in the size of their lymph nodes at one year (Table 10.4) (Nishimoto et al. 2005). This study represents the largest clinical trial on MCD to date and makes IL-6 receptor antagonists a promising choice for the treatment of MCD. Another humanized anti-IL-6R (rhPM-1) antibody was used earlier in seven MCD patients and produced good response rates. However, patients relapsed directly after treatment discontinuation (Nishimoto et al. 2000).

*10.4.1.4.1.3 Targeting IL-1 or IL-1R* IL-1 activates the NF-κB pathway once it is bound to its receptor. This, in turn, upregulates the expression of multiple inflammatory proteins implicated in the pathogenesis of MCD, particularly IL-6 (Dinarello 2009). Anakinra, a recombinant IL-1R antagonist, was used with success for the first time in a 13-year-old boy with refractory MCD (Galeotti et al. 2008). Furthermore, El-Osta et al. showed that treatment of a 62-year-old patient with refractory MCD using anakinra led to a rapid and durable resolution of clinical symptoms and laboratory abnormalities (Table 10.4) (El-Osta et al. 2010).

##### *10.4.1.4.2 Targeting the Proteasome by Using Bortezomib*

Bortezomib was approved in 2003 for the treatment of relapsed and refractory multiple myeloma. Encouraging outcomes with the use of bortezomib in MCD patients were illustrated in two case reports but both patients were HHV-8 negative. One had a complete remission for more than one year after treatment with bortezomib (Hess et al. 2006), and the other patient sustained complete remission for more than four years after discontinuation of the treatment (a combination of bortezomib and dexamethasone) (Table 10.4) (Sobas et al. 2010).

### 10.4.2 Treatment of PEL

PEL is a rare lymphoma accounting for 2% or less of HIV-associated lymphomas, and it is even more rarely encountered in HIV-seronegative patients. This disease carries a dismal prognosis and a median survival of around six months (Boulanger et al. 2005). Given its rarity, there are very few observational series of patients with PEL and very few prospective trials testing chemotherapy antiviral therapy or targeted therapy. Therefore, there is no clear standard of care established in the treatment of PEL.

#### 10.4.2.1 Preclinical Models of HHV-8-Infected PEL

In preclinical studies, a number of drugs are effective on HHV-8-infected PEL cells (Aoki et al. 2003; Wang et al. 2004; Uddin et al. 2005; Petre et al. 2007; Sarosiek et al. 2010; Goto et al. 2012). Indeed, rapamycin (sirolimus) as well as the combination of IFN-α and zidovudine (AZT) induce apoptosis in PEL cell lines and in nonobese diabetic severe combined immunodeficient (NOD/SCID) mice xenografts (Lee et al. 1999; Gosh et al. 2003; Wu et al. 2005; Sin et al. 2007). The current and most promising treatment strategies in PEL patients are directed toward combining the available antiviral treatments with other agents including chemicals and cytokines.

Arsenic trioxide (arsenic) is a very effective treatment for acute promyelocytic leukemia (APL) (Shen et al. 2004; Lallemand-Breitenbach et al. 2008; Nasr et al. 2008, 2009; Hu et al. 2009; Kogan 2009; Powell et al. 2010; Ghavamzadeh et al. 2011). In human T-cell leukemia virus type 1 (HTLV-1)-associated adult T-cell leukemia (ATL) (Bazarbachi et al. 2004), the combination of arsenic and IFN degrades the viral oncoprotein Tax and cures murine and human ATL (Bazarbachi et al. 1999; El-Sabban et al. 2000; Mahieux et al. 2001; Nasr et al. 2003; Hermine et al. 2004; Kchour et al. 2009; El Hajj et al. 2010). Similarly, in HHV-8 positive PEL cell lines, the combination of arsenic and IFN inhibits growth and NF-κB activation and induces caspase-dependent apoptosis (Abou-Merhi et al. 2007). In a preclinical NOD/SCID mouse model, this combination downregulates the latent viral transcripts LANA-1, vFLIP, and v-Cyc in PEL cells derived from malignant ascites, decreases the peritoneal volume, and synergistically increases survival of PEL mice. These results provide a promising rational for the therapeutic use of arsenic/IFN in PEL patients (El Hajj unpublished data).

#### 10.4.2.2 Chemotherapy

Despite the improvement in therapeutic strategies during the past few years, there is no evidence for the cure of PEL patients with conventional systemic chemotherapy. Moreover, survival with conventional chemotherapy is very poor (Nador et al. 1996; Simonelli et al. 2003). Furthermore, p53 mutations, which are rare in PEL, accumulate in PEL cell lines after chemotherapy (Petre et al. 2007; Chen et al. 2010). High-dose methotrexate with cyclophosphamide, doxorubicin, vincristine, prednisone (CHOP)-derived chemotherapy rescued three complete response (CR) out of seven patients (Boulanger et al. 2003). Two case reports of PEL-HIV patients documented the successful use of high-dose chemotherapy with autologous stem cell transplant (Boulanger et al. 2003; Waddington and Aboulafia 2004; Won et al. 2006; Riva et al. 2012). Furthermore, one rare case of HIV-negative PEL in an elderly HHV-8-negative patient, who developed cardiac tamponade due to pericardial effusion,

**TABLE 10.5**
**Treatment Strategies Used for Primary Effusion Lymphoma**

| Treatment Strategies | Primary Effusion lymphoma |
|---|---|
| Chemotherapy | **Combinations:** |
| | CHOP |
| | CHOP with methotrexate |
| | CHOP with rituximab and cyclophosphamide |
| | High-dose chemotherapy followed by autologous stem cell transplantation |
| Antiviral therapy | |
| Anti-HHV-8 therapy | Cidofovir |
| | Cidofovir with ganciclovir |
| | Valproate followed by the treatment with antiviral agents |
| HAART | **Nucleoside or nucleotide reverse transcriptase inhibitors:** |
| | Efavirenz, tenofovir, emtricitabine, zidovudine, lamivudine, abacavir, stavudine, or tenofovir disoproxil fumarate |
| | **Non nucleoside reverse transcriptase inhibitors:** |
| | Cidofovir, efavirenz, nevirapine, or etravirine |
| | **Protease inhibitors:** |
| | Atazanavir, saquinavir, ritonavir, indinavir, nelfinavir, fosamprenavir, lopinavir, darunavir, or tipranavir |
| Targeted therapy | |
| Monoclonal antibodies | Rituximab |
| | Rituximab combined with pirarubicin/cyclophosphamide/vincristine/prednisolone |
| mTOR-signaling inhibitors | Rapamycin (Sirolimus) |
| Proteasome inhibitors | Bortezomib |
| | Bortezomib and Tyrphostin AG 490 |
| | Bortezomib and HDAC inhibitors |

was treated with rituximab and CHOP showing complete remission without signs of disease 30 months after the last treatment (Table 10.5) (Suzuki et al. 2008).

#### 10.4.2.3 Antiviral Therapy

Antitumor activity of antiviral therapy against HHV-8-associated PEL is based on only case report. Patients with a diagnosis of PEL, related to HIV infection or not, experienced prolonged complete remission after the intracavitary administration of cidofovir (Hocqueloux et al. 2001; Luppi et al. 2005; Halfdanarson et al. 2006), which has a broad activity against multiple DNA viruses, specifically targeting their replication. Another report was published on the use of the combination of cidofovir with ganciclovir, which prompted the adjunctive use of these drugs in PEL (Table 10.5) (Casper and Wald 2007). The same study showed that the use of valproate induces the lytic replication of HHV-8, and when followed by the treatment with antiviral agents, HHV-8-infected cells undergo apoptosis (Casper and Wald 2007).

##### *10.4.2.3.1 HIghly Active Antiretroviral Therapy*

Before introduction of HAART, the therapeutic results of the use of aggressive polychemotherapy regimens—including anthracyclines—in the cohorts of HIV-positive patients were unsatisfactory. The administration of HAART was also applied to the PEL setting. However, the improvement in patients with PEL has been low compared to that reported in patients with HIV (Weiss et al. 2006). Clinical studies reported the prognostic impact of HAART in association with chemotherapy on PEL outcome (Simonelli et al. 2003; Boulanger et al. 2005). Complete remissions of PEL have been reported after implementation of only HAART without any antimitotic drug (Spina et al. 1998; Chen et al. 2007). Thus, initiating or continuing HAART as part of supportive therapy was recommended when commencing treatment for HIV-positive patients with PEL (Table 10.5).

#### 10.4.2.4 Targeted Therapy

##### *10.4.2.4.1 Monoclonal Antibodies*

*10.4.2.4.1.1 Rituximab (Rituxan™)* is a humanized monoclonal antibody against CD20. Individual case reports documented the responses to rituximab-containing chemotherapy regimens (Oksenhendler et al. 1998; Boulanger et al. 2001; Hocqueloux et al. 2001; Ghosh et al. 2003; Lim et al. 2005; Siddiqi and Joyce 2008), or pleurodesis with bleomycin (Yiakoumis et al. 2010). A recent study by Kumode et al. reports the cases of two elderly HHV-8-related HIV-negative PEL patients, where rituximab-based chemotherapy (rituximab, pirarubicin, cyclophosphamide, vincristine, prednisolone, or R-THPCOP) was administered to the two patients and both achieved complete remission (Table 10.5) (Kumode et al. 2013).

##### *10.4.2.4.2 Targeting mTOR Signaling by Rapamycin*

Rapamycin (Sirolimus™) is an inhibitor of mTOR signaling. Its use has been documented in two reports where the patients underwent renal transplantation and later developed PEL (Table 10.5). The first study described the cases of two HIV-1-negative males of African origin who developed PEL while receiving rapamycin as an immunosuppressive treatment. Both patients were retrospectively found to be HHV-8 seropositive before renal transplantation. This case report suggested that rapamycin might not protect HHV-8-infected renal transplant recipients from the occurrence of PEL or the progression of pre-existing PEL (Boulanger et al. 2008). Another more recent study reinforced the previous finding because PEL was poorly controlled by sirolimus and chemotherapy in two male kidney transplant recipients (Régnier-Rosencher et al. 2010).

*10.4.2.4.3 Targeting the Proteosome by Bortezomib*

Bortezomib's (Velcade™) primary mechanism of action is inhibition of the 26S proteasome and nuclear factor B (NF-κB) activity. This approach has been documented in very few cases of PEL. It has been shown to induce apoptosis of PEL cell lines in vitro (An et al. 2004; Abou-Merhi et al. 2007) alone or in combination with chemotherapy (Siddiqi and Joyce 2008). A more recent study compared the immunogenicity of PEL cell death induced by bortezomib with that induced by Tyrphostin AG 490, an inhibitor of janus-activated kinase 2/signal transducer and activator of transcription-3 (JAK2/STAT3). These studies showed that both treatments were able to induce PEL apoptosis with similar kinetics and promote dendritic cells maturation. However, PEL cell death induced by bortezomib alone was more effective in inducing uptake by dendritic cells as compared to AG 490 or a combination of both drugs. These studies concluded that the activation of dendritic cells by bortezomib and AG 490-treated PEL cells might have important implications for a combined chemo and immunotherapy in such patients (Cirone et al. 2012). Very recently, bortezomib was combined with the histone deacetylase (HDAC) inhibitor, suberoylanilide hydroxamic acid (SAHA, also known as vorinostat), in a PEL xenografted mouse model, resulting in a significantly prolonged survival (Table 10.5) (Bhatt et al. 2013).

### 10.4.3 Treatment of KS

With the wide spectrum of KS subtypes and their persistent global burden, continuous efforts are exerted on prevention and therapy. Many treatment strategies have been used depending on the KS subtype. The widest strategies were used in the case of the most aggressive form, AIDS-associated or epidemic KS (Table 10.6).

#### 10.4.3.1 Systemic Chemotherapy

Although early lesions of epidemic KS are highly responsive to antiretroviral therapy, systemic chemotherapy is generally indicated for advanced disease with a poor prognostic index (Lynen et al. 2005; Martellotta et al. 2009; Fatahzadeh 2012). The current first-line systemic therapy for advanced, progressive AIDS-KS includes liposomal anthracyclines (Table 10.6) (Von Roenn 2003; Lynen et al. 2005; Martellotta et al. 2009; Fatahzadeh 2012). Indeed, in patients with moderate to severe AIDS-KS, the addition of liposomal anthracyclines to HAART led to a significantly better response rate (76%) compared to HAART alone (20%) (Martín-Carbonero et al. 2004). The second-line systemic U.S. Food and Drug Administration (FDA)-approved drug in patients, refractory to or intolerant to liposomal anthracycline, is paclitaxel (Table 10.6) (Welles et al. 1998; Von Roenn 2003; Lynen et al. 2005;

**TABLE 10.6**
**Treatment Strategies Used for Kaposi Sarcoma**

| | Kaposi Sarcoma | | | |
|---|---|---|---|---|
| **Treatment Strategies** | **AIDS-Associated KS (Epidemic)** | **Classic KS** | **Transplant-Associated KS (Iatrogenic)** | **African (Endemic)** |
| Chemotherapy | **Single agent:**<br>Paclitaxel<br>Liposomal anthracyclines<br>Oral etoposide<br>Vinorebline<br>**Combinations:**<br>Doxorubicin/bleomycin/vincristine<br>Doxorubicin/bleomycin/vinblastine | Oral etoposide | – | vincristine<br>vinblastine |
| Antiviral therapy | | | | |
| Anti-HHV-8 therapy | | | Ganciclovir/foscarnet<br>Ganciclovir/foscarnet/HAART | |
| HAART | **Nucleoside or nucleotide reverse transcriptase inhibitors:**<br>Efavirenz, tenofovir, emtricitabine, zidovudine, lamivudine, abacavir, stavudine, or tenofovir disoproxil fumarate<br>**Non nucleoside reverse transcriptase inhibitors:**<br>Cidofovir, efavirenz, nevirapine, or etravirine | **Nucleoside or nucleotide reverse transcriptase inhibitors:**<br>Efavirenz, tenofovir, emtricitabine, zidovudine, lamivudine, abacavir, stavudine, or tenofovir disoproxil fumarate<br>**Non nucleoside reverse transcriptase inhibitors:**<br>Cidofovir, efavirenz, nevirapine, or etravirine | – | – |

*Continued*

**TABLE 10.6** (*Continued*)
**Treatment Strategies used for Kaposi Sarcoma**

| | Kaposi Sarcoma | | | |
|---|---|---|---|---|
| **Treatment Strategies** | **AIDS-Associated KS (Epidemic)** | **Classic KS** | **Transplant-Associated KS (Iatrogenic)** | **African (Endemic)** |
| | **Protease inhibitors:** Atazanavir, saquinavir, ritonavir, indinavir, nelfinavir, fosamprenavir, lopinavir, darunavir, or tipranavir | **Protease inhibitors:** Atazanavir, saquinavir, ritonavir, indinavir, nelfinavir, fosamprenavir, lopinavir, darunavir, or tipranavir | | |
| Immunomodulators | INF-alpha bleomycin and interferon alpha | Bleomycin and interferon alpha | – | – |
| Local therapy | + | + | – | – |
| Radiotherapy | + | + | – | – |
| Cryotherapy | + | – | – | – |
| Topical therapy | Alitretinoin or imiquimod | – | – | – |
| Targeted therapy | | | | |
| VEGFI | IM862, Bevacizumab | – | – | – |
| TKI | imatinib | – | – | – |
| MMPI | + | – | – | |

*Note:* + used; – not used.

Martellotta et al. 2009; Fatahzadeh 2012). The response rate of advanced AIDS-KS to paclitaxel was found to vary in different studies but was reported as high as 71% (Welles et al. 1998; Lynen et al. 2005). Another second-line drug for progressive severe classic KS and advanced AIDS-KS is oral etoposide (Laubenstein et al. 1984; Brambilla et al. 1994; Evans et al. 2002; Lynen et al. 2005). The combination regimens, including doxorubicin, bleomycin, and either vincristine or vinblastine, have been widely studied in AIDS-KS with response rates varying from 25% to 88% (Table 10.5) (Laubenstein et al. 1984; Gill et al. 1990, 1996; Northfelt et al. 1998). Nasti et al. reported the efficacy of the vinca alkaloid vinorebline in the treatment of AIDS-KS (Nasti et al. 2000). IFN-alpha, an immunomodulatory agent with antiviral and antiangiogenic properties, has dose-dependent efficacy in the treatment of AIDS-KS when administered systemically (Table 10.5) (Krown 2001; Lynen et al. 2005). Drawbacks, however, include hepatotoxicity, flulike symptoms, myelosuppression, and predisposition to opportunistic infections (Lynen et al. 2005; Martellotta et al. 2009). The delayed therapeutic response renders IFN-x inappropriate for rapidly progressive or symptomotic visceral KS (Lynen et al. 2005; Martellotta et al. 2009).

### 10.4.3.2 Antiviral Therapy

#### *10.4.3.2.1 Anti-HHV-8 Therapy*

To improve efficacy and overcome chemotherapy associated toxicity, new therapeutic approaches have been adopted to directly control or prevent HHV-8 infection. For example, antiherpes medications were shown to reduce plasma viral load of HHV-8 and prevent KS in HHV-8-seropositive transplant recipients (Schwartz 2004). In addition, the antiherpes drugs ganciclovir and foscarnet were found to reduce the risk of KS as much as 62% among HIV-positive subjects (Table 10.5) (Mocroft et al. 1996). Finally, the combination of antiherpes medications with HAART offers a promise of diminishing replication of both viruses, preventing new lesions, and helping to regress lesions already present (Feller et al. 2007).

#### *10.4.3.2.2 Highly Active Antiretroviral Therapy*

HAART induces clinical responses in epidemic (Diz Dios et al. 1998; Tirelli and Bernardi 2001; Murdaca et al. 2002; Spano et al. 2008) and HIV-associated KS (Rohrmus et al. 2000; Nasti et al. 2003a, 2003b; Epstein et al. 2005). Indeed, potent antiretroviral medications have led to a dramatic decline in the incidence of KS among HIV-infected individuals (Mocroft et al. 2004; Vanni et al. 2006). Moreover, the advent of HAART led to a change in the criteria of staging for epidemic KS (Lynen et al. 2005). Nasti et al. proposed new prognostic indicators for AIDS-KS staging based on tumor extension (T) and HIV-related systemic illness (S) (Nasti et al. 2003). Thus, the severity of immunosuppression reflected in CD4 count is no longer an independent prognostic indicator for staging AIDS-KS. Accordingly, treated HIV-positive patients with the combination of poor tumor stage (e.g., tumor-associated edema) and constitutional symptoms (T1S1) have an unfavorable prognosis with a three-year survival rate of 53%, whereas patients with none or only one prognostic criteria—T0S0, T0S1, or T1S0—have a good prognosis with three-year survival rates of 88%, 80%, and 81%, respectively (Nasti et al. 2003; Krown 2004;

Lynen et al. 2005; Spano et al. 2008; Martellotta et al. 2009). Therefore, HAART alone is not sufficient for advanced epidemic KS with poor prognostic index (T1S1), which requires additional interventions. For instance, AIDS-KS with unfavorable prognosis (e.g., pulmonary involvement or rapidly progressive disease) is best managed with a combination of HAART and systemic chemotherapy (Tirelli and Bernardi 2001; Von Roenn 2003; Martellotta et al. 2009; Bihl et al. 2007; Di Lorenzo et al. 2007; Henderson 2009). Although protease inhibitors are thought to have specific antiangiogenic effects, the choice of HAART regimen does not appear to influence protection against epidemic KS (Table 10.6) (Portsmouth et al. 2003; Vanni et al. 2006; Di Lorenzo et al. 2007; Henderson 2009).

#### 10.4.3.3 Local Therapy

HAART is usually used as a palliative measure in patients with rapidly progressive mucocutaneous lesions causing pain, esthetic concerns, or with interference with oral function. Conversely, local therapy remains a safe and efficacious measure for limited, asymptomatic mucocutaneous lesions of AIDS-KS (Lynen et al. 2005; Martellotta et al. 2009; Fatahzadeh 2012). Local therapy is indeed widely used in classic KS, preventing pain, ulceration, and bleeding of indolent lesions (Brambilla et al. 2010).

Radiotherapy is also a highly effective modality for local control of KS with remission rates exceeding 90% of AIDS-KS (Table 10.6) (Lynen et al. 2005; Martellotta et al. 2009; Fatahzadeh 2012). Furthermore, in classic KS, radiotherapy was fond to achieve response rates higher than 80% (Bottler et al. 2007; Mohanna et al. 2007). Similarly, the focal, superficial, mucocutaneous KS is amenable to surgical excision.

Intralesional injection with vinca alkaloids, such as vincristine and vinblastine, and bleomycin or IFN-alpha has been efficacious for local treatment of mucocutaneous lesions of classic and epidemic KS (Table 10.6) (Boudreaux et al. 1993; Trattner et al. 1993; Poignonec et al. 1995; Brambilla et al. 2010). Focal skin lesions of AIDS-KS were also treated with liquid nitrogen, called cryotherapy, and 80% of cases showed full resolution of the KS lesions (Tappero et al. 1991; Hengge et al. 2002). Retinoids were also used to treat KS and were found to have an inhibitory effect on IL-6, implicated in KS pathogenesis (Vanni et al. 2006). Application of 0.1% alitretinoin gel, the only self-administered FDA-approved topical agent for cutaneous AIDS-KS, has shown efficacy for skin lesions of classic and HIV-KS (Table 10.6) (Lynen et al. 2005; Martellotta et al. 2009). In a study of cutaneous KS in HIV-seronegative patients, the overall response rate to topical 5% imiquimod cream was 47%, although 53% of subjects experienced local erythema and pruritis (Table 10.6) (Célestin-Schartz et al. 2008). However, compared with topical drugs, such as alitretinoin or imiquimod, intralesional injections are more efficacious because of faster and more precise delivery of therapeutic agents (Brambilla et al. 2010).

#### 10.4.3.4 New Targeted Therapeutic Approaches

##### *10.4.3.4.1 Vascular Endothelial Growth Factor Inhibitors*

KS is among the most vascular tumors with abnormally dense and irregular blood vessels. It highly interacts with the human VEGF pathway. The importance of angiogenesis in KS tumor development is further highlighted by the fact that early-stage KS might not yet be a true tumor but a neoplasm of proliferative spindle-shaped cells driven by angiogenesis and inflammation (Folkman 2006). For instance, VEGF receptors 2 and 3 are known to be expressed in KS lesions (Masood et al. 2002). Furthermore, HHV-8 glycoproteins K1 and B are shown to increase the expression of VEGF and activate VEGF-R3, respectively (Bais et al. 2003; Wang et al. 2004). Transgenic mice expressing vGPCR display multifocal and angioproliferative KS-like lesions (Guo et al. 2004; Jensen et al. 2005). In a SCID mouse model with human skin grafts, VEGF is essential for the inoculated early-stage KS cells to grow into KS-like tumors (Samaniego et al. 2002). Many of the cytokines induced by HHV-8 infection, including VEGF, bFGF, IL-6, IL-8, GRO-α, TNF-β, and ephrin B2, are angiogenic (Masood et al. 2002, 2005; Naranatt et al. 2004; Wang et al. 2004; Xie et al. 2005). Higher levels of serum VEGF were also seen in HHV-1-infected persons with KS compared with HHV-1-infected persons without KS (Weindel et al. 1992), and higher mRNA levels of VEGF and angiopoietins, including Ang-1 and Ang-2, were also detected at higher levels in KS lesions compared to the adjacent normal tissues (Brown et al. 2000). Hence, VEGF inhibitors were used to treat KS. Clinical trials using pentosan and tecogalan showed moderate toxicity but minimal efficacy (Table 10.6) (Pluda et al. 1993; Eckhardt et al. 1996; Schwartsmann et al. 1996). Another VEGFI, IM862, reached a phase II clinical study and induced partial response in 36% of 44 patients with epidemic KS (Table 10.6) (Tulpule et al. 2000). A phase I clinical study using an antisense oligonucleotide to VEGF led to dramatic decreases in VEGF plasma levels and a complete remission of KS in a refractory patient to chemotherapy and HAART (Table 10.6) (Levine et al. 2006). Finally, bevacizumab (Avastin) resulted in a significant increase in the long-term responsiveness of KS tumors in a xenografted nude mouse model (Table 10.6) (Ferrario and Gomer 2006). Furthermore, a phase II study investigated the efficacy of bevacizumab in patients with HIV-KS (Table 10.6). Best tumor responses were observed in 5 out of 17 enrolled patients with CR in 3 patients (19%), partial response (PR) in two patients (12%), stable disease in nine patients (56%), and progressive disease in two patients (12%). Four of five responders had received prior chemotherapy for KS (Uldrick et al. 2012).

#### *10.4.3.4.2 Tyrosine Kinase Inhibitors*

KS expresses the tyrosine kinase receptor for stem cell factor, *c-kit* (Moses et al. 2002). This observation led to propose *c-kit* inhibitors to treat HHV-8-associated malignancies (Moses et al. 2002). In addition, the TKI, imatinib, was used on ten epidemic KS patients (Table 10.6). Only five patients displayed a PR to the drug. In fact, diarrhea limited the dose of imatinib; it could be administered in only six of ten participants (Koon et al. 2005).

#### *10.4.3.4.3 Matrix Metalloproteinases Inhibitors*

As with most malignant cancers, KS tumors overexpress matrix metalloproteinases (MMP), which might confer a role in invasion and extravasation through the basement membrane in different tissues (Dezube et al. 2006). A phase II clinical trial investigated the effect of 6-demethyl-6-deoxy-4-dimethylamino tetracycline (COL)-3 in KS, a topical MMPI structurally similar to tetracycline (Table 10.6) (Dezube et al. 2006); 41% of 75 patients with AIDS-associated KS responded to this treatment strategy. However, participants frequently discontinued using the drug, and photosensitivity was common (Dezube et al. 2006).

## 10.5 CONCLUSION

The prognosis for HHV-8 associated diseases remains poor, and currently used treatment modalities are far from achieving complete and sustainable remission. Indeed, the persistent global burden of KS and the few successful strategies of treatment in MCD and PEL suggest that novel effective drugs are badly needed. Based on recent preclinical data and translational studies, several new targeted therapies have been found promising. These include inhibitors of viral replication (zidovudine and interferon, HAART), modulators of cell signaling and inflammation (interferon), immunomodulators (thalidomide), monoclonal antibodies (rituximab) and inhibitors of angiogenesis (bevacizumab), NF-κB inhibitors such as arsenic trioxide or proteasome inhibitors (bortezomib), histone deacetylase inhibitors (SAHA), metalloprotease inhibitors (COL-3), and TKI (imatinib). However, the efficacy of these strategies can only be validated through careful controlled clinical trials. Some of these targeted therapies have not yet reached clinical studies, although others were used in a few individual case reports with low numbers of patients.

## REFERENCES


Aaron L, Olivier L, Cherine Y, et al. (2002). Human herpesvirus 8–positive Castleman disease in human immunodeficiency virus–infected patients: The impact of highly active antiretroviral therapy. *Clinical Infectious Diseases* 35: 880–882.

Abou-Merhi R, Khoriaty R, Arnoult D, et al. (2007). PS-341 or a combination of arsenic trioxide and interferon-α inhibit growth and induce caspase-dependent apoptosis in KSHV/HHV-8-infected primary effusion lymphoma cells. *Leukemia* 21: 1792–1801.

Adams J. (2001). Proteasome inhibition in cancer: Development of PS-341. *Semin Oncol.* 28: 613–619.

Alexander L, Lynn D, Amanda K, et al. (2000). The primary sequence of rhesus monkey rhadinovirus isolate 26–95: Sequence similarities to Kaposi's sarcoma-associated herpesvirus and rhesus monkey rhadinovirus isolate 17577. *Journal of Virology* 74: 3388–3398.

An J, Sun Y, Fisher M, et al. (2004). Antitumor effects of bortezomib (PS-341) on primary effusion lymphomas. *Leukemia* 18: 1699–1704.

Antman K, Chang Y. (2000). Kaposi's sarcoma. *New England Journal of Medicine* 342: 1027–1038.

Aoki Y, Feldman GM, Tosato G. (2003). Inhibition of STAT3 signaling induces apoptosis and decreases survivin expression in primary effusion lymphoma. *Blood* 101: 1535–1542.

Aoki Y, Jaffe ES, Chang Y, et al. (1999). Angiogenesis and hematopoiesis induced by Kaposi's sarcoma-associated herpesvirus-encoded interleukin-6. Presented in Part at the 40th Annual American Society of Hematology Meeting, December 7, 1998 (Miami Beach, FL). *Blood* 93: 4034–4043.

Bais C, Santomasso B, Coso O, et al. (1998). G-protein-coupled receptor of Kaposi's sarcoma-associated herpesvirus is a viral oncogene and angiogenesis activator. *Nature* 391: 86–89.

Bais C, Van Geelen A, Eroles P, et al. (2003). Kaposi's sarcoma associated herpesvirus G protein-coupled receptor immortalizes human endothelial cells by activation of the VEGF receptor-2/KDR. *Cancer Cell* 3: 131–143.

Ballestas ME, Chatis PA, Kaye KM. (1999). Efficient persistence of extrachromosomal KSHV DNA mediated by latency-associated nuclear antigen. *Science* 284: 641–644.

Banks PM, Warnke RA. (2001). Pathology and genetics of tumours of haematopoietic and lymphoid tissues. In: Jaffe ES, Harris NL, Stein H, Vardiman JW (eds.), *World Health Organization Classification of Tumours*. IARC Press, Lyon, France, 178–182.

Bazarbachi A, El-Sabban M, Nasr R, et al. (1999). Arsenic trioxide and interferon-α synergize to induce cell cycle arrest and apoptosis in human T-cell lymphotropic virus type I–transformed cells. *Blood* 93: 278–283.

Bazarbachi A, Ghez D, Lepelletier Y, et al. (2004). New therapeutic approaches for adult T-cell leukaemia. *The Lancet Oncology* 5: 664–672.

Beck JT, Hsu SM, Wijdenes J, et al. (1994). Alleviation of systemic manifestations of Castleman' disease by monoclonal anti-interleukin-6 antibody. *New England Journal of Medicine* 330: 602–605.

Bhatt S, Ashlock B, Toomey N, et al. (2013). Efficacious proteasome/HDAC inhibitor combination therapy for primary effusion lymphoma. *The Journal of Clinical Investigation* 123: 2616–2628.

Bi X, Yang L, Mancl M, et al. (2011). Modulation of interferon regulatory factor 5 activities by the Kaposi sarcoma-associated herpesvirus-encoded viral interferon regulatory factor 3 contributes to immune evasion and lytic induction. *Journal of Interferon and Cytokine Research* 31: 373–382.

Bihl F, Mosam A, Henry LN, et al. (2007). Kaposi's sarcoma-associated herpesvirus-specific immune reconstitution and antiviral effect of combined HAART/chemotherapy in HIV clade C-infected individuals with Kaposi's sarcoma. *AIDS* 21: 1245–1252.

Boshoff C, Weiss R. (2002). AIDS-related malignancies. *Nature Reviews Cancer* 2: 373–382.

Bottler T, Hardt K, Oehen HP, et al. (2007). Non-HIV-associated Kaposi's sarcoma of the tongue: Case report and review of the literature. *International Journal of Oral and Maxillofacial Surgery* 36: 1218–1220.

Boudreaux A, Smith L, Cosby C, et al. (1993). Intralesional vinblastine for cutaneous Kaposi's sarcoma associated with acquired immunodeficiency syndrome: A clinical trial to evaluate efficacy and discomfort associated with injection. *Journal of the American Academy of Dermatology* 28: 61–65.

Boulanger E, Afonso PV, Yahiaoui Y, et al. (2008). Human herpes virus-8 (HHV-8)-associated primary effusion lymphoma in two renal transplant recipients receiving rapamycin. *American Journal of Transplantation* 8: 707–710.

Boulanger E, Agbalika F, Maarek O, et al. (2001). A clinical, molecular and cytogenetic study of 12 cases of human herpes virus 8 associated primary effusion lymphoma in HIV-infected patients. *The Hematology Journal* 2: 172–179.

Boulanger E, Daniel MT, Agbalika F, et al. (2003). Combined chemotherapy including high-dose methotrexate in KSHV/HHV8-associated primary effusion lymphoma. *American Journal of Hematology* 73: 143–148.

Boulanger E, Gérard L, Gabarre J, et al. (2005). Prognostic factors and outcome of human herpes virus 8–associated primary effusion lymphoma in patients with AIDS. *Journal of Clinical Oncology* 23: 4372–4380.

Bower M, Powles T, Williams S, et al. (2007). Brief communication: Rituximab in HIV-associated multicentric Castleman disease. *Annals of Internal Medicine* 147: 836–839.

Bowne B, Lewis J, Filippa D, et al. (1999). The management of unicentric and multicentric Castleman's disease. *Cancer* 85: 706–717.

Brambilla L, Bellinvia M, Tourlaki A, et al. (2010). Intralesional vincristine as first-line therapy for nodular lesions in classic Kaposi sarcoma: A prospective study in 151 patients. *British Journal of Dermatology* 162: 854–859.

Brambilla L, Labianca R, Boneschi V, et al. (1994). Mediterranean Kaposi's sarcoma in the elderly. A randomized study of oral etoposide versus vinblastine. *Cancer* 74: 2873–2878.

Brandt J, Bodine D, Dunbar C, et al. (1990). Dysregulated interleukin 6 expression produces a syndrome resembling Castleman's disease in mice. *Journal of Clinical Investigation* 86: 592–599.

Brinkmann M, Glenn M, Rainbow L, et al. (2003). Activation of mitogen-activated protein kinase and NF-κB pathways by a Kaposi's sarcoma-associated herpes virus k15 membrane protein. *Journal of Virology* 77: 9346–9358.

Brown F, Dezube B, Tognazzi K, et al. (2000). Expression of Tie1, Tie2, and angiopoietins 1, 2, and 4 in Kaposi's sarcoma and cutaneous angiosarcoma. *The American Journal of Pathology* 156: 2179–2183.

Burysek L, Yeow WS, Pitha PM. (1999). Unique properties of a second human herpes virus 8-encoded interferon regulatory factor (vIRF-2). *Journal of Human Virology* 2: 19–32.

Cai QL, Knight J, Verma S, et al. (2006). EC5S ubiquitin complex is recruited by KSHV latent antigen LANA for degradation of the VHL and p53 tumor suppressors. *PLoS Pathogens* 2: 1002–1012.

Cannon J, Ciufo D, Hawkins A, et al. (2000). A new primary effusion lymphoma-derived cell line yields a highly infectious Kaposi's sarcoma herpesvirus-containing supernatant. *Journal of Virology* 74: 10187–10193.

Cannon M, Cesarman E. (2004). The KSHV G protein-coupled receptor signals via multiple pathways to induce transcription factor activation in primary effusion lymphoma cells. *Oncogene* 23: 514–523.

Carbone A, Gaidano G. (1997). HHV-8-positive body-cavity-based lymphoma: A novel lymphoma entity. *British Journal of Haematology* 97: 515–522.

Carbone A, Gloghini A. (2008). PEL and HHV8-unrelated effusion lymphomas. *Cancer Cytopathology* 114: 225–227.

Carbone A, Gloghini A, Capello D, et al. (2001). Genetic pathways and histogenetic models of AIDS-related lymphomas. *Eur J Cancer* 37: 1270–1275.

Carbone A, Gloghini A, Vaccher E, et al. (1996). Kaposi's sarcoma-associated Herpesvirus DNA Sequences in AIDS-related and AIDS-unrelated lymphomatous effusions. *British Journal of Haematology* 94: 533–543.

Casper C. (2008). New approaches to the treatment of human herpesvirus 8-associated disease. *Reviews in Medical Virology* 18: 321–329.

Casper C, Nichols WG, Huang ML, et al. (2004). Remission of HHV-8 and HIV-associated multicentric Castleman disease with ganciclovir treatment. *Blood* 103: 1632–1634.

Casper C, Wald A. (2007). The use of antiviral drugs in the prevention and treatment of Kaposi sarcoma, multicentric Castleman disease and primary effusion lymphoma. In: Boshoff C, Weiss RA (eds.), *Kaposi Sarcoma Herpesvirus: New Perspectives*. Springer, Heidelberg, Berlin, 289–307.

Castleman B, Towne VW. (1954). Case records of the Massachusetts General Hospital; Weekly clinicopathological exercises; Founded by Richard C. Cabot. *The New England Journal of Medicine* 251: 396–400.

Célestin-Schartz NE, Chevret S, Paz C, et al. (2008). Imiquimod 5% cream for treatment of HIV-negative Kaposi's sarcoma skin lesions: A phase I to II, open-label trial in 17 patients. *Journal of the American Academy of Dermatology* 58: 585–591.

Cesarman E, Chang Y, Moore PS, et al. (1995a). Kaposi's sarcoma–associated herpesvirus-like DNA sequences in AIDS-related body-cavity–based lymphomas. *New England Journal of Medicine* 332: 1186–1191.

Cesarman E, Moore PS, Rao PH, et al. (1995b). In vitro establishment and characterization of two acquired immunodeficiency syndrome-related lymphoma cell lines (BC-1 and BC-2) containing Kaposi's sarcoma-associated herpesvirus-like (KSHV) DNA sequences. *Blood* 86: 2708–2714.

Chadburn A, Cesarman E, Nador RG, et al. (1997). Kaposi's sarcoma-associated herpesvirus sequences in benign lymphoid proliferations not associated with human immunodeficiency virus. *Cancer* 80: 788–797.

Chadburn A, Hyjek E, Mathew S, et al. (2004). KSHV-positive solid lymphomas represent an extra-cavitary variant of primary effusion lymphoma. *The American Journal of Surgical Pathology* 28: 1401–1416.

Chadburn A, Hyjek EM, Tam W, et al. (2008). Immunophenotypic analysis of the Kaposi sarcoma herpesvirus (KSHV; HHV-8)-infected B Cells in HIV multicentric Castleman disease (MCD). *Histopathology* 53: 513–524.

Chang Y, Cesarman E, Pessin MS, et al. (1994). Identification of herpesvirus-like DNA sequences in AIDS-associated Kaposi's sarcoma. *Science* 266: 1865–1869.

Chatterjee M, Osborne J, Bestetti G, et al. (2002). Viral IL-6-induced cell proliferation and immune evasion of interferon activity. *Science Signaling* 298: 1432–1435.

Chen W, Hilton IB, Staudt MR, et al. (2010). Distinct p53, p53: LANA, and LANA complexes in Kaposi's sarcoma-associated herpesvirus lymphomas. *Journal of Virology* 84: 3898–3908.

Chen YB, Rahemtullah A, Hochberg E. (2007). Primary effusion lymphoma. *The Oncologist* 12: 569–576.

Choi JK, Lee BS, Shim S, et al. (2000). Identification of the novel K15 gene at the rightmost end of the Kaposi's sarcoma-associated herpesvirus genome. *Journal of Virology* 74: 436–446.

Chronowski GM, Ha CS, Wilder RB, et al. (2001). Treatment of unicentric and multicentric castleman disease and the role of radiotherapy. *Cancer* 92: 670–676.

Chugh P, Matta H, Schamus S, et al. (2005). Constitutive NF-κB activation, normal fas-induced apoptosis, and increased incidence of lymphoma in human herpes virus 8 K13 transgenic mice. *Proceedings of the National Academy of Sciences of the United States of America* 102: 12885–12890.

Chung YH, Means RE, Choi JK, et al. (2002). Kaposi's sarcoma-associated herpesvirus OX2 glycoprotein activates myeloid-lineage cells to induce inflammatory cytokine production. *Journal of Virology* 76: 4688–4698.

Cirone M, Di Renzo L, Lotti L, et al. (2012). Primary effusion lymphoma cell death induced by bortezomib and AG 490 activates dendritic cells through CD91. *PLoS One* 7: e31732.

Coscoy L, Ganem D. (2000). Kaposi's sarcoma-associated herpesvirus encodes two proteins that block cell surface display of MHC class I chains by enhancing their endocytosis. *Proceedings of the National Academy of Sciences* 97: 8051–8056.

Coscoy L, Ganem D. (2001). A viral protein that selectively downregulates ICAM-1 and B7–2 and modulates T cell costimulation. *Journal of Clinical Investigation* 107: 1599–1606.

Cunningham C, Barnard S, Blackbourn DJ, et al. (2003). Transcription mapping of human herpesvirus 8 genes encoding viral interferon regulatory factors. *Journal of General Virology* 84: 1471–1483.

DePond W, Said JW, Tasaka T, et al. (1997). Kaposi's sarcoma-associated herpesvirus and human herpesvirus 8 (KSHV/HHV8)-associated lymphoma of the bowel: Report of two cases in HIV-positive men with secondary effusion lymphomas. *The American Journal of Surgical Pathology* 21: 719–724.

Dezube B, Krown S, Lee J, et al. (2006). Randomized phase II trial of matrix metalloproteinase inhibitor COL-3 in AIDS-related Kaposi's sarcoma: An aids malignancy consortium study. *Journal of Clinical Oncology* 24: 1389–1394.

Di Lorenzo G, Konstantinopoulos P, Pantanowitz L, et al. (2007). Management of AIDS-related Kaposi's sarcoma. *The Lancet Oncology* 8: 167–176.

Dinarello CA. (2009). Immunological and inflammatory functions of the interleukin-1 family. *Annual Review of Immunology* 27: 519–550.

Dittmer D. (2003). Transcription profile of Kaposi's sarcoma-associated herpesvirus in primary Kaposi's sarcoma lesions as determined by real-time PCR arrays. *Cancer Research* 63: 2010–2015.

Dittmer D, Lagunoff M, Renne R, et al. (1998). A cluster of latently expressed genes in Kaposi's sarcoma-associated herpesvirus. *Journal of Virology* 72: 8309–8315.

Diz Dios P, Oscampo A, Hermida C, et al. (1998). Regression of AIDS-related Kaposi's sarcoma following ritonavir therapy. *Oral Oncology* 34: 236–238.

Djerbi M, Screpanti V, Catrina AI, et al. (1999). The inhibitor of death receptor signaling, FLICE-inhibitory protein defines a new class of tumor progression factors. *The Journal of Experimental Medicine* 190: 1025–1032.

Dourmishev LA, Dourmishev A, Palmeri D, et al. (2003). Molecular genetics of Kaposi's sarcoma-associated herpesvirus (human herpesvirus 8) epidemiology and pathogenesis. *Microbiology and Molecular Biology Reviews* 67: 175–212.

Du MQ, Liu H, Diss T, et al. (2001). Kaposi sarcoma–associated herpesvirus infects monotypic (IgMλ) but polyclonal naive B cells in Castleman disease and associated lymphoproliferative disorders. *Blood* 97: 2130–2136.

Dupin N, Diss T, Kellam P, et al. (2000). HHV-8 is associated with a plasmablastic variant of Castleman disease that is linked to HHV-8–positive plasmablastic lymphoma. *Blood* 95: 1406–1412.

Eckhardt SG, Burris HA, Eckardt JR, et al. (1996). A phase I clinical and pharmacokinetic study of the angiogenesis inhibitor, tecogalan sodium. *Annals of Oncology* 7: 491–496.

Edelman D. (2005). Human herpesvirus 8–a novel human pathogen. *Virology Journal* 2: 78.

El Hajj H, El-Sabban M, Hasegawa H, et al. (2010). Therapy-induced selective loss of leukemia-initiating activity in murine adult T cell leukemia. *The Journal of Experimental Medicine* 207: 2785–2792.

El Hajj H, Ali J, Ghantous A, et al. (2013). Combination of arsenic and interferon-α inhibits expression of KSHV latent transcripts and synergistically improves survival of mice with primary effusion lymphomas. *PLoS* One 8:e79474.

El-Osta H, Janku F, Kurzrock R. (2010). Successful treatment of Castleman's disease with interleukin-1 receptor antagonist (anakinra). *Molecular Cancer Therapeutics* 9: 1485–1488.

El-Osta H, Kurzrock R. (2011). Castleman's disease: From basic mechanisms to molecular therapeutics. *The Oncologist* 16: 497–511.

El-Sabban M, Nasr R, Dbaibo G, et al. (2000). Arsenic-interferon-α–triggered apoptosis in HTLV-I transformed cells is associated with tax down-regulation and reversal of NF-κB activation. *Blood* 96: 2849–2855.

Epstein J, Cabay R, Glick M. (2005). Oral malignancies in HIV disease: Changes in disease presentation, increasing understanding of molecular pathogenesis, and current management. *Oral Surgery, Oral Medicine, Oral Pathology, Oral Radiology, and Endodontology* 100: 571–578.

Estephan F, Elghetany M, Berry M, et al. (2005). Complete remission with anti-CD20 therapy for unicentric, non-HIV-associated, Hyaline-vascular type, Castleman's disease. *Cancer Investigation* 23: 191.

Evans SR, Krown S, Testa M, et al. (2002). Phase II evaluation of low-dose oral etoposide for the treatment of relapsed or progressive AIDS-related Kaposi's sarcoma: An AIDS clinical trials group clinical study. *Journal of Clinical Oncology* 20: 3236–3241.

Fakhari F, Dittmer D. (2002). charting latency transcripts in Kaposi's sarcoma-associated herpesvirus by whole-genome real-time quantitative PCR. *Journal of Virology* 76: 6213–6223.

Fakhari F, Jeong J, Kanan Y, et al. (2006). The Latency-associated nuclear antigen of Kaposi sarcoma–associated herpesvirus induces B cell hyperplasia and lymphoma. *Journal of Clinical Investigation* 116: 735–742.

Fatahzadeh M. (2012). Kaposi sarcoma: Review and medical management update. *Oral Surgery, Oral Medicine, Oral Pathology and Oral Radiology* 113: 2–16.

Feller L, Wood NH, Lemmer J. (2007). Herpes zoster infection as an immune reconstitution inflammatory syndrome in HIV-seropositive subjects: A review. *Oral Surg Oral Med Oral Pathol Oral Radiol Endod* 104: 455–460.

Feng P, Park J, Lee BS, et al. (2002). Kaposi's sarcoma-associated herpesvirus mitochondrial K7 protein targets a cellular calcium-modulating cyclophilin ligand to modulate intracellular calcium concentration and inhibit apoptosis. *Journal of Virology* 76: 11491–11504.

Ferrario A, Gomer C. (2006). Avastin enhances photodynamic therapy treatment of Kaposi's sarcoma in a mouse tumor model. *Journal of Environmental Pathology, Toxicology and Oncology* 25: 251–259.

Folkman J. (2006). Angiogenesis. *Annual Reviews in Medicine* 57: 1–18.

Friborg J, Kong WP, Flowers C, et al. (1998). Distinct biology of Kaposi's sarcoma-associated herpesvirus from primary lesions and body cavity lymphomas. *Journal of Virology* 72: 10073–10082.

Friborg J, Kong WP, Hottiger M, et al. (1999). P53 inhibition by the LANA protein of KSHV protects against cell death. *Nature* 402: 889–894.

Friedrichs C, Neyts J, Gaspar G, et al. (2004). Evaluation of antiviral activity against human herpesvirus 8 (HHV-8) and Epstein–Barr virus (EBV) by a quantitative real-time PCR assay. *Antiviral Research* 62: 121–123.

Fukumoto H, Kanno T, Hasegawa H, et al. (2011). Pathology of Kaposi's sarcoma-associated herpesvirus infection. *Frontiers in Microbiology* 2: 175.

Galeotti C, Tran TA, Franchi-Abella S, et al. (2008). IL-1RA agonist (anakinra) in the treatment of multifocal Castleman disease: Case report. *Journal of Pediatric Hematology/ Oncology* 30: 920–924.

Gao SJ, Boshoff C, Jayachandra S, et al. (1997). KSHV ORF K9 (vIRF) is an oncogene which inhibits the interferon signaling pathway. *Oncogene* 15: 1979–1985.

Gao SJ, Kingsley L, Hoover DR, et al. (1996). Seroconversion to antibodies against Kaposi's sarcoma–associated herpesvirus–related latent nuclear antigens before the development of Kaposi's sarcoma. *New England Journal of Medicine* 335: 233–241.

Gessain A. (1997). Human herpesvirus 8 and associated diseases: Kaposi's sarcoma, body cavity based lymphoma and multicentric Castleman disease: Clinical and molecular epidemiology. *Bulletin De l'Academie Nationale De Medecine* 181: 1023–1034.

Ghavamzadeh A, Alimoghaddam K, Rostami S, et al. (2011). Phase II study of single-agent arsenic trioxide for the front-line therapy of acute promyelocytic leukemia. *Journal of Clinical Oncology* 29: 2753–2757.

Ghosh SK, Wood C, Boise L, et al. (2003). Potentiation of TRAIL-induced apoptosis in primary effusion lymphoma through azidothymidine-mediated inhibition of NF-κB. *Blood* 101: 2321–2327.

Gill P, Rarick M, Bernstein-Singer M, et al. (1990). Treatment of advanced Kaposi's sarcoma using a combination of bleomycin and vincristine. *American Journal of Clinical Oncology* 13: 315–319.

Gill P, Wernz J, Scadden DT, et al. (1996). Randomized phase III trial of liposomal daunorubicin versus doxorubicin, bleomycin, and vincristine in AIDS-related Kaposi's sarcoma. *Journal of Clinical Oncology* 14: 2353–2364.

Godden-Kent D, Talbot S, Boshoff C, et al. (1997). The cyclin encoded by Kaposi's sarcoma-associated herpesvirus stimulates cdk6 to phosphorylate the retinoblastoma protein and histone H1. *Journal of Virology* 71: 4193–4198.

Golks A, Brenner D, Krammer PH, et al. (2006). The c-FLIP–NH2 terminus (p22-FLIP) induces NF-κB activation. *The Journal of Experimental Medicine* 203: 1295–1305.

Goto H, Kariya R, Shimamoto M, et al. (2012). Antitumor effect of berberine against primary effusion lymphoma via inhibition of NF-κB pathway. *Cancer Science* 103: 775–81.

Grandadam M, Dupin N, Calvez V, et al. (1997). Exacerbations of clinical symptoms in human immunodeficiency virus type 1—infected patients with multicentric Castleman's disease are associated with a high increase in Kaposi's sarcoma herpesvirus DNA load in peripheral blood mononuclear cells. *Journal of Infectious Diseases* 175: 1198–1201.

Greene W, Kuhne K, Ye F, et al. (2007). Molecular biology of KSHV in relation to AIDS-associated oncogenesis. *Cancer Treatment Research* 133: 69–127.

Grisotto MG, Garin A, Martin A, et al. (2006). The human herpesvirus 8 chemokine receptor vGPCR triggers autonomous proliferation of endothelial cells. *Journal of Clinical Investigation* 116: 1264–1273.

Groves AK, Cotter MA, Subramanian C, et al. (2001). The latency-associated nuclear antigen encoded by Kaposi's sarcoma-associated herpesvirus activates two major essential Epstein–Barr virus latent promoters. *Journal of Virology* 75: 9446–9457.

Grundhoff A, Ganem D. (2004). Inefficient establishment of KSHV latency suggests an additional role for continued lytic replication in Kaposi sarcoma pathogenesis. *Journal of Clinical Investigation* 113: 124–136.

Guasparri I, Keller S, Cesarman E. (2004). KSHV vFLIP is essential for the survival of infected lymphoma cells. *The Journal of Experimental Medicine* 199: 993–1003.

Guo HG, Pati S, Sadowska M, et al. (2004). Tumorigenesis by human herpesvirus 8 vGPCR is accelerated by human immuodeficiency virus type 1 Tat. *Journal of Virology* 78: 9336–9342.

Halfdanarson TR, Markovic SN, Kalokhe U, et al. (2006). A non-chemotherapy treatment of a primary effusion lymphoma: Durable remission after intracavitary cidofovir in HIV negative PEL refractory to chemotherapy. *Annals of Oncology* 17: 1849–1850.

Henderson H. (2009). Kaposi sarcoma is the most common cancer diagnosed in HIV-infected persons. *HIV Clinician* 21: 1–2.

Hengge U, Ruzicka T, Tyring SK, et al. (2002). Update on Kaposi's sarcoma and other HHV8 associated diseases. Part 2: Pathogenesis, Castleman's disease, and pleural effusion lymphoma. *The Lancet Infectious Diseases* 2: 344–352.

Hermine O, Dombret H, Poupon J, et al. (2004). Phase II trial of arsenic trioxide and alpha interferon in patients with relapsed/refractory adult T-cell leukemia/lymphoma. *The Hematology Journal* 5: 130–134.

Hess G, Wagner V, Kreft A, et al. (2006). Effects of bortezomib on pro-inflammatory cytokine levels and transfusion dependency in a patient with multicentric Castleman disease. *British Journal of Haematology* 134: 544–545.

Hocqueloux L, Agbalika F, Oksenhendler E, et al. (2001). Long-term remission of an AIDS-related primary effusion lymphoma with antiviral therapy. *AIDS* 15: 280–282.

Hu J, Garber A, Renne R. (2002). The latency-associated nuclear antigen of Kaposi's sarcoma-associated herpesvirus supports latent DNA replication in dividing cells. *Journal of Virology* 76: 11677–11687.

Hu J, Liu YF, Wu CF, et al. (2009). Long-term efficacy and safety of all-trans retinoic acid/arsenic trioxide-based therapy in newly diagnosed acute promyelocytic leukemia. *Proceedings of the National Academy of Sciences* 106: 3342–3347.

Huang Q, Chang KL, Gaal K, et al. (2004). KSHV/HHV8-associated lymphoma simulating anaplastic large cell lymphoma. *The American Journal of Surgical Pathology* 28: 693–697.

Hyun TS, Subramanian C, Cotter M, et al. (2001). Latency-associated nuclear antigen encoded by Kaposi's sarcoma-associated herpesvirus interacts with Tat and activates the long terminal repeat of human immunodeficiency virus type 1 in human cells. *Journal of Virology* 75: 8761–8771.

Irmler M, Thome M, Hahne M, et al. (1997). Inhibition of death receptor signals by cellular FLIP. *Nature* 388: 190–195.

Ishido S, Wang C, Lee BS, et al. (2000). Downregulation of major histocompatibility complex class I molecules by Kaposi's sarcoma-associated herpesvirus K3 and K5 proteins. *Journal of Virology* 74: 5300–5309.

Järviluoma A, Koopal S, Räsänen S, et al. (2004). KSHV viral cyclin binds to p27KIP1 in primary effusion lymphomas. *Blood* 104: 3349–3354.
Jenner R, Mar Albà M, Boshoff C, et al. (2001). Kaposi's sarcoma-associated herpesvirus latent and lytic gene expression as revealed by DNA arrays. *Journal of Virology* 75: 891–902.
Jenner RG, Boshoff C. (2002). The molecular pathology of Kaposi's sarcoma-associated herpesvirus. *Biochimica Et Biophysica Acta (BBA)-Reviews on Cancer* 1602: 1–22.
Jenner RG, Maillard K, Cattini N, et al. (2003). Kaposi's sarcoma-associated herpesvirus-infected primary effusion lymphoma has a plasma cell gene expression profile. *Proceedings of the National Academy of Sciences* 100: 10399–10404.
Jensen K, Manfra D, Grisotto M, et al. (2005). The human herpes virus 8-encoded chemokine receptor is required for angioproliferation in a murine model of Kaposi's sarcoma. *The Journal of Immunology* 174: 3686–3694.
Jeong J, Orvis J, Kim JW, et al. (2004). Regulation and autoregulation of the promoter for the latency-associated nuclear antigen of Kaposi's sarcoma-associated herpesvirus. *Journal of Biological Chemistry* 279: 16822–16831.
Jeong J, Papin J, Dittmer D. (2001). Differential regulation of the overlapping Kaposi's sarcoma-associated herpesvirus vGCR (orf74) and LANA (orf73) promoters. *Journal of Virology* 75: 1798–1807.
Judde JG, Lacoste V, Brière J, et al. (2000). Monoclonality or oligoclonality of human herpesvirus 8 terminal repeat sequences in Kaposi's sarcoma and other diseases. *Journal of the National Cancer Institute* 92: 729–736.
Kaposi M. (1872). Idiopathic multiple pigmented sarcoma of the skin. *Arch Dermatol Syphil* 4: 265–273.
Kchour G, Tarhini M, Kooshyar MM, et al. (2009). Phase 2 study of the efficacy and safety of the combination of arsenic trioxide, interferon alpha, and zidovudine in newly diagnosed chronic adult T-cell leukemia/lymphoma (ATL). *Blood* 113: 6528–6532.
Kellam PC, Boshoff D, Whitby S, et al. (1997). Identification of a major latent nuclear antigen, LNA-1, in the human herpesvirus 8 genome. *Journal of Human Virology* 1: 19–29.
Klein U, Gloghini A, Gaidano G, et al. (2003). Gene expression profile analysis of AIDS-related primary effusion lymphoma (PEL) suggests a plasmablastic derivation and identifies PEL-specific transcripts. *Blood* 101: 4115–4121.
Kliche S, Nagel W, Kremmer E, et al. (2001). Signaling by human herpesvirus 8 kaposin A through direct membrane recruitment of cytohesin-1. *Molecular Cell* 7: 833–843.
Kogan SC. (2009). Curing APL: Differentiation or destruction? *Cancer Cell* 15: 7–8.
Komatsu T, Ballestas ME, Barbera A, et al. (2002). The KSHV latency-associated nuclear antigen: A multifunctional protein. *Frontiers in Bioscience* 7: d726–730.
Koon H, Bubley G, Pantanowitz L, et al. (2005). Imatinib-induced regression of AIDS-related Kaposi's sarcoma. *Journal of Clinical Oncology* 23: 982–989.
Koopal S, Furuhjelm JH, Järviluoma A, et al. (2007). Viral oncogene–induced DNA damage response is activated in Kaposi sarcoma tumorigenesis. *PLoS Pathogens* 3: e140.
Krown SE. (2001). Management of Kaposi sarcoma: The role of interferon and thalidomide. *Current Opinion in Oncology* 13: 374–381.
Krown SE. (2004). Highly active antiretroviral therapy in AIDS-associated Kaposi's sarcoma: Implications for the design of therapeutic trials in patients with advanced, symptomatic Kaposi's sarcoma. *Journal of Clinical Oncology* 22: 399–402.
Kumode T, Ohyama Y, Kawauchi M, et al. (2013). Clinical importance of human herpes virus-8 and human immunodeficiency virus infection in primary effusion lymphoma. *Leukemia & Lymphoma* 54: 1947–1952.
Lagunoff M, Ganem D. (1997). The structure and coding organization of the genomic termini of Kaposi's sarcoma-associated herpesvirus (human herpesvirus 8). *Virology* 236: 147–154.

Lallemand-Breitenbach V, Jeanne M, Benhenda S, et al. (2008). Arsenic degrades PML or PML–RARα through a SUMO-triggered RNF4/ubiquitin-mediated pathway. *Nature Cell Biology* 10: 547–555.

Laubenstein L, Krigel RL, Odajnyk CM, et al. (1984). Treatment of epidemic Kaposi's sarcoma with etoposide or a combination of doxorubicin, bleomycin, and vinblastine. *Journal of Clinical Oncology* 2: 1115–1120.

Lee BS, Alvarez X, Ishido S, et al. (2000). Inhibition of intracellular transport of B cell antigen receptor complexes by Kaposi's sarcoma–associated herpesvirus K1. *The Journal of Experimental Medicine* 192: 11–22.

Lee BS, Lee SH, Feng P, et al. (2005). Characterization of the Kaposi's sarcoma-associated herpesvirus K1 signalosome. *Journal of Virology* 79: 12173–12184.

Lee RK, Cai JP, Deyev V, et al. (1999). Azidothymidine and interferon-α induce apoptosis in herpesvirus-associated lymphomas. *Cancer Research* 59: 5514–5520.

Levine A, Tulpule A, Quinn D, et al. (2006). Phase I study of antisense oligonucleotide against vascular endothelial growth factor: Decrease in plasma vascular endothelial growth factor with potential clinical efficacy. *Journal of Clinical Oncology* 24: 1712–1719.

Li CF, Ye H, Liu H, et al. (2006). Fatal HHV-8-associated hemophagocytic syndrome in an HIV-negative immunocompetent patient with plasmablastic variant of multicentric Castleman disease (plasmablastic microlymphoma). *The American Journal of Surgical Pathology* 30: 123–127.

Li M, Lee H, Yoon DW, et al. (1997). Kaposi's sarcoma-associated herpesvirus encodes a functional cyclin. *Journal of Virology* 71: 1984–1991.

Lim C, Gwack Y, Hwang S, et al. (2001). The transcriptional activity of cAMP response element-binding protein-binding protein is modulated by the latency associated nuclear antigen of Kaposi's sarcoma-associated herpesvirus. *Journal of Biological Chemistry* 276: 31016–31022.

Lim ST, Rubin N, Said J, et al. (2005). Primary effusion lymphoma: Successful treatment with highly active antiretroviral therapy and rituximab. *Annals of Hematology* 84: 551–552.

Loi S, Goldstein D, Clezy K, et al. (2004). Castleman's disease and HIV infection in Australia. *HIV Medicine* 5: 157–162.

Lubyova B, Kellum MJ, Frisancho AJ, et al. (2004). Kaposi's sarcoma-associated herpesvirus-encoded vIRF-3 stimulates the transcriptional activity of cellular IRF-3 and IRF-7. *Journal of Biological Chemistry* 279: 7643–7654.

Lubyova B, Kellum MJ, Frisancho JA, et al. (2007). Stimulation of c-Myc transcriptional activity by vIRF-3 of Kaposi sarcoma-associated herpesvirus. *Journal of Biological Chemistry* 282: 31944–31953.

Luppi M, Barozzi P, Maiorana A, et al. (1996). Human herpesvirus-8 DNA sequences in human immunodeficiency virus-negative angioimmunoblastic lymphadenopathy and benign lymphadenopathy with giant germinal center hyperplasia and increased vascularity. *Blood* 87: 3903–3909.

Luppi M, Trovato R, Barozzi P, et al. (2005). Treatment of herpesvirus associated primary effusion lymphoma with intracavity cidofovir. *Leukemia* 19: 473–476.

Lynen L, Zolfo M, Huyst V, et al. (2005). Management of Kaposi's sarcoma in resource-limited settings in the era of HAART. *AIDS Reviews* 7: 13–21.

Mahieux R, Pise-Masison C, Gessain A, et al. (2001). Arsenic trioxide induces apoptosis in human T-cell leukemia virus type 1– and type 2–infected cells by a caspase-3–dependent mechanism involving Bcl-2 cleavage. *Blood* 98: 3762–3769.

Marcelin AG, Aaron L, Mateus C, et al. (2003). Rituximab therapy for HIV-associated Castleman disease. *Blood* 102: 2786–2788.

Martellotta F, Berretta M, Vaccher E, et al. (2009). AIDS-related Kaposis sarcoma: State of the art and therapeutic strategies. *Current HIV Research* 7: 634–638.

Martín-Carbonero L, Barrios A, Saballs P, et al. (2004). Pegylated liposomal doxorubicin plus highly active antiretroviral therapy versus highly active antiretroviral therapy alone in HIV patients with Kaposi's sarcoma. *AIDS* 18: 1737–1740.

Masood R, Cesarman E, Smith LD, et al. (2002). Human herpesvirus-8-transformed endothelial cells have functionally activated vascular endothelial growth factor/vascular endothelial growth factor receptor. *The American Journal of Pathology* 160: 23–29.

Masood R, Xia G, Smith DL, et al. (2005). Ephrin B2 expression in Kaposi sarcoma is induced by human herpesvirus type 8: Phenotype switch from venous to arterial endothelium. *Blood* 105: 1310–1318.

McCormick C, Ganem D. (2005). The kaposin B protein of KSHV activates the p38/MK2 pathway and stabilizes cytokine mRNAs. *Science Signaling* 307: 739–741.

Miltenyi Z, Toth J, Gonda A, et al. (2009). Successful immunomodulatory therapy in Castleman disease with paraneoplastic pemphigus vulgaris. *Pathology & Oncology Research* 15: 375–381.

Mocroft A, Kirk O, Clumeck N, et al. (2004). The changing pattern of Kaposi sarcoma in patients with HIV, 1994–2003. *Cancer* 100: 2644–2654.

Mocroft A, Youle M, Gazzard B, et al. (1996). Anti-herpesvirus treatment and risk of Kaposi's sarcoma in HIV infection. *AIDS* 10: 1101–1105.

Mohanna S, Bravo F, Ferrufino JC, et al. (2007). Classic Kaposi s sarcoma presenting in the oral cavity of two HIV-negative quechua patients. *Medicina Oral, Patología Oral y Cirugía Bucal (Internet)* 12: 365–368.

Moore PS, Boshoff C, Weiss RA, et al. (1996a). Molecular mimicry of human cytokine and cytokine response pathway genes by KSHV. *Science* 274: 1739–1744.

Moore PS, Kingsley L, Holmberg S, et al. (1996b). Kaposi's sarcoma-associated herpesvirus infection prior to onset of Kaposi's sarcoma. *AIDS* 10: 175–180.

Moses AV, Jarvis MA, Raggo C, et al. (2002). Kaposi's sarcoma-associated herpesvirus-induced upregulation of the c-kit proto-oncogene, as identified by gene expression profiling, is essential for the transformation of endothelial cells. *Journal of Virology* 76: 8383–8399.

Muralidhar S, Veytsmann G, Chandran B, et al. (2000). Characterization of the human herpesvirus 8 (Kaposi's sarcoma-associated herpesvirus) oncogene, kaposin (ORF K12). *Journal of Clinical Virology* 16: 203–213.

Murdaca G, Campelli A, Setti M, et al. (2002). Complete remission of AIDS/Kaposi's sarcoma after treatment with a combination of two nucleoside reverse transcriptase inhibitors and one non-nucleoside reverse transcriptase inhibitor. *AIDS* 16: 304–305.

Nador R, Cesarman E, Chadburn A, et al. (1996). Primary effusion lymphoma: A distinct clinicopathologic entity associated with the Kaposi's sarcoma-associated herpes virus. *Blood* 88: 645–656.

Nador R, Tsang P, Reed J, et al. (1997). Expression of KSHV/HHV 8 cyclin in primary effusion lymphomas and Kaposi's sarcoma: 37. *Journal of Acquired Immune Deficiency Syndromes* 14: A25.

Nakamura H, Lu M, Gwack Y, et al. (2003). Global changes in Kaposi's sarcoma-associated virus gene expression patterns following expression of a tetracycline-inducible Rta transactivator. *Journal of Virology* 77: 4205–4220.

Naranatt P, Krishnan H, Svojanovsky S, et al. (2004). Host gene induction and transcriptional reprogramming in Kaposi's sarcoma-associated herpesvirus (KSHV/HHV-8)-infected endothelial, fibroblast, and B cells insights into modulation events early during infection. *Cancer Research* 64: 72–84.

Nasr R, Guillemin MC, Ferhi O, et al. (2008). Eradication of acute promyelocytic leukemia-initiating cells through PML-RARA degradation. *Nature Medicine* 14: 1333–1342.

Nasr R, Lallemand-Breitenbach V, Zhu J, et al. (2009). Therapy-induced PML/RARA proteolysis and acute promyelocytic leukemia cure. *Clinical Cancer Research* 15: 6321–6326.

Nasr R, Rosenwald A, El-Sabban M, et al. (2003). Arsenic/interferon specifically reverses 2 distinct gene networks critical for the survival of HTLV-1–infected leukemic cells. *Blood* 101: 4576–4582.

Nasti G, Errante D, Talamini R, et al. (2000). Vinorelbine is an effective and safe drug for AIDS-related Kaposi's sarcoma: Results of a phase II study. *Journal of Clinical Oncology* 18: 1550–1557.

Nasti G, Martellotta F, Berretta M, et al. (2003a). Impact of highly active antiretroviral therapy on the presenting features and outcome of patients with acquired immunodeficiency syndrome–related Kaposi sarcoma. *Cancer* 98: 2440–2446.

Nasti G, Talamini R, Antinori A, et al. (2003b). AIDS-related Kaposi's sarcoma: Evaluation of potential new prognostic factors and assessment of the AIDS clinical trial group staging system in the HAART era—the Italian Cooperative Group on AIDS and Tumors and the Italian Cohort of Patients Naive from Antiretrovirals. *Journal of Clinical Oncology* 21: 2876–2882.

Neyts J, De Clercq E. (1997). Antiviral drug susceptibility of human herpesvirus 8. *Antimicrobial Agents and Chemotherapy* 41: 2754–2756.

Nishimoto N, Kanakura Y, Aozasa K, et al. (2005). Humanized anti–interleukin-6 receptor antibody treatment of multicentric Castleman disease. *Blood* 106: 2627–2632.

Nishimoto N, Sasai M, Shima Y, et al. (2000). Improvement in Castleman's disease by humanized anti-interleukin-6 receptor antibody therapy. *Blood* 95: 56–61.

Northfelt DW, Dezube B, Thommes J, et al. (1998). Pegylated-liposomal doxorubicin versus doxorubicin, bleomycin, and vincristine in the treatment of AIDS-related Kaposi's sarcoma: Results of a randomized phase III clinical trial. *Journal of Clinical Oncology* 16: 2445–2451.

Ogita M, Hoshino J, Sogawa Y, et al. (2007). Multicentric Castleman disease with secondary AA renal amyloidosis, nephrotic syndrome and chronic renal failure, remission after high-dose melphalan and autologous stem cell transplantation. *Clinical Nephrology* 68: 171–176.

Oksenhendler E, Clauvel JP, Jouveshomme S, et al. (1998). Complete remission of a primary effusion lymphoma with antiretroviral therapy. *Am J Hematol.* 57: 266.

Oksenhendler E. (2009). HIV-associated multicentric castleman disease. *Current Opinion in HIV and AIDS* 4: 16–21.

Oksenhendler E, Boulanger E, Galicier L, et al. (2002). High incidence of Kaposi sarcoma–associated herpesvirus–related non-Hodgkin lymphoma in patients with HIV infection and multicentric Castleman disease. *Blood* 99: 2331–2336.

Oksenhendler E, Duarte M, Soulier J, et al. (1996). Multicentric Castleman's disease in HIV infection: A clinical and pathological study of 20 patients. *AIDS* 10: 61–68.

Osborne J, Moore PS, Chang Y. (1999). KSHV-encoded viral IL-6 activates multiple human IL-6 signaling pathways. *Human Immunology* 60: 921–927.

Otsuki T, Kumar S, Ensoli B, et al. (1996). Detection of HHV-8/KSHV DNA sequences in AIDS-associated extranodal lymphoid malignancies. *Leukemia* 10: 1358–1364.

Parravicini C, Chandran B, Corbellino M, et al. (2000). Differential viral protein expression in Kaposi's sarcoma-associated herpesvirus-infected diseases: Kaposi's sarcoma, primary effusion lymphoma, and multicentric Castleman's disease. *The American Journal of Pathology* 156: 743–749.

Paulose-Murphy M, Ha NK, Xiang C, et al. (2001). Transcription program of human herpesvirus 8 (Kaposi's sarcoma-associated herpesvirus). *Journal of Virology* 75: 4843–4853.

Petre C, Sin SH, Dittmer D. (2007). Functional p53 signaling in Kaposi's sarcoma-associated herpesvirus lymphomas: Implications for therapy. *Journal of Virology* 81: 1912–1922.

Pluda J, Shay L, Foli A, et al. (1993). Administration of pentosan polysulfate to patients with human immunodeficiency virus-associated Kaposi's sarcoma. *Journal of the National Cancer Institute* 85: 1585–1592.

Poignonec S, Lachiver LD, Lamas G, et al. (1995). Intralesional bleomycin for acquired immunodeficiency syndrome-associated cutaneous Kaposi's sarcoma. *Archives of Dermatology* 131: 228.

Portsmouth S, Stebbing J, Gill J, et al. (2003). A comparison of regimens based on non-nucleoside reverse transcriptase inhibitors or protease inhibitors in preventing Kaposi's sarcoma. *AIDS* 17: F17–F22.

Powell B, Moser B, Stock W, et al. (2010). Arsenic trioxide improves event-free and overall survival for adults with acute promyelocytic leukemia: North American Leukemia Intergroup Study C9710. *Blood* 116: 3751–3757.

Prakash O, Swamy R, Peng X, et al. (2005). Activation of Src kinase Lyn by the Kaposi sarcoma–associated herpesvirus K1 protein: Implications for lymphomagenesis. *Blood* 105: 3987–3994.

Radkov S, Kellam P, Boshoff C. (2000). The latent nuclear antigen of Kaposi sarcoma-associated herpesvirus targets the retinoblastoma–E2F pathway and with the oncogene Hras transforms primary rat cells. *Nature Medicine* 6: 1121–1127.

Radu O, Pantanowitz L. (2013). Kaposi sarcoma. *Archives of Pathology and Laboratory Medicine* 137: 289–294.

Régnier-Rosencher E, Barrou B, Marcelin AG, et al. (2010). Primary effusion lymphoma in two kidney transplant recipients. *Annales De Dermatologie Et De Venereologie* 137: 285–289.

Renne R, Barry C, Dittmer D, et al. (2001). Modulation of cellular and viral gene expression by the latency-associated nuclear antigen of Kaposi's sarcoma-associated herpesvirus. *Journal of Virology* 75: 458–468.

Renne R, Lagunoff M, Zhong W, et al. (1996a). The size and conformation of Kaposi's sarcoma-associated herpesvirus (human herpesvirus 8) DNA in infected cells and virions. *Journal of Virology* 70: 8151–8154.

Renne R, Zhong W, Herndier B, et al. (1996b). Lytic growth of Kaposi's sarcoma–associated herpesvirus (human herpesvirus 8) in culture. *Nature Medicine* 2: 342–346.

Riva G, Luppi M, Barozzi P, et al. (2012). How I treat HHV8/KSHV-related diseases in post-transplant patients. *Blood* 120: 4150–4159.

Rohrmus B, Thoma-Greber E, Bogner J, et al. (2000). Outlook in oral and cutaneous Kaposi's sarcoma. *The Lancet* 356(9248): 2160.

Russo J, Bohenzky R, Chien MC, et al. (1996). Nucleotide sequence of the Kaposi sarcoma-associated herpesvirus (HHV8). *Proceedings of the National Academy of Sciences* 93: 14862–14867.

Sadler R, Wu L, Forghani B, et al. (1999). A complex translational program generates multiple novel proteins from the latently expressed kaposin (K12) locus of Kaposi's sarcoma-associated herpesvirus. *Journal of Virology* 73: 5722–5730.

Samaniego F, Pati S, Karp J, et al. (2001). Human herpesvirus 8 K1-associated nuclear factor-kappa B-dependent promoter activity: Role in Kaposi's sarcoma inflammation? *Journal of the National Cancer Institute Monographs* 28: 15–23.

Samaniego F, Young D, Grimes C, et al. (2002). Vascular endothelial growth factor and Kaposi's sarcoma cells in human skin grafts. *Cell Growth and Differentiation* 13: 387–395.

Sarid R, Flore O, Bohenzky RA, et al. (1998). Transcription mapping of the Kaposi's sarcoma-associated herpesvirus (Human Herpesvirus 8) genome in a body cavity-based lymphoma cell line (BC-1). *Journal of Virology* 72: 1005–1012.

Sarid R, Sato T, Bohenzky RA, et al. (1997). Kaposi's sarcoma-associated herpesvirus encodes a functional Bcl-2 homologue. *Nature Medicine* 3: 293–298.

Sarosiek K, Cavallin L, Bhatt S, et al. (2010). Efficacy of bortezomib in a direct xenograft model of primary effusion lymphoma. *Proceedings of the National Academy of Sciences* 107: 13069–13074.

Schwam D, Luciano R, Mahajan S, et al. (2000). Carboxy terminus of human herpesvirus 8 latency-associated nuclear antigen mediates dimerization, transcriptional repression, and targeting to nuclear bodies. *Journal of Virology* 74: 8532–8540.

Schwartsmann S, Kalakun L, Yamagushi N. (1996). Phase II study of pentosan polysulfate (PPS) in patients with AIDS-related Kaposi's sarcoma. *Tumori* 82: 360–363.

Schwartz RA. (2004). Kaposi's sarcoma: An update. *Journal of Surgical Oncology* 87: 146–151.

Scott D, Cabral L, Harrington WJ. (2001). Treatment of HIV-associated multicentric Castleman's disease with oral etoposide. *American Journal of Hematology* 66: 148–150.

Shen ZX, Shi ZZ, Fang J, et al. (2004). All-trans retinoic acid/As2O3 combination yields a high quality remission and survival in newly diagnosed acute promyelocytic leukemia. *Proceedings of the National Academy of Sciences of the United States of America* 101: 5328–5335.

Shin YC, Joo CH, Gack M, et al. (2008). Kaposi's sarcoma–associated herpesvirus viral IFN regulatory factor 3 stabilizes hypoxia-inducible factor-1$\alpha$ to induce vascular endothelial growth factor expression. *Cancer Research* 68: 1751–1759.

Si H, Robertson ES. (2006). Kaposi's sarcoma-associated herpesvirus-encoded latency-associated nuclear antigen induces chromosomal instability through inhibition of p53 function. *Journal of Virology* 80: 697–709.

Siddiqi T, Joyce RM. (2008). A case of HIV-negative primary effusion lymphoma treated with bortezomib, pegylated liposomal doxorubicin, and rituximab. *Clinical Lymphoma and Myeloma* 8: 300–304.

Simonelli C, Spina M, Cinelli R, et al. (2003). Clinical features and outcome of primary effusion lymphoma in HIV-infected patients: A single-institution study. *Journal of Clinical Oncology* 21: 3948–3954.

Sin SH, Roy D, Wang L, et al. (2007). Rapamycin is efficacious against primary effusion lymphoma (PEL) cell lines in vivo by inhibiting autocrine signaling. *Blood* 109: 2165–2173.

Sobas MA, Vence N, Arias JD, et al. (2010). Efficacy of bortezomib in refractory form of multicentric castleman disease associated to poems syndrome (MCD-POEMS Variant). *Annals of Hematology* 89: 217–219.

Soulier J, Grollet L, Oksenhendler E, et al. (1995). Kaposi's sarcoma-associated herpesvirus-like DNA sequences in multicentric Castleman's disease. *Blood* 86: 1276–1280.

Spano JP, Costagliola D, Katlama C, et al. (2008). AIDS-related malignancies: State of the art and therapeutic challenges. *Journal of Clinical Oncology* 26: 4834–4842.

Spiller OB, Blackbourn DJ, Mark L, et al. (2003a). Functional activity of the complement regulator encoded by Kaposi's sarcoma-associated herpesvirus. *Journal of Biological Chemistry* 278: 9283–9289.

Spiller OB, Mark L, Blue CE, et al. (2006). Dissecting the regions of virion-associated Kaposi's sarcoma-associated herpesvirus complement control protein required for complement regulation and cell binding. *Journal of Virology* 80: 4068–4078.

Spiller OB, Robinson M, O'Donnell E, et al. (2003b). Complement regulation by Kaposi's sarcoma-associated herpesvirus ORF4 protein. *Journal of Virology* 77: 592–599.

Spina M, Gaidano G, Carbone A, et al. (1998). Highly active antiretroviral therapy in human herpesvirus-8-related body-cavity-based lymphoma. *AIDS (London, England)* 12: 955–956.

Stary G, Kohrgruber N, Herneth MA, et al. (2008). Complete regression of HIV-associated multicentric Castleman disease treated with rituximab and thalidomide. *AIDS* 22: 1232–1234.

Staskus KA, Zhong W, Gebhard K, et al. (1997). Kaposi's sarcoma-associated herpesvirus gene expression in endothelial (spindle) tumor cells. *Journal of Virology* 71: 715–719.

Staudt MR, Dittmer DP. (2006). Promoter switching allows simultaneous transcription of LANA and K14/vGPCR of Kaposi's sarcoma-associated herpesvirus. *Virology* 350: 192–205.

Sullivan R, Pantanowitz L, Casper C, et al. (2008). Epidemiology, pathophysiology, and treatment of Kaposi sarcoma—associated herpesvirus disease: Kaposi sarcoma, primary effusion lymphoma, and multicentric castleman disease. *Clinical Infectious Diseases* 47: 1209–1215.

Suzuki K, Ino K, Sugawara Y, et al. (2008). Prolonged survival in a patient with human herpesvirus-8-negative primary effusion lymphoma after combination chemotherapy with rituximab. *Gan to Kagaku Ryoho. Cancer & Chemotherapy* 35: 691–694.

Swanton C, Mann DJ, Fleckenstein B, et al. (1997). Herpes viral cyclin/Cdk6 complexes evade inhibition by CDK inhibitor proteins. *Nature* 390: 184–187.

Tamayo M, Gonzalez C, Juliana M, et al. (1995). Long-term complete remission after interferon treatment in a case of multicentric Castelman's disease. *American Journal of Hematology* 49: 359–360.

Tappero JW, Berger TG, Kaplan LD, et al. (1991). Cryotherapy for cutaneous Kaposi's sarcoma (KS) associated with acquired immune deficiency syndrome (AIDS): A phase II trial. *Journal of Acquired Immune Deficiency Syndromes* 4: 839–846.

Tirelli U, Bernardi D. (2001). Impact of HAART on the clinical management of AIDS-related cancers. *European Journal of Cancer* 37: 1320–1324.

Tomkowicz B, Singh SP, Cartas M, et al. (2002). Human herpesvirus-8 encoded Kaposin: Subcellular localization using immunofluorescence and biochemical approaches. *DNA and Cell Biology* 21: 151–162.

Trattner A, Reizis Z, David M, et al. (1993). The therapeutic effect of intralesional interferon in classical Kaposi's sarcoma. *British Journal of Dermatology* 129: 590–593.

Trus B, Heymann JB, Nealon K, et al. (2001). Capsid structure of Kaposi's sarcoma-associated herpesvirus, a gammaherpesvirus, compared to those of an alphaherpesvirus, herpes simplex virus type 1, and a betaherpesvirus, cytomegalovirus. *Journal of Virology* 75: 2879–2890.

Tulpule A, Scadden DT, Espina BM, et al. (2000). Results of a randomized study of IM862 nasal solution in the treatment of AIDS-related Kaposi's sarcoma. *Journal of Clinical Oncology* 18: 716–723.

Uddin S, Hussain AR, Al-Hussein KA, et al. (2005). Inhibition of phosphatidylinositol 3′-kinase/AKT signaling promotes apoptosis of primary effusion lymphoma cells. *Clinical Cancer Research* 11: 3102–3108.

Uldrick TS, Wyvill KM, Kumar P, et al. (2012). Phase ii study of bevacizumab in patients with HIV-associated Kaposi's sarcoma receiving antiretroviral therapy. *Journal of Clinical Oncology* 30: 1476–1483.

Vanni T, Sprinz E, Machado MW, et al. (2006). Systemic treatment of AIDS-related Kaposi sarcoma: Current status and perspectives. *Cancer Treatment Reviews* 32: 445–455.

Verma SC, Lan K, Robertson E. (2007). Structure and function of latency-associated nuclear antigen. In: Boshoff C, Weiss RA (eds.), *Kaposi Sarcoma Herpesvirus: New Perspectives.* Springer, Heidelberg, Berlin, 101–136.

Verschuren EW, Hodgson G, Gray J, et al. (2004). The role of p53 in suppression of KSHV cyclin-induced lymphomagenesis. *Cancer Research* 64: 581–589.

Verschuren EW, Klefstrom J, Evan GI, et al. (2002). The oncogenic potential of Kaposi's sarcoma-associated herpesvirus cyclin is exposed by p53 loss in vitro and in vivo. *Cancer Cell* 2: 229–241.

Viejo-Borbolla A, Schulz TF. (2003). Kaposi's sarcoma-associated herpesvirus (KSHV/HHV8): Key aspects of epidemiology and pathogenesis. *AIDS Reviews* 5: 222–229.

Virgin HW, Latreille P, Wamsley P, et al. (1997). Complete sequence and genomic analysis of murine gammaherpesvirus 68. *Journal of Virology* 71: 5894–5904.

Von Roenn JH. (2003). Clinical presentations and standard therapy of AIDS-associated Kaposi's sarcoma. *Hematology/Oncology Clinics of North America* 17: 747–762.

Waddington TW, Aboulafia DM. (2004). Failure to eradicate AIDS-associated primary effusion lymphoma with high-dose chemotherapy and autologous stem cell reinfusion: Case report and literature review. *AIDS Patient Care and STDs* 18: 67–73.

Wang CY, Sugden B. (2004). New viruses shake old paradigms. *Journal of Clinical Investigation* 113(1): 21–23.

Wang HW, Sharp TV, Koumi A, et al. (2002). Characterization of an anti-apoptotic glycoprotein encoded by Kaposi's sarcoma-associated herpesvirus which resembles a spliced variant of human survivin. *The EMBO Journal* 21: 2602–2615.

Wang HW, Trotter M, Lagos D, et al. (2004a). Kaposi sarcoma herpesvirus–induced cellular reprogramming contributes to the lymphatic endothelial gene expression in Kaposi sarcoma. *Nature Genetics* 36: 687–693.

Wang L, Dittmer DP, Tomlinson CC, et al. (2006). Immortalization of primary endothelial cells by the K1 protein of Kaposi's sarcoma–associated herpesvirus. *Cancer Research* 66: 3658–3666.

Wang L, Wakisaka N, Tomlinson C, et al. (2004b). The Kaposi's sarcoma-associated herpesvirus (KSHV/HHV-8) K1 protein induces expression of angiogenic and invasion factors. *Cancer Research* 64: 2774–2781.

Wang XP, Zhang YJ, Deng JH, et al. (2001). Characterization of the promoter region of the viral interferon regulatory factor encoded by Kaposi's sarcoma-associated herpesvirus. *Oncogene* 20: 523–530.

Wang YF, Hsieh YF, Lin CL, et al. (2004c). Staurosporine-induced G2/M arrest in primary effusion lymphoma BCBL-1 cells. *Annals of Hematology* 83: 739–744.

Weber K, Gröne HJ, Röcken M, et al. (2001). Selective recruitment of Th2-type cells and evasion from a cytotoxic immune response mediated by viral macrophage inhibitory protein-II. *European Journal of Immunology* 31: 2458–2466.

Weindel K, Marmé D, Weich HA. (1992). AIDS-associated Kaposi's sarcoma cells in culture express vascular endothelial growth factor. *Biochemical and Biophysical Research Communications* 183: 1167–1174.

Weiss R, Mitrou P, Arasteh K, et al. (2006). Acquired immunodeficiency syndrome-related lymphoma. *Cancer* 106: 1560–1568.

Welles L, Saville W, Lietzau J, et al. (1998). Phase II trial with dose titration of paclitaxel for the therapy of human immunodeficiency virus-associated Kaposi's sarcoma. *Journal of Clinical Oncology* 16: 1112–1121.

Wen KW, Damania B. (2010). Kaposi sarcoma-associated herpesvirus (KSHV): Molecular biology and oncogenesis. *Cancer Letters* 289: 140–150.

Wies E, Mori Y, Hahn A, et al. (2008). The viral interferon-regulatory factor-3 is required for the survival of KSHV-infected primary effusion lymphoma cells. *Blood* 111: 320–327.

Won JH, Han SH, Bae SB, et al. (2006). Successful eradication of relapsed primary effusion lymphoma with high-dose chemotherapy and autologous stem cell transplantation in a patient seronegative for human immunodeficiency virus. *International Journal of Hematology* 83: 328–330.

Wong E, Damania B. (2006). Transcriptional regulation of the Kaposi's sarcoma-associated herpesvirus K15 gene. *Journal of Virology* 80: 1385–1392.

Wu W, Rochford R, Toomey L, et al. (2005). Inhibition of HHV-8/KSHV infected primary effusion lymphomas in NOD/SCID mice by azidothymidine and interferon-α. *Leukemia Research* 29: 545–555.

Xie J, Pan H, Yoo S, et al. (2005). Kaposi's sarcoma-associated herpesvirus induction of AP-1 and interleukin 6 during primary infection mediated by multiple mitogen-activated protein kinase pathways. *Journal of Virology* 79: 15027–15037.

Yang TY, Chen SC, Leach MW, et al. (2000). Transgenic expression of the chemokine receptor encoded by human herpesvirus 8 induces an angioproliferative disease resembling Kaposi's sarcoma. *The Journal of Experimental Medicine* 191: 445–454.

Ye FC, Zhou FC, Xie JP, et al. (2008). Kaposi's sarcoma-associated herpesvirus latent gene vFLIP inhibits viral lytic replication through NF-κB-mediated suppression of the AP-1 pathway: A novel mechanism of virus control of latency. *Journal of Virology* 82: 4235–4249.

Yiakoumis X, Pangalis GA, Kyrtsonis MC. (2010). Primary effusion lymphoma in two HIV-negative patients successfully treated with pleurodesis as first-line therapy. *Anticancer Research* 30: 271–276.

Zhong W, Wang H, Herndier B. (1996). Restricted expression of Kaposi sarcoma-associated herpesvirus (human herpesvirus 8) genes in Kaposi sarcoma. *Proceedings of the National Academy of Sciences* 93: 6641–6646.

Zietz C, Bogner JR, Goebel FD, et al. (1999). An unusual cluster of cases of Castleman's disease during highly active antiretroviral therapy for AIDS. *New England Journal of Medicine* 340: 1923–1924.

# 11 Tea Polyphenolic Compounds against Herpes Simplex Viruses

*Tin-Chun Chu, Sandra D. Adams, and Lee H. Lee*

## 11.1 NATURAL PRODUCTS, TEA, AND TEA POLYPHENOLS

The constant need for novel antimicrobial treatments is driving pharmaceutical companies to scavenge many natural products. Natural compounds from plants have traditionally been known to treat microbial infections. Plant extracts have been used for centuries to heal cuts and bruises and for other medical emergencies. Several natural compounds such as cerulenin and thiolactomycin are antibacterial agents (Zhang and Rock 2004). Plant extracts have a number of low-molecular-weight metabolites. These metabolites could have arisen as an evolutionary response to microbial attacks on the plants. They serve as an important defense mechanism against pathogens and animal herbivores. If plants can prevent microbial attack, then humans could potentially use these metabolites against human pathogenic microorganisms.

Many studies have reported that natural antiviral compounds from plants can inhibit different viruses. Several alkaloids reported activity on herpes simplex virus type 1 (HSV-1) and influenza virus (Jassim and Naji 2003; Orhana et al. 2007; Ozcelik et al. 2011; Rajbhandari et al. 2001; Serkedjieva and Velcheva 2003) and furyl compounds (Hudson 1989; Jassim and Naji 2003) and hypericin as anti-Sindbis and against other viruses (Hudson 1989; Hudson et al. 1997; Lopez-Bazzocchi et al. 1991); polyacetylenes and thiophenes can damage viral membrane integrity or inhibit viral replication (Hudson 1989). Some plants contain terpenoids that have been demonstrated to have antiviral activities against herpes simplex virus (HSV) and human immunodeficiency virus (HIV) (Ito et al. 1988; Kashiwada et al. 1996; Li et al. 2003; Min et al. 1999a, 1999b; Rukachaisirikul et al. 2003; Yu et al. 2005, 2007; Zhang et al. 2003). Essential oils from various plants including *Eugenia brasiliensis*, *Rosmarinus officinalis*, *Origanum syriacum*, and *Foeniculum vulgare* also showed antiviral activities on HSV-1, HSV-2 (Allahverdiyev et al. 2004; Armaka et al. 1999; Benencia and Courreges 1999; De Logu et al. 2000; Farag et al. 2004; Hayashi et al. 1995; Primo et al. 2001; Schnitzler et al. 2001; Schuhmacher et al. 2003; Sinico et al. 2005; Sivropoulou et al. 1997; Vijayan et al. 2004), and the yellow fever virus (Meneses et al. 2009). Protein and peptides extracted from different plants have shown strong antiviral activities against influenza virus H1N1, HIV (D'Cruz and

Uckun 2001; Rajamohan et al. 1999; Uckun et al. 1998, 1999; Wang and Tumer 1999), and HSV-1 infection (Camargo Filho et al. 2008). Some studies on the polysaccharide extracted from plants and red marine algae have been reported to have inhibitory activity on some strains of HSV-1 (Thompson and Dragar 2004) and influenza virus (Huang et al. 1991; Serkedjieva 2004).

The most common plant products that illustrate antiviral activities are phenolic compounds. Some phenolic compounds have been shown to exhibit antiviral activity against poliomyelitis viruses (Konowalchuk and Speirs 1976), HSVs, coxsackie viruses (Baldé et al. 1990; Fukuchi et al. 1989), parainfluenza viruses (Ozcelik et al. 2011), and HIV (Heredia et al. 2000). The most studied polyphenols are those antimicrobial metabolites found in plants such as green tea (Dixon 2001). In the past several decades, an immense amount of studies have been conducted on the valuable green tea extract. Nature's most powerful antioxidant is green tea, and it is one of the most consumed beverages around the globe (Mak 2012). Polyphenols are found in a variety of plants including green tea leaves and grape seeds. Along with tea polyphenols, other types of polyphenols such as tannins are also found in many plant species. These tannins behave similarly to the catechins found in tea (Taguri et al. 2004).

Three main types of tea—green, oolong, and black—are produced from the *Camellia sinensis* plant. Cultivation of tea from this plant can occur in many places with the following conditions: acidic soil, high humidity, and fair temperature, although the majority of tea is grown in subtropical and tropical zones (Dufresne and Farnworth 2001; Gupta et al. 2002). Consumption of tea averages about 120 ml per person each day, making it the second most popular drink in the world, overshadowed only by water (Mak 2012). Of the 2.5 million metric tons of dried tea that is produced annually, approximately 78% is black tea, 20% is green tea, and only 2% is oolong tea. These different types of tea are favored in various parts of the world. Whereas green tea is mainly consumed in Asia, North Africa, and the Middle East, oolong tea is confined mainly to China, and black tea is preferred mainly in western Europe and the United States (Cheng 2006; Gupta et al. 2002; Khan and Mukhtar 2007; Luczaj and Skrzydlewska 2005).

The three types of tea result from differences in the manufacturing fermentation process. Fermentation in the tea industry often refers to exposing the tea leaves to air in order to dry, which results in oxidation. The amount of fermentation that tea leaves undergo determines which tea type is produced; fermentation of the tea leaves is often halted by heating and dehydration, which inactivates certain enzymes in the tea leaves. Generally, green tea does not undergo fermentation, while black tea undergoes a complete fermentation or 100% oxidation; oolong tea undergoes a partial fermentation and is thus only partially oxidized (Cheng 2006; Luczaj and Skrzydlewska 2005). The amount of fermentation that tea undergoes to produce the three different varieties is responsible for the characteristics of the tea; these include color, aroma, and taste, and are due to the different compounds that are present (Babich et al. 2006; Wang et al. 2000).

On average, tea leaves contain about 36% polyphenolic compounds, 25% carbohydrates, 15% proteins, 6.5% lignin, 5% ash, 4% amino acids, 2% lipids, 1.5% organic acids, 0.5% carotenoids, and about 0.1% volatile substances, though the age of the tea leaves can affect their composition (Luczaj and Skrzydlewska 2005). Because tea is often consumed as a hot water extract (the tea leaves are steeped in hot water), it is usually considered an extract of a plant (Gupta et al. 2002). Dry tea extracts or tea extract powders are made by first creating an infusion by soaking tea leaves in water and/or alcohol, then spray-drying the infusion after it has been concentrated (Wang et al. 2000). Although the different tea extracts vary in the amount and nature of their compounds, they all contain polyphenolic compounds that fall into the class of flavonoids and are, more specifically, flavanols. However, it is the type and amount of flavanols originally produced during the fermentation process that gives the characteristics to each type of tea extract (Dufresne and Farnworth 2001; Wang et al. 2000).

### 11.1.1 Catechins

The main type of flavanol in green tea extract is the catechins, which include the following compounds: (-)-epigallocatechin gallate (EGCG), (-)-epigallocatechin (EGC), (-)-epicatechin gallate (ECG), and (-)-epicatechin (EC) (Dufresne and Farnworth 2001; Wang et al. 2000). According to an high-performance liquid chromatography-liquid chromatography/mass spectrometry (HPLC-LC/MS) study, the major constituent of green tea is EGCG (Vodnar and Socaciu 2012). It is the most abundant green tea polyphenol (GTP) that can accumulate in concentrations up to 1 mg/mL (Sakanaka et al. 1989; Sharangi 2009). A previous study that conducted a bioassay-guided fractionation to observe the composition of green tea revealed that out of all the catechins, the two strongest antimicrobials are ECG and EGCG (Si et al. 2006). EGCG is an isomer of the catechin ECG (Shimamura et al. 2007), and its molecular weight is 290. The molecular weight of EGCG is 458 (Shimamura et al. 2007). EGCG has a four-ring structure and also has eight hydroxyl groups (Zhong et al. 2012). Epicatechins have three hydroxyl groups in the B ring. Antioxidant activity is known by structures that have the most hydroxyl groups (Macedo et al. 2011). The U.S. Food and Drug Administration has classified EGCG as a safe compound (Isaacs et al. 2008; Paterson and Anderson 2005), and it has been shown to have a number of medicinal benefits for humans (Shimamura et al. 2007), including anticancer, antiviral, antibacterial, and anti-inflammatory activity (Imai et al. 1997; Mukhtar and Ahmad 2000; Sueoka et al. 2001; Zu et al. 2012).

Green tea contains a higher amount of catechins compared to black tea because it is unfermented. Fermentation causes the catechins to polymerize and create theaflavins and thearubigens in black tea, but these are not present in green tea. The four theaflavins are as follows: theaflavin (TF1), theaflavin-3-monogallate (TF2A), theaflavin-3′-monogallate (TF2B), and theaflavin-3,3′-digallate (TF3), which are formed by the combination of the following catechins: EC with EGC, EC with ECG, EGC with EGCG, and ECG with EGCG, respectively (Bonnely et al. 2003; Kuroda and Hara 1999). Thearubigins are more complex; they can be joined with various theaflavins and catechin dimer products called theasinensins. Combination of the catechin monomers occurs with the addition of oxygen via an enzyme in the tea leaves (polyphenol oxidase). This happens during the fermentation process creating black tea (Ferruzzi 2010). As a result, in terms of dry weight, black tea contains a lower percentage of catechins, only 10%–12%, as compared to 30%–42% in green tea; however, black tea contains 3%–6% theaflavins and 12%–18% thearubigens, which are responsible for many of the distinctive properties of black tea (Babich et al. 2006; Dufresne and Farnworth 2001; Khan and Mukhtar 2007; Luczaj and Skrzydlewska 2005).

## 11.2 MODIFICATION OF GTPS

EGCG cannot be dissolved in hydrophobic media and is only dissolved in water (Zhong et al. 2012). EGCG is hydrophilic in nature and does not penetrate within a lipophilic environment (Zhong et al. 2012). EGCG has poor membrane permeability and low chemical stability, and it is usually metabolized rapidly (Mori et al. 2008). It loses its abilities long before one would be able to apply it. Freshly prepared EGCG has to be used in order to show its potent antimicrobial activity (Chen et al. 2009). The GTP needs to be modified to form a lipophilic tea polyphenol (LTP), which is soluble in any lipid medium (Chen et al. 2003). This allows more uses of it as a topical application. LTP is usually prepared by catalytic esterification of GTP with hexadeconoyl chloride (Chen et al. 2003). The structure of EGCG-ester was purified previously by a team of researchers in China through a catalytic esterification between GTPs and $C_{16}$-fatty acid (Chen et al. 2003). The LTP product was purified using high current chromatography separation. A new long-chain acyl derivative of EGCG was isolated from this LTP by high-speed countercurrent chromatography (Chen et al. 2003).

The incorporation of fatty acids has been shown to increase antioxidant activities (Zhong et al. 2012). These molecules also have higher antiviral activities (Zhong et al. 2012). The ester form of EGCG has increased beneficial properties compared to the parent molecule. It has been proposed that fatty acid esters of the polyphenols (EGCG-ester) can be used as an effective antiviral agent and as an ingredient in lipophilic preparations for HSV prevention (de Oliveira et al. 2013; Mori et al. 2008). Esterization of EGCG can be achieved either enzymatically or chemically (Chen et al. 2003). Data generated from experiments using this influenza virus showed that the esters of EGCG are 24 times more effective in inhibiting the infection and in inactivating influenza virus (Mori et al. 2008). Because a high concentration of EGCG is needed for an antiviral effect to be seen in influenza, researchers thought of a way to increase its lipid membrane permeability and its chemical stability, and to slow down its metabolism (Mori et al. 2008).

## 11.3 PHARMACOLOGICAL ACTIVITIES

### 11.3.1 ANTIOXIDANT

Green tea extract has high concentrations of catechins that have been found to have more antioxidant activity than black tea extract (Cheng 2006). Epicatechins have a high number of hydroxyl (OH) groups that are responsible for antioxidant activity (Luczaj and Skrzydlewska 2005; Zhang and Rock 2004). GTPs have the ability to chelate redox active transition metals, interrupt chain oxidation reactions and accept free radicals, and they also can provide hydrogen atoms (Chen et al. 2003). All of these properties make GTPs the ideal antioxidant. Recent studies have shown that certain extracts, specifically concentrated theaflavin extracts made from black tea, can be just as effective as catechins in terms of antioxidant properties. In fact, theaflavins have been shown to protect cells against oxidative damage, thus confirming its antioxidant capabilities (Luczaj and Skrzydlewska 2005; Yang et al. 2007).

### 11.3.2 Antiviral

GTPs (especially EGCG) have been reported to be broad-spectrum antiviral agents. Many reports have suggested that they can work on many families of viruses including some DNA viruses such as Adenoviridae (adenovirus), Herpesviridae (HSV-1/HSV-2 and Epstein–Barr virus), and Hepadnaviridae (hepatitis B) and some RNA viruses such as Flaviviridae (HCV), Orthomyxoviridae (influenza virus), Picornaviridae (enterovirus), and Retroviridae (HIV) (Steinmann et al. 2013).

The mode of action of EGCG varies in different viruses. In DNA viruses, EGCG works on adenovirus by inactivating viral particles, inhibiting viral growth and viral protease activity (Weber et al. 2003). In HSV, EGCG damages and inactivates glycoproteins on the envelope, thus inhibiting binding of the virus to the host cells (de Oliveira et al. 2013; Isaacs et al. 2008, 2011). In Epstein–Barr virus (EBV), EGCG can inhibit immediate early gene products, *Rta*, *Zta*, and EA-D (Chang et al. 2003). Recent study on HBV suggested that EGCG interferes with viral DNA and nuclear covalent closed circular DNA (cccDNA) synthesis, thus stopping viral replication (He et al. 2011; Xu et al. 2008). For RNA viruses, it inhibits HCV viral binding on the host cells (Calland et al. 2012a, 2012b; Chen et al. 2012; Ciesek et al. 2011).

The mode of action on the influenza virus is to bind to hemagglutinin and therefore inhibit the virus's ability to bind and to enter the target cells (Imanishi et al. 2002; Nakayama et al. 1993; Song et al. 2005). EGCG suppresses enterovirus replication by modulation of cellular redox milieu (Ho et al. 2009).

Studies of EGCG on HIV suggested that EGCG inhibits the binding of HIV glycoprotein gp120 to the CD4 receptor on the host (Williamson et al. 2006). Reverse transcriptase and integrase inhibition has been suggested as mechanisms on HIV (Fassina et al. 2002; Jiang et al. 2010; Kawai et al. 2003; Li et al. 2011; Nakane and Ono 1989; Nance et al. 2009; Williamson et al. 2006; Yamaguchi et al. 2002).

Many studies have been done regarding the antiviral effect of black tea polyphenols. Clark's group (1998) has reported the black tea polyphenols, mainly theaflavins (TF1, TF2A, TF2B, and TF3), are able to neutralize bovine rotavirus and bovine coronavirus. Their results indicated that the combination of all theaflavins possesses higher antiviral activities than each compound individually (Clark et al. 1998). Another study showed that one of the major components of theaflavin, TF3, is able to inhibit various enteric viral infections including poliovirus 1 and rotaviruses (strains Wa and 69M), coxsackie virus A16 and B3, and enteric cytopathic human orphan virus (echovirus) 11 (Mukoyama et al. 1991).

In 1993, Nakayama's group found that 1 mM TF3 was able to bind to the glycoprotein spikes—hemagglutinin of influenza A and B viruses—leading to the adsorption inhibition, which eventually blocks the viral infection (Nakayama et al. 1993). In addition, 80% of TF derivatives (a combination of all derivatives) showed not only hemagglutinin inhibition but also the neuraminidase inhibition for three subtypes of human influenza viruses (A/H1N1, A/H3N2, and B/2003) (Zu et al. 2012).

Black tea polyphenols have also been found to have the ability to bind to the hydrophobic pocket, which is on the surface of glycoprotein gp41 of HIV (Liu et al. 2005). This suggests that theaflavins may be used as lead compounds for developing HIV-1 entry inhibitors (Liu et al. 2005). Black tea extract, consisting primarily of theaflavins, was found to inhibit HSV-1 infection in cultured cells (Cantatore et al. 2013). Furthermore, TFmix, a topical microbicide containing 90% of TF derivatives, was recently developed by the Yingshili Natural Plant Company in Zhejiang, China. It displayed potent anti-HIV activities while exhibiting low cytotoxicity on human vaginal and cervical epithelial cells in vitro (Yang et al. 2012).

EGCG and other catechins found in green tea have been shown to be effective antiviral agents against other viruses, capable of inhibiting the hepatitis B virus (Xu et al. 2008) and the influenza virus (Song et al. 2005). Although the benefits of drinking green tea and the antiviral capability of its compounds have been studied and well documented (Cheng 2006; Dufresne and Farnworth 2001), those of black tea have remained comparatively unstudied. This is surprising given the fact that black tea is one of the most consumed beverages in the world. Further still, black tea contains many of the same compounds and similar types of compounds that are found in green tea. Thus, although the health benefits of black tea compounds may be just as effective as those of green tea, the former has gone largely unnoticed and has remained unstudied until recently (Gupta et al. 2002).

### 11.3.3 Anticancer

Catechins are known to inhibit enzyme activity of ornithine decarboxylase, urokinase, and protein kinase. Some of these enzymes are involved in the progression of cancerous cells (Rani et al. 2012). Catechins have been shown to inhibit the growth of cancerous human colon and hepatic epithelial cells (Shimamura et al. 2007).

In 2000, a study showed that theaflavin-3,3′-digallate (TF3) was able to induce apoptosis in Ras-transformed human bronchial cell lines (Yang et al. 2000). The result also suggested that the gallate structure of theaflavin plays a key role when it comes to cell growth inhibition (Yang et al. 2000). Singh's group indicated that theaflavins were able to block phosphorylation and therefore inhibited Akt and nuclear factor-κB (NF-κB) activation and induced apoptosis in HeLa cells (Singh et al. 2011). Theaflavins are capable of inhibiting certain types of cancer including lung cancers, skin tumors, and leukemia (Kundu et al. 2005; Kuroda and Hara 1999; Wang and Li 2006). Reports suggested intracellular and extracellular mechanisms for tumor inhibition, including DNA replication or repair modulation, reactive molecule blocking, metastasis inhibition, metabolism modulation, and apoptosis induction (Ebata et al. 1998; Kuroda 1996; Kuroda and Hara 1999; Sazuka et al. 1997; Shiraki et al. 1994; Wang et al. 1988; Yamane et al. 1996).

### 11.3.4 Antibacterial

Many studies have shown that GTPs have antibacterial activities and the concentrations required are usually 10–100 times higher (Steinmann et al. 2013) than when they are used as antiviral agents. There are many reports suggesting that EGCG can work on different bacteria as an antibacterial agent. The modes of action are not fully understood; EGCG is suggested to be able to disrupt the membrane of pathogenic bacteria and may also affect the folic acid metabolism by inhibiting the enzyme dihydrofolate reductase (Steinmann et al. 2013). EGCG has different affinities toward the different cell wall components of Gram-negative and Gram-positive bacteria (Blanco et al. 2003). Reports have shown that EGCG activity against *Enterobacteriaceae* is less active and effective compared to Gram-positive microorganisms (Matsumoto et al. 2012). Gram-positive bacteria are more susceptible to EGCG than Gram-negative bacteria due to better adherence and the composition of the cell peptidoglycan layer (Cui et al. 2012). For example, EGCG can directly bind to the peptidoglycan layer of *Staphylococcus aureus* (Cui et al. 2012). It can disrupt the structures of peptides through hydrogen bonding determined by its hydroxyl group. This disruption can lead to the degradation of the peptidoglycan layer and eventually to the rupture of the cell (Cui et al. 2012).

Some studies showed that the mode of action of EGCG on enterohemorrhagic *Escherichia coli* (Gram-negative bacteria) is Shiga toxin release inhibition (Sugita-Konishi et al. 1999) and biofilm formation reduction (Lee et al. 2009). Alteration of the cell wall was also observed (Cui et al. 2012). The antibactericidal effect of EGCG was also reported on *Helicobacter pylori*, *Salmonella typhi*, and *Bacillus cereus* (Kim and Kim 2007; Si et al. 2006; Stoicov et al. 2009).

There are also reports indicating that EGCG can work on the heat-resistant spores in *Bacillus stearothermophilus* and *Clostridium thermoaceticum* (Sakanaka et al. 2000). Reports indicated that EGCG works in combination with antibiotics against some antibiotic-resistant bacteria; the antibacterial effects of EGCG alone and in combination with different antibiotics have been intensively analyzed against a number of bacteria including multidrug-resistant strains such as methicillin-resistant *Staphylococcus aureus* (MRSA) or *Stenotrophomonas* (Cui et al. 2012; Hu et al. 2001, 2002; Stapleton et al. 2004). EGCG can inhibit the penicillin-binding protein 2a (PBP2a) in MRSA and thus combat MRSA (Zhao et al. 2002).

In addition, several studies indicated that the inhibition of biofilm formation by EGCG in several Gram-positive bacteria such as *S. mutans* (Anderson et al. 2011; Taylor et al. 2005; Xu et al. 2012) and *S. pyogenes* (Hull Vance et al. 2011).

## 11.4 HERPES SIMPLEX VIRUSES

### 11.4.1 Classification

The Herpesviridae family contains more than 100 different herpesviruses that infect a multitude of host organisms, including fish, birds, horses, and humans. Herpesviruses are further classified into three subfamilies: Alphaherpesvirinae, Betaherpesvirinae, and Gammaherpesvirinae. HSV types 1 and 2 (human herpesvirus 1 and 2, or HHV-1 and -2) are members of the Alphaherpesvirinae subfamily, *Simplexvirus* genus.

The other genus in the *Alphaherpesvirinae* subfamily is the *Varicellovirus* genus that also includes varicella zoster virus (HHV-3) that causes chicken pox and shingles. This subfamily is distinguished by its short reproductive cycle, rapid spread, destruction of host cells, and by establishing its latent cycle (Mettenleiter et al. 2009; Roizman and Baines 1991).

Although some herpesviruses can infect multiple types of host organisms, most herpesviruses, including HSV-1 and HSV-2, are restricted to just one host type. Aside from host range, other differences among the herpesviruses include genetic content, reproductive cycle duration, and mechanism of latency, yet all herpesviruses are united by some basic properties: DNA genome, virion structure, and latent infection cycle (Mettenleiter et al. 2009; Roizman and Baines 1991).

Although HSV-1 and HSV-2 are closely related, they differ in several aspects of their pathology. HSV-1 is primarily associated with throat and mouth diseases, as well as ocular and genital infections, while HSV-2 is the leading cause of recurring genital herpes cases worldwide (Haddow et al. 2006; Pereira et al. 2012; Stanberry 2006; Wheeler 1988).

### 11.4.2 Genome

Herpesvirus particles can range from 120 to 300 nm in diameter, though many are approximately 200 nm in diameter. All herpesviruses contain several distinct morphological features, including core, capsid, tegument, and envelope. The core of a herpes virion consists of a linear, double-stranded DNA (dsDNA) genome arranged in toroid form, ranging from 120 to 230 kilobase pairs (kbp) in length; about 30–35 different proteins reside with the genome in the core. The core is protected by an icosahedral capsid composed of 150 hexons and 12 pentons, which has a diameter ranging from 100 to 110 nm. A tegument separates the inner capsid from the outer envelope; the tegument contains several proteins, some of which are present at up to 2000 copies. The viral envelope is composed of a lipid bilayer and contains various glycoprotein spikes (Kelly et al. 2009; Roizman and Baines 1991).

Genomes of HSV-1 and HSV-2 are substantially diverged, although the HSV-1 and HSV-2 gene sets have very close correspondence with each other and include 74 genes that encode distinct proteins. HSV-1 and HSV-2 linear dsDNA genomes are 152 and 154 kb in length, respectively, and consist of two main protein-coding components—the unique long ($U_L$) and unique short ($U_S$) sequences—which can be inverted to produce four isomers (Dolan et al. 1998). The difference in length is due to the longer $U_S$ region of HSV-2. The $U_L$ and $U_S$ sequences are flanked by a pair of inverted repeat regions, $TR_L$-$IR_L$ and $TR_S$-$IR_S$, respectively, which have functions in gene regulation and genome replication. Residing between two of the inverted repeat regions, which connect the $U_L$ and $U_S$ sequences, and at the ends of these repeat regions, which serve as the terminal ends of the genome, are “a” sequences that contain packaging signals. The $U_L$ sequence contains 58 genes, while the $U_S$ sequence contains 13 genes, which are activated at different times during its 18-hour life cycle, yet it has been shown that HSV-1 can produce more than 80 different proteins, some of which have been found to be the same protein but with various post-translational modifications. Gene expression and function that has been extensively studied for HSV-1 can serve as a model for these processes in other herpesviruses, too. Although many proteins serve a structural purpose, some are solely for viral DNA replication and some have multiple functions (Bataille and Epstein 1995; Dolan et al. 1998; Garner 2003; Kelly et al. 2009; Mettenleiter et al. 2009; Watanabe 2010).

### 11.4.3 Prevalence

It is estimated that HSV-1 and HSV-2 persist in approximately 45%–98% and 7% of the world population, respectively. In the United States, HSV-1 persists in about 40%–63% of the population, although HSV-2 prevalence remains low (16.2%). Of these infected individuals, about 15%–40% experience symptomatic recurrent infections. Although there is no seasonal variation to the spread of these herpesviruses, demographic factors seem to affect their infections. For example, the infection rate in less developed/industrialized countries is higher than that in more developed/industrialized nations, 70%–80% versus 40%–60%, respectively. Race is another factor, with infection rates ranging from 35% in African American children under five years old to only 18% for their white counterparts in the United States. Socioeconomic conditions also influence HSV-1 infection, with 70%–80% of adults belonging to a lower socioeconomic condition being infected as compared to only 40%–60% of adults belonging to an improved socioeconomic condition. HSV-2 continues to disproportionately burden African Americans (39.2% prevalence), particularly women (48.0% prevalence). Women of all races, however, have a greater incidence of infection than men (Fatahzadeh and Schwartz 2007; Glick 2002; Smith and Robinson 2002; Whitley and Roizman 2001).

Because HSV primarily affects skin, mucous membranes, and neurons, this herpesvirus often results in oral, facial, pharyngeal, ocular, and central nervous system infections. Primary and recurrent infections generally produce the same symptoms with the exception that recurrent infections are often milder and persist for a shorter amount of time as compared to primary infections. HSV infections can result in primary herpetic gingivostomatitis (PHGS), herpes simplex labialis (HSL), recurrent intraoral herpes (RIH), genital herpes, Kaposi's varicelliform eruption (KVE), herpes gladitorum, herpetic whitlow, ocular herpes, encephalitis, and neonatal herpes. An individual's symptoms of these various HSV manifestations may vary according to one's genetic makeup, immune status, site of infection, and dose of inoculum (Fatahzadeh and Schwartz 2007; Huber 2003; Lin et al. 2011).

PHGS, HSL (also known as fever blisters or cold sores), and RIH are all types of orofacial herpes, which are the most common types of HSV-1 infections. Lesions that develop in and/or around the oral cavity characterize each type of infection. These lesions are often preceded or accompanied by a burning sensation, pain, and discomfort; lesions form when vesicles at the site of infection rupture to form erosions that coalesce into ulcerations. In each case, these symptoms persist for approximately two weeks after infection, while viral shedding can continue for up to several weeks after resolution of the symptoms (Corey 1988; Corey et al. 1983). PHGS generally affects the oral mucosa and can also cause gingivitis. HSL lesions, most commonly associated with HSV-1 infections, are located primarily on the outer vermilion border (the border marking the transition from one's lip to skin) and generally crust over. RIH typically affects the hard palate, surrounding gingival, and other keratinized tissue of the oral cavity. HSV-1 can also cause genital herpes, which produces symptoms similar to those of orofacial herpes; however, genital herpes is localized to the labia minora and urethra meatus in women and the shaft and glans of the penis in men. Genital infections of HSV-1 have increased in recent years (Brady and Bernstein 2004).

Genital HSV-2 infections result in pain at the site of the lesion and swelling in local lymph nodes as a result of the cytopathology of the virus. Primary infection results in the appearance of macules and papules and leads to vesicles and pustules. Most people infected with HSV-2 are unaware that they have been infected because they have mild or atypical symptoms. Subclinical virus shedding results in the majority of transmissions (Corey 1988; Corey et al. 1983; Wald 2004).

### 11.4.4 Infection Cycle

HSV infects epithelial cells during lytic infection and travels to sensory neurons in latent infections. Viruses stay in a dormant state within nerve cells during the latent infection until they are activated into the lytic cycle. This permits HSV to survive and replicate permanently for the HSV-infected patient (Kang et al. 2003) The lytic infection cycle of HSV begins when the virus attaches to, and then fuses with, a susceptible cell. Attachment and fusion take place when glycoproteins on the virus particle bind to suitable receptors on the plasma membrane of the host cell. Cell surface receptors bind specifically to the viral glycoproteins. Important HSV-1 glycoproteins (gs) extruding from the viral envelope include glycoproteins gB, gC, gD, gH, and gL, as well as gE, gI, and gK. Cell receptors to which these glycoproteins bind include heparan sulfate (HS), nectin-1, nectin-2, and herpesvirus entry mediator (HVEM). Although HS is found on many types of cells, nectin-1 and nectin-2 are found primarily at junctions of epithelial cells, as well as the synaptic junctions of neurons; HVEM is located on T lymphocytes and trabecular meshwork cells. The location of the receptors influences the tropism of HSV-1 and places a limit on the types of cells to which it can attach and thus infect (Akhtar and Shukla 2009; Garner 2003; O'Donnell et al. 2010).

Initial attachment of HSV-1 to its host cell occurs when gB with or without gC on the viral envelope binds to HS on the host cell's membrane. HS is a glycosaminoglycan, in which the polysaccharides, composed of repeating disaccharide units of uronic (either glucuronic or iduronic) acid and (either *N*-acetylated or *N*-sulfo-) glucosamine, are covalently linked to a protein core, which is located on the plasma membrane of most cells in humans. Binding of the glycoprotein to HS is only facilitated by gC; gB alone can successfully bind to the HS cell receptor to allow HSV-1 attachment to a host cell. Although attachment of HSV-1 gB to its HS receptor on a host cell can occur anywhere along the plasma membrane of the host cell, it has been found to often occur along filopodia or protrusions of the plasma membrane that contain an abundant actin network. Viral particles attached to HS on filopodia have been observed to be transported to the cell body for subsequent fusion in a process called viral surfing. Viral surfing is a likely result of the reorganization of the host cell's actin cytoskeleton that occurs when HSV-1 gB binds to HS, which functions as a signaling pathway for the aforementioned process. In addition to HS, HSV-1 gB has also recently been found to bind to paired immunoglobulin-like type 2 receptor-alpha (PILR-α) in order to achieve attachment to monocytes, macrophages, and dendritic cells. Binding of gB to PILR-α has also been implicated in the subsequent process of viral fusion to the host cell, although this is still poorly understood (Akhtar and Shukla 2009; O'Donnell et al. 2010; Watanabe 2010).

Once the viral particle has attached itself to a suitable host cell, the viral envelope must fuse with the plasma membrane of the host cell; this process too, like that of viral attachment, is dependent upon viral glycoproteins and host cell receptors. Fusion begins when gD of HSV-1 binds to a cellular receptor such as nectin-1, nectin-2, HVEM, or a modified HS molecule called 3-O-sulfated HS (3-OS HS), which is found on corneal cells; binding of gD to any of these receptors results in a conformational change in the gD molecule that recruits gH and gL, which form a heterodimer, as well as gB to form a multiprotein complex. Fusion then proceeds in a two-step process that begins by bringing the HSV-1 envelope close to the host cell's plasma membrane; close contact is achieved by the gD/gH/gL portion of the complex and receptor binding and results in the mixing of lipids between the viral envelope and the host cell's plasma membrane, which creates a fusion intermediate. The presence of gB in the glycoprotein complex allows for the subsequent fusion pore to form, which allows for mixing of the host cell's contents with that of the virus particle; specifically, the viral core, surrounded by its capsid and tegument, is released into the host cell's cytoplasm (Akhtar and Shukla 2009; Nakano et al. 2011; O'Donnell et al. 2010; Watanabe 2010).

Viral entry can also be achieved by cell-to-cell spread, in which an infected cell transmits viral particles to an adjacent noninfected cell. This process still requires the interaction of HSV-1 gD with a suitable cell receptor; however, for cell-to-cell spread, gD is complexed with a heterodimer formed by gE and gI, which is transported to cell junctions via the trans-Golgi network (TGN) of cells infected with HSV-1. Fusion then proceeds in a manner similar to that previously described for viral adsorption. In addition, gK has recently been implicated in the cell-to-cell spread of specific cell types, including corneal cells and trigeminal ganglia cells. Regardless of how HSV-1 enters a host cell, viral adsorption is followed by transport of the capsid to the nucleus (Akhtar and Shukla 2009; Meckes et al. 2010; Watanabe 2010).

The HSV-1 genome can undergo replication and transcription only when it is in the nucleus of the host cell. Thus, once a virion has gained entry into a host cell, transport of the viral particle to the nucleus must ensue. Upon entry into a cell, many viral proteins that make up the tegument dissociate from the viral capsid, while some tegument proteins still remain bound; disassociation of tegument proteins is aided by pUS3 and pUL13, which are protein kinases that are part of the outer tegument of the viral particle. (The "p" designates a protein, while "US" and "UL" signify that the protein is coded for by $U_S$ and $U_L$ segments in the viral genome, respectively.) Viral transport to the nucleus soon follows when the capsid protein, pUL35, interacts with the motor protein, dynein. Thus, HSV-1 uses the available microtubule network in host cells to engage in dynein-dependent transport in order to move to the nucleus. As transport occurs, the dissociated tegument proteins begin to alter cellular function to favor viral propagation; two of these proteins include the previously mentioned pUS3, which hinders apoptosis by inactivating members of the Bcl-2 pro-apoptotic family through phosphorylation, and pUL41, which is responsible for the downregulation of Major Histocompatibility Complexes Class II (MHCII) cell receptors, thereby avoiding an immune response, and degradation of host mRNA, allowing for efficient viral protein synthesis (Kelly et al. 2009; Watanabe 2010).

Once at the nucleus, HSV-1 will employ pUL36 to transport the viral genome and associated viral proteins from the capsid into the host nucleus by interacting with the nuclear pore complex. In the nucleus, HSV-1 DNA is transcribed in a regulated process that proceeds in three main stages, which is separated temporally into the following during an active or lytic infection: immediate early (IE), early (E), and late (L). IE genes generally contain promoters with many binding motifs and, thus, are initially activated by a tegument protein, pUL48, which is a transcriptional activator and recruits host cell transcription factors such as HCF-1 and Oct-1. The five IE genes code for transcription factors, which serve to regulate HSV-1 gene expression. Specifically, infected cell protein 4 (ICP4) has been found to form a transcriptional complex on the promoter of viral E genes, thereby transitioning viral transcription to the second stage. Activation of E genes proceeds by ICP0 via the blocking interferon, thereby preventing E gene silencing. The 12 E genes mainly function to replicate the viral genome; some protein products of E genes include pUL23, thymidine kinase (which is involved in nucleic acid metabolism), pUL5, DNA helicase, and pUL30 and pUL42, which are subunits of DNA polymerase. The 56 L genes function mainly to produce structural proteins for the formation of new virions, including capsid proteins such as pUL19 and pUL26.5 (which are structural and scaffolding proteins, respectively), tegument proteins such as pUL46 and pUL36 (involved in viral envelopment and transport, respectively), and envelope proteins such as pUL27 and pUS6—gB and gD, respectively (Bloom et al. 2010; Kelly et al. 2009; Nakabayashi and Sasaki 2009; Watanabe 2010).

In its latent state, the HSV genome is markedly different than in its lytic stage. HSV is able to achieve latency by infecting neuronal cells. Not only does the HSV genome circularize in its latent stage, but it also produces one major transcript, latency-associated transcript (LAT). LAT is usually spliced to produce a stable 2 Kb RNA intron, which is thought to repress lytic genes by recruiting histone modification enzymes that interact with the viral DNA genome (Kang et al. 2003). When exposed to certain stresses—physical, radioactive, hyperthermic, and so on—this repression can, however, be reversed in order to reactivate lytic genes, causing thereby viral propagation to resume (Bloom et al. 2010; Kelly et al. 2009; Nakabayashi and Sasaki 2009; Watanabe 2010).

A lytic HSV-1 infection will result in the production of new viral DNA and proteins, which must then assemble to produce new, infectious virions. Viral assembly is a complex process, and different hypotheses have been proposed including the luminal and nuclear pore pathways; however, one model, the envelopment–deenvelopment–reenvelopment model, is the one most supported by a variety of studies.

HSV-1 assembly begins with capsid formation in the infected host cell's nucleus, where the capsid proteins pUL19, pUL18, and pUL38 assemble around pUL26.5, a scaffolding protein. Two other proteins, pUL17 and pUL32, are involved in cleaving the newly synthesized viral DNA and transporting it into the newly formed capsid. Tegument proteins may also begin to associate with the capsid; however, this process still remains unclear. Consequently, three separate models have been proposed in which some tegument proteins, such as pUL16, are attached to the capsid in the nucleus (nuclear loading model), cytoplasm (capsid loading model) or TGN (TGN loading model); the most recent studies have supported the capsid loading model, where tegument proteins are added and the capsid travels through the host cell's cytoplasm. Regardless, the viral capsid must first exit the nucleus in a process called primary envelopment. In order for the capsid to be attached to the inner nuclear membrane and bud into the perinuclear space, the nuclear lamin network must be disassembled; this occurs when pUL31 and pUL34 form a complex to recruit pUS3, a protein kinase, which causes changes in the nuclear lamin by phosphorylating lamins A and C, which are involved in maintaining nuclear integrity. Once the newly enveloped capsid buds into the perinuclear space, the virus soon loses this primary envelope as it fuses with the outer nuclear membrane, in which pUS3 is again thought to play a significant role; this deenvelopment of the virus releases it into the cytoplasm of the host cell (Kelly et al. 2009; Meckes et al. 2010; Mettenleiter et al. 2009; Watanabe 2010).

In the cytoplasm, several viral proteins associate with the capsid, including pUL36, pUL37, and pUS3, which constitute an inner tegument. Exit of the virus from the host cell proceeds as the viral particle moves along the microtubule network via kinesin-dependent transport with the aid of pUL36 and pUL37. While in transport, additional viral proteins are recruited by pUL48, which interacts with pUL41, pUL46, pUL47, and pUL49 to form an outer tegument; pUL16 also associates with the outer tegument. Secondary envelopment or reenvelopment occurs at the TGN, where glycoproteins are also added. Glycoproteins such as gB and gD are found at the TGN, bound to pUL11, which in turn binds to the tegument of the virus via interactions with pUL16. Similar interactions occur between pUL48 on the tegument and gB, gD, and gH at the TGN. At this stage, the viral capsid with its tegument becomes enveloped and then proceeds to exit the cell via exocytosis. Though this process is not well understood, pUL20 and gK have been implicated in viral egress or exit from the cell. Once released, these new virions are capable of infecting new host cells and starting the lytic cycle once again (Kelly et al. 2009; Meckes et al. 2010; Mettenleiter et al. 2009; Watanabe 2010).

## 11.5 TREATMENT AND PREVENTION

### 11.5.1 Treatment

There is no cure for HSV infection. Thus, antiviral treatments—whether intravenous, oral, or topical—focus on limiting the symptoms caused by HSV, usually by inhibiting viral replication. Generally, a topical antiviral agent treats accessible lesions while inaccessible lesions are treated with systemic antiviral agents, either oral or intravenous. The current antiviral drugs of preference for the treatment of oral or genital HSV infections are acyclovir, valaciclovir, penciclovir, and famciclovir (Brady and Bernstein 2004; Snoeck 2000). These drugs are often administered orally for a period of seven to ten days. They are analogs of nucleosides and have a similar mode of action to shut off viral replication (Morfin and Thouvenot 2003). The most common form of treatment is with acyclovir, a guanosine analog. Acyclovir can be phosphorylated by HSV thymidine kinase and further phosphorylated by the host enzyme into an active acyclovir triphosphate. Active acyclovir can prevent viral DNA elongation by inhibiting viral DNA polymerase (Brady and Bernstein 2004). Acyclovir has only a rare side effect of causing irreversible nephropathy. The problems with this drug are low bioavailability and short half-life in blood, requiring frequent doses to remain effective. In addition, acyclovir is expensive, and patients who suffer from frequent recurrence of HSV-1 infections may not be able to afford this medication (Fatahzadeh and Schwartz 2007; Khan et al. 2005; Thompson 2006; Whitley and Roizman 2001). Famiciclovir, another anti-HIV drug, has a longer half-life and can be given less frequently than acyclovir (Stanberry 2006). Other drugs such as cidofovir, an acyclic nucleoside 5′-monophosphate, can inhibit the viral DNA polymerase. This is often used on HSV-1 strains resistant to acyclovir and other TK-dependent drugs. However, HSV may develop resistance to cidofovir when the DNA polymerase gene is mutated (Brady and Bernstein 2004).

Although acyclovir remains the current standard treatment for HSV-1 infections, new treatments are being developed to help combat resistant strains. Some of these treatments still focus on inhibition of viral replication but through a different process. One such class of compounds is called helicase-primase inhibitors (HPIs); these inhibit either pUL5 (helicase) or pUL52 (primase), both of which are vital to viral replication. However, mutant HSV-1 strains resistant to some of these drugs have already been detected, though a strategy of employing a combination of drugs that inhibit helicase and primase have proved successful thus far (Field and Biswas 2011; Sukla et al. 2010). Another strategy to inhibit HSV-1 propagation is to inhibit protein synthesis; trichosanthin (TCS) is a compound extracted from the root of a plant that is capable of inactivating the 60s subunit of ribosomes, thereby preventing protein synthesis. The action of TCS is believed to induce cellular apoptosis, thereby inhibiting further HSV-1 propagation (He and Tam 2010). Increased resistance resulted from taking these different drugs; novel and more effective medications need to be developed in order to prevent HSV infection (Morfin and Thouvenot 2003).

### 11.5.2 Vaccine Development

Researchers have been trying to develop an effective vaccine against HSV, but it has proven to be extremely challenging because HSV establishes latency and an infection may arise even if the immune system becomes activated (Awasthi et al. 2008). Two vaccines are being tested that allow the immune system to identify and aid in the eradication of HSV-1. One method employed plasmids that coded for Bax and gB. Bax is capable of inducing apoptosis. Thus, when these transfected cells died, they released large amounts of gB, which could be presented to T cells by antigen-presenting cells, leading to an increase in the immune response (Parsania et al. 2010). Another vaccination method employed a phage particle expressing another HSV-1 glycoprotein, gD. These phages were injected into mice, which induced a cellular and a humoral response (Hashemi et al. 2010). More efforts are needed to develop vaccines for HSV.

### 11.5.3 The Search for Natural Products with Anti-HSV Activities as a Novel Approach for Anti-HSV Treatment

In an effort to find low-cost treatments for HSV that can be taken more conveniently with fewer side effects and a lower chance of developing resistant strains, researchers have examined a variety of compounds. A number of these treatments are focusing on compounds derived from plants because many of them have been used throughout history to treat various human diseases (Khan et al. 2005). One study found that herbal extracts and compounds from a variety of medicinal plants, such as *Geum japonicum* and *Syzygium aromaticum*, were quite effective as anti-HSV agents. Compounds identified from these extracts had a wide range of structures, including polyphenols, polysaccharides, tannins, and flavonoids (Thompson 2006). Another study confirmed that tannins inhibit HSV-1 and revealed that these compounds block virus adsorption into cells (Fukuchi et al. 1989), whereas similarly structured polyphenolic compounds inhibited attachment and penetration of HSV-1 into cells (Khan et al. 2005). The effectiveness of flavonoids, especially when used together instead of separately, was also confirmed to be due to the inhibition of HSV-1, though the mechanism still remained unclear (Amoros et al. 1992).

### 11.5.4 Natural and Modified GTP as Anti-HSV Therapeutic Agents

GTPs EGCG and other catechins have been shown to effectively inhibit a diverse group of viruses such as HIV, hepatitis B, influenza, and other pathogenic viruses. There are many studies that have focused on their effect on HSV-1 and HSV-2. Previous reports indicated that when Vero cells were infected with HSV-1 and HSV-2, EGCG successfully inhibited HSV-1 infection by 3000-fold and HSV-2 infection by 10,000–fold, respectively, in a dosage-dependent manner. Among all tested green tea catechins, only EGCG produced the inhibitory effect (Isaacs et al. 2008). The report suggested that EGCG inhibits viral adsorption. However, EGCG does not affect viral reproduction if it is applied after virus entry. Also, after treatment of HSV with 100 μl of 100 μM EGCG, the envelopes of virions were damaged. This suggested that EGCG can inhibit HSV directly (Isaacs et al. 2008). Further study using immunogold-labeled antibodies for gB, gD, and a capsid protein in EGCG-treated HSV-1, resulted in a 30%–40% decrease as compared to untreated HSV-1. This suggested that EGCG-treated HSV affects the binding of the monoclonal antibodies on its envelope glycoproteins (Isaacs et al. 2008).

It is promising that the EGCG compound in green tea inhibits HSV, but it is unstable in aqueous solution and readily oxidizes, resulting in loss of activity (Chen et al. 2003, 2009). This is a major problem with EGCG, and it makes EGCG less favored as a potential therapeutic agent. It has been proposed that fatty acid esters of the polyphenols (EGCG-ester) can be used as an effective antiviral agent in the form of an ingredient in lipophilic preparations for HSV prevention (Mori et al. 2008). Esterization of EGCG can be synthesized either enzymatically or chemically

(Chen et al. 2003). Because an ester of EGCG worked so well on the influenza virus, similar procedures may be potentially applied to HSV (Mori et al. 2008).

It has been proposed that fatty acid-modified polyphenols could be effective HSV antiviral agents that could be formulated in lipophilic preparations (Chen et al. 2009; Mori et al. 2008). EGCG lipid esters are 24-fold more effective than EGCG as inhibitors and inactivators of the influenza virus (Mori et al. 2008) and are, therefore, candidate HSV antiviral agents for topical application. Recently, palmitoyl-epigallocatechin gallate (p-EGCG), a lipid ester of EGCG, has shown more effectiveness than EGCG in inhibiting HSV-1 (de Oliveira et al. 2013). Palmitoylation increases the affinity of EGCG for the viral envelope (de Oliveira et al. 2013; Lebel and Boivin 2006; Sakanaka et al. 1989) and is nontoxic. This may provide a novel treatment for HSV. In fact, a case study using lipophilic EGCG with glycerin showed reduced duration of symptoms of HSV infection; p-EGCG is suggested to have future promise for HSV control (Zhao et al. 2012).

## 11.6 CONCLUSION

For many centuries, it was believed that tea possessed therapeutic properties. There is significant evidence that polyphenols from tea, specifically EGCG and theaflavins, have pharmacological activities. They have been shown to be effective antioxidant, antibacterial, anticancer, and antiviral agents through various modes of action. HSV infections are among the most common infections in humans, and there is no cure. Tea polyphenols have shown significant promise for inhibiting and reducing the infectivity of HSV. Specifically, lipid esters of EGCG have the potential to be developed as effective topical therapeutic agents to limit the spread and reduce the infection of HSV.

## REFERENCES

Akhtar J, Shukla D. (2009). Viral entry mechanisms: Cellular and viral mediators of herpes simplex virus entry. *FEBS J* 276: 7228–7236.

Allahverdiyev A, Duran N, Ozguven M, et al. (2004). Antiviral activity of the volatile oils of *Melissa officinalis* L. against herpes simplex virus type-2. *Phytomedicine* 11: 657–661.

Amoros M, Simoes CM, Girre L, et al. (1992). Synergistic effect of flavones and flavonols against herpes simplex virus type 1 in cell culture. Comparison with the antiviral activity of propolis. *J Nat Prod* 55: 1732–1740.

Anderson JC, McCarthy RA, Paulin S, et al. (2011). Anti-staphylococcal activity and beta-lactam resistance attenuating capacity of structural analogues of (-)-epicatechin gallate. *Bioorg Med Chem Lett* 21: 6996–7000.

Armaka M, Papanikolaou E, Sivropoulou A, et al. (1999). Antiviral properties of isoborneol, a potent inhibitor of herpes simplex virus type 1. *Antiviral Res* 43: 79–92.

Awasthi S, Lubinski JM, Eisenberg RJ, et al. (2008). An HSV-1 gD mutant virus as an entry-impaired live virus vaccine. *Vaccine* 26: 1195–1203.

Babich H, Pinsky SM, Muskin ET, et al. (2006). In vitro cytotoxicity of a theaflavin mixture from black tea to malignant, immortalized, and normal cells from the human oral cavity. *Toxicol In Vitro* 20: 677–688.

Baldé AM, van Hoof L, Pieters LA, et al. (1990). Plant antiviral agents. VII. Antiviral and antibacterial proanthocyanidins from the bark of *Pavetta owariensis*. *Phytother Res* 4: 182–188.

Bataille D, Epstein AL. (1995). Herpes simplex virus type 1 replication and recombination. *Biochimie* 77: 787–795.

Benencia F, Courreges MC. (1999). Antiviral activity of sandalwood oil against herpes simplex viruses-1 and -2. *Phytomedicine* 6: 119–123.

Blanco AR, La Terra Mule S, Babini G, et al. (2003). (-)Epigallocatechin-3-gallate inhibits gelatinase activity of some bacterial isolates from ocular infection, and limits their invasion through gelatine. *Biochim Biophys Acta* 1620: 273–281.

Bloom DC, Giordani NV, Kwiatkowski DL. (2010). Epigenetic regulation of latent HSV-1 gene expression. *Biochim Biophys Acta* 1799: 246–256.

Bonnely S, Davis AL, Lewis JR, et al. (2003). A model oxidation system to study oxidised phenolic compounds present in black tea. *Food Chem* 83: 485–492.

Brady RC, Bernstein DI. (2004). Treatment of herpes simplex virus infections. *Antiviral Res* 61: 73–81.

Calland N, Albecka A, Belouzard S, et al. (2012a). (-)-Epigallocatechin-3-gallate is a new inhibitor of hepatitis C virus entry. *Hepatology* 55: 720–729.

Calland N, Dubuisson J, Rouille Y, et al. (2012b). Hepatitis C virus and natural compounds: A new antiviral approach? *Viruses* 4: 2197–2217.

Camargo Filho I, Cortez DA, Ueda-Nakamura T, et al. (2008). Antiviral activity and mode of action of a peptide isolated from Sorghum bicolor. *Phytomedicine* 15: 202–208.

Cantatore A, Randall SD, Traum D, et al. (2013). Effect of black tea extract on herpes simplex virus-1 infection of cultured cells. *BMC Complement Altern Med* 13: 139.

Chang LK, Wei TT, Chiu YF, et al. (2003). Inhibition of Epstein-Barr virus lytic cycle by (-)-epigallocatechin gallate. *Biochem Biophys Res Commun* 301: 1062–1068.

Chen C, Qiu H, Gong J, et al. (2012). (-)-Epigallocatechin-3-gallate inhibits the replication cycle of hepatitis C virus. *Arch Virol* 157: 1301–1312.

Chen P, Dickinson D, Hsu SD. (2009). Lipid-soluble green tea polyphenols: Stabilized for effective formulation. In: McKinley H, Jamieson M (eds.), *Handbook of Green Tea and Health Research*. Hauppauge, NY: NOVA Publishers, pp. 45–61.

Chen P, Tan Y, Sun D, et al. (2003). A novel long-chain acyl-derivative of epigallocatechin-3-O-gallate prepared and purified from green tea polyphenols. *J Zhejiang Univ Sci* 4: 714–718.

Cheng TO. (2006). All teas are not created equal: The Chinese green tea and cardiovascular health. *Int J Cardiol* 108: 301–308.

Ciesek S, von Hahn T, Colpitts CC, et al. (2011). The green tea polyphenol, epigallocatechin-3-gallate, inhibits hepatitis C virus entry. *Hepatology* 54: 1947–1955.

Clark KJ, Grant PG, Sarr AB, et al. (1998). An in vitro study of theaflavins extracted from black tea to neutralize bovine rotavirus and bovine coronavirus infections. *Vet Microbiol* 63: 147–157.

Corey L. (1988). First-episode, recurrent, and asymptomatic herpes simplex infections. *J Am Acad Dermatol* 18: 169–172.

Corey L, Adams HG, Brown ZA, et al. (1983). Genital herpes simplex virus infections: Clinical manifestations, course, and complications. *Ann Intern Med* 98: 958–972.

Cui Y, Oh YJ, Lim J, et al. (2012). AFM study of the differential inhibitory effects of the green tea polyphenol (-)-epigallocatechin-3-gallate (EGCG) against Gram-positive and Gram-negative bacteria. *Food Microbiol* 29: 80–87.

D'Cruz OJ, Uckun FM. (2001). Pokeweed antiviral protein: A potential nonspermicidal prophylactic antiviral agent. *Fertil Steril* 75: 106–114.

De Logu A, Loy G, Pellerano ML, et al. (2000). Inactivation of HSV-1 and HSV-2 and prevention of cell-to-cell virus spread by *Santolina insularis* essential oil. *Antiviral Res* 48: 177–185.

de Oliveira A, Adams SD, Lee LH, et al. (2013). Inhibition of herpes simplex virus type 1 with the modified green tea polyphenol palmitoyl-epigallocatechin gallate. *Food Chem Toxicol* 52: 207–215.

Dixon RA. (2001). Natural products and plant disease resistance. *Nature* 411: 843–847.

Dolan A, Jamieson FE, Cunningham C, et al. (1998). The genome sequence of herpes simplex virus type 2. *J Virol* 72: 2010–2021.

Dufresne CJ, Farnworth ER. (2001). A review of latest research findings on the health promotion properties of tea. *J Nutr Biochem* 12: 404–421.

Ebata J, Fukagai N, Furukawa H. (1998). Mechanism of antimutagenesis by catechins towards *N*-nitrosodimethylamine. *Environ Mutagen Res* 20: 45–50.

Farag RS, Shalaby AS, El-Baroty GA, et al. (2004). Chemical and biological evaluation of the essential oils of different *Melaleuca* species. *Phytother Res* 18: 30–35.

Fassina G, Buffa A, Benelli R, et al. (2002). Polyphenolic antioxidant (-)-epigallocatechin-3-gallate from green tea as a candidate anti-HIV agent. *AIDS* 16: 939–941.

Fatahzadeh M, Schwartz RA. (2007). Human herpes simplex virus infections: Epidemiology, pathogenesis, symptomatology, diagnosis, and management. *J Am Acad Dermatol* 57: 737–763; quiz 764–736.

Ferruzzi MG. (2010). The influence of beverage composition on delivery of phenolic compounds from coffee and tea. *Physiol Behav* 100: 33–41.

Field HJ, Biswas S. (2011). Antiviral drug resistance and helicase-primase inhibitors of herpes simplex virus. *Drug Resist Updat* 14: 45–51.

Fukuchi K, Sakagami H, Okuda T, et al. (1989). Inhibition of herpes simplex virus infection by tannins and related compounds. *Antiviral Res* 11: 285–297.

Garner JA. (2003). Herpes simplex virion entry into and intracellular transport within mammalian cells. *Adv Drug Deliv Rev* 55: 1497–1513.

Glick M. (2002). Clinical aspects of recurrent oral herpes simplex virus infection. *Compend Contin Educ Dent* 23: 4–8.

Gupta S, Saha B, Giri AK. (2002). Comparative antimutagenic and anticlastogenic effects of green tea and black tea: A review. *Mutat Res* 512: 37–65.

Haddow LJ, Dave B, Mindel A, et al. (2006). Increase in rates of herpes simplex virus type 1 as a cause of anogenital herpes in western Sydney, Australia, between 1979 and 2003. *Sex Transm Infect* 82: 255–259.

Hashemi H, Bamdad T, Jamali A, et al. (2010). Evaluation of humoral and cellular immune responses against HSV-1 using genetic immunization by filamentous phage particles: A comparative approach to conventional DNA vaccine. *J Virol Methods* 163: 440–444.

Hayashi K, Kamiya M, Hayashi T. (1995). Virucidal effects of the steam distillate from *Houttuynia cordata* and its components on HSV-1, influenza virus, and HIV. *Planta Med* 61: 237–241.

He DX, Tam SC. (2010). Trichosanthin affects HSV-1 replication in Hep-2 cells. *Biochem Biophys Res Commun* 402: 670–675.

He W, Li LX, Liao QJ, et al. (2011). Epigallocatechin gallate inhibits HBV DNA synthesis in a viral replication – inducible cell line. *World J Gastroenterol* 17: 1507–1514.

Heredia A, Davis C, Redfield R. (2000). Synergistic inhibition of HIV-1 in activated and resting peripheral blood mononuclear cells, monocyte-derived macrophages, and selected drug-resistant isolates with nucleoside analogues combined with a natural product, resveratrol. *J Acquir Immune Defic Syndr* 25: 246–255.

Ho HY, Cheng ML, Weng SF, et al. (2009). Antiviral effect of epigallocatechin gallate on enterovirus 71. *J Agric Food Chem* 57: 6140–6147.

Hu ZQ, Zhao WH, Asano N, et al. (2002). Epigallocatechin gallate synergistically enhances the activity of carbapenems against methicillin-resistant *Staphylococcus aureus*. *Antimicrob Agents Chemother* 46: 558–560.

Hu ZQ, Zhao WH, Hara Y, et al. (2001). Epigallocatechin gallate synergy with ampicillin/sulbactam against 28 clinical isolates of methicillin-resistant *Staphylococcus aureus*. *J Antimicrob Chemother* 48: 361–364.

Huang R, Dietsch E, Lockhoff O, et al. (1991). Antiviral activity of some natural and synthetic sugar analogues. *FEBS Lett* 291: 199–202.
Huber MA. (2003). Herpes simplex type-1 virus infection. *Quintessence Int* 34: 453–467.
Hudson JB. (1989). Plant photosensitizers with antiviral properties. *Antiviral Res* 12: 55–74.
Hudson JB, Imperial V, Haugland RP, et al. (1997). Antiviral activities of photoactive perylenequinones. *Photochem Photobiol* 65: 352–354.
Hull Vance S, Tucci M, Benghuzzi H. (2011). Evaluation of the antimicrobial efficacy of green tea extract (egcg) against streptococcus pyogenes in vitro – biomed 2011. *Biomed Sci Instrum* 47: 177–182.
Imai K, Suga K, Nakachi K. (1997). Cancer-preventive effects of drinking green tea among a Japanese population. *Prev Med* 26: 769–775.
Imanishi N, Tuji Y, Katada Y, et al. (2002). Additional inhibitory effect of tea extract on the growth of influenza A and B viruses in MDCK cells. *Microbiol Immunol* 46: 491–494.
Isaacs CE, Wen GY, Xu W, et al. (2008). Epigallocatechin gallate inactivates clinical isolates of herpes simplex virus. *Antimicrob Agents Chemother* 52: 962–970.
Isaacs CE, Xu W, Merz G, et al. (2011). Digallate dimers of (-)-epigallocatechin gallate inactivate herpes simplex virus. *Antimicrob Agents Chemother* 55: 5646–5653.
Ito M, Sato A, Hirabayashi K, et al. (1988). Mechanism of inhibitory effect of glycyrrhizin on replication of human immunodeficiency virus (HIV). *Antiviral Res* 10: 289–298.
Jassim SA, Naji MA. (2003). Novel antiviral agents: A medicinal plant perspective. *J Appl Microbiol* 95: 412–427.
Jiang F, Chen W, Yi K, et al. (2010). The evaluation of catechins that contain a galloyl moiety as potential HIV-1 integrase inhibitors. *Clin Immunol* 137: 347–356.
Kang W, Mukerjee R, Fraser NW. (2003). Establishment and maintenance of HSV latent infection is mediated through correct splicing of the LAT primary transcript. *Virology* 312: 233–244.
Kashiwada Y, Hashimoto F, Cosentino LM, et al. (1996). Betulinic acid and dihydrobetulinic acid derivatives as potent anti-HIV agents. *J Med Chem* 39: 1016–1017.
Kawai K, Tsuno NH, Kitayama J, et al. (2003). Epigallocatechin gallate, the main component of tea polyphenol, binds to CD4 and interferes with gp120 binding. *J Allergy Clin Immunol* 112: 951–957.
Kelly BJ, Fraefel C, Cunningham AL, et al. (2009). Functional roles of the tegument proteins of herpes simplex virus type 1. *Virus Res* 145: 173–186.
Khan MT, Ather A, Thompson KD, et al. (2005). Extracts and molecules from medicinal plants against herpes simplex viruses. *Antiviral Res* 67: 107–119.
Khan N, Mukhtar H. (2007). Tea polyphenols for health promotion. *Life Sci* 81: 519–533.
Kim JS, Kim Y. (2007). The inhibitory effect of natural bioactives on the growth of pathogenic bacteria. *Nutr Res Pract* 1: 273–278.
Konowalchuk J, Speirs JI. (1976). Virus inactivation by grapes and wines. *Appl Environ Microbiol* 32: 757–763.
Kundu T, Dey S, Roy M, et al. (2005). Induction of apoptosis in human leukemia cells by black tea and its polyphenol theaflavin. *Cancer Lett* 230: 111–121.
Kuroda Y. (1996). Bio-antimutagenic activity of green tea catechins in cultured Chinese hamster V79 cells. *Mutat Res* 361: 179–186.
Kuroda Y, Hara Y. (1999). Antimutagenic and anticarcinogenic activity of tea polyphenols. *Mutat Res* 436: 69–97.
Lebel A, Boivin G. (2006). Pathogenicity and response to topical antiviral therapy in a murine model of acyclovir-sensitive and acyclovir-resistant herpes simplex viruses isolated from the same patient. *J Clin Virol* 37: 34–37.
Lee KM, Kim WS, Lim J, et al. (2009). Antipathogenic properties of green tea polyphenol epigallocatechin gallate at concentrations below the MIC against enterohemorrhagic *Escherichia coli* O157:H7. *J Food Prot* 72: 325–331.

Li F, Goila-Gaur R, Salzwedel K, et al. (2003). PA-457: A potent HIV inhibitor that disrupts core condensation by targeting a late step in Gag processing. *Proc Natl Acad Sci U S A* 100: 13555–13560.

Li S, Hattori T, Kodama EN. (2011). Epigallocatechin gallate inhibits the HIV reverse transcription step. *Antivir Chem Chemother* 21: 239–243.

Lin LT, Chen TY, Chung CY, et al. (2011). Hydrolyzable tannins (chebulagic acid and punicalagin) target viral glycoprotein-glycosaminoglycan interactions to inhibit herpes simplex virus 1 entry and cell-to-cell spread. *J Virol* 85: 4386–4398.

Liu S, Lu H, Zhao Q, et al. (2005). Theaflavin derivatives in black tea and catechin derivatives in green tea inhibit HIV-1 entry by targeting gp41. *Biochim Biophys Acta* 1723: 270–281.

Lopez-Bazzocchi I, Hudson JB, Towers GH. (1991). Antiviral activity of the photoactive plant pigment hypericin. *Photochem Photobiol* 54: 95–98.

Luczaj W, Skrzydlewska E. (2005). Antioxidative properties of black tea. *Prev Med* 40: 910–918.

Macedo JA, Battestin V, Ribeiro ML, et al. (2011). Increasing the antioxidant power of tea extracts by biotransformation of polyphenols. *Food Chem* 126: 491–497.

Mak JC. (2012). Potential role of green tea catechins in various disease therapies: Progress and promise. *Clin Exp Pharmacol Physiol* 39: 265–273.

Matsumoto Y, Kaihatsu K, Nishino K, et al. (2012). Antibacterial and antifungal activities of new acylated derivatives of epigallocatechin gallate. *Front Microbiol* 3: 53.

Meckes DG, Jr, Marsh JA, Wills JW. (2010). Complex mechanisms for the packaging of the UL16 tegument protein into herpes simplex virus. *Virology* 398: 208–213.

Meneses R, Ocazionez RE, Martinez JR, et al. (2009). Inhibitory effect of essential oils obtained from plants grown in Colombia on yellow fever virus replication in vitro. *Ann Clin Microbiol Antimicrob* 8: 8.

Mettenleiter TC, Klupp BG, Granzow H. (2009). Herpesvirus assembly: An update. *Virus Res* 143: 222–234.

Min BS, Hattori M, Lee HK, et al. (1999a). Inhibitory constituents against HIV-1 protease from *Agastache rugosa*. *Arch Pharm Res* 22: 75–77.

Min BS, Jung HJ, Lee JS, et al. (1999b). Inhibitory effect of triterpenes from *Crataegus pinatifida* on HIV-I protease. *Planta Med* 65: 374–375.

Morfin F, Thouvenot D. (2003). Herpes simplex virus resistance to antiviral drugs. *J Clin Virol* 26: 29–37.

Mori S, Miyake S, Kobe T, et al. (2008). Enhanced anti-influenza A virus activity of (-)-epig allocatechin-3-O-gallate fatty acid monoester derivatives: Effect of alkyl chain length. *Bioorg Med Chem Lett* 18: 4249–4252.

Mukhtar H, Ahmad N. (2000). Tea polyphenols: Prevention of cancer and optimizing health. *Am J Clin Nutr* 71: 1698S–1702S; discussion 1703S–1694S.

Mukoyama A, Ushijima H, Nishimura S, et al. (1991). Inhibition of rotavirus and enterovirus infections by tea extracts. *Jpn J Med Sci Biol* 44: 181–186.

Nakabayashi J, Sasaki A. (2009). The function of temporally ordered viral gene expression in the intracellular replication of herpes simplex virus type 1 (HSV-1). *J Theor Biol* 261: 156–164.

Nakane H, Ono K. (1989). Differential inhibition of HIV-reverse transcriptase and various DNA and RNA polymerases by some catechin derivatives. *Nucleic Acids Symp Ser* 21: 115–116.

Nakano K, Kobayashi M, Nakamura K, et al. (2011). Mechanism of HSV infection through soluble adapter-mediated virus bridging to the EGF receptor. *Virology* 413: 12–18.

Nakayama M, Suzuki K, Toda M, et al. (1993). Inhibition of the infectivity of influenza virus by tea polyphenols. *Antiviral Res* 21: 289–299.

Nance CL, Siwak EB, Shearer WT. (2009). Preclinical development of the green tea catechin, epigallocatechin gallate, as an HIV-1 therapy. *J Allergy Clin Immunol* 123: 459–465.

O'Donnell CD, Kovacs M, Akhtar J, et al. (2010). Expanding the role of 3-O sulfated heparan sulfate in herpes simplex virus type-1 entry. *Virology* 397: 389–398.

Orhana I, Ozcelik B, Karaoglu T, et al. (2007). Antiviral and antimicrobial profiles of selected isoquinoline alkaloids from *Fumaria* and *Corydalis* species. *Z Naturforsch C* 62: 19–26.

Ozcelik B, Kartal M, Orhan I. (2011). Cytotoxicity, antiviral and antimicrobial activities of alkaloids, flavonoids, and phenolic acids. *Pharm Biol* 49: 396–402.

Parsania M, Bamdad T, Hassan ZM, et al. (2010). Evaluation of apoptotic and anti-apoptotic genes on efficacy of DNA vaccine encoding glycoprotein B of Herpes Simplex Virus type 1. *Immunol Lett* 128: 137–142.

Paterson I, Anderson EA. (2005). Chemistry. The renaissance of natural products as drug candidates. *Science* 310: 451–453.

Pereira VS, Moizeis RN, Fernandes TA, et al. (2012). Herpes simplex virus type 1 is the main cause of genital herpes in women of Natal, Brazil. *Eur J Obstet Gynecol Reprod Biol* 161: 190–193.

Primo V, Rovera M, Zanon S, et al. (2001). [Determination of the antibacterial and antiviral activity of the essential oil from *Minthostachys verticillata* (Griseb.) Epling]. *Rev Argent Microbiol* 33: 113–117.

Rajamohan F, Venkatachalam TK, Irvin JD, et al. (1999). Pokeweed antiviral protein isoforms PAP-I, PAP-II, and PAP-III depurinate RNA of human immunodeficiency virus (HIV)-1. *Biochem Biophys Res Commun* 260: 453–458.

Rajbhandari M, Wegner U, Julich M, et al. (2001). Screening of Nepalese medicinal plants for antiviral activity. *J Ethnopharmacol* 74: 251–255.

Rani A, Singh K, Ahuja PS, et al. (2012). Molecular regulation of catechins biosynthesis in tea [*Camellia sinensis* (L.) O. Kuntze]. *Gene* 495: 205–210.

Roizman B, Baines J. (1991). The diversity and unity of Herpesviridae. *Comp Immunol Microbiol Infect Dis* 14: 63–79.

Rukachaisirikul V, Pailee P, Hiranrat A, et al. (2003). Anti-HIV-1 protostane triterpenes and digeranylbenzophenone from trunk bark and stems of *Garcinia speciosa. Planta Med* 69: 1141–1146.

Sakanaka S, Juneja LR, Taniguchi M. (2000). Antimicrobial effects of green tea polyphenols on thermophilic spore-forming bacteria. *J Biosci Bioeng* 90: 81–85.

Sakanaka S, Kim M, Taniguchi M, et al. (1989). Antibacterial substances in Japanese green tea extract against *Streptococcus mutans*, a cariogenic bacterium. *Agric Biol Chem* 53: 2307–2311.

Sazuka M, Imazawa H, Shoji Y, et al. (1997). Inhibition of collagenases from mouse lung carcinoma cells by green tea catechins and black tea theaflavins. *Biosci Biotechnol Biochem* 61: 1504–1506.

Schnitzler P, Schon K, Reichling J. (2001). Antiviral activity of Australian tea tree oil and eucalyptus oil against herpes simplex virus in cell culture. *Pharmazie* 56: 343–347.

Schuhmacher A, Reichling J, Schnitzler P. (2003). Virucidal effect of peppermint oil on the enveloped viruses herpes simplex virus type 1 and type 2 in vitro. *Phytomedicine* 10: 504–510.

Serkedjieva J. (2004). Antiviral activity of the red marine alga *Ceramium rubrum. Phytother Res* 18: 480–483.

Serkedjieva J, Velcheva M. (2003). In vitro anti-influenza virus activity of the pavine alkaloid (-)-thalimonine isolated from *Thalictrum simplex* L. *Antivir Chem Chemother* 14: 75–80.

Sharangi AB. (2009). Medicinal and therapeutic potentialities of tea (Camellia sinensis L.) – A review. *Food Res Int* 42: 529–535.

Shimamura T, Zhao W-H, Hu Z-Q. (2007). Mechanism of action and potential for use of tea catechin as an anti-infective agent. *Anti Infect Agents Med Chem* 6: 57–62.

Shiraki M, Hara Y, Osawa T, et al. (1994). Antioxidative and antimutagenic effects of theaflavins from black tea. *Mutat Res* 323: 29–34.

Si W, Gong J, Tsao R, et al. (2006). Bioassay-guided purification and identification of antimicrobial components in Chinese green tea extract. *J Chromatogr A* 1125: 204–210.

Singh M, Singh R, Bhui K, et al. (2011). Tea polyphenols induce apoptosis through mitochondrial pathway and by inhibiting nuclear factor-kappaB and Akt activation in human cervical cancer cells. *Oncol Res* 19: 245–257.

Sinico C, De Logu A, Lai F, et al. (2005). Liposomal incorporation of *Artemisia arborescens* L. essential oil and in vitro antiviral activity. *Eur J Pharm Biopharm* 59: 161–168.

Sivropoulou A, Nikolaou C, Papanikolaou E, et al. (1997). Antimicrobial, cytotoxic, and antiviral activities of *Salvia fruticosa* essential oil. *J Agric Food Chem* 45: 3197–3201.

Smith JS, Robinson NJ. (2002). Age-specific prevalence of infection with herpes simplex virus types 2 and 1: A global review. *J Infect Dis* 186(Suppl 1): S3–28.

Snoeck R. (2000). Antiviral therapy of herpes simplex. *Int J Antimicrob Agents* 16: 157–159.

Song JM, Lee KH, Seong BL. (2005). Antiviral effect of catechins in green tea on influenza virus. *Antiviral Res* 68: 66–74.

Stanberry LR. (2006). *Understanding Herpes*. Mississippi: University Press of Mississippi.

Stapleton PD, Shah S, Anderson JC, et al. (2004). Modulation of beta-lactam resistance in *Staphylococcus aureus* by catechins and gallates. *Int J Antimicrob Agents* 23: 462–467.

Steinmann J, Buer J, Pietschmann T, et al. (2013). Anti-infective properties of epigallocatechin-3-gallate (EGCG), a component of green tea. *Br J Pharmacol* 168: 1059–1073.

Stoicov C, Saffari R, Houghton J. (2009). Green tea inhibits Helicobacter growth in vitro and in vitro. *Int J Antimicrob Agents* 33: 473–478.

Sueoka N, Suganuma M, Sueoka E, et al. (2001). A new function of green tea: Prevention of lifestyle-related diseases. *Ann N Y Acad Sci* 928: 274–280.

Sugita-Konishi Y, Hara-Kudo Y, Amano F, et al. (1999). Epigallocatechin gallate and gallocatechin gallate in green tea catechins inhibit extracellular release of Vero toxin from enterohemorrhagic *Escherichia coli* O157:H7. *Biochim Biophys Acta* 1472: 42–50.

Sukla S, Biswas S, Birkmann A, et al. (2010). Effects of therapy using a helicase-primase inhibitor (HPI) in mice infected with deliberate mixtures of wild-type HSV-1 and an HPI-resistant UL5 mutant. *Antiviral Res* 87: 67–73.

Taguri T, Tanaka T, Kouno I. (2004). Antimicrobial activity of 10 different plant polyphenols against bacteria causing food-borne disease. *Biol Pharm Bull* 27: 1965–1969.

Taylor PW, Hamilton-Miller JM, Stapleton PD. (2005). Antimicrobial properties of green tea catechins. *Food Sci Technol Bull* 2: 71–81.

Thompson KD. (2006). Herbal extracts and compounds active against herpes simplex virus. *Adv Phytomedicine* 2: 65–86.

Thompson KD, Dragar C. (2004). Antiviral activity of *Undaria pinnatifida* against herpes simplex virus. *Phytother Res* 18: 551–555.

Uckun FM, Bellomy K, O'Neill K, et al. (1999). Toxicity, biological activity, and pharmacokinetics of TXU (anti-CD7)-pokeweed antiviral protein in chimpanzees and adult patients infected with human immunodeficiency virus. *J Pharmacol Exp Ther* 291: 1301–1307.

Uckun FM, Chelstrom LM, Tuel-Ahlgren L, et al. (1998). TXU (anti-CD7)-pokeweed antiviral protein as a potent inhibitor of human immunodeficiency virus. *Antimicrob Agents Chemother* 42: 383–388.

Vijayan P, Raghu C, Ashok G, et al. (2004). Antiviral activity of medicinal plants of Nilgiris. *Indian J Med Res* 120: 24–29.

Vodnar DC, Socaciu C. (2012). Green tea increases the survival yield of Bifidobacteria in simulated gastrointestinal environment and during refrigerated conditions. *Chem Cent J* 6: 61.

Wald A. (2004). Herpes simplex virus type 2 transmission: Risk factors and virus shedding. *Herpes* 11(Suppl 3): 130A–137A.

Wang C, Li Y. (2006). Research progress on property and application of theaflavins. *Afr J Biotechnol* 5: 213–218.

Wang H, Provan GJ, Helliwell K. (2000). Tea flavonoids: Their function, utilisation and analysis. *Trends Food Sci Technol* 11: 152–160.

Wang P, Tumer NE. (1999). Pokeweed antiviral protein cleaves double-stranded supercoiled DNA using the same active site required to depurinate rRNA. *Nucleic Acids Res* 27: 1900–1905.

Wang ZY, Das M, Bickers DR, et al. (1988). Interaction of epicatechins derived from green tea with rat hepatic cytochrome P-450. *Drug Metab Dispos* 16: 98–103.

Watanabe D. (2010). Medical application of herpes simplex virus. *J Dermatol Sci* 57: 75–82.

Weber JM, Ruzindana-Umunyana A, Imbeault L, et al. (2003). Inhibition of adenovirus infection and adenain by green tea catechins. *Antiviral Res* 58: 167–173.

Wheeler CE, Jr. (1988). The herpes simplex problem. *J Am Acad Dermatol* 18: 163–168.

Whitley RJ, Roizman B. (2001). Herpes simplex virus infections. *Lancet* 357: 1513–1518.

Williamson MP, McCormick TG, Nance CL, et al. (2006). Epigallocatechin gallate, the main polyphenol in green tea, binds to the T-cell receptor, CD4: Potential for HIV-1 therapy. *J Allergy Clin Immunol* 118: 1369–1374.

Xu J, Wang J, Deng F, et al. (2008). Green tea extract and its major component epigallocatechin gallate inhibits hepatitis B virus in vitro. *Antiviral Res* 78: 242–249.

Xu X, Zhou XD, Wu CD. (2012). Tea catechin epigallocatechin gallate inhibits *Streptococcus mutans* biofilm formation by suppressing gtf genes. *Arch Oral Biol* 57: 678–683.

Yamaguchi K, Honda M, Ikigai H, et al. (2002). Inhibitory effects of (-)-epigallocatechin gallate on the life cycle of human immunodeficiency virus type 1 (HIV-1). *Antiviral Res* 53: 19–34.

Yamane T, Nakatani H, Kikuoka N, et al. (1996). Inhibitory effects and toxicity of green tea polyphenols for gastrointestinal carcinogenesis. *Cancer* 77: 1662–1667.

Yang GY, Liao J, Li C, et al. (2000). Effect of black and green tea polyphenols on c-jun phosphorylation and $H_2O_2$ production in transformed and non-transformed human bronchial cell lines: Possible mechanisms of cell growth inhibition and apoptosis induction. *Carcinogenesis* 21: 2035–2039.

Yang J, Li L, Jin H, et al. (2012). Vaginal gel formulation based on theaflavin derivatives as a microbicide to prevent HIV sexual transmission. *AIDS Res Hum Retroviruses* 28: 1498–1508.

Yang Z, Tu Y, Xia H, et al. (2007). Suppression of free-radicals and protection against $H_2O_2$-induced oxidative damage in HPF-1 cell by oxidized phenolic compounds present in black tea. *Food Chem* 105: 1349–1356.

Yu D, Morris-Natschke SL, Lee KH. (2007). New developments in natural products-based anti-AIDS research. *Med Res Rev* 27: 108–132.

Yu D, Wild CT, Martin DE, et al. (2005). The discovery of a class of novel HIV-1 maturation inhibitors and their potential in the therapy of HIV. *Expert Opin Investig Drugs* 14: 681–693.

Zhang HJ, Tan GT, Hoang VD, et al. (2003). Natural anti-HIV agents. Part IV. Anti-HIV constituents from *Vatica cinerea*. *J Nat Prod* 66: 263–268.

Zhang YM, Rock CO. (2004). Evaluation of epigallocatechin gallate and related plant polyphenols as inhibitors of the FabG and FabI reductases of bacterial type II fatty-acid synthase. *J Biol Chem* 279: 30994–31001.

Zhao M, Jiang J, Zheng R, et al. (2012). A proprietary topical preparation containing EGCG-stearate and glycerin with inhibitory effects on herpes simplex virus: Case study. *Inflamm Allergy Drug Targets* 11: 364–368.

Zhao WH, Hu ZQ, Hara Y, et al. (2002). Inhibition of penicillinase by epigallocatechin gallate resulting in restoration of antibacterial activity of penicillin against penicillinase-producing *Staphylococcus aureus*. *Antimicrob Agents Chemother* 46: 2266–2268.

Zhong Y, Ma CM, Shahidi F. (2012). Antioxidant and antiviral activities of lipophilic epigallocatechin gallate (EGCG) derivatives. *J Funct Foods* 4: 87–93.

Zu M, Yang F, Zhou W, et al. (2012). In vitro anti-influenza virus and anti-inflammatory activities of theaflavin derivatives. *Antiviral Res* 94: 217–224.

# 12 HIV Integrase Inhibitors
## *Qualitative and Quantitative Structure–Activity Relationship Studies*

*Rajni Garg and Gene M. Ko*

## 12.1 INTRODUCTION

The human immunodeficiency virus type-1 (HIV-1) is the causative agent of the acquired immunodeficiency syndrome (AIDS) and its related disorders (Barré-Sinoussi et al. 1983). HIV-1 is commonly referred to as "HIV," and HIV infection is a major health concern worldwide. In 2011 alone, an estimated 34 million people were infected with HIV and 1.7 million deaths were attributed to AIDS (UNAIDS 2012).

Studies to understand critical events required for viral replication revealed several key targets for developing anti-HIV drugs, including three crucial viral enzymes: integrase (IN), protease, and reverse transcriptase (De Clercq 1995; Fauci 1988). Currently, 26 drugs have been approved by the U.S. Food and Drug Administration (FDA) for the treatment of HIV infection, and many new drugs are in clinical trials (Belavic 2013; De Clercq 2009; Neamati 2011). The approved drugs belong to the following categories: reverse transcriptase inhibitors (12 drugs), protease inhibitors (10 drugs), fusion inhibitor (1 drug), cell entry (CCR-5 co-receptor) inhibitors (1 drug), and IN inhibitors (2 drugs). The most effective approach for the treatment of HIV-infected patients is called highly active antiretroviral therapy (HAART), in which a combination of drugs targeting two or more key steps in the viral life cycle are administered together (De Clercq 2009). However, mutations arising in the HIV genome that confer resistance toward existing drugs in a large number of patients, physiological side effects, and drug toxicity limit the effectiveness of HAART (Mouscadet and Desmaële 2010; Shafer and Schapiro 2008). These issues drive the need to develop new, effective anti-HIV drugs with acceptable tolerability in terms of side effects, toxicity, and mutation profile. Discovery of novel inhibitors that are structurally and/or mechanistically different is required to inhibit viral strains resistant to the approved drugs.

People with AIDS have a weak immune system and thus are at an increased risk of developing infections, lymphoma, and other types of cancer. Thus, HIV is also known as an oncovirus. The most common types of AIDS-related cancers are Kaposi sarcoma and non-Hodgkin's lymphoma. Other AIDS-related cancers include Hodgkin's disease and cancers of the lung, mouth, cervix, and digestive system (Chow et al. 2009; Palefsky 2012). A study conducted by the International Collaboration on HIV and Cancer (2000) reported a sharp decline in the incidence of Kaposi sarcoma and non-Hodgkin lymphoma in HIV-infected people on HAART treatment. Anti-HIV drugs target multiple pathways in the life cycle of HIV and provide a large source of potential anticancer drugs. Several FDA-approved anti-HIV drugs are being considered for repositioning as anticancer agents; among them, HIV protease inhibitor drugs seem to be more promising (Chow et al. 2009, Oprea et al. 2011). Recently, two classes of HIV IN inhibitors were also studied for repositioning as anticancer agents (Oprea and Mestres 2012; Zeng et al. 2012). Development of new drugs is a lengthy and costly process. Drug repositioning (i.e., repurposing or finding a new use for an existing drug) has gained considerable attention (Collins 2011), and the use of computational methods has significantly helped in these efforts (Andronis et al. 2011; Keiser et al. 2009). The repositioning of an approved drug has financial and other benefits, such as knowledge of its toxicity profile, drug metabolism, pharmacokinetics, and drug interactions.

In the past 20 years, a large number of structure–activity relationship (SAR) studies on HIV IN inhibitors have been published. Subsequently, quantitative SAR (QSAR) studies were undertaken for biological structure–activity data gathered by various research groups to determine the full potential of inhibitors belonging to various classes. In this chapter, a comprehensive overview of QSAR studies published since 2000 is presented with a discussion of relevant SAR. The aim of this chapter is not to discuss each individual SAR study published but rather to provide insight on QSAR studies performed on anti-HIV IN inhibitors and present a historical perspective on the progression of QSAR modeling efforts. As per the scope of this chapter, emphasis has been placed on second-generation IN inhibitors targeting drug-resistant mutant viruses, inhibitors targeting the interaction between IN and the newly discovered human cellular cofactor protein lens epithelium-derived growth factor/p75 (LEDGF/p75), and dual inhibitors targeting the inhibition of two or more steps during the integration process. A discussion is also included about HIV-related cancers and repositioning efforts to redevelop some HIV IN inhibitors as anticancer agents.

## 12.2 HIV INTEGRASE: A NOVEL ANTIVIRAL TARGET

HIV IN, one of the three crucial viral enzymes that drive the HIV replication process, is an important target for the discovery of new anti-HIV drugs (De Clercq 1995; Gupta and Nagappa 2003; Pommier et al. 2005). IN is responsible for the integration of the viral DNA into the host DNA, which results in a functional proviral formation, ensuring chronic infection and continuous transmission of the viral genes from the parent host cell to all new generations of daughter cells. The integration process occurs in two key steps. The first step, 3′-processing, is the cleavage of dinucleotides from each end of the reverse-transcribed viral DNA. This reaction takes place in the cytoplasm and results in the formation of two free 3′-OH groups, which are used for a nucleophilic attack in the second reaction. The second step, strand transfer, is the integration of the 3′-processed ends of the viral DNA into the host cell's DNA, which occurs in the nucleus (Engelman et al. 1991). It has been proposed that multimeric forms of IN are involved in the integration process. Monomeric, dimeric, tetrameric, and higher order oligomeric forms of IN coexist in equilibrium; however, only the active dimeric form is able to bind to the viral DNA. Dimers are supposed to be involved in 3′-processing, whereas tetrameric structures are required for strand transfer (Craigie 2011; Neamati 2011).

IN is a 32 kDa oligomeric viral protein consisting of three domains with distinct structures and function: the N-terminal domain (NTD) (amino acids 1–49), the catalytic core domain (CCD) (amino acids 50–212), and the C-terminal domain (CTD) (amino acids 213–288) (Figure 12.1) (Engelman et al. 1993). The NTD contains a zinc-binding motif (H12, H16, C40, C43), which is known to promote protein multimerization. The CCD contains a conserved amino acid motif (D64, D116, E152, also known as DDE), which is involved in DNA substrate recognition for the 3′-processing and strand transfer reactions. The presence of a divalent metallic cofactor, $Mg^{2+}$ or $Mn^{2+}$, which binds at the DDE motif, is necessary for the strand transfer reaction to proceed. The CTD contains a SH3-like motif and is involved in nonspecific DNA binding, which helps to stabilize the HIV IN-DNA complex (Craigie 2011).

Several human cellular protein cofactors that have distinct interactions with IN throughout the viral life cycle have been reported; among these, LEDGF/p75 has been identified as one of the important cellular proteins involved in the interactions between the viral genome and host DNA. LEDGF/p75 plays an important role in the tetramerization of IN-DNA dimers in the nucleus (Al-Mawsawi and Neamati 2007; Tintori et al. 2010). LEDGF/p75 has an N-terminal chromosomal DNA binding region and a well-characterized small (~80 residues) C-terminal IN-binding domain (IBD). It has been demonstrated that the interaction between the CCD of IN and IBD of LEDGF/p75 is crucial for integration to take place (Al-Mawsawi and Neamati 2011). Disruption/inhibition of this interaction between IN and LEDGF/p75 is currently the most promising approach for the development of novel IN inhibitors. Because the host factors are generally conserved in relevant host–virus interaction, resistance is less likely to occur, increasing the clinical potential of these drugs.

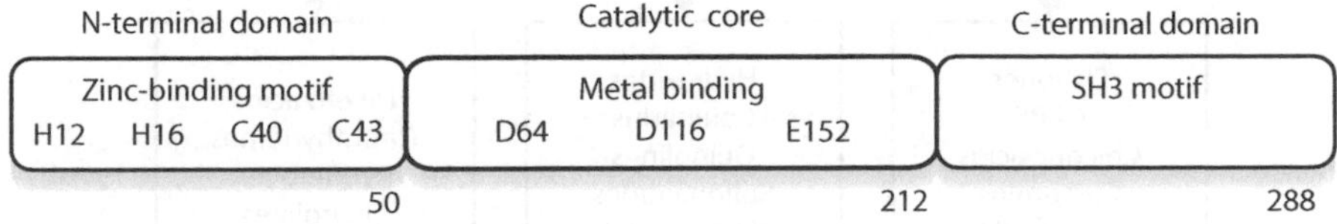


**FIGURE 12.1** Schematic representation of the three structural domains of HIV integrase. Amino acids: histidine (H), cysteine (C), aspartic acid (D), and glutamic acid (E). (Reprinted with permission from Ko GM et al., *Curr Comput Aided Drug Des* 8, 255–270, 2012. Copyright 2012 Bentham Science Publishers.)

The IN enzyme is considered an important target for HIV inhibition for several reasons: It is an essential enzyme in the life cycle of the virus, and integration of the proviral DNA into transcriptional active sites of the host DNA represents a point of no return. It has no cellular homolog; therefore, IN inhibitor drugs are expected to show minimum toxicity in the host. And, a mutation in any of its conserved residues in the CCD reduces the ability of the virus to replicate. In addition, inclusion of approved IN inhibitor drugs in the HAART regimen has increased the life expectancy of HIV-infected patients.

## 12.3 HIV INTEGRASE INHIBITORS

Unlike the successful development of HIV protease and reverse transcriptase inhibitors, IN inhibitor design has been a slow process. Major hindrances in the design of IN inhibitors include the complexity of the integration process and a lack of understanding about the role of multimeric forms of IN in the preintegration complex, the presence of a shallow substrate-binding site on the IN surface, and the lack of a complete full-length x-ray crystal structure (Liao and Nicklaus 2011). IN was found to be the most difficult and "undruggable" target among 27 target binding sites studied (Cheng et al. 2007).

A major breakthrough in HIV IN inhibitor design came in the early 1990s when Fesen et al. (1993) investigated several natural products and topoisomerase inhibitors as potential HIV IN inhibitors. It was found that compounds possessing a dihydroxynaphthoquinone moiety and keto-enol motif inhibit IN activity. Most of the promising IN inhibitors reported to date have either an exact keto-enol motif or a slight variation of this motif. These findings led to the development of hydroxylated aromatic (catechol) compounds. Soon it was realized that a key chemical feature missing in all of these early leads was the lack of a hydrophobic group. Intense research efforts directed to find a potent IN inhibitor drug later on identified several classes of catechol- and non-catechol-based IN inhibitors (Figure 12.2) (Neamati 2011).

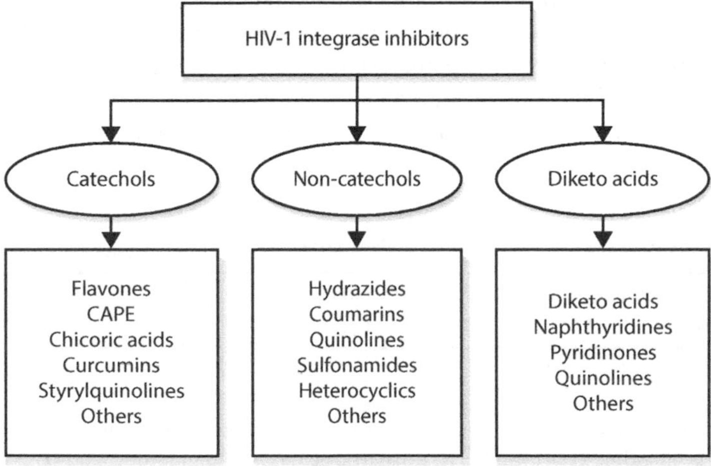


**FIGURE 12.2** Various classes of HIV integrase inhibitors.

Significant research on developing select IN inhibitors has so far resulted in the FDA approval of two-strand transfer IN inhibitor drugs, raltegravir (MK-0518) in 2007 and elvitegravir (GS-9137) in 2012, with a few others in various stages of clinical trial (Figure 12.3). However, mutations conferring resistance to these inhibitors highlight the urgent need to discover the next generation of IN inhibitor molecules (Hombrouck et al. 2011). Most of the drugs in clinical trial and inhibitor classes being developed are similar to the already approved drugs and may have similar interaction patterns, which may result in a low success rate against drug-resistant strains. To counteract the high mutation rate of HIV, identification of drugs with new

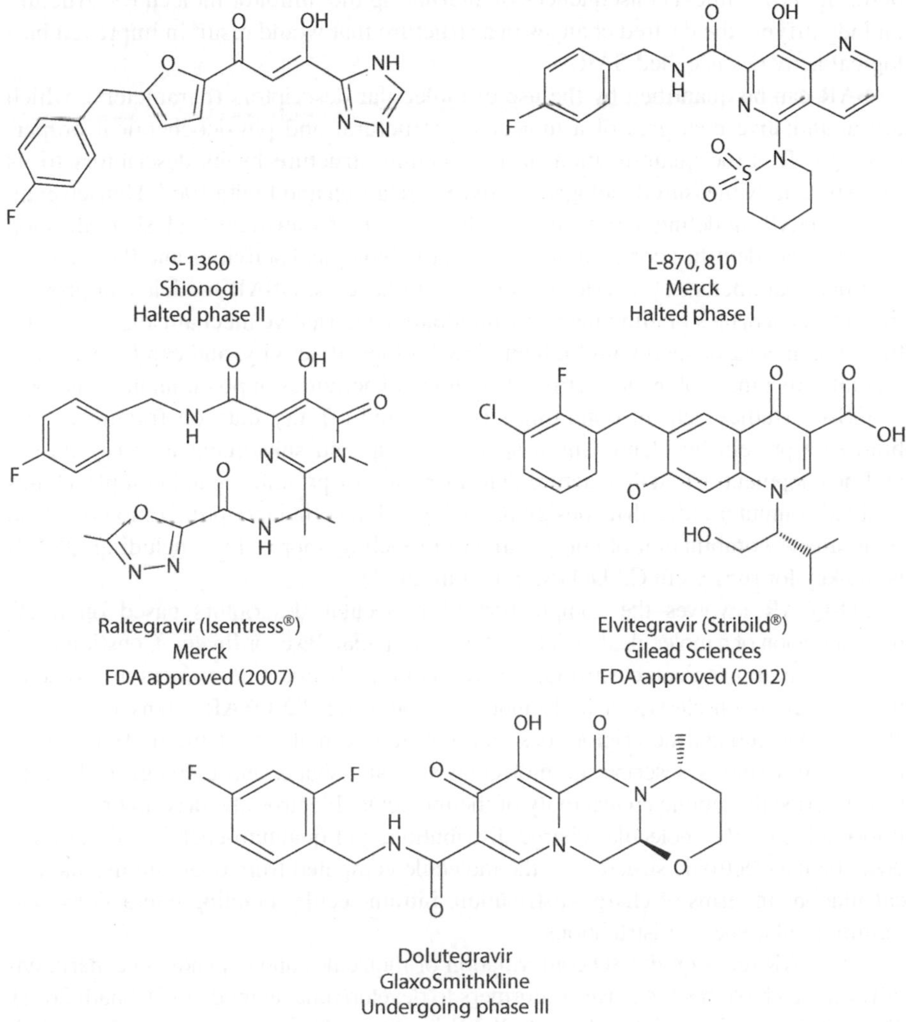

S-1360
Shionogi
Halted phase II

L-870, 810
Merck
Halted phase I

Raltegravir (Isentress®)
Merck
FDA approved (2007)

Elvitegravir (Stribild®)
Gilead Sciences
FDA approved (2012)

Dolutegravir
GlaxoSmithKline
Undergoing phase III

**FIGURE 12.3** Select HIV integrase inhibitors that have gone through clinical trials. Stribild® is the brand name of a combination drug containing elvitegravir, cobicistat, emtricitabine, and tenofovir disoproxil fumarate.

active substituents and new mechanisms of action is required for the development of next generation inhibitors.

## 12.4 QUANTITATIVE STRUCTURE–ACTIVITY RELATIONSHIP

The premise of the drug discovery process involves synthesizing analogs of lead compounds to determine their biological activity toward a specific target such that one would obtain progressively better molecules in terms of potency, efficacy, pharmacokinetics, and a favorable adsorption, distribution, metabolism, excretion, and toxicity (ADMET) profile. The systematic study of a cause and effect relationship in defining the chemical consequences of alternating the inhibitor molecule's structure and identifying the desired changes in a structure that would result in improved biological activities is called SAR.

SAR can be quantified by the use of molecular descriptors (parameters), which are quantitative measures of a molecule's structural and physicochemical properties. QSAR is the quantification of a molecular structure by its descriptors to its experimentally measured biological activities (Hansch and Fujita 1964; Hansch et al. 1962). QSAR modeling approaches involve the use of statistical analysis to develop a model that identifies correlations between the biological activities and the physicochemical parameters of a series of inhibitor molecules. QSAR models can provide insight into a drug's binding mechanism, identify alternative mechanisms and structural features associated with high and low biological activity, and can be used as a virtual screening tool by predicting the biological activities of potential new analogs.

QSAR studies help in obtaining structure and affinity data on fragments that bind to a protein by identifying a specific group and substituent for drug design and development. Crystal structure determination of protein–ligand complexes has enabled computer-aided drug design (CADD) and the efficient optimization of a lead structure. A combination of computational modeling approaches, including QSAR, is the key for success in CADD (Serrao et al. 2011).

2D-QSAR involves the computation of molecular descriptors based on a 2D representation of a molecule that does not rely on spatial 3D coordinates. Constitutional, geometrical, topological, electrostatic, and quantum-chemical descriptors are some of the most common classes of molecular descriptors for 2D-QSAR (Hansch and Leo 1995). Constitutional descriptors describe the atomic makeup of the molecule, geometrical descriptors describe the molecular size and shape, and topological descriptors describe the atomic connectivity of the molecule. Electrostatic descriptors provide information on the molecular charge distribution, and quantum-chemical descriptors describe the electronic structure of the molecule computed from quantum mechanical calculations in terms of charge distribution, intramolecular bonding interactions, and intramolecular energy distributions.

3D-QSAR relies on the 3D conformation of molecules and depends on experimentally derived biologically active conformers to develop reliable models (Debnath 2001). The generation of descriptors for 3D-QSAR is obtained by several techniques, including comparative molecular field analysis (CoMFA), comparative molecular surface analysis (CoMSA), comparative molecular similarity indices analysis (CoMSIA), comparative residue interaction analysis (CoRIA), molecular field analysis (MFA), and molecular shape analysis (MSA). These methods use force field grid calculations to compute descriptors that provide information on steric, hydrophobic, and electrostatic interactions. Analysis of these interactions provides information on structural regions, which positively or negatively affect the biological activity of the compounds studied. 4D-QSAR extends upon 3D-QSAR and uses 4D descriptors derived from the 4D molecular similarity method, which provides molecular conformational information and captures the molecular size and shape (Hopfinger et al. 1997).

Several statistical modeling techniques are available for developing QSAR models but not all have been used in anti-HIV IN inhibitor design. The methods that have been used are shown in Figure 12.4 and discussed here briefly. Readers are encouraged to refer to Dudek et al. (2006) for more details. 2D-QSAR classification models were developed using hierarchical clustering (HC), linear discriminant analysis (LDA), and principle component analysis (PCA) algorithms. For 2D-QSAR regression models, linear modeling methods such as multiple linear regression (MLR), principle component regression (PCR), and partial least squares (PLS) were explored, whereas nonlinear 2D-QSAR models were developed using back propagation neural networks (BPNN) and support vector machine (SVM) techniques. 3D-QSAR models (MFA, MSA, CoMFA, CoMSA, CoMSIA, and CoRIA) were developed using genetic function approximation (GFA), genetic/partial least squares (G/PLS), K-nearest neighbor (KNN), MLR, and PLS techniques. The development of 4D-QSAR models involved the use of GFA and PLS.

Many examples of the successful use of QSAR modeling to understand drug–receptor interactions in the discovery of anti-HIV drugs have been reported in the literature (Bhhatarai and Garg 2008; Debnath 2005; Garg et al. 1999; Hansch and Leo 1995; Ko et al. 2010, 2012; Kurup et al. 2003; Reddy et al. 2010, 2011; Srivastava et al. 2012). In this chapter, a comprehensive overview of QSAR studies published

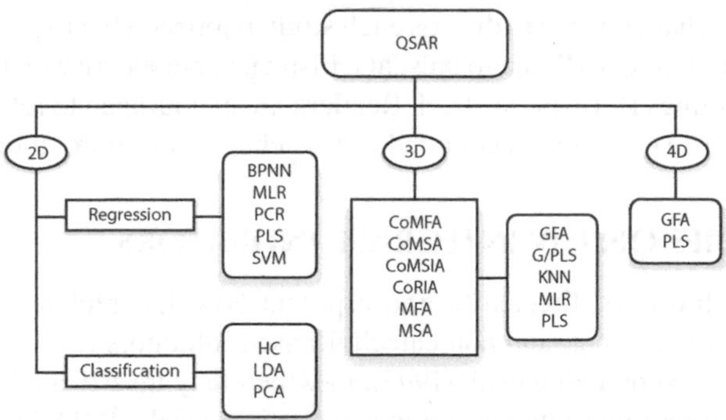


**FIGURE 12.4** Common techniques applied to develop QSAR models for HIV integrase inhibitors. BPNN: back propagation neural networks; MLR: multiple linear regression; PCR: principle component regression; PLS: partial least squares; SVM: support vector machines; HC: hierarchical clustering; LDA: linear discriminant analysis; PCA: principle component analysis; GFA: genetic function analysis; G/PLS: genetic partial least squares; and KNN: k-nearest neighbor. (Reprinted with permission from Ko GM et al., *Curr Comput Aided Drug Des* 8, 255–270, 2012. Copyright 2012 Bentham Science Publishers.)

**TABLE 12.1**
**SAR and QSAR Review Articles on HIV Integrase Inhibitors**

| Title | SAR/QSAR | Reference |
|---|---|---|
| HIV-1 integrase inhibitors: past, present, and future | SAR | Neamati et al. 2000 |
| Retroviral integrase inhibitors year 2000: update and perspectives | SAR | Pommier et al. 2000 |
| Small-molecule HIV-1 integrase inhibitors: the 2001–2002 update | SAR | Dayam and Neamati 2003 |
| HIV-1 integrase inhibitors: a decade of research and two drugs in clinical trials | SAR | Johnson et al. 2004 |
| HIV-1 integrase inhibitors: 2003–2004 update | SAR | Dayam et al. 2006a |
| HIV-1 integrase inhibitors: 2005–2006 update | SAR | Dayam et al. 2008b |
| Anti-HIV drugs: 25 compounds approved within 25 years after the discovery of HIV | SAR | De Clercq 2009 |
| HIV-1 integrase inhibitors: 2007–2008 update | SAR | Ramkumar et al. 2010a |
| HIV-1 integrase inhibitor design: overview and historical perspectives | SAR | Neamati 2011 |
| Application of 3D-QSAR techniques in anti-HIV-1 drug design—an overview | QSAR | Debnath 2005 |
| Computational modeling methods for QSAR studies on HIV-1 integrase inhibitors between 2005 and 2010 | QSAR | Ko et al. 2012 |
| Modeling anti-HIV compounds: the role of analog-based approaches | QSAR | Srivastava et al. 2012 |

on anti-HIV IN inhibitors since 2000 is presented with a discussion of relevant SAR. The aim of this chapter is not to discuss each study published but to provide an insight on the application of QSAR and to present a historical perspective on the progression of QSAR modeling efforts in this field. Readers are encouraged to refer to the review articles (Table 12.1) and references cited in this chapter for more information.

## 12.5 STUDIES ON HIV INTEGRASE INHIBITORS

Intense research efforts directed to find a potent HIV IN inhibitor drug identified several classes of catechol- and non-catechol-based inhibitors (Neamati 2011). Early leads were reported by Fesen et al. (1993), in which they investigated several natural products, including topoisomerase inhibitors, antimalarials, DNA binders, naphthoquinones, and flavones, as potential IN inhibitors. It was found that the topoisomerase I inhibitor, camptothecin, was inactive against IN, whereas topoisomerase II inhibitors, doxorubicin and mitoxantrone, inhibited IN activities (Figure 12.5). This study established that there is a significant difference in the mode of action of topoisomerase I and II inhibitors to block IN-mediated 3′-processing and strand transfer activities. An important finding of this work was that a dihydroxynaphthoquinone motif, found in some of the topoisomerase II inhibitors as well as in some natural products, was able

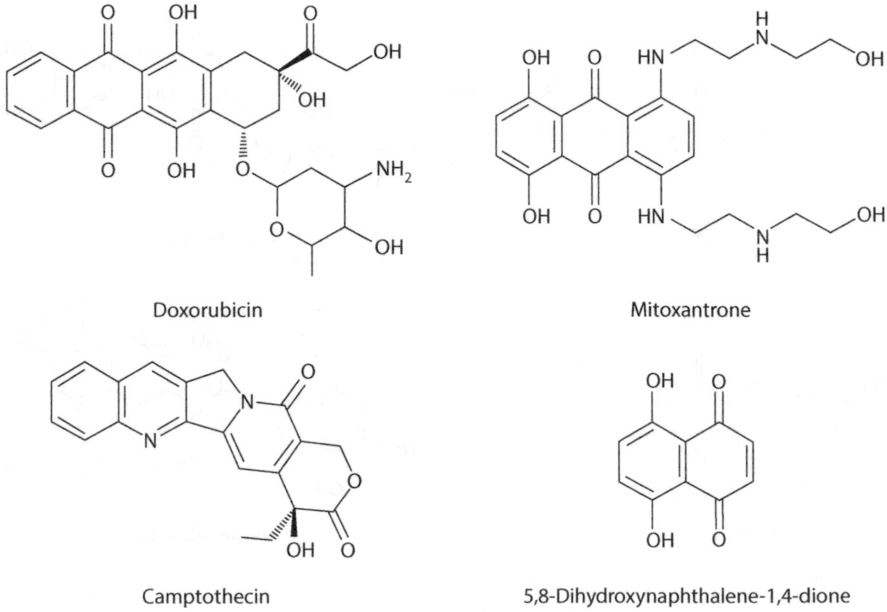

Doxorubicin

Mitoxantrone

Camptothecin

5,8-Dihydroxynaphthalene-1,4-dione

**FIGURE 12.5** Early lead compounds as HIV integrase inhibitors.

to inhibit IN. This work also concluded for the first time that a keto-enol motif, capable of chelating divalent metals, can also inhibit IN activity. Most of the promising IN inhibitors reported to date have either an exact keto-enol motif or a slight variation. These findings later on led to the development of hydroxylated aromatic (catechol) compounds. Soon it was realized that a key chemical feature missing in all of these early leads was the lack of a hydrophobic group required for selectivity toward the IN strand transfer activity. Various hydrophobic groups were subsequently explored in different non-catechol scaffolds, and a fluorobenzyl group, commonly found in many drugs, was found to be an ideal functional group for inhibition.

All of these research efforts resulted in the development of two-strand transfer IN inhibitor drugs, raltegravir and elvitegravir, with a few others in clinical trial (Figure 12.3). These inhibitors were identified using structural information gathered from extensive synthetic work and a combination of computational approaches, including QSAR, pharmacophore modeling, docking studies, and high-throughput virtual screening. Many QSAR modeling studies were performed, which provided deeper insight for understanding the binding interactions of these compounds with the IN active site and helped in further structure optimization to improve potency. The results of these studies are presented here.

### 12.5.1 Catechol (Hydroxylated Aromatic) Derivatives

Polyhydroxylated aromatic compounds possessing one or two catechol units, separated by different aromatic or aliphatic linker groups, were the first compounds found active against HIV IN (Fesen et al. 1993, 1994). Some of the natural products, caffeic

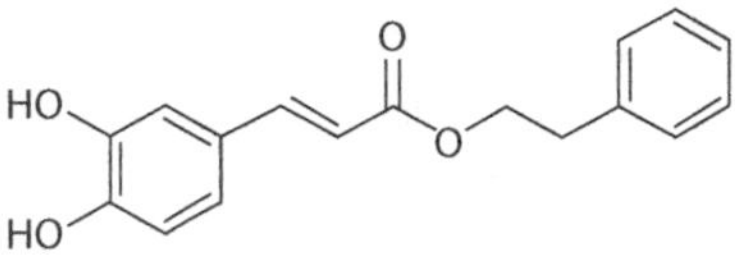

CAPE

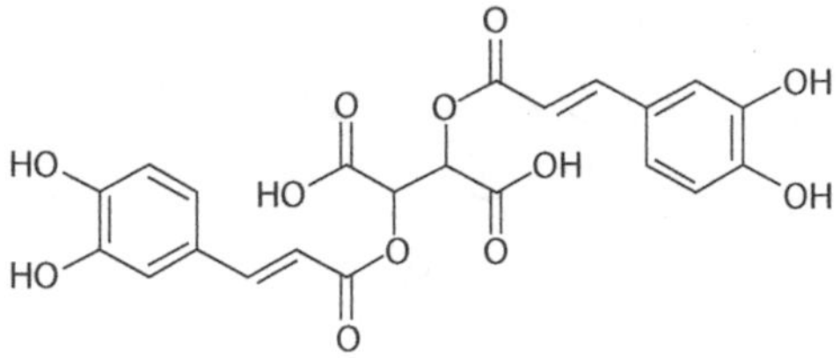

(L,D)-Chicoric acids

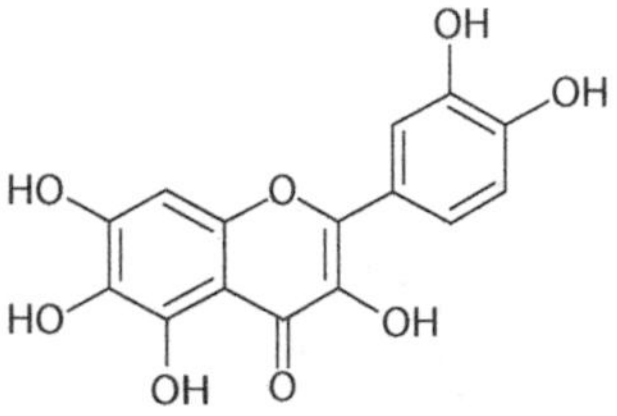

Quercetagetin

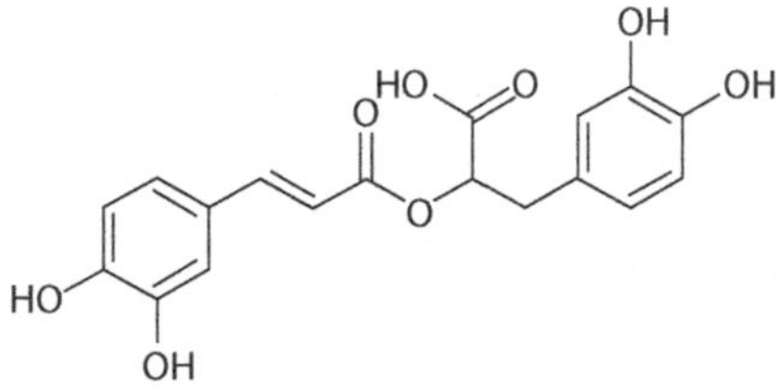

Rosmarinic acid

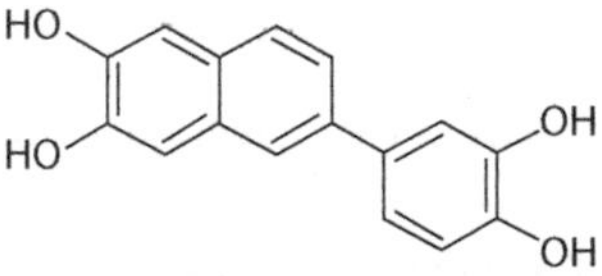

Naphthalene-diol

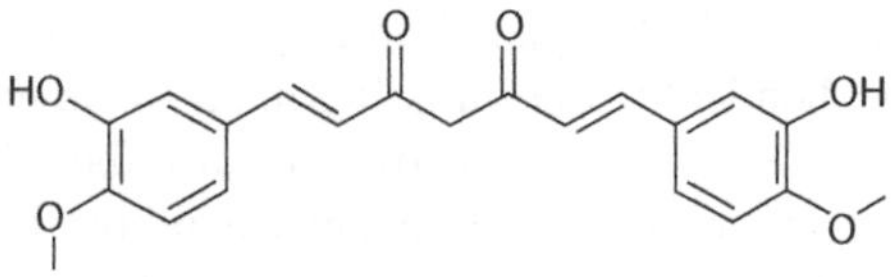

Curcumin

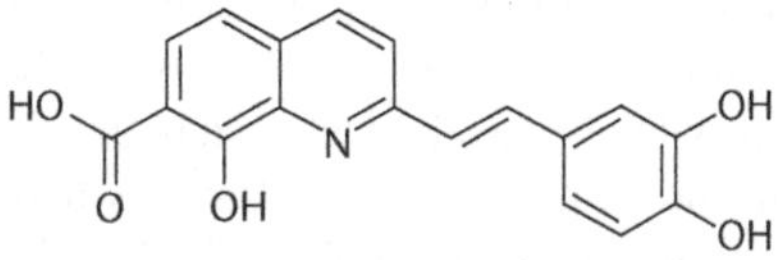

Styrylquinoline

**FIGURE 12.6** Catechol-containing polyhydroxylated aromatic compounds studied as HIV integrase inhibitors (CAPE: caffeic acid phenethyl ester).

acid phenethyl ester (CAPE), flavonoids, curcumin, chicoric acid, and styrylquinoline derivatives, were studied extensively, which resulted in some of these inhibitors being selected for clinical trials (Figure 12.6).

#### 12.5.1.1 Flavones, Curcumins, and Cinnamoyls

Early work on catechol derivatives focused on developing naturally found flavones, curcumin, and cinnamoyl derivatives as potential IN inhibitors (Figure 12.7). QSAR studies on these compounds are discussed here.

A few research groups analyzed SAR data reported by Fesen et al. (1993, 1994) using various QSAR approaches. First, Raghavan et al. (1995) performed 3D-QSAR studies using CoMFA and PLS methods for a series of 15 flavones and related

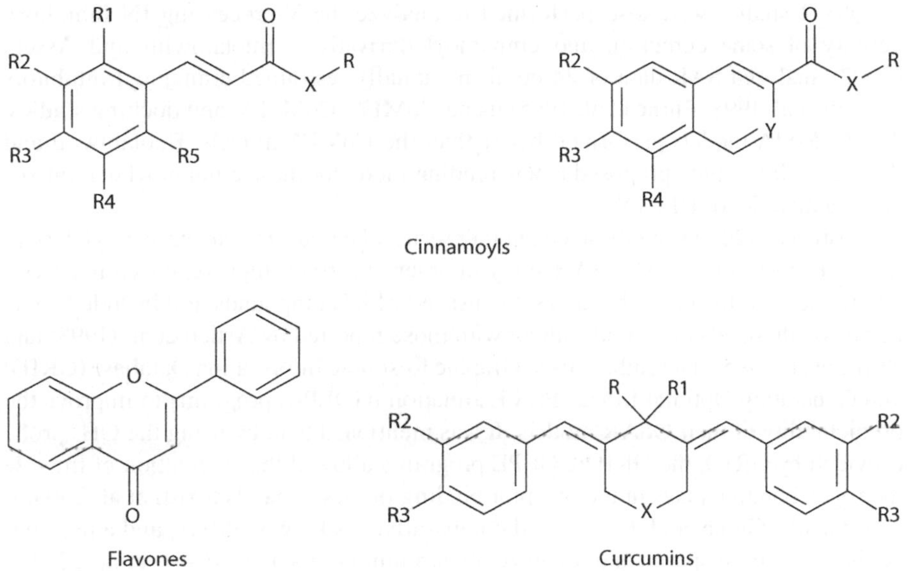

FIGURE 12.7 Substituted flavones, curcumins, and cinnamoyls studied by QSAR modeling.

polyhydroxylated aromatic compounds (Fesen et al. 1994) for their 3′-processing and strand transfer inhibitory activity. The model by Raghavan et al. (1995) indicated a strong correlation between the inhibitory activity of these compounds and the steric and electrostatic fields around them.

Buolamwini et al. (1996) studied the same series of 15 flavones (Fesen et al. 1994) using electrotopological state (E-state) indices. E-state indices are Kier and Hall descriptors that encode electronic and topological information as nonempirical atomic level structural descriptors. This study suggested that polyhydroxylation of the aromatic ring is required for potency and substitution of an electronegative group, such as hydroxyl at C-6, will enhance activity. E-state indices at C-6, C-3′, C-5′, C-5, and O-4 were found to be more important for the prediction of activity than those for any of the other 12 flavone skeletal atoms that were common to the molecules in the dataset.

Marrero-Ponce (2004) proposed a new methodology to analyze the same series of flavones (Fesen et al. 1994). They used linear indices of the molecular pseudograph's atom adjacency matrix for discovery and optimization of new lead compounds.

Lameira et al. (2006) published a 2D-QSAR study on a series of 32 flavone compounds (Fesen et al. 1993) using density function theory. They developed classification models for determining active or inactive molecules for the inhibition of HIV IN. HC, LDA, and PCA with stepwise variable selection being used to develop the classification models. The LDA model correctly classified all of the inactive compounds, while correctly classifying 93.33% of the active molecules. In all three models, low hydrophobicity, electrophilic index value, and a greater negative charge on oxygen at C-11 led to a greater inhibitory activity of flavonoid compounds.

QSAR studies were also performed to analyze the 3′-processing IN inhibitory activity of some curcumin and cinnamoyl derivatives. Buolamwini and Assefa (2002) analyzed SAR data of 24 conformationally restrained cinnamoyl inhibitors (Artico et al. 1998; Burke et al. 1995) using CoMFA, CoMSIA, and docking studies. The CoMSIA models performed better than the CoMFA models. Buolamwini and Assefa's (2002) study proposed a new binding mode for these cinnamoyl derivatives at the active site of HIV IN.

Costi et al. (2004a) synthesized another series of curcumin-like cinnamoyl derivatives and performed a 3D-QSAR study in order to further improve antiviral activity and reduce cytotoxicity. This dataset consisted of 87 compounds and included some newly synthesized compounds, along with those reported by Artico et al. (1998) and Burke et al. (1995). The authors used Graphic Resource Information Database (GRID) and Generating Optimal Linear PLS Estimation (GOLPE) programs to improve the predictability of their QSAR models. It was mentioned that, by using the $OH_2$ probe provided by GRID, the GRID/GOLPE programs allowed the elucidation of the role hydrogen bonding plays in the receptor binding of these ligands (Costi et al. 2004a).

Recently, Gupta et al. (2011) used comparative docking, CoMFA, and a pharmacophore mapping approach to analyze 39 curcumin derivatives (Costi et al. 2004a). The high contribution of polar interactions in pharmacophore mapping was well supported by docking and CoMFA results. This comparative study provided insight on important structural and binding features of these derivatives for the rational design of novel inhibitors.

#### 12.5.1.2 Styrylquinolines

Styrylquinolines were developed as another class of catechol derivatives possessing good IN enzyme inhibition and antiviral activities as well as reduced cellular cytotoxicity (Mekouar et al. 1998). These molecules link a quinoline substructure to an aryl nucleus (possessing various hydroxyl substituents) by an ethylenic spacer. Computational docking and spectroscopic studies soon established styrylquinolines as among the few IN inhibitors showing selectivity for the 3′-processing step during IN inhibition (Deprez et al. 2004). Many derivatives of styrylquinolines, including quinoline ring replacement by quinazolines and benzofuran, were synthesized. QSAR studies on these derivatives are presented here (Figure 12.8).

Polanski et al. (2002) analyzed SAR data of 19 styrylquinoline derivatives (Mekouar et al. 1998; Zouhiri et al. 2000) using Kohonen neural network (NN) to model the ex vivo activities of these compounds. These maps of electrostatic potential represent a powerful tool for analyzing structure–activity data. Through an extensive study of the comparative Kohonen maps of different structural motifs of the molecules, they found an interesting pattern that clearly differentiated active from inactive molecules. Their analysis revealed a key structural factor defining the ex vivo activity of the series that helped in the design of additional analogs.

In another paper, Polanski et al. (2004) used SAR data of the same styrylquinolines (Mekouar et al. 1998) containing 20 compounds for 3′-processing and addressed the probability issues in molecular design using CoMFA and CoMSIA methodologies. Ma et al. (2004) also analyzed 3′-processing inhibitory data of 38 styrylquinolines (Mekouar et al. 1998; Zouhiri et al. 2000) using CoMFA and

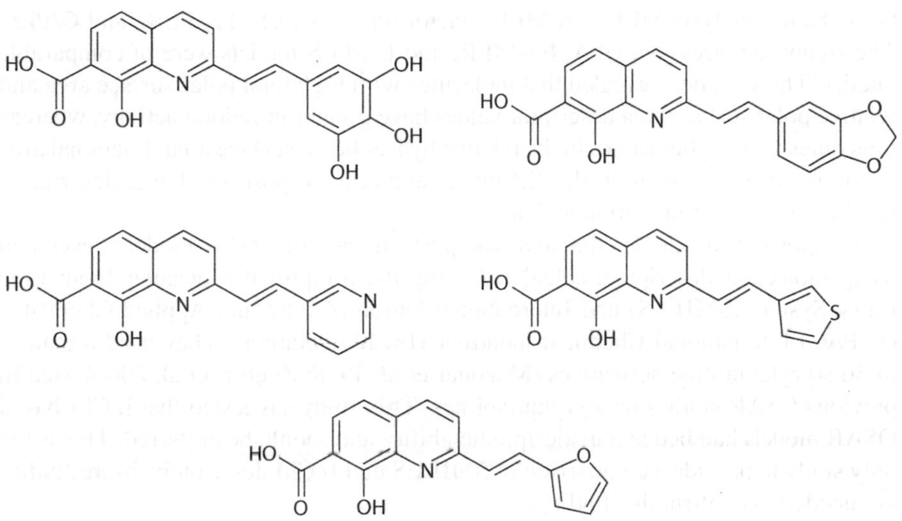

**FIGURE 12.8** Examples of styrylquinolines studied by QSAR modeling.

docking methods. Their results showed that a carboxyl group at C-7 and a hydroxyl group at C-8 (in the quinolone subunit) are closely bound to the divalent metal cofactor ($Mg^{2+}$) around the IN catalytic site. They also reported a linear correlation between the binding energy of the inhibitors with IN and their inhibitory effect.

Niedbala et al. (2006) published a 3D-QSAR model using CoMSA on a series of 29 styrylquinoline derivatives (Deprez et al. 2004; Mekouar et al. 1998; Ouali et al. 2000) for virtual combinatorial library screening. They developed several models using CoMFA, CoMSA, 4D-QSAR, and self-organizing map-4D-QSAR to generate the chemical descriptors to find a correlation between the inhibitory activities and the molecular descriptors derived from each of the ligand structures. The models were developed using PLS for the inhibition of the strand transfer reaction, 3′-processing reaction, and the inhibitory concentration at which half of the viruses were killed in an ex vivo experiment. The CoMSA models were found to perform better than all other models. The CoMSA model was used as a virtual screening tool for the identification of new promising inhibitor molecules. A virtual combinatorial library containing 26,784 molecules based on a quinazoline skeleton was created. PCA and Kohenen NN were used to identify potentially active compounds. Imine analogs of styrylquinolines were identified as potential drug candidates; 11 were selected for in vitro experimentation to measure inhibitory activities for the 3′-processing mechanism, of which nine were found active.

An MSA based 3D-QSAR study on a set of 36 styrylquinoline derivatives was reported by Leonard and Roy (2008) using 3′-processing inhibitory SAR data of these compounds (Mekouar et al. 1998; Zouhiri et al. 2000). K-means clustering was used to classify the molecules based on their topological and structural descriptors in order to divide the dataset into a training and test set. Spatial, MSA, thermodynamic, and structural descriptors were considered for the MSA model development. Several statistical techniques were used to develop the models: stepwise regression,

GFA, factor analysis-MLR (FA-MLR), factor analysis-PLS (FA-PLS), and G/PLS. The stepwise regression, GFA, FA-MLR, and FA-PLS models were of comparable quality. These models revealed that molecules with high total polar surface area and relative polar surface area descriptor values have greater inhibition activity, whereas molecules with higher values in the relative hydrophobic surface area, fractional area of the molecular shadow in the XZ plane, and relative positive charge descriptors resulted in reduced inhibition activity.

Toropova et al. (2011) compared the performance of QSAR models developed using molecular descriptors calculated using the Simplified Molecular Input Line Entry System (SMILES) and International Union of Pure and Applied Chemistry (IUPAC) International Chemical Identifier (InChI) notations. They used a dataset of 36 styrylquinoline derivatives (Mekouar et al. 1998; Zouhiri et al. 2000) used in previous QSAR studies on styrylquinolines. This study suggested that InChI-based QSAR models had better statistics/predictability and should be preferred. This is the only study to provide a comparison of SMILES and InChI descriptors. More studies are needed to confirm this finding.

Goudarzi et al. (2012) also analyzed inhibitory data of 36 styrylquinoline derivatives (Mekouar et al. 1998; Zouhiri et al. 2000) using GA coupled with MLR 2D-QSAR approach. This study demonstrated that GA-MLR is a simple and fast methodology for modeling these compounds. Recently, Sun et al. (2012) performed a 3D-QSAR study using SAR data of 77 quinoline ring derivatives (Mekouar et al. 1998; Mouscadet and Desmaële 2010; Normand-Bayle et al. 2005; Ouali et al. 2000; Polanski et al. 2002; Zouhiri et al. 2000). They compared the performances of CoMFA, CoMSIA, and Topomer CoMFA and found that CoMFA-based QSAR models have better predictability.

### 12.5.2 Non-Catechols

It was soon realized that polyhydroxylated aromatic compounds (catechol) are not selective inhibitors against IN 3′-processing or strand transfer reactions, show significant cytotoxicity, have low oral bioavailability, and are not metabolically stable. It could be because the cellular oxidation of catechol to semiquinone or orthoquinone results in reactive intermediate capable of interacting with other cellular targets. Attempts to improve the biological activity of catechol derivatives by substitution or removal of the hydroxyl group resulted in the complete loss of IN inhibitory activity. The National Cancer Institute (NCI) 3D database of pharmacophore-based searches and computational molecular modeling approaches identified several active structures devoid of catechol moiety (Neamati 2011). Among them, coumarins, hydrazides, quinones, sulfones, sulfonamides, diketo acids (DKAs), naphthyridines, and pyrimidinones were explored extensively. Raltegravir, the first FDA-approved IN inhibitor drug, was the result of research on pyrimidinone carboxamides, whereas elvitegravir, the second FDA-approved IN inhibitor drug, resulted from research on quinolinoyl carboxylic acids. Several other diverse compounds uniquely different from many examples mentioned earlier have been reported to inhibit IN. Some of them, such as thiazolothiazepines, chalcones, dihydropyridine carboxylic acids, benzoxazoles, tetrazoles, and benzene tricarboxamides, were extensively optimized in search of potent and selective IN inhibitors. Many QSAR studies on these inhibitors were reported and are discussed in this section.

#### 12.5.2.1 Hydrazides, Coumarins, Quinones, and Thiazozepines

Hydrazides and coumarins were among the first promising class of inhibitors synthesized without a catechol ring. *N,N′*-Bissalicylhydrazide was considered an interesting lead due to its ease of synthesis and availability in the NCI database. The presence of a salicyl group known to chelate divalent metal such as $Mg^{2+}$, $Mn^{2+}$, and $Zn^{2+}$ was essential for inhibitory activity of these compounds. Soon, it was noticed that these hydrazides possessed more cytotoxicity than the hydroxylated aromatics and were not suitable to be developed as HIV IN inhibitors (Zhao et al. 1997c). An important outcome of this research was repurposing studies on these compounds as novel anticancer agents due to their cellular cytotoxicity. Many coumarin derivatives were also studied to inhibit various targets in the HIV life cycle. Some of the select compounds of this class explored in detail were tetrameric coumarin (NSC 158393), biscoumarins, benzophenone-linked coumarins, caffeoyl-coumarins, and many conjugated coumarins. A mass spectroscopy study identified that these coumarins bind at a novel binding site at the dimeric interface of IN (Al-Mawsawi and Neamati 2011). The same binding site was also shown to be responsible for the interaction between IN and LEDGF/p75. Biscoumarins were also predicted to disrupt the formation of catalytically active IN tetramers by binding at this allosteric dimeric interface. QSAR studies on hydrazides, coumarins, and some other non-catechol derivatives reported in the literature are discussed here (Figure 12.9).

In a series of papers, Makhija and Kulkarni (2001a, 2001b, 2002a, 2002b) used 3D-QSAR methods to examine 3′-processing and strand transfer inhibitory activities of structurally diverse classes of IN inhibitors. In the first paper, Makhija and

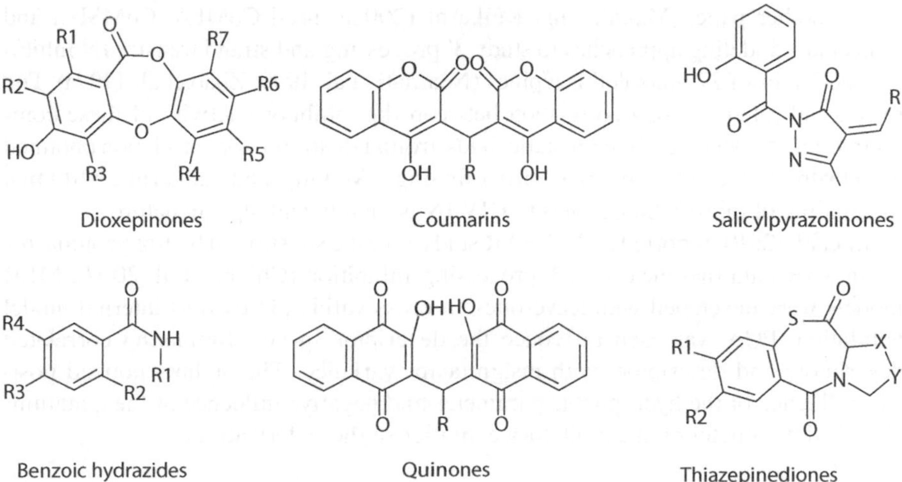


**FIGURE 12.9** Substituted hydrazides, coumarins, quinones, and thiazolothiazepines studied by QSAR modeling.

Kulkarni (2001b) used CoMSIA to study a training set of 34 compounds belonging to these classes: salicylpyrazolinones (Neamati et al. 1998), dioxepinones (Neamati et al. 1997a), coumarins (Zhao et al. 1997a), quinones (Mazumder et al. 1996), and benzoic hydrazides (Zhao et al. 1997c). In addition, seven molecules belonging to a different structural class, thiazepinediones (Neamati et al. 1999), were used as the test set. A CoMFA QSAR model was developed from an alignment strategy based on molecular electrostatic potentials (MEPs). This study noticed that there is not much difference between the steric and electrostatic contours for 3′-processing and strand transfer activities, indicating similar steric and electrostatic requirements in terms of substituents on the inhibition of both these activities. Steric bulk at two phenyl rings was found to be favorable for inhibition, and a hydroxyl group and carbonyl function of the pyrazole ring indicated the involvement of these groups in electrostatic interactions with IN. The authors concluded that the binding of these inhibitors to IN seems mainly enthalpic in nature and, during binding, the areas that get desolvated were polar, not hydrophobic.

Next, Makhija and Kulkarni (2001a) used CoMSIA 3D-QSAR and eigenvalue analysis (EVA) to analyze 35 molecules from five diverse classes: salicylhydrazines (Neamati et al. 1998), lichen acids (Neamati et al. 1997a), coumarins (Zhao et al. 1997a), quinones (Mazumder et al. 1996), and thiazolothiazepines (Neamati et al. 1999). In this work, they used a test set of six molecules from a different structural class, hydrazides (Zhao et al. 1997c), to validate their models. EVA descriptors are alignment-free values, derived from calculated infrared vibrational frequencies. EVA profiles indicated prominent peaks in the hydrogen bond stretching frequency region for potent salicylhydrazine derivatives that possess two hydrogen bond donor groups. These peaks were absent or attenuated in less active thiazolothiazepine analogs, which do not possess any free hydrogen bond donor group. These results agreed well with SAR studies documenting that hydrogen bond donor groups such as OH, COOH, $SO_3H$, and so on are required for potent inhibitory activity of salicylhydrazines.

In another paper, Makhija and Kulkarni (2002a) used CoMFA, CoMSIA, and molecular modeling approaches to study 3′-processing and strand transfer IN inhibitory activities of 27 thiazolothiazepines (Neamati et al. 1999; Zhao et al. 1997a). The results indicated a strong correlation between the inhibitory activity of these compounds and the steric and electrostatic fields around them. A poor correlation obtained with hydrophobic field supported earlier findings (Makhija and Kulkarni 2001b) that the binding of thiazolothiazepines to HIV IN is mainly enthalpic in nature.

Li et al. (2010) reported a 2D-QSAR study on nine substituted hydroxycoumarins using SAR data that measured 3′-processing inhibition (Chiang et al. 2007). MLR models were developed with leave-one-out cross validation used as internal model validation. PCA was used to reduce the descriptor space, eliminating correlated descriptors and descriptors with insignificant variance. The authors noticed positive influence of the hydrophobic parameter and negative influence of the quantum-chemical parameter on the inhibitory activities of these derivatives.

#### 12.5.2.2 Sulfonamides

Continued research in a quest for finding novel non-catechol derivatives identified several sulfone, sulfonamide, sulfide, and benzodithiazine derivatives as promising lead compounds. The first x-ray crystal structure of a naphthalene disulfone bound to avian sarcoma virus (ASV) IN facilitated further optimization of these compounds (Chen et al. 2000; Neamati et al. 1997a). Many potential compounds interacting with the IN active site residues were identified. Among them, mercaptobezesulfonamides emerged as an important class, and a free mercapto group in these compounds was found important for antiviral as well as IN inhibitory activity (Brzozowski et al. 2004). Compounds bearing a hydroxyl instead of a mercapto group were found to be inactive. Mercaptobenzesulfonamides were also found to possess antiviral activity against mutant strains resistant to reverse transcriptase and viral entry inhibitors and hold great promise to be developed as dual inhibitors. Geminal disulfones bearing polyhydroxylated aromatic groups displayed potent inhibitory activity against 3′-processing and strand transfer mechanisms. This study indicated that these disulfones are capable of inhibiting different stages in the viral life cycle and have the potential to be developed as dual inhibitors. Benzodithiazines were found active against IN mutants and were studied in detail. Results of QSAR studies on these derivatives are presented here (Figure 12.10).

Kuo et al. (2004) examined 66 mercaptobenzenesulfonamides (Chen et al. 2000; Neamati et al. 1997c; Nicklaus et al. 1997) using CoMFA- and CoMSIA-based 3D-QSAR to analyze 3′-processing and strand transfer inhibitory activities of these compounds. A significant hydrophobic contribution at the R-substituent and the edge of the common substituted phenyl ring around the chlorine substituent was reflected by the presence of a hydrophobic descriptor in the CoMSIA QSAR model.

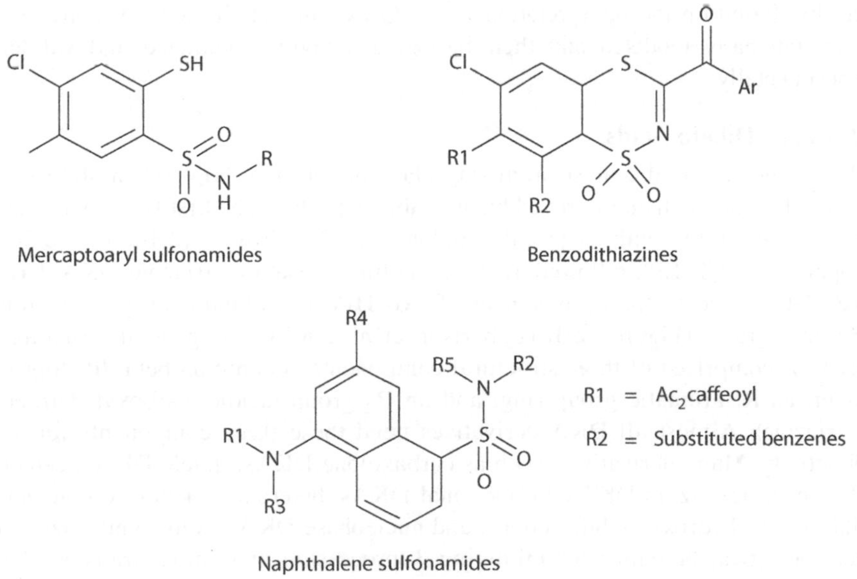

**FIGURE 12.10** Substituted sulfonamides studied by QSAR modeling.

The authors noted that the presence of steric fields in the CoMFA model also reflect this hydrophobic contribution.

Kaushik et al. (2011b) modeled another series of 22 mercaptobenzenesulfonamides (Brzozowski et al. 2008) and reported 2D-QSAR models for 3′-processing and strand transfer inhibitory activities. They noticed that the 3′-processing activity is significantly correlated with hydrophobic and surface tension properties of the molecules, whereas strand transfer activity is correlated with only the hydrophobic property of the molecules, and an indicator variable pointed toward the steric role of aryl substituents with the enzyme.

A 3D-QSAR study using CoMFA and CoMSIA was reported by Gupta et al. (2009). They used the 3′-processing inhibitory SAR data of benzodithiazine (Brzozowski et al. 2004) and mercaptobenzenesulfonamide (Brzozowski et al. 2008) derivatives for this study. PLS was used to correlate CoMFA and CoMSIA fields to the inhibitory activity values. The models were internally validated using leave-one-out cross validation. The best models indicated the significance of steric and hydrophobic properties of the inhibitors for binding to the active site.

In continuation of their work, Gupta et al. (2012) used shape-based screening, QSAR, and docking to further analyze SAR data of 22 benzodithiazines (Brzozowski et al. 2004). They used GFA and shape-based screening methods to develop 2D-QSAR models. The authors reported that the most important descriptors were related to the topology and 3D arrangement of atoms of these molecules.

Sahu et al. (2007) reported a 2D-QSAR study on 20 caffeoyl naphthalene sulfonamide derivatives (Xu et al. 2003). The selected descriptors in the optimal MLR model indicated that an increase in the partition coefficient, total energy, and shape index, and a decrease in the connectivity index, would result in increased biological activity. Based on the interpretation of the QSAR model, 30 new structures were computationally modeled and their biological activities predicted and validated experimentally.

#### 12.5.2.3 Diketo Acids

DKAs emerged as the most promising class of selective HIV IN inhibitors and antiviral agents with an improved bioavailability profile. The first HIV IN inhibitor crystal structure with 1-(5-chloroindole-3-yl)-3-hydroxy-3-(2H-tetrazole-5-yl)-propenone (5-CITEP), followed by the first clinical trial of a triazole-DKA derivative, S-1360, led to the development of two HIV IN inhibitor drugs, raltegravir and elvitegravir (Figure 12.3) (Egbertson et al. 2007). The general structure of DKAs is comprised of three structural components, a common beta-diketopropyl linker, an $R_1$ aromatic group ring, and an $R_2$ group (acidic, carboxyl, tetrazole, or triazole). Almost all DKA derivatives need these three components for anti-IN activity. Many derivatives, such as carbazolone DKAs, indole DKAs, catechol DKA hydrides, azido DKAs, bifunctional DKAs, heteroaryldiketohexanoic acids, triketoacids, hydroxynaphthyridines, and nucleobase DKAs, were synthesized and their biological activities for inhibiting 3′-processing and strand transfer of the IN mechanism were reported. Select examples of DKA derivatives are shown in Figure 12.11 (Johnson et al. 2004).

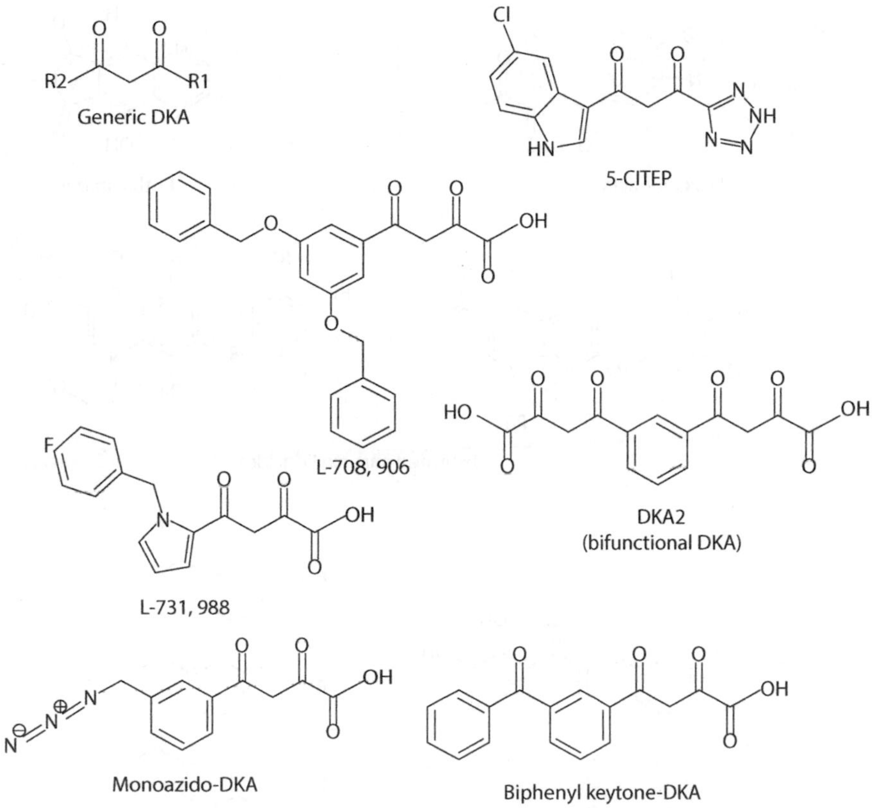

Generic DKA

5-ClTEP

L-708, 906

DKA2
(bifunctional DKA)

L-731, 988

Monoazido-DKA

Biphenyl keytone-DKA

**FIGURE 12.11** Diketo acid (DKA) lead compounds as inhibitors of HIV integrase (5-ClTEP: 1-(5-chloroindol-3-yl)-3-hydroxy-3-(2H-tetrazol-5-yl)-propenone).

Exhaustive SAR and molecular modeling efforts at two pharmaceutical companies, Shionogi & Co., Ltd. and Merck Research Laboratory, resulted in the identification of many DKA-derived lead compounds inhibiting the strand transfer mechanism of IN. Among them, a triazole derivative (S-1360) and a naphthyridine carboxamide derivative (L-870,810) progressed to various stages of clinical trials. A number of QSAR studies using various approaches on several classes of DKA derivatives supported the pharmacopore mapping, docking, and molecular modeling efforts and expedited the discovery of novel IN inhibitors. A discussion of these QSAR studies is presented in this section (Figure 12.12).

Barreca et al. (2005) published a 3D-QSAR study for 33 DKA derivatives for the inhibition of the strand transfer mechanism (Grobler et al. 2002; Hazuda et al. 2004a; Johnson et al. 2004; Pais et al. 2002; Wai et al. 2000; Zhuang et al. 2003) in the development of SAR pharmacophore hypotheses. The best 3D pharmacophore model consisted of four features: a hydrophobic aromatic region, two hydrogen-bond acceptors, and one-hydrogen bond donor site. One molecule in the dataset failed to map to the hydrophobic aromatic region, which prompted further molecular modeling studies to add hydrophobic functional groups. The predicted enzyme inhibitory

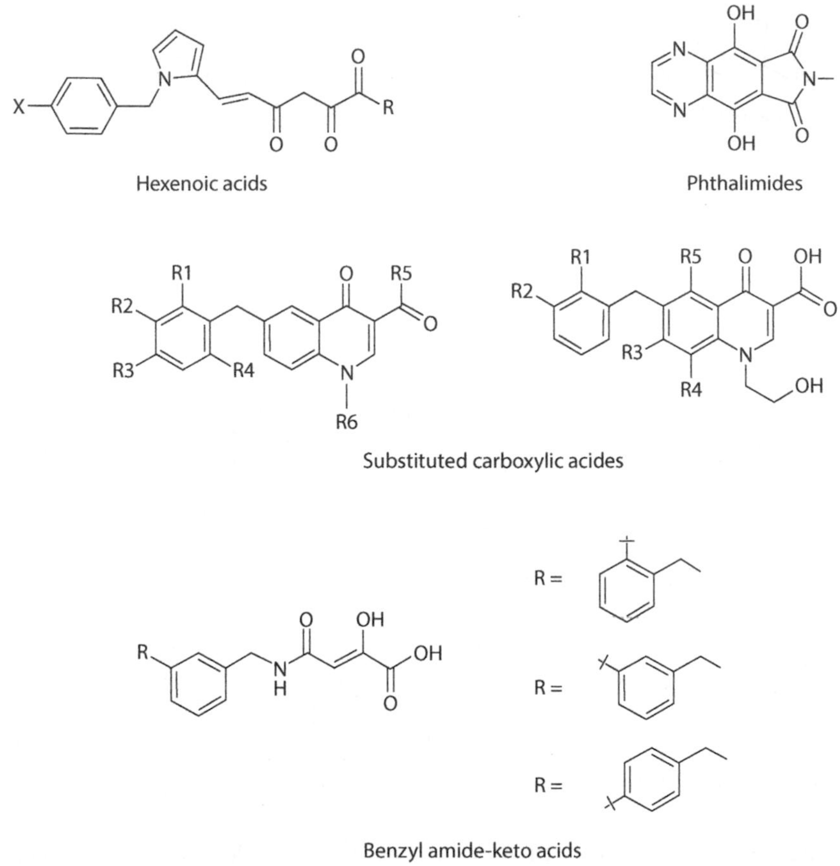

Hexenoic acids

Phthalimides

Substituted carboxylic acides

Benzyl amide-keto acids

**FIGURE 12.12** Substituted diketo acids studied by QSAR modeling.

activity of a new N-benzyl derivative closely matched its experimentally measured activity. This molecular modeling success led to the design and synthesis of 12 new benzyl-indole derivatives. The two most potent indole derivatives had experimentally measured strand transfer inhibition activities (0.01 μM and 0.004 μM). This study is a good example of the application of QSAR in drug design.

A 2D-QSAR study on 12 aryl dioxohexenoic acids was reported by Ravichandran et al. (2008). The biological activities were collected from SAR data, which measured 3′-processing and strand transfer inhibition (Costi et al. 2004b). MLR models revealed that an increase in hydrophobicity results in a negative inhibitory effect, whereas high electropositive chemical groups and heat of formation are positive contributions to the biological activity.

Dessalew (2009) published a 2D-QSAR study on 37 phthalimide-based DKA derivatives using strand transfer inhibitory activities (Verschueren et al. 2005). 2D-QSAR models were derived using MLR with pairwise correlation analysis. It was noted that the total dipole moment, molar refractivity, and Balaban topological index parameters of the R-group substituents have a positive effect on the inhibitory activity.

The Verloop's length and width steric parameter and Weiner's topological index descriptors were found to have a negative effect on the biological activity of these derivatives.

Cheng et al. (2010) reported a 2D-QSAR study of 62 carboxylic acid derivatives using SAR data that measured inhibition of the strand transfer reaction (Pasquini et al. 2008; Sato et al. 2006, 2009; Sechi et al. 2009a, 2009b). In this comprehensive study, the authors developed MLR, BPNN, and SVM models using 895 chemical descriptors such as 2D autocorrelations; topological, geometrical, radial distribution function (RDF); geometry, topology, and atoms-weighted assembly (GETAWAY); weighted holistic invariant molecular (WHIM); and 3D molecular representation of structure based on electron diffraction (3D-MoRSE) descriptors. The best predictive QSAR model using 3D-MoRSE descriptors was based on a combination of SVM with replacement method, indicating the importance of atomic van der Waals volumes and polarizabilities for improving the inhibitory activity of these compounds.

Vengurlekar et al. (2010) reported a 2D- and 3D-QSAR study on 24 benzyl amide-keto acids using a SAR dataset that measured strand transfer inhibition (Walker et al. 2007). These 2D-QSAR models were developed using MLR, PLS, and PCR with stepwise variable selection. The models were validated using leave-one-out cross validation. Of the three 2D-QSAR models, the MLR model was the most optimal. The descriptors in the 2D-QSAR model indicate high dipole, polarizability, and hydrogen count contribute positively to the biological activity. An MFA-based 3D-QSAR study on this dataset was also performed by these authors. Contour plots of the KNN-MFA 3D-QSAR model indicated that freedom of an amide C-N bond as opposed to an acyclic C-C bond is required for biological activity. In addition, the authors mentioned that the presence of less bulky groups on the benzene ring would lead to favorable activity. These results complement the results of the 2D-QSAR study.

A 3D-QSAR study of 48 quinolone carboxylic acid derivatives using CoMFA and CoMSIA was reported by Lu et al. (2010) for the inhibition of the strand transfer mechanism (Sato et al. 2009). The best PLS-based CoMFA and CoMSIA models provided insight as to the regions of potential atomic substituents to increase inhibitory activities of quinolone carboxylic acid derivatives.

Satpathy and Ghosh (2011) analyzed 26 quinolone carboxylic acids (elvitegravir analogs) reported by Serrao et al. (2009). The 2D-QSAR models highlighted the importance of topological descriptors for improved inhibitory activity of these compounds.

Recently, Sharma et al. (2011) reported SAR data for strand transfer enzyme inhibitory activity of a novel series of 3-keto salicylic acid chalcones and related amides as a diketo isostere. These second-generation IN inhibitors were developed to design new drugs active against raltegravir-resistant mutant viral strains. The same authors also performed QSAR studies on this dataset using 53 synthesized compounds. Statistically significant CoMFA and CoMSIA 3D-QSAR models were obtained using alignment based on pharmacophore mapping. Most of these novel compounds inhibited HIV replication in cell culture with moderate antiviral activity and could be further developed.

Dubey et al. (2011) analyzed 3′-processing and strand transfer inhibitory activities of 25 novel bifunctional quinolinoyl diketo acids (Bona et al. 2006; Di Santo et al. 2006; Sato et al. 2006). 2D-QSAR models developed using MLR, PCA, PCR, and PLS analyses were reported. It was found that electrostatic descriptors played an important role in regulating the activity of these derivatives. In continuation of their work, Dubey et al. (2012) also analyzed the performance of various types of descriptors (WHIM, topological, RDF, 2D autocorrelation, E-state indices, GETAWAY, and 3D-MoRSE) in the same dataset (Bona et al. 2006; Di Santo et al. 2006; Sato et al. 2006). It was noted that 3D-MoRSE descriptors were particularly significant in the PCA and MLR models. These studies suggested that the inclusion of electronegative groups such as halogens and nitrogen-, sulfur-, and oxygen-containing functional groups attached to aromatic rings would be helpful in designing novel potent inhibitors belonging to this class.

A dataset of 50 substituted quinolone carboxylic acids (Sato et al. 2009) was analyzed by Gupta and Madan (2012). 2D-QSAR models were developed using classification (DT and RF) and regression (MLR, PLS, and PCR) methods. Alignment-independent descriptors were found to be the most important descriptors by all methods.

Recently, de Melo and Ferreira (2012) developed a 4D-QSAR model for 85 IN strand transfer inhibitors with a DKA substructure (de Melo and Ferreira 2009; Egbertson et al. 2007; Guare et al. 2006; Hazuda et al. 2000; Petrocchi et al. 2007; Summa et al. 2006; Wai et al. 2000; Zhuang et al. 2003). They used a new approach named LQTA-QSAR (LQTA, Laboratório de Quimiometria Teórica e Aplicada) and derived robust QSAR models with good external and internal validation. The model was comparable to other 2D- and 3D-QSAR models developed for other classes of DKA derivatives and was also able to provide mechanistic interpretation.

#### 12.5.2.4 Naphthyridines and Pyrimidinone Carboxamides

The 1,3-DKA moiety essential for inhibitory activity of DKAs is biologically not very stable, and the need for its suitable replacement led to the discovery of naphthyridine derivatives. The nitrogen atom of naphthyridine is thought to mimic the carboxylate anion of DKAs, while the enol tautomer of one of the ketones is represented by an isostereic phenolic hydroxyl group in naphthyridine derivatives. Among all of the DKA derivatives, compounds containing nucleobase (pyrimidine and purine) scaffolds were also recognized as potent inhibitors for the 3′-processing and strand transfer mechanisms (Gardelli et al. 2007). Extensive research on these inhibitors resulted in the development of raltegravir (a pyrimidinone carboxamide derivative) and elvitegravir (a naphthyridine carboxamide derivative). QSAR modeling efforts discussed here supported the synthetic efforts (Figure 12.13).

A 2D-QSAR study was reported by Ravichandran et al. (2010) using SAR data of 67 oxadiazole substituted naphthyridine derivatives, inhibitors of the IN strand transfer mechanism (Johns et al. 2009a). The best MLR-based 2D-QSAR model indicated that high electronegativity, low dielectric constant, and highly branched, unsaturated, and long chain groups are conducive for high enzyme inhibitory activity of these derivatives. Veerasamy et al. (2011) developed a 2D-QSAR model to analyze the antiviral inhibitory activity of 63 compounds from the same series of

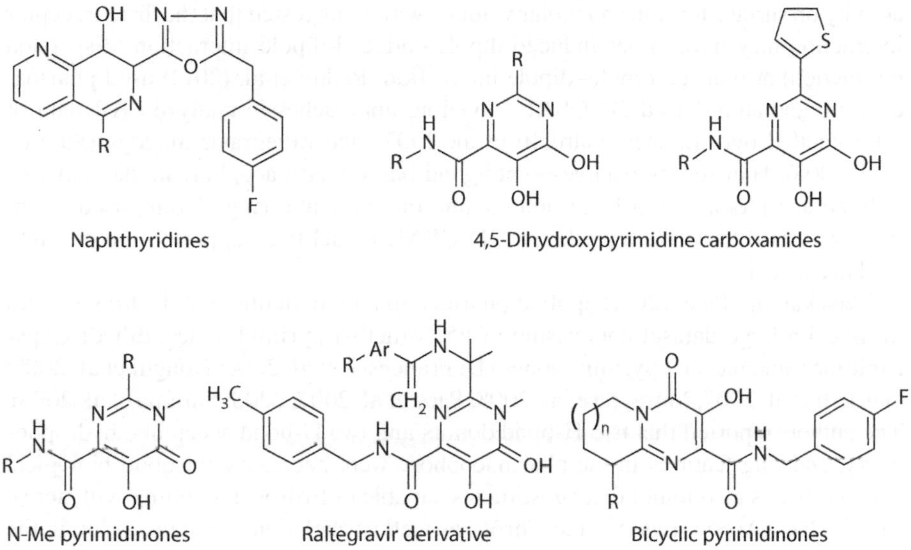

Naphthyridines 4,5-Dihydroxypyrimidine carboxamides

N-Me pyrimidinones Raltegravir derivative Bicyclic pyrimidinones

**FIGURE 12.13** Substituted naphthyridines and pyrimidinone carboxamides studied by QSAR modeling.

oxadiazole-substituted naphthyridines (Johns et al. 2009b) for which Ravichandran et al. (2010) also analyzed strand transfer enzyme inhibitory activity. The authors reported that the shape index, partition coefficient, and solvent accessible surface are related descriptors and play an important role in governing the anti-HIV activities of these derivatives. These two QSAR studies (Ravichandran et al. 2010; Veerasamy et al. 2011) indicated the importance of different physicochemical parameters for improving the enzyme inhibitory activity and anti-HIV activity of these derivatives. More studies comparing two different biological end points are required to delineate the similarities and differences in these two biological activities.

Jalali-Heravia and Ebrahimi-Najafabadia (2011) studied 84 oxadiazole- and triazole-substituted naphthyridines (Johns et al. 2009a, 2009b) using a different approach. Artificial neural network (ANN) was combined with ladder particle swarm optimization (LPSO) to develop 2D-QSAR models. Their results indicated superiority of LPSO-ANN models over LPSO-MLR and LPSO-MLR-ANN. The best model indicated that the atomic mass, molecular size, and electronic structure of the molecules play a significant role in inhibitory behavior of these compounds.

A 2D-QSAR study on 33 dihydroxypyrimidine carboxamide derivatives was reported by de Melo and Ferreira (2009). The SAR data for strand transfer inhibitory activity was taken from Petrocchi et al. (2007). The descriptors selected by the optimal PLS model that have a positive influence on the biological activity values were related to E-HOMO and E-state indices, whereas the total energy descriptor has a negative influence on the biological activity of these compounds.

Kaushik et al. (2011a) studied 51 N-methyl pyrimidones (Gardelli et al. 2007) to derive 2D-QSAR models. The authors noted positive dependence of the inhibitory activity on surface tension and molar volume, which suggested that the drug–receptor interaction may have either induced-dipole–induced-dipole interaction (dispersion interaction) or induced dipole–dipole interaction. Reddy et al. (2012) used pharmacophore and atom-based 3D-QSAR modeling approaches to analyze SAR data of 50 N-methyl pyrimidones (Gardelli et al. 2007) and raltegravir analogs (Summa et al. 2008). They reported a five-point ligand-based pharmacophore model with two H-bond acceptors, two H-bond donors, and one aromatic ring. Atom-based alignment was used to derive a predictive 3D-QSAR model that supported the pharmacophore model.

Telvekar and Patel (2011) applied pharmacophore modeling and docking studies to model a large dataset comprising of 65 N-methyl pyrimidinones, dihydroxypyrimidines, and bicyclic pyrimidinones (Di Francesco et al. 2008; Donghi et al. 2009; Gardelli et al. 2007; Muraglia et al. 2008; Pace et al. 2007, 2008; Summa et al. 2008). The authors reported that two H-bond donors and two H-bond acceptors, hydrophobicity, and ring features in the pharmacophore were necessary for good biological activity. It was also found that substituents capable of hydrogen bonding with active site residues (Asp64, Lys159, and Thr66) as well as chelation of the metal ion in the active site of IN are important features for developing a good pyrimidinone-based inhibitor.

### 12.5.3 Diverse Chemical Classes

SAR data of structurally diverse classes of catechol and non-catechol inhibitors (such as tyrphostins, coumarins, sulfonamides, chicoric acids, tetracyclines, arylamides, naphthalenes, thiazolothiazepines, curcumins, salicylhydrazines, styrylquinolines, DKAs, quinoline carboxamides, and pyrimidinones) were combined into large datasets. Many QSAR studies using different 2D-, 3D-, and 4D-QSAR approaches were performed. Results of these studies discussed here suggested that the compounds in different groups might bind at different binding sites. These results were in agreement with experimental studies.

IN inhibitory activity and antiviral data of two structurally diverse classes—14 salicylhydrazines (Neamati et al. 1998) and 28 tyrphostins (Mazumder et al. 1995a)—were analyzed by Yuan and Parrill (2000). The results of this study suggested that salicylhydrazines and tyrphostins might have different binding sites in HIV IN. QSAR models for the two different classes of inhibitors have common descriptors but different coefficients and signs, which indicated that the structural features characterized by these descriptors have different effects on the interaction between the inhibitor and protein for the two different structural groups. A QSAR equation for the combined classes resulted in poor correlation for tyrphostins, indicating a possible difference in the way this structural class interacted with the IN enzyme. This difference can be manifested either through interactions with different amino acid residues that line a common inhibitor-binding pocket, or it could arise from these structurally different inhibitors interacting with two completely overlapping sites. Two of the crystallographic studies on the interactions of IN proteins with HIV IN inhibitors provided support for the latter possibility (Yuan and Parrill 2000).

3′-processing inhibitory activity data of 11 structurally diverse classes (174 compounds) were analyzed by Yuan and Parrill (2002). 2D-QSAR models were derived using topological, geometrical, and electronic descriptors. GFA was used to select the best descriptors for QSAR modeling. Descriptor-based clustering analysis indicated that the 11 structural classes studied belong to two groups, group I consisting of five classes: tyrphostins (Mazumder et al. 1995a), coumarins (Zhao et al. 1997a), sulfonamides (Nicklaus et al. 1997), chicoric acids (Lin et al. 1999), and tetracyclines (Neamati et al. 1997b); and group II consisting of six classes: arylamides and naphthalenes (Zhao et al. 1997b), thiazolothiazepines (Neamati et al. 1999), curcumins (Mazumder et al. 1997), salicylhydrazines (Neamati et al. 1998), styrylquinolines (Zouhiri et al. 2000), and depsides and depsidones (Neamati et al. 1997a). The compounds in each group were believed to interact at two different binding sites of HIV IN. These observations supported the results of an earlier QSAR study (Yuan and Parrill 2000) in which it was hypothesized that styrylquinolines and tyrphostins bind at different sites. It was suggested that the six structural classes comprising the larger cluster (group 2) may bind near the metal ion in a fashion similar to that observed in one publicly available cocrystal structure of an inhibitor bound to HIV IN (Yuan and Parrill 2002).

Makhija and Kulkarni (2002b) analyzed 3′-processing and strand transfer inhibitory activities of 81 molecules from diverse classes of catechols and non-catechols: catechols consisting of chicoric acids (Lin et al. 1999; Mazumder et al. 1997; Zhao et al. 1997b) and non-catechols consisting of salicylhydrazines (Neamati et al. 1998), lichen acids (Neamati et al. 1997a), coumarins (Zhao et al. 1997a), quinones (Mazumder et al. 1996), hydrazides (Zhao et al. 1997c), and thiazolothiazepines (Neamati et al. 1999). QSAR models were generated separately for catechols and non-catechols. Models generated for catechols showed that electronic, shape-related, and thermodynamic parameters were important for their activity, whereas those developed for non-catechols showed that spatial, structural, and thermodynamic properties played an important role. The results also indicated that for non-catechols, these properties were mostly important for 3′-processing inhibition. For strand transfer, apart from spatial and thermodynamic descriptors, structural parameters such as number of H-bond donor groups were also important. Molecules containing a catechol substructure act mainly through chelation of divalent metal ions. Hence, apart from electronic descriptors, shape-related physicochemical properties based on MSA and thermodynamic descriptors describing desolvation phenomenon during binding were also found to contribute significantly toward inhibition of 3′-processing and strand transfer activities.

An MFA-based 3D-QSAR study on 11 classes of HIV IN inhibitors (60 compounds) was reported by Yuan and Parrill (2005). The biological activities of all the structures were experimentally derived for the inhibition of the 3′-processing mechanism. HC analysis was performed on the 11 classes of inhibitors using a set of approximately 200 topological and geometric descriptors that resulted in two distinct groups of inhibitor molecules: group I consisting of tyrphostins (Mazumder et al. 1995a), coumarins (Zhao et al. 1997a), sulfonamides (Nicklaus et al. 1997), chicoric acids (Lin et al. 1999), and tetracyclines (Neamati et al. 1997b); and group II consisting of arylamides and naphthalenes (Zhao et al. 1997b), thiazolothiazepines

(Neamati et al. 1999), curcumins (Mazumder et al. 1997), styrylquinolines (Zouhiri et al. 2000), salicylhydrazines (Neamati et al. 1998), and depsides and depsidones (Neamati et al. 1997a). The selected descriptors in the models indicated that the two groups of inhibitors have different binding modes. These models were used to predict the activities of the test molecules, which showed that the predicted activities were in good agreement with their observed ones.

3D-QSAR models developed using CoMFA and CoMSIA techniques in 11 classes (89 compounds) of HIV IN inhibitors were reported by Nunthaboot et al. (2006). These 11 classes of compounds known to inhibit 3′-processing were coumarins (Mao et al. 2002; Zhao et al. 1997a), chicoric acids (Lin et al. 1999), hydrazides (Zhao et al. 1997c), carbonyl derivatives (Maurer et al. 2000), styrylquinolines (Bénard et al. 2004; Zouhiri et al. 2000), curcumins (Mazumder et al. 1997), cinnamoyls (Artico et al. 1998), 2,6-bis(3,4,5-trihydroxybenzylydene) derivatives of cyclohexanones (Costi et al. 2004a), salicylhydrazines (Neamati et al. 1998), tetracyclines (Neamati et al. 1997b), and 3-aryl-1,3-diketo-containing compounds (Pais et al. 2002). Based on their models, Nunthaboot et al. (2006) concluded that larger substituents in a plane of the indole ring and small bulky groups at the tetrazole ring with respect to the 5-CITEP inhibitor may improve the inhibition of 3′-processing, and that functional groups with a higher electron density may result in better interaction with the metal ion in the IN active site. The CoMSIA models indicated the significance of hydrogen bond interactions between the ligand and the IN protein.

Iyer and Hopfinger (2007) reported a 4D-QSAR analysis for a set of 213 molecules belonging to 12 structurally diverse classes of IN inhibitors: cinnamoyl derivatives (Artico et al. 1998; Burke et al. 1995), tyrphostins (Mazumder et al. 1995a), coumarins (Zhao et al. 1997a), aromatic sulfonamides (Nicklaus et al. 1997), chicoric acids (Lin et al. 1999), tetracyclines (Neamati et al. 1997b), arylamides and naphthalenes (Zhao et al. 1997b), thiazolothiazepines (Neamati et al. 1999), curcumins (Mazumder et al. 1997), salicylhydrazines (Neamati et al. 1998), styrylquinolines (Zouhiri et al. 2000), and depsides and depsidones (Neamati et al. 1997a), which are known to inhibit the 3′-processing mechanism of IN. The molecules were divided into a training set comprising 148 molecules and a test set comprising 65 molecules across all the 12 classes. The test set molecules were selected such that the biological activities were well represented. The 4D-fingerprint descriptors and intramolecular solute, dissolution, solvation, and ClogP descriptors were calculated for these molecules. To determine groups of molecules that share key structural features, partitioning around medoids and divisive HC was performed using the two most potent molecules in each structural class with their 4D-fingerprint descriptors. Clustering analysis revealed three main clusters, of which cluster I was further subdivided into three smaller clusters: 1A, 1B, and 1C. Cluster 1A consisted of cinnamoyl derivatives, chicoric acids, and curcumins; cluster 1B of tyrphostins and salicylhydrazines; cluster 1C of depsides and depsidones; cluster II of coumarins; and cluster III of aromatic sulfonamides, tetracyclines, arylamides, thiazolothiazepines, and styrylquinolines. GFA was used to develop linear QSAR models for each of the clusters, which were internally validated using cross validation. Each of the models exhibited high correlation statistics and contained primarily 4D-fingerprint descriptors.

Saíz-Urra et al. (2007) performed a 2D-QSAR study on 172 compounds that all have experimentally derived biological activities for the inhibition of the 3′-processing mechanism. These compounds belong to 11 classes: tyrphostins (Mazumder et al. 1995a), coumarins (Zhao et al. 1997a), aromatic sulfonamides (Nicklaus et al. 1997), chicoric acids (Lin et al. 1999), tetracyclines (Neamati et al. 1997b), arylamide- and naphthalene-based compounds (Zhao et al. 1997b), thiazolothiazepines (Neamati et al. 1999), curcumins (Mazumder et al. 1997), salicylhydrazines (Neamati et al. 1998), styrylquinolines (Zouhiri et al. 2000), and depsides and depsidones (Neamati et al. 1997a). The authors calculated GETAWAY, Randic molecular profiles, geometrical, WHIM, RDF, and 3D-MoRSE descriptors. QSAR models were developed using each of the aforementioned classes of descriptors, and in each case, the best model that had the fewest parameters and the highest statistical significance was identified. The QSAR model involving the use of the GETAWAY descriptors explained more than 72.5% of the biological activity after the removal of outliers. The descriptors in the GETAWAY QSAR model were related to the polarizabilities, van der Waals volume, electronegativities, and the atomic mass; thus, the model suggested that a large van der Waals volume would be favorable and high electronegativity would be unfavorable for the inhibitory activity. Aromatic moiety common to many inhibitors has been proposed to interact with the divalent cation in a cation-π type interaction, and there is also a possibility of a typical charge–charge interaction between the metal ion and ionic or partial charges of the ligands.

Dhaked et al. (2009) reported a 3D-QSAR CoRIA study of 81 HIV IN inhibitors belonging to 13 classes that inhibit 3′-processing: arylamides and naphthalenes (Zhao et al. 1997b), geminal disulfones (Meadows et al. 2005), coumarins (Zhao et al. 1997a), salicylhydrazines (Neamati et al. 1998) and hydrazides (Zhao et al. 1997c), indole β-DKAs (Sechi et al. 2004), thiazolothiazepines (Neamati et al. 1999), quinolinone-3-carboxamides (Dayam et al. 2006b), curcumins (Mazumder et al. 1997), mercaptobenzenesulfonamides (Kuo et al. 2004), sulfonamides (Nicklaus et al. 1997), tyrphostins (Mazumder et al. 1995a), diarylsulfones (Neamati et al. 1997c), and depsides and depsidones (Neamati et al. 1997a). All QSAR models were generated using GA/PLS. Three QSAR models were developed using different descriptor combinations: model 1 was based on the Coulombic and van der Waals interaction energies between the ligands and amino acid residues in the active site, model 2 added descriptors to model one that reflected ligand-receptor binding, and model 3 further added the total nonbonded (van der Waals and Coulombic) interaction energy descriptor to produce the best statistics. In each model, almost the same set of amino acid residues appeared, indicating the significance of these residues in protein–ligand interactions.

Liu et al. (2011) performed multitarget QSAR modeling on the same dataset of HIV IN inhibitors analyzed by Iyer and Hopfinger (2007). They used 2D descriptors and a multitask learning approach to design HIV–HCV co-inhibitors. This approach provided an efficient way to identify and design inhibitors that simultaneously and selectively could bind to multiple targets from multiple viruses with high affinity.

Sharma et al. (2012) used homology-model-guided 3D-QSAR (CoMFA and CoMSIA) to study strand transfer inhibitory activity of a large dataset consisting of 301 compounds. These compounds were 3-keto salicylic acid chalcones/amides

(Patil et al. 2007; Sharma et al. 2011), N-methyl-4-hydroxypyrimidinone carboxamides (Gardelli et al. 2007; Summa et al. 2008), 4-quinolone-3-carboxylic acid carboxamides (Pasquini et al. 2008; Sato et al. 2006), dihydroxypyrimidine carboxamides (Pace et al. 2007; Petrocchi et al. 2007; Summa et al. 2006), bicyclic pyrimidinones (Muraglia et al. 2008), tert-butyl-N-methyl pyrimidones (Di Francesco et al. 2008), hydroxypyrrolinone carboxamides (Pace et al. 2008), pyridopyrimidinone carboxamides (Donghi et al. 2009), pyrazinopyrimidine carboxamides (Petrocchi et al. 2009), N-methyl pyrimidones (Nizi et al. 2009), and 2-pyrrolidinyl N-methyl pyrimidones (Ferrara et al. 2010). Their study showed that docking-based alignment was better than atom-based alignment for developing good predictive 3D-QSAR models.

Yan et al. (2012) used SVM to classify 1,336 active and weakly active IN strand transfer inhibitors collected from various literature sources. Their results indicated that electrostatic properties, van der Waals surface area, H-bonds, and number of fluorine atoms based molecular descriptors play an important role in governing IN inhibitory activity of compounds.

### 12.5.4 Overview of QSAR Studies

A comprehensive overview of QSAR studies on HIV IN inhibitors illustrates that QSAR models are classified according to the generation of molecular descriptors: 2D-, 3D-, and 4D-QSAR. Linear and nonlinear modeling methods have been applied to develop these QSAR models, with the majority of the models derived from linear statistical methods such as MLR and PLS. Each of the published QSAR models provided an insight into the distinct chemical features of HIV IN inhibitors crucial for their biological activity.

2D-QSAR models developed for HIV IN inhibitors fall into two categories: classification models and regression models, where the former are developed using HC, LDA, and PCA, and the latter are developed using MLR, PCR, and PLS. The nonlinear regression models are based on BPNN and SVM. In most of the 2D-QSAR models, electronegativit- and hydrophobicity-related descriptors have been found to be influential in determining HIV-1 IN inhibition activity. Most of the 2D-QSAR models derived involved MLR analysis. Cheng et al. (2010) compared MLR with the nonlinear methods of BPNN and SVM in their study and found that the QSAR models involving the use of SVM had the highest predictive ability compared to MLR and BPNN. However, in one instance, the 2D-QSAR model by Sahu et al. (2007) was found to be very useful to rationalize the design of new structures of caffeoyl naphthalene sulfonamide derivatives.

Relatively few 3D-QSAR models have been reported for HIV IN inhibitors as compared to 2D-QSAR models. This could be due to the lack of a complete crystal structure and the unavailability of the experimental binding pocket description; thus, the binding conformation of the inhibitors are unknown (Liao and Nicklaus 2011). 3D-QSAR relies on the 3D conformation of molecules and depends on experimentally derived, biologically active conformers to develop reliable models. Analysis of these interactions provided information on structural regions that positively or negatively affect the biological activity of the compounds studied.

Many of the 3D-QSAR studies involved the use of PLS-based CoMFA and CoMSIA to construct models to understand the steric and electrostatic interactions between the inhibitors and IN.

Only two studies have been reported using the 4D-QSAR approach. In the first study, Iyer and Hopfinger (2007) used 4D fingerprints to capture the 3D shape and conformation of molecules without the constraints of alignment, as well as information about molecular flexibility and conformational entropy. The authors mentioned that 4D descriptors are abstract and do not lead to precise chemical interpretation. However, it seemed that even if 4D descriptors could not provide any mechanistic interpretations, these models could be used as a virtual screening tool to identify potentially active novel compounds for synthesis. In the second study, de Melo and Ferreira (2012) derived a 4D-QSAR model for 85 IN strand transfer inhibitors with a DKA substructure using a new approach named LQTA-QSAR. This was comparable to some 2D- and 3D-QSAR models developed for other classes of DKA derivatives and was able to provide mechanistic interpretation.

## 12.6 STUDIES ON IN–LEDGF/p75 INHIBITORS

Recently, some studies have been reported on compounds that inhibit the interaction between the viral protein IN and host protein LEDGF/p75. These compounds belong to the following classes: peptide derivatives, thiazolidinone benzoic acids, quinoline acetic acids (LEDGIN), hydroxyquinolines, indolebutenoic acids, indolebenzamides (CAB), hydroxybenzamides, acylhydrazones, and pyridyl dibenzylideneacetone (pyr-dba). SAR studies performed on these IN–LEDGF/p75 inhibitors are discussed here (Figures 12.14 and 12.15).

Hayouka et al. (2007, 2010) used rational drug design and combinatorial library screening to synthesize a few cyclic peptides derived from the amino acid sequence (residues 361–370) of the LEDGF/p75 cellular cofactor. These compounds showed anti-HIV IN activity and disrupted the interactions between IN and LEDGF/p75. It was shown that the ring size and linker structure of cyclic peptides have a huge effect on the conformation, binding, and activity of the peptides. The NMR structure of one of the cyclic peptides showed that it mimicked the bioactive conformation of the IN-binding loop in LEDGF/p75, suggesting that restricting linear peptides to their bioactive confirmation could be a useful strategy for improving their stability and activity.

Peptides and natural products are used to model protein–protein interactions in several therapeutic areas, but their physicochemical properties are less suitable for drug development. Therefore, a few research groups have focused on developing small-molecule inhibitors for disrupting interactions between IN and LEDGF/p75.

Benzoic acid derivatives were the first small-molecule inhibitors found to disrupt IN–LEDGF/p75 interactions (Du et al. 2008). The most active compound of this series, a thiazolidinone-based benzoic acid derivative (D77), was reported to inhibit IN–LEDGF/p75 binding, disturb the nuclear distribution of IN, and exhibit antiviral activity. Molecular docking studies and site-directed mutagenesis identified IN-binding site residues (Gln95, Thr125, Trp131, and Thr174) that are involved in the interaction.

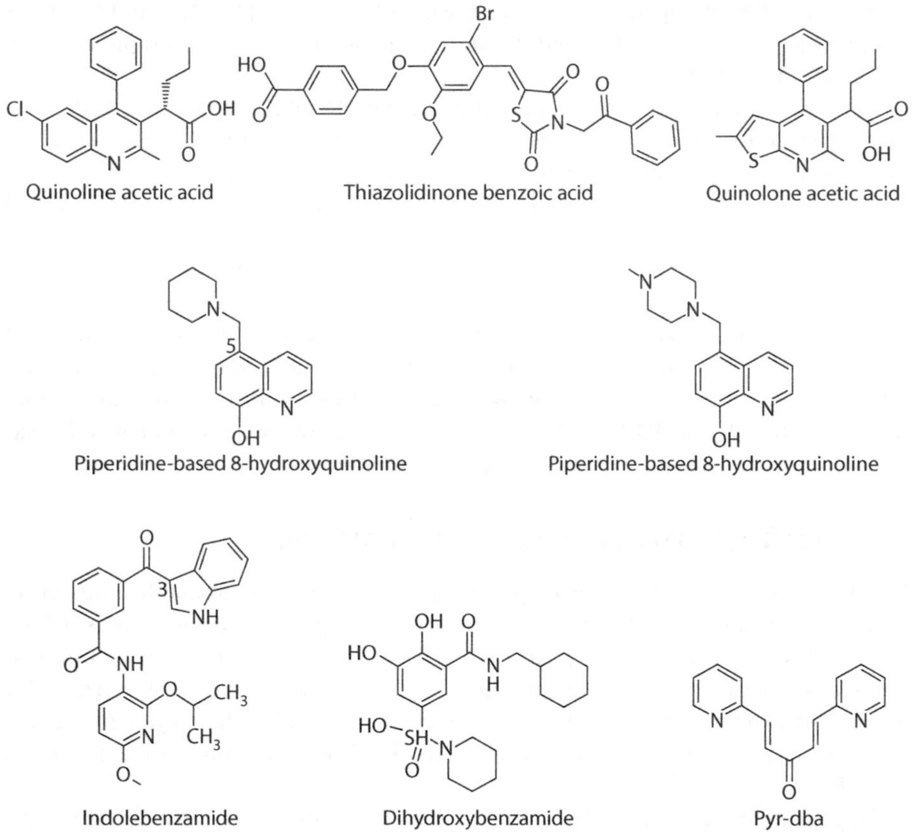

Quinoline acetic acid Thiazolidinone benzoic acid Quinolone acetic acid

Piperidine-based 8-hydroxyquinoline Piperidine-based 8-hydroxyquinoline

Indolebenzamide Dihydroxybenzamide Pyr-dba

**FIGURE 12.14** Examples of inhibitors of the IN–LEDGF/p75 interaction (Pyr-dba: pyridinyl analog of dibenzylideneacetone).

To further identify new leads, Christ et al. (2010) reported quinoline-based acetic acid derivatives (LEDGIN) as potent inhibitors of the IN–LEDGF/p75 interaction. Crystal structures of LEDGIN showed that these inhibitors bind in the LEDGF/p75-binding pocket. Nitrogen in the main chain residues, Glu170 and His171, of the IN enzyme were found to form H-bonds with the carboxyl moiety of these compounds. These interactions were similar to those found in the IN–LEDGF/p75 active site. LEDGIN are the first class of allosteric inhibitors possessing dual mode of action (i.e., interruption of the interactions between IN–LEDGF/p75 and inhibition of IN). Desimmie et al. (2012) used structure-based drug design to further improve the biological activity of LEDGIN derivatives. They reported that LEDGIN do not show any cross-resistance with currently approved IN inhibitors and might serve as second-generation IN strand transfer inhibitors or dual inhibitors of strand transfer and disruption of the LEDGF/p75–IN interaction.

Serrao et al. (2013) used a fragment-based drug design approach to develop a series of 8-hydroxyquinoline derivatives that are the part of several FDA-approved drugs. Over the past decade, several quinoline-related scaffolds have been explored

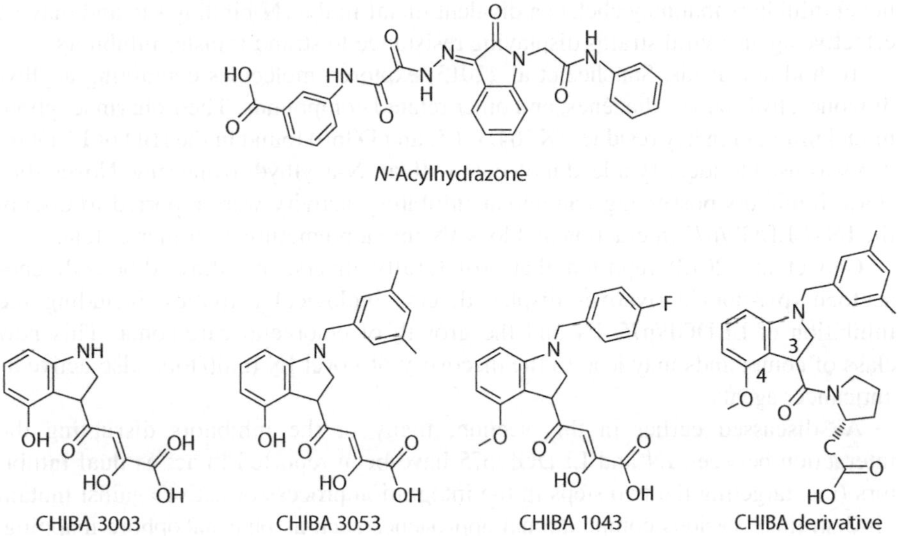

**FIGURE 12.15** N-acylhydrazone and indolebutenoic acids (CHIBA) as inhibitors of the IN–LEDGF/p75 interaction.

for the development of IN inhibitors. Modifications at the C-5 position yielded potent inhibitors with high potency and low cytotoxicity. The lead compounds of this series were reported to be drug-like, have low molecular weights, and were suitable for substituent modifications to enhance the potency and selectivity.

In a series of studies, De Luca et al. (2008, 2009, 2010, 2011a, 2011b, 2013) used various computational approaches such as pharmacophore mapping, virtual screening, docking, and fragment-based methods to study indolebutenoic acid derivatives (CHIBA) as potent dual inhibitors disrupting the IN–LEDGF/p75 interaction. These studies described the main contact points between Ile365-Ile366 of IN–LEDGF/p75 and key amino acid residues of IN. Replacement of the 1,3-diketo moiety in CHIBA-3003 and some other key modifications resulted in the most potent compound in this series: CHIBA-3053 (1-[3,5-dimethylbenzyl)-4-methoxy-1H-indole-3-carbonyl] pyrrolidine-2-carboxylic acid).

Cavalluzzo et al. (2012) reported a new class of indolebenzamides (CAB) inhibiting LEDGF/p75. These benzamide derivatives mimicked the alpha-3 helix of IN, a major contributor for the interactions between the IBD of LEDGF/p75 and CCD of IN.

In search of structurally and mechanistically novel dual inhibitors targeting the IN–LEDGF/p75 interaction, Fan et al. (2011) used a scaffold hopping approach. They merged the salicylic acid and catechol pharmacophores to generate a new 2,3-dihydroxy benzamide scaffold. It was found that a heteroaromatic functional group on the carboxamide portion and the piperidinyl sulfonyl substituent at the phenyl ring improve the biological activity of the lead compound. Molecular modeling studies showed that these new inhibitors inhibited the catalytic domain of IN and interactions between IN and LEDGF/p75 proteins. This scaffold may produce novel inhibitors that may chelate a divalent metal in the IN-binding site and may be effective against viral strains displaying resistance to strand transfer inhibitors.

To find new leads, Sanchez et al. (2013) explored molecules containing acylhydrazones, hydrazines, diazenes, and other related compounds. Their pharmacophore model mimicking key residues (K364, I365, and D366) found in the IBD of LEDGF/p75 was used to identify a lead molecule with an N-acylhydrazone ring. Novel allosteric inhibitors possessing nanomolar inhibitory activity were reported to disrupt the IN–LEDGF/p75 interaction and lock IN into a premature multimeric state.

Cao et al. (2012) reported that structurally diverse pyridinyl dibenzylideneacetone (pyr-dba) derivatives display diverse biological activities, including the inhibition of LEDGF/p75–IN and the growth of colorectal carcinoma. This new class of compounds may lead to the discovery of novel IN inhibitors, also active as anticancer agents.

As discussed earlier in this section, many of the inhibitors disrupting the interaction between IN and LEDGF/p75 have been reported to act as dual inhibitors (i.e., targeting the two steps in the integration process or active against mutant viral strains). Various computational approaches such as pharmacophore mapping, virtual screening, docking, and fragment-based methods have been used in developing these molecules; however, results of QSAR studies on IN–LEDGF/p75 inhibitors have not been reported in the literature, which could help in further optimizing these molecules.

## 12.7 HIV INTEGRASE INHIBITORS AS ANTICANCER AGENTS

As discussed earlier in this chapter, a study conducted by the International Collaboration on HIV and Cancer (2000) reported a sharp decline in the incidence of Kaposi sarcoma and non-Hodgkin lymphoma in HIV-infected people under HAART treatment. Anti-HIV drugs targeting different pathways in the life cycle of HIV provide a large source of potential anticancer drugs. Thus, several FDA-approved anti-HIV drugs and other HIV inhibitors abandoned early in clinical trials are being considered for repositioning as anticancer agents (Oprea and Mestres 2012). Two studies on IN inhibitors (Oprea and Mestres 2012; Zeng et al. 2012) are discussed here (Figure 12.16).

Raltegravir, the first FDA-approved HIV IN inhibitor drug, is being considered as a potential drug for treating certain types of cancers. Recent studies identified the DNA repair enzyme, mentase, as a potential target for cancer therapy (Allen et al. 2011; Lee et al. 2005). Structure-based virtual screening studies confirmed raltegravir as a mentase inhibitor at a dosage ten times greater than the currently approved maximum dose. A pilot study is ongoing to evaluate its use as an anticancer drug (Oprea et al. 2011).

Zeng et al. (2012) reported that the 1,6-naphthyridine scaffold has the potential to be developed as anti-HIV IN inhibitors as well as anticancer agents. 8-Hydroxy-1,6-naphthyridine-7 carboxamides were initially developed as potent antiviral agents (Hazuda et al. 2004a). One of the compounds of this class, L-870,810, showed good antiviral activity but exhibited an unfavorable pharmacological profile during initial clinical studies and hence was not pursued further. Zeng et al. (2012) noticed that

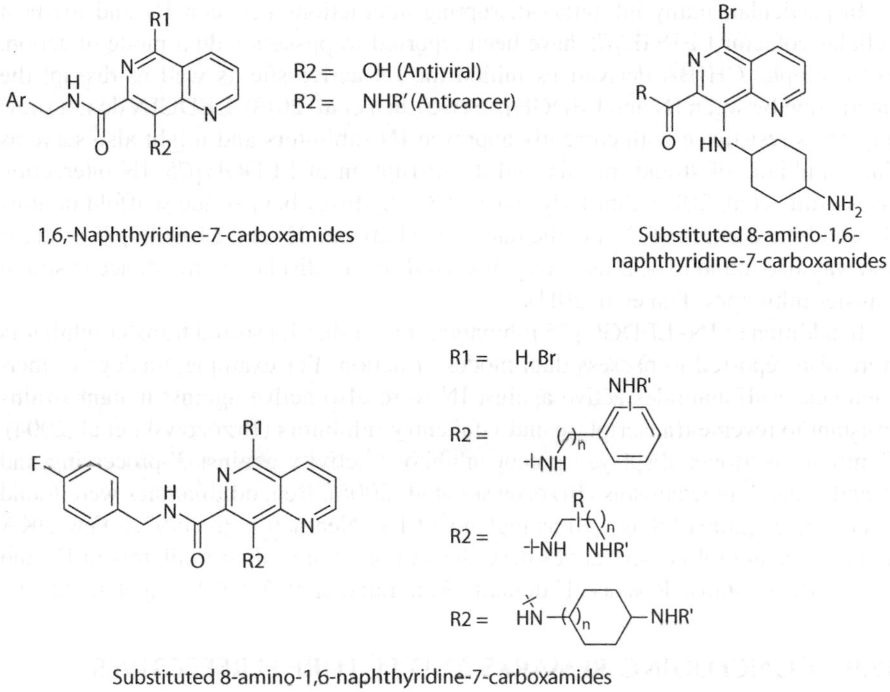

**FIGURE 12.16** Substituted naphthyridine carboxamide-based HIV integrase inhibitors repurposed as anticancer agents.

variations at the 5- and 8-positions of 1,6-naphthyridine-7-carboxamide resulted in a promising scaffold for repositioning anti-HIV IN inhibitors as anticancer agents. It was found that although 8-hydroxy substituted naphthyridines act as potent anti-HIV inhibitors, 8-amino derivatives exhibit significant cytotoxicity against several cancer cell lines and are effective inhibitors against select oncogenic kinases. Several of these compounds are under study to further understand their mechanism of action. This study showed that development of anti-HIV IN inhibitors, abandoned in early clinical phase trials, could lead to new anticancer agents.

Some hydrazines, coumarins, and pyr-dba derivatives were found to possess cellular toxicity and could be of interest for development as anticancer drugs (Cao et al. 2012).

## 12.8 DUAL INHIBITORS DESIGN: A NEW APPROACH

Designing a single drug to inhibit multiple targets is an increasingly important strategy to develop mechanistically and structurally novel HIV inhibitors (De Luca et al. 2010). Many examples discussed in this chapter show that several of the IN inhibitor molecules could act as dual inhibitors. It has been observed that, most of the time, modest modifications in an inhibitor molecule led to selectivity against one target over the other.

In particular, many inhibitors disrupting interactions between IN and the host cellular cofactor LEDGF/p75 have been reported to possess a dual mode of action. For example, CHIBA derivatives inhibit the IN active site as well as disrupt the interaction between IN and LEDGF/p75 (De Luca et al. 2013). LEDGIN do not show any cross-resistance with currently approved IN inhibitors and might also serve as dual inhibitors of strand transfer and the disruption of LEDGF/p75–IN interaction (Desimmie et al. 2012). Similarly, a new 2,3-dihydroxy benzamide scaffold inhibits the catalytic domain of IN and the interaction between IN and LEDGF/p75 proteins and was also found to be effective against viral strains displaying resistance to strand transfer inhibitors (Fan et al. 2011).

In addition to IN–LEDGF/p75 inhibitors, some other IN strand transfer inhibitors were also reported to possess dual modes of action. For example, analogs of mercaptobenzesulfonamides active against IN were also active against mutant strains resistant to reverse transcriptase and viral entry inhibitors (Brzozowski et al. 2004). Geminal disulfones displayed potent inhibitory activity against 3′-processing and strand transfer mechanisms (Brzozowski et al. 2008). Benzodithiazines were found to be active against IN as well as mutants of IN (Neamati et al. 1997c). Few DKA and hydroxyquinoline derivatives have shown potential as dual inhibitors of IN and reverse transcriptase RNAase H domain (Marchand et al. 2008; Wang et al. 2010).

## 12.9 CONCLUDING REMARKS AND FUTURE PERSPECTIVES

In this chapter, a comprehensive overview of QSAR studies published on HIV IN inhibitors since 2000 has been presented with a discussion of relevant SAR. Table 12.2 lists a summary of SAR data published on different types of inhibitors and QSAR studies (if any) performed on that data. Table 12.3 provides a summary of the modeling methods used by different research groups for developing QSAR models. A comparison of SAR and QSAR studies published between 1993 and 2012 shows that, although SAR studies on HIV IN inhibition started in 1993, it took almost seven years to begin reporting QSAR studies on any HIV IN inhibition data, with the exception of two studies (Figure 12.17). Thus, currently only about 40% of SAR data reported in the literature has been analyzed by QSAR studies (Table 12.2; Figure 12.17). Each of the published QSAR models reviewed in this chapter has provided a novel insight into the distinct chemical features of the HIV IN inhibitors. This highlights the need to analyze the remaining SAR data by using QSAR modeling approaches.

The results of QSAR modeling supported SAR studies and suggested a common mode of binding. QSAR models stressed the involvement of hydrophobic, steric, and electrostatic interactions. In many QSAR models, hydrophobic descriptors have not been found to be important, although there are certainly hydrophobic binding sites among the active sites of IN. It is possible that ligand-induced conformational changes (allosteric binding sites) modify the size of the hydrophobic cavity at the receptor binding site.

QSAR models are developed to better understand the physicochemical interactions between the ligand and drug target and to help improve existing molecules. These models can also be used to predict the biological activities of proposed compounds.

**TABLE 12.2**
**SAR Studies on HIV Integrase Inhibitors**

| Class | Sub-Class | Inhibitor Type | Inhibitory Activity | SAR Reference | QSAR Reference |
|---|---|---|---|---|---|
| Catechol | CAPE | Caffeic acids | Antiviral | Xia et al. 2008 | |
| Catechol | CAPE | Dicaffeoylquinic acids | 3′, ST | Robinson et al. 1996 | |
| Catechol | CAPE | Dicaffeoylquinic acids | Antiviral, cytotoxicity | Bodiwala et al. 2011 | |
| Catechol | CAPE | Mercapto sulfonamides | 3′, ST | Nicklaus et al. 1997 | Dhaked et al. 2009; Iyer and Hopfinger 2007; Kuo et al. 2004; Liu et al. 2011; Saíz-Urra et al. 2007; Yuan and Parrill 2002, 2005 |
| Catechol | CAPE | Polyhydroxylated aromatics | 3′, ST | Wang et al. 2009 | |
| Catechol | Chicoric acids | Chicoric acids | 3′, ST | Lin et al. 1999 | Iyer and Hopfinger 2007; Liu et al. 2011; Makhija and Kulkarni 2002b; Nunthaboot et al. 2006; Saíz-Urra et al. 2007; Yuan and Parrill 2002, 2005 |
| Catechol | Curcumins | Cinnamoyls | 3′, ST | Burke et al. 1995 | Buolamwini and Assefa 2002; Costi et al. 2004a; Iyer and Hopfinger 2007; Liu et al. 2011 |
| Catechol | Curcumins | Cinnamoyls | 3′ | Artico et al. 1998 | Buolamwini and Assefa 2002; Costi et al. 2004a; Iyer and Hopfinger 2007; Liu et al. 2011; Nunthaboot et al. 2006 |
| Catechol | Curcumins | Cinnamoyls | 3′, ST | Costi et al. 2004a | Gupta et al. 2011; Nunthaboot et al. 2006 |
| Catechol | Curcumins | Curcumins | 3′, ST | Mazumder et al. 1995b | |

*Continued*

**TABLE 12.2 (*Continued*)**
**SAR Studies on HIV Integrase Inhibitors**

| Class | Sub-Class | Inhibitor Type | Inhibitory Activity | SAR Reference | QSAR Reference |
|---|---|---|---|---|---|
| Catechol | Curcumins | Curcumins | 3′, ST | Mazumder et al. 1997 | Dhaked et al. 2009; Iyer and Hopfinger 2007; Liu et al. 2011; Makhija and Kulkarni 2002b; Nunthaboot et al. 2006; Saíz-Urra et al. 2007; Yuan and Parrill 2002, 2005 |
| Catechol | Curcumins | Rosmarinic acids | 3′, ST | Dubois et al. 2008 | |
| Catechol | Flavones | Flavones | 3′, ST | Fesen et al. 1993 | Lameira et al. 2006 |
| Catechol | Flavones | Flavones | 3′, ST | Fesen et al. 1994 | Buolamwini et al. 1996; Marrero-Ponce 2004; Raghavan et al. 1995 |
| Catechol | Styrylquinolines | Styrylbenzofurans | 3′ | Yoo et al. 2003 | |
| Catechol | Styrylquinolines | Styrylquinazolines | 3′ | Lee et al. 2002 | |
| Catechol | Styrylquinolines | Styrylquinolines | 3′, ST | Mekouar et al. 1998 | Dubey et al. 2012; Leonard and Roy 2008; Ma et al. 2004; Niedbala et al. 2006; Polanski et al. 2002, 2004; Sun et al. 2012; Telvekar and Patel 2011 |
| Catechol | Styrylquinolines | Styrylquinolines | 3′, ST | Ouali et al. 2000 | Niedbala et al. 2006; Sun et al. 2012 |
| Catechol | Styrylquinolines | Styrylquinolines | 3′, ST | Zouhiri et al. 2000 | Goudarzi et al. 2012; Iyer and Hopfinger 2007; Leonard and Roy 2008; Liu et al. 2011; Ma et al. 2004; Nunthaboot et al. 2006; Polanski et al. 2002; Saíz-Urra et al. 2007; Sun et al. 2012; Toropova et al. 2011; Yuan and Parrill 2002, 2005 |
| Catechol | Styrylquinolines | Styrylquinolines | 3′ | Bénard et al. 2004 | Nunthaboot et al. 2006; Sun et al. 2012 |
| Catechol | Styrylquinolines | Styrylquinolines | Antiviral | Normand-Bayle et al. 2005 | Sun et al. 2012 |

| | | | | | |
|---|---|---|---|---|---|
| Catechol | Others | Tyrphostins | 3′, ST | Mazumder et al. 1995a | Dhaked et al. 2009; Iyer and Hopfinger 2007; Liu et al. 2011; Saíz-Urra et al. 2007; Yuan and Parrill 2000, 2002, 2005 |
| Non-catechol | Coumarins | Coumarins | 3′, ST | Mazumder et al. 1996 | Makhija and Kulkarni 2001a, 2001b, 2002b |
| Non-catechol | Coumarins | Coumarins | 3′, ST | Zhao et al. 1997a | Dhaked et al. 2009; Iyer and Hopfinger 2007; Liu et al. 2011; Makhija and Kulkarni 2001a, 2001b, 2002a, 2002b; Nunthaboot et al. 2006; Saíz-Urra et al. 2007; Yuan and Parrill 2002, 2005 |
| Non-catechol | Coumarins | Coumarins | 3′ | Mao et al. 2002 | Nunthaboot et al. 2006 |
| Non-catechol | Coumarins | Coumarins | 3′ | Chiang et al. 2007 | Li et al. 2010 |
| Non-catechol | Curcumins | Pyridines | LEDGF/p75 | Cao et al. 2012 | |
| Non-catechol | Heterocyclic | Benzothiazines | 3′, ST | Brzozowski et al. 2009 | |
| Non-catechol | Heterocyclic | Benzyl indoles | ST | Ferro et al. 2010 | |
| Non-catechol | Heterocyclic | Naphthoquinones | 3′, ST | Crosby et al. 2010 | |
| Non-catechol | Heterocyclic | Piperazines | ST | Yang et al. 2010 | |
| Non-catechol | Heterocyclic | Pyranones | 3′, ST, cytotoxicity | Ramkumar et al. 2008 | |
| Non-catechol | Heterocyclic | Pyrazolones | 3′, ST | Hadi et al. 2010 | |
| Non-catechol | Heterocyclic | Pyridazines | Antiviral | Wang et al. 2012a | |
| Non-catechol | Heterocyclic | Pyrimidines | ST | Summa et al. 2006 | de Melo and Ferreira 2012; Sharma et al. 2012 |
| Non-catechol | Heterocyclic | Pyrimidines | ST | Gardelli et al. 2007 | Kaushik et al. 2011a; Reddy et al. 2012; Sharma et al. 2012; Telvekar and Patel 2011 |
| Non-catechol | Heterocyclic | Pyrimidines | ST | Pace et al. 2007 | Sharma et al. 2012; Telvekar et al. 2011 |

*Continued*

**TABLE 12.2 (*Continued*)**
**SAR Studies on HIV Integrase Inhibitors**

| Class | Sub-Class | Inhibitor Type | Inhibitory Activity | SAR Reference | QSAR Reference |
|---|---|---|---|---|---|
| Non-catechol | Heterocyclic | Pyrimidines | ST | Petrocchi et al. 2007 | de Melo and Ferreira 2009, 2012; Sharma et al. 2012 |
| Non-catechol | Heterocyclic | Pyrimidines | ST | Di Francesco et al. 2008 | Sharma et al. 2012; Telvekar et al. 2011 |
| Non-catechol | Heterocyclic | Pyrimidines | ST | Summa et al. 2008 | Reddy et al. 2012; Sharma et al. 2012; Telvekar et al. 2011 |
| Non-catechol | Heterocyclic | Pyrimidines | 3′ | Donghi et al. 2009 | Sharma et al. 2012; Telvekar et al. 2011 |
| Non-catechol | Heterocyclic | Pyrimidines | ST | Petrocchi et al. 2009 | Sharma et al. 2012 |
| Non-catechol | Heterocyclic | Pyrimidines | Antiviral | Ferrara et al. 2010 | Sharma et al. 2012 |
| Non-catechol | Heterocyclic | Pyrimidines | 3′, ST | Jones et al. 2010 | |
| Non-catechol | Heterocyclic | Pyrimidines | Antiviral | Wang et al. 2012c | |
| Non-catechol | Heterocyclic | Pyrimidines | Antiviral | Hajimahdi et al. 2013 | |
| Non-catechol | Heterocyclic | Pyrimidinones | ST | Muraglia et al. 2008 | Sharma et al. 2012; Telvekar et al. 2011 |
| Non-catechol | Heterocyclic | Pyrimidinones | ST | Nizi et al. 2009 | Sharma et al. 2012 |
| Non-catechol | Heterocyclic | Pyrimidinones | 3′, ST | Ramajayam et al. 2009 | |
| Non-catechol | Heterocyclic | Pyrimidinones | ST | De Luca et al. 2011a | |
| Non-catechol | Heterocyclic | Pyrimidinones | 3′, ST | Tang et al. 2011 | |
| Non-catechol | Heterocyclic | Pyrimidinones | ST, cytotoxicity | Kawasuji et al. 2012 | |
| Non-catechol | Heterocyclic | Pyrrolinones | 3′, ST | Dayam et al. 2007 | |
| Non-catechol | Heterocyclic | Pyrrolinones | ST | Pace et al. 2008 | Sharma et al. 2012; Telvekar et al. 2011 |

| | | | | | |
|---|---|---|---|---|---|
| Non-catechol | Heterocyclic | Pyrrolinones | Antiviral | Pendri et al. 2010 | |
| Non-catechol | Heterocyclic | Pyrrolinones | 3′, ST | Ma et al. 2011 | |
| Non-catechol | Heterocyclic | Quinazolines | Antiviral | Wang et al. 2012b | |
| Non-catechol | Heterocyclic | Quinolines | 3′, ST | Sechi et al. 2009b | Cheng et al. 2010 |
| Non-catechol | Heterocyclic | Quinolines | Antiviral, cytotoxicity | He et al. 2011 | |
| Non-catechol | Heterocyclic | Quinolines | ST | Hu et al. 2012a | |
| Non-catechol | Heterocyclic | Quinolones | 3′, ST, cytotoxicity | Sato et al. 2009 | Cheng et al. 2010; Gupta and Madan 2012; Lu et al. 2010 |
| Non-catechol | Heterocyclic | Quinolones | Antiviral, cytotoxicity | He et al. 2013 | |
| Non-catechol | Hydrazides | Hydrazides | 3′, ST | Zhao et al. 1997c | Dhaked et al. 2009; Makhija and Kulkarni 2001a, 2001b, 2002b |
| Non-catechol | Hydrazides | Hydrazones | 3′, ST, LEDGF/p75 | Sanchez et al. 2013 | |
| Non-catechol | Hydrazides | Salicylhydrazides | 3′, ST | Al-Mawsawi et al. 2007 | |
| Non-catechol | Hydrazides | Salicylhydrazines | 3′, ST | Neamati et al. 1998 | Dhaked et al. 2009; Iyer and Hopfinger 2007; Liu et al. 2011; Makhija and Kulkarni 2001a, 2001b, 2002b; Nunthaboot et al. 2006; Saíz-Urra et al. 2007; Yuan and Parrill 2000, 2002, 2005 |
| Non-catechol | Naphthyridine | Naphthyridinones | ST | Johns et al. 2013 | |
| Non-catechol | Naphthyridine | Oxadiazoles, triazoles | ST | Johns et al. 2009a | Jalali-Heravia and Ebrahimi-Najafabadia 2011; Ravichandran et al. 2010 |
| Non-catechol | Naphthyridine | Oxadiazoles, triazoles | ST | Johns et al. 2009b | Jalali-Heravia and Ebrahimi-Najafabadia 2011; Veerasamy et al. 2011 |

*Continued*

**TABLE 12.2 (*Continued*)**
**SAR Studies on HIV Integrase Inhibitors**

| Class | Sub-Class | Inhibitor Type | Inhibitory Activity | SAR Reference | QSAR Reference |
|---|---|---|---|---|---|
| Non-catechol | Naphthyridine | Oxazepines | 3′, ST | Garofalo et al. 2008 | |
| Non-catechol | Naphthyridine | Quinolines | Cytotoxicity | Nagasawa et al. 2011 | |
| Non-catechol | Naphthyridines | Naphthyridines | ST, cytotoxicity | Zhuang et al. 2003 | Barreca et al. 2005; de Melo and Ferreira 2012 |
| Non-catechol | Naphthyridines | Naphthyridines | ST | Hazuda et al. 2004a | |
| Non-catechol | Naphthyridines | Naphthyridines | ST, cytotoxicity | Guare et al. 2006 | de Melo and Ferreira 2012 |
| Non-catechol | Naphthyridines | Naphthyridines | ST | Egbertson et al. 2007 | de Melo and Ferreira 2012 |
| Non-catechol | Naphthyridines | Naphthyridinones | ST | Johns et al. 2011 | |
| Non-catechol | Others | Arylamides | 3′, ST | Zhao et al. 1997b | Dhaked et al. 2009; Iyer and Hopfinger 2007; Liu et al. 2011; Makhija and Kulkarni 2002b; Saíz-Urra et al. 2007; Yuan and Parrill 2002, 2005 |
| Non-catechol | Others | Carbonyls | 3′, ST | Maurer et al. 2000 | Nunthaboot et al. 2006 |
| Non-catechol | Others | Cyclic peptides | LEDGF/p75 | Hayouka et al. 2010 | |
| Non-catechol | Others | Depsides, depsidones | 3′, ST | Neamati et al. 1997a | Dhaked et al. 2009; Iyer and Hopfinger 2007; Liu et al. 2011; Makhija and Kulkarni 2001a, 2001b, 2002b; Saíz-Urra et al. 2007; Yuan and Parrill 2002, 2005 |
| Non-catechol | Others | Diverse | 3′, ST | Deng et al. 2005 | |
| Non-catechol | Others | Quinolines | LEDGF/p75 | Christ et al. 2010 | |
| Non-catechol | Others | Quinolines | LEDGF/p75 | Desimmie et al. 2013 | |
| Non-catechol | Others | Quinolines | 3′, ST, LEDGF/p75 | Serrao et al. 2013 | |

| | | | | | |
|---|---|---|---|---|---|
| Non-catechol | Others | Tetracyclines | 3′, ST | Neamati et al. 1997b | Iyer and Hopfinger 2007; Liu et al. 2011; Nunthaboot et al. 2006; Saíz-Urra et al. 2007; Yuan and Parrill 2002, 2005 |
| Non-catechol | Others | Thiazolothiazepines | 3′, ST | Neamati et al. 1999 | Dhaked et al. 2009; Iyer and Hopfinger 2007; Liu et al. 2011; Makhija and Kulkarni 2001a, 2001b, 2002a, 2002b; Saíz-Urra et al. 2007; Yuan and Parrill 2002, 2005 |
| Non-catechol | Others | Thiazolothiazepines | 3′, ST | Aiello et al. 2004 | |
| Non-catechol | Others | Thiazolothiazepines | 3′, ST | Aiello et al. 2006 | |
| Non-catechol | Sulfonamides | Aryl sulfones | 3′, ST | Neamati et al. 1997c | Dhaked et al. 2009; Kuo et al. 2004 |
| Non-catechol | Sulfonamides | Benzo thiazines | 3′, ST | Brzozowski et al. 2004 | Gupta et al. 2009, 2012 |
| Non-catechol | Sulfonamides | Geminal disulfones | 3′, ST | Meadows et al. 2005 | Dhaked et al. 2009 |
| Non-catechol | Sulfonamides | Mercaptobenzenesulfonamides | 3′, ST | Chen et al. 2000 | Kuo et al. 2004 |
| Non-catechol | Sulfonamides | Mercaptobenzenesulfonamides | 3′, ST | Kuo et al. 2004 | Dhaked et al. 2009 |
| Non-catechol | Sulfonamides | Mercaptobenzenesulfonamides | 3′, ST | Brzozowski et al. 2008 | Gupta et al. 2009, 2012; Kaushik et al. 2011b |
| Non-catechol | Sulfonamides | Naphthalene sulfonamides | Antiviral | Xu et al. 2003 | Sahu et al. 2007 |
| Non-catechol | Sulfonamides | Vinyl sulfones | 3′, ST, cytotoxicity | Meadows et al. 2007 | |
| Others | Diketo acids | Amide diketo acids | ST | Walker et al. 2007 | Vengurlekar et al. 2010 |
| Others | Diketo acids | Amide diketo acids | 3′, ST | Li et al. 2009 | |
| Others | Diketo acids | Aryl diketo acids | ST | Wai et al. 2000 | Barreca et al. 2005; de Melo and Ferreira 2012 |
| Others | Diketo acids | Aryl diketo acids | 3′, ST | Pais et al. 2002 | Barreca et al. 2005; Nunthaboot et al. 2006 |
| Others | Diketo acids | Aryl diketo acids | 3′, ST, cytotoxicity | Zhang et al. 2003 | |

*Continued*

**TABLE 12.2 (*Continued*)**
**SAR Studies on HIV Integrase Inhibitors**

| Class | Sub-Class | Inhibitor Type | Inhibitory Activity | SAR Reference | QSAR Reference |
|---|---|---|---|---|---|
| Others | Diketo acids | Aryl diketo acids | 3′, ST | Costi et al. 2004b | Ravichandran et al. 2008 |
| Others | Diketo acids | Aryl diketo acids | ST | Zhang et al. 2004 | |
| Others | Diketo acids | Aryl diketo acids | ST | Kehlenbeck et al. 2006 | |
| Others | Diketo acids | Aryl diketo acids | 3′, ST | Zeng et al. 2008a | |
| Others | Diketo acids | Aryl diketo acids | 3′, ST | Zeng et al. 2008b | |
| Others | Diketo acids | Benzoic acids | LEDGF/p75 | Du et al. 2008 | |
| Others | Diketo acids | Benzylamides | 3′, ST, LEDGF/p75 | Fan et al. 2011 | |
| Others | Diketo acids | Benzylindoles | ST | De Luca et al. 2008 | |
| Others | Diketo acids | Benzylindoles | LEDGF/p75 | De Luca et al. 2009 | |
| Others | Diketo acids | Benzylindoles | LEDGF/p75 | De Luca et al. 2011b | |
| Others | Diketo acids | Benzylindoles | LEDGF/p75 | De Luca et al. 2013 | |
| Others | Diketo acids | Chalcones | 3′, ST, cytotoxicity | Deng et al. 2006 | |
| Others | Diketo acids | Chalcones | 3′, ST | Deng et al. 2007 | |
| Others | Diketo acids | Chalcones | ST | Sharma et al. 2011 | Sharma et al. 2011, 2012 |
| Others | Diketo acids | Chromones | ST | Park et al. 2008 | |
| Others | Diketo acids | Diketo acids | 3′, ST | Hazuda et al. 2000 | de Melo and Ferreira 2012 |
| Others | Diketo acids | Diketo acids | ST | Grobler et al. 2002 | Barreca et al. 2005 |
| Others | Diketo acids | Diketo acids | 3′, ST | Long et al. 2004 | |
| Others | Diketo acids | Diketo acids | 3′, ST | Dayam et al. 2005a | |
| Others | Diketo acids | Diketo acids | 3′, ST | Dayam et al. 2005b | |
| Others | Diketo acids | Diketo acids | ST | Sechi et al. 2005 | |
| Others | Diketo acids | Diketo acids | 3′, ST | Sechi et al. 2009a | |

| | | | | | |
|---|---|---|---|---|---|
| Others | Diketo acids | Diketo acids | ST, cytotoxicity | Hu et al. 2012b | |
| Others | Diketo acids | Indole diketo acids | 3′, ST | Sechi et al. 2004 | Dhaked et al. 2009 |
| Others | Diketo acids | Indole diketo acids | LEDGF/p75 | De Luca et al. 2010 | |
| Others | Diketo acids | Indole diketo acids | LEDGF/p75 | Cavalluzzo et al. 2012 | |
| Others | Diketo acids | Metal complexed | 3′, ST | Sechi et al. 2006 | |
| Others | Diketo acids | Metal complexed | 3′, ST | Bacchi et al. 2008 | |
| Others | Diketo acids | Nucleobases | 3′, ST | Nair et al. 2006a | |
| Others | Diketo acids | Nucleobases | 3′, ST | Nair et al. 2006b | |
| Others | Diketo acids | Phthalimide | ST | Verschueren et al. 2005 | Dessalew 2009 |
| Others | Diketo acids | Polycyclic diketo acids | 3′, ST | Patil et al. 2007 | Sharma et al. 2012 |
| Others | Diketo acids | Quinolones | 3′, ST | Bona et al. 2006 | Dubey et al. 2011, 2012 |
| Others | Diketo acids | Quinolones | 3′, ST | Dayam et al. 2006b | Dhaked et al. 2009 |
| Others | Diketo acids | Quinolones | ST, 3′ | Di Santo et al. 2006 | Dubey et al. 2011, 2012 |
| Others | Diketo acids | Quinolones | ST | Sato et al. 2006 | Cheng et al. 2010; Dubey et al. 2011, 2012; Sharma et al. 2012 |
| Others | Diketo acids | Quinolones | 3′, ST | Dayam et al. 2008a | |
| Others | Diketo acids | Quinolones | ST | Di Santo et al. 2008 | |
| Others | Diketo acids | Quinolones | 3′, ST | Pasquini et al. 2008 | Cheng et al. 2010; Sharma et al. 2012 |
| Others | Diketo acids | Rhodanines | 3′, ST | Ramkumar et al. 2010b | |
| Others | Diverse | Diverse | 3′, ST | Carlson et al. 2000 | |
| Others | Others | Salicylic acids | 3′, ST, LEDGF/p75 | Fan et al. 2011 | |

*Note:* ST = strand transfer.

**TABLE 12.3**
**QSAR Studies on HIV Integrase Inhibitors**

| Class | Sub-Class | Inhibitor Type | # of Compounds | Inhibitory Activity | QSAR Method | Modeling Method | References |
|---|---|---|---|---|---|---|---|
| Catechol | Curcumins | Cinnamoyls | 24 | 3′ | 3D: CoMFA, CoMSIA | PLS | Buolamwini and Assefa 2002 |
| Catechol | Curcumins | Cinnamoyls | 87 | 3′ | 3D: GRID/GOLPE | PLS | Costi et al. 2004b |
| Catechol | Curcumins | Curcumins | 39 | 3′ | 3D: CoMFA | PLS | Gupta et al. 2011 |
| Catechol | Flavones | Flavones | 15 | 3′, ST | 3D: CoMFA | PLS | Raghavan et al. 1995 |
| Catechol | Flavones | Flavones | 15 | 3′, ST | 2D | PLS | Buolamwini et al. 1996 |
| Catechol | Flavones | Flavones | 15 | 3′ | 2D | GFA | Marrero-Ponce 2004 |
| Catechol | Flavones | Flavones | 32 | Active/inactive | 2D | HC, LDA, PCA | Lameira et al. 2006 |
| Catechol | Styrylquinolines | Styrylquinolines | 19 | Antiviral, cytotoxicity | 2D | KNN | Polanski et al. 2002 |
| Catechol | Styrylquinolines | Styrylquinolines | 20 | 3′ | 3D: CoMFA, CoMSIA | PLS | Polanski et al. 2004 |
| Catechol | Styrylquinolines | Styrylquinolines | 38 | 3′ | 3D: CoMFA | PLS | Ma et al. 2004 |
| Catechol | Styrylquinolines | Styrylquinolines | 29 | 3′, ST | 3D: CoMFA, CoMSA; 4D | PLS | Niedbala et al. 2006 |
| Catechol | Styrylquinolines | Styrylquinolines | 36 | 3′ | 3D: MSA | GFA, MLR, PLS | Leonard and Roy 2008 |
| Catechol | Styrylquinolines | Styrylquinolines | 36 | Antiviral | SMILES and InChI | PLS | Toropova et al. 2011 |

| | | | | | | | |
|---|---|---|---|---|---|---|---|
| Catechol | Styrylquinolines | Styrylquinolines | 77 | 3′ | 3D: CoMFA, CoMSIA | PLS | Sun et al. 2012 |
| Catechol | Styrylquinolines | Styrylquinolines | 36 | 3′ | 2D: GA, MLR | MLR | Goudarzi et al. 2012 |
| Non-catechol | Coumarins | Hydroxycoumarins | 9 | 3′ | 2D | MLR | Li et al. 2010 |
| Non-catechol | Diverse | Diverse | 41 | 3′, ST | 3D: CoMSIA | PLS | Makhija and Kulkarni 2001b |
| Non-catechol | Diverse | Diverse | 41 | 3′, ST | 2D: Eigenvalue analysis; 3D: CoMSIA | PLS | Makhija and Kulkarni 2001a |
| Non-catechol | Diverse | Thiazolothiazepines | 27 | 3′, ST | 3D: CoMFA, CoMSIA | PLS | Makhija and Kulkarni 2002a |
| Non-catechol | Sulfonamides | Benzodithiazine | 22 | Unspecified | 2D | GFA | Gupta et al. 2012 |
| Nón-catechol | Sulfonamides | Diverse | 41 | 3′ | 3D: CoMFA, CoMSIA | PLS | Gupta et al. 2009 |
| Non-catechol | Sulfonamides | Mercaptobenzenesulfonamides | 66 | 3′, ST | 3D: CoMFA, CoMSIA | PLS | Kuo et al. 2004 |
| Non-catechol | Sulfonamides | Mercaptobenzenesulfonamides | 22 | 3′, ST | 2D | MLR | Kaushik et al. 2011b |
| Non-catechol | Sulfonamides | Naphthalene sulfonamide | 20 | Unspecified | 2D | MLR | Sahu et al. 2007 |
| Others | Diketo acids | Benzyl amide-ketoacids | 24 | ST | 2D; 3D: MFA | MLR, PCR, PLS, KNN | Vengurlekar et al. 2010 |
| Others | Diketo acids | Carboxylic acids | 62 | ST | 2D | MLR, BPNN, SVM | Cheng et al. 2010 |
| Others | Diketo acids | Chalcones | 53 | ST | 3D: CoMFA, CoMSIA | PLS | Sharma et al. 2011 |
| Others | Diketo acids | Diketo acids | 33 | ST | 3D | Pharmacophore | Barreca et al. 2005 |

*Continued*

**TABLE 12.3 (*Continued*)**
**QSAR Studies on HIV Integrase Inhibitors**

| Class | Sub-Class | Inhibitor Type | # of Compounds | Inhibitory Activity | QSAR Method | Modeling Method | References |
|---|---|---|---|---|---|---|---|
| Others | Diketo acids | Diverse | 65 | ST | 3D | PLS, Pharmacophore | Telvekar and Patel 2011 |
| Others | Diketo acids | Diverse | 85 | ST | 4D LQTA-QSAR | PLS | de Melo and Ferreira 2012 |
| Others | Diketo acids | Hexenoic acids | 12 | 3′, ST | 2D | MLR | Ravichandran et al. 2008 |
| Others | Diketo acids | Oxadiozole naphthyridine | 67 | ST | 2D | MLR | Ravichandran et al. 2010 |
| Others | Diketo acids | Oxadiozole naphthyridine | 84 | ST | 2D | MLR, ANN | Jalali-Heravia and Ebrahimi-Najafabadi 2011 |
| Others | Diketo acids | Oxadiozole naphthyridine | 63 | Cytotoxicity | 2D | MLR | Veerasamy et al. 2011 |
| Others | Diketo acids | Phthalimides | 37 | ST | 2D | MLR | Dessalew 2009 |
| Others | Diketo acids | Pyrimidinones | 33 | ST | 2D | PLS | de Melo and Ferreira 2009 |
| Others | Diketo acids | Pyrimidinones | 51 | ST | 2D | MLR | Kaushik et al. 2011a |
| Others | Diketo acids | Pyrimidinones | 49 | ST | 3D | Pharmacophore | Reddy et al. 2012 |
| Others | Diketo acids | Quinolones | 48 | ST | 3D: CoMFA, CoMSIA | PLS | Lu et al. 2010 |
| Others | Diketo acids | Quinolones | 26 | Unspecified | 2D | MLR | Satpathy and Ghosh 2011 |

| | | | | | | | |
|---|---|---|---|---|---|---|---|
| Others | Diketo acids | Quinolones | 25 | 3′, ST, cytotoxicity | 2D | MLR, PCR, PLS | Dubey et al. 2011 |
| Others | Diketo acids | Quinolones | 50 | Unspecified | 2D | MLR, PLS, PCR, DT, RF | Gupta and Madan 2012 |
| Others | Diketo acids | Quinolones | 25 | 3′, ST | 2D | MLR | Dubey et al. 2012 |
| Others | Diverse | Diverse | 42 | Antiviral | 2D | MLR, GFA | Yuan and Parrill 2000 |
| Others | Diverse | Diverse | 81 | 3′, ST | 2D | GFA | Makhija and Kulkarni 2002b |
| Others | Diverse | Diverse | 174 | 3′ | 2D | GFA | Yuan and Parrill 2002 |
| Others | Diverse | Diverse | 60 | 3′ | 3D: MFA | GFA | Yuan and Parrrill 2005 |
| Others | Diverse | Diverse | 89 | 3′ | 3D: CoMFA, CoMSIA | PLS | Nunthaboot et al. 2006 |
| Others | Diverse | Diverse | 172 | 3′ | 2D | MLR | Saíz-Urra et al. 2007 |
| Others | Diverse | Diverse | 213 | 3′ | 4D | GFA | Iyer and Hopfinger 2007 |
| Others | Diverse | Diverse | 81 | 3′ | 3D: CoRIA | G/PLS | Dhaked et al. 2009 |
| Others | Diverse | Diverse | 213 | 3′ | 2D | Multitask learning | Liu et al. 2011 |
| Others | Diverse | Diverse | 103 | ST | 3D: CoMFA, CoMSIA | PLS | Sharma et al. 2012 |
| Others | Diverse | Diverse | 1336 | Active/ inactive | 2D | SVM | Yan et al. 2012 |

*Note:* ST = strand transfer.

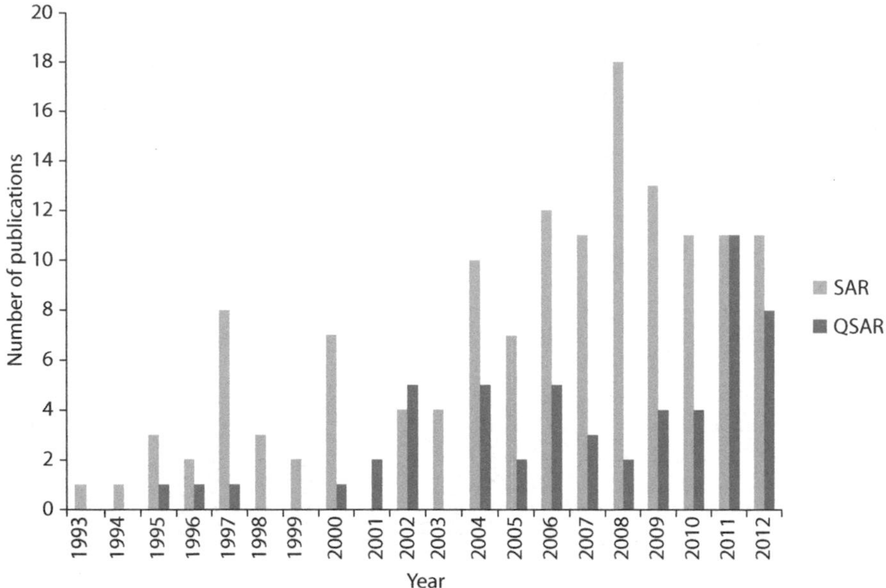


**FIGURE 12.17** A comparison of SAR and QSAR studies published between 1993 and 2012.

A statistically and mechanistically robust QSAR model can be used to screen chemical compounds for potential lead compounds that may be further optimized to become viable drug candidate molecules crucial for biological activity.

A review of different QSAR models used in HIV IN inhibitor design shows that the long-established 2D-QSAR techniques have been dominant in these studies. The biggest advantage of 2D-QSAR methods is that they permit modeling on a wide assortment of ligands even in the absence of the 3D crystal structure of the target receptor. Due to the lack of a complete crystal structure, relatively few 3D and 4D-QSAR models have been reported. In the absence of a complete crystal structure, 3D-QSAR models resort to methods such as ligand alignment, energy minimized conformers, and molecular docking. The 4D-QSAR models used molecular dynamics simulations to obtain a conformation ensemble required for their method.

In most of the QSAR studies, the models were internally validated using cross validation or externally validated by dividing the dataset into a training and test set. However, in some studies, it was observed that model validation methods were lacking, providing lower confidence in the models for use in mechanistic interpretation and biological activity prediction. Model validation is an important aspect of QSAR modeling for ensuring the robustness of the model (Tropsha et al. 2003). Validation ensures that the QSAR model is not overfitting to the dataset and has an acceptable level of predictivity.

A review of SAR data has shown that research efforts from different groups have resulted in the development of several second-generation inhibitors with nanomolar potency. Most of these inhibitors are the result of bioisosteric replacement of the carboxylic group (e.g., tetrazole and triazole), rearrangement of the keto-enol moiety for metal chelation, modification of the fluorobenzyl hydrophobic group, and other suitable modifications to improve their pharmacokinetic profile. Select examples of these second-generation inhibitors include benzylamide and benzylindole DKA, 1,6-naphthyridine carboxamides, dihydropyrimidine 4-carboxamides, N-methylpyrimidinones, pyrrolinones, and quinolone 3-carboxylic acids. Optimization of these inhibitors may lead to new drugs active against IN drug-resistant and mutant viruses.

Inhibition of the interaction between viral protein IN and host protein LEDGF/p75 is emerging as an important approach to develop new inhibitors. Many of the inhibitors disrupting these interactions either have been reported to target two steps in the integration process or are active against mutant viral strains. Results of QSAR studies on these inhibitors have not been reported in the literature, which could help in further optimization of these molecules.

Development of new drugs is a lengthy and costly process and, in recent years, drug repositioning (using old drugs on new targets) has gained significant attention for addressing these issues. Due to the dual inhibitory nature of many anti-HIV inhibitors, they constitute a large source for the development of new anticancer drugs. Recent studies show that the HIV inhibitors that were abandoned in early clinical trials could be redeveloped as potential anticancer agents. The use of various CADD approaches, including QSAR modeling, can significantly help in these efforts.

IN inhibitor drugs act selectively on HIV IN and have no risk of developing cross-resistance against other classes of HIV inhibitor drugs. However, these inhibitors are still susceptible to developing drug-resistant mutant viral strains. There is also a high probability of cross-resistance against other mechanistically similar IN inhibitors. The studies discussed in this chapter show several examples where an inhibitor molecule has been found to act as a dual inhibitor and could be further optimized as a novel HIV inhibitor. Future efforts for developing mechanistically and structurally distinct next generation novel IN inhibitors, targeting the interaction between viral protein IN and host cellular cofactor protein LEDGF/p75 and/or acting as dual inhibitor inhibiting two or more steps during the integration process, seem to be the most important approach for addressing these challenges.

## ACKNOWLEDGMENTS

The authors thank the editor, Professor S.P. Gupta, for motivating them to write this chapter. The authors are also thankful to the Computational Science Research Center at San Diego State University for infrastructure support. The authors also thank Paras Kumar for proofreading the manuscript. Every attempt has been made to make this work as complete as possible. However, sincere apology is extended if any pertinent work(s) is not included in this chapter.

## REFERENCES

Aiello F, Brizzi A, De Grazia O, et al. (2006). An approach to the stereo controlled synthesis of polycyclic derivatives of 1-4-thiazolidinecarboxylic acid active against HIV-1 integrase. *Eur J Med Chem* 41: 914–917.

Aiello F, Brizzi A, Garofalo A, et al. (2004). Synthesis of novel thiazolothiazepine based HIV-1 integrase inhibitors. *Bioorg Med Chem* 12: 4459–4466.

Allen C, Ashley AK, Hromas R, Nickoloff JA. (2011). More forks on the road to replication stress recovery. *J Mol Cell Biol* 3: 4–12.

Al-Mawsawi LQ, Dayam R, Taheri L, et al. (2007). Discovery of novel non-cytotoxic salicylhydrazide containing HIV-1 integrase inhibitors. *Bioorg Med Chem Lett* 17: 6472–6475.

Al-Mawsawi LQ, Neamati N. (2007). Blocking interactions between HIV-1 integrase and cellular cofactors: An emerging anti-retroviral strategy. *Trends Pharmacol Sci* 28: 526–535.

Al-Mawsawi LQ, Neamati N. (2011). Allosteric inhibitor development targeting HIV-1 integrase. *Chem Med Chem* 6: 228–241.

Andronis C, Sharma A, Virvilis V, Deftereos S, Persidis A. (2011). Literature mining, ontologies and information visualization for drug repurposing. *Brief Bioinform* 12: 357–368.

Artico M, Di Santo R, Costi R, et al. (1998). Geometrically and conformationally restrained cinnamoyl compounds as inhibitors of HIV-1 integrase: Synthesis, biological evaluation, and molecular modeling. *J Med Chem* 41: 3948–3960.

Bacchi A, Biemmi M, Carcelli M, et al. (2008). From ligand to complexes. Part 2. Remarks on human immunodeficiency virus type 1 integrase inhibition by β-diketo acid metal complexes. *J Med Chem* 51: 7253–7264.

Barreca ML, Ferro S, Rao A, et al. (2005). Pharmacophore-based design of HIV-1 integrase strand-transfer inhibitors. *J Med Chem* 48: 7084–7088.

Barré-Sinoussi F, Chermann JC, Rey F, et al. (1983). Isolation of a T-lymphotropic retrovirus from a patient at risk for acquired immune deficiency syndrome (AIDS). *Science* 220: 868–871.

Belavic JM. (2013). Drug updates and approvals: 2012 in review. *Nurse Pract* 28: 24–42.

Bénard C, Zouhiri F, Normand-Bayle M, et al. (2004). Linker-modified quinoline derivatives targeting HIV-1 integrase: Synthesis and biological activity. *Bioorg Med Chem Lett* 14: 2473–2476.

Bhhatarai B, Garg R. (2008). Comparative QSAR as a cheminformatics tool in the design of dihydro-pyranone based HIV-1 protease inhibitors. *Curr Comput Aided Drug Des* 4: 283–310.

Bodiwala HS, Sabde S, Gupta P, et al. (2011). Design and synthesis of caffeoyl-anilides as portmanteau inhibitors of HIV-1 integrase and CCR5. *Bioorg Med Chem* 19: 1256–1263.

Bona R, Andreotti M, Buffa V, et al. (2006). Development of a human immunodeficiency virus vector-based, single-cycle assay for evaluation of anti-integrase compounds. *Antimicrob Agents Chemother* 50: 3407–3417.

Brzozowski Z, Saczewski F, Sanchez T, et al. (2004). Synthesis, antiviral, and anti-HIV-1 integrase activities of 3-aroyl-1,1-dioxo-1,4,2-benzodithiazines. *Bioorg Med Chem* 12: 3663–3672.

Brzozowski Z, Sączewski F, Sławiński J, Sanchez T, Neamati N. (2009). Synthesis and anti-HIV-1 integrase activities of 3-aroyl-2,3-dihydro-1,1-dioxo-1,4,2-benzodithiazines. *Eur J Med Chem* 44: 190–196.

Brzozowski Z, Sławiński J, Sączewski F, Sanchez T, Neamati N. (2008). Synthesis, anti-HIV-1 integrase, and cytotoxic activities of 4-chloro-N-(4-oxopyrimidin-2-yl)-2-mercaptobenzenesulfonamide derivatives. *Eur J Med Chem* 43: 1188–1198.

Buolamwini JK, Assefa H. (2002). CoMFA and CoMSIA 3D QSAR and docking studies on conformationally-restrained cinnamoyl HIV-1 integrase inhibitors: Exploration of a binding mode at the active site. *J Med Chem* 45: 841–852.

Buolamwini JK, Raghavan K, Fesen MR, et al. (1996). Application of the electrotopological state index to QSAR analysis of flavone derivatives as HIV-1 integrase inhibitors. *Pharm Res* 13: 1892–1895.

Burke TR, Jr, Fesen MR, Mazumder A, et al. (1995). Hydroxylated aromatic inhibitors of HIV-1 integrase. *J Med Chem* 38: 4171–4178.

Cao B, Wang Y, Ding K, Neamati N, Long YQ. (2012). Synthesis of the pyridinyl analogues of dibenzylideneacetone (pyr-dba) via an improved Claisen-Schmidt condensation, displaying diverse biological activities as curcumin analogues. *Org Biomol Chem* 10: 1239–1245.

Carlson HA, Masukawa KM, Rubins K, et al. (2000). Developing a dynamic pharmacophore model for HIV-1 integrase. *J Med Chem* 43: 2100–2114.

Cavalluzzo C, Voet A, Christ F, et al. (2012). De novo design of small molecule inhibitors targeting the LEDGF/p75-HIV integrase interaction. *RSC Adv* 2: 974–984.

Chen IJ, Neamati N, Nicklaus MC, et al. (2000). Identification of HIV-1 integrase inhibitors via three-dimensional database searching using ASV and HIV-1 integrases as targets. *Bioorg Med Chem* 8: 2385–2398.

Cheng AC, Coleman RG, Smyth KT, et al. (2007). Structure-based maximal affinity model predicts small-molecule druggability. *Nat Biotechnol* 25: 71–75.

Cheng Z, Zhang Y, Fu W. (2010). QSAR study of carboxylic acid derivatives as HIV-1 integrase inhibitors. *Eur J Med Chem* 45: 3970–3980.

Chiang CC, Mouscadet JF, Tsai HJ, Liu CT, Hsu LY. (2007). Synthesis and HIV-1 integrase inhibition of novel bis- or tetra-coumarin analogues. *Chem Pharm Bull* 55: 1740–1743.

Chow WA, Jiang C, Guan M. (2009). Anti-HIV drugs for cancer therapeutics: Back to the future? *Lancet Oncol* 10: 61–71.

Christ F, Voet A, Marchand A, et al. (2010). Rational design of small-molecule inhibitors of the LEDGF/p75-integrase interaction and HIV replication. *Nat Chem Biol* 6: 442–448.

Collins FS. (2011). Mining for therapeutic gold. *Nat Rev Drug Discov* 10: 397.

Costi R, Di Santo R, Artico M, et al. (2004a). 2,6-Bis(3,4,5-trihydroxybenzylydene) derivatives of cyclohexanone: Novel potent HIV-1 integrase inhibitors that prevent HIV-1 multiplication in cell-based assays. *Bioorg Med Chem* 12: 199–215.

Costi R, Di Santo R, Artico M, et al. (2004b). 6-Aryl-2,4-dioxo-5-hexenoic acids, novel integrase inhibitors active against HIV-1 multiplication in cell-based assays. *Bioorg Med Chem Lett* 14: 1745–1749.

Craigie R. (2011). Integrase mechanism and function. In: Neamati N (ed.), *HIV-1 Integrase: Mechanism and Inhibitor Design*. Hokoben, NJ: John Wiley & Sons, pp. 23–33.

Crosby IT, Bourke DG, Jones ED, et al. (2010). Antiviral agents 2. Synthesis of trimeric naphthoquinone analogues of conocurvone and their antiviral evaluation against HIV. *Bioorg Med Chem* 18: 6442–6450.

Dayam R, Al-Mawsawi LQ, Neamati N. (2007). Substituted 2-pyrrolinone inhibitors of HIV-1 integrase. *Bioorg Med Chem Lett* 17: 6155–6159.

Dayam R, Al-Mawsawi LQ, Zawahir Z, et al. (2008a). Quinolone 3-carboxylic acid pharmacophore: Design of second generation HIV-1 integrase inhibitors. *J Med Chem* 51: 1136–1144.

Dayam R, Deng J, Neamati N. (2006a). HIV-1 integrase inhibitors: 2003–2004 update. *Med Res Rev* 26: 271–309.

Dayam R, Gundla R, Al-Mawsawi LQ, Neamati N. (2008b). HIV-1 integrase inhibitors: 2005–2006 update. *Med Res Rev* 28: 118–154.

Dayam R, Neamati N. (2003). Small-molecule HIV-1 integrase inhibitors: The 2001–2002 update. *Curr Pharm Des* 9: 1789–1802.

Dayam R, Sanchez T, Clement O, et al. (2005a). Beta-diketo acid pharmacophore hypothesis. 1. Discovery of a novel class of HIV-1 integrase inhibitors. *J Med Chem* 48: 111–120.

Dayam R, Sanchez T, Neamati N. (2005b). Diketo acid pharmacophore. 2. Discovery of structurally diverse inhibitors of HIV-1 integrase. *J Med Chem* 48: 8009–8015.

Dayam R, Sanchez T, Neamati N. (2006b). Discovery and structure-activity relationship studies of a unique class of HIV-1 integrase inhibitors. *Chem Med Chem* 1: 238–244.

Debnath AK. (2001). Quantitative structure-activity relationship (QSAR): A versatile tool in drug design. In: Ghose AK, Viswanadhan VN (ed.), *Combinatorial Library Design and Evaluation: Principles, Software Tools, and Applications in Drug Discovery*. New York: Marcel Dekker, pp. 73–129.

Debnath AK. (2005). Application of 3D-QSAR techniques in anti-HIV-1 drug design—An overview. *Curr Pharm Des* 11: 3091–3110.

De Clercq E. (1995). Antiviral therapy for human immunodeficiency virus infections. *Clin Microbiol Rev* 8: 200–239.

De Clercq E. (2009). Anti-HIV drugs: 25 compounds approved within 25 years after the discovery of HIV. *Int J Antimicrob Agents* 33: 307–320.

De Luca L, Barreca ML, Ferro S, et al. (2008). A refined pharmacophore model for HIV-1 integrase inhibitors: Optimization of potency in the 1H-benzylindole series. *Bioorg Med Chem Lett* 18: 2891–2895.

De Luca L, Barreca ML, Ferro S, et al. (2009). Pharmacophore-based discovery of small-molecule inhibitors of protein-protein interactions between HIV-1 integrase and cellular cofactor LEDGF/p75. *Chem Med Chem* 4: 1311–1316.

De Luca L, De Grazia S, Ferro S, et al. (2011a). HIV-1 integrase strand-transfer inhibitors: Design, synthesis and molecular modeling investigation. *Eur J Med Chem* 46: 756–764.

De Luca L, Ferro S, Gitto R, et al. (2010). Small molecules targeting the interaction between HIV-1 integrase and LEDGF/p75 cofactor. *Bioorg Med Chem* 18: 7515–7521.

De Luca L, Ferro S, Morreale F, et al. (2013). Fragment hopping approach directed at design of HIV IN-LEDGF/p75 interaction inhibitors. *J Enzyme Inhib Med Chem* 28: 1002–1009.

De Luca L, Gitto R, Christ F, et al. (2011b). 4-[1-(4-Fluorobenzyl)-4-hydroxy-1H-indol-3-yl]-2-hydroxy-4-oxobut-2-enoic acid as a prototype to develop dual inhibitors of HIV-1 integration process. *Antiviral Res* 92: 102–107.

de Melo EB, Ferreira MMC. (2009). Multivariate QSAR study of 4,5-dihydroxypyrimidine carboxamides as HIV-1 integrase inhibitors. *Eur J Med Chem* 44: 3577–3583.

de Melo EB, Ferreira MMC. (2012). Four-dimensional structure-activity relationship model to predict HIV-1 integrase strand transfer inhibition using LQTA-QSAR methodology. *J Chem Inf Model* 52: 1722–1732.

Deng J, Kelley JA, Barchi JJ, et al. (2006). Mining the NCI antiviral compounds for HIV-1 integrase inhibitors. *Bioorg Med Chem* 14: 3785–3792.

Deng J, Lee KW, Sanchez T, et al. (2005). Dynamic receptor-based pharmacophore model development and its application in designing novel HIV-1 integrase inhibitors. *J Med Chem* 48: 1496–1505.

Deng J, Sanchez T, Al-Mawsawi LQ, et al. (2007). Discovery of structurally diverse HIV-1 integrase inhibitors based on a chalcone pharmacophore. *Bioorg Med Chem* 15: 4985–5002.

Deprez E, Barbe S, Kolaski M, et al. (2004). Mechanism of HIV-1 integrase inhibition by styrylquinoline derivatives in vitro. *Mol Pharmacol* 65: 85–98.

Desimmie BA, Demeulemeester J, Christ F, Debyser Z. (2013). Rational design of LEDGINs as first allosteric integrase inhibitors for the treatment of HIV infection. *Drug Discovery Today: Technol.* 10: e517–e522.

Dessalew N. (2009). Investigation of the structural requirement for inhibiting HIV integrase: QSAR study. *Acta Pharm* 59: 31–43.

Dhaked DK, Verma J, Saran A, Coutinho EC. (2009). Exploring the binding of HIV-1 integrase inhibitors by comparative residue interaction analysis (CoRIA). *J Mol Model* 15: 233–245.

Di Francesco ME, Pace P, Fiore F, et al. (2008). Development of 2-t butyl-N-methyl pyrimidones as potent inhibitors of HIV integrase. *Bioorg Med Chem Lett* 18: 2709–2713.

Di Santo R, Costi R, Roux A, et al. (2006). Novel bifunctional quinolonyl diketo acid derivatives as HIV-1 integrase inhibitors: Design, synthesis, biological activities, and mechanism of action. *J Med Chem* 49: 1939–1945.

Di Santo R, Costi R, Roux A, et al. (2008). Novel quinolinonyl diketo acid derivatives as HIV-1 integrase inhibitors: Design, synthesis, and biological activities. *J Med Chem* 51: 4744–4750.

Donghi M, Kinzel OD, Summa V. (2009). 3-Hydroxy-4-oxo-4H-pyrido[1,2-a]pyrimidine-2-carboxylates-a new class of HIV-1 integrase inhibitors. *Bioorg Med Chem Lett* 19: 1930–1934.

Du L, Zhao Y, Chen J, et al. (2008). D77, one benzoic acid derivative, functions as a novel anti-HIV-1 inhibitor targeting the interaction between integrase and cellular LEDGF/p75. *Biochem Biophys Res Commun* 375: 139–144.

Dubey S, Abbas N, Goutham G, Bhosle PA. (2012). QSAR modeling of bifunctional quinolonyl diketo acid derivatives as integrase inhibitors. *Med Chem Res* 21: 964–973.

Dubey S, Abbas NK, Goutham G, Bhosle PA. (2011). QSAR studies on quinolonyl diketo acid derivatives. *NSHM J Pharm Healthcare Manage* 2: 66–73.

Dubois M, Bailly F, Mbemba G, et al. (2008). Reaction of rosmarinic acid with nitrite ions in acidic conditions: Discovery of nitro- and dinitrorosmarinic acids as new anti-HIV-1 agents. *J Med Chem* 51: 2575–2579.

Dudek AZ, Arodz T, Gálvez J. (2006). Computational methods in developing quantitative structure-activity relationships (QSAR): A review. *Comb Chem High Throughput Screen* 9: 213–228.

Egbertson MS, Moritz HM, Melamed JY, et al. (2007). A potent and orally active HIV-1 integrase inhibitor. *Bioorg Med Chem Lett* 17: 1392–1398.

Engelman A, Bushman FD, Craigie R. (1993). Identification of discrete functional domains of HIV-1 integrase and their organization within an active multimeric complex. *EMBO J* 12: 3269–3275.

Engelman A, Mizuuchi K, Craigie R. (1991). HIV-1 DNA integration: Mechanism of viral DNA cleavage and DNA strand transfer. *Cell* 67: 1211–1221.

Fan X, Zhang FH, Al-Safi RI, et al. (2011). Design of HIV-1 integrase inhibitors targeting the catalytic domain as well as its interaction with LEDGF/p75: A scaffold hopping approach using salicylate and catechol groups. *Bioorg Med Chem* 19: 4935–4952.

Fauci AS. (1988). The human immunodeficiency virus: Infectivity and mechanisms of pathogenesis. *Science* 239: 617–622.

Ferrara M, Fiore F, Summa V, Gardelli C. (2010). Development of 2-pyrrolidinyl-N-methyl pyrimidones as potent and orally bioavailable HIV integrase inhibitors. *Bioorg Med Chem Lett* 20: 5031–5034.

Ferro S, De Luca L, Barreca ML, et al. (2010). New chloro, fluorobenzylindole derivatives as integrase strand-transfer inhibitors (INSTIs) and their mode of action. *Bioorg Med Chem* 18: 5510–5518.

Fesen MR, Kohn KW, Leteurtre F, Pommier Y. (1993). Inhibitors of human immunodeficiency virus integrase. *Proc Natl Acad Sci U S A* 90: 2399–2403.

Fesen MR, Pommier Y, Leteurtre F, et al. (1994). Inhibition of HIV-1 integrase by flavones, caffeic acid phenethyl ester (CAPE) and related compounds. *Biochem Pharmacol* 48: 595–608.

Gardelli C, Nizi E, Muraglia E, et al. (2007). Discovery and synthesis of HIV integrase inhibitors: Development of potent and orally bioavailable N-methyl pyrimidones. *J Med Chem* 50: 4957–4975.

Garg R, Gupta SP, Gao H, et al. (1999). Comparative quantitative structure-activity relationship studies on anti-HIV drugs. *Chem Rev* 99: 3525–3602.

Garofalo A, Grande F, Brizzi A, et al. (2008). Naphthoxazepine inhibitors of HIV-1 integrase: Synthesis and biological evaluation. *Chem Med Chem* 3: 986–990.

Goudarzi N, Goodarzi M, Chen T. (2012). QSAR prediction of HIV inhibition activity of styrylquinoline derivatives by genetic algorithm coupled with multiple linear regressions. *Med Chem Res* 21: 437–443.

Grobler JA, Stillmock K, Hu B, et al. (2002). Diketo acid inhibitor mechanism and HIV-1 integrase: Implications for metal binding in the active site of phosphotransferase enzymes. *Proc Natl Acad Sci U S A* 99: 6661–6666.

Guare JP, Wai JS, Gomez RP, et al. (2006). A series of 5-aminosubstituted 4-fluorobenzyl-8-hydroxy-[1,6]naphthyridine-7-carboxamide HIV-1 integrase inhibitors. *Bioorg Med Chem Lett* 16: 2900–2904.

Gupta M, Madan AK. (2012). Diverse models for the prediction of HIV integrase inhibitory activity of substituted quinolone carboxylic acids. *Arch Pharm (Weinheim)* 345: 989–1000.

Gupta P, Garg P, Roy N. (2011). Comparative docking and CoMFA analysis of curcumine derivatives as HIV-1 integrase inhibitors. *Mol Diversity* 15: 733–750.

Gupta P, Garg P, Roy N. (2012). Identification of novel HIV-1 integrase inhibitors using shape-based screening, QSAR, and docking approach. *Chem Biol Drug Des* 79: 835–849.

Gupta P, Roy N, Garg P. (2009). Docking-based 3D-QSAR study of HIV-1 integrase inhibitors. *Eur J Med Chem* 44: 4276–4287.

Gupta SP, Nagappa AN. (2003). Design and development of integrase inhibitors as anti-HIV agents. *Curr Med Chem* 10: 1779–1794.

Hadi V, Koh YH, Sanchez TW, et al. (2010). Development of the next generation of HIV-1 integrase inhibitors: Pyrazolone as a novel inhibitor scaffold. *Bioorg Med Chem Lett* 20: 6854–6857.

Hajimahdi Z, Zarghi A, Zabihollahi R, Aghasadeghi MR. (2013). Synthesis, biological evaluation, and molecular modeling studies of new 1,3,4-oxadiazole- and 1,3,4-thiadiazole-substituted 4-oxo-4H-pyrido[1,2-a]pyrimidines as anti-HIV-1 agents. *Med Chem Res* 22: 2467–2475.

Hansch C, Fujita T. (1964). p-$\sigma$-$\pi$ analysis. A method for the correlation of biological activity and chemical structure. *J Am Chem Soc* 86: 1616–1626.

Hansch C, Leo A. (1995). *Exploring QSAR: Volume 1: Fundamentals and Applications in Chemistry and Biology.* Washington, DC: American Chemical Society.

Hansch C, Maloney PP, Fujita T, Muir RM. (1962). Correlation of biological activity of phenoxyacetic acids with Hammett substituent constants and partition coefficients. *Nature* 194: 178–180.

Hayouka Z, Hurevich M, Levin A, et al. (2010). Cyclic peptide inhibitors of HIV-1 integrase derived from the LEDGF/p75 protein. *Bioorg Med Chem* 18: 8388–8395.

Hayouka Z, Rosenbluh J, Levin A, et al. (2007). Inhibiting HIV-1 integrase by shifting its oligomerization equilibrium. *Proc Natl Acad Sci U S A* 104: 8316–8321.

Hazuda DJ, Anthony NJ, Gomez RP, et al. (2004a). A naphthyridine carboxamide provides evidence for discordant resistance between mechanistically identical inhibitors of HIV-1 integrase. *Proc Natl Acad Sci U S A* 101: 11233–11238.

Hazuda DJ, Felock P, Witmer M, et al. (2000). Inhibitors of strand transfer that prevent integration and inhibit HIV-1 replication in cells. *Science* 287: 646–650.

Hazuda DJ, Young SD, Guare JP, et al. (2004b). Integrase inhibitors and cellular immunity suppress retroviral replication in rhesus macaques. *Science* 305: 528–532.

He QQ, Zhang X, Wu HQ, et al. (2011). Synthesis and biological evaluation of HQCAs with aryl or benzyl substituents on N-1 position as potential HIV-1 integrase inhibitors. *Bioorg Med Chem* 19: 5553–5558.

He QQ, Zhang X, Yang LM, Zheng YT, Chen F. (2013). Synthesis and biological evaluation of 5-fluoroquinolone-3-carboxylic acids as potential HIV-1 integrase inhibitors. *J Enzyme Inhib Med Chem* 28: 671–676.

Hombrouck A, Clayton R, Voet A, Witvrouw M, Debyser Z. (2011). Resistance to inhibitors of HIV-1 integrase. In: Neamati N (ed.), *HIV-1 Integrase: Mechanism and Inhibitor Design*. Hokoben, NJ: John Wiley & Sons, pp. 477–498.

Hopfinger AJ, Wang S, Tokarski JS, et al. (1997). Construction of 3D-QSAR models using the 4D-QSAR analysis formalism. *J Am Chem Soc* 119: 10509–10524.

Hu L, Yan S, Luo Z, et al. (2012a). Design, practical synthesis, and biological evaluation of novel 6-(pyrazolylmethyl)-4-quinoline-3-carboxylic acid derivatives as HIV-1 integrase inhibitors. *Molecules* 17: 10652–10666.

Hu L, Zhang S, He X, et al. (2012b). Design and synthesis of novel β-diketo derivatives as HIV-1 integrase inhibitors. *Bioorg Med Chem* 20: 177–182.

International Collaboration on HIV and Cancer. (2000). Highly active antiretroviral therapy and incidence of cancer in human immunodeficiency virus-infected adults. *J Natl Cancer Inst* 92: 1823–1830.

Iyer M, Hopfinger AJ. (2007). Treating chemical diversity in QSAR analysis: Modeling diverse HIV-1 integrase inhibitors using 4D fingerprints. *J Chem Inf Model* 47: 1945–1960.

Jalali-Heravia M, Ebrahimi-Najafabadia H. (2011). The use of ladder particle swarm optimisation for quantitative structure-activity relationship analysis of human immunodeficiency virus-1 integrase inhibitors. *Mol Simul* 37: 1221–1233.

Johns BA, Kawasuji T, Weatherhead JG, et al. (2011). Combining symmetry elements results in potent naphthyridinone (NTD) HIV-1 integrase inhibitors. *Bioorg Med Chem Lett* 21: 6461–6464.

Johns BA, Kawasuji T, Weatherhead JG, et al. (2013). Naphthyridinone (NTD) integrase inhibitors: N1 protio and methyl combination substituent effects with C3 amide groups. *Bioorg Med Chem Lett* 23: 422–425.

Johns BA, Weatherhead JG, Allen SH, et al. (2009a). 1,3,4-Oxadiazole substituted naphthyridines as HIV-1 integrase inhibitors. Part 2: SAR of the C5 position. *Bioorg Med Chem Lett* 19: 1807–1810.

Johns BA, Weatherhead JG, Allen SH, et al. (2009b). The use of oxadiazole and triazole substituted naphthyridines as HIV-1 integrase inhibitors. Part 1: Establishing the pharmacophore. *Bioorg Med Chem Lett* 19: 1802–1806.

Johnson AA, Marchand C, Pommier Y. (2004). HIV-1 integrase inhibitors: A decade of research and two drugs in clinical trial. *Curr Top Med Chem* 4: 1059–1077.

Jones ED, Vandegraaff N, Le G, et al. (2010). Design of a series of bicyclic HIV-1 integrase inhibitors. Part 1: Selection of the scaffold. *Bioorg Med Chem Lett* 20: 5913–5917.

Kaushik S, Gupta SP, Sharma PK, Anwer Z. (2011a). A QSAR study on a series of N-methyl pyrimidones acting as HIV integrase inhibitors. *Indian J Biochem Biophys* 48: 427–434.

Kaushik S, Gupta SP, Sharma PK, Anwer Z. (2011b). A QSAR study on some series of HIV-1 integrase inhibitors. *Med Chem* 7: 553–560.

Kawasuji T, Johns BA, Yoshida H, et al. (2012). Carbamoyl pyridone HIV-1 integrase inhibitors. 1. Molecular design and establishment of an advanced two-metal binding pharmacophore. *J Med Chem* 55: 8735–8744.

Kehlenbeck S, Betz U, Birkmann A, et al. (2006). Dihydroxythiophenes are novel potent inhibitors of human immunodeficiency virus integrase with a diketo acid-like pharmacophore. *J Virol* 80: 6883–6894.

Keiser MJ, Setola V, Irwin JJ, et al. (2009). Predicting new molecular targets for known drugs. *Nature* 462: 175–181.

Ko GM, Reddy AS, Garg R, Kumar S, Hadaegh AR. (2012). Computational modeling methods for QSAR studies on HIV-1 integrase inhibitors (2005–2010). *Curr Comput Aided Drug Des* 8: 255–270.

Ko GM, Reddy AS, Kumar S, Bailey BA, Garg R. (2010). Computational analysis of HIV-1 protease protein binding pockets. *J Chem Inf Model* 50: 1759–1771.

Kuo CL, Assefa H, Kamath S, et al. (2004). Application of CoMFA and CoMSIA 3D-QSAR and docking studies in optimization of mercaptobenzenesulfonamides as HIV-1 integrase inhibitors. *J Med Chem* 47: 385–399.

Kurup A, Mekapati SB, Garg R, Hansch C. (2003). HIV-1 protease inhibitors: A comparative QSAR analysis. *Curr Med Chem* 10: 1679–1688.

Lameira J, Medeiros IG, Reis M, Santos AS, Alves CN. (2006). Structure-activity relationship study of flavone compounds with anti-HIV-1 integrase activity: A density functional theory study. *Bioorg Med Chem* 14: 7105–7112.

Lee JY, Park JH, Lee SJ, Park H, Lee YS. (2002). Styrylquinazoline derivatives as HIV-1 integrase inhibitors. *Arch Pharm (Weinheim)* 335: 277–282.

Lee SH, Oshige M, Durant ST, et al. (2005). The SET domain protein Metnase mediates foreign DNA integration and links integration to nonhomologous end-joining repair. *Proc Natl Acad Sci U S A* 102: 18075–18080.

Leonard JT, Roy K. (2008). Exploring molecular shape analysis of styrylquinoline derivatives as HIV-1 integrase inhibitors. *Eur J Med Chem* 43: 81–92.

Li BJ, Chiang CC, Hsu LY. (2010). QSAR studies of 3,3'-(substituted-benzylidene)-bis-4-hydroxycoumarin, potential HIV-1 integrase inhibitor. *J Chin Chem Soc* 57: 742–749.

Li H, Wang C, Sanchez T, et al. (2009). Amide-containing diketoacids as HIV-1 integrase inhibitors: Synthesis, structure-activity relationship analysis, and biological activity. *Bioorg Med Chem* 17: 2913–2919.

Liao C, Nicklaus MC. (2011). HIV-1 integrase-DNA models. In: Neamati N (ed.), *HIV-1 Integrase: Mechanism and Inhibitor Design*. Hokoben, NJ: John Wiley & Sons, pp. 429–455.

Lin Z, Neamati N, Zhao H, et al. (1999). Chicoric acid analogues as HIV-1 integrase inhibitors. *J Med Chem* 42: 1401–1414.

Liu Q, Zhou H, Liu L, et al. (2011). Multi-target QSAR modelling in the analysis and design of HIV-HCV co-inhibitors: An in-silico study. *BMC Bioinformatics* 12: 294.

Long YQ, Jiang XH, Dayam R, et al. (2004). Rational design and synthesis of novel dimeric diketoacid-containing inhibitors of HIV-1 integrase: Implication for binding to two metal ions on the active site of integrase. *J Med Chem* 47: 2561–2573.

Lu P, Wei X, Zhang R. (2010). CoMFA and CoMSIA 3D-QSAR studies on quionolone caroxylic acid derivatives inhibitors of HIV-1 integrase. *Eur J Med Chem* 45: 3413–3419.

Ma K, Wang P, Fu W, et al. (2011). Rational design of 2-pyrrolinones as inhibitors of HIV-1 integrase. *Bioorg Med Chem Lett* 21: 6724–6727.

Ma XH, Zhang XY, Tan JJ, Chen WZ, Wang CX. (2004). Exploring binding mode for styrylquinoline HIV-1 integrase inhibitors using comparative molecular field analysis and docking studies. *Acta Pharmacol Sin* 25: 950–958.

Makhija MT, Kulkarni VM. (2001a). Eigen value analysis of HIV-1 integrase inhibitors. *J Chem Inf Model* 41: 1569–1577.

Makhija MT, Kulkarni VM. (2001b). Molecular electrostatic potentials as input for the alignment of HIV-1 integrase inhibitors in 3D QSAR. *J Comput Aided Mol Des* 15: 961–978.

Makhija MT, Kulkarni VM. (2002a). 3D-QSAR and molecular modeling of HIV-1 integrase inhibitors. *J Comput Aided Mol Des* 16: 181–200.

Makhija MT, Kulkarni VM. (2002b). QSAR of HIV-1 integrase inhibitors by genetic function approximation method. *Bioorg Med Chem* 10: 1483–1497.

Mao PCM, Mouscadet JF, Leh H, Auclair C, Hsu LY. (2002). Chemical modification of coumarin dimer and HIV-1 integrase inhibitory activity. *Chem Pharm Bull* 50: 1634–1637.

Marchand C, Beutler JA, Wamiru A, et al. (2008). Madurahydroxylactone derivatives as dual inhibitors of human immunodeficiency virus type 1 integrase and RNase H. *Antimicrob Agents Chemother* 52: 361–364.

Marrero-Ponce Y. (2004). Linear indices of the "molecular pseudograph's atom adjacency matrix": Definition, significance-interpretation, and application to QSAR analysis of flavone derivatives as HIV-1 integrase inhibitors. *J Chem Inf Comput Sci* 44: 2010–2026.

Maurer K, Tang AH, Kenyon GL, Leavitt AD. (2000). Carbonyl J derivatives: A new class of HIV-1 integrase inhibitors. *Bioorg Chem* 28: 140–155.

Mazumder A, Gazit A, Levitzki A, et al. (1995a). Effects of tyrphostins, protein kinase inhibitors, on human immunodeficiency virus type 1 integrase. *Biochem* 34: 15111–15122.

Mazumder A, Neamati N, Sunder S, et al. (1997). Curcumin analogs with altered potencies against HIV-1 integrase as probes for biochemical mechanisms of drug action. *J Med Chem* 40: 3057–3063.

Mazumder A, Raghavan K, Weinstein J, Kohn KW, Pommier Y. (1995b). Inhibition of human immunodeficiency virus type-1 integrase by curcumin. *Biochem Pharmacol* 18: 1165–1170.

Mazumder A, Wang S, Neamati N, et al. (1996). Antiretroviral agents as inhibitors of both human immunodeficiency virus type 1 integrase and protease. *J Med Chem* 39: 2472–2481.

Meadows DC, Mathews TB, North TW, et al. (2005). Synthesis and biological evaluation of geminal disulfones as HIV-1 integrase inhibitors. *J Med Chem* 48: 4526–4534.

Meadows DC, Sanchez T, Neamati N, North TW, Gervay-Hague J. (2007). Ring substituent effects on biological activity of vinyl sulfones as inhibitors of HIV-1. *Bioorg Med Chem* 15: 1127–1137.

Mekouar K, Mouscadet JF, Desmaële D, et al. (1998). Styrylquinoline derivatives: A new class of potent HIV-1 integrase inhibitors that block HIV-1 replication in CEM cells. *J Med Chem* 41: 2846–2857.

Mouscadet JF, Desmaële D. (2010). Chemistry and structure-activity relationship of the styrylquinoline-type HIV integrase inhibitors. *Molecules* 15: 3048–3078.

Muraglia E, Kinzel O, Gardelli C, et al. (2008). Design and synthesis of bicyclic pyrimidinones as potent and orally bioavailable HIV-1 integrase inhibitors. *J Med Chem* 51: 861–874.

Nagasawa JY, Song J, Chen H, et al. (2011). 6-Benzylamino 4-oxo-1,4-dihydro-1,8-naphthyridines and 4-oxo-1,4-dihydroquinolines as HIV integrase inhibitors. *Bioorg Med Chem Lett* 21: 760–763.

Nair V, Chi G, Ptak R, Neamati N. (2006a). HIV integrase inhibitors with nucleobase scaffolds: Discovery of a highly potent anti-HIV agent. *J Med Chem* 49: 445–447.

Nair V, Uchil V, Neamati N. (2006b). β-diketo acids with purine nucleobase scaffolds: Novel, selective inhibitors of the strand transfer step of HIV integrase. *Bioorg Med Chem Lett* 16: 1920–1923.

Neamati N. (2011). HIV-1 integrase inhibitor design: Overview and historical perspectives. In: Neamati N (ed.), *HIV-1 Integrase: Mechanism and Inhibitor Design*. Hokoben, NJ: John Wiley & Sons, pp. 165–196.

Neamati N, Hong H, Mazumder A, et al. (1997a). Depsides and depsidones as inhibitors of HIV-1 integrase: Discovery of novel inhibitors through 3D database searching. *J Med Chem* 40: 942–951.

Neamati N, Hong H, Owen JM, et al. (1998). Salicylhydrazine-containing inhibitors of HIV-1 integrase: Implication for a selective chelation in the integrase active site. *J Med Chem* 41: 3202–3209.

Neamati N, Hong H, Sunder S, Milne GWA, Pommier Y. (1997b). Potent inhibitors of human immunodeficiency virus type 1 integrase: Identification of a novel four-point pharmacophore and tetracyclines as novel inhibitors. *Mol Pharmacol* 52: 1041–1055.

Neamati N, Marchand C, Pommier Y. (2000). HIV-1 integrase inhibitors: Past, present, and future. *Adv Pharmacol* 49: 147–165.

Neamati N, Mazumder A, Zhao H, et al. (1997c). Diarylsulfones, a novel class of human immunodeficiency virus type 1 integrase inhibitors. *Antimicrob Agents Chemother* 41: 385–393.

Neamati N, Turpin JA, Winslow HE, et al. (1999). Thiazolothiazepine inhibitors of HIV-1 integrase. *J Med Chem* 42: 3334–3341.

Nicklaus MC, Neamati N, Hong H, et al. (1997). HIV-1 integrase pharmacophore: Discovery of inhibitors through three-dimensional database searching. *J Med Chem* 40: 920–929.

Niedbala H, Polanski J, Gieleciak R, et al. (2006). Comparative molecular surface analysis (CoMSA) for virtual combinatorial library screening of styrylquinoline HIV-1 blocking agents. *Comb Chem High Throughput Screen* 9: 753–770.

Nizi E, Orsale MV, Crescenzi B, et al. (2009). Dihydroxy-pyrimidine and N-methylpyrimidone HIV-integrase inhibitors: Improving cell based activity by the quaternarization of a chiral center. *Bioorg Med Chem Lett* 19: 4617–4621.

Normand-Bayle M, Bénard C, Zouhiri F, et al. (2005). New HIV-1 replication inhibitors of the styryquinoline class bearing aroyl/acyl groups at the C-7 position: Synthesis and biological activity. *Bioorg Med Chem Lett* 15: 4019–4022.

Nunthaboot N, Tonmunphean S, Parasuk V, Wolschann P, Kokpol S. (2006). Three-dimensional quantitative structure: Activity relationship studies on diverse structural classes of HIV-1 integrase inhibitors using CoMFA and CoMSIA. *Eur J Med Chem* 41: 1359–1372.

Oprea TI, Bauman JE, Bologa CG, et al. (2011). Drug repurposing from an academic perspective. *Drug Discov Today Ther Strateg* 8: 61–69.

Oprea TI, Mestres J. (2012). Drug repurposing: Far beyond new targets for old drugs. *AAPS J* 14: 759–763.

Ouali M, Laboulais C, Leh H, et al. (2000). Modeling of the inhibition of retroviral integrases by styrylquinoline derivatives. *J Med Chem* 43: 1949–1957.

Pace P, Di Francesco ME, Gardelli C, et al. (2007). Dihydroxypyrimidine-4-carboxamides as novel potent and selective HIV integrase inhibitors. *J Med Chem* 50: 2225–2239.

Pace P, Spieser SAH, Summa V. (2008). 4-Hydroxy-5-pyrrolinone-3-carboxamide HIV-1 integrase inhibitors. *Bioorg Med Chem Lett* 18: 3865–3869.

Pais GCG, Zhang X, Marchand C, et al. (2002). Structure activity of 3-aryl-1,3-diketo-containing compounds as HIV-1 integrase inhibitors. *J Med Chem* 45: 3184–3194.

Palefsky J. (2012). HIV protease inhibitors to prevent progression of cervical intraepithelial neoplasia to cervical cancer: Therapeutic opportunities and challenges. *AIDS* 26: 1035–1036.

Park JH, Lee SU, Kim SH, et al. (2008). Chromone and chromanone derivatives as strand transfer inhibitors of HIV-1 integrase. *Arch Pharm Res* 31: 1–5.

Pasquini S, Mugnaini C, Tintori C, et al. (2008). Investigations on the 4-quinolone-3-carboxylic acid motif. 1. Synthesis and structure-activity relationship of a class of human immunodeficiency virus type 1 integrase inhibitors. *J Med Chem* 51: 5125–5129.

Patil S, Kamath S, Sanchez T, et al. (2007). Synthesis and biological evaluation of novel 5(H)-phenanthridin-6-ones, 5(H)-phenanthridin-6-one diketo acid, and polycyclic aromatic diketo acid analogs as new HIV-1 integrase inhibitors. *Bioorg Med Chem* 15: 1212–1228.

Pendri A, Troyer TL, Sofia MJ, et al. (2010). Solid phase synthesis of novel pyrrolidinedione analogs as potent HIV-1 integrase inhibitors. *J Comb Chem* 12: 84–90.

Petrocchi A, Jones P, Rowley M, Fiore F, Summa V. (2009). N-(4-Fluorobenzyl)-3-hydroxy-9,9-dimethyl-4-oxo-6,7,8,9-tetrahydro-4H-pyrazino[1,2-a]pyrimidine-2-carboxamides a novel class of potent HIV-1 integrase inhibitors. *Bioorg Med Chem Lett* 19: 4245–4249.

Petrocchi A, Koch U, Matassa VG, et al. (2007). From dihydroxypyrimidine carboxylic acids to carboxamide HIV-1 integrase inhibitors: SAR around the amide moiety. *Bioorg Med Chem Lett* 17: 350–353.

Polanski J, Gieleciak R, Bak A. (2004). Probability issues in molecular design: Predictive and modeling ability in 3D-QSAR schemes. *Comb Chem High Throughput Screen* 7: 793–807.

Polanski J, Zouhiri F, Jeanson L, et al. (2002). Use of the Kohonen neural network for rapid screening of ex vivo anti-HIV activity of styrylquinolines. *J Med Chem* 45: 4647–4654.

Pommier Y, Johnson AA, Marchand C. (2005). Integrase inhibitors to treat HIV/AIDS. *Nat Rev Drug Discov* 4: 236–248.

Pommier Y, Marchand C, Neamati N. (2000). Retroviral integrase inhibitors year 2000: Update and perspectives. *Antiviral Res* 49: 139–148.

Raghavan K, Buolamwini JK, Fesen MR, et al. (1995). Three-dimensional quantitative structure-activity relationship (QSAR) of HIV integrase inhibitors: A comparative molecular field analysis (CoMFA) study. *J Med Chem* 38: 890–897.

Ramajayam R, Mahera NB, Neamati N, Yadav MR, Giridhar R. (2009). Synthesis and anti-HIV-1 integrase activity of cyano pyrimidinones. *Arch Pharm (Weinheim)* 342: 710–715.

Ramkumar K, Serrao E, Odde S, Neamati N. (2010a). HIV-1 integrase inhibitors: 2007–2008 update. *Med Res Rev* 30: 890–954.

Ramkumar K, Tambov KV, Gundla R, et al. (2008). Discovery of 3-acetyl-4-hydroxy-2-pyranone derivatives and their difluoridoborate complexes as a novel class of HIV-1 integrase inhibitors. *Bioorg Med Chem* 16: 8988–8998.

Ramkumar K, Yarovenko VN, Nikitina AS, et al. (2010b). Design, synthesis and structure-activity studies of rhodanine derivatives as HIV-1 integrase inhibitors. *Molecules* 15: 3958–3992.

Ravichandran V, Mourya VK, Agrawal RK. (2008). QSAR analysis of 6-aryl-2,4-dioxo-5-hexenoic acids as HIV-1 integrase inhibitors. *Indian J Pharm Educ Res* 42: 133–140.

Ravichandran V, Shalini S, Sundram K, Sokkalingam AD. (2010). QSAR study of substituted 1,3,4-oxadiazole naphthyridines as HIV-1 integrase inhibitors. *Eur J Med Chem* 45: 2791–2797.

Reddy AS, Jalahalli V, Kumar S, et al. (2011). Analysis of HIV protease binding pockets based on 3D shape and electrostatic potential descriptors. *Chem Biol Drug Des* 77: 137–151.

Reddy AS, Kumar S, Garg R. (2010). Hybrid-genetic algorithm based descriptor optimization and QSAR models for predicting the biological activity of tipranavir analogs for HIV protease inhibition. *J Mol Graph Model* 28: 852–862.

Reddy KK, Singh SK, Dessalew N, Tripathi SK, Selvaraj C. (2012). Pharmacophore modelling and atom-based 3D-QSAR studies on N-methyl pyrimidones as HIV-1 integrase inhibitors. *J Enzyme Inhib Med Chem* 27: 339–347.

Robinson WE, Jr, Reinecke MG, Abdel-Malek S, Jia Q, Chow SA. (1996). Inhibitors of HIV-1 replication that inhibit HIV integrase. *Proc Natl Acad Sci U S A* 93: 6326–6331.

Sahu KK, Ravichandran V, Mourya VK, Agrawal RK. (2007). QSAR analysis of caffeoyl naphthalene sulfonamide derivatives as HIV-1 integrase inhibitors. *Med Chem Res* 15: 418–430.

Saíz-Urra L, González MP, Fall Y, Gómez G. (2007). Quantitative structure-activity relationship studies of HIV-1 integrase inhibition. 1. GETAWAY descriptors. *Eur J Med Chem* 42: 64–70.

Sanchez TW, Debnath B, Christ F, et al. (2013). Discovery of novel inhibitors of LEDGF/p75-IN protein-protein interactions. *Bioorg Med Chem* 21: 957–963.

Sato M, Kawakami H, Motomura T, et al. (2009). Quinolone carboxylic acids as a novel monoketo acid class of human immunodeficiency virus type 1 integrase inhibitors. *J Med Chem* 52: 4869–4882.

Sato M, Motomura T, Aramaki H, et al. (2006). Novel HIV-1 integrase inhibitors derived from quinolone antibiotics. *J Med Chem* 49: 1506–1508.

Satpathy R, Ghosh S. (2011). In-silico comparative study and quantitative structure-activity relationship analysis of some structural and physiochemical descriptors of elvitegravir analogs. *J Young Pharm* 3: 246–249.

Sechi M, Bacchi A, Carcelli M, et al. (2006). From ligand to complexes: Inhibition of human immunodeficiency virus type 1 integrase by beta-diketo acid metal complexes. *J Med Chem* 49: 4248–4260.

Sechi M, Carta F, Sannia L, et al. (2009a). Design, synthesis, molecular modeling, and anti-HIV-1 integrase activity of a series of photoactivatable diketo acid-containing inhibitors as affinity probes. *Antiviral Res* 81: 267–276.

Sechi M, Derudas M, Dallocchio R, et al. (2004). Design and synthesis of novel indole β-diketo acid derivatives as HIV-1 integrase inhibitors. *J Med Chem* 47: 5298–5310.

Sechi M, Rizzi G, Bacchi A, et al. (2009b). Design and synthesis of novel dihydroquinoline-3-carboxylic acids as HIV-1 integrase inhibitors. *Bioorg Med Chem* 17: 2925–2935.

Sechi M, Sannia L, Carta F, et al. (2005). Design of novel bioisosteres of beta-diketo acid inhibitors of HIV-1 integrase. *Antiviral Chem Chemother* 16: 41–61.

Serrao E, Debnath B, Otake H, et al. (2013). Fragment-based discovery of 8-hydroxyquinoline inhibitors of the HIV-1 integrase-lens epithelium-derived growth factor/p75 (IN-LEDGF/p75) interaction. *J Med Chem* 56: 2311–2322.

Serrao E, Gundla R, Deng J, Odde S, Neamati N. (2011). Computer-aided techniques in design of HIV-1 integrase inhibitors. In: Neamati N (ed.), *HIV-1 Integrase: Mechanism and Inhibitor Design*. Hokoben, NJ: John Wiley & Sons, pp. 477–498.

Serrao E, Odde S, Ramkumar K, Neamati N. (2009). Raltegravir, elvitegravir, and metoogravir: The birth of "me-too" HIV-1 integrase inhibitors. *Retrovirology* 6: 25.

Shafer RW, Schapiro JM. (2008). HIV-1 drug resistance mutations: An updated framework for the second decade of HAART. *AIDS Rev* 10: 67–84.

Sharma H, Cheng X, Buolamwini JK. (2012). Homology model-guided 3D-QSAR studies of HIV-1 integrase inhibitors. *J Chem Inf Model* 52: 515–544.

Sharma H, Patil S, Sanchez TW, et al. (2011). Synthesis, biological evaluation and 3D-QSAR studies of 3-keto salicylic acid chalcones and related amides as novel HIV-1 integrase inhibitors. *Bioorg Med Chem* 19: 2030–2045.

Srivastava HK, Bohari MH, Sastry GN. (2012). Modeling anti-HIV compounds: The role of analogue-based approaches. *Curr Comput Aided Drug Des* 8: 224–248.

Summa V, Petrocchi A, Bonelli F, et al. (2008). Discovery of raltegravir, a potent, selective orally bioavailable HIV-integrase inhibitor for the treatment of HIV-AIDS infection. *J Med Chem* 51: 5843–5855.

Summa V, Petrocchi A, Matassa VG, et al. (2006). 4,5-Dihydroxypyrimidine carboxamides and N-alkyl-5-hydroxypyrimidinone carboxamides are potent, selective HIV integrase inhibitors with good pharmacokinetic profiles in preclinical species. *J Med Chem* 49: 6646–6649.

Sun XH, Guan JQ, Tan JJ, Liu C, Wang CX. (2012). 3D-QSAR studies of quinoline ring derivatives as HIV-1 integrase inhibitors. *SAR QSAR Environ Res* 23: 683–703.

Tang J, Maddali K, Metifiot M, et al. (2011). 3-Hydroxypyrimidine-2,4-diones as an inhibitor scaffold of HIV integrase. *J Med Chem* 54: 2282–2292.

Telvekar VN, Patel KN. (2011). Pharmacophore development and docking studies of the HIV-1 integrase inhibitors derived from N-methylpyrimidones, dihydroxypyrimidines, and bicyclic pyrimidinones. *Chem Biol Drug Des* 78: 150–160.

Tintori C, Veljkovic N, Veljkovic V, Botta M. (2010). Computational studies of the interaction between the HIV-1 integrase tetramer and the cofactor LEDGF/p75: Insights from molecular dynamics simulations and the informational spectrum method. *Proteins* 78: 3396–3408.

Toropova AP, Toropov AA, Benfenati E, Gini G. (2011). Simplified molecular input-line entry system and international chemical identifier in the QSAR analysis of styrylquinoline derivatives as HIV-1 integrase inhibitors. *Chem Biol Drug Des* 77: 343–360.

Tropsha A, Gramatica P, Gombar VK. (2003). The importance of being earnest: Validation is the absolute essential for successful application and interpretation of QSPR models. *QSAR Comb Sci* 22: 69–77.

UNAIDS. (2012). *UNAIDS Report on the Global AIDS Epidemic 2012*. Geneva: Joint United Nations Programme on HIV/AIDS.

Veerasamy R, Ying KH, Reneeraj JS, Hui LS. (2011). Designing hypothesis of substituted naphthyridines as anti-HIV agent: A QSAR approach. *Int J Drug Des Discovery* 2: 591–604.

Vengurlekar S, Sharma R, Trivedi P. (2010). Two- and three-dimensional QSAR studies on benzyl amide-ketoacid inhibitors of HIV integrase and their reduced analogues. *Med Chem Res* 19: 1106–1120.

Verschueren WG, Dierynck I, Amssoms KIE, et al. (2005). Design and optimization of tricyclic phthalimide analogues as novel inhibitors of HIV-1 integrase. *J Med Chem* 48: 1930–1940.

Wai JS, Egbertson MS, Payne LS, et al. (2000). 4-Aryl-2,4-dioxobutanoic acid inhibitors of HIV-1 integrase and viral replication in cells. *J Med Chem* 43: 4923–4926.

Walker MA, Johnson T, Naidu BN, et al. (2007). Benzyl amide-ketoacid inhibitors of HIV-integrase. *Bioorg Med Chem Lett* 27: 4886–4890.

Wang P, Liu C, Sanches T, et al. (2009). Design and synthesis of novel nitrogen-containing polyhydroxylated aromatics as HIV-1 integrase inhibitors from caffeic acid phenethyl ester. *Bioorg Med Chem Lett* 19: 4574–4578.

Wang Z, Tang J, Salomon CE, Dreis CD, Vince R. (2010). Pharmacophore and structure-activity relationships of integrase inhibition within a dual inhibitor scaffold of HIV reverse transcriptase and integrase. *Bioorg Med Chem* 18: 4202–4211.

Wang Z, Wang M, Yao X, et al. (2012a). Design, synthesis and antiviral activity of novel pyridazines. *Eur J Med Chem* 54: 33–41.

Wang Z, Wang M, Yao X, et al. (2012b). Design, synthesis and antiviral activity of novel quinazolinones. *Eur J Med Chem* 53: 275–282.

Wang Z, Wang M, Yao X, et al. (2012c). Hydroxyl may not be indispensable for raltegravir: Design, synthesis and SAR studies of raltegravir derivatives as HIV-1 inhibitors. *Eur J Med Chem* 50: 361–369.

Xia CN, Li HB, Liu F, Hu WX. (2008). Synthesis of trans-caffeate analogues and their bioactivities against HIV-1 integrase and cancer cell lines. *Bioorg Med Chem Lett* 18: 6553–6557.

Xu YW, Zhao GS, Shin CG, et al. (2003). Caffeoyl naphthalene sulfonamide derivatives as HIV integrase inhibitors. *Bioorg Med Chem* 11: 3589–3593.

Yan A, Xuan S, Hu X. (2012). Classification of active and weakly active ST inhibitors of HIV-1 integrase using a support vector machine. *Comb Chem High Through Screen*, 12: 792–805.

Yang L, Xu X, Huang Y, et al. (2010). Synthesis of polyhydroxylated aromatics having amidation of piperazine nitrogen as HIV-1 integrase inhibitor. *Bioorg Med Chem Lett* 20: 5469–5471.

Yoo H, Lee JY, Park JH, Chung BY, Lee YS. (2003). Synthesis of styrylbenzofuran derivatives as styrylquinoline analogues for HIV-1 integrase inhibitor. *Farmaco* 58: 1243–1250.

Yuan H, Parrill AL. (2000). QSAR development to describe HIV-1 integrase inhibition. *J Mol Struct* 529: 273–282.

Yuan H, Parrill AL. (2002). QSAR studies of HIV-1 integrase inhibition. *Bioorg Med Chem* 10: 4169–4183.

Yuan H, Parrill AL. (2005). Cluster analysis and three-dimensional QSAR studies of HIV-1 integrase inhibitors. *J Mol Graph Model* 23: 317–328.

Zeng LF, Jiang XH, Sanchez T, et al. (2008a). Novel dimeric aryldiketo containing inhibitors of HIV-1 integrase: Effects of the phenyl substituent and the linker orientation. *Bioorg Med Chem* 16: 7777–7787.

Zeng LF, Wang Y, Kazemi R. (2012). Repositioning HIV-1 integrase inhibitors for cancer therapeutics: 1,6-naphthyridine-7-carboxamide as a promising scaffold with drug-like properties. *J Med Chem* 55: 9492–9509.

Zeng LF, Zhang HS, Wang YH, et al. (2008b). Efficient synthesis and utilization of phenyl-substituted heteroaromatic carboxylic acids as aryl diketo acid isosteres in the design of novel HIV-1 integrase inhibitors. *Bioorg Med Chem Lett* 18: 4521–4524.

Zhang X, Marchand C, Pommier Y, Burke TR, Jr. (2004). Design and synthesis of photoactivatable aryl diketo acid-containing HIV-1 integrase inhibitors as potential affinity probes. *Bioorg Med Chem Lett* 14: 1205–1207.

Zhang X, Pais GCG, Svarovskaia ES, et al. (2003). Azido-containing aryl β-diketo acid HIV-1 integrase inhibitors. *Bioorg Med Chem Lett* 13: 1215–1219.

Zhao H, Neamati N, Hong H, et al. (1997a). Coumarin-based inhibitors of HIV integrase. *J Med Chem* 40: 242–249.

Zhao H, Neamati N, Mazumder A, et al. (1997b). Arylamide inhibitors of HIV-1 integrase. *J Med Chem* 40: 1186–1194.

Zhao H, Neamati N, Sunder S, et al. (1997c). Hydrazide-containing inhibitors of HIV-1 integrase. *J Med Chem* 40: 937–941.

Zhuang L, Wai JS, Embrey MW, et al. (2003). Design and synthesis of 8-hydroxy-[1,6]-naphthyridines as novel inhibitors of HIV-1 integrase in vitro and in infected cells. *J Med Chem* 46: 453–456.

Zouhiri F, Mouscadet JF, Mekouar K, et al. (2000). Structure-activity relationships and binding mode of styrylquinolines as potent inhibitors of HIV-1 integrase and replication of HIV-1 in cell culture. *J Med Chem* 43: 1533–1540.

# 13 Virus Wars

## *Oncolytic versus Oncogenic Viruses*

*Markus Vähä-Koskela, Fabrice Le Boeuf, and Vincenzo Cerullo*

## 13.1 INTRODUCTION

In this chapter, after a brief introduction to oncoviruses, we aim to present tumor-killing (oncolytic) viruses as therapeutic tools against oncovirus-induced cancers. We attempt to provide an overview of the mechanistic aspects of oncolysis of oncovirus tumors without going too deeply into the oncolytic viruses themselves; these have been described in detail elsewhere (Harrington et al. 2008; Sinkovics and Horvath 2005). Also in this chapter, we do not describe the use of viruses as vaccine platforms despite the fact that it is entirely possible to "arm" oncolytic viruses with antigens from oncoviruses. Rather, we provide hypothetical and actual examples of rational exploitation of oncovirus-enforced tumor biology to enable/promote destructive replication of oncolytic viruses. We believe that the concept of pitting virus against virus, which we have affectionately labeled "virus wars," has matured to the stage of formal testing in the clinics. As we discuss, there are several good reasons to use oncolytic viruses against oncovirus-associated tumors, and efficacy might be achieved even in the absence of direct virus–virus interference within the same cell. Indeed, a shifting view in the field—largely influenced by an accumulating wealth of data from cancer patients—is that although replication is necessary for therapeutic efficacy with oncolytic viruses, induction of antitumor immune responses is the principal determinant of long-term benefit.

## 13.2 ONCOVIRUSES

The World Health Organization International Agency for Research on Cancer estimated that in 2002, 12% of all human cancers worldwide had a causal link to oncovirus infection (Parkin 2006). Typically, among these cancers we find cervical cancer, Burkitt's lymphoma, hepatocellular carcinoma, and Kaposi's sarcoma. Several viruses are now known to possess tumor-inducing capacity. Viruses that have a confirmed link to oncogenesis in humans include hepatitis B and C viruses (HBV and HCV, 5% of all cancers), Epstein–Barr virus (EBV, 1%), human papillomavirus (HPV, 5%), human T-cell leukemia virus type I (HTLV-1, below 1%), Kaposi's sarcoma-associated herpesvirus (HHV-8, below 1%), and Merkel cell polyomavirus (below 1%). Most of these are DNA viruses; only hepatitis C virus is an RNA virus, and HTLV-1 is a retrovirus.

Retroviruses were discovered at the beginning of 20th century when it was shown by Vilhelm Ellermann, Oluf Bang, and Peyton Rous that chicken leucosis and sarcomas were caused by a virus, the Rous sarcoma virus (RSV). In 1964, the EBV was the first for which a direct oncogenic potential was demonstrated (Saha et al. 2010). Murine leukemia viruses (MuLVs) include the Moloney, Rauscher, Abelson, and Friend viruses, named for their discoverers. These are gammaretroviruses that induce a wide spectrum of leukemias having lymphoid, myeloid, or stem-like characteristics with varying penetrance depending on virus strain and mouse background (Eckner and Steeves 1972). The Friend and Raucher spleen focus-forming viruses (SFFVs; named for their ability to induce agglomeration—"foci"—of splenocytes upon intravenous injection in mice) induce erythroleukemia in all susceptible mouse strains in a biphasic pattern associated with molecular and retrovirus insertional events (initially EpoR and sf-STK activation by the viral gp55, followed by Spi-1/PU.1 activation due to retrovirus insertion), and that is recapitulated in acute myeloid leukemia (AML) in humans (Moreau-Gachelin 2008). Thus, the murine retroviruses have served as models for human retrovirus-associated cancers.

After the discovery of RSV, new retroviruses were found infecting multiple vertebrate species (Levine and Enquist 2007; Peyton 1911, 1911b). Discovery of the first human retrovirus, the human T-cell leukemia virus 1 (HTLV-1), became possible only after the 1976 and 1977 publications on what is now known as interleukin-2 (IL-2). (See review by main/co-discoverer of HTLV-1 and IL-2 [Gallo 2005].) Subsequently, in 1979, T-cell malignancy-associated HTLV-1 was found from a T-cell line that originated from a patient with cutaneous T-cell lymphoma (Poiesz et al. 1980; Yoshida et al. 1982). A noteworthy anecdote describing the aforementioned predispositions toward the existence of human retroviruses is that when Gallo first submitted the HTLV-1 article into the *Journal of Virology* it was rejected, and the editor was "insisting that they should cease, and not continue to perpetuate the controversy, strongly implying that we all know human retroviruses do not exist" (Vahlne 2009).

HTLV-1 typically causes a chronic type of eczematous disease, HTLV-1-associated infective dermatitis (HAID), with a prevalence of 5%–10%, characterized by exudative eruptions and persistent infection with *Staphylococcus aureus* and β-hemolytic streptococci. The more serious diseases caused by HTLV-1 include adult T-cell leukemia/lymphoma (ATLL) and tropical spastic paraparesis/HTLV-1-associated myelopathy (TSP/HAM) (McGill et al. 2012).

## 13.3 ONCOLYTIC VIRUSES

### 13.3.1 A Brief History

Interestingly, many observations dating back at least to the 18th century describe beneficial effects of infections on the progression of cancer (Hoption Cann et al. 2002). Sometimes these were attributed to divine intervention, but by the latter part of the 19th century, formal testing (deliberate induction of fever/inflammation) had begun. Work pioneered by William Coley was followed by more incidental observations confirming the therapeutic effect of induced inflammatory events, until in 1912 when spurred by regression of cervical cancer following rabies vaccination, several other patients were given the rabies vaccine with the purpose of inducing tumor regression (De Pace 1912). Next, concurrent with the explosion of virology as a field of research and the plethora of new viruses being shipped in from all around the world, viruses isolated from the wild were tested directly in cancer patients. Some of the viruses were obviously pathogenic and several patients incurred viral disease, sometimes with a fatal outcome (e.g., West Nile virus, Langat- and Bunyamwera virus encephalitis, hepatitis in 14/22 patients, and the death of one inoculated with hepatitis B virus) (Kelly and Russell 2007). By the 1970s, these seemingly cavalier ad hoc experiments had been terminated and some of the associated clinicians dismissed. However, several clinical trials continued throughout these trying times, perhaps fortuitously, with sufficiently mild viruses, and the findings from these trials constitute the foundation of today's more regulated oncolytic virus research. Of the more notable studies from this era, a 1974 publication describes a remission rate of approximately 90% (2/3 partial, 1/3 full) in a cohort of 90 patients with terminal cancer treated with the mumps virus (Asada 1974).

Today, tumor cell specificity is a prerequisite for the development of virus platforms against cancer. With the advent of genetic engineering and the increased selection of stably attenuated viruses, oncolytic virus research has experienced a resurgence. Modern oncolytic viruses are an emerging form of biological cancer therapy and have been selected or engineered for their capacity to infect and kill tumor cells without causing damage to normal cells.

### 13.3.2 Mechanism of Action

Oncolytic viruses are either attenuated viruses, such as natural vaccine strains or serially passaged viruses, or genetically engineered recombinant viruses. Both exploit for their replication molecular pathways over- or underrepresented in cancer cells compared to normal cells. As an example, most tumors display overactivated Ras pathway, which increases cell proliferation and intracellular pools of nucleotides, which are important for the life cycle of most viruses. Indeed, because of the inherent characteristics of cancer cells, some of the viral homologs of cellular functions that enforce virus replication in normal cells can be deleted—the loss of function is compensated by the activated pathways in cancer cells, but now the oncolytic viruses are unable to replicate in normal cells. As an example, the adenovirus E1A gene product features a domain that sequesters Rb tumor suppressor. When sequestered, Rb is unable to inactivate eukaryotic initiation factor 2, which in turn increases translation of virus transcripts. Modern oncolytic adenoviruses feature a deletion in the Rb-binding domain, rendering them virtually unable to replicate in normal cells and instead restrict them to cells in which Rb is inactivated via other mechanisms (most cancer cells).

Oncolytic viruses based on natural unmodified virus strains, particularly RNA viruses such as reovirus type 3 strain Dearing (Reolysin®, Oncolytics Biotech Inc.), rely on defective antiviral defenses in cancer cells (Vähä-Koskela et al. 2007). Antiviral defenses overlap with cellular DNA repair and suicide systems, and it is not surprising that activation of oncogenic pathways (such as increased Ras signaling) and loss of tumor suppressors (such as Rb and p53) result in a severe inability of cancer cells to respond to prototypical antiviral and apoptotic defense mediator, type I interferon (IFN). Such cells become fertile growing ground for viruses that in normal cells induce a self-limiting type I IFN response. One particularly strong type I IFN-inducing oncolytic virus is M mutated vesicular stomatitis virus (VSV), which harbors a mutation or deletion in the matrix (M) gene that renders the M protein unable to prevent nuclear export of interferon transcripts, resulting in copious amounts of type I IFN secreted into the cell surroundings, rendering IFN-responsive normal cells refractory to infection (Stojdl et al. 2000 and 2003).

Oncolytic viruses are able to replicate in tumor cells, turning them into factories of new viral progenies that propagate the infection throughout the tumor and possibly to distant tumor nests via the circulation. Being derived from human or animal pathogens, oncolytic viruses trigger several innate intruder alarm systems, which in turn instigate inflammation and adaptive antiviral immune responses (Thompson et al. 2011). As the oncolytic viruses are cleared (even from highly immunosuppressive tumors), the immune-evasion mechanisms of the tumor are disrupted and tumors may also become targets for antitumor immune responses. Such responses may be augmented by engineering the oncolytic viruses to express immunostimulatory transgenes or tumor antigens. Understanding of the immunological aspects of oncolytic viruses has encouraged development of oncolytic viruses as de facto immunotherapeutics (in situ cancer vaccines). Consequently, combinations of virotherapy with other immunotherapeutic modalities have been quite successful in preclinical settings, and clinical translation is ongoing (Cerullo et al. 2012; Melcher et al. 2011). For an overview of oncolytic virus action, see Figure 13.1.

### 13.3.3 Clinical Development of Oncolytic Viruses

So far, two oncolytic viruses have been approved for use in cancer treatment in China: Gendicine (Ad5-ΔE1B55K-p53) by Shenzhen SiBiono GeneTech in 2003 and Oncorine (H101, Ad5-ΔE1B55K-TK) by Sunway Biotech in 2005. In other parts of the world, some private clinical centers provide oncolytic viruses on an experimental or hospital exemption basis (medical tourism, e.g., www.virotherapy.eu/rigvir/#). In June 2013, positive interim results showing a significantly greater durable response rate and a trend toward increased survival compared to recombinant granulocyte/macrophage colony-stimulating factor injections (GM-CSF) of a multicenter phase III study with oncolytic herpes simplex virus expressing GM-CSF, T-Vec® (Amgen Inc.), herald the long-sought marketing approval in Western countries as well. Other ongoing phase III studies include Reolysin® (wild type reovirus type 3, strain Dearing) in solid tumors (SCCHN) by Oncolytic Biotech Inc., Pexa-Vec® (thymidine kinase-deleted Wyeth strain vaccinia virus expressing GM-CSF, also known as JX-594) in hepatocellular carcinoma by Jennerex Inc., USA, and a conditionally replicating adenovirus-expressing yeast cytosine deaminase and adenovirus death protein (Ad5ΔE1B55K-yCD/mutTKSR39rep/ADP) by the National Institutes

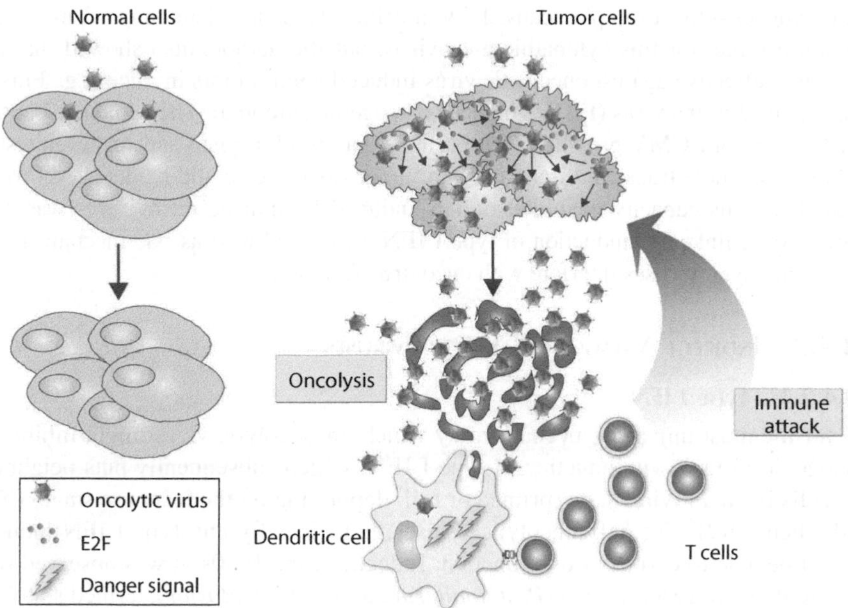


**FIGURE 13.1 (See color insert.)** Schematic representation of oncolytic virus mechanisms. Oncolytic adenovirus can only replicate in tumor cells due to abundance of eukaryotic initiation factor 2F. In normal cells, tumor suppressor Rb binds and inactivates E2F. Recruited to the site of inflammation, antigen-presenting cells (dendritic cells) phagocytize tumor material, and virus-triggered danger signals also drive adaptive T-cell immune responses against tumor antigens.

of Health (NIH). In addition, there is an ongoing phase III clinical trial with a replication-defective adenovirus vector expressing herpesvirus thymidine kinase (Ad-TK) for treatment of prostate cancer by Advantagene Inc., USA.

## 13.4 VIRUS WARS—INHIBITION OF ONCOVIRUSES BY ONCOYTIC VIRUSES

### 13.4.1 Virus Wars—The Early Days

During the first surge in oncolytic virus research in the 1940s to 1960s, seminal works into harnessing oncolytic viruses to combat tumors caused by oncogenic viruses were also published. Although it took Harald zur Hausen several years (after his original hypothesis in 1976) to confirm the link between human papillomavirus and cervical cancer, the disease had, much before that, already been targeted by oncolytic viruses on more than one occasion. Notably, injection of adenovirus (called APC virus at the time) by various routes in cervical carcinoma patients resulted in localized tumor necrosis in 26/40 cases, albeit without apparent impact on survival (Huebner et al. 1956). Two other pioneering papers from this era published in *Nature* by Molomut et al. (1964) and Molomut and Padnos (1965) described the use of lymphocytic choriomeningitis virus (LCMV) to destroy tumors caused by oncogenic murine viruses. Because LCMV displays selective tropism for lymphoid cells, lymphoproliferative malignancies—for example, caused by murine (Raucher) leukemia virus—were a natural target for this cytopathic arenavirus, but the authors also showed that the virus was effective against oncogenic virus-induced solid tumors in mice (e.g., Finkel, Biskis, and Jinkins virus (FBJV)-induced osteogenic sarcoma). The Molomut-Padnos (M-P) strain of LCMV grown in HeLa cells had a broad tropism and was even tested in humans, where transient viremia but no tumor responses could be detected (Webb et al. 1975). Its capacity to inhibit tumors induced by murine retroviruses was presumed to be linked to induction of type I IFN, discussed next as one mechanism by which oncolytic viruses interfere with oncovirus function.

## 13.4.2 Indirect Antagonism of Oncoviruses

### 13.4.2.1 Type I IFN

By far the most universal mechanism by which an oncolytic virus might inhibit an oncogenic virus is via induction of type I IFN, which subsequently puts neighboring cells in an antiviral state (primed or full, depending on the response stage) (Yan and Chen 2012). Not only oncolytic viruses, but virtually any type I IFN inducer should be able to carry out this function. Indeed, in the 1960s, it was observed that culture fluids of certain fungi (*P. stoloniferum* and *P. funiculosum*), termed statolon, inhibited plaque formation by VSV, mengovirus, and later also the oncogenic Friend murine leukemia retrovirus (Kleinschmidt et al. 1968). Intraperitoneal statolon injection protected chickens against the lethal neoplastic Marek's disease virus (MDV), even when given 72 hours after the virus, with the degree of protection correlating with type I IFN levels in the sera of the chickens (Vengris and Mare 1973). Notably, in this model Polyinosinic: polycytidylic acid had no protective effect against MDV, arguing that the quality of antiviral signaling influences the ability to control oncovirus activity—an aspect not studied systematically with oncolytic viruses.

Today, aside from recombinant type I IFNs (Multiferon and PEGintron/PEGasys), at least toll-like receptor (TLR)-7 agonist and type I IFN inducer Aldara (Imiquimod) have been approved for clinical use in the United States and in Europe. Imiquimod is applied topically to combat certain skin conditions (keratosis)—superficial basal cell carcinoma as well as morbidities of oncogenic viral etiology—including papilloma and genital warts. It must be noted, however, that although Imiquimod acts through TLR7, it may inhibit cancer cell proliferation through induction of opioid growth factor receptor expression rather than via type I IFN (Zagon et al. 2008). In the case of papilloma, the respective roles of type I IFN, opioid growth factor receptor, or other hitherto undiscovered mechanisms of virus inhibition by Imiquimod have not been elucidated in detail.

Several oncolytic viruses induce secretion of type I IFN, and exogenous IFN is known to inhibit at least RSV and lymphoid leukemia virus replication in vitro, with protective effects seen in vivo when given before virus challenge (Lampson et al. 1963; Strube et al. 1982). Several other integrating retroviruses are also controlled by the type I IFN system, as demonstrated by Nobel laureate and oncovirus pioneer Luc Montagnier and his colleagues in the 1970s and 1980s (Barre-Sinoussi et al. 1979). Intriguingly, superinfection of chicks or chicken cells by IFN-inducing rabies virus or VSV prevented, in a temporal and dose-dependent fashion, the cellular transformation induced by RSV (Semmel and Sathasivam 1983). Persistence of rhabdovirus proteins in chicken cells was associated with reduced capacity of RSV to replicate in those cells, and supernatant from rhabdovirus-infected cells also abrogated the transforming potential of Rous virus. Superinfection of VSV-infected cells by RSV was reported to lead to reduction of mRNA and protein synthesis of both viruses (Semmel et al. 1988). Taken together, soluble factors elicited by rhabdovirus infection, likely type I IFN, interfere with RSV infection and transforming capacity, but also intracellular heterologous virus interference seems to occur in cells displaying rhabdovirus persistence and/or rhabdovirus protein. Conversely, RSV does not abrogate cellular responsiveness to type I IFN, and cells superinfected with RSV are still able to limit superinfection by VSV when pretreated with type I IFN (Strube et al. 1982). This is in contrast to HPV, which clearly renders cells unresponsive to type I IFN (Le Boeuf et al. 2012).

However, type I IFN responses on their own may not be sufficient to provide long-term control of retroviruses or their transforming capacity. Namely, several retrovirus countermeasures of the IFN system have been identified, and coupled with extrinsic signals from the host, such as stress and other ongoing conditions, the innate antiviral defenses may ultimately fail (Douville and Hiscott 2010). Other oncoviruses also employ anti-IFN countermeasures. But if antiviral signaling via type I IFN is suppressed by the oncovirus, such tumors can become excellent targets for IFN-sensitive oncolytic viruses, as in the case of HPV-transformed cells becoming extremely prone to destruction by VSV (Le Boeuf et al. 2012).

#### 13.4.2.2 Transactivation of Oncolytic Viruses by Oncovirus Replication Factories

Some oncolytic viruses, such as minute virus of mice (MVM), have been observed to capitalize on replication machineries or strategies employed by oncoviruses. As an example, replication of MVM is normally restricted in many human cells, but when coinfected with oncogenic adenovirus (Ad2), MVM DNA was found to colocalize with adenovirus DNA—in particular, nucleolar structures where adenovirus replicates (Walton et al. 1989). There was a concomitant increase in MVM replication. Although the role of adenoviruses in human oncogenesis has been fiercely debated (Sanchez-Prieto et al. 1999), it could be worth investigating whether MVM could capitalize on similar nucleolar replication mechanisms of other oncoviruses. Nevertheless, it was also shown that despite effective MVM DNA recompartmentalization to adenovirus replication factories, there was significant variation in the overall replication of MVM in several ad-transformed cell lines, arguing that other factors of permissiveness will still impose limitations to MVM as an oncolytic agent (Fox et al. 1990). In comparison to MVM, another parvovirus under development as an oncolytic agent, the Lull rat parvovirus, shows a much broader capacity to infect and kill human cancer cells (Paglino and Tattersall 2011). This difference was shown at least in part to be due to VP2 capsid protein, interacting with hitherto undiscovered host cell proteins. Similarly to MVM, the Lull virus could potentially exploit nuclear oncovirus replication factories to further its own replication.

Overall, hijacking or exploitation of oncovirus replication machineries may be characterized as a form of *heterologous virus interaction*, which can result in either inhibition or enhancement of one or the other virus. The phenomenon is dependent on the viruses involved, and because it may also occur through soluble factors, its relevance/validity as an exclusive virus-specific intracellular event can be difficult to establish. For more on this topic, see sections that follow.

#### 13.4.2.3 Oncolytic Virus Infection May Expose Oncoviruses to Immune Purging

Expanding tumors in immunocompetent hosts have undergone a process termed *immunoediting*, whereby tumor cells capable of growing have been selected for their capacity to avoid eradication by the immune system (Dunn et al. 2006). Cellular transformation by oncoviruses and subsequent oncogenesis entails that tumors survive/evade the negative pressure exerted by the host immune system. Thus, with the possible exception of tumors that manifest exclusively in immunosuppressed patients, oncovirus tumors have undergone immunoediting and are capable of either actively subverting immune responses (e.g., via Treg/myeloid-derived suppressor cells secreting transforming growth factor beta (TGF-β), IL-10, or indoleamine 2,3-dioxygenase (IDO)-derived kynurenine) or evading recognition (e.g., loss of HLA expression coupled with natural killer (NK)-inhibitory receptor expression). In simian virus 40 (SV40) preimmunized animals, rejection of in vitro SV40-transformed tumor cells occurred 100 times more effectively than rejection of SV40-transformed tumor cells that had been growing as tumors in other animals (Deichman et al. 1999). In other words, the capacity of oncovirus-transformed cells to resist rejection by innate immune effectors (macrophages and NK cells) upon transplantation increased gradually as the cells were growing as tumors in vivo (Deichman et al. 2001). In this regard, oncolytic viruses may prove particularly attractive to treat oncovirus-associated malignancies because they possess the capacity to overcome tumor immunosuppression/evasion. For instance, reovirus infection of tumors in mice was shown to induce expression of proinflammatory cytokines—including IFN gamma—to cause upregulation of MHC-I molecules in tumors, and to increase immune cell infiltration of the tumors, with induction of tumor-specific CD8 T cells that could protect naïve recipient mice from tumor challenge following adoptive transfer (Gujar et al. 2011). Antitumor effects of reovirus may not be dependent on virus replication/oncolysis (Prestwich et al. 2009). Still, virus-induced immunogenicity of preinfected cells in the HPV-transformed TC-1 mouse model cells was weaker than with irradiated cells, arguing that virus infection per se does not guarantee greater immunogenicity (Figova et al. 2010).

In summary, oncolytic virus infection triggers a robust innate response typically involving at least type I and type II IFNs, which increase chemokine secretion and HLA expression on tumors, and when antigen-presenting cells enter the tumor and process-infected tumor cells, viruses trigger several pattern-recognition receptors within them, causing maturation and activation. This results in adaptive antivirus immune responses, but because tumor immunosuppression/evasion is at least temporarily lifted during oncolytic virus infection, antitumor immune responses also have a chance to develop. It is thus perhaps not surprising that antigen-specific antitumor T-cell responses can be detected in animals and cancer patients treated with oncolytic viruses (Cerullo et al. 2012).

On a macroscopic level, the human body is under a constant barrage from potential pathogens and must be capable of handling multiple insults simultaneously. Still, some virus and/or bacterial infections influence susceptibility to subsequent infections, and immunity to a previously encountered pathogen may also change as a result of immune responses against a heterologous infection (DaPalma et al. 2010). If double infections occur during ongoing/chronic infection, T-cell exhaustion may ensue and predispose to postinfectious complications (Salek-Ardakani and Schoenberger 2013). Otherwise, the T-cell memory pools against previously encountered antigens may shift in quality and quantity, in ways that remain poorly understood. These aspects with regard to oncolytic–oncovirus-interactions remain largely unstudied.

#### 13.4.2.4 Bystander Targeting of Oncovirus-Transformed Cells

Another way by which oncolytic viruses could be used to target oncovirus-associated tumors is to equip them with capacity to destroy uninfected neighboring cells, for instance, by arming them with thymidine kinase (TK) enzyme. As TK converts non-toxic prodrugs into toxic metabolites locally, such metabolites are capable of diffusing out from the oncolytic virus-infected TK-expressing cells. In such cases, it is possible to extend the anti-oncovirus influence of the oncolytic virus beyond infected cells, which otherwise might limit overall efficacy in large and heterogeneous tumor nests.

Another intriguing approach is to extend the zone of infection in tumors beyond the original infected cell without relying on virus spread via replication. In one study, oncolytic adenovirus was engineered to express the membrane-fusogenic GALV gene (from Gibbon ape leukemia virus), resulting in the capacity of adenovirus to spread via normal healthy cells (Guedan et al. 2012). In an adaptation of this strategy, Human Embryonic Kidney (HEK) 293 cells expressing adenovirus E1A protein were injected into tumors before the oncolytic adenovirus; through syncytium formation between these and normal cells and non-HEK tumor cells—resulting in transcomplementation of oncolytic adenovirus replication—therapeutic efficacy was enhanced (Chen et al. 2011). As early as the 1960s and 1970s, observations were made that VSV induced syncytiosomes in cultures of cells preinfected with MLV and RSV (Sinkovics and Horvath 2008). The effect was later attributed to the VSV glycoprotein (G). At physiological conditions, however, fusogenic activity of G may be suboptimal, which has at least in part also spurred arming VSV with membrane-fusogenic transgenes. In place of G, VSV expressing reptile reovirus small fusogenic glycoprotein p14FAST showed increased oncolytic efficacy over parental virus, forming whole-well syncytia in cell culture and increasing therapeutic benefit in several in vivo mouse tumor models (Brown et al. 2009). VSV-fusion-associated small transmembrane (FAST) has not been tested against oncovirus-induced tumors, perhaps in part because of its increased neurotoxicity compared to the parental virus. Nevertheless, the cell-fusion concept is understudied in combating oncovirus-induced tumors and deserves systematic probing.

## 13.5 VIRUS WARS—EXPLOITING ONCOVIRUS SIGNALING CASCADES

In this section, we highlight some molecular loopholes left open by oncoviruses that can be exploited/usurped by oncolytic viruses. It must be noted that because signaling pathways in cells are overlapping, it is unlikely that defects in single cascades will fully predict predisposition for oncolysis. As such, the pathways discussed here function as conceptual examples of molecular exploitation of oncovirus tumors by oncolytic viruses— not as predictors of actual therapeutic efficacy.

### 13.5.1 Antiviral Defenses—Role of Rat sarcoma (Ras) and c-Jun NH(2)-terminal kinase (JNK)

Influenza virus was shown to reduce RSV infection in chickens (Kravchenko et al. 1965). The first oncolytic influenza virus was engineered to have a defect in NS1, a multifunctional protein that is responsible for antagonizing cellular antiviral defenses at least by blocking nuclear export of messenger RNA (mRNA) and by inhibiting activity of 2′-5′ oligoadenylate synthetase (OAS), retinoic acid-inducible gene 1 (RIG-I), double-stranded RNA-activated kinase (PKR) (Bergmann et al. 2001; Hale et al. 2008). Thus, oncolytic influenza virus is restricted to cells in which antiviral mechanisms are impaired. This is often the case in tumors, where oncogenic Ras activation may induce an inhibitor of PKR. It was observed that enforced Ras expression resulted in reduction of a particular subset of IFN-regulated gene expression, including signal transducer and transcription activator-2 (STAT-2), as well as a reduction of phosphorylation of other key IFN-signaling mediators, such as STAT-1 (Christian et al. 2009, 2012). Oncolytic reovirus and VSV display a preference for cells with an activated Ras-MAPK pathway, associated with defects in type I IFN-mediated antiviral signaling (Noser et al. 2007). An example of an oncovirus that induces Ras activation is HHV-8.

Replication of oncolytic HSV-1 vector G207 was enhanced in SV40 T-expressing cells rather than in nontransduced counterparts, owing to increased levels/activity of Ras/Raf (Farassati et al. 2008). Conversely, Ras/Raf inhibitors reduced HSV replication in tumor cells. Overall, the authors provide clues that high Ras activity favors replication of HSV because extracellular signal-regulated kinase (ERK) and JNK mitogenic pathways are activated in preference over pro-apoptotic Phosphoinositide 3-kinase (PI3K) pathways. On the other hand, replication of oncolytic vaccinia virus was increased in JNK-deficient cells compared to JNK-expressing cells, associated with reduced activation of PKR in JNK –/– cells (Hu et al. 2008). Thus, depending on virus or context (the viruses in these studies were not compared head-to-head in the same model systems), JNK may display either pro- or antiviral effects, likely through PKR activation during oncolytic virus infection. Because Kaposi's sarcoma-associated herpesvirus (KSHV) viral Fas-associated death domain-like interleukin-1β-converting enzyme-inhibitory protein (vFLIP) also activates the JNK/AP1 pathway (An et al. 2003), it is difficult to predict on a general level the net effect of oncovirus-induced JNK pathway activation on oncolytic virus replication/efficacy.

As a hypothetical example of targeting veterinary oncovirus tumors in the context of Ras activation, ovine pulmonary adenocarcinomas and enzootic nasal adenocarcinomas, two contagious respiratory cancers in sheep and goats caused by the Jaagsiekte sheep retrovirus and enzootic nasal tumor virus, respectively, display activated Ras pathway, marked by ETS domain-containing protein (Elk-1) phosphorylation (De Las Heras et al. 2006). Such tumors could be favorably targeted (tropism willing) by oncolytic herpesviruses in which ICP4 expression is regulated by Elk-response elements (Pan et al. 2009), and possibly by oncolytic reovirus or VSV, provided that Ras activation results in abrogation of IFN-mediated antiviral defenses.

### 13.5.2 Exploiting Oncovirus Cell Cycle Deregulation

The HTLV-1 Tax protein induces cell cycle checkpoint proteins p21/waf1 and cyclin D2, which consequently frees cyclin E/cdk2 complex to promote cell cycle (Kehn et al. 2004). Because adenovirus E1B55K carries out a very similar function to HTLV-1 Tax protein (Cheng et al. 2013), oncolytic adenoviruses deleted for E1B55K may be able to replicate in HTLV-1-induced tumors.

### 13.5.3 Retinoblastoma Pathway

Adenovirus E1A mediates cell transformation, in part, through E1A CR1 domain-mediated dislodging of transcription factor E2F from the retinoblastoma (Rb) tumor suppressor protein (Liu and Marmorstein 2007). In typical second-generation oncolytic adenoviruses, the CR1 Rb-binding domain of E1A has been disrupted, rendering the virus dependent on activated Rb pathway, which is frequently encountered in many cancers. Oncovirus HPV and SV40 also display Rb-binding properties (Chellappan et al. 1992), and it was shown that HPV16 E7 protein but not SV40 T-antigen could complement replication of an adenovirus with disrupted CR1 region (Wong and Ziff 1996). The inability of SV40 T-antigen to complement adenovirus E1A CR1 defects reveals additional factors contributing to activation of adenovirus early gene transcription.

### 13.5.4 The Wnt Pathway

Latency-associated nuclear antigen of Kaposi's sarcoma virus, large T-antigen of the human polyomavirus JC, and latent membrane protein 2A of EBV activate the Wnt pathway (Shackelford and Pagano 2004). In addition, latent membrane protein 1 (LMP1) expressed by EBV induces expression of β-catenin, the principal signaling mediator of the Wnt pathway (Jang et al. 2005). In Wnt-activating oncovirus-transformed cells, β-catenin translocates into the nucleus and forms a complex with T-cell factor (TCF)/Lef that activates transcription via the TCF response element. Although not displaying a direct oncovirus causality, Wnt pathway, measured as transcriptional increase of the catenin (cadherin-associated protein), beta 1 (CTNNB1) gene that encodes β-catenin, is prominently activated in at least breast cancer, colorectal cancer, melanoma, prostate cancer, and lung cancer. Several of these types of tumors have been effectively targeted with oncolytic herpes-, adeno-, and parvoviruses engineered with TCF response elements controlling vital oncolytic virus replicase components (Fuerer and Iggo 2002; Kuroda et al. 2006; Malerba et al. 2003). Interestingly, the E1B55K and E4orf6 gene products of adenovirus participate to form an E3 ubiquitin ligase complex that causes degradation of p53 (Berk 2005). Because Wnt signaling is directly regulated by p53-dependent proteasomal control of β-catenin levels in the cells (Shackelford and Pagano 2004), oncolytic adenovirus with intact E1B55K and E4orf6, such as the modern E1A Δ24-Rb pathway restricted viruses (which feature a CR1 domain deletion in E1A that abrogates the Rb-binding capacity of the protein), could be particularly well suited to target oncovirus-afflicted cells with activated Wnt signaling by reducing p53-mediated control of β-catenin levels. (In Δ24 viruses, only the Rb binding is lost, not p53 binding.)

Hypothetically, by placing virus replicase components of oncolytic E1B55K/E4orf6-competent adenoviruses under control of TCF response elements, specificity to Wnt-dependent tumor could be increased further. An oncolytic adenovirus engineered to express E1A under control of a β-catenin-responsive promoter, AdTOP-PUMA, replicated and killed human cancer cells in a β-catenin-dependent fashion (Dvory-Sobol et al. 2006).

### 13.5.5 Oncolytic Viruses against Src-STAT3-Activating Oncoviruses

The concept of cancer induction by viruses was introduced by Nobel laureate Francis Peyton Rous in the early 1900s. Rous sarcoma virus, named after its discoverer, causes cellular transformation via a single protein, v-Src, which functions as a homolog of the cellular tyrosine kinase c-Src but which is constitutively active due to a missing inhibitory phosphorylation site at its C-terminus. Src kinases direct signaling events mediated by the CD4 and CD8 T-cell receptors, platelet-derived and other growth factors, and several interleukins, as well as integrins, and serves to promote cellular motility, survival, and division (Martin 2001). Importantly, signaling by Src occurs via its immediate downstream substrate STAT-3, which in turn is a target for proteasomal degradation by rubulavirus V protein (family: Paramyxoviridae). Indeed, mumps virus V protein alone was shown to induce apoptosis in multiple myeloma cells and v-Src-transformed murine fibroblasts in vitro (Ulane et al. 2003). Targeting Src or STAT-3 is also appealing from the standpoint of combination with chemotherapy; STAT-3 activation confers resistance against cisplatin, and accordingly, an oncolytic adenovirus engineered to express antisense DNA against STAT-3 showed therapeutic synergy with cisplatin in several cisplatin-resistant preclinical models (Han et al. 2009, 2012). Such an adenovirus could potentially also be used to target oncovirus-induced tumors with an activated src pathway.

In addition to proteasomal degradation, several signaling components of antiviral and proinflammatory pathways are targeted for inhibition by the V protein of other paramyxovirus members. Measles virus V protein binds to and inhibits nuclear translocation of STAT-2 and STAT-1 in a STAT-2-dependent manner (Chinnakannan et al. 2013; Palosaari et al. 2003; Ramachandran et al. 2008). In doing so, it inhibits signaling via type I and type II IFN receptors. In addition to STAT-2, V protein of measles virus specifically binds to the Rel homology domain of the NF-κB subunit p65, as well as to interferon regulatory factor 7 (IRF7), and melanoma differentiation-associated protein 5 (MDA5) significantly inhibiting type I IFN-independent interferon-stimulated gene (ISG) activation and nuclear factor kappa-light-chain-enhancer of activated B cells (NfκB)-dependent proinflammatory cytokine expression (Schuhmann et al. 2011). Another oncolytic virus featuring multiple anti-inflammatory mechanisms is vaccinia virus (e.g., scavenger receptors against type I and type II IFNs, IL-1β, and several chemokines) (Turner and Moyer 2002); overall, the role of targeting the inflammatory cascades by oncolytic viruses in treatment of oncovirus-associated tumors (where the oncovirus itself may drive similar anti-inflammatory functions) is unclear.

### 13.5.6 Notch Pathway and Epithelial-to-Mesenchymal Transition

Many tumor-forming viruses, including KSHV and HHV-8, activate the cellular Notch kinase, which has central roles in regulating cell proliferation, DNA-damage responses, and resistance to chemotherapeutics (Cheng et al. 2012). Notch has recently been targeted by small interfering ribonucleotidic acid (siRNA) in combination with an oncolytic adenovirus, H101, forming an interesting avenue to explore further, such as by encoding microRNAs in the vector itself and testing the virus on Kaposi's sarcoma herpesvirus (KSHV)-induced tumors (Yao et al. 2012). Notch signaling by KSHV is activated by the virus-expressed analog of cellular FLIP oncogene, vFLIP, as well as viral G protein-coupled receptor (vGPCR), and it promotes endothelial-to-mesenchymal transition (EndMT) of KSHV-infected cells in a process similar to epithelial-to-mesenchymal transition (EMT). Such cells are also induced to secrete membrane-type-1 matrix metalloproteinase (MT1-MMP), which increases the invasive properties of KSHV-infected cells (Cheng et al. 2011). Interestingly, oncolytic viruses engineered to express natural tissue inhibitors of metalloproteases (TIMPs), such as Ad5/3-CXCR4-TIMP2 (Yang et al. 2011), could potentially be used to target Notch-activated (KSHV-induced) tumors. Putatively, not only could the KSHV-infected cells be destroyed by oncolytic virus replication but the KSHV-driven MMP-dependent tumor invasiveness could also be reduced through TIMP-secretion by the oncolytic virus infected cells.

EMT is known to occur following infection, also with oncolytic viruses, including several members of the adenovirus serogroup (Strauss et al. 2009), as well as vaccinia virus (Wang et al. 2012a). These studies show that oncolytic virus replication is favored in cells with mesenchymal phenotype, and conversely that OV replication is limited in cells displaying epithelial characteristics. Thus, it would be reasonable to assume that oncolytic viruses could replicate favorably in KSHV-infected cells in which vFLIP and vGPCR have induced EndMT.

## 13.6 VIRUS WARS—SPECIFIC EXAMPLES

Although several potential molecular and immunological oncovirus complementation mechanisms of oncolytic viruses have been presented here, the sum of oncolytic virus replication/antitumor efficacy in oncovirus-associated cancers is difficult to predict due to multiple overlapping and, in many cases, promoting and inhibitory pathways. Therefore, empirical testing remains the most viable way of establishing whether an oncolytic virus would be suitable to target a particular type of oncovirus-induced tumor. Here we provide examples of such testing, where the aim has specifically been to combat oncovirus tumors through molecular exploitation by the oncolytic virus. We also highlight some unwitting targeting of oncovirus-induced tumors (unwitting because oncovirus etiology and/or presence was not known or not addressed).

### 13.6.1 Targeting HPV-Associated Cancer

#### 13.6.1.1 VSV against HPV

HPVs induce benign warts and papillomas; however, infection with high-risk types (HPV-16, -18, -31, and -45) is a major risk factor for the development of 100% of cervical carcinomas, 50% of other anogenital carcinomas, and 20% of head and

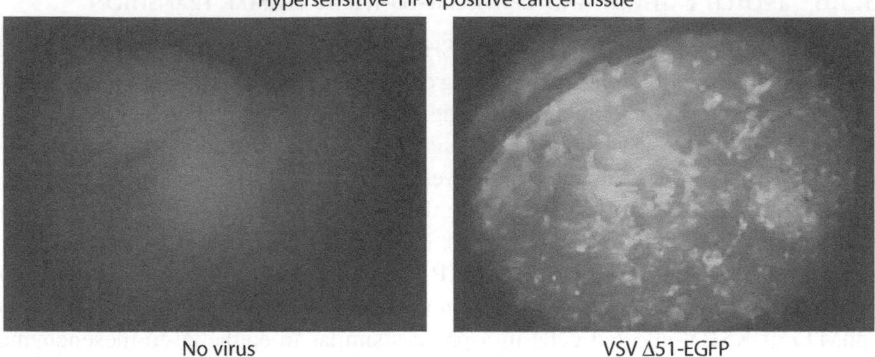


**FIGURE 13.2 (See color insert.)** Targeting of HPV by oncolytic vesicular stomatitis virus. Fresh surgical cervical cancer tissue was infected ex vivo with IFN-sensitive oncolytic VSV. Infection was established only in IFN-unresponsive tissue.

neck squamous cell carcinomas (HNSCCs); this may also play a role in oropharyngeal cancer. Recurrent and metastatic cervical carcinomas and HNSCCs are associated with a very poor prognosis with limited treatment options. A key event of HPV-induced carcinogenesis is the integration of the HPV genome into a host chromosome, usually occurring near a common fragile site of the human genome. The main HPV oncogenes are E6 and E7. It is important to note that E6 protein has been recently shown to inhibit the cellular antiviral IFN response as Tyk2, which binds to the type I interferon receptor (IFNAR) and facilitates IFN-induced JAK-STAT pathway activation by this receptor, a binding partner of E6 that inhibits its function. We observed that HPV infection, through the expression of its E6 protein, inhibits IFN receptor signaling. We evaluated several cell lines for their sensitivity to VSV IFN-sensitive OV and found three groups of cells: high resistant, semiresistant, and highly permissive to VSV oncolysis. The resistant cell lines readily induced the IFN pathway and IFNα secretion. The sensitive cells lines were sensitive to VSV-induced oncolysis but showed an attenuated response to IFN-sensitive VSV strains; the hypersensitive group showed significant cytotoxicity with VSV OV and failed to induce an IFN response (Figure 13.2).

Thus, a direct correlation between HPV-positive cell lines and their sensitivity to VSV oncolysis was established. A critical observation was that if we express HPV-E6 in the VSV-resistant HNSCC cell line SCC25 inhibiting IFN signaling, the cytotoxicity of VSV was significantly enhanced compared to parental controls. These data were confirmed in vivo as well as using human explant tumor tissues from CC patients. In conclusion, this study suggests that HPV-infected cells are susceptible to oncolytic virus therapy and that this approach may represent a novel therapeutic tool in HPV-positive CC and HNSCC.

#### 13.6.1.2 Reovirus against HPV

Oncolytic reovirus was also shown to replicate in and kill HPV-transformed cells (Sobotkova et al. 2008). The virus eradicated HPV16-transformed TC-1 cells in culture and displayed modest oncolytic effect with tumors established using the same cells. Interestingly, whereas cytokines IL-2, IL-12, and GM-CSF have, in other contexts (such as when expressed from other oncolytic virus vectors), been shown to mediate potent anticancer properties, in this study, when nontransformed TC-1 cells engineered to express these cytokines were co-administered with oncolytic reovirus into HPV16-transformed TC-1 tumors, the net therapeutic effect was reduced. The authors speculated that if stronger antitumor immune responses were generated using cytokine combination treatment, they were masked by simultaneous/preferential increase of antiviral immune responses. A synergistic therapeutic effect was achieved when reovirus was combined with cyclophosphamide, favoring regimens where the chemotherapeutic was administered up to two weeks after the virus.

Because cyclophosphamide affects tumor cells and cells of the immune system, and also may affect virus replication, it remains a matter of speculation as to how/why combination yielded synergy. One explanation offered by the authors was that antiviral immune responses may have been reduced. Although this is plausible, it must be kept in mind that in mice IgM is detectable against reovirus as early as four days post inoculation, IgG is detectable beginning from day five, and virus-specific T-cell responses likely peak around one week post infection. Therefore, somewhat contradictory to this thinking, it was found that cyclophosphamide given two weeks after virus yielded the best overall therapeutic efficacy in the HPV16-TC1 tumor model. Instead, another possible explanation includes reduction of immunosuppressive elements in the tumor by cyclophosphamide, such as regulatory T cells (Tongu et al. 2010), which could hypothetically allow reovirus-induced antitumor responses to exert a greater effect than when only virus was administered. Also, although CPA does reduce regulatory T cells, it may during chronic inflammation and perhaps over time increase influx of myeloid-derived suppressor cells, which counteract its antitumor immune effects (Sevko et al. 2013). Also, reovirus-induced antitumor immune responses in two HPV-transformed mouse models, TC-1 and MK16, appear to be very different, despite the virus replication in both being similar (Figova et al. 2010). Thus, the presence of HPV does not automatically guarantee therapeutic success; other biological and/or immunological factors contribute. Nevertheless, it would be interesting to compare cytokine secretion in these two models with reovirus in order to assess impact on immunogenicity.

#### 13.6.1.3 Oncolytic Adenovirus against HPV-Associated Cancer

At the beginning of the 20th century, virus research exploded and a veritable stream of new viruses flowed from the tropics to various research institutes around the world. Spurred by the preceding anecdotal clinical studies with cancer patients (Coley's fluid, rabies, and smallpox vaccinations, influenza spats) and by cytotoxic virus characteristics revealed in cultured cells and in some cases in the Erlich ascites model, many viruses were then systematically tested as oncolytics in cancer patients. One of the viruses that was tested on more than one occasion was adenovirus. In a study from 1956, out of 30 patients with cervical carcinoma treated with the virus via intratumoral, intra-arterial, or intravascular routes, roughly one-third displayed signs of tumor necrosis within ten days post virus inoculation (Smith et al. 1956). However, survival was not increased.

As a modern day example, an oncolytic E1A Rb-binding defective adenovirus engineered to express HPV E2, a negative regulator of the HPV oncogenes E6 and E7, showed increased oncolytic activity against HPV-positive cancer cells compared to parental backbone virus (Wang et al. 2011). The E2-expressing adenovirus also conferred tumor cell sensitivity to ionizing radiation, which is a hallmark of HPV-associated cervical cancer, and it synergized with external beam irradiation in treating animals bearing HPV-positive xenografts. This strategy potentially also benefited from the capacity of HPV E6 protein to transactivate heterologous virus promoters, including the herpes simplex TK and adenovirus E2-promoter (Shirasawa et al. 1994). Moreover, HPV E7 protein counters the Rb-activating properties of HPV E2 protein, and constitutive E7 expression is required to maintain cell cycle in HPV-transformed cells (Psyrri et al. 2004). Therefore, oncolytic adenovirus in which the Rb-inactivating features have been removed could benefit from HPV-E7-mediated Rb silencing. Although not oncolytic, a replication-defective adenovirus vector expressing p53 homolog p73beta could overcome HPV E6-mediated loss of p53, inducing cell cycle arrest and apoptosis in HPV-infected tumor cells (Das et al. 2003). Overall, oncolytic adenovirus may be an excellent oncolytic candidate for treatment of HPV-induced cancer.

### 13.6.2 Targeting Hepatitis Virus-Associated Cancer

Both HBV and HCV are causative agents of hepatocellular carcinoma. Although HBV may cause insertional mutagenesis, both viruses may establish chronic infection that drives oncogenesis on a phenotypical level—whereas the role of specific gene products of these viruses in cellular transformation is still under debate (Banerjee et al. 2010). Independent of the potential viral etiology, liver cancer has been successfully targeted with oncolytic viruses. Perhaps the most striking contemporary example is provided by a recent phase II dose-finding clinical trial where Jennerex Biotherapeutics' oncolytic thymidine kinase-deleted Wyeth strain vaccinia virus expressing GM-CSF, JX-594 (trade name Pexa-Vec), was injected intratumorally in 30 patients with advanced treatment-refractory liver cancer (Heo et al. 2013). Results from this study revealed that although low-dose ($10^8$ plaque-forming unit (PFU)) and high-dose ($10^9$ PFU) virus resulted in equal tumor response rates, measured by modified Response Evaluation Criteria in Solid Tumors (mRECIST), (15%) and Choi's criteria (62%), objective survival was significantly related to dose (median survival of 14.1 months compared to 6.7 months on the high and low dose, respectively). However, with regard to attacking the underlying oncovirus etiology of liver cancer, the phase II trial was preceded by a report describing virotherapy of three patients with HCC, all of whom were confirmed HBV-positive and two of whom had been on lamivudine before oncolytic virus infusion (Liu et al. 2008). In these patients, intratumoral injection of JX-594 not only caused antitumor effects, marked by tumor shrinkage and vascular collapse, but also reduction in HBV replication in all three patients (70%–91% of baseline). The mechanism of HBV suppression by vaccinia virus was not studied further, but cytokines (IFN-$\gamma$, TNF-$\alpha$) known to display antiviral activity against HBV were induced by JX-594, possibly contributing to HBV suppression. In any case, this is a bona fide example of pitting virus against virus in clinical settings and demonstrates the feasibility of the virus wars concept.

Apart from vaccinia virus, herpesvirus has been tested in targeting HCC, where a selection of gene promoters specific for HCC was shown to direct HCC-specific gene expression (Foka et al. 2010). HBV and HCV-associated HCC have been targeted by other gene therapy modalities besides oncolytic viruses (Smerdou et al. 2010).

### 13.6.3 Targeting HTLV

Oncolytic VSV has successfully been used to target and destroy HTLV-1-infected human T cells (Cesaire et al. 2006). In this case, virus replication and killing of the HTLV-preinfected T cells required an active cell cycle, wherein activation of MEK1/2, c-Jun NH(2)-terminal kinase, or phosphatidylinositol 3-kinase but not p38 kinase was critical for cell death induced by VSV (Oliere et al. 2008).

### 13.6.4 Oncolytic Viruses against Oncogenic Herpesviruses

Because measles and mumps viruses display a natural tropism for leukocytes lymphocytes, they may be particularly well suited to target leukocytic cancers induced by oncoviruses. Other oncovirus-induced cancers that may be well suited for targeting by V-expressing paramyxoviruses include Kaposi's sarcoma, primary effusion lymphoma (PEL), and plasma cell-type multicentric Castleman's disease (MCD), which are caused by human herpesvirus 8 (HHV-8) or KSHV. These viruses encode for several viral analogs of human cytokines that have potent effects on host cell defenses and proliferation (Moore et al. 1996; Nicholas 2007). Primary among these is vIL-6, which displays many (most) of the biological effects of human IL-6, except that vIL-6 has a much broader range of cells it affects due to direct interaction with the ubiquitous IL-6 receptor subunit gp130, which bypasses the need for other receptor subunits and for the secretion and extracellular receptor-binding (Hu and Nicholas 2006; Molden et al. 1997; Sakakibara and Tosato 2011). Nevertheless, vIL-6 signaling is dependent on gp130-associated Janus or PI3 kinases, both of which phosphorylate STAT-3 (on different tyrosine residues), which may render KSHV-induced tumors sensitive to destruction by STAT-3-degrading (V-expressing) paramyxoviruses. In parallel, however, vIL-6 interaction with gp130 also leads to activation of SHP2 (SH2 domain-containing phosphatase 2), which in turn activates a complex cascade through Grb2, Sos, and Ras, which activates together with a number of other kinases c-Raf, which in turn activates the MEK/mitogen-activated protein kinase (MAPK)/ERK cascade; the net effect of MEK/MAPK/ERK signaling on replication of paramyxoviruses is poorly known and may influence the permissiveness of vIL-6-expressing cells to oncolytic paramyxoviruses. Moreover, even in the presence of cytokine signaling/cell proliferation inhibitors, such as p16 INK4a, p27Kip1, or STAT-3 inhibitor, the KSHV analog of cellular cyclin D1, cyclin K, is able to activate cell proliferation (Lundquist et al. 2003). KSHV-enforced cell division would complement for engineered defects in conditionally replicating (oncolytic) adenoviruses deleted for the proliferation-promoting Rb-binding domain of E1A. On a molecular level, KSHV LANA and EBV EBNA1 proteins cause degradation of p53 and pRb tumor suppressors, which renders these cells ideal growing ground for Rb-pathway-targeted E1A Δ24 oncolytic adenoviruses (Cerullo et al. 2012).

Oncogenic herpesvirus vFLIP activates the NfκB pathway, which induces expression/secretion of IL-6, Cyclooxygenase, and PGE2, essential for the generation of a favorable tumor microenvironment (An et al. 2003; Sharma-Walia et al. 2012). At least the COX-2-activating property of KSHV vFLIP could be exploited for specific targeting by oncolytic viruses with COX-2-dependent promoters, as described (Ono et al. 2005). On the other hand, COX-2 is highly expressed in B-lymphocytes where it is important for antibody production, and administering COX-2 inhibitors was shown to greatly reduce antibody generation against oncolytic vaccinia virus in mouse tumor models, allowing repeated administration of the virus where otherwise not possible (Chang et al. 2009). Thus, although COX-2-promoter-selective oncolytic viruses could benefit from KSHV-induced COX-2 activity, it is conceivable that pharmaceutical COX-2 inhibition in combination with oncolytic virotherapy would still prove a superior therapeutic option because it would simultaneously antagonize oncovirus activity while inhibiting neutralizing antibody development against the oncolytic virus.

On the other hand, for other oncoviruses that do not rely on or induce COX-2, such as HPV-16, using COX-2 inhibitors may be detrimental for the generation of anti-oncovirus antibodies and may hamper anti-oncovirus vaccine efficacy (Ryan et al. 2006). It is likely that oncolytic virotherapy of oncovirus-associated cancers also generates a vaccine effect against the oncovirus (as is the case for other tumor-associated antigens); therefore, the net effect of COX-2 inhibition during oncolytic virotherapy would have to be gauged on a case-by-case basis depending on the type of oncovirus-driven cancer and on the desired antitumor effect (oncolysis vs vaccine effect).

Finally, oncoherpesvirus EBV has been targeted by oncolytic viruses. An oncolytic adenovirus deleted for VA (viral associated) RNA I displayed excellent selectivity to EBV-transformed tumor cells as well as oncolytic potency in a mouse EBV-positive xenograft model (Wang et al. 2005). This was possible because EBV virus RNA1 can functionally complement adenovirus VAI RNA. EBV-induced cancers can also be targeted by oncolytic herpesvirus, but here efficacy was dependent on the levels of herpesvirus receptor, nectin-1 as well as intracellular PKR, both of which vary significantly in EBV-associated tumors (Wang et al. 2012b).

### 13.6.5 Oncolytic Viruses against Oncovirus-Induced Osseous Tumors?

Tumors induced by the FBJ virus in mice were characterized histologically to loosely resemble osteosarcomas, displaying regions of cartilage and bone as well as minimal vascularity and calcification but with some important histological and biochemical differences compared to human cancer—such as lack of type II collagen (Price et al. 1972; Yamagata and Yamagata 1984). Although not of confirmed viral etiology, osteosarcomas have successfully been targeted in animal models by Semliki Forest virus (SFV) (Ketola et al. 2008), VSV (Kubo et al. 2011), as well as human and canine adenoviruses (Hemminki et al. 2003; Li et al. 2011). In addition, both VSV and influenza virus replicate productively in osteoblasts, suggesting potential by these viruses to target tumors of osseous oncovirus origin (Ilvesaro et al. 1999). If, on the other hand, the oncolytic virus displays a broad receptor-tropism, osteoblastic tumors may still be refractory if cellular mechanisms of tropism, which render normal osteoblasts refractory to viruses, remain intact. Further, because of minimal blood vessel invasion in the tumor, oncolytic viruses may not be able to cause vascular shutdown (Breitbach et al. 2007) or even enter the tumors effectively when administered systemically.

## 13.7 ONCOVIRUS DEFENSES

In this section, we discuss potential defense mechanisms of oncoviruses. Although typically speculative, such defenses may contribute to ensuring oncovirus survival and may in part explain why oncovirus tumors emerge despite the fact that humans are constantly exposed to viral infections.

### 13.7.1 Chronic Inflammation

In the book *Infections Causing Human Cancer*, Nobel laureate Harald zur Hausen discusses in detail the two aspects of oncovirus-induced transformation: virus-products/intracellular disruptions and chronic inflammation (zur Hausen 2006). With regard to opposing/thwarting oncolytic viruses, chronic inflammation may under some circumstances also activate antiviral defenses (presumably also affecting the oncovirus to some extent). However, whether oncolytic virus-induced inflammation exacerbates the already inflamed tumor microenvironment or whether oncolytic viruses may infect tumor tissue under such circumstances remains poorly understood. The classical view has been that the protypical antiviral STAT-1 mediator functions as the counterpart to STAT-3, overactivated in many cancers, and that the two signaling mediators carry out reciprocal functions (Pensa et al. 2009). However, emerging evidence points toward STAT-1 in promoting tumor survival during challenge because STAT-1 transcriptional activity increases the DNA repair mechanisms required to withstand genotoxic stress (Khodarev et al. 2012). If so, it is possible that in some tumors, STAT-1-mediated antiviral signaling is active and such tumors may be refractory to oncolytic virus replication. As discussed earlier, although many oncoviruses activate STAT-3, less in known about their interactions with STAT-1.

### 13.7.2 Heterologous Virus Interference

As an example of oncogenic virus-induced defense against heterologous (oncolytic) virus superinfection, RSV-transformed Armenian hamster cells were more resistant to superinfection by vaccinia virus than nontransformed or transformed but non-Rous-virus producing cells (Nadzharian and Kamalian 1975). The effect was shown by antibody depletion being independent of type I IFN, which was also corroborated by superinfection with Newcastle disease virus (which, unlike vaccinia virus, is very sensitive to type I IFN-induced antiviral defenses). The nature of this heterologous interference on vaccinia virus replication remains unknown.

However, because replication of oncogenic viruses is not required to maintain tumor growth after transformation, targeting such tumors by oncolytic viruses may relate more to the current molecular makeup of the tumor than to specific virus–virus interference among the heterologous viruses. In other words, even if a tumor were of oncovirus origin, there may not be actively replicating oncovirus in the cells of that tumor (certainly not in all cells). Therefore, oncolytic viruses are expected to target/exploit whatever molecular characteristics are present in the cells at the time of infection, irrespective of the presence of the initiating oncovirus. The phenomenon of real-time coinfection and simultaneous replication of the oncovirus and the superinfecting oncolytic virus may not occur, or may be irrelevant for antitumor efficacy of the oncolytic virus. Nevertheless, oncovirus presence has, in experimental models, been shown to affect replication of superinfecting viruses. As an example, Shope fibroma virus typically induces self-limiting and spontaneously regressing tumor-like lesions in rabbits. Although Shope virus is of negligible societal or economic import, it has still facilitated discovery of a number of important aspects of heterologous virus interactions. Even before the 1970s, it was observed that the presence of Shope fibroma virus enhanced replication of several superinfecting RNA viruses, including VSV, encephalomyocarditis virus, Sindbis virus, and Newcastle disease virus, whereas replication of heterologous poxiviruses was suppressed (Padgett and Walker 1970). The phenomenon of heterologous superinfection exclusion or heterologous virus enhancement is complex and dependent on which viruses are involved.

### 13.7.3 Immunological Cross-Inhibition of Oncolytic Viruses

A potential mechanism of inhibition of oncolytic virus efficacy by an oncovirus may involve immunological cross-reactivity through phenotypic particle intermixing. Oncolytic VSV budding from cells superinfected by RSV was shown to contain antigens from both viruses (Ogura and Bauer 1976), and such particles may be prone to neutralization by antibodies against either virus. Indeed, it was shown that VSV particles bearing glycoproteins from RSV-infected cells were neutralized by serum from chickens infected with RSV alone (Mandeville et al. 1979). To what extent such cross-neutralization occurs in a therapeutic scenario, however, remains to be studied.

### 13.7.4 Oncovirus Interference with Oncolytic Virus Cell Entry

Gross murine retrovirus has been detected in L4946 cells, and it was proposed to underlie the reduced infection capacity of the encephalomyocarditis virus in freshly explanted L4946 cells compared to long-term cultured cells (Burrin and MacLennan 1969). MuLVs are known to alter the surface properties of the cells they infect, including exposing retrovirus-associated T-cell neo-antigens (Green 1980); this was proposed to hinder Encephalomyocarditis Virus (EMCV) attachment to freshly explanted L4946 cells as opposed to the cells in established cultures. The authors also provided evidence that EMCV attachment onto rather than replication in L4946 cells was reduced, arguing against a role of type I IFN in restricting infection capacity. Similarly, BHK-21 cells transformed by SV40 polyomavirus are nonpermissive to foot-and-mouth disease virus (FMDV), putatively due to lack of virus attachment as opposed to replication (Diderholm and Dinter 1965).

Although L4946 cells were also refractory to bovine enterovirus (BEV) 1, the virus displayed remarkable oncolytic potency against intramuscular Erlich ascites and SA-1 ascites tumors in mice, "curing" most virus-treated animals following a single intratumoral virus injection, with virus attachment onto these types of cancer cells occurring with much greater efficiency than onto normal mouse embryo or kidney cells (Taylor et al. 1971). A variety of other cancer cell lines were tested for permissiveness to the virus, but critically it was shown that BEV absorption onto 3T3 mouse fibroblasts occurred effectively only when the cells had been transformed with SV40 polyomavirus. These results are in contrast to the findings with EMCV and FMDV in putatively oncovirus-associated cells and would instead argue in favor of a positive influence of oncoviruses on oncolytic virus attachment.

## 13.8 VIRUS WARS

### 13.8.1 Oncoviruses versus Oncoviruses?

Although targeting oncovirus-associated cancers with oncolytic viruses that exploit molecular vulnerabilities conferred by oncovirus presence is an appealing concept, infection and productive replication of the oncolytic virus in all oncovirus-transformed cells is not automatically guaranteed. Indeed, most oncolytic viruses retain their parental cell tropism and may therefore be poorly suited to target "off-target" cells. One possibility for overcoming this problem is to use the oncovirus against itself. One of the earliest clinical studies involved the paradoxical use of extracts from the lymph nodes of patients with glandular fever (infectious mononucleosis = Epstein–Barr virus, which itself is a B-cell transforming oncovirus) to treat acute monoblastic leukemia (Kelly and Russell 2007). The etiology of leukemia in the patients of this study was not known, but it is interesting to note that an oncogenic virus may also serve to deplete malignant cells, highlighting a somewhat counterintuitive scenario of fighting fire with fire.

As an intriguing contemporary example, an EBV vector deleted for three oncogenes still transduced its primary target cells, including EBV-associated B-cell lymphoma cells, and when engineered to express GM-CSF, this vector conferred significant EBV-induced anti-B-cell lymphoma efficacy (Hellebrand et al. 2006).

AIDS patients with PEL display frequent coinfection with KSHV and EBV. Interestingly, it was recently shown that the latency-controlling transcription factors of each of these two oncoviruses, K-RTA and EBV-Z, respectively, interact physically and inhibit each other's functions (maintaining latency of both viruses) (Jiang et al. 2008). KSHV infection also induces EBV latency-inducing LMP-1 expression (Xu et al. 2007), which likely keeps EBV latent and contributes to PEL pathogenesis.

Heterologous superinfection of rat tumors with Friend leukemia virus resulted in immunological recognition of the tumors, suggesting that such superinfection could be harnessed in immune-mediated control of oncovirus-induced cancer (Sinkovics and Horvath 2008). Intriguingly, although homologous superinfection with lentivirus (HIV-1) in humans may result in increased humoral immune responses against the virus (Cortez et al. 2012), such responses may not correlate with decreased risk/capacity to be infected by serologically overlapping strains, possibly due to insufficient levels of neutralizing antibodies (Blish et al. 2008).

It is thus likely that homologous oncovirus immune-mediated inhibition of oncovirus spread would be inefficient, unless antibody responses can be significantly enhanced. Therefore, heterologous superinfection would appear more feasible as a form of therapy. Perhaps the inherent smoldering nature of oncovirus infection could prove more effective than aggressive oncolysis if it would avoid premature clearance but still result in tumors becoming immunologically recognizable. In this regard, the replication-competent mouse retrovirus promoted by Tocagen Inc. (www.tocagen.com) is a tantalizing therapeutic in targeting human oncovirus-associated tumors, not only because of its gene therapy potential but also because of this putative capacity of heterologous (xenogenic) oncoviruses to unveil other oncoviruses.

### 13.8.2 Virus Wars in the "-omics" Era

Because biomolecular research is accelerated by high-throughput and high-content screening, the patterns of genetic, epigenetic, transcriptional, translational, and cellular changes afforded by oncolytic and oncoviruses are becoming clearer (Mahoney et al. 2011; Simonis et al. 2012). Because connectivity analyses may predict drug efficacy of specific cancers or identify putative antiviral drugs based on transcriptional signatures (Josset et al. 2010; Lamb et al. 2006), so could they help with identifying the best oncolytic virus(es) to use against particular oncovirus-induced cancers when signatures are known for both. So far, no systematic multivirus in silico screenings have been performed in this regard, but it is likely only a matter of time before such testings are published.

On a macroscopic scale, however, biomolecular screening methods based on in vitro data are inadequate to predict effects on the immune system or other whole-host functions that govern tumor growth and resistance. This implies that although a particular oncolytic virus may replicate well in a "bioinformatically matched" oncovirus tumor, net therapeutic benefit may be difficult to predict solely based on in silico approaches. Instead, a shifting view in the oncolytic virus field is that oncolytic virus replication per se is less important for therapeutic efficacy than antitumor immune responses and that oncolytic viruses are in fact a form of immunotherapy. This concept is being translated into the next generation of clinical trials where oncolytic viruses are administered as prime-boost tumor vaccines. A level of research that is still understudied is how oncovirus-induced tumors and immune-evasion mechanisms can be countered by oncolytic-virus-induced inflammation. Indeed, in virus wars, the final battle is likely to be decided not by ground battles in single cells or even in the tumor bed but rather systemwide by the third player in the war—the host immune system.

## 13.9 CONCLUSIONS

In this chapter, we have given an overview of the concept of targeting virus-induced cancer by cancer-killing viruses and have highlighted a few but important specific examples of using oncolytic viruses against oncovirus-associated tumors. We believe oncoviruses and oncovirus-induced tumors may be a particularly attractive target for oncolytic virotherapy because many of the host defenses against viruses have already been compromised in such tumors and may therefore provide fertile ground for oncolytic virus replication. Thus, we expect an increase in studies that specifically address this issue and that highlight yet undiscovered molecular vulnerabilities that oncolytic viruses may exploit. A challenge for upcoming studies will be not only to elucidate the complex interactions of the viruses with each other but also their interactions with the host.

For further discussion on virus interactions with host cells and the immune system from a comprehensive evolutionary and mechanistic view, readers may refer to Sinkovics (2008a, 2008b) and Sinkovics and Horvath (2008).

## REFERENCES


An J, Sun Y, Sun R, et al. (2003). Kaposi's sarcoma-associated herpesvirus encoded vFLIP induces cellular IL-6 expression: The role of the NF-kappaB and JNK/AP1 pathways. *Oncogene* 22: 3371–3385.

Asada T. (1974). Treatment of human cancer with mumps virus. *Cancer* 34: 1907–1928.

Banerjee A, Ray RB, Ray R. (2010). Oncogenic potential of hepatitis C virus proteins. *Viruses* 2: 2108–2133.

Barre-Sinoussi F, Montagnier L, Lidereau R, et al. (1979). Enhancement of retrovirus production by anti-interferon serum. *Ann Microbiol (Paris)* 130B: 349–362.

Bergmann M, Romirer I, Sachet M, et al. (2001). A genetically engineered influenza A virus with ras-dependent oncolytic properties. *Cancer Res* 61: 8188–8193.

Berk AJ. (2005). Recent lessons in gene expression, cell cycle control, and cell biology from adenovirus. *Oncogene* 24: 7673–7685.

Blish CA, Dogan OC, Derby NR, et al. (2008). Human immunodeficiency virus type 1 superinfection occurs despite relatively robust neutralizing antibody responses. *J Virol* 82: 12094–12103.

Breitbach CJ, Paterson JM, Lemay CG, et al. (2007). Targeted inflammation during oncolytic virus therapy severely compromises tumor blood flow. *Mol Ther* 15: 1686–1693.

Brown CW, Stephenson KB, Hanson S, et al. (2009). The p14 FAST protein of reptilian reovirus increases vesicular stomatitis virus neuropathogenesis. *J Virol* 83: 552–561.

Burrin DH, MacLennan AP. (1969). Changes in virus attachment and other surface properties of L4946 mouse leukemia cells on adaptation of growth in tissue culture. *Cancer Res* 29: 435–441.

Cerullo V, Koski A, Vähä-Koskela M, et al. (2012). Chapter eight—Oncolytic adenoviruses for cancer immunotherapy: Data from mice, hamsters, and humans. *Adv Cancer Res* 115: 265–318.

Cesaire R, Oliere S, Sharif-Askari E, et al. (2006). Oncolytic activity of vesicular stomatitis virus in primary adult T-cell leukemia. *Oncogene* 25: 349–358.

Chang CL, Ma B, Pang X, et al. (2009). Treatment with cyclooxygenase-2 inhibitors enables repeated administration of vaccinia virus for control of ovarian cancer. *Mol Ther* 17: 1365–1372.

Chellappan S, Kraus VB, Kroger B, et al. (1992). Adenovirus E1A, simian virus 40 tumor antigen, and human papillomavirus E7 protein share the capacity to disrupt the interaction between transcription factor E2F and the retinoblastoma gene product. *Proc Natl Acad Sci U S A* 89: 4549–4553.

Chen HH, Cawood R, El-Sherbini Y, et al. (2011). Active adenoviral vascular penetration by targeted formation of heterocellular endothelial-epithelial syncytia. *Mol Ther* 19: 67–75.

Cheng F, Pekkonen P, Laurinavicius S, et al. (2011). KSHV-initiated notch activation leads to membrane-type-1 matrix metalloproteinase-dependent lymphatic endothelial-to-mesenchymal transition. *Cell Host Microbe* 10: 577–590.

Cheng F, Pekkonen P, Ojala PM. (2012). Instigation of Notch signaling in the pathogenesis of Kaposi's sarcoma-associated herpesvirus and other human tumor viruses. *Future Microbiol* 7: 1191–1205.

Cheng PH, Rao XM, McMasters KM, et al. (2013). Molecular basis for viral selective replication in cancer cells: Activation of CDK2 by adenovirus-induced cyclin E. *PLoS One* 8: e57340.

Chinnakannan SK, Nanda SK, Baron MD. (2013). Morbillivirus v proteins exhibit multiple mechanisms to block type 1 and type 2 interferon signalling pathways. *PLoS One* 8: e57063.

Christian SL, Collier TW, Zu D, et al. (2009). Activated Ras/MEK inhibits the antiviral response of alpha interferon by reducing STAT2 levels. *J Virol* 83: 6717–6726.

Christian SL, Zu D, Licursi M, et al. (2012). Suppression of IFN-induced transcription underlies IFN defects generated by activated Ras/MEK in human cancer cells. *PLoS One* 7: e44267.

Cortez V, Odem-Davis K, McClelland RS, et al. (2012). HIV-1 superinfection in women broadens and strengthens the neutralizing antibody response. *PLoS Pathog* 8: e1002611.

DaPalma T, Doonan BP, Trager NM, et al. (2010). A systematic approach to virus-virus interactions. *Virus Res* 149: 1–9.

Das S, El-Deiry WS, Somasundaram K. (2003). Efficient growth inhibition of HPV 16 E6-expressing cells by an adenovirus-expressing p53 homologue p73beta. *Oncogene* 22: 8394–8402.

Deichman G, Dyakova N, Kashkina L, et al. (1999). In vivo acquired mechanisms of tumor cells local defense against the host innate immunity effectors: Implication in specific antitumor immunity. *Immunol Lett* 70: 37–42.

Deichman G, Dyakova N, Kashkina L, et al. (2001). The fingerprints of the host innate immunity on the cells of primary virus-induced tumors. *Immunol Lett* 75: 209–214.

De Las Heras M, Ortin A, Benito A, et al. (2006). In-situ demonstration of mitogen-activated protein kinase Erk 1/2 signalling pathway in contagious respiratory tumours of sheep and goats. *J Comp Pathol* 135: 1–10.

De Pace N. (1912). Sulla scomparsa di un enorme cancro vegetante del callo dell'utero senza cura chirurgica. *Ginecologia* 9: 82–88.

Diderholm H, Dinter Z. (1965). A line of polyoma-transformed Bhk21 cells insusceptible to foot-and-mouth disease virus. *Virology* 26: 369–372.

Douville RN, Hiscott J. (2010). The interface between the innate interferon response and expression of host retroviral restriction factors. *Cytokine* 52: 108–115.

Dunn GP, Koebel CM, Schreiber RD. (2006). Interferons, immunity and cancer immunoediting. *Nat Rev Immunol* 6: 836–848.

Dvory-Sobol H, Sagiv E, Kazanov D, et al. (2006). Targeting the active beta-catenin pathway to treat cancer cells. *Mol Cancer Ther* 5: 2861–2871.

Eckner RJ, Steeves RA. (1972). A classification of the murine leukemia viruses. Neutralization of pseudotypes of Friend spleen focus-forming virus by type-specific murine antisera. *J Exp Med* 136: 832–850.

Farassati F, Pan W, Yamoutpour F, et al. (2008). Ras signaling influences permissiveness of malignant peripheral nerve sheath tumor cells to oncolytic herpes. *Am J Pathol* 173: 1861–1872.

Figova K, Sobotkova E, Duskova M, et al. (2010). In vitro and in vivo effects of reovirus on HPV16-transformed mice cells. *Neoplasma* 57: 207–214.

Foka P, Pourchet A, Hernandez-Alcoceba R, et al. (2010). Novel tumour-specific promoters for transcriptional targeting of hepatocellular carcinoma by herpes simplex virus vectors. *J Gene Med* 12: 956–967.

Fox E, Moen PT, Jr, Bodnar JW. (1990). Replication of minute virus of mice DNA in adenovirus-infected or adenovirus-transformed cells. *Virology* 176: 403–412.

Fuerer C, Iggo R. (2002). Adenoviruses with Tcf binding sites in multiple early promoters show enhanced selectivity for tumour cells with constitutive activation of the wnt signalling pathway. *Gene Ther* 9: 270–281.

Gallo RC. (2005). The discovery of the first human retrovirus: HTLV-1 and HTLV-2. *Retrovirology* 2: 17.

Green WR. (1980). H-2-restricted cytolytic T lymphocytes specific for a subclass of AKR endogenous leukemia virus-induced tumors: Correlation of tumor cell susceptibility with expression of the gross cell surface antigen. *J Immunol* 125: 2584–2590.

Guedan S, Grases D, Rojas JJ, et al. (2012). GALV expression enhances the therapeutic efficacy of an oncolytic adenovirus by inducing cell fusion and enhancing virus distribution. *Gene Ther* 19: 1048–1057.

Gujar SA, Pan DA, Marcato P, et al. (2011). Oncolytic virus-initiated protective immunity against prostate cancer. *Mol Ther* 19: 797–804.

Hale BG, Randall RE, Ortin J, et al. (2008). The multifunctional NS1 protein of influenza A viruses. *J Gen Virol* 89: 2359–2376.

Han Z, Hong Z, Chen C, et al. (2009). A novel oncolytic adenovirus selectively silences the expression of tumor-associated STAT3 and exhibits potent antitumoral activity. *Carcinogenesis* 30: 2014–2022.

Han Z, Hong Z, Gao Q, et al. (2012). A potent oncolytic adenovirus selectively blocks the STAT3 signaling pathway and potentiates cisplatin antitumor activity in ovarian cancer. *Hum Gene Ther* 23: 32–45.

Harrington KJ, Pandha H, Vile RG. (2008). *Viral Therapy of Cancer*. Hoboken, NJ: John Wiley & Sons.

Hellebrand E, Mautner J, Reisbach G, et al. (2006). Epstein-Barr virus vector-mediated gene transfer into human B cells: Potential for antitumor vaccination. *Gene Ther* 13: 150–162.

Hemminki A, Kanerva A, Kremer EJ, et al. (2003). A canine conditionally replicating adenovirus for evaluating oncolytic virotherapy in a syngeneic animal model. *Mol Ther* 7: 163–173.

Heo J, Reid T, Ruo L, et al. (2013). Randomized dose-finding clinical trial of oncolytic immunotherapeutic vaccinia JX-594 in liver cancer. *Nat Med* 19: 329–336.

Hoption Cann SA, van Netten JP, van Netten C, et al. (2002). Spontaneous regression: A hidden treasure buried in time. *Med Hypotheses* 58: 115–119.

Hu F, Nicholas J. (2006). Signal transduction by human herpesvirus 8 viral interleukin-6 (vIL-6) is modulated by the nonsignaling gp80 subunit of the IL-6 receptor complex and is distinct from signaling induced by human IL-6. *J Virol* 80: 10874–10878.

Hu W, Hofstetter W, Guo W, et al. (2008). JNK-deficiency enhanced oncolytic vaccinia virus replication and blocked activation of double-stranded RNA-dependent protein kinase. *Cancer Gene Ther* 15: 616–624.

Huebner RJ, Rowe WP, Schatten WE, et al. (1956). Studies on the use of viruses in the treatment of carcinoma of the cervix. *Cancer* 9: 1211–1218.

Ilvesaro J, Metsikko K, Vaananen K, et al. (1999). Polarity of osteoblasts and osteoblast-like UMR-108 cells. *J Bone Miner Res* 14: 1338–1344.

Jang KL, Shackelford J, Seo SY, et al. (2005). Up-regulation of beta-catenin by a viral oncogene correlates with inhibition of the seven in absentia homolog 1 in B lymphoma cells. *Proc Natl Acad Sci U S A* 102: 18431–18436.

Jiang Y, Xu D, Zhao Y, et al. (2008). Mutual inhibition between Kaposi's sarcoma-associated herpesvirus and Epstein-Barr virus lytic replication initiators in dually-infected primary effusion lymphoma. *PLoS One* 3: e1569.

Josset L, Textoris J, Loriod B, et al. (2010). Gene expression signature-based screening identifies new broadly effective influenza a antivirals. *PLoS One* 5: e13169.

Kehn K, Deng L, de la Fuente C, et al. (2004). The role of cyclin D2 and p21/waf1 in human T-cell leukemia virus type 1 infected cells. *Retrovirology* 1: 6.

Kelly E, Russell SJ. (2007). History of oncolytic viruses: Genesis to genetic engineering. *Mol Ther* 15: 651–659.

Ketola A, Hinkkanen A, Yongabi F, et al. (2008). Oncolytic Semliki forest virus vector as a novel candidate against unresectable osteosarcoma. *Cancer Res* 68: 8342–8350.

Khodarev NN, Roizman B, Weichselbaum RR. (2012). Molecular pathways: Interferon/stat1 pathway: Role in the tumor resistance to genotoxic stress and aggressive growth. *Clin Cancer Res* 18: 3015–3021.

Kleinschmidt WJ, Ellis LF, Van Frank RM, et al. (1968). Interferon stimulation by a double stranded RNA of a mycophage in statolon preparations. *Nature* 220: 167–168.

Kravchenko AT, Altstein AD, Voronin ES. (1965). Interference between influenza and Rous sarcoma viruses in chicks. *Acta Virol* 9: 130–136.

Kubo T, Shimose S, Matsuo T, et al. (2011). Oncolytic vesicular stomatitis virus administered by isolated limb perfusion suppresses osteosarcoma growth. *J Orthop Res* 29: 795–800.

Kuroda T, Rabkin SD, Martuza RL. (2006). Effective treatment of tumors with strong beta-catenin/T-cell factor activity by transcriptionally targeted oncolytic herpes simplex virus vector. *Cancer Res* 66: 10127–10135.

Lamb J, Crawford ED, Peck D, et al. (2006). The Connectivity Map: Using gene-expression signatures to connect small molecules, genes, and disease. *Science* 313: 1929–1935.

Lampson GP, Tytell AA, Nemes MM, et al. (1963). Purification and characterization of chick embryo interferon. *Proc Soc Exp Biol Med* 112: 468–478.

Le Boeuf F, Niknejad N, Wang J, et al. (2012). Sensitivity of cervical carcinoma cells to vesicular stomatitis virus-induced oncolysis: Potential role of human papilloma virus infection. *Int J Cancer* 131: E204–215.

Levine AJ, Enquist LW., (2007). "History of virology," in Fields Virology, 5th Edn, eds Knipe D. M., Howley P. M., Griffin D. E., Lamb R. A., Martin M. A., Roizman B., Straus S. E., editors. (Philadelphia, PA: Lippincott Williams & Wilkins), 3–23.

Li G, Kawashima H, Ogose A, et al. (2011). Efficient virotherapy for osteosarcoma by telomerase-specific oncolytic adenovirus. *J Cancer Res Clin Oncol* 137: 1037–1051.

Liu TC, Hwang T, Park BH, et al. (2008). The targeted oncolytic poxvirus JX-594 demonstrates antitumoral, antivascular, and anti-HBV activities in patients with hepatocellular carcinoma. *Mol Ther* 16: 1637–1642.

Liu X, Marmorstein R. (2007). Structure of the retinoblastoma protein bound to adenovirus E1A reveals the molecular basis for viral oncoprotein inactivation of a tumor suppressor. *Genes Dev* 21: 2711–2716.

Lundquist A, Barre B, Bienvenu F, et al. (2003). Kaposi sarcoma-associated viral cyclin K overrides cell growth inhibition mediated by oncostatin M through STAT3 inhibition. *Blood* 101: 4070–4077.

Mahoney DJ, Lefebvre C, Allan K, et al. (2011). Virus-tumor interactome screen reveals ER stress response can reprogram resistant cancers for oncolytic virus-triggered caspase-2 cell death. *Cancer Cell* 20: 443–456.

Malerba M, Daeffler L, Rommelaere J, et al. (2003). Replicating parvoviruses that target colon cancer cells. *J Virol* 77: 6683–6691.

Mandeville R, Rohan P, Wainberg MA. (1979). Neutralization of pseudotypes of vesicular stomatitis virus by sera from avian retrovirus-infected hosts. *Int J Cancer* 23: 415–423.

Martin GS. (2001). The hunting of the Src. *Nat Rev Mol Cell Biol* 2: 467–475.

McGill NK, Vyas J, Shimauchi T, et al. (2012). HTLV-1-associated infective dermatitis: Updates on the pathogenesis. *Exp Dermatol* 21: 815–821.

Melcher A, Parato K, Rooney CM, et al. (2011). Thunder and lightning: Immunotherapy and oncolytic viruses collide. *Mol Ther* 19: 1008–1016.

Molden J, Chang Y, You Y, et al. (1997). A Kaposi's sarcoma-associated herpesvirus-encoded cytokine homolog (vIL-6) activates signaling through the shared gp130 receptor subunit. *J Biol Chem* 272: 19625–19631.

Molomut N, Padnos M. (1965). Inhibition of transplantable and spontaneous murine tumours by the M-P virus. *Nature* 208: 948–950.

Molomut N, Padnos M, Gross L, et al. (1964). Inhibition of a transplantable murine leukaemia by a lymphocytopenic virus. *Nature* 204: 1003–1004.

Moore PS, Boshoff C, Weiss RA, et al. (1996). Molecular mimicry of human cytokine and cytokine response pathway genes by KSHV. *Science* 274: 1739–1744.

Moreau-Gachelin F. (2008). Multi-stage Friend murine erythroleukemia: Molecular insights into oncogenic cooperation. *Retrovirology* 5: 99.

Nadzharian NU, Kamalian LA. (1975). Sensitivity of normal and Rous virus-transformed lines of Armenian hamster cells to infectious viruses. *Vopr Virusol* 167–171.

Nicholas J. (2007). Human herpesvirus 8-encoded proteins with potential roles in virus-associated neoplasia. *Front Biosci* 12: 265–281.

Noser JA, Mael AA, Sakuma R, et al. (2007). The RAS/Raf1/MEK/ERK signaling pathway facilitates VSV-mediated oncolysis: Implication for the defective interferon response in cancer cells. *Mol Ther* 15: 1531–1536.

Ogura H, Bauer H. (1976). Biological and electron microscopic studies on the phenotypic mixing of the thermolabile mutant of vesicular stomatitis virus, tl-17, with avian RNA tumor viruses. *Arch Virol* 52: 233–242.

Oliere S, Arguello M, Mesplede T, et al. (2008). Vesicular stomatitis virus oncolysis of T lymphocytes requires cell cycle entry and translation initiation. *J Virol* 82: 5735–5749.

Ono HA, Davydova JG, Adachi Y, et al. (2005). Promoter-controlled infectivity-enhanced conditionally replicative adenoviral vectors for the treatment of gastric cancer. *J Gastroenterol* 40: 31–42.

Padgett BL, Walker DL. (1970). Effect of persistent fibroma virus infection on susceptibility of cells to other viruses. *J Virol* 5: 199–204.

Paglino J, Tattersall P. (2011). The parvoviral capsid controls an intracellular phase of infection essential for efficient killing of stepwise-transformed human fibroblasts. *Virology* 416: 32–41.

Palosaari H, Parisien JP, Rodriguez JJ, et al. (2003). STAT protein interference and suppression of cytokine signal transduction by measles virus V protein. *J Virol* 77: 7635–7644.

Pan W, Bodempudi V, Esfandyari T, et al. (2009). Utilizing ras signaling pathway to direct selective replication of herpes simplex virus-1. *PLoS One* 4: e6514.

Parkin DM. (2006). The global health burden of infection-associated cancers in the year 2002. *Int J Cancer* 118: 3030–3044.

Pensa S, Regis G, Boselli D, et al. (2009). STAT1 and STAT3 in tumorigenesis: Two sides of the same coin? In: Stephanou A (ed.), *JAK-STAT Pathway in Disease*. Austin, TX: Landes Bioscience, pp. 110–121.

Peyton R. (1911). A sarcoma of the fowl transmissible by an agent separable from the tumor cells. *J Exp Med.* 13: 397–411.

Poiesz BJ, Ruscetti FW, Gazdar AF, et al. (1980). Detection and isolation of type C retrovirus particles from fresh and cultured lymphocytes of a patient with cutaneous T-cell lymphoma. *Proc Natl Acad Sci U S A* 77: 7415–7419.

Prestwich RJ, Ilett EJ, Errington F, et al. (2009). Immune-mediated antitumor activity of reovirus is required for therapy and is independent of direct viral oncolysis and replication. *Clin Cancer Res* 15: 4374–4381.

Price CH, Moore M, Jones DB. (1972). FBJ virus-induced tumours in mice. A histopathological study of FBJ virus tumours and their relevance to murine and human osteosarcoma arising in bone. *Br J Cancer* 26: 15–27.

Psyrri A, DeFilippis RA, Edwards AP, et al. (2004). Role of the retinoblastoma pathway in senescence triggered by repression of the human papillomavirus E7 protein in cervical carcinoma cells. *Cancer Res* 64: 3079–3086.

Ramachandran A, Parisien JP, Horvath CM. (2008). STAT2 is a primary target for measles virus V protein-mediated alpha/beta interferon signaling inhibition. *J Virol* 82: 8330–8338.

Rous P. (1911). Transmission of a malignant new growth by means of a cell-free filtrate. *JAMA* 56: 198.

Ryan EP, Malboeuf CM, Bernard M, et al. (2006). Cyclooxygenase-2 inhibition attenuates antibody responses against human papillomavirus-like particles. *J Immunol* 177: 7811–7819.

Saha A, Kaul R, Murakami M, et al. (2010). Tumor viruses and cancer biology: Modulating signaling pathways for therapeutic intervention. *Cancer Biol Ther* 10: 961–978.

Sakakibara S, Tosato G. (2011). Viral interleukin-6: Role in Kaposi's sarcoma-associated herpesvirus: Associated malignancies. *J Interferon Cytokine Res* 31: 791–801.

Salek-Ardakani S, Schoenberger SP. (2013). T cell exhaustion: A means or an end? *Nat Immunol* 14: 531–533.

Sanchez-Prieto R, de Alava E, Palomino T, et al. (1999). An association between viral genes and human oncogenic alterations: The adenovirus E1A induces the Ewing tumor fusion transcript EWS-FLI1. *Nat Med* 5: 1076–1079.

Schuhmann KM, Pfaller CK, Conzelmann KK. (2011). The measles virus V protein binds to p65 (RelA) to suppress NF-kappaB activity. *J Virol* 85: 3162–3171.

Semmel M, Mercier G, Pavloff N, et al. (1988). Viral products in cells infected with vesicular stomatitis virus and superinfected with Rous sarcoma virus. *Arch Virol* 100: 121–129.

Semmel M, Sathasivam A. (1983). Inhibition of Rous sarcoma virus-induced transformation by preinfection with rhabdoviruses. *J Gen Virol* 64(Pt 2): 275–284.

Sevko A, Sade-Feldman M, Kanterman J, et al. (2013). Cyclophosphamide promotes chronic inflammation-dependent immunosuppression and prevents antitumor response in melanoma. *J Invest Dermatol* 133: 1610–1619.

Shackelford J, Pagano JS. (2004). Tumor viruses and cell signaling pathways: Deubiquitination versus ubiquitination. *Mol Cell Biol* 24: 5089–5093.

Sharma-Walia N, Patel K, Chandran K, et al. (2012). COX-2/PGE2: Molecular ambassadors of Kaposi's sarcoma-associated herpes virus oncoprotein-v-FLIP. *Oncogenesis* 1: e5.

Shirasawa H, Jin MH, Shimizu K, et al. (1994). Transcription-modulatory activity of full-length E6 and E6*I proteins of human papillomavirus type 16. *Virology* 203: 36–42.

Simonis N, Rual JF, Lemmens I, et al. (2012). Host-pathogen interactome mapping for HTLV-1 and -2 retroviruses. *Retrovirology* 9: 26.

Sinkovics JG. (2008a). Cytolytic immune lymphocytes in the armamentarium of the human host. *Acta Microbiol Immunol Hung* 55: 371–382.

Sinkovics JG. (2008b). *Cytolytic Immune Lymphocytes: In the Armamentarium of the Human Host Products of the Evolving Universal Immune System*. Germany: Schenk Verlag Gmbh.

Sinkovics JG, Horvath JC. (2005). *Viral Therapy of Human Cancers*. New York: Marcel Dekker.

Sinkovics JG, Horvath JC. (2008). Natural and genetically engineered viral agents for oncolysis and gene therapy of human cancers. *Arch Immunol Ther Exp (Warsz)* 56(Suppl 1): 1–59.

Smerdou C, Menne S, Hernandez-Alcoceba R, et al. (2010). Gene therapy for HCV/HBV-induced hepatocellular carcinoma. *Curr Opin Investig Drugs* 11: 1368–1377.

Smith RR, Huebner RJ, Rowe WP, et al. (1956). Studies on the use of viruses in the treatment of carcinoma of the cervix. *Cancer* 9: 1211–1218.

Sobotkova E, Duskova M, Eckschlager T, et al. (2008). Efficacy of reovirus therapy combined with cyclophosphamide and gene-modified cell vaccines on tumors induced in mice by HPV16-transformed cells. *Int J Oncol* 33: 421–426.

Stojdl DF, Lichty B, Knowles S, et al. (2000). Exploiting tumor-specific defects in the interferon pathway with a previously unknown oncolytic virus. *Nat Med* 6: 821–825.

Stojdl DF, Lichty BD, tenOever BR, et al. (2003). VSV strains with defects in their ability to shutdown innate immunity are potent systemic anti-cancer agents. *Cancer Cell* 4: 263–275.

Strauss R, Sova P, Liu Y, et al. (2009). Epithelial phenotype confers resistance of ovarian cancer cells to oncolytic adenoviruses. *Cancer Res* 69: 5115–5125.

Strube W, Strube M, Kroath H, et al. (1982). Interferon inhibits establishment of fibroblast infection with avian retroviruses. *J Interferon Res* 2: 37–49.

Taylor MW, Cordell B, Souhrada M, et al. (1971). Viruses as an aid to cancer therapy: Regression of solid and ascites tumors in rodents after treatment with bovine enterovirus. *Proc Natl Acad Sci U S A* 68: 836–840.

Thompson MR, Kaminski JJ, Kurt-Jones EA, et al. (2011). Pattern recognition receptors and the innate immune response to viral infection. *Viruses* 3: 920–940.

Tongu M, Harashima N, Yamada T, et al. (2010). Immunogenic chemotherapy with cyclophosphamide and doxorubicin against established murine carcinoma. *Cancer Immunol Immunother* 59: 769–777.

Turner PC, Moyer RW. (2002). Poxvirus immune modulators: Functional insights from animal models. *Virus Res* 88: 35–53.

Ulane CM, Rodriguez JJ, Parisien JP, et al. (2003). STAT3 ubiquitylation and degradation by mumps virus suppress cytokine and oncogene signaling. *J Virol* 77: 6385–6393.

Vähä-Koskela MJ, Heikkilä JE, Hinkkanen AE. (2007). Oncolytic viruses in cancer therapy. *Cancer Lett* 254: 178–216.

Vahlne A. (2009). A historical reflection on the discovery of human retroviruses. *Retrovirology* 6: 40.

Vengris VE, Mare CJ. (1973). Protection of chickens against Marek's disease virus JM-V strain with statolon and exogenous interferon. *Avian Dis* 17: 758–767.

Walton TH, Moen PT, Jr, Fox E, et al. (1989). Interactions of minute virus of mice and adenovirus with host nucleoli. *J Virol* 63: 3651–3660.

Wang H, Chen NG, Minev BR, et al. (2012a). Oncolytic vaccinia virus GLV-1h68 strain shows enhanced replication in human breast cancer stem-like cells in comparison to breast cancer cells. *J Transl Med* 10: 167.

Wang PY, Currier MA, Hansford L, et al. (2012b). Expression of HSV-1 receptors in EBV-associated lymphoproliferative disease determines susceptibility to oncolytic HSV. *Gene Ther* 20: 761–769.

Wang W, Xia X, Wang S, et al. (2011). Oncolytic adenovirus armed with human papillomavirus E2 gene in combination with radiation demonstrates synergistic enhancements of antitumor efficacy. *Cancer Gene Ther* 18: 825–836.

Wang Y, Xue SA, Hallden G, et al. (2005). Virus-associated RNA I-deleted adenovirus, a potential oncolytic agent targeting EBV-associated tumors. *Cancer Res* 65: 1523–1531.

Webb HE, Molomut N, Padnos M, et al. (1975). The treatment of 18 cases of malignant disease with an arenavirus. *Clin Oncol* 1: 157–169.

Wong HK, Ziff EB. (1996). The human papillomavirus type 16 E7 protein complements adenovirus type 5 E1A amino-terminus-dependent transactivation of adenovirus type 5 early genes and increases ATF and Oct-1 DNA binding activity. *J Virol* 70: 332–340.

Xu D, Coleman T, Zhang J, et al. (2007). Epstein-Barr virus inhibits Kaposi's sarcoma-associated herpesvirus lytic replication in primary effusion lymphomas. *J Virol* 81: 6068–6078.

Yamagata S, Yamagata T. (1984). FBJ virus-induced osteosarcoma contains type I, type I trimer, type III as well as type V collagens. *J Biochem* 96: 17–26.

Yan N, Chen ZJ. (2012). Intrinsic antiviral immunity. *Nat Immunol* 13: 214–222.

Yang SW, Cody JJ, Rivera AA, et al. (2011). Conditionally replicating adenovirus expressing TIMP2 for ovarian cancer therapy. *Clin Cancer Res* 17: 538–549.

Yao Y, Wang L, Zhang H, et al. (2012). A novel anticancer therapy that simultaneously targets aberrant p53 and Notch activities in tumors. *PLoS One* 7: e46627.

Yoshida M, Miyoshi I, Hinuma Y. (1982). Isolation and characterization of retrovirus from cell lines of human adult T-cell leukemia and its implication in the disease. *Proc Natl Acad Sci U S A* 79:2031–2035.

Zagon IS, Donahue RN, Rogosnitzky M, et al. (2008). Imiquimod upregulates the opioid growth factor receptor to inhibit cell proliferation independent of immune function. *Exp Biol Med (Maywood)* 233: 968–979.

zur Hausen H. (2006). *Infections Causing Human Cancer*. Weinheim: Wiley-VCH.

# 14 Oncolytic Viruses and Histone Deacetylase Inhibitors

*Vaishali M. Patil and Satya P. Gupta*

## 14.1 INTRODUCTION

Over the past decades, viruses have been used as potential anticancer therapeutics for the treatment of cancer. A variety of agents including rabies virus, adenovirus, paramyxovirus, Newcastle disease virus, and mumps virus are well reported (Sinkovics and Horvath 1993). The newer technologies have helped in the understanding of the cancer biology and signaling. It includes characterization of the selectivity of viruses such as reovirus for cells with an activated Ras signaling pathway (Strong and Lee 1996; Strong et al. 1998). The role of tumor suppressor p53 in cell cycle control has been elaborated and, based on this, E1B-deleted mutants were developed as potential cancer therapeutics (Yew and Berk 1992; Bischoff et al. 1996). The herpesvirus biology has allowed engineering of nonneurovirulent mutants lacking genes as essential elements for replication of most of the cells including tumor cells. Therefore, development of oncolytic viruses has been an attractive field of study and has led to the transition of many novel therapeutics into the clinic, such as the mutant adenovirus ONYX-015, herpes simplex virus type 1 (HSV-1) mutant G207 (related mutant R3616), the naturally occurring human reovirus (laboratory strain type 3 Dearing), and vesicular stomatitis virus (VSV). Among them, the adenovirus was clinically tested for cervical carcinoma in the 1950s (Smith et al. 1956). In the 20th century, it emerged as a potential cancer therapeutic, and new generations of mutant adenoviruses are under development for improved specificity and efficacy (Tollefson et al. 1996).

In 1977, Hashiro et al. reported certain cell lines of murine origin susceptible to cytotoxic induction by reovirus (Hashiro et al. 1977). Later Duncan et al. (1978) found different sensitivities to reovirus infection by normal human and subhuman primate cells and cytopathology only in transformed cells. Also the virus–receptor interaction studies have explored the molecular basis of reovirus oncolysis. Further, because the immune system is largely suppressed in tumors, the microorganism distributed in a tumor escapes the immune surveillance system of the host. The host immune system clears the remaining circulating viral particles shortly after intravenous administration (Yu et al. 2004).

Various cellular pathways are involved when viruses kill cells directly and the appearance of drug resistance is unlikely. The antitumor toxicity of therapeutic genes can be increased by combining with oncolytic viruses.

## 14.2 ONCOLYTIC AGENTS

The studies conducted in the 1950s and 1960s failed to establish oncolytic viral therapy as a viable anticancer modality. Until the advances of genetic engineering, this field remained a medical curiosity. In the 19th century, oncolytic agents came to medical prominence with the regression of some forms of hematological malignancies in certain viral infections. Similarly, rabies virus inoculation also was shown to cause advanced cervical carcinoma to regress (Kelly and Russell 2007). Rapid advancements have been observed and have been intensively studied for nearly two decades. Some naturally occurring oncolytic viruses such as the vaccinia virus (VV), reovirus, and Newcastle disease virus have been discovered that spare normal tissues and infect only tumor cells. Other viruses such as HSV-1 and adenovirus need attenuation to act as successful oncolytic agents. The attenuated viruses have been found to be more tumor specific and less pathogenic to normal tissues; various modifications have been carried out to achieve this profile (Wong et al. 2010).

## 14.3 TYPES OF ONCOLYTIC VIRUSES AND THEIR THERAPEUTIC ROLE IN CANCER MANAGEMENT

### 14.3.1 ADENOVIRUS

Adenovirus-based cancer therapy forms an intensive area of viral oncolytic research. In the 1950s, it was clinically tested as a therapy for cervical carcinoma (Smith et al. 1956). A series of proteins (E1A, E1B-55KD, E3-11.6KD) cooperate to promote viral replication of adenovirus after its infection in cells (Tollefson et al. 1996).

Study of adenovirus mutants has developed a variety that can specifically replicate within tumor cells, for example, dl1520 (E1BSSK-deficient adenovirus) (Barker and Berk 1987), having oncolytic potential (Bischoff et al. 1996). In most of the cancers, p53 function gets disrupted (Hollstein et al. 1991; Levine et al. 1991), and this proves the significance of a mutant p53 selective oncovirus (OV). In vitro studies proved the role of dl1520 in tumor regression and in improved survival rate (Bischoff et al. 1996; Heise et al. 1997; Heise and Kirn 2000). Various reports that proposed selectivity of dl1520 for cells in which p53 was inactive appeared to be inconsistent (Goodrum and Ornelles 1998; Rothmann et al. 1998). Clinical trials have supported safety and thus selectivity of viral replication (Ganly et al. 2000; Heise et al. 2000). Recent studies with intratumoral viral administration and chemotherapeutics in recurrent squamous cell carcinoma have produced promising results (Heise et al. 1997; Khuri et al. 2000), and no virus was detected in normal tissues. Further modifications to improve efficacy through restriction of adenovirus E1A expression (Rodriguez et al. 1997; Chen et al. 2000) and incorporation of cytotoxic genes (Wildner et al. 1999; Wildner and Morris 2000) have been found to promote more effective and sustained tumor regression (Kurihara et al. 2000).

The mutated viruses have shown selectivity of tumor cells over normal cells and improved therapeutic efficacy over dl1520 (Fueyo et al. 2000; Heise et al. 2000). The adenovirus proteins can modulate the activity of tumor suppressors (O53, pRb) to circumvent cellular responses to infection and thus rationalize the resurgence of interest in viral oncolysis. These identified mutants have shown replication within human tumors with significant response rates in combination with other treatments as compared to conventional therapeutics. A variety of replicant-competent adenovirus mutants have been developed to improve specificity and potency. The results of clinical trials provide a promising novel alternative in the management of cancer.

### 14.3.2 Reovirus

Reovirus is an orphan virus not associated with a significant diseased state (Sabin 1959). Reovirus is a nonenveloped, small, icosahedral virus with double-stranded RNA genome (Yue and Shatkin 1998). These are three serotypes (Weiner and Fields 1977), and its field isolates are useful models for viral infection. Reovirus infection in humans is characteristically benign. Due to this intrinsically benign nature and endogenous oncocytic properties, no modifications are required. It has entered clinical trials as in unengineered viral therapeutic (Sabin 1959).

Reovirus has innate specificity for cells having activated Rat sarcoma (Ras) pathway and thus can be proven as a possible oncocytic agent against a broad variety of cancers. Its therapeutic potential as a Ras-selective OV is important/enticing because in at least 30% of human cancers, Ras-signaling plays an important role in tumorigenesis. More than 80% of cell lines from various tumor origins have been effectively lysed by this virus. To evaluate the in vitro efficacy of reovirus, it was injected into severe combined immunodeficiency (SCID) mouse U87 (glioblastoma) or THC-11 (v-erb B-transformed NH4-373) hind flank tumor xenografts and has shown complete tumor regression in up to 80% of animals within four weeks. The tumor regression ability of reovirus was also evaluated in animals with competent immune systems (Coffey et al. 1998). The multiple intratumoral infections of reovirus in Ras-transformed C3H tumor allografts resulted in complete responses (66%). Additionally, virus induced toxicity was not observed, and the mice remained tumor-free for at least six months after termination of the study. Reovirus has emerged as a potential cancer therapeutic. Also, the broad host range of reovirus permits the testing of its oncocytic potential in vivo (in human xenografts) as well as evaluating the effect of host immune responses to viral therapy in syngeneic tumor modals. These results provide the basis for future studies including the evaluation of concurrent chemotherapy and radiation with reovirus administration.

### 14.3.3 Herpes Simplex Virus

Herpes simplex virus (HSV) has been known for its pathogenicity in humans. HSVs can also cause latent infection of neurons, followed by severe damage to the central nervous system (CNS), and ultimately meningitis or encephalitis (Whitley et al. 1998).

HSV genes are divided into three classes: alpha (α), beta (β), and gama (γ). They are transcribed sequentially (Roizman 1996) as well as encoding viral structural proteins. Mutant versions of HSV have been developed with the help of recent advances in HSV molecular biology and are made oncolytic by deleting major genes ($\gamma_1$ 34:5 and $U_2$ 39) (Martuza 2000; Rampling et al. 2000).

Similar to reovirus, in a case of HSV-1 infection, the ras signaling pathway is the major determinant of host cell permissiveness. High ras activity inhibits or reverses RNA-activated protein kinase (PKR) phosphorylation and allows for viral protein synthesis. Therefore, cancer cells with an activated ras signaling pathway would be suitable targets. The ras pathway plays an important role in deciding host cell susceptibility to HSV-1 and can be explained using drugs (farnesyl transferase inhibitors, PD98059) that can block the extracellular regulated (erk) pathway (Dudley et al. 1995; Gibbs et al. 1997).

Further deletion of the $U_L$39 gene compels the G207 virus replication in proliferating cancer cells having high endogenous ribonucleotide reductase activity (Boviatsis et al. 1994; Mineta et al. 1994). G207 has been tested on animal models of various types of cancer, such as colon (Kooby et al. 1999), ovarian (Coukos et al. 1999), breast (Toda et al. 1998), prostate (Walker et al. 1999), head and neck squamous cell (Chalavi et al. 1999), melanoma (Randazzo et al. 1997), and bladder (Oyama et al. 2000). The efficacy of G207 with chemotherapy or radiotherapy has also been studied (Bradley et al. 1999; Chalavi et al. 1999; Coukos et al. 2000). This includes the combination of G207 with cisplatin for the treatment of squamous cell carcinoma (Chalavi et al. 1999), which has shown promising results compared to either of them individually. Combination with radiation therapy in human U-87 malignant glioma xenografts, reduced tumor size and longer survivals were observed (Bradley et al. 1999). Herpes therapy has also been found effective against cancers that are resistant to chemotherapy (Coukos et al. 2000). The results of phase I clinical trials of HSV mutants shows no toxicity or major adverse clinical symptoms (Homa and Brown 1997; Markert et al. 2000; Rampling et al. 2000) (i.e., G207 and 1716 tumor progression) as well as showing tumor shrinkage.

### 14.3.4 Vesicular Stomatitis Virus

Vesicular stomatitis virus (VSV) is an RNA virus with oncolytic properties (Stojdl et al. 2000a); studies have reported its ability to replicate in carcinoma and melanoma cells. It has shown inhibition of tumor growth in nude mice on intravenous administration. The role of VSV in cancer therapy is based on the fact that interferon-responsive antiviral pathways are mostly defective in tumors; thus, VSV can replicate within these cells regardless of interferon treatment. It is supported by the reduced viral output in case of in vitro interferon treatment of primary cells and high levels of virus in interferon-treated tumor derived cells. Also it is observed that PKR status in the host cells dictates permissiveness to OVs (Stojdl et al. 2000b). Due to inability of PKR, defects arise in the interferon-responsive antiviral pathway and thus improve the susceptibility to virus infection. Various approaches have been applied to control side effects and to improve the therapeutic index of this oncolytic agent.

### 14.3.5 Myxoma Virus

Myxoma virus is a double-stranded DNA genome that does not infect normal human cells. It is, however, highly replicative and lytic for many different human cancer cells (Stanford and MacFadden 2007). The infection of neuroblastoma cancer-initiating cell (CIC) cultures by myxoma virus has been reported (Redding et al. 2009). It also focuses toward altered Akt signaling in neuroblastoma CICs, which is an important survival signal in neuroblastoma cells (Sartelet et al. 2008) and thus suggests an effective virus tumor combination.

## 14.4 HISTONE DEACETYLASES AS POTENTIAL CANCER TARGETS

For various cellular functions, it is fundamental to regulate cellular processes by reversible phosphorylation of key regulatory proteins. In the past decade, research has focused toward the importance of acetylation and deacetylation reactions of histone lysine residues and other cellular factors that regulate gene expression. Studies have addressed the role of histone acetylation in controlling gene transcription, and reversible non-histone proteins' acetylation regulates various cellular functions (Sun and Hampsey 1999). They can be termed as a molecular "on-off" switch (Sun and Hampsey 1999; Kouzarides 2000). The structure of histone deacetylase (HDAC) has been characterized with the cloning of HDAC-1 enhancing gene (Taunton et al. 1996). The class I HDACs are comprised of HDAC-1, -2, -3, and -8 having similar domain organization and high sequence identity (Rundlett et al. 1996). Class II HAD-1-like proteins include HDAC-4, -5, -6, -7, -9 (Grozinger et al. 1999; Gray and Ekstrom 2001; Zhou et al. 2001), and class III sirtuins (in mammals) and silent information regulator-2 (SIR2) (in yeast).

Among the 18 identified HDACs, class I, II, and IV are zinc-dependent metalloproteins and class III is an Nicotinamide adenine dinucleotide (NAD) $^{+}$-dependent protein deacetylase. Class I (i.e., HDAC-1, -2, -3, and -8) is mainly implied in cancer progression as well as other diseases (Minucci and Pelicci 2006). HDACs have been successfully implied in various types of cancer progression as well as other diseases (Minucci and Pelicci 2006). This includes examples of HDAC-1 as a possible prognostic indicator for lung cancer and as overexpressed in prostate, gastric, and colorectal cancers. Similarly, HDAC-2 has been overexpressed in colorectal and gastric cancers, and HDAC-3 is overexpressed in lung cancer and several solid tumors. HDAC-8 has been found to be correlated with the inhibition of cell growth in various human cancerous cells. These studies help to provide specific biomarkers to enhance the available therapeutic options. HDAC-5 and HDAC-6 have been correlated with colon cancer, acute myeloid leukemia, and breast cancer.

Certain complex molecular modifications (i.e., DNA methylation, histone acetylation, phosphorylation, ubiquitylation, and ADP ribosylation) lead to development and progression of cancer. The p53 protein plays an important role in regulation of gene expression and cell signaling pathways; therefore, mutation of p53 is correlated with many forms of cancer (Kaghad et al. 1997). Low levels of p53 and increased production of Mdm2 protein is found to be associated with increased cell growth and proliferation (Rich et al. 2000). Downregulation of p53 is found to be responsible for keratinocyte hyperproliferation (Michel et al. 1996). In the case of cisplatin-treated tumors, overexpression of p53 is found to be associated with cell resistance to apoptosis (Nakayama et al. 2003).

As shown in Figure 14.1, in the absence of stress, p53 does not get phosphorylated. While in stress conditions phosphorylation of p53 is followed by release from Mdm2, and acetylations lead to activation of p53 (Prives and Manley 2001) and, finally, cell division cycle, permanent cell division arrest, and cell death. Various studies have emphasized the role of deacetylases in the regulation of p53 activity. Also, it is well reported that the class III deacetylases interact to modify the functions and biological properties of p53 (Luo et al. 2001; Vaziri et al. 2001) and tubulin (North et al. 2003). The activation/repression of p 53 transcription is controlled by the selective binding of HDACs; hence, HDACs have gained attention from a large number of researchers.

Various studies have supported the potential role of HDAC in tumorigenesis (Yang et al. 1997; Hu et al. 2000; Kristeleit et al. 2004). HDACs have been found overexpressed in certain conditions (i.e., hypoxia, hypoglycemia, and serum deprivation) (Kim et al. 2001). Some reports highlight NAD-dependent deacetylase sirtuin-2 (SIRT2) as a tumor suppressor gene in glioma cells (Morales et al. 1999; Hiratsuka et al. 2003). Further nuclear factor-kappa B (NF-κB) is also an important target of HDACs (Karin et al. 2002). It is reported to be active in pancreatic adenocarcinoma (Wang et al. 1999), T leukemia lymphocytes (Mori et al. 1999), and breast cancer (Cogswell et al. 2000) and thus facilitates cell survival and tumor progression. It is also observed that the antiapoptotic activity of NF-κB is regulated by deacetylation of RelA/p65 lysine310 (Yeung et al. 2004).

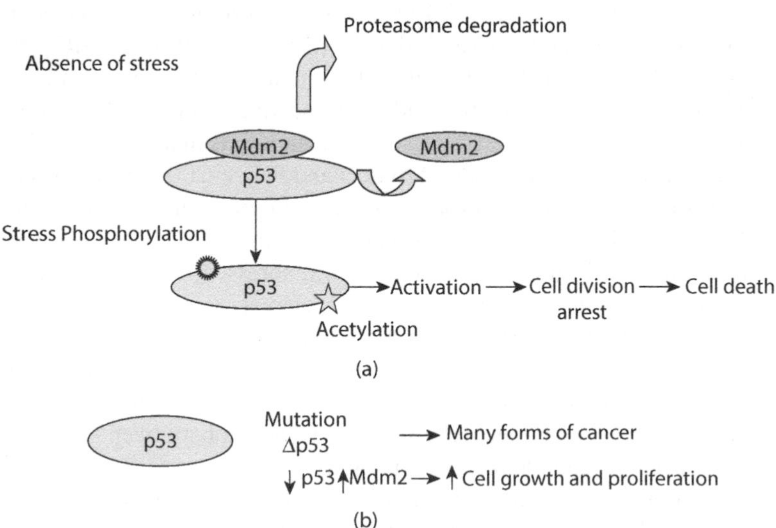


**FIGURE 14.1** Schematic representation of some modifications of p53 leading to cell growth and proliferation or death. (a) Under stress conditions, phosphorylation of p53 and release from Mdm2 and further acetylation led to p53 activation; (b) mutations of p53 and the modifications in the levels of p53/Mdm2 might affect cell growth and proliferation. (Reprinted with permission from Ouaissi M, Ouaissi A., *J Biomed Biotech* 1–10, 2006, Copyright 2006 PubMed Central.)

The multitarget action mechanism of valproic acid (VPA) has been an area of considerable importance. Alvarez-Breckenridge et al. (2012) attempted to improve OV therapy through application of synergism mechanism. To achieve this, oncolytic agents were combined with pharmacological modulators. In the previous studies (Liu et al. 2008a; Nguyen et al. 2008, 2010; Otsuki et al. 2008), VPA and histone deacetylase inhibitors (HDIs) were hypothesized to improve OV replication by downregulating type I interferon (IFN) response. VPA functions were explored by combining with oncolytic herpes simplex virus (oHSV) and using glioblastoma xenograft model. The synergistic mechanism was multifaceted because it resulted in decreased immune cell activation along with suppression of Natural killer (NK)-cell-mediated cytotoxicity against virally infected glioblastoma cells but was related to the detrimental antiviral response to oHSV. Decreased production of IFN-γ was correlated with signal transducer and activator of transcription 5 (STAT5) activity and T-bet, a T-cell-associated transcription factor expression. All the results indicated VPA's pleiotropic action mechanism by suppressing intracellular and innate cellular antiviral activity to enhance viral oncolysis. VPA has been marked as a U.S. Food and Drug Administration (FDA) approved drug for the treatment of epilepsy. Recently, VPA has been established as an effective adjuvant for the treatment of oHSV (Otsuki et al. 2008). Currently cyclophosphamide (CPA) has been utilized in combination with cancer virotherapeutic trials using measles virus (NCT00450814), reovirus (NCT01240538), and adenovirus. CPA has proven its efficacy by circumventing complement-mediated viral depletion and attenuation of antiviral response (Ikeda et al 1999; Fulci et al. 2006).

Thus, HDACs are the key elements in regulating gene expression, differentiation, and the development and maintenance of homeostasis. HDACs are associated with a number of cellular oncogenes and tumor suppressor genes. This makes HDACs an attractive target for anticancer drugs and therapies.

## 14.5 HDAC INHIBITORS

In the discovery and development of anticancer therapy, the study of histone deacetylation/acetylation has offered a detailed insight into tumor initiation and progression. The enzymes histone aetyl transferases (HATs) and histone deacetylases (HDACs) act as both histone substrates and nonhistone proteins. The 18 members of the HDAC family are grouped into four classes: class I (HDAC-1, -2, -3, and -8); class IIa (HDAC-4, -5, -7, and -9) and class IIb (HDAC-6 and -10); class III (sirtuins); and class IV (HDAC-11) (Balasubramanian et al. 2009). Class I HDACs are involved in genetic regulation and are important transcriptional repressors. Class I and II HDACs exist as multi-enzyme complexes, while class II HDACs are important for differentiation as well as function as individual proteins. Class III HDACs are NAD dependent and are important regulators of metabolism and transcription.

Various studies have observed high-level expression of HDACs and hypoacetylation of histones in cancerous cells (Weichert 2009). Thus, alteration of HDACs will play an active role in tumor onset and progression making HDAC an attractive target for therapeutic interventions. Based on this, various classes of histone deacetylase inhibitors (HDIs/HDACIs) have been identified (such as natural and synthetic analogs).

HDIs/HDACIs also includes hydroxamic acid cyclic tetrapeptides, benzamides, and short-chain aliphatic acids (among the latter, the hydroxamate analogs of suberoylanilide hydroxamic acid (SAHA) are the most important). The structure of HDIs has been elaborated in three parts: a head group to coordinate the $Zn^{2+}$ ion in the active pocket of the HDAC enzyme, a hydrophobic spacer to fill out the narrow channel, and a capping group to interact with the rim surrounding the entrance into the pocket.

### 14.5.1 HDIs in Cancer Therapy

In the development of anticancer therapeutics, one of the important strategies is inhibition of HDACs. HDIs which act through transcriptional repression-induced cell cycle arrest, differentiation, and apoptosis of tumor suggests interesting strategy for antitumor chemotherapy development (Pandolfi 2001). In HDI-mediated growth inhibition, p53 plays an important role; cell death due to apoptosis was found to be reduced in the absence of p53 (Joseph et al. 2005). Some other mechanisms have been reported that do not support the role of p53 in HDI functioning including the examples of trichostatin A (TSA), which arrests G2/M phase transition, and cell death by apoptosis (Hirose et al. 2003). TSA has been reported as the most potent HDAC class I and II inhibitors (Grignani et al. 1998; He et al. 1998; Lin et al. 1998), and various synthetic drugs were designed based on its structure. The structural features include a "cap" group, a central aliphatic chain, and a functional group that can chelate the zinc cation in the HDAC active site. Several compounds have successfully entered phase I/II clinical trials (Acharya et al. 2005).

Among the various reported molecules, hydroxamic acid derivatives are the efficient inhibitors of HDAC class I and II enzymes at nanomolar to micromolar concentrations, but their reversible activity is the major problem. TSA, the first HDI with a hydroxamic functional group, inhibits the growth of pancreatic adenocarcinoma cells and the renal carcinoma cell cycle. Suberoylanilide hydroxamic acid and pyroxamide were reported as inducers of growth arrest and in differentiation and apoptosis in various cancer cells. The natural product, apicidin, is a cyclic tetrapeptide that inhibits HDACs and the growth of tumor cells as well as blocking tumor cell proliferation. The cyclin-dependent kinase (CDK) inhibitor $p21^{WAFI}$ arrests G1 phase by blocking CDK activity. Most of the HDIs induce expression of p21 and block tumor cell proliferation (Acharya et al. 2005). The bicyclic depsipeptide, FK 228, inhibits HDACs at nanomolar concentration and induces apoptosis and finally cell death. It induces HDAC inhibition by increasing expression of the angiogenic-inhibitory factor encoding gene and suppressing the gene encoding angiogenic-stimulating factor. Thus, it interferes with the neovascularization required for certain tumor growth (Ouaissi and Ouaissi 2006).

Most of the HDIs induce expression of p21 and block tumor cell proliferation (Acharya et al. 2005). FK-228, bicyclic depsipeptide, inhibits HDACs at nanomolar concentration with induced apoptosis and thus cell death. It also induces HDAC inhibition with increased expression of angiogenic-inhibitory factor encoding gene and suppression of gene encoding angiogenic-stimulating factor. Thus, it interferes with neovascularization required for certain tumor growth.

HDIs' action is generally through interference with cell cycle differentiation and apoptosis pathways and depends on the nature of tumor cells (Figure 14.2).

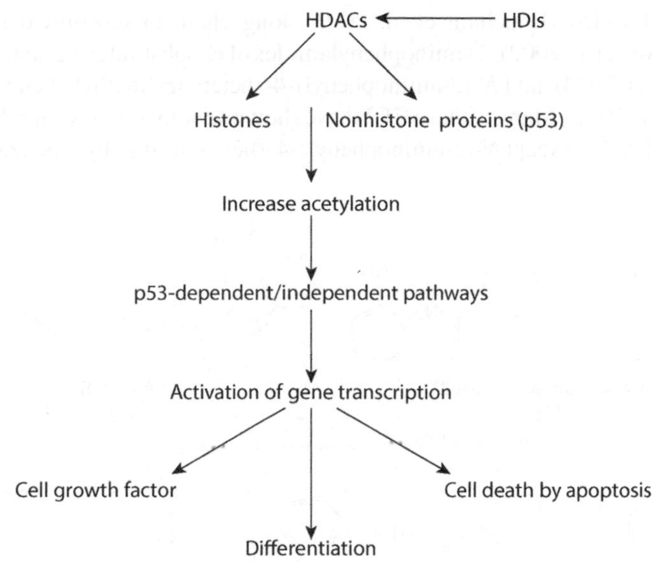


**FIGURE 14.2** A schematic diagram showing mechanism of action of HDIs.

The NF-κB is one of the targets of HDACs. In precancerous cells, it improves pro inflammatory and survival gene expression and inhibits the death-promoting pathway with enhanced premalignant potential. HDIs prevent NF-κB transactivation and enable re expression of silenced genes in tumor cells, leading to reversion of malignant phenotypes and showing their therapeutic effect.

The effect of anticancer drugs can be synergized by HDIs as observed by co-incubation of human leukemia cells (K562) with a proteasome inhibitor, bortezomid, and suberoylanilide hydroxamic acid or sodium butyrate as HDIs. This study has shown increased mitochondrial injury and apoptosis (Yu et al. 2003). Similarly, fludarabine has shown activity in B-cell malignancies, and pretreatment of human leukemic cells with MS-275 has enhanced the activity of cytotoxic drugs (Maggio et al. 2004).

The enzymatic reaction of sirtuins results in the final deacetylated product, such as nicotinamide (Nam) and 2′-O-acetyl-ADP-ribose. Nam exerts feedback regulation of sirtuin activity and is a physiological inhibitor of certain deacetylase SIR2 enzymes (yeast SIR2 and human SIRT1) (Bitterman et al. 2002). Nam is in anticancer and anti-diabetic trials (Kaanders et al. 2002). Various small molecules from different classes are undergoing clinical trials as potent inhibitors of HDAC and as effective in vivo antitumor agents (Figure 14.3) (Saito et al. 1999; Suzuki et al. 1999; Plumb et al. 2003; Hansen et al. 2005; Ryan et al. 2005; Buggy et al. 2006; Maiso et al. 2006; Vaisburg 2006; Zhou et al. 2006). Two such HDIs, suberoylanilide hydroxamic acid Zolinza™ (vorinostat; SAHA, Merck) and the cyclic depsipeptide romidepsin (FK228), have been approved for the treatment of advanced cutaneous T-cell lymphoma (CTCL) (Kim et al. 1999; Meinke and Liberator 2001; Kelly et al. 2005; Bates et al. 2010).

Vaisburg and his research group reported some novel HDIs (Figure 14.4) such as arylsulfonamide-based hydroxamates (**14**) (Lavoie et al. 2001), arylsulfonamide-based amino-anilides (**15**) (Bouchain et al. 2003), long chain ω-substituted hydroxamic acids (**16**) (Woo et al. 2002), 2-aminophenylamides of ω-substituted alkanoic acids (**17**) (Vaisburg et al. 2004), and *N*-(2-aminophenyl)-4-(heteroarylmethyl)-benzamides (**18**) (Vaisburg et al. 2007). Many of these HDIs have shown a comparatively short half-life and poor bioavailability, except *N*-(2-aminophenyl)-4-(heteroarylmethyl)-benzamides (**18**),

Mocetinostat (MGCD0103)
**(1)**

CRA-024781
**(2)**

Belinostat (PXD-101)
**(3)**

Panobinostat (LBH-589)
**(4)**

Entinostat (SNDX-275; MS-275)
**(5)**

Vorinostat (SAHA)
**(6)**

Givinostat (ITF2357)
**(7)**

PCI-24781
**(8)**

Dacinostat (LAQ824)
**(9)**

Romidepsin (FK228)
**(10)**

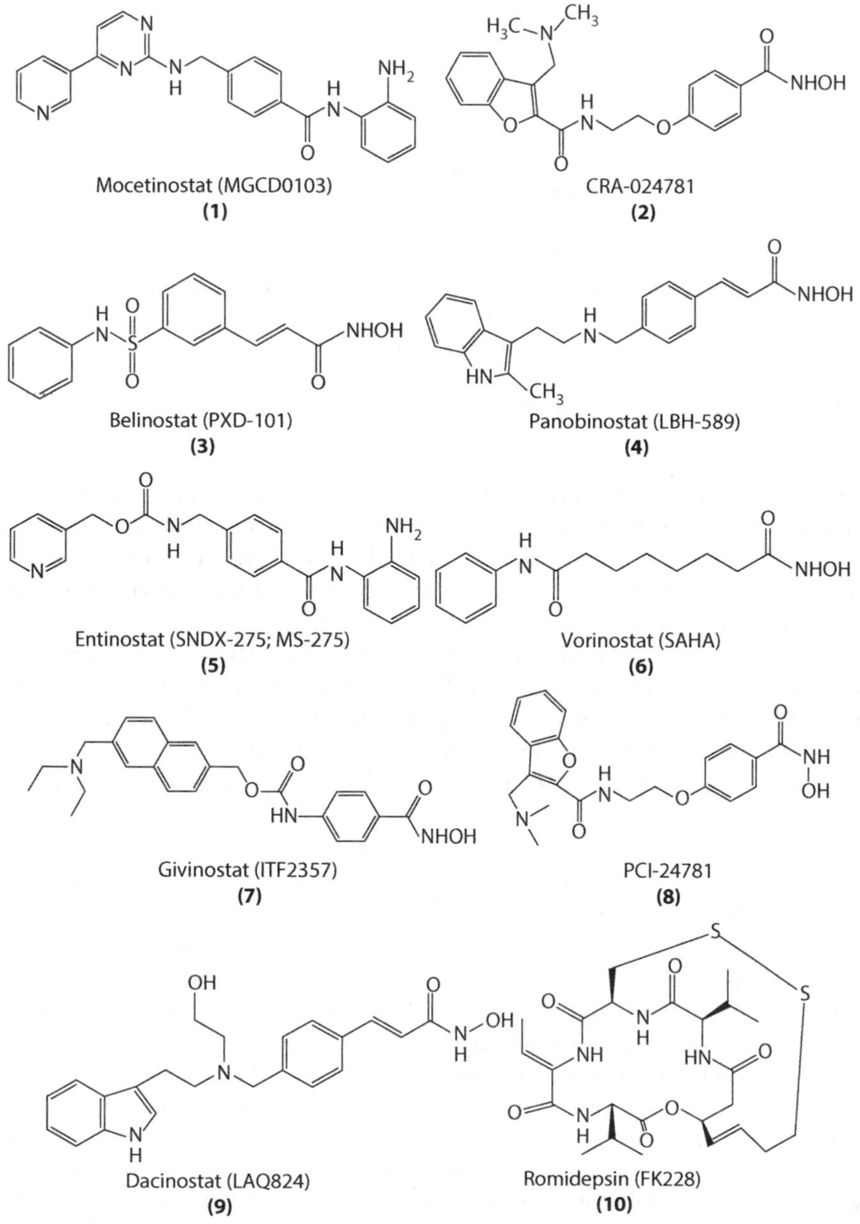

**FIGURE 14.3** Structures of some antitumor HDIs in clinical trials.

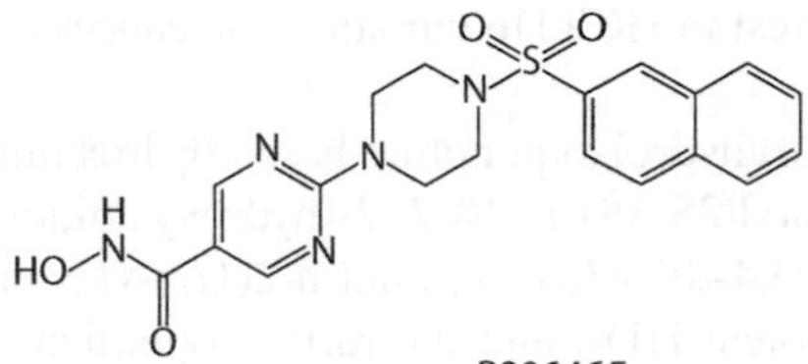

R306465
(11)

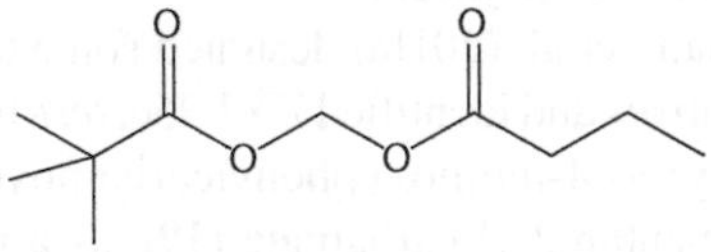

Pivanex
(12)

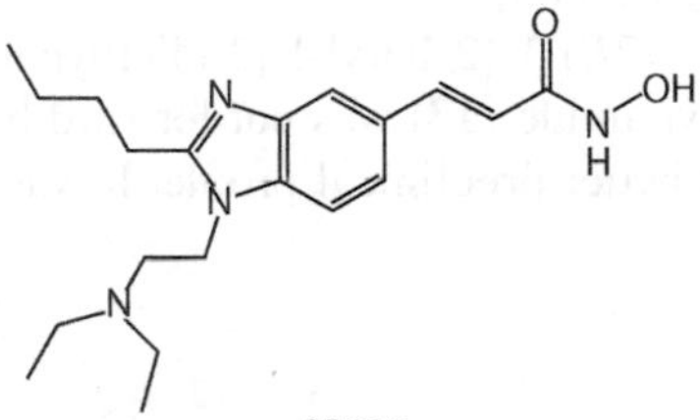

SB939
(13)

**FIGURE 14.3 (*Continued*)** Structures of some antitumor HDIs in clinical trials.

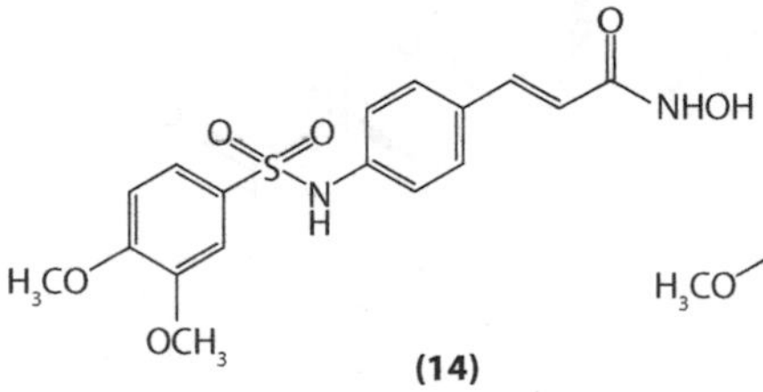

(14)

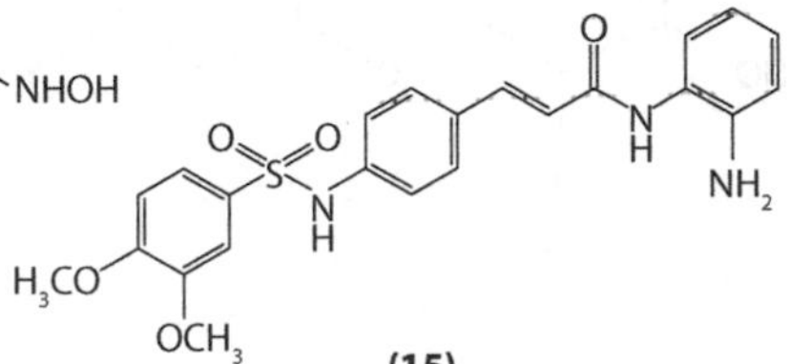

(15)

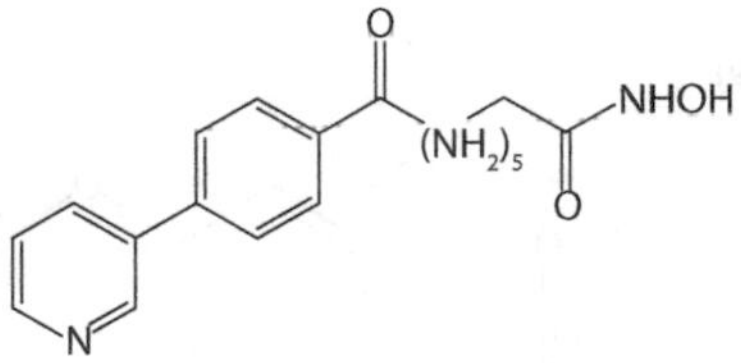

(16)

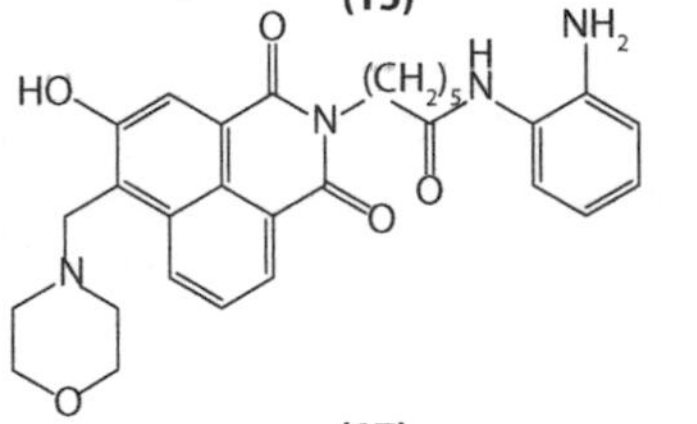

(17)

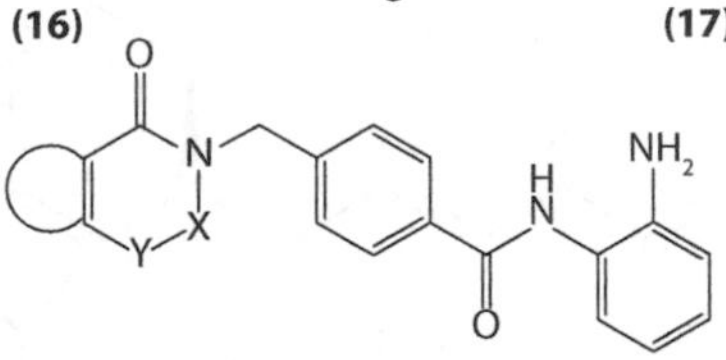

(18)

**FIGURE 14.4** Structures of some reported HDIs.

which have significant in vivo antitumor activity. The reported series of compounds having one/two carbonyl groups on the left-hand side heterocyclic moiety were found to inhibit recombinant HDAC-1 with an $IC_{50}$ of 0.1–1.0 μM with good in vitro antiproliferative potency in human colon cancer cell line. These compounds also induced the expression of the CDK inhibitor p21 $^{WAF1/Cip1}$. Some of the compounds caused apoptosis and G2/M cell cycle arrest in HCT116 human colon cancer cell line (Vaisburg et al. 2007).

Zhang et al. (2011a) designed some tetrahydroisoquinoline-based hydroxamic acid derivatives and identified ZYJ-25e, *tert*-butyl(2S,3S)-1-((S)-7-(2-(hydroxyamino)-2-oxo ethoxy)-3-(4-methoxyphenylcarbamoyl)3,4-dihydroisoquinolin-2(1*H*)-yl-3-methyl-1-oxopentan-2-yl carbamate (**19**), as a potent HDI, and the further optimization led to more a potent, orally active HDI, ZYJ-34c (**20**), which exhibited in vivo antitumor activity (Zhang et al. 2011a, 2011b) (Figure 14.5).

A compound SB-939 such as (2*E*)-3-{2-butyl-1-[2-(diethylamino) ethyl]-1*H*-benzimidazol-5-yl}-*N*-hydroxyacryl amide (**13**) was put forward by Wang et al. (2011) as an orally active HDI with a better preclinical profile. It was identified by

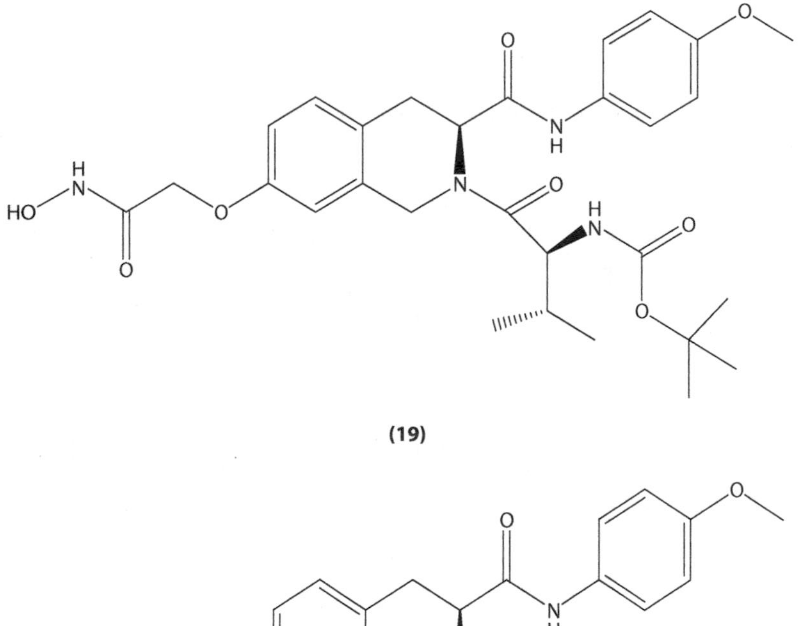

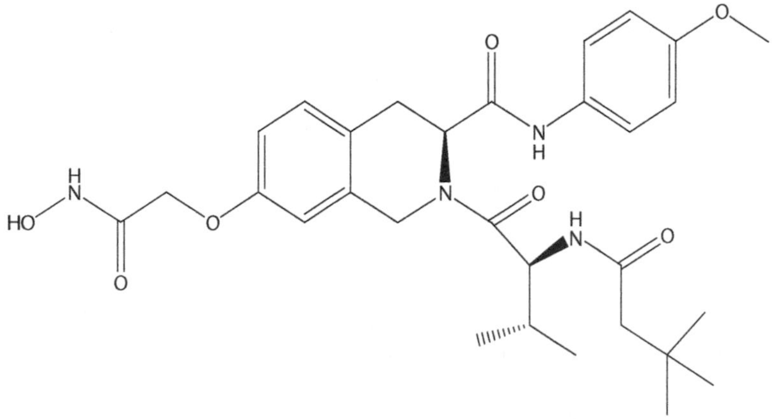

**FIGURE 14.5** Structures of some tetrahydroisoquinoline-based hydroxamic acid derivatives.

extensive SAR evaluation of a series of 3-(1,2-disubstituted-1*H*-benzimidazol-5-yl)-*N*-hydroxyacrylamides. SB-939 showed comprehensive in vitro and in vivo profiles and excellent drug-like properties in tumor models (HCT-116, PC-3, A2780, MV4-11, and Ramos). Currently, it is being tested in phase I and II clinical trials.

A series of novel spiropiperidine hydroxamic acid-based derivatives as HDIs were discovered and evaluated by Varasi et al. (2011). The study concluded identification of spirocycle **21** with good oral bioavailability and tumor growth inhibition in an HCT-116 murine xenograft model.

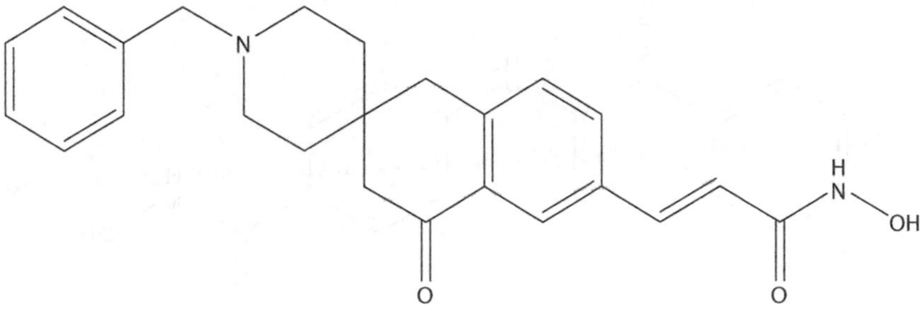

N H N OH O O

**(21)**

Thaler et al. (2010) proposed 4-phenylpiperazinyl analogs with substituted phenylacrylate, 5- or 6-substituted pyridine-2-yl acrylates as HDIs. Compounds **22** and **23** have shown notable antiproliferative activity in leukemic K562 and human colon cancer HCT116 cells ($IC_{50}$ ~1 μM).

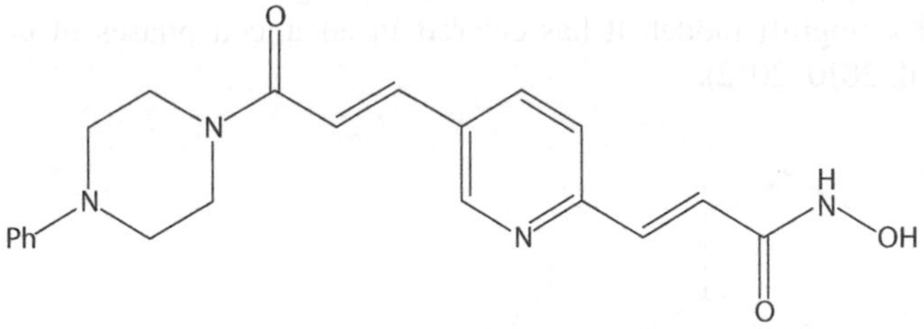

O N N Ph N H N OH O

**(22)**

O N N Ph N H N OH O

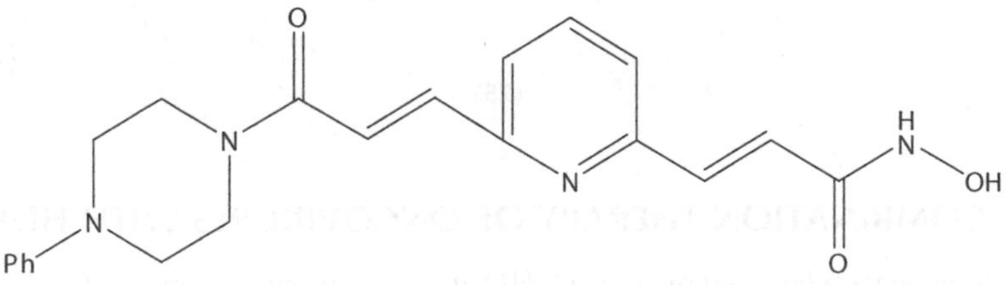

**(23)**

Further investigations identified three representative compounds with high clearance rates in rats and human hepatocytes AR-42. The phenyl hydroxamic acid derivative (**24**) was shown to have antiproliferative activities in submicromolar range against three myeloma cell lines, IM-9 (Lu et al. 2005), RPMI-8226 (Bai et al. 2011), and U226 (Cheng et al. 2006a, 2006b). Due to good in vitro potency and pharmacokinetic (PK) profile of these compounds, their phase I/IIa trials for recurrent hematological malignancies and solid tumors are ongoing (Hofmeister 2012).

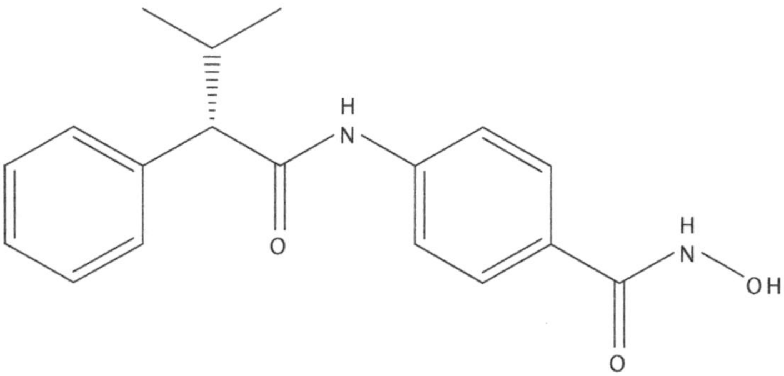

*(S)*AR-42
**(24)**

Moffat et al. (2010) carried out a project on pyrimidine hydroxamates derivatives bound to bicyclic hexahydropyrrolo[3,4-*c*]pyrrole and azabicyclo[3,1,0]hexane linkers and observed their antiproliferative response at micro- or submicromolar range. The study put forward CHR-3996 (**25**) with good in vivo antitumor activity in HCT-116 xenograft model. It has entered in advanced phases of clinical trials (Banerji et al. 2010, 2012).

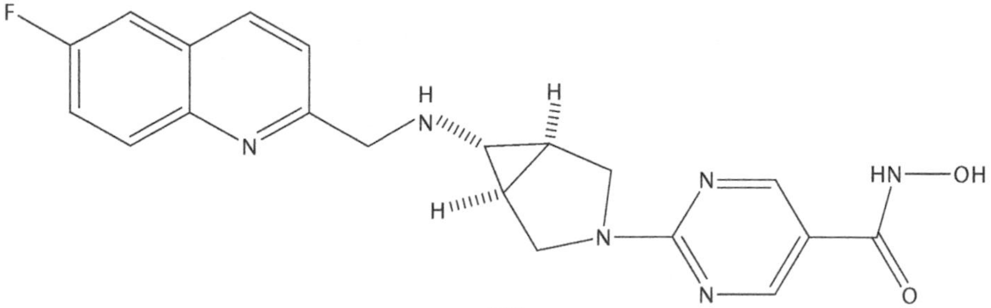

**(25)**

## 14.6 COMBINATION THERAPY OF ONCOVIRUSES WITH HDIs

The histone deacetylase and oncovirus (HDI/OV) combination approach is based on the fact that the addition of a compound that can reversibly compromise host antiviral genetic programs could enhance OV growth in tumor cells. HDIs are known to prevent the transcriptional activation of antiviral genes after IFN stimulation or virus infection and thus have potent antitumor activity (Genin et al. 2003; Chang et al. 2004; Joseph et al. 2004; Kelly and Marks 2005; Minucci and Pelicci 2006; Nusinzon and Horvath 2006; Mehnert and Kelly 2007; Vlasakova et al. 2007). Because HDIs are tumor-specific viral sensitizers, they have considerable potential to enhance the spectrum of malignancies required for OV therapy.

Various studies have been performed to evaluate whether HDIs can be combined with oncoviruses to enhance potency. Some of the examples are included in Table 14.1 and also are discussed in detail. The combination therapy approach has helped to establish rational, multipronged therapies against cancer. The HDIs have been well established as enhancers of tumor-specific replication and oncolysis for mainly adenovirus, HSV, VSV, VV, and Semliki Forest virus (SFV) (Sachs et al. 2004; Bieler et al. 2006; Nguyen et al. 2008; Katsura et al. 2009). Various categories of HDIs suggest a broad range of effects of these compounds on viral oncolysis. Thus OV/HDI combination is a rational approach and is expected to offer exceptional treatment benefits (Nguyen et al. 2008; Otsuki et al. 2008). Inhibition of therapeutic spread of OVs is correlated with intratumoral innate immunity. All OVs have to exert selective targeting and must overcome the cells' innate resistance to viral infection.

OVs have been used with CPA as a modulator of immune response to improve therapeutic efficacy of the virus within tumors and to decrease antiviral cytokines that facilitate OV intraneoplastic propagation (Wakimoto et al. 2004). When immune modulators were evaluated, there was increased susceptibility to infection or undue propagation of replication-competent OVs. Some of the interesting advantages that emphasize the implementation of combination HDI/OV therapy are as follows:

- Specificity of action to transformed cells and not rendering normal tissues sensitive to OV replication.
- Reversibility of HDI treatment (i.e., when therapy is discontinued, OV replication stops and resumes reinitiation of the treatment).
- HDIs can be used as chemical switches to regulate growth of therapeutic viruses.
- HDIs have their effect on angiogenesis (i.e., they downregulate the expression of angiogenic proteins).

One possible mechanism by which HDIs act as oncolytics is in downregulation of the cellular antiviral response. Some studies highlighted the involvement of deacetylation in regulating the early interferon response. When the effects of TSA on NF-κB and IRF activity were evaluated, it prevented induction of interferon regulatory factor (IRF) genes without affecting translocation of IRF3 to the nucleus and DNA-binding activity. TSA concludes requirement of HDIs' activity to initiate transcription from IFN-β promoter (Nusinzon and Horvath 2006). The antiviral response may be regulated by STAT signaling through acetylation and deacetylation (Tang et al. 2007). In the case of virus infection as well as treatment with IFNαt, there is a decrease in IFNα, ISG15 (ubiquitin-like modifier), and ISG54 (interferon-stimulated gene 54) activity, and impaired interferon-stimulated gene factor 3 (ISGF3) complex

## TABLE 14.1
## HDIs and OV Combination Therapy and their Mechanism

| OV | HDI | Mechanism | Cancer Model |
|---|---|---|---|
| Adenoviral gene transfer | Trichostatin A (TSA) and sodium phenylbutyrate | Coxackie-adenovirus receptor (CAR) upregulation | Bladder cancer |
| Adenoviral gene transfer | Depsipeptide (FR901228, FK228) | CAR upregulation | Melanoma |
| Ad5-TRAIL | TSA, sodium butyrate, MS-275, oxamflatin, and depsipeptide | CAR upregulation, TRAIL-R1 and -R2 upregulation, and decreased PCKC2 activity | Renal cell carcinoma, prostate cancer |
| Adenovirus OBP-301 (Telomelysin) | Depsipeptide (FR901228, FK228) | CAR upregulation | Non-small-cell lung cancer |
| Adenovirus d1520 (E1A 13S deleted) | TSA and irinotecan | CAR upregulation (and enhanced nuclear localization of YB-1) | Glioblastomas |
| Herpes simplex virus (HSV) (rQNestin34.5) | Valproic acid | Inhibition of IFN-responsive antiviral genes, enhancement of viral genes, enhancement of viral transcription, and improved viral propagation | Gliomas |
| HSV (G47Δ) | TSA | Reduced level of cyclin D1, enhanced antiangiogenic effects through vascular endothelial growth factor inhibition | Colon, lung, and brain cancers |
| HSV | TSA | Enhanced NF-κB activation | Oral squamous cell carcinoma |
| Vesicular stomatitis virus (Δ51) | MS-275 and suberoylanilide hydroxamic acid (SAHA, Vorinostat, Zolinza) | Dampening of IFN antiviral response, synergistic activation of apoptosis | Colon, breast, prostate, and ovarian cancers |
| Vaccinia virus (double deleted) | MS-275 | Enhanced viral replication | Breast and lung cancers and melanoma |
| Semliki Forest virus | MS-275 and suberoylanilide hydroxamic acid (SAHA, Vorinostat, Zolinza) | Enhanced viral replication | Breast cancer |

*Source:* Nguyen et al., *Cytokine Growth Factor Rev* 21, 153–159, 2010.

**TABLE 14.2**
**Some Clinical Successes**

| OV | Brand Name | Anticancer Drug | Type of Cancer | Phase |
|---|---|---|---|---|
| Reovirus | REOLYSIN | Paclitaxel and Carboplatin | Head and neck carcinoma; | II |
| | | | non-small-cell lung cancer; | II |
| | | | metastatic melanoma; | II |
| | | | squamous cell carcinoma of lungs | II |
| | | Gemcitabine | Advanced pancreatic adenocarcinoma | II |

formation and interferon regulatory factor 7 (IRF7) activation were sustained (Genin et al. 2003; Cheng et al. 2004).

Also the mechanism of "vascular shutdown" that controls the tumor progression can be regulated by synergistic action of HDIs and OV (Liu et al. 2008; Nguyen et al. 2008). Some of the promising results include examples of HSV and VSV showing decreased tumor growth, increased survival, and no signs of toxicity. To establish effective combination of OV/HDI, various studies have been carried out (Table 14.1). A brief discussion of them is given here. Some of them have entered in clinical phase of development (Table 14.2). The combination of HDIs and OV also helps to distinguish between antitumoral activities and enhanced efficacy of HDIs in combination treatment compared to a single treatment regimen.

Some studies were carried out by Alvarez-Breckenridge et al. (2012) to support the preclinical success of HDI/OV combination therapy and the synergistic mechanism. VPA and other HDIs have been very well reported to enhance OV replication through downregulating type I IFN responses (Liu et al. 2008; Nguyen et al. 2008, 2010; Otsuki et al. 2008) and to lead to additional downregulation of antiviral innate cell responses. VPA has been combined with oHSV by a glioblastoma xenograft model and has been found to decrease immune cell activation at the early stages of infection and to suppress STAT5 activity and Th1-specific T box transcription factor (T-BET) expression (Alvarez-Breckenridge et al. 2012). The results highlighted the multifaceted mechanism of VPA and oHSV synergism, mediated through NK cells linked to a detrimental antiviral response to oHSV (Fulci et al. 2006). The mechanism of enhanced viral oncolysis by VPA is achieved by suppressing intracellular and innate cellular antiviral responses.

### 14.6.1 HDIs and Vaccinia Virus

Vaccinia virus (VV) is an ideal and safe OV with various properties such as large cloning capacity, systemic activity, immune-modulating genes, and no genotoxic activity (Kim and Thorne 2009). VV achieves tumor selectivity by attenuating deletion of viral genes or gene fragments. VV entered the preclinical phase of development in 2004 (Zeh and Bertlett 2002). Some of the major advantages of using VV include attenuation based on general tumor biology, targeting signaling pathways, and no requirement of foreign DNA elements or genes (Gromeier et al. 2000; Ries and Korn 2002; Russell 2002; Varghese and Rabkin 2002; Zeh and Bartlett 2002;

Dobbelstein 2004). JX-594 (thymidine kinase deleted and GM-CSF) and JX-929 (double deleted and cytosine deaminase) from Jennerex Biopharmaceutics have successfully entered phase I/II and I of clinical trials, respectively (Nguyen et al. 2010). In cases of advanced liver tumors, JX-594 has shown acceptable safety and promising anticancer activity (Kirn and Thorne 2009).

Some of the genes encoded by vaccinia are reported to be reductant for growth of tumor cells (McCart et al. 2001; Shen and Nemunaitis 2005; Kirn and Thorne 2009). VV mutants with B18R, TK, and/or the VGF gene manipulation have entered preclinical and clinical studies (Colamonici et al. 1995; Symons et al. 1995; Alcami et al. 2000; Kirn et al. 2007; Thorne et al. 2007; Liu et al. 2008; Park et al. 2008; Kirn and Thorne 2009). The B18R gene in JX-594 has shown reduced antagonization of interferon activity and makes it acceptable in humans due to a better safety profile (Alcami et al. 2000; Park et al. 2008). Studies have highlighted the defects in HDIs as well as their effects on the growth of OVs in tumors. It is observed that HDIs can enhance the growth of vasicular stomatitis virus (VSVΔ51) in tumor cells. Here, HDIs have shown increased acetylation of histones and other proteins (Yoshida et al. 1990; Marks et al. 2001; Bolden et al. 2006; Minucci and Pelicci 2006). They can inhibit the ability of tumor cells to block antiviral responses (Chang et al. 2004; Nguyen et al. 2008; Otsuki et al. 2008).

MacTavish et al. (2010) examined various HDIs to augment oncolytic activity of the VV. In the first step, a panel of HDIs comprised of SAHA, MS-275, Oxamflatin, Apicidin, SBHA, Scriptaid, CHAHA, VPA, M344, and TSA challenged with a green fluorescent protein (GFP)-expressing VV (VVdd) were evaluated (McCart et al. 2001). Mac Tavish et al. concluded that TSA was the most potent enhancer of VVdd replication. Therefore, they evaluated it further using various in vitro and in vivo models (MacTavish et al. 2010) such as (a) spread of VV strains specifically in cancer cells; (b) VVdd efficacy in a syngeneic lung metastasis; and (c) attenuated B18R-deleted vaccinia strain. They found that TSA could enhance activity in all three models but was less effective against in vitro VV spread.

TSA has shown a synergistic effect with VV and thus better cell killing. Therefore, its dose could be reduced but still retain efficient antitumor activity. The use of a TSA/VV combination having increased efficacy has been therefore been suggested. Increased activity was also assumed with TSA and B18R-deleted VV strains. The vaccinia B18R gene is found to be enhanced by TSA. It was concluded that the anti-interferon activities of B18R fail to impair the cellular interferon response and/or that the effects of TSA are beyond interferon-induced antiviral responses.

Some studies focused on the development of small molecule, OV enhancers with key properties such as a restricted effect on tumor cells and a minimum impact on the antiviral programs of normal tissues as well as higher levels of HDAs in tumor cells emphasis dependency of such epigenetic modifications to control gene expression (Mehnert and Kelly 2007). In vitro and in vivo studies, TSA is a potent enhancer of VV, and further clinical evaluations are under way (MacTavish et al. 2010).

### 14.6.2 HDIs and VSV

The spread of OVs can be effectively blocked by modulating the intratumoral innate immunity; the HDIs can blunt the cellular antiviral response through epigenetic modifications of chromatin. Nguyen et al. (2008) have attempted to show that HDIs can enhance the spread of VSV in a variety of cancer cells in vitro.

VSV is sensitive to the IFN response; tumors that lack components of this pathway provide an environment that is conducive to its replication. This is the basis of VSV tumor selectivity and oncolytic activity. Many tumor cells retain components of the IFN response and limit VSV tumor spread. This diminishes their activity in vivo and restricts VSV clinical utility. To facilitate VSV replication and oncolysis, HDIs that can decrease antiviral gene expression can be combined. Some studies were carried out with a few HDIs (such as MS275 and SAHA) to find enhanced VSV replication and cell death in resistant cell lines treated with these HDIs. Pretreatment with HDIs induced VSV replication within 24 hours post infection and inhibited transcriptional activation of IFN-β, MXA, and IRF7 without affecting the IFN signaling cascade of IRF3. These results concluded that HDIs act to inhibit IFN and IFN-inducible gene expression. This combination treatment synergistically kills cancer cells by activation of the mitochondrial apoptosis pathway. In vivo, this combination shows no adverse toxicity or enhancement of off-target viral replication, and HDIs do not render primary normal cells sensitive to VSV. HDIs act as a reversible chemical switch that can regulate the extent of virus replication within tumors. Sensitivity of VSV to IFN response and effects of HDIs on antiviral immune response are the major contributing factors. HDI sensitization also enhances the oncolytic activity of SFV and VV (Nguyen et al. 2008). Thus, HDI treatment proves a promising way to overcome resistance to OV therapy.

It has been concluded that there can be enhanced tumor killing by the combination of VSV and HDIs, but this combination may not bring any significant improvement in VSV infection of normal human/mouse tissues. The results were consistent with the fact that HDIs blunt the cellular IFN response in tumor cells. In the case of PC3 cells treated with combination therapy, some important antiviral genes were expressed at low levels and were poorly inducible by virus infection (Genin et al. 2003; Chang et al. 2004; Joseph et al. 2004; Kelly and Marks 2005; Minucci and Pelicci 2006; Nusinzon and Horvath 2006; Mehnert and Kelly 2007; Vlasakova et al. 2007; Nguyen et al. 2008). HDIs effectively alter IFN activity in tumor cells along with their additional stress responses, which impacts virus replication. Because HDIs target different classes of HDACs, they may have variable effects depending on the tumor or the OV (Hess-Stumpp et al. 2007; Marks 2007; Marks and Breslow 2007). Continuous systemic administration of HDI is required to maintain robust virus replication within the tumor, and it may be possible to regulate the magnitude of OV therapy by applying or withdrawing HDI. This approach is applicable when OV therapy harbors gene facilitating imaging of the infection. In the presence of HDIs, VSV, SFV, and VV have shown increased oncolytic activity. The effect of HDIs on the rate of oncolysis can be correlated with a wide spectrum of OVs that are under development (Nguyen et al. 2008).

### 14.6.3 HDIs and HSV

Various studies have found pretreatment of oncolytic HSV therapy with HDIs as a beneficial option. One of the examples includes pretreatment with VPA showing enhanced viral replication and spread only in tumor cells where the innate antiviral response was found to be correlated with several IFN-responsive genes. Pretreatment with VPA decreased expression levels of STAT1, PKR, and promyelocytic leukemia protein (PML). Thus, STAT1-deficient cells resulted in significant enhancement in HSV replication, and treatment with VPA does not further increase HSV replication. Therefore, decreased STAT1 activity is partially responsible for VPA-treated HSV replication. Also, VPA counteracts inhibition of OV replication by IFN-β treatment and thus highlights the inhibitory mechanism of the IFN antiviral response. The timing of HDI treatment is also an important factor in HDI/OV combination therapy. Concurrent treatment with VPA in HSV infection failed to show enhancing effect compared to pretreatment. Further, HSV/TSA combination resulted in synergistic enhancement of anti-angiogenesis and inhibition of the angiogenic factor, vascular endothelial growth factor (VEGF), to a better extent (Cinatl et al. 2004). In the case of HSV or TSA-treated cells, decreased VGEF was observed.

Katsura et al. (2009) observed that pretreatment of oral squamous cells carcinoma (SCC) cells with TSA shows enhanced replication of oncolytic HSV and modified gene expression. The mechanism of activity showed induced p65 translocation by HSV infection and enhanced acetylation level of p65 by TSA. Also, the effect of TSA on HSV replication was found to be diminished due to inhibition of NF-κB nuclear translocation with p50 subunit (SN50); this indicated the role of NF-κB in increased replication. The oncolytic activity of HSV NF-κB is supposed to exert an effect on late viral gene expression and the suppression of apoptosis required for HSV replication (Katsura et al. 2009).

The results of in vivo HSV/VPA and HSV/TSA combination were observed to be parallel to each other (Katsura et al. 2009), showing enhanced angiogenic activity and reduced vascularity (Liu et al. 2008). No significant toxicity was found with VPA and TSA when used alone at a concentration effective for enhancement of HSV replication.

### 14.6.4 HDIs and Adenoviruses

Kitazono et al. (2001, 2002) was the first to report increased efficiency of adenoviral infection by high levels of Coxackie-adenovirus receptor (CAR), and αv integrin mRNA was observed in different cancer cell lines by some HDIs (namely FR901228, TSA and VPA). The results were confirmed using other cancer types (Sachs et al. 2004). Now the potential benefits of HDIs/adenovirus combination therapy are being explored (Fan et al. 2005).

Studies were carried out to further explore the effects of HDI on CAR expression in normal tissues. It was beneficial to evaluate the safety before translating the approach into clinical applications (Goldsmith et al. 2003). One study suggested increased efficiency of adenovirus therapies in in vivo xenograft model systems (Goldsmith et al. 2007).

Some exceptional results were observed when HDIs were used in combination with engineered oncolytic adenoviruses (Ad5-TRAIL adenoviral vector), such as enhanced transgene expression, increased Ad5-TRAIL infection, induced CAR expression, and enhanced cancer cell sensitivity to TRAIL-induced apoptosis

(Vanoosten et al 2005, 2006). Some mechanistic studies focused on HDAC inhibition, which caused decreased PCKC2 activity, activation of caspase-2, and partial cleavage of caspase-8 to sensitize the tumor cells to TRAIL (Vanoosten et al. 2007).

The HDIs, such as FR901228, and attenuated replication-competent adenoviruses (OBP-301; Telomelysin) were proved synergistic in non-small-cell lung cancer cell lines (A-549, H460) (Watanabe et al. 2006). FR901228 did not affect CAR expression and infection efficiency, indicating its specificity to transformed cells. The studies also rationalized the multitherapy or tritherapy concept to enhance the oncolytic effect of adenoviruses. The adenovirus dl520 lacks the E1A135 protein and selectively replicates in cells with nuclear localization of the transcription factor γB-1. Irinotecan and TSA were used in this combination therapy and found highly targeted and suggests use of combination therapy to enhance the oncolytic efficacy of dl520 (Bieler et al. 2006). In this combination, irinotecan increases the replication efficacy of dl520, and TSA augments the low infection rate by upregulation of CAR expression. Together, these reports suggested that HDIs enhance the activity of adenoviruses by multiple mechanisms and confirmed the importance of the rational design of HDI/adenovirus combination therapy.

## 14.7 CONCLUSION

Recent advances in virology and cancer biology have helped to provide a better understanding of the molecular basis of viral oncolysis. All OVs (natural/engineered) take advantage of certain unique features of target cancer cells. Various studies have been carried out that focus on combining viruses with classical treatments (chemotherapy, radiation, kinase inhibitors, immunosuppressants, etc.) (Bauzon and Hermiston 2008). In tumor cells, replication of OV appears to be inhibited by antiviral defense, but in comparison to this, combination therapy aims to provide a possible solution. One of the ways is exogenous administration of HDIs as they downregulate antiviral defense in cancer cells and enhance oncolytic efficacy using IFN-sensitive OVs (Nguyen et al. 2008). A major advantage of combining HDIs and OVs is that HDIs do not automatically lead to increased replication in normal cells, and this suggests crucial cellular differences. These results rationalize further exploration of HDIs and OV combination therapies. Thus, it seems clear that tumor regression is enhanced when OVs are used in combination with HDIs. Based on the results obtained, it can be concluded that OV/HDIs will be an important part of cancer treatment in the near future.

## REFERENCES

Acharya MR, Sparreboom A, Venitz J, et al. (2005). Rational development of histone deacetylase inhibitors as anticancer agents: A review. *Mol Pharmacol* 68: 917–932.

Alcami A, Symons JA, Smith GL. (2000). The vaccinia virus soluble alpha/beta interferon (IFN) receptor binds to the cell surface and protects cells from the antiviral effects of IFN. *J Virol* 74: 11230–11239.

Alvarez-Breckenridge CA, Yu J, Price R, et al. (2012). The HDAC inhibitor valproic acid lessens NK cell action against oncolytic virus-infected glioblastoma cells with inhibition of STAT5/T-BET signaling and IFNγ generation. *J Virol* 86: 4566–4577.

Bai LY, Omar HA, Chiu CF, et al. (2011). Antitumor effects of (S)-HDAC42, a phenylbutyrate-derived histone deacetylase inhibitor, in multiple myeloma cells. *Cancer Chemother Pharmacol* 68: 489–496.

Balasubramanian S, Verner E, Buggy JJ. (2009). Isoform-specific histone deacetylase inhibitors: The next step? *Cancer Lett* 280: 211–221.

Banerji U, van Doorn L, Papadatos-Pastos D, et al. (2010). A phase I pharmacokinetic (PK) and pharmacodynamic (PD) study of CHR-3996, a class 1 selective histone deacetylase inhibitor (HDACi), in patients with advanced solid tumors. *J Clin Oncol* 28(Suppl):2552.

Banerji U, van Doorn L, Papadatos-Pastos D, et al. (2012). A phase I pharmacokinetic and pharmacodynamic study of CHR-3996, an oral class I selective histone deacetylase inhibitor in refractory solid tumors. *Clin Cancer Res* 18: 2687–2694.

Barker DD, Berk AJ. (1987). Adenovirus proteins from both E1B reading frames are required for transformation of rodent cells by viral infection and DNA transfection. *Virology* 156: 107–121.

Bates SE, Zhan Z, Steadman K, et al. (2010). Laboratory correlates for a phase II trial of romidepsin in cutaneous and peripheral T-cell lymphoma. *Br J Haematol* 148: 256–267.

Bauzon M, Hermiston TW. (2008). Exploiting diversity: Genetic approaches to creating highly potent and efficacious oncolytic viruses. *Curr Opin Mol Ther* 10: 350–355.

Bieler A, Mantwill K, Dravits T, et al. (2006). Novel three-pronged strategy to enhance cancer cell killing in glioblastoma cell lines: Histone deacetylase inhibitor, chemotherapy, and oncolytic adenovirus dl520. *Hum Gene Ther* 17: 55–70.

Bischoff JR, Kirn DH, Williams A, et al. (1996). An adenovirus mutant that replicates selectively in p53-deficient human tumor cells. *Science* 274: 373–376.

Bitterman KJ, Anderson RM, Cohen HY, et al. (2002). Inhibition of silencing and accelerated aging by nicotinamide, a putative negative regulator of yeast Sir2 and human SIRT1. *J Bio Chem* 277: 45099–45107.

Bolden JE, Peart MJ, Johnstone RW. (2006). Anticancer activities of histone deacetylase inhibitors. *Nat Rev Drug Discov* 5: 769–784.

Bouchain G, Leit S, Frechette S, et al. (2003). Development of potential antitumor agents with cell-based phenotypic screens and array-based gene expression analysis. *J Med Chem* 46: 820–830.

Boviatsis EJ, Scharf JM, Chase M, et al. (1994). Antitumor activity and reporter gene transfer into rat brain neoplasms inoculated with herpes simplex virus vectors defective in thymidine kinase or ribonucleotide reductase. *Gene Ther* 1: 323–331.

Bradley JD, Kataoka Y, Advani S, et al. (1999). Ionizing radiation improves survival in mice bearing intracranial high-grade gliomas injected with genetically modified herpes simplex virus. *Clin Cancer Res* 5: 1517–1522.

Buggy JJ, Cao ZA, Bass KE, et al. (2006). CRA-024781: A novel synthetic inhibitor of histone deacetylase enzymes with antitumor activity in vitro and in vivo. *Mol Cancer Ther* 5: 1309–1317.

Chalavi A, Todo T, Martuza RL, et al. (1999). Replication-competent herpes simplex virus vector G207/cisplatin combination therapy for head and neck squamous cell carcinoma. *Neoplasia* 1: 162–169.

Chang HM, Paulson M, Holko M, et al. (2004). Induction of interferon-stimulated gene expression and antiviral responses require protein deacetylase activity. *Proc Natl Acad Sci U S A* 101: 9578–9583.

Cheng H, Jones W, Wei X, et al. (2006a). Preclinical pharmacokinetics studies of R- and S-enantiomers of the histone deacetylase inhibitor, HDAC-42 (NSC 731438), in the rat. *Proc Amer Assoc Cancer Res* 47 (abstr 686).

Cheng H, Liu Z, Kulp SK, et al. (2006b). Preclinical pharmacokinetic studies with s-HDAC-42 (NSC 736012), an inhibitor of histone deacetylase, by LC-MS/MS. *Proc Amer Assoc Cancer Res* 47 (abstr 3091).

Chen Y, Yu DC, Charlton D, et al. (2000). Pre-existent adenovirus antibody inhibits systemic toxicity and antitumor activity of CN706 in the nude mouse LNCaP xenograft model: Implications and proposals for human therapy. *Hum Gene Ther* 11: 1553–1567.

Cinatl J, Jr, Michaelis M, Driever PH, et al. (2004). Multimutated herpes simplex virus g207 is a potent inhibitor of angiogenesis. *Neoplasia* 6: 725–735.

Coffey MC, Strong JE, Forsyth PA, Lee PW. (1998). Reovirus therapy of tumors with activated Ras pathway. *Science* 282: 1332–1334.

Cogswell PC, Guttridge DC, Funkhouser WK, et al. (2000). Selective activation of NF-κB subunits in human breast cancer: Potential roles for NF-κB2/p52 and for Bcl-3. *Oncogene* 19: 1123–1131.

Colamonici OR, Domanski P, Sweitzer SM, et al. (1995). Vaccinia virus B18R gene encodes a type I interferon-binding protein that blocks interferon alpha transmembrane signaling. *J Biol Chem* 270: 15974–15978.

Coukos G, Makrigiannakis A, Kang EH, et al. (1999). Multi-attenuated herpes simplex virus-1 mutant G207 exerts cytotoxicity against epithelial ovarian cancer but not normal mesothelium, and is suitable for intraperitoneal oncolytic therapy. *Cancer Gene Ther* 5: 1523–1537.

Coukos G, Makrigiannakis A, Kang EH, et al. (2000). Oncolytic herpes simplex virus-1 lacking ICP34.5 induces p53-independent death and is efficacious against chemotherapy-resistant ovarian cancer. *Clin Cancer Res* 6: 3342–3353.

Dobbelstein, M. (2004). Replicating adenoviruses in cancer therapy. *Curr Top Microbiol Immunol* 273: 291–334.

Dudley DT, Pang L, Decker SJ, et al. (1995). A synthetic inhibitor of the mitogen-activated protein kinase cascade. *Proc Natl Acad Sci U S A* 92: 7686–7689.

Duncan MR, Stanish SM, Cox DC. (1978). Differential sensitivity of normal and transformed human cells to reovirus infection. *J Virol* 28: 444–449.

Fan S, Maguire CA, Ramirez SH, et al. (2005). Valproic acid enhances gene expression from viral gene transfer vectors. *J Virol Methods* 125: 23–33.

Fueyo J, Gomez-Manzano C, Alemany R, et al. (2000). A mutant oncolytic adenovirus targeting the Rb pathway produces anti-glioma effect in vivo. *Oncogene* 19: 2–12.

Fulci G, Breymann L, Gianni D, et al. (2006). Cyclophosphamide enhances glioma virotherapy by inhibiting innate immune responses. *Proc Natl Acad Sci U S A* 103: 12873–12878.

Ganly I, Eckhardt SG, Rodriguez GI, et al. (2000). A phase I study of Onyx-015, an E1B attenuated adenovirus, administered intratumorally to patients with recurrent head and neck cancer. *Clin Cancer Res* 6: 798–806, erratum May 6: 2120.

Genin P, Morin P, Civas A. (2003). Impairment of interferon-induced IRF-7 gene expression due to inhibition of ISGF3 formation by trichostatin A. *J Virol* 77: 7113–7119.

Gibbs JB, Graham SL, Hartman GD, et al. (1997). Farnesyltransferase inhibitors versus Ras inhibitors. *Curr Opin Chem Biol* 1: 197–203.

Goldsmith ME, Aguila A, Steadman K, et al. (2007). The histone deacetylase inhibitor FK228 given prior to adenovirus infection can boost infection in melanoma xenograft model systems. *Mol Cancer Ther* 6: 496–505.

Goldsmith ME, Kitazono M, Fok P, et al. (2003). The histone deacetylase inhibitor FK228 preferentially enhances adenovirus transgene expression in malignant cells. *Clin Cancer Res* 9: 5394–5401.

Goodrum FD, Ornelles DA. (1998). p53 status does not determine outcome of E1B 55-kilodalton mutant adenovirus lytic infection. *J Virol* 72: 9479–9490.

Gray SG, Ekstrom TJ. (2001). The human histone deacetylase family. *Exp Cell Res* 262: 75–83.

Grignani F, DeMatteis S, Nervi C, et al. (1998). Fusion proteins of the retinoic acid receptor-α recruit histone deacetylase in promyelocytic leukaemia. *Nature* 391: 815–818.

Gromeier M, Lachmann S, Rosenfeld MR, et al. (2000). Intergeneric poliovirus recombinants for the treatment of malignant glioma. *Proc Natl Acad Sci U S A* 97: 6803–6808.

Grozinger CM, Hassig CA, Schreiber SL. (1999). Three proteins define a class of human histone deacetylases related to yeast Hda1p. *Proc Natl Acad Sci U S A* 96: 4868–4873.

Hansen M, Gimsing P, Rasmussen A, et al. (2005). A phase I study of the histone deacetylase (HDAC) inhibitor PXD101 in patients with advanced hematological tumors. *J Clin Oncol* 23: 3137.

Hashiro G, Loh PC, Yau JT. (1977). The preferential cytotoxicity of reovirus for certain transformed cell lines. *Arch Virol* 54: 307–315.

He L-Z, Guidez F, Tribioli C, et al. (1998). Distinct interactions of PML-RARα and PLZF-RARα with co-repressors determine differential responses to RA in APL. *Nat Genet* 18: 126–135.

Heise C, Hermiston T, Johnson L, et al. (2000). An adenovirus E1A mutant that demonstrates potent and selective systemic anti-tumoral efficacy. *Nat Med* 6: 1134–1139.

Heise C, Kirn DH. (2000). Replication-selective adenoviruses as oncolytic agents. *J Clin Invest* 105: 847–851.

Heise C, Sampson-Johannes A, Williams A, et al. (1997). ONYX-015, an E1B gene-attenuated adenovirus, causes tumor-specific cytolysis and antitumoral efficacy that can be augmented by standard chemotherapeutic agents. *Nat Med* 3: 639–645.

Hess-Stumpp H, Bracker TU, Henderson D, et al. (2007). MS-275, a potent orally available inhibitor of histone deacetylases—the development of an anticancer agent. *Int J Biochem Cell Biol* 39: 1388–1405.

Hiratsuka M, Inoue T, Toda T, et al. (2003). Proteomics-based identification of differentially expressed genes in human gliomas: Down-regulation of SIRT2 gene. *Biochem Biophys Res Comm* 309: 558–566.

Hirose T, Sowa Y, Takahashi S, et al. (2003). p53-independent induction of Gadd45 by histone deacetylase inhibitor: Coordinate regulation by transcription factors Oct-1 and NF-Y. *Oncogene* 22: 7762–7773.

Hofmeister C. (2012). AR-42 in treating patients with advanced or relapsed multiple myeloma, chronic lymphocytic leukemia or lymphoma. ClinicalTrials.gov Identifier: NCT01129193, Craig Hofmeister, Ohio State University Comprehensive Cancer Center.

Hollstein M, Sidransky D, Vogelstein B, et al. (1991). p53 mutations in human cancers. *Science* 253: 49–53.

Homa FL, Brown JC. (1997). Capsid assembly and DNA packaging in herpes simplex virus. *Rev Med Virol* 7: 107–122.

Hu E, Chen Z, Fredrickson T, et al. (2000). Cloning and characterization of a novel human class I histone deacetylase that functions as a transcription repressor. *J Bio Chem* 275: 15254–15264.

Ikeda K, Ichikawa T, Wakimoto H, et al. (1999). Oncolytic virus therapy of multiple tumors in the brain requires suppression of innate and elicited antiviral responses. *Nat Med* 5: 881–887.

Joseph J, Mudduluru G, Antony S, et al. (2004). Expression profiling of sodium butyrate (NaB)-treated cells: Identification of regulation of genes related to cytokine signaling and cancer metastasis by NaB. *Oncogene* 23: 6304–6315.

Joseph J, Wajapeyee N, Somasundaram K. (2005). Role of p53 status in chemosensitivity determination of cancer cells against histone deacetylase inhibitor sodium butyrate. *Int J Cancer* 115: 11–18.

Kaanders JHAM, Pop LAM, Marres HAM, et al. (2002). ARCON: Experience in 215 patients with advanced head-and-neck cancer. *Int J Radiat Oncol Biol Phys* 52: 769–778.

Kaghad M, Bonnet H, Yang A, et al. (1997). Monoallelically expressed gene related to p53 at 1p36, a region frequently deleted in neuroblastoma and other human cancers. *Cell* 90: 809–819.

Karin M, Cao Y, Greten FR, et al. (2002). NF-κB in cancer: From innocent bystander to major culprit. *Nat Rev Cancer* 2: 301–310.

Katsura T, Iwai S, Ota Y, et al. (2009). The effects of trichostatin A on the oncolytic ability of herpes simplex virus for oral squamous cell carcinoma cells. *Cancer Gene Ther* 16: 237–245.

Kelly E, Russell SJ. (2007). History of oncolytic viruses: Genesis to genetic engineering. *Mol Ther* 155: 651–659.

Kelly WK, Marks PA. (2005). Drug insight: Histone deacetylase inhibitors-development of the new targeted anticancer agent suberoylanilide hydroxamic acid. *Nat Clin Pract Oncol* 2: 150–157.

Kelly WK, Richon VM, O'Connor O, et al. (2005). Phase I study of an oral histone deacetylase inhibitor, suberoylanilide hydroxamic acid, in patients with advanced cancer. *J Clin Oncol* 23: 3923–3931.

Khuri FR, Nemunaitis J, Ganly I, et al. (2000). A controlled trial of intratumoral ONYX-015, a selectively-replicating adenovirus, in combination with cisplatin and 5-fluorouracil in patients with recurrent head and neck cancer. *Nat Med* 6: 879–885.

Kim MS, Kwon HJ, Lee YM, et al. (2001). Histone deacetylases induce angiogenesis by negative regulation of tumor suppressor genes. *Nat Med* 7: 437–443.

Kim YB, Lee KH, Sugita K, et al. (1999). Oxamflatin is a novel antitumor compound that inhibits mammalian histone deacetylase. *Oncogene* 18: 2461–2470.

Kirn DH, Thorne SH. (2009). Targeted and armed oncolytic poxviruses: A novel multi-mechanistic therapeutic class for cancer. *Nat Rev Cancer* 9: 64–71.

Kirn DH, Wang Y, Le Boeuf F, et al. (2007). Targeting of interferon-beta to produce a specific, multi-mechanistic oncolytic vaccinia virus. *PLoS Med* 4: e353.

Kitazono M, Goldsmith ME, Aikou T, et al. (2001). Enhanced adenovirus transgene expression in malignant cells treated with the histone deacetylase inhibitor FR901228. *Cancer Res* 61: 6328–6330.

Kitazono M, Rao VK, Robey R, et al. (2002). Histone deacetylase inhibitor FR901228 enhances adenovirus infection of hematopoietic cells. *Blood* 99: 2248–2251.

Kooby DA, Carew JF, Halterman MW, et al. (1999). Oncolytic viral therapy for human colorectal cancer and liver metastases using a multi-mutated herpes simplex virus type-1 (G207). *FASEB J* 13: 1325–1334.

Kouzarides T. (2000). Acetylation: A regulatory modification to rival phosphorylation? *EMBO J* 19: 1176–1179.

Kristeleit R, Stimson L, Workman P, et al. (2004). Histone modification enzymes: Novel targets for cancer drugs. *Expet Opin Emerg Drugs* 9: 135–154.

Kurihara T, Brough DE, Kovesdi I, et al. (2000). Selectivity of a replication-competent adenovirus for human breast carcinoma cells expressing the MUC1 antigen. *J Clin Invest* 106: 763–771.

Lavoie R, Bouchain G, Frechette S, et al. (2001). Design and synthesis of a novel class of histone deacetylase inhibitors. *Bioorg Med Chem Lett* 11: 2847–2850.

Levine AJ, Momand J, Finlay CA. (1991). The p53 tumour suppressor gene. *Nature* 351: 453–456.

Lin RJ, Nagy L, Inoue S, et al. (1998). Role of the histone deacetylase complex in acute promyelocytic leukaemia. *Nature* 391: 811–814.

Liu TC, Castelo-Branco P, Rabkin SD, et al. (2008a). Trichostatin A and oncolytic HSV combination therapy shows enhanced antitumoral and antiangiogenic effects. *Mol Ther* 16: 1041–1047.

Liu TC, Hwang T, Park BH, et al. (2008b). The targeted oncolytic poxvirus JX-594 demonstrates antitumoral, antivascular, and anti-HBV activities in patients with hepatocellular carcinoma. *Mol Ther* 16: 1637–1642.

Lu Q, Wang DS, Chen CS, et al. (2005). Structure-based optimization of phenylbutyrate-derived histone deacetylase inhibitors. *J Med Chem* 48: 5530–5535.

Luo J, Nikolaev AY, Imai S-I, et al. (2001). Negative control of p53 by Sir2α promotes cell survival under stress. *Cell* 107: 137–148.

MacTavish H, Diallo JS, Huang B, et al. (2010). Enhancement of vaccinia virus based oncolysis with histone deacetylase inhibitors. *PLoS One* 5: e14462.

Maggio SC, Rosato RR, Kramer LB, et al. (2004). The histone deacetylase inhibitor MS-275 interacts synergistically with fludarabine to induce apoptosis in human leukemia cells. *Cancer Res* 64: 2590–2600.

Maiso P, Carvajal-Vergara X, Ocio EM, et al. (2006). The histone deacetylase inhibitor LBH589 is a potent antimyeloma agent that overcomes drug resistance. *Cancer Res* 66: 5781–5789.

Markert JM, Medlock MD, Rabkin SD, et al. (2000). Conditionally replicating herpes simplex virus mutant, G207 for the treatment of malignant glioma: Results of a phase I trial. *Gene Ther* 7: 867–874.

Marks P, Rifkind RA, Richon VM, et al. (2001). Histone deacetylases and cancer: Causes and therapies. *Nat Rev Cancer* 1: 194–202.

Marks PA. (2007). Discovery and development of SAHA as an anticancer agent. *Oncogene* 26: 1351–1356.

Marks PA, Breslow R. (2007). Dimethyl sulfoxide to vorinostat: Development of this histone deacetylase inhibitor as an anticancer drug. *Nat Biotechnol* 25: 84–90.

Martuza RL. (2000). Conditionally replicating herpes vectors for cancer therapy. *J Clin Invest* 105: 841–846.

McCart JA, Ward JM, Lee J, et al. (2001). Systemic cancer therapy with a tumor-selective vaccinia virus mutant lacking thymidine kinase and vaccinia growth factor genes. *Cancer Res* 61: 8751–8757.

Mehnert JM, Kelly WK. (2007). Histone deacetylase inhibitors: Biology and mechanism of action. *Cancer J* 13: 23–29.

Meinke PT, Liberator P. (2001). Histone deacetylase: A target for antiproliferative and antiprotozoal agents. *Curr Med Chem* 8: 211–235.

Michel G, Auer H, Keményny L, et al. (1996). Antioncogene P53 and mitogenic cytokine interleukin-8 aberrantly expressed in psoriatic skin are inversely regulated by the antipsoriatic drug tacrolimus (FK506). *Biochem Pharmacol* 51: 1315–1320.

Mineta T, Rabkin SD, Martuza RL. (1994). Treatment of malignant gliomas using ganciclovir-hypersensitive, ribonucleotide reductase-deficient herpes simplex viral mutant. *Cancer Res* 54: 3963–3966.

Minucci S, Pelicci PG. (2006). Histone deacetylase inhibitors and the promise of epigenetic (and more) treatments for cancer. *Nat Rev Cancer* 6: 38–51.

Moffat D, Patel S, Day F, et al. (2010). Discovery of 2-(6-{[(6-fluoroquinolin-2-yl) methyl] amino}bicyclo[3.1.0]hex-3-yl)-n-hydroxypyrimidine-5-carboxamide (chr-3996), a class I selective orally active histone deacetylase inhibitor. *J Med Chem* 53: 8663–8678.

Morales CP, Holt SE, Ouellette M, et al. (1999). Absence of cancer associated changes in human fibroblasts immortalized with telomerase. *Nat Genet* 21: 115–118.

Mori N, Nunokawa Y, Yamada Y, et al. (1999). Expression of human inducible nitric oxide synthase gene in T-cell lines infected with human T-cell leukemia virus type-I and primary adult T-cell leukemia cells. *Blood* 94: 2862–2870.

Nakayama K, Takebayashi Y, Nakayama S, et al. (2003). Prognostic value of overexpression of p53 in human ovarian carcinoma patients receiving cisplatin. *Cancer Lett* 192: 227–235.

Nguyen TL, Abdelbary H, Arguello M, et al. (2008). Chemical targeting of the innate antiviral response by histone deacetylase inhibitors renders refractory cancers sensitive to viral oncolysis. *Proc Natl Acad Sci U S A* 105: 14981–14986.

Nguyen TLA, Wilson MG, Hiscott J. (2010). Oncolytic viruses and histone deacetylase inhibitors-A multi-pronged strategy to target tumor cells. *Cytokine Growth Factor Rev* 21: 153–159.

North BJ, Marshall BL, Borra MT, et al. (2003). The human Sir2 ortholog, SIRT2, is an NAD+-dependent tubulin deacetylase. *Mol Cell* 11: 437–444.

Nusinzon I, Horvath CM. (2006). Positive and negative regulation of the innate antiviral response and beta interferon gene expression by deacetylation. *Mol Cell Biol* 26: 3106–3113.

Otsuki A, Patel A, Kasai K, et al. (2008). Histone deacetylase inhibitors augment antitumor efficacy of herpes-based oncolytic viruses. *Mol Ther* 16: 1546–1555.

Ouaissi M, Ouaissi A. (2006). Histone deacetylase enzymes as potential drug targets in cancer and parasitic diseases. *J Biomed Biotechnol* 2006: 13474.

Oyama M, Ohigashi T, Hoshi M, et al. (2000). Intravesical and intravenous therapy of human bladder cancer by the herpes vector G207. *Hum Gene Ther* 11: 1683–1693.

Pandolfi PP. (2001). Histone deacetylases and transcriptional therapy with their inhibitors. *Cancer Chemother Pharmacol* 48(Suppl 1): S17–S19.

Park BH, Hwang T, Liu TC, et al. (2008). Use of a targeted oncolytic poxvirus, JX-594, in patients with refractory primary or metastatic liver cancer: A phase I trial. *Lancet Oncol* 9: 533–542.

Plumb JA, Finn PW, Williams RJ, et al. (2003). Pharmacodynamic response and inhibition of growth of human tumor xenografts by the novel histone deacetylase inhibitor PXD101. *Mol Cancer Ther* 2: 721–728.

Prives C, Manley JL. (2001). Why is p53 acetylated? *Cell* 107: 815–818.

Rampling R, Cruickshank G, Papanastassiou V, et al. (2000). Toxicity evaluation of replication-competent herpes simplex virus (ICP 34.5 null mutant 1716) in patients with recurrent malignant glioma. *Gene Ther* 7: 859–866.

Randazzo BP, Bhat MG, Kesari S, et al. (1997). Treatment of experimental subcutaneous human melanoma with a replication-restricted herpes simplex virus mutant. *J Invest Dermatol* 108: 933–937.

Redding N, Zhou H-Y, Lun X, et al. (2009). The utility of oncolytic viruses against neuroblastoma. In The 5th International Meeting on Replicating Oncolytic Virus Therapeutics, Banff, Canada.

Rich T, Allen RL, Wyllie AH. (2000). Defying death after DNA damage. *Nature* 407: 777–783.

Ries S, Korn WM. (2002). ONYX-015: Mechanisms of action and clinical potential of a replication-selective adenovirus. *Br J Cancer* 86: 5–11.

Rodriguez R, Schuur ER, Lim HY, et al. (1997). Prostate attenuated replication competent adenovirus (ARCA) CN706: A selective cytotoxic for prostate-specific antigen-positive prostate cancer cells. *Cancer Res* 57: 2559–2563.

Roizman B. (1996). The function of herpes simplex virus genes: A primer for genetic engineering of novel vectors. *Proc Natl Acad Sci U S A* 93: 11307–11312.

Rothmann T, Hengstermann A, Whitaker NJ, et al. (1998). Replication of ONYX-015, a potential anticancer adenovirus, is independent of p53 status in tumor cells. *J Virol* 72: 9470–9478.

Rundlett SE, Carmen AA, Kobayashi R, et al. (1996). HDA1 and RPD3 are members of distinct yeast histone deacetylase complexes that regulate silencing and transcription. *Proc Natl Acad Sci U S A* 93: 14503–14508.

Russell SJ. (2002). RNA viruses as virotherapy agents. *Cancer Gene Ther* 9: 961–996.

Ryan QC, Headlee D, Acharya M, et al. (2005). Phase I and pharmacokinetic study of MS-275, a histone deacetylase inhibitor, in patients with advanced and refractory solid tumors or lymphoma. *J Clin Oncol* 23: 3912–3922.

Sabin AB. (1959). Reoviruses: A new group of respiratory and enteric viruses formerly classified as ECHO type 10 is described. *Science* 130: 1387–1389.

Sachs MD, Ramamurthy M, Poel H, et al. (2004). Histone deacetylase inhibitors upregulate expression of the coxsackie adenovirus receptor (CAR) preferentially in bladder cancer cells. *Cancer Gene Ther* 11: 477–486.

Saito A, Yamashita T, Mariko Y, et al. (1999). A synthetic inhibitor of histone deacetylase, MS-27-275, with marked in vivo antitumor activity against human tumors. *Proc Natl Acad Sci U S A* 96: 4592–4597.

Sartelet H, Oligny LL, Vassal G. (2008). AKT pathway in neuroblastoma and its therapeutic implication. *Expert Rev Anticancer Ther* 8: 757–769.

Shen Y, Nemunaitis J. (2005). Fighting cancer with vaccinia virus: Teaching new tricks to an old dog. *Mol Ther* 11: 180–195.

Sinkovics J, Horvath J. (1993). New developments in the virus therapy of cancer: A historical review. *Intervirology* 36: 193–214.

Smith RR, Huebner RJ, Rowe WP, et al. (1956). Studies on the use of viruses in the treatment of carcinoma of the cervix. *Cancer* 9: 1211–1218.

Stanford MM, McFadden G. (2007). Myxoma virus and oncolytic virotherapy: A new biologic weapon in the war against cancer. *Expert Opin Biol Ther* 7: 1415–1425.

Stojdl DF, Abraham N, Knowles S, et al. (2000b). The murine double-stranded RNA-dependent protein kinase PKR is required for resistance to vesicular stomatitis virus. *J Virol* 74: 9580–9585.

Stojdl DF, Lighty B, Knowles S, et al. (2000a). Exploiting tumor-specific defects in the interferon pathway with a previously unknown oncolytic virus. *Nat Med* 6: 821–825.

Strong JE, Coffey MC, Tang D, et al. (1998). The molecular basis of viral oncolysis: Usurpation of the Ras signaling pathway by reovirus. *EMBO J* 17: 3351–3362.

Strong JE, Lee PW. (1996). The v-erbB oncogene confers enhanced cellular susceptibility to reovirus infection. *J Virol* 70: 612–616.

Sun ZW, Hampsey M. (1999). A general requirement for the Sin3-Rpd3 histone deacetylase complex in regulating silencing in *Saccharomyces cerevisiae*. *Genetics* 152: 921–932.

Suzuki T, Ando T, Tsuchiya K, et al. (1999). Synthesis and histone deacetylase inhibitory activity of new benzamide derivatives. *J Med Chem* 42: 3001–3003.

Symons JA, Alcami A, Smith GL. (1995). Vaccinia virus encodes a soluble type I interferon receptor of novel structure and broad species specificity. *Cell* 81: 551–560.

Tang X, Gao JS, Guan YJ, et al. (2007). Acetylation-dependent signal transduction for type I interferon receptor. *Cell* 131: 93–105.

Taunton J, Hassig CA, Schreiber SL. (1996). A mammalian histone deacetylase related to the yeast transcriptional regulator Rpd3p. *Science* 272: 408–411.

Thaler F, Colombo A, Mai A, et al. (2010). Synthesis and biological evaluation of N-hydroxyphenylacrylamides and N-hydroxypyridin-2-ylacrylamides as novel histone deacetylase inhibitors. *J Med Chem* 53: 822–829.

Thorne SH, Hwang TH, O'Gorman WE, et al. (2007). Rational strain selection and engineering creates a broad-spectrum, systemically effective oncolytic poxvirus, JX-963. *J Clin Invest* 117: 3350–3358.

Toda M, Rabkin SD, Martuza RL. (1998). Treatment of human breast cancer in a brain metastatic model by G207, a replication-competent multimutated herpes simplex virus 1. *Hum Gene Ther* 9: 2177–2185.

Tollefson AE, Ryerse JS, Scaria A, et al. (1996). The E3-11.6-kDa adenovirus death protein (ADP) is required for efficient cell death: Characterization of cells infected with adp mutants. *Virology* 220: 152–162.

Vaisburg A. (2006). Discovery and development of MGCD0103 – an orally active HDAC inhibitor in human clinical trials. Presented at the XIXth International Symposium on Medicinal Chemistry, Istanbul, August 2006; Paper L57.

Vaisburg A, Bernstein N, Frechette S, et al. (2004). (2-amino-phenyl)-amides of omega-substituted alkanoic acids as new histone deacetylase inhibitors. *Bioorg Med Chem Lett* 14: 283–287.

Vaisburg A, Paquin I, Bernstein N, et al. (2007). N-(2-amino-phenyl)-4-(heteroarylmethyl)-benzamides as new histone deacetylase inhibitors. *Bioorg Med Chem Lett* 17: 6729–6733.

Vanoosten RL, Earel JK, Jr, Griffith TS. (2006). Enhancement of Ad5-TRAIL cytotoxicity against renal cell carcinoma with histone deacetylase inhibitors. *Cancer Gene Ther* 13: 628–632.

Vanoosten RL, Earel JK, Jr, Griffith TS. (2007). Histone deacetylase inhibitors enhance Ad5-TRAIL killing of TRAIL-resistant prostate tumor cells through increased caspase-2 activity. *Apoptosis* 12: 561–571.

Vanoosten RL, Moore JM, Ludwig AT, et al. (2005). Depsipeptide (FR901228) enhances the cytotoxic activity of TRAIL by redistributing TRAIL receptor to membrane lipid rafts. *Mol Ther* 11: 542–552.

Varasi M, Thaler F, Abate A, et al. (2011). Discovery, synthesis and pharmacological evaluation of spiropiperidine hydroxamic acid based derivatives as structurally novel histone deacetylase (HDAC) inhibitors. *J Med Chem* 54: 3051–3064.

Varghese S, Rabkin SD. (2002). Oncolytic herpes simplex virus vectors for cancer virotherapy. *Cancer Gene Ther* 9: 967–978.

Vaziri H, Dessain SK, Eaton EN, et al. (2001). hSIR2SIRT1 functions as an NAD-dependent p53 deacetylase. *Cell* 107: 149–159.

Vlasakova J, Novakova Z, Rossmeislova L, et al. (2007). Histone deacetylase inhibitors suppress IFNalpha-induced up-regulation of promyelocytic leukemia protein. *Blood* 109: 1373–1380.

Wakimoto H, Fulci G, Tyminski E, et al. (2004). Altered expression of antiviral cytokine mRNAs associated with cyclophosphamide's enhancement of viral oncolysis. *Gene Ther* 11: 214–223.

Walker JR, McGeagh KG, Sundaresan P, et al. (1999). Local and systemic therapy of human prostate cancer with the conditionally replicating herpes simplex virus vector G207. *Hum Gene Ther* 10: 2237–2245.

Wang W, Abbruzzese JL, Evans DB, et al. (1999). Overexpression of urokinase-type plasminogen activator in pancreatic adenocarcinoma is regulated by constitutively activated RelA. *Oncogene* 18: 4554–4563.

Wang H, Yu N, Chen D, et al. (2011) Discovery of (2E)-3-{2-butyl-1-[2-(diethylamino) ethyl]-1H-benzimidazol-5-yl}-N-hydroxyacrylami de (SB939), an orally active histone deacetylase inhibitor with a superior preclinical profile. *J Med Chem* 54: 4694–4720.

Watanabe T, Hioki M, Fujiwara T, et al. (2006). Histone deacetylase inhibitor FR901228 enhances the antitumor effect of telomerase-specific replication-selective adenoviral agent OBP-301 in human lung cancer cells. *Exp Cell Res* 312: 256–265.

Weichert W. (2009). HDAC expression and clinical prognosis in human malignancies. *Cancer Lett* 280: 168–176.

Weiner HL, Fields BN. (1977). Neutralization of reovirus: The gene responsible for the neutralization antigen. *J Exp Med* 146: 1305–1310.

Whitley RJ, Kimberlin DW, Roizman B. (1998). Herpes simplex viruses. *Clin Infect Dis* 26: 541–553.

Wildner O, Blaese RM, Morris JC. (1999). Therapy of colon cancer with oncolytic adenovirus is enhanced by the addition of herpes simplex virus-thymidine kinase. *Cancer Res* 59: 410–413.

Wildner O, Morris JC. (2000). The role of the E1B 55 kDa gene product in oncolytic adenoviral vectors expressing herpes simplex virus-tk: Assessment of antitumor efficacy and toxicity. *Cancer Res* 60: 4167–4174.

Wong HH, Lemoine NR, Wang Y. (2010). Oncolytic viruses for cancer therapy: Overcoming the obstacles. *Viruses* 2: 78–106.

Woo SH, Frechette S, Abou-Khalil E, et al. (2002). Structurally simple trichostatin A-like straight chain hydroxamates as potent histone deacetylase inhibitors. *J Med Chem* 45: 2877–2885.

Yang WM, Yao YL, Sun JM, et al. (1997). Isolation and characterization of cDNAs corresponding to an additional member of the human histone deacetylase gene family. *J Bio Chem* 272: 28001–28007.

Yeung F, Hoberg JE, Ramsey CS, et al. (2004). Modulation of NF-κB-dependent transcription and cell survival by the SIRT1 deacetylase. *EMBO J* 23: 2369–2380.

Yew PR, Berk AJ. (1992). Inhibition of p53 transactivation required for transformation by adenovirus early 1B protein. *Nature* 357: 82–85.

Yoshida M, Kijima M, Akita M, et al. (1990). Potent and specific inhibition of mammalian histone deacetylase both in vivo and in vitro by trichostatin A. *J Biol Chem* 265: 17174–17179.

Yu C, Rahmani M, Conrad D, Subler M, et al. (2003). The proteasome inhibitor bortezomib interacts synergistically with histone deacetylase inhibitors to induce apoptosis in Bcr/Abl+ cells sensitive and resistant to STI571. *Blood* 102: 3765–3774.

Yu, YA, Shabahang S, Timiryasova TM, et al. (2004). Visualization of tumors and metastases in live animals with bacteria and vaccinia virus encoding light-emitting proteins. *Nat Biotechnol* 22: 313–320.

Yue Z, Shatkin AJ. (1998). Enzymatic and control functions of reovirus structural proteins. *Curr Top Microbiol Immunol* 233: 31–56.

Zeh HJ, Bartlett DL. (2002). Development of a replication selective, oncolytic poxvirus for the treatment of human cancers. *Cancer Gene Ther* 9: 1001–1012.

Zhang Y, Fang H, Feng J, et al. (2011b). Discovery of a tetrahydroisoquinoline-based hydroxamic acid derivatives (ZYJ-34c) as histone deacetylase inhibitor with potent oral antitumor activities. *J Med Chem* 54: 5532–5539.

Zhang Y, Feng J, Jia Y, et al. (2011a). Development of tetrahydroisoquinoline-based hydroxamic acid derivatives: Potent histone deacetylase inhibitors with marked in vitro and in vivo antitumor activities. *J Med Chem* 54: 2823–2838.

Zhou N, Moradei O, Raeppel S, et al. (2006). In novel benzamide derivatives as potent histone deacetylase (HDAC) inhibitors: Synthesis and antiproliferative evaluation. Presented at the 232nd American Chemical Society National Meeting, San Francisco, CA, September 2006; Abstract MEDI 147.

Zhou X, Marks PA, Rifkind RA, et al. (2001). Cloning and characterization of a histone deacetylase, HDAC9. *Proc Natl Acad Sci U S A* 98: 10572–10577.

# Index

Index page numbers in italics refer to figures and tables

## G



## H

## I

## J



## K



## L

## O

## P



## Q



## R



## S

## T



## U



## V

## W



## X



## Z

T - #0354 - 101024 - C16 - 234/156/27 - PB - 9781138374836 - Gloss Lamination